RESOURCES AND DYNAMICS OF THE BOREAL ZONE

Proceedings of a conference held at Thunder Bay, Ontario, August, 1982

Ross W. Wein, Roderick R. Riewe and Ian R. Methven
(Editors)

Association of Canadian Universities for Northern Studies

The organizers gratefully acknowledge generous financial support received from the following organizations:

Department of Environment, Government of Canada
Department of Fisheries and Oceans, Government of Canada
Department of Indian Affairs and Northern Development, Government of Canada
Government of the Northwest Territories
Great Lakes Forest Products
NOVA, An Alberta Corporation
Petro-Canada
Polar Gas Project
Province of Alberta
Province of Newfoundland-Labrador
Province of Ontario
Province of Saskatchewan

Cover photo by Roderick R. Riewe

130 Albert Street,
Ottawa, Canada K1P 5G4
Tel: (613) 238-3525

Deposited at the National Library of Canada
Ottawa
1983

ISBN 0-9690987-2-3

Table of Contents

Dynamics – Succession

Dynamics – Fire

PREFACE

One of the means that the Association of Canadian Universities for Northern Studies employs to achieve its purpose of advancing northern scholarship is the periodic holding of symposia on topics of special interest to northern researchers as well as policy makers, managers, and the Canadian public. This is the second major symposium held by the Association, the first on Renewable Resources and the Economy of the North having been held in Banff, Alberta in May 1981, with the Proceedings appearing in September 1981. In both symposia participants were invited from relevant sectors of Canadian society as well as from other countries. The aim was to promote as much interchange as possible among producers and users of northern research and scholarship, in the wider public interest.

The three day International Symposium on the Dynamics of Boreal Forest Ecosystems: Future Research and Management Requirements was held at Lakehead University in Thunder Bay, Ontario, between August 23 and 26, 1982. The opening plenary sessions provided an overview of current thinking on the boreal forest. In the second session four papers were presented under the heading ‘‘Potential of Renewable and Non-Renewable Resources’’.

The participants were divided into six workshops on the second day and considered the following topics: Forest Management Issues in Northern Forests, Fire in Boreal Ecosystems, Northern Communities, Food and Energy for Northern Forests, Classification, Succession and Nutrient Cycling in Northern Forests, and Toxic Substance Releases in the North. About forty-five papers were presented at the six workshop sessions. On the final day of the Symposium, summary papers on environmental and management issues discussed during the workshops were presented during a plenary session entitled ‘‘Resource Conflicts and Compatibilities’’.

A number of major concerns emerged during the Symposium. Many participants saw a communications gap to be bridged between researchers and managers. Concern was expressed about lack of understanding of the forest as a whole, and in particular of key processes such as succession. Recommendations were made on manager-scientist relations, fire, boreal forest ecosystem dynamics, and social issues, including forest food potential.

Readers of these Proceedings should be aware, as I am, of their debt to a number of people who made the Symposium possible. Included are the members of the ACUNS Board of Directors, who worked hard on the concept and organization, the Chairmen of the sessions and workshops, the faculty and staff at Lakehead University who hosted the symposium and worked effectively in organizing sessions and field trips, Cam Mackie and David Judd, past and current ACUNS Executive Directors who provided thrust for the whole effort, and especially Eileen van Heyst whose conceptual and organizational skills and energy made the Symposium work.

Special thanks are due to Dr. Ross Wein of the University of New Brunswick who undertook the major task of organizing the design and content of the Symposium. He and his Organizing Committee shared the effort involved in putting together a meeting of this scope. The job of editing these Proceedings fell to Ross Wein and to his colleagues Professor Rick Riewe and Professor Ian Methven.

Finally we are very grateful to the government departments, agencies and corporations which provided funding. They and the many participants involved ensured the Symposium's ultimate success.

These Proceedings should contribute to the dissemination of their ideas to a much wider audience.

J. Gordon Nelson
President
Association of Canadian
Universities for Northern Studies

Introduction

In the past decade, there has been a burst of research and other scholarly activity in the north due to the location and development of mineral and petroleum resources. In 1969, in response to these pressures, a conference on the ''Productivity and conservation in circumpolar lands'' at the University of Alberta, in Edmonton, Alberta, brought together scientists to examine the functioning of tundra ecosystems (Fuller and Keven 1970). Out of that conference came a number of research activities best exemplified by the High Arctic ecosystem study on Truelove Lowland, Devon Island (Bliss 1977). Unfortunately, a similar ecosystem study was not initiated in the North American boreal zone.

Relatively little research and scholarly activity has taken place recently in the circumpolar boreal zone yet this area is 1000 km wide in parts of North America and Eurasia. Roughly two-thirds of the boreal zone is located in Eurasia. In Canada the zone occupies one-third of the land mass and involves both provincial and federal jurisdictions.

In August 1969, a conference in Thunder Bay, Ontario attracted a wide range of individuals interested in the development potential of the boreal zone (Mid-Canada Development Foundation, Inc. 1970). Since that time, there has been no major interdisciplinary conference to stimulate research, training and other scholarly activities. There has however, been recognition that information about the functioning and the management of this huge geographic area should be summarized. Synthesis volumes such as Larsen (1980), Nilsson and Kullman (1981), Persson (1980), Pruitt (1978) and Tamm (1976) have made admirable strides in that direction.

On August 23 to 26, 1982, the Association of Canadian Universities for Northern Studies sponsored an international symposium on the ''Dynamics of boreal forest ecosystems: Future research and management requirements'' at Lakehead University, Thunder Bay, Ontario. The objective of this symposium was to bring together scientists, decision-makers, and managers to stimulate information exchange on how the boreal zone functions, to identify what research information is required and to determine how policy and management programs should operate within the constraints of the ecosystem.

Papers presented in this volume represent, to a large extent, the presentations at the conference although a number of significant changes have been made. We wish to recognize the extremely valuable summary account of the conference, prepared by Dr. J.M. Harrison and Dr. J.E. Lillycrop (Harrison and Lillycrop 1982). Many of the comments below are based on their report.

One of the major concerns of the conference participants was the gap which exists between the research scientists and the managers who are responsible for the resource harvesting. There are many reasons for this gap. One is that the published information about northern resources is largely stored in libraries located in southern Canada. There are also too few communicators of scientific knowledge. Technology transfer should be strongly encouraged. It was recognized that the distinction between management and research becomes blurred and at times disappears, and this, in a way was considered as a positive goal.

As already indicated, the boreal region is an extremely large circumpolar area. It is not surprising therefore that there is a lack of understanding of the total resource or, more importantly, of the productivity potential of the various areas of the region. There is a continuous requirement for the use of the zone's resources, whether it be pressures from the south for recreation and fibre or from northern peoples for food, fuel and shelter. To gain some understanding of the resources of this huge geographic area, and to transfer knowledge from one similar geographic area to another, a more coordinated approach is needed to develop better classification systems. New technology, such as satellite imagery, is readily available to aid in these approaches.

Under the topic, dynamics of boreal ecosystems, it is recognized that more research is needed of theories and hypotheses using biologically-sound input. If biologically-sound models can be developed to capture the dynamics of ecosystems, then managers will have a powerful tool at their disposal to make decisions after modelling the potential effects of a range of management strategies. Nutrient cycling is obviously a key process in ecosystem dynamics in that it is one of the main limiting factors to production. It is also recognized that in studying the dynamics of the boreal ecosystem, a better understanding of succession is needed.

One of the driving variables that influences the boreal ecosystem is that of fire which traditionally has been viewed in terms of total fire suppression. It is now recognized that the more inclusive term 'fire management' is gaining increasing support. This does not mean that fire suppression approaches have been weakened; indeed, they must be strengthened so that the manager has the capability of achieving objectives such as suppressing a fire, simply monitoring a fire or conducting prescribed burns.

It is also recognized that the effect of pollutants such as heavy metals and acid rain will vary across the boreal zone, depending on surficial deposits. Baseline data, as with all other studies, are required but accelerated simulation experiments are needed to predict long-term changes in the quality of waters and soils.

Finally, but certainly not least important, there are social issues that relate to management and research issues. The gap between researchers and managers has already been emphasized but that gap is just as large as the gap between the communities and scientists or the local people and managers. It is recognized that a better assessment is needed of the economic potential of using more food from the boreal region and more intensive investigations are required on the uses of bio-energy by small communities. There are as well many social issues that revolve around large-scale development projects. Baseline biological and social data are required but just as important, if not more so, are follow-up studies during and after the project. They are not only necessary but could be seen as an obligation of the developers. Better methods of involving northern peoples in planning projects that will ultimately affect their livelihood must be given more serious attention.

Readers will recognize that the present volume presents the views of the boreal zone of individuals with a wide diversity of backgrounds. However great the diversity of views, the common goals are to seek information about how the boreal systems function and to use this information for the betterment of peoples who rely on the resources found in the boreal zone.

Literature Cited

Bliss, L.C. (Ed.). 1977. Truelove Lowland, Devon Island, Canada: A high Arctic ecosystem. The University of Alberta Press, Edmonton, Alberta.

Fuller, W.A. and P.G. Keven (Eds.). Productivity and conservation in northern circumpolar lands. IUCN Publ., New Series #16. Morges, Switzerland.

Harrison, J.M. and J.E. Lillycrop. 1982. Summary account: International symposium on the "Dynamics of boreal forest ecosystems: Future research and management requirements". Association of Canadian Universities for Northern Studies, Ottawa, Ontario.

Larsen, J.A. 1980. The boreal ecosystem. Academic Press, New York.

Mid-Canada Development Foundation, Inc. 1970. Essays on Mid-Canada. Maclean-Hunter, Ltd., Toronto, Ontario.

Nilsson, C. and L. Kullman. 1981. Studies in boreal plant ecology. Wahlenbergia, Vol. 7, University of Umeå, Umeå, Sweden.

Persson, T. 1980. Structure and function of northern coniferous forests: an ecosystem study. Ecol. Bull. No. 32, Swedish Nat. Sci. Res. Council (NFR), Stockholm, Sweden.

Pruitt, W.O., Jr. 1978. Boreal ecology. The Institute of Biology's Studies in Biology No. 91. Edward Arnold, London, England.

Tamm, C.O. (Ed.). 1976. Man and the boreal forest. Ecol. Bull. No. 21, Swedish Nat. Sci. Res. Council (NFR), Stockholm, Sweden.

R.W.W.
R.R.R.
I.R.M.

Acknowledgements

In preparing the volume, the Editors wish first to recognize the authors who revised their symposium contributions in light of the external and internal reviews and editorial requirements. We also recognize the contributions of anonymous peer reviewers. Janice M. Moore deserves a special thanks for spending many days organizing the uniform format of the volume. Patricia Ovenden was largely responsible for producing the uniform format figures. Production of the volume was coordinated by David A.W. Judd, Executive Director, Association of Canadian Universities for Northern Studies (A.C.U.N.S.) and Eileen van Heyst, Administrative Officer of the Association.

Since the present volume is based on the conference, the Editors wish to acknowledge the efforts of the Organizing Committee (J.K. Jeglum, P.C. Mackie, I.R. Methven, B.A.M. Phillips, R.R. Riewe and R.W. Wein) and the Workshop Coordinators (G.M. Courtin, A. Gill, A.G. Gordon, J.K. Jeglum, A.J. Kayll, I.R. Methven and R.R. Riewe). The Local Arrangements Committee, initially chaired by B.A.M. Phillips at Lakehead University, provided an excellent environment for discussion and exchange of ideas during the conference. P.C. (Cam) Mackie (former Executive Director of A.C.U.N.S.) and Eileen van Heyst deserve special thanks for coordinating much of the symposium development.

R.W.W.
R.R.R.
I.R.M.

List of Contributors

Addison, P.A.
Environment Canada, Canadian Forestry Service,
Northern Forest Research Centre, 5320–122nd St.,
Edmonton, Alberta T6H 3S5

Baskerville, G.
Faculty of Forestry, University of New Brunswick,
P.O. Box 4400, Fredericton, New Brunswick E3B 5A3

Battson, R.A.
Department of Geography, University of Western Ontario,
London, Ontario N6A 5B8

Bendell, J.F.
Faculty of Forestry, University of Toronto,
203 College St., Toronto, Ontario M5S 1A1

Benson, E.A.H.
Department of Botany, Erindale College,
Mississauga, Ontario L5L 1C6

Berkes, Fikret
Institute of Urban and Environmental Studies,
Brock University, St. Catharines, Ontario L2S 3A1

Boothroyd, P.N.
Environment Canada, Canadian Wildlife Service,
501 University Crescent, Winnipeg, Manitoba R3T 2N6

Boyle, Margaret
Department of Zoology, University of Manitoba,
Winnipeg, Manitoba R3T 2N2

Bustillos, Franklin
Department of Zoology, University of Manitoba,
Winnipeg, Manitoba R3T 2N2

Carleton, T.J.
Faculty of Forestry, University of Toronto,
203 College St., Toronto, Ontario M5S 1A1

Cawker, K.B.
Department of Social Sciences, University of Toronto,
1265 Military Trail, West Hill, Ontario M1C 1A4

Ducruc, Jean-Pierre
Service des inventaires écologiques, Environnement Québec,
2360 Chemin Sainte-Foy, Sainte-Foy, Québec G1V 4H2

Dyrness, C.T.
Institute of Northern Forestry, Fairbanks, Alaska, U.S.A. 99701

El-Bayoumi, M.A.
Department of Biology and Fire Science Centre,

University of New Brunswick, P.O. Box 4400,
Fredericton, New Brunswick E3B 5A3

Farmer, R.E.
School of Forestry, Lakehead University,
Thunder Bay, Ontario P7B 5E1

George, P.J.
Department of Economics, McMaster University,
Hamilton, Ontario L8S 4M4

Gerardin, Vincent
Service des inventaires écologiques, Environnement Québec,
2360 Chemin Sainte-Foy, Sainte-Foy, Québec G1V 4H2

Gill, Alison
Department of Geography, Brandon University,
Brandon, Manitoba R7A 6A9

Gordon, Alan G.
Ontario Tree Improvement and Forest Biomass Institute,
Ontario Ministry of Natural Resources,
P.O. Box 490, Sault Ste. Marie, Ontario P6A 5M7

Gordon, Andrew M.
Forest Soils Laboratory, University of Alaska,
Fairbanks, Alaska, U.S.A. 99701

Guse, Lorna W.
Department of Anthropology, 452 University College,
University of Manitoba, Winnipeg, Manitoba R3T 2N2

Häkkilä, Matti
University of Oulu, Research Institute of Northern Finland,
Subdepartment of Kainuu, Kauppakatu 25A, 87100 Kajaani 10, Finland.

Hawkes, Brad C.
Environment Canada, Canadian Forestry Service, Pacific Forest Research Centre,
506 West Burnside Road, Victoria, British Columbia V8Z 1M5

Jeglum, J.K.
Environment Canada, Canadian Forestry Service, Great Lakes Forest Research Centre, P.O. Box 490, Sault Ste. Marie, Ontario P6A 5M7

Jones, R.K.
Ontario Institute of Pedology, Agriculture Canada, Blackwood Hall,
University of Guelph, Guelph, Ontario N1G 2W1

Kaufert, Joseph M.
Department of Anthropology, 452 University College, University of Manitoba,
Winnipeg, Manitoba R3T 2N2

Kennedy, K.A.
Environment Canada, Canadian Forestry Service, Northern Forest Research Centre, 5320–122nd St., Edmonton, Alberta T6H 3S5

Knowles, P.
School of Forestry, Lakehead University, Thunder Bay, Ontario P7B 5E1

Koolage, William W.
Department of Anthropology, 452 University College, University of Manitoba, Winnipeg, Manitoba R3T 2N2

Lei, T.T.
Department of Botany, Erindale College, Mississauga, Ontario L5L 1C6

Lopoukhine, N.
Natural Resources Division, Parks Canada, Les Terrasses de la Chaudière, Ottawa, Ontario K1A 1G2

Marek, G.T.
Ontario Ministry of Natural Resources, P.O. Box 97, Nipigon, Ontario P0T 2J0

Martell, David L.
Faculty of Forestry, University of Toronto, 203 College St., Toronto, Ontario M5S 1A1

Martin, Antonio M.
Department of Biochemistry, Memorial University, St. John's, Newfoundland A1B 3X9

Maynard, D.G.
Environment Canada, Canadian Forestry Service, Northern Forest Research Centre, 5320 – 122nd St., Edmonton, Alberta T6H 3S5

Methven, Ian R.
Department of Forest Resources, University of New Brunswick, P.O. Box 4400, Fredericton, New Brunswick E3B 5A3

Muller, R.A.
Department of Economics, McMaster University, Hamilton, Ontario L8S 4M4

Naylor, B.J.
Faculty of Forestry, University of Toronto, 203 College St., Toronto, Ontario M5S 1A1

Norum, R.
Institute of Northern Forestry, Fairbanks, Alaska, U.S.A. 99701

Overend, Ralph P.
Renewable Energy, Division of Energy, National Research Council of Canada, Montreal Road, Ottawa, Ontario K1A 0R6

Parker, W.H.
School of Forestry, Lakehead University, Thunder Bay, Ontario P7B 5E1

Pierpoint, G.
Ontario Tree Improvement and Biomass Institute, Ontario Ministry of Natural Resources, Maple, Ontario L0J 1E0

Pringle, William L.
Agriculture Canada, Experimental Farm, R.R. #8, RMD #6, Prince George, British Columbia V2N 4M6

Reader, Brian
Department of Zoology, University of Manitoba, Winnipeg, Manitoba R3T 2N2

Riewe, Roderick R.
Department of Zoology, University of Manitoba, Winnipeg, Manitoba R3T 2N2

Saastamoinen, Olli
The Finnish Forest Research Institute, Rovaniemi Research Station,
Etelaranta 55, SF-96300 Rovaniemi 30, Finland.

Savinsky, H.
Faculty of Forestry, University of Toronto, 203 College St.,
Toronto, Ontario M5S 1A1

Senyk, John P.
Environment Canada, Lands Directorate, 506 West Burnside Road,
Victoria, British Columbia V8Z 1M5

Silversides, C.R.
R.R. #1, Prescott, Ontario K0E 1T0

Siuruainen, Eino
Department of Geography, University of Oulu, Linnanmaa, SF-90570,
Oulu 57, Finland

Slaughter, C.W.
Institute of Northern Forestry, Fairbanks, Alaska, U.S.A. 99701

Spires, S.
Faculty of Forestry, University of Toronto, 203 College St.,
Toronto, Ontario M5S 1A1

Street, R.B.
Atmospheric Environment Service, 4905 Dufferin St.,
Downsview, Ontario M3H 5T4

Stocks, B.J.
Environment Canada, Canadian Forestry Service, Great Lakes Forest Research Centre, P.O. Box 490, Sault Ste. Marie, Ontario P6A 5M7

Svoboda, J.
Department of Botany, Erindale College, Mississauga, Ontario L5L 1C6

Szuba, K.J.
Faculty of Forestry, University of Toronto, 203 College St.,
Toronto, Ontario M5S 1A1

Taylor, H.W.
Department of Physics, Erindale College, Mississauga, Ontario L5L 1C6

Thomas, P.A.
Department of Biology and Fire Science Centre, University of New Brunswick,
P.O. Box 4400, Fredericton, New Brunswick E3B 5A3

Timmer, V.R.
Faculty of Forestry, University of Toronto, 203 College St.,
Toronto, Ontario M5S 1A1

Van Ballenberghe, V.
Institute of Northern Forestry, Fairbanks, Alaska, U.S.A. 99701

Van Cleve, Keith
Forest Soils Laboratory, University of Alaska, Fairbanks, Alaska, U.S.A. 99701

Viereck, Leslie A.
Institute of Northern Forestry, Fairbanks, Alaska, U.S.A. 99701

Wannamaker, B.A.
Faculty of Forestry, University of Toronto, 203 College St.,
Toronto, Ontario M5S 1A1

Weetman, G.F.
Faculty of Forestry, MacMillan Building, University of British Columbia,
2075 Wesbrook Mall, Vancouver, British Columbia V6T 1W5

Wein, Ross W.
Department of Biology and Fire Science Centre, University of New Brunswick,
P.O. Box 4400, Fredericton, New Brunswick E3B 5A3

Weller, Geoffrey R.
Department of Political Studies, Lakehead University,
Thunder Bay, Ontario P7B 5E1

Werner, R.A.
Institute of Northern Forestry, Fairbanks, Alaska, U.S.A. 99701

Wickware, G.M.
Environment Canada, Lands Directorate, P.O. Box 5050,
Burlington, Ontario L7N 4A6

Wilkinson, Paul F.
Centre for Northern Studies and Research, McGill University,
1020 Pine Ave., W., Montreal, Quebec H3A 1A2

Winterhalder, Bruce
Department of Anthropology, The University of North Carolina, Alumni Bldg.
004A, Chapel Hill, North Carolina, U.S.A. 27514

Zasada, J.
Institute of Northern Forestry, Fairbanks, Alaska, U.S.A. 99701

PROBLEMS OF MULTIPLE-USE FORESTRY IN THE NORTHERN BOREAL ZONE OF FINLAND

Olli Saastamoinen
The Finnish Forest Research Institute

Abstract: Some multiple-uses of forestry land are encountered in Finnish Lapland. This area contains less than 5 per cent of the population but more than one-quarter of the total land area of Finland. Lapland is no longer a frontier economy in that it is linked to southern Finland although it retains some unique economic activities. The basis for the regional economy is timber production and associated industries. Outdoor recreation and tourism continue to expand rapidly. There are conflicts between timber production and conservation, between timber production and reindeer production and to a certain extent, between timber production and subsistence collecting of berries and fungi, or hunting and fishing. The author argues that the economic value of hunting and fishing, reindeer grazing, gathering berries and mushrooms, and outdoor recreation may exceed that of timber production and suggests that there are a number of complementary relationships between forest uses.

Finnish Lapland covers more than a quarter (9.4 million hectares) of the total land area of Finland but its population (196,000 persons) is less than five per cent of the whole country. Most of the Finnish Lapland area is located north of the Polar Circle but most of its people live south of it.

Due to the northern location, the climate is severe, but thanks to the proximity of the Gulf Stream it is milder than at corresponding latitudes in Canada or Siberia. The annual mean temperature ranges from $-1°$ to $+2°C$; the mean temperature in July ranges from $+14°$ to $+16°C$, and in February, from $-13°$ to $-10°C$. The average number of degree days (d.d.) exceeding the threshold value of $+5°C$ is 500 to 900. The coniferous timber line approximately follows the isogram of 550 d.d. Annual precipitation ranges from 400 to 500 mm, of which 40–50 per cent is snow. The lasting snow arrives in Lapland between October 25 and November 20 and disappears between May 5 and 25 (Pohtila 1977). Climatically, as can be seen from the above figures, Lapland is not a homogenous area. Sometimes, for example, the differences between the southernmost and northernmost parts of Lapland are greater than the average differences between it and central parts of Finland.

In Finland, Finnish Lapland is clearly regarded as a region with its own character. It not only has a different nature but also some peculiar economic and social features. For example, it has some unique economic activities the best known of which is reindeer husbandry. Others include the collecting of Arctic raspberries, salmon fishing in rivers not developed for hydropower and panning for gold. The last two activities are also important tourist activities.

Lapland is no longer a frontier economy. The post-war re-building activities, the development of agricultural settlements, road construction, water-power plant construction and large scale timber harvesting in virgin forests are mainly over. With these activities a kind of regional boom which lasted from the late 1940s to the mid-1960s also ceased. In the post-boom period high unemployment rates and emigration have been the most serious economic problems of Finnish Lapland. An additional social-economic feature of Lapland which needs to be mentioned in this context, is the encountering of traditional culture of Lappish ethnic minority with the mainstream Finnish culture. This background information helps to understand the economic and social situation in Finnish Lapland.

The Lapland Forest

One may recognize that the role of forests has always been important for the people of Lapland. As a renewable resource and as a source of not only many traditional but also some new (e.g. recreational) or newly discovered (e.g. fuelwood) benefits, its importance cannot be overemphasized. However, due to the increased demand for many forest benefits, the impacts and conflicts between different forest uses have become more and more common.

The purpose of this paper is to describe briefly the most important forest uses in Finnish Lapland and to indicate some of the most serious impacts and conflicts between them.

The forests of Finnish Lapland belong mainly to the northern boreal coniferous zone. The most northern coniferous forests in the timber line region of Finnish Lapland are almost pure Scots pine *(Pinus silvestris)* stands, in contrast to most other parts of the polar timber-line region, where spruce or larch constitute the timber line (e.g., Pohtila 1977). Single Norway spruce *(Picea abies)* trees or stands also grow sparsely near the boreal timber line of pine, however, most spruce forests are found in the central and southern parts of Finnish Lapland (Norokorpi 1979). North of the coniferous timber line, and usually also in mountains above the coniferous forests, a birch zone occurs. The pubescent birch *(Betula pubescens)* is very common throughout the whole of Finnish Lapland. Silver birch *(Betula verrucosa)* is much more productive but is rare in the area and, in fact, attempts to increase it by plantings have proven to be difficult because of browsing semi-domestic reindeer *(Rangifer tarandus tarandus)*. This is the only apparent economic harm caused by reindeer to forests. The most productive tree species, the Scots pine, clearly exceeds the growth of spruce and birch. Therefore, in forest management, the goal has been to favour pine at the expense of spruce and birch wherever possible. This development, even if well-grounded from the point of view of timber production, has caused some problems for reindeer husbandry.

Forest Use Conflicts — Timber vs Conservation

Timber growing and harvesting is the major land use in Finnish Lapland. Cutting is carried out almost through the whole of Lapland and present high stumpage prices have made even the most distant tracts of forests economically accessible. Although timber is harvested not far from the timber line, the special Protection Forest Zone which includes polar and alpine timber line areas, regulates the scale and intensity of cuttings in order to preserve the timber line forests. Even in the areas outside the Protection Forest Zone the cuttings are strictly limited in so-called summit area forests (forests above the altitude of 300 m, in some cases above 250 m). The northern Protection Forest Zone and summit area forests are examples of forest protection, even if their establishment was stimulated mainly by the need to avoid risks of poor regeneration. The altitude and geographical limits and other restrictions for cutting intensity in the northernmost and summit area forests are at present topical questions. These questions are also closely related to the recent establishment of new national parks and strict nature reserves (the difference between the two types of conservation areas is the restricted access of the public to the latter) and the proposed creation of some wilderness areas, where, as in national parks and strict nature reserves, timber harvesting is not allowed.

The forest industry and forest authorities are afraid that new and proposed conservation areas will cause a severe shortage in timber supply in the short term (i.e., 10–20 years). Because there are somewhat different ways and bases for calculating the reduction of allowable removals caused by the established and planned conservation areas, one can only state that the range of estimated reduction varies from five to ten per cent of the total allowable removals, which at present are 4.0 million m^3 in Finnish Lapland. The balance between the value of nature conservation areas and the need for industrial wood is perhaps the most obvious conflict but it is not the only one among the different forest uses.

Forest Use Conflicts — Timber vs Reindeer

Great public concern is devoted to the relationships between timber production and reindeer husbandry. Three-fourths of the approximate total 200,000 reindeer graze in the coniferous zone, which as previously mentioned, is also intensively used for timber growing and harvesting. There are numerous effects of forestry operations on the ranges of reindeer. Silvicultural cuttings in pine forests, for example, decrease the productivity of ground lichen stands and make it more difficult for reindeer to dig out lichens because of the harder snow cover. Cuttings in old spruce forests, which from the point of view of timber are unproductive, substantially decrease the crop of arboreal lichens in the long term but increase their short-term availability when the trees are felled in winter (Saastamoinen 1978). As examples of beneficiary effects of forestry on reindeer husbandry, one may mention the increased growth of wavy hairgrass *(Deschampsia flexuosa)* after clearcutting. The building of forest roads also facilitates reindeer management.

The relationships between reindeer husbandry and forestry are variable and complicated. At present we have few quantitative data, although in recent years these problems have been studied (e.g., Helle and Saastamoinen 1979) and a reindeer range inventory has been taken in connection with the National Forest Inventory (Mattila 1981). A difficult problem seems to be the separation of the effects of heavy grazing from those of forestry operations (Saastamoinen 1982).

It has been concluded that even if intensive timber production decreases the reindeer production it only does so partially. In fact reindeer grazing and timber production may be conducted simultaneously in the same area.

Forest Use Conflicts — Timber vs Other Uses

Collecting of forest and peatland berries and edible fungi is carried out widely not only for personal use but also for sale as a cash crop. The must utilized forest berries are red whortleberry *(Vaccinium vitis-idaea)* and blueberry *(Vaccinium myrtillus)* which usually increase in production under various kinds of timber harvests. Soil treatments, especially ploughing, decrease the berry vegetation although access for picking is improved.

The most valuable edible fungi, the false morel *(Gyromitra esculenta)*, thrives very well in ploughed areas and in places where the mineral soil is exposed by forestry operations. Even if the most appreciated peatland berry, Arctic raspberry *(Rubus chamaemorus)*, has suffered from widespread peatland drainage to increase timber production, one can say that timber production and berry and mushroom picking can be regarded as complementary forest uses.

Hunting and fishing at one time were primary occupations but now they are carried out only for secondary income. To some extent they have importance for domestic consumption, but primarily they are popular recreational activities. It is commonly held, that forestry operations make the composition of game species less valuable rather than decreasing their quantities. Moose *(Alces alces)* have benefitted from the increase of young pine stands but at the same time cause damage to them.

Fishing has suffered from forest and peatland drainage and from floating of timber. However, due to the much more decisive damage caused by water-power plants to salmon, these impacts have attracted little public concern.

Among the uses of forestry land, outdoor recreation is a newcomer. It has developed rapidly during recent decades (Saastamoinen 1982) and tourism is now regarded as one of the main industries in Finnish Lapland. It is also one of the few industries with significant potential for expansion.

Because tourism in Lapland is closely nature-oriented, it is understandable that conflicts arise between timber production and outdoor touristic activities. Of course, outdoor recreation also includes activities by local inhabitants which are important. The conflicts between recreation and timber harvesting and/or related silvicultural operations (such as forest ploughing) have been mainly local but may be growing. Many times the most significant outdoor recreation pressures occur in national parks or other conservation areas, or in proposed wilderness areas. Therefore it often is difficult to separate the role of outdoor recreation from the closely related arguments for nature conservation. It seems probable, especially in those timber-line forest areas with heavy recreational demand, that income from tourism may well exceed the value of timber production (Saastamoinen 1982).

On the other hand, even nature conservation and outdoor recreation may have contradictory aims. Similarly, relationships among other non-timber uses may conflict, for example, outdoor recreation with reindeer grazing (deterioration of range) or hunting with picking forest berries (the beginning of the hunting season is sometimes considered to be too early from the point of view of red whortleberry pickers).

Economics of Multiple-Use Forestry

Finally, some data are supplied to show the economic importance of different uses of forestry land (Table 1). It must be pointed out that the figures are approximate and not quite comparable; most are based on market prices except in the case of game for which mainly estimated prices are used. The value of output is that of the harvested product, e.g. in the case of timber it is the value of timber removed from forests and in the case of berries and mushrooms it is the value of natural product collected from forests and peatlands including both the marketed harvest as well as that used for domestic consumption. In addition it must be emphasized that the income from tourism is not

Table 1. Tentative figures showing the economic importance of some forestry land uses in Finnish Lapland, current prices (not deflated) and relative weights are compared to timber production (Can. $1 = 3.7 FIM, in 1982).

	Value					
	1970		1974		1980	
Forestry land use	Mill.FIM	%	Mill.FIM	%	Mill.FIM	%
Timber production	153	(100)	270	(100)	520	(2)
Hunting	2	(1)	8	(3)	10	(2)
Reindeer grazing	9	(6)	11	(4)	30	(6)
Berries and mushrooms	16	(10)	6	(2)	40	(8)
Outdoor recreation (tourism)[1]	—	(—)	100	(37)	500[2]	(96)[2]

[1] Not directly comparable with other figures. Data for 1970 not available.
[2] Estimate of 1981.

wholly comparable to the output values of other forest uses. The data for 1970 and 1974 are from my previous article (Saastamoinen 1977) where the material and calculation bases are described in more detail. The estimate of timber value for 1980 is from Kuusela (1982) and it is, as for other years, the value of timber at the forest edge.

One may add that the total turnover of the forest industry in Finnish Lapland was about FIM 2100 million (Can. $1 = 3.7 FIM in 1982) in 1980 (Kuusela 1982). The tourism income is a preliminary 1981 estimate by tourism authorities; the figure for 1980 was not available. The values of hunting and picked berries and mushrooms are my tentative estimates, and indicate only the size class. All figures express current prices and include inflationary increases.

Harvesting timber generates the major part of the economic benefit from forests of Finnish Lapland. As earlier stated, the forest industry is the backbone of the regional economy. In spite of direct comparison difficulties, one can see that tourism takes second place and its regional importance is clearly growing. Reindeer husbandry, collecting berries and mushrooms, and hunting have less economic importance than the other two, mainly because of the extensive character of these industries. However, each of them by their own weight increases economic benefits from forestry land, and together they produce a substantial output of about 10–15 per cent of the value of timber production. One can conclude that, although all forest benefits (e.g. protective functions of forests or nature conservation areas) may not be well represented by monetary measurements, the direct economic benefits from these forest uses are so real and substantial that the principle of multiple-use forestry is well supported economically.

Multiple-use of forestry land in the northern boreal zone of Finnish Lapland provides numerous examples of dual or one-way impacts and effects among different forest uses. Many conflicts occur but there are also examples of complementary effects. There are many problems but few are insurmountable. As it is customary to state, more research is needed, and it should be more systematic and better carried out. According to our experience, what is still needed, is a better balance between practical and scientifically rewarding approaches in research projects.

Acknowledgements

I would like to express my appreciation to Professor J. Harry G. Smith, Faculty of Forestry, University of British Columbia, who kindly made suggestions on the content of the paper. The paper was produced while the author was a Post-doctoral Fellow in the Faculty of Forestry at the University of British Columbia.

Literature Cited

Helle, T. and O. Saastamoinen. 1979. The winter use of food resources of semi-domestic reindeer in Northern Finland. Commun. Inst. For. Fenn. 95: 1-27.

Kuusela, K. 1982. Pohjois-Suomen metsa-ja puutalous. Metsa ja Puu, 4: 4-12.

Mattila, E. 1981. Survey of reindeer winter ranges as a part of the Finnish National Forest Inventory in 1976-1978. Commun. Inst. For. Fenn. 99: 1-74.

Norokorpi, Y. 1979. Old Norway spruce stands, amount of decay and decay-causing microbes in northern Finland. Commun. Inst. For. Fenn. 97: 1-77.

Pohtila, E. 1977. Reforestation of ploughed sites in Finnish Lapland. Commun. Inst. For. Fenn. 91: 1-98.

Saastamoinen, O. 1977. Economics of forest uses in Finnish Lapland. Silva Fenn. 11: 162-168.

Saastamoinen, O. 1978. Cutting areas as reindeer pasturage. Commun. Inst. For. Fenn. 95: 1-28.

Saastamoinen, O. 1982. Economics of multiple-use forestry in the Saariselka forest and fell area. Commun. Inst. For. Fenn. 104: 1-102.

AREAL VARIATIONS IN INCREMENT AND SPECIES AND TIMBER ASSORTMENT COMPOSITION OF THE FINNISH FORESTS

Matti Häkkilä
University of Oulu

Abstract: There are only three main tree species in the Finnish forests; pine, spruce and birch. As a result of the great extent of the country from south to north, from 60° to 70° N latitude, considerable areal differences are found between different parts of Finland in volume of growing stock, species and timber assortment composition, increment per unit area, etc. An east-west axis of variation is also noted, in that the forests in the western part of the country are in a less advantageous position with respect to certain growing stock properties than those of central and eastern Finland. This is chiefly due to the predominantly flat terrain and the resulting high incidence of paludification, and also in part to historical factors connected with forest utilization and its consequences.

The aim of the study was to achieve areal typologies of timber yield in Finland. The yield is a result of the amount of the increment and its species and timber assortment composition. Typologies are made based on the results of the 6th National Forest Inventory and the structures of the 'taxable cubic metre' by communes or commune assemblages. The areal types were constructed by means of triangular graphs, factor and grouping analyses and cartographical methods. On maps these give fairly logically organized divisions of areal timber yield types, in which a very obvious regional pattern is revealed.

Finland is one of the most heavily forested countries in the world, with 64.6 per cent of its total land area of 305 500 square kilometres taken up by actual forest land and a further 11.7 per cent by poorly productive forest land and 11.0 per cent by waste land. Thus, over 87 per cent of the land area falls within the scope of forestry operations to some extent. The country has approximately 4.2 hectares of actual forest land per head of population.

Finland lies almost entirely within the boreal coniferous forest zone, with forests of this kind extending across its relatively flat surface in a zone some 900–1000 km in width from the south coast up into the north of Lapland. The dominant trees are only three in number, pine *(Pinus silvestris)*, spruce *(Picea abies)* and birch *(Betula pubescens* and *B. verrucosa)*. The greatest proportion of forest land, 58.2 per cent, is dominated by pine, with spruce-dominated forests accounting for 30.1 per cent and predominantly deciduous forests for 7.7 per cent. The remaining 4.0 per cent of forest land consists of treeless areas awaiting reforestation. The proportions of the species are less striking when examined in terms of growing stock, however, with 45.1 per cent of the total stock consisting of pine, 37.4 per cent of spruce and 17.5 per cent of deciduous species (Yearbook of Forest Statistics 1980).

Since Finland is also an extensive country in a north-south direction, stretching from 60° to 70°N latitude and with a third of its land area beyond the Arctic Circle, it is obvious that the conditions for forestry production will vary widely from one region to another. The most significant factor affecting these conditions is the decline in temperatures from south to north (Koivisto 1970; Heikurainen and Seppälä 1973; Kuusela 1977), which is fairly regular due to the low-lying nature of the terrain by comparison with the Scandinavian countries, for example. Even so, the isotherms do not run in an east-west direction, but instead slope from southwest to northeast, reflecting the country's intermediate position between the more oceanic climates in the west and the more continental climates in the east. The decline in mean annual temperatures with latitude within Finland is from +5.5°C on the islands off the southwest coast to 0°C around the Arctic Circle and −2°C in the extreme north of Lapland.

The length of the growing season and the effective temperature sum are nevertheless more important factors as far as the vegetation is concerned. The thermal growing season varies in length from 180 days on the Åland Islands to 100 days in Northern Lapland, and the temperature sum for this same period ranges from over 1300° in the southwestern and southeastern corners of the country to only 500° at Utsjoki in Lapland and only 400° in the 'arm' of Kilpisjärvi.

Other sources of areal differences in timber production involve soils, topographical factors and the structure of the growing stock (Alho 1972). Most of these also give advantage to the southern parts of the country, although some east-west variation is noted, with the eastern regions generally more favoured than the west, the Gulf of Bothnia coast in particular being extremely flat, with a high incidence of mires. Similarly, the history of forest exploitation places the coastal region in a disadvantageous position, with the greater density of settlement near the coast and the high proportions of agricultural land. The inland areas, with their more variable topography are better placed with respect to all the above factors, and this is reflected in both the amounts of forest land and its quality.

The aim of this work was to examine the areal differences existing in Finland with respect to the distribution of forests, annual increment per hectare, tree species relations and timber assortments. These aspects are examined both individually and in terms of an areal typology which attempts to provide an overall impression of the areal organization within the forests of Finland from the point of view of the factors most relevant to timber production. Thus, the work is to a great extent an exercise in methodology, an experiment in the application of the areal typology approach common elsewhere in geographical research to the question of the areal distribution of forest resource data.

Materials and Methods

The material used here is based for the most part on the detailed areal data provided in the Sixth National Forest Inventory. Certain values are calculated for relatively small areal units, normally for taxation purposes, e.g. the structure of the taxable cubic metre, which is indicated for almost 160 areas. This unit is theoretical and represents a cubic metre of stemwood distributed by timber assortment according to a felling structure conforming to the annual increment in the local forests, and may thus be used in practice to denote the local timber assortment structure with reasonable accuracy. The same detailed data also permit calculation of the species composition of the forests for similar areal units. There are also some important measures such as the percentage of forest land and the mean increment on forest land which are even available at the commune level. The areal units employed here are depicted in Figure 1.

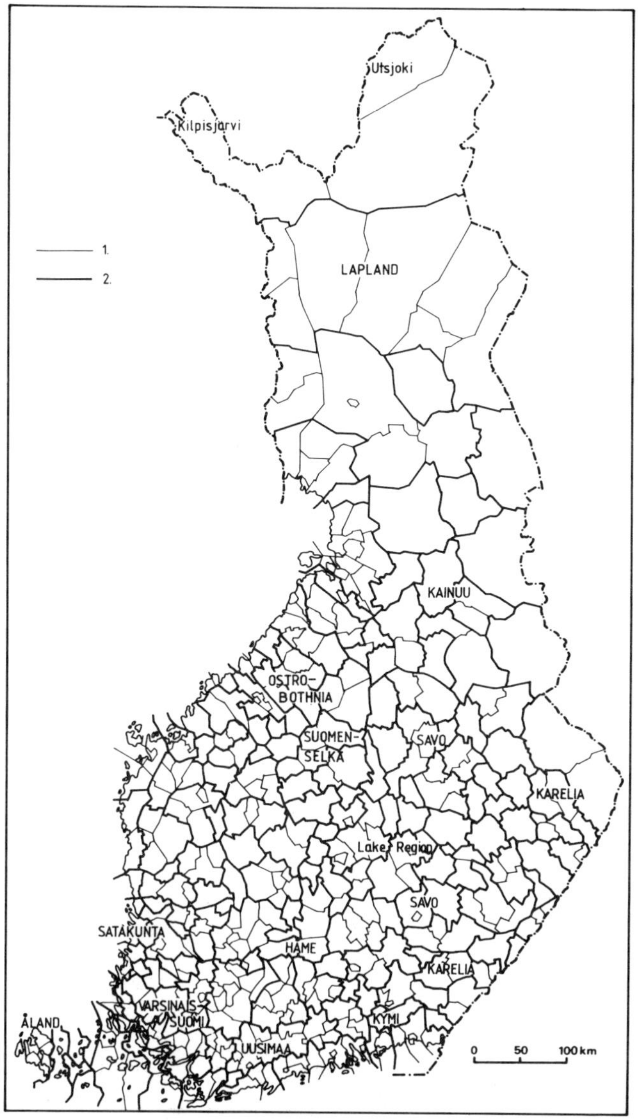

Figure 1. The communes and commune assemblages. (1) communal boundary, (2) commune assemblage boundary.

The areal distributions of the contributory factors are presented first, after which triangular graphs are constructed to obtain areal typologies for species composition and timber assortments. Attempts are then made to apply factor and group analysis techniques to achieve an overall synthesis including incidence of forest land, forest increment per hectare and the species composition and assortment structure of the timber produced.

Incidence of Forest Land

As mentioned above, practically two-thirds of the surface area of Finland is composed of forest land, i.e. land on which a forest of full density containing species appropriate to the growing site will produce an annual increment of at least one cubic metre (solid measure) of stemwood with bark per hectare given a rotation interval of 100 years. The maps produced from the National Forest Inventory (Figure 2, see Atlas of Finland 1976) show the proportion of the land area occupied by such forest land to vary from 5 per cent to over 90 per cent. The range is somewhat more restricted, however, when the data are examined at the commune level, since the communes of northern Finland in particular are so large that they normally encompass both areas with a high incidence of forest land and without trees.

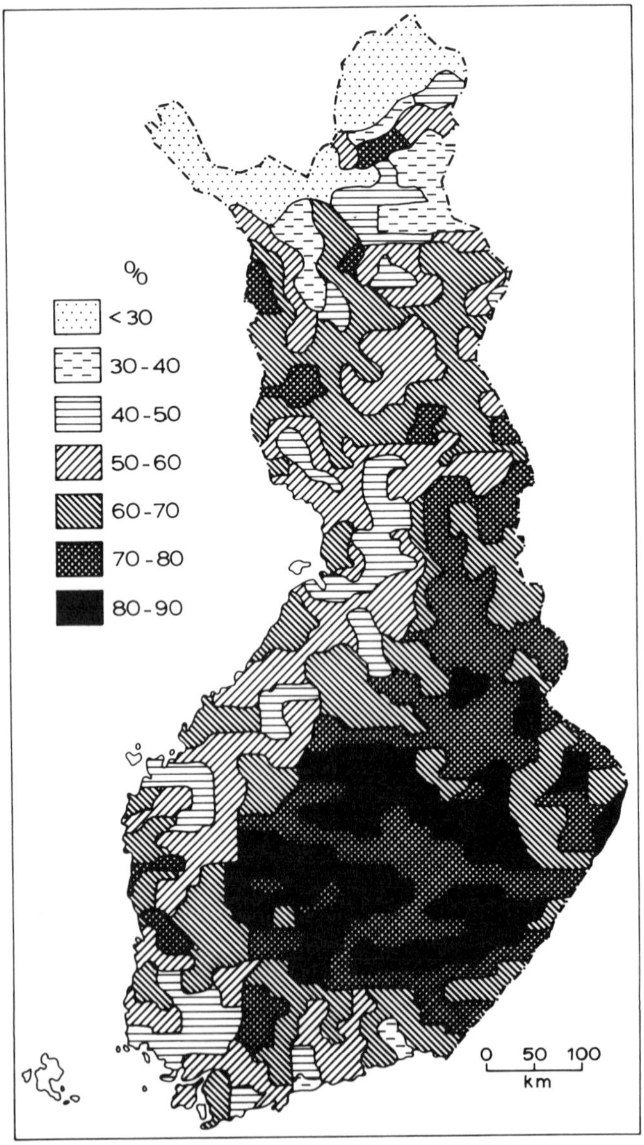

Figure 2. Proportion of total land area occupied by forest land.

The incidence of forest land as a proportion of total land area is lowest in the extreme north of Lapland, on account of the extent of the areas remaining treeless for climatic reasons. In Ostrobothnia the proportion is low on account of the areas of sparsely forested mire and the density of agriculturally-based settlement. Similarly, the incidence of forest land in the southern part of the country is reduced by the presence of urban settlements, industry and a dense road network and the frequent occurrence of bedrock outcrops, especially on the southwestern coasts, the off-shore islands and the Åland Islands. The Lake Region, with its variable topography and favourable climate, is the most heavily forested part of the country.

The area of forest land is steadily increasing, principally due to the ditching of mires and fertilization of the land. Such measures are gradually converting areas of poorly productive forest and waste land into forest land, although some increase is also due to the reforestation of unwanted fields and pasture land. It has been estimated that by the year 2000, forest land will account for as much as 70 to 75 per cent of the total land area of Finland (Kuusela 1976).

Mean Increment Growth on Forest Land

The mean increment per hectare of the growing stock in Finland decreases gradually from the south and southwest of the country to the north and northeast (Koivisto 1970; Häkkilä 1977; Kuusela 1977). Thus, while it exceeds 5.5 m^3 (solid measure) in numerous communes in southern Finland, there are some in Lapland in which it remains below 1 m /ha/year (Figure 3). The mean increment per hectare is particularly low in relation to latitude in the Suomenselkä area (Figure 1), where soils are known to be poor.

A number of studies (Koivisto 1970; Heikurainen and Seppälä 1973) show the length of the growing season and the temperature sum to be the chief factors explaining areal variations in mean increment, even though other factors are also implicated to some extent. Thus topography, soils and the structure of the growing stock can be seen to cause deviations in the general climatically determined zones.

Species Composition

Although the Finnish forests possess only a small number of tree species, the species relations can vary considerably from one area to another. Since there are really only two major species, pine and spruce, their distribution maps are broadly the inverse one of the other. The areas which are pine dominant normally contain little spruce and vice versa. The high incidence of pine forests in northern Finland, Suomenselkä and Karelia (Figure 4) reflects the prevalence of dry heath forests and poor pine bogs in these regions. Comparable results for the southwest coast and its islands are due to the common occurrence of bedrock outcrops. The highest proportions of spruce forest (Figure 5) are correspondingly found in southwestern Finland and southern Ostrobothnia, where there are extensive amounts of fresh, rich forests and spruce mires. Spruce is similarly prevalent in the hilly regions of eastern Finland with their moist climate and fresh forest biotopes (Kalliola 1973). The modern trend in forest management has been to replace poorly productive spruce forest on less fertile land with pine forest, and partly for this reason the proportion of pine-dominated forests is slowly increasing.

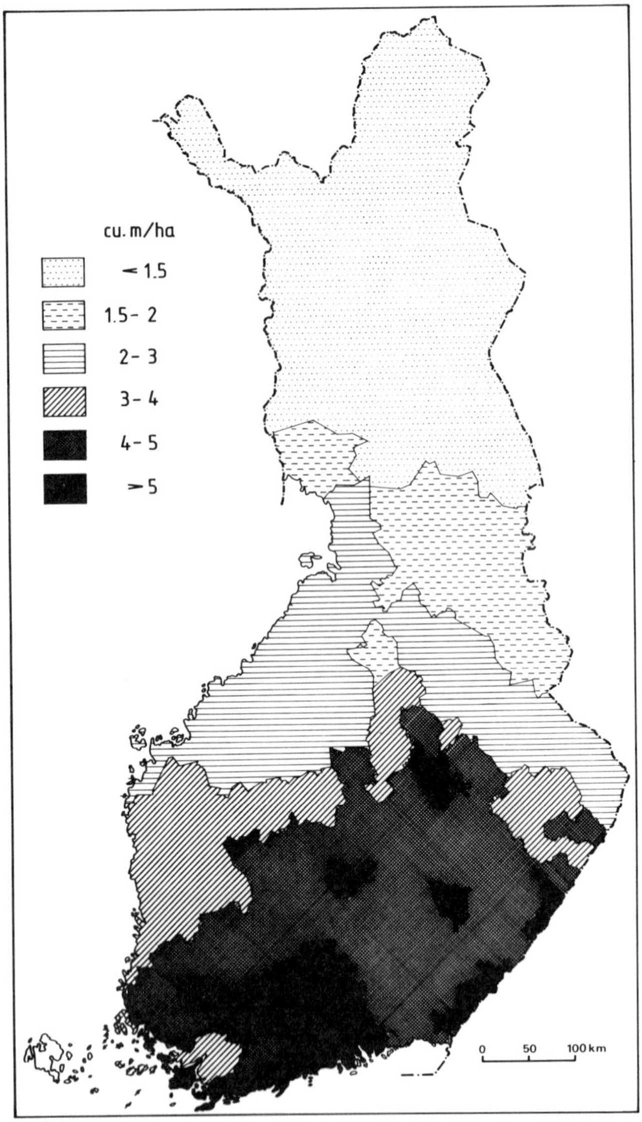

Figure 3. Mean annual increment per hectare on forest land.

The proportion of predominantly deciduous forests in Finland has declined markedly since the 1950s. Such forests are now most common in the eastern parts of southern Finland, the last areas in which 'slash and burn' crop cultivation was practised, and in northern Ostrobothnia and the Peräpohjola district (Figure 6). The extensive programme of mire drainage undertaken in the latter regions has caused an increase in the proportion of birch within the growing stock.

A summary of the situation regarding the species composition of the growing stock in the Finnish forests was obtained by constructing an areal typology based on a triangular graph

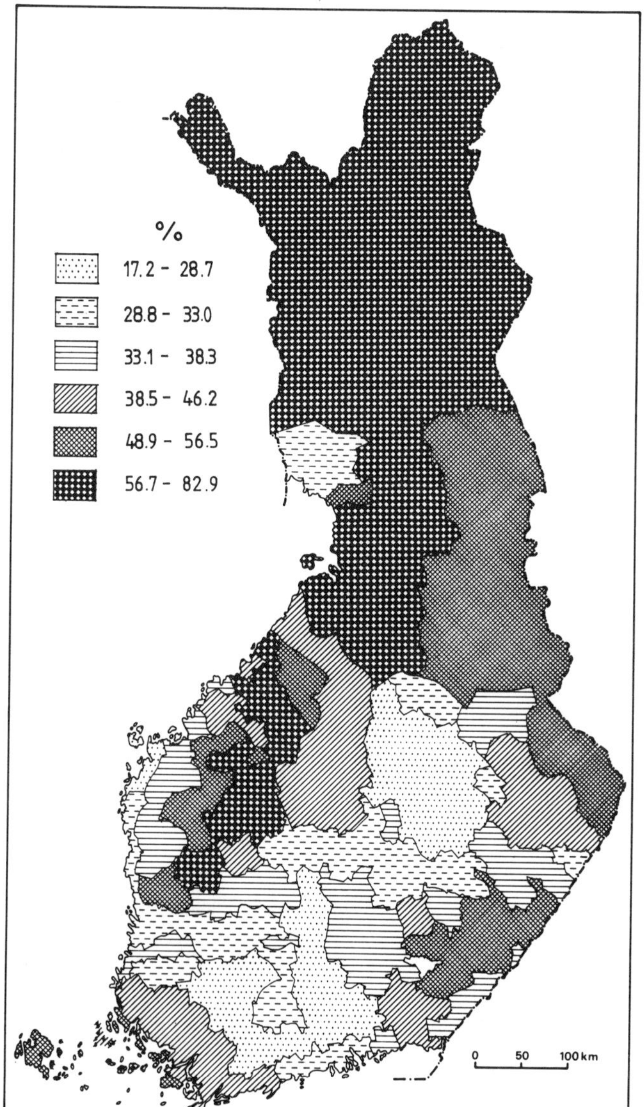

Figure 4. Proportion of pine in the cut conforming to the annual increment (according to the structure of the taxable cubic metre by commune assemblages).

(Figure 7). With only three species to deal with, such a typology is relatively easy to construct once one knows their percentage distributions within the areal units.

The usual way of obtaining a typology of this kind would be to divide the area of the triangle into six sectors on the basis of the mean species composition of the growing stock for the whole country (Häkkilä 1977). In the present case, however, such a large number of areal units fall close to the mean value that one would be justified in setting aside a seventh category for these alone (Talman 1979). The starting point here consists of the

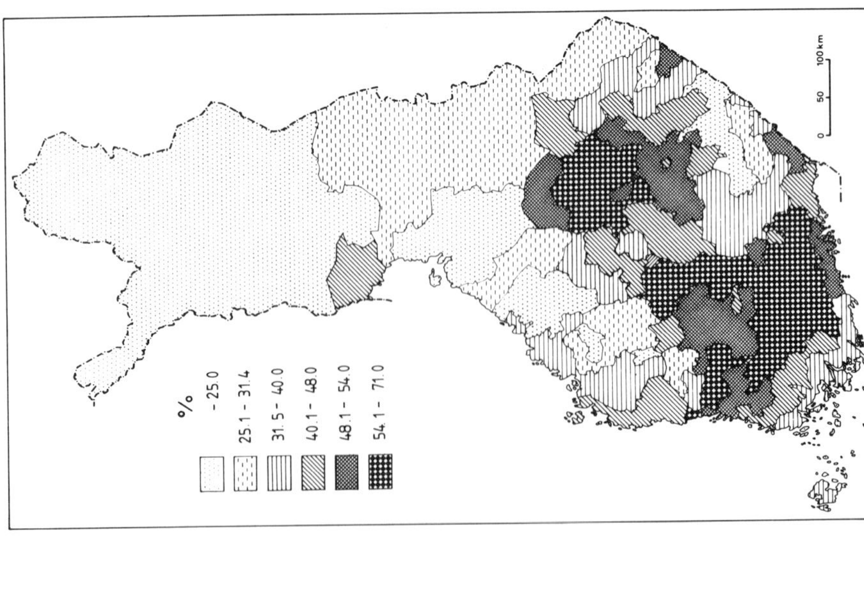

Figure 5. Proportion of spruce in the cut conforming to the annual increment (according to the structure of the taxable cubic metre by commune assemblages).

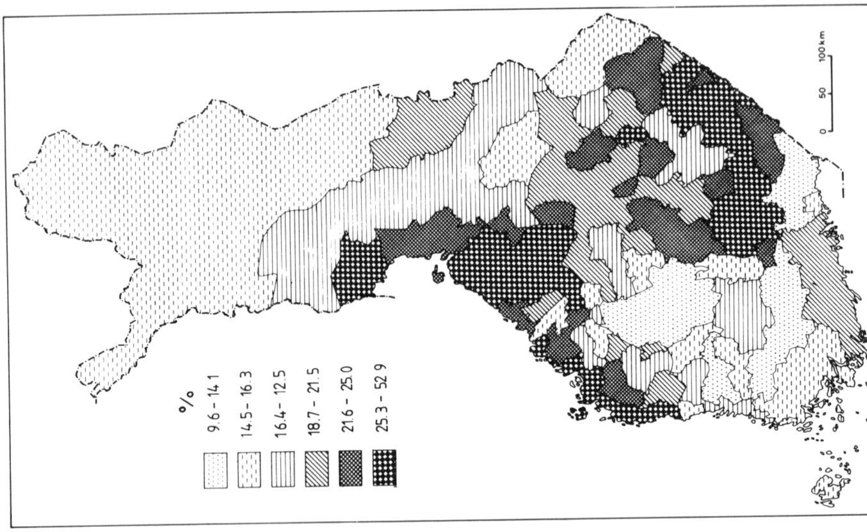

Figure 6. Proportion of hardwoods in the cut conforming to the annual increment (according to the structure of the taxable cubic metre by commune assemblages).

mean proportions of pine, spruce and deciduous species for the total set of commune assemblages, increased in each case by an amount equivalent to half the standard deviation. These values, when drawn on the triangular graph, produce a set of divisions representing commune assemblages whose tree species composition is dominated by each of these species, together with intermediate types (Figure 7).

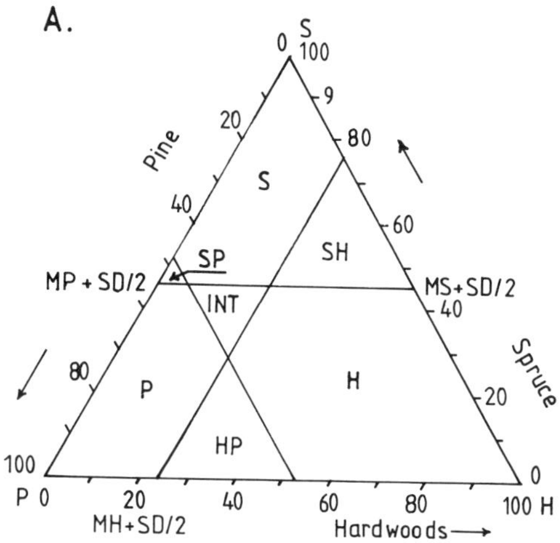

Figure 7. A: Triangular graph for the typology of the communes by species composition of the forests (in cubic metres of timber). MP, MS, MH – means for pine (P), spruce (S) and hardwoods (H), SD – standard deviation. Types: S – spruce type, SH – spruce-hardwoods type, H – hardwoods type, HP – hardwoods-pine type, P – pine type, SP – spruce-pine type, INT – intermediate type.

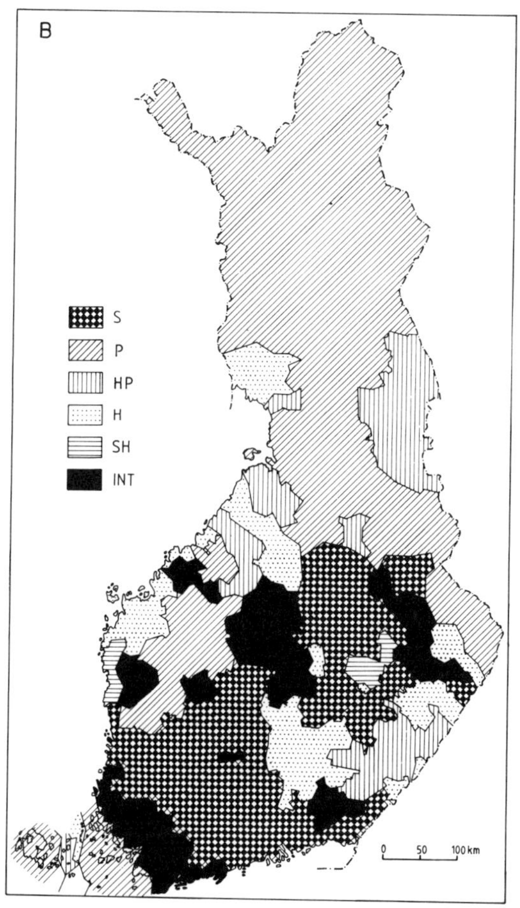

B: Distribution of the species composition types.

We have now obtained one single diagram which depicts the distribution of the tree species in terms of areal types defined according to mathematically specified criteria. The resulting map has much the same areal features as were described above. It should be noted that there is one theoretical type (type SP) defined by the triangular graph which in fact does not occur at all. The type comprising distributions lying close to the mean for the whole country, type INT, is distributed in an irregular manner throughout the southern half of the country.

Timber Assortment Structure

The timber assortments recognized here consist of sawlogs (as raw material for sawmills and plywood factories), pulpwood and logging waste. The areal distribution of the proportions of these within a cut equivalent to the annual increment in the growing stock will be examined here.

The areal differences in the timber assortment structure of the forests correspond relatively closely to those in the diameter of the timber, a factor which is in turn dependent on the fertility of the forest land and the age structure of the growing stock. The stock reaches its greatest diameter, and thus its greatest proportions of sawlogs, in the Uusimaa and Häme areas (Figure 8), largely on account of the fertility of the soils. There are also some more remote parts of the country, e.g. northeastern Ostrobothnia, where large amounts of extremely old forest are still standing, so that diameters well in excess of the national mean are found even in such areas of slow growth. The smallest mean stem size is recorded in Ostrobothnia, due to the high incidence of mires and the long tradition of selective felling (Alho 1968, 1975; Kuusela 1976, 1977).

The map depicting the proportion of pulpwood (Figure 9) is of course, practically the inverse of that for sawlogs. Along the Gulf of Bothnia coast in particular, but also in a narrow zone along the Gulf of Finland, over a half of the growing stock does not meet the dimensions required for sawlogs. The higher the proportion of sapling and pole-stage forests, and the higher the proportion of deciduous species, particularly alder and *Betula pubescens,* the fewer sawlogs the forests will contain on average and the greater the quantity of logging waste (Figure 10; Kuusela 1977). The proportion of the latter is also affected by geographical factors connected with settlement and transportation, since denser settlements, with their increased demand for wood for home use and other purposes will tend to promote more efficient logging, as also will a dense road network and shorter transportation distances. At the same time the greater natural wastage in excessively old forests which have been left unexploited for a long time, and the loss of timber through rotting and other defects, will increase the proportion of unused wood in such cases (Alho 1975). Consequently, the proportion of logging waste increases for the most part from south to north across the country. The fact that the proportion of waste is also higher in the west than in the east may be attributed largely to the incidence of mires in the former area. Areas which have been intensively drained tend to show a higher wastage than usual in connection with felling and the management of forests with a low growing stock. Calculated in terms of the commune assemblages used here (Figure 1), the lowest proportion of logging waste for any area in Finland is 4 per cent, while in some communes in Lapland it accounts for as much as 24 per cent of the cut.

Since the timber assortments can also be arranged in three groups, it is easy to obtain an areal typology for these on the same principle as for the species composition, employing a

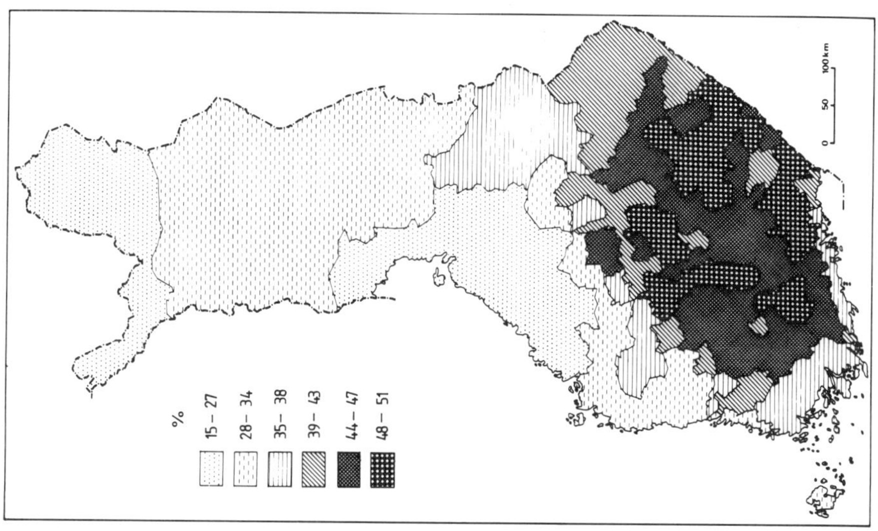

Figure 8. Proportion of sawlogs in the cut conforming to the annual increment (according to the structure of the taxable cubic metre by commune assemblages).

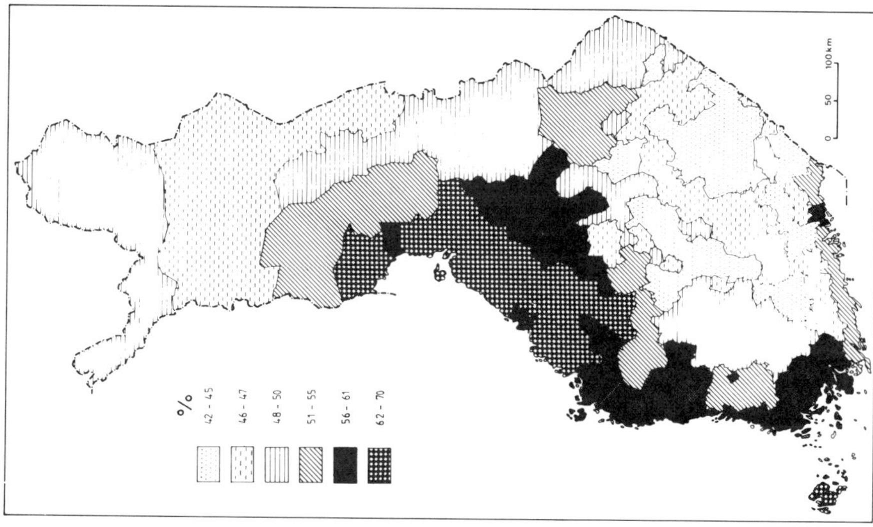

Figure 9. Proportion of pulpwood in the cut conforming to the annual increment (according to the structure of the taxable cubic metre by commune assemblages).

triangular graph. Here only five out of the seven theoretically possible categories are actually realized (Figure 11). The categories recognized show a fairly clear-cut areal distribution.

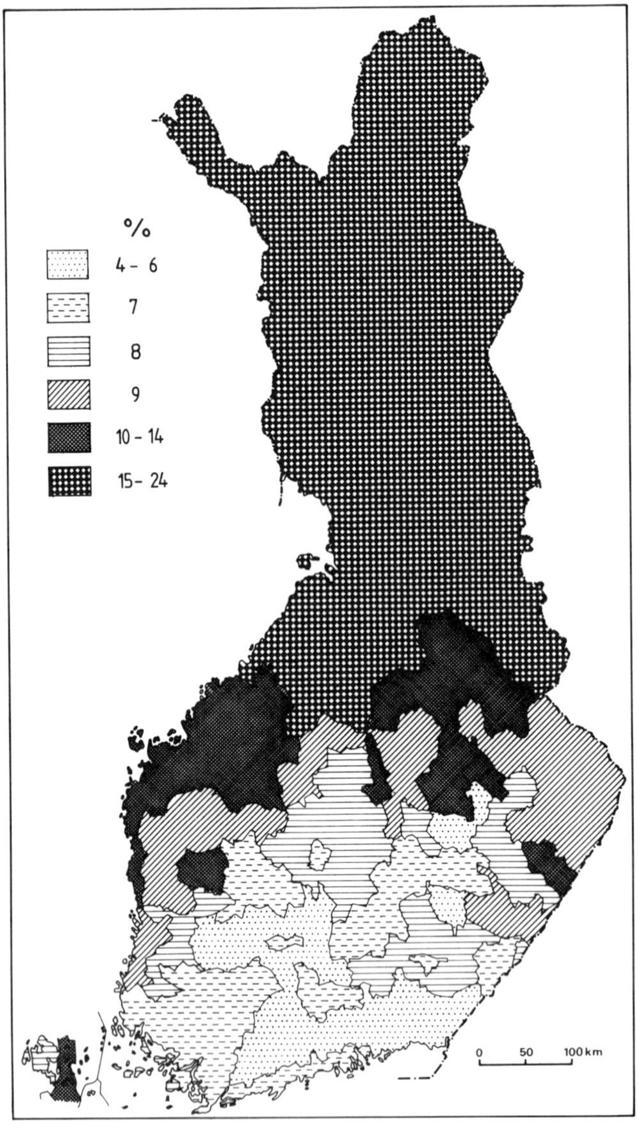

Figure 10. Proportion of logging waste in the cut conforming to the annual increment (according to the structure of the taxable cubic metre by commune assemblages).

Types of Timber Yield and Their Areal Patterns

The above is an areal examination of the various individual factors used here to account for variations in timber yields, namely the incidence of forest land as a proportion of total land area, mean increment per hectare on forest land and the species and timber assortment composition of a volume of timber equivalent to the annual increment. We shall now attempt to construct an areal synthesis of these factors, i.e. to construct areal types for timber yields. Since there are a number of factors involved, we must first

A.

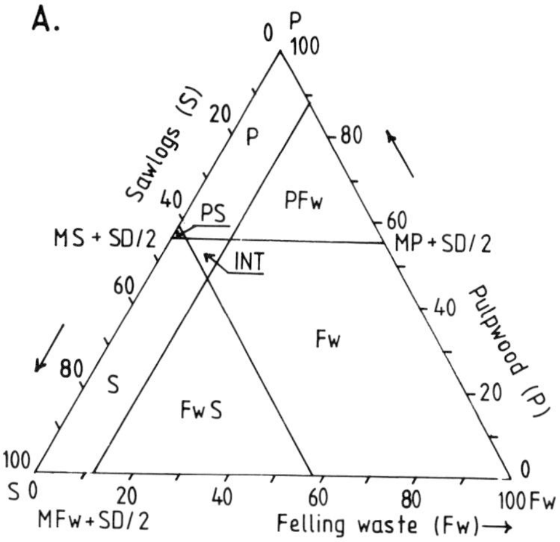

Figure 11. A: Triangular graph for the typology of the commune assemblages by timber assortment structure in the mean cut. MS, MP, MFw – means for sawlogs (S), pulpwood (P) and felling waste (Fw), SD – standard deviation.
Types: P – pulpwood type, PFw – pulpwood-felling waste type, Fw – felling waste type, FwS – felling waste-sawlogs type, S – sawlogs type, PS – pulpwood-sawlogs type, INT – intermediate type.

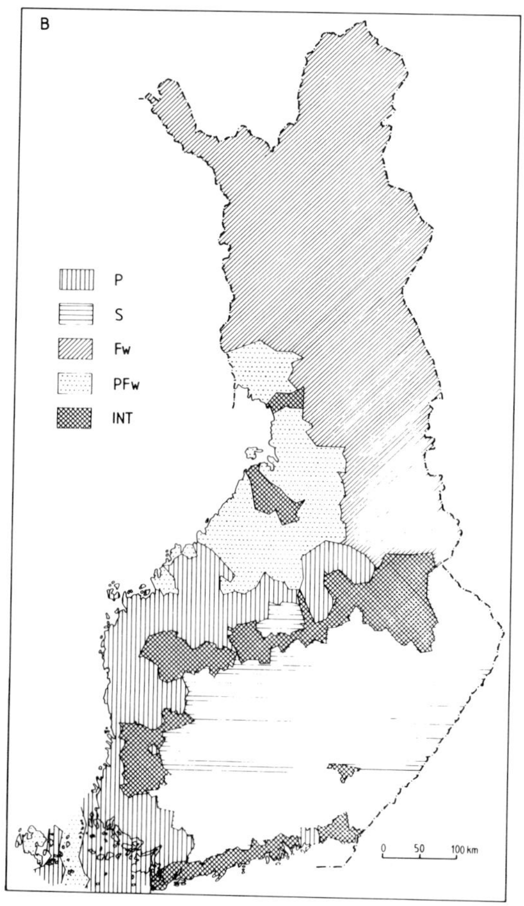

B: Distribution of the timber assortment composition types.

compress the information contained in these into a more manageable form. One possibility for doing this is offered by factor analysis, the aim of which is to gather the information offered by a set of variables into a small number of basic dimensions which depict as accurately as possible the co-variance existing between all original variables. The set of basic dimensions obtained in this way and the factor score map constructed from these may be regarded in a sense as a partial solution to eventual areal synthesis which frequently serves as the true aim in geographical research.

The areal synthesis itself is achieved subsequent to the factor analysis by assembling the areal units into suitable groups on the basis of their factor scores and depicting the results cartographically. The use of factor scores for this purpose is justified on the grounds that the grouping analysis programme developed by the Helsinki University Computing Centre (HYLPS/GA), which is that most commonly used in Finland, requires that the variables should not correlate one with another. This means that in practice we are more or less obliged to use factor scores, since variables of any other kind will almost inevitably show autocorrelation to some extent (Sänkiaho 1974).

The variable selected to form the basis for constructing the areal synthesis were:

X_2 forest land as a proportion of total land area,
X_3 mean increment on forest land,
X_4 proportion of pine in the cut conforming to the annual increment,
X_5 proportion of spruce in the cut conforming to the annual increment,
X_6 proportion of deciduous species in the cut conforming to the annual increment,
X_7 proportion of sawlogs in the cut conforming to the annual increment,
X_8 proportion of pulpwood in the cut conforming to the annual increment,
X_9 proportion of logging waste in the cut conforming to the annual increment.

Factor analysis is based on the correlations existing between the selected variables. The correlation matrix for the present analysis is depicted in Table 1. In its broad outlines this matrix is logical in form, the only surprising features are the powerful negative correlations between the proportion of pine in the growing stock and the mean increment and proportion of sawlogs. This may be due to the fact that the areas where pine is predominant in the forests, particularly Ostrobothnia, Suomenselkä and northern Finland, are precisely those in which growth is relatively slow and the proportion of sawlogs frequently low. The opposite situation prevails in those areas where spruce is dominant. This state of affairs nevertheless does not arise from the nature of these species as such, but from the growth factors involved, since pine will also grow well when given favourable sites.

Factorization was performed in three ways, yielding two, three and four-factor solutions. The last of these appears to be the best for the subsequent steps in the analysis (Table 2). Factor I is dominant in the principal axis solution, possessing high weightings on all of the variables except for the proportion of deciduous species (X_6). As has become apparent in a number of previous studies (Alho 1975; Kuusela 1977; Häkkilä 1977, 1980, 1982; Salminen 1981) many of the measures showing the quality of forests in Finland prove to be very much alike in their areal distribution. Consequently these consistently high weightings on the same factor would seem only natural. The second factor features relatively high weightings on only a few variables, while the weightings on the third and fourth factors are fairly low throughout. The first factor alone explains 55.7 per cent of the total variance, the second factor 12.3 per cent and the third and fourth only 7.5 per cent and 6.9 per cent respectively.

Table 1. Matrix of correlation coefficients. Those which are at least significant at the 1 per cent risk level are underlined

	Variables							
Variables	2	3	4	5	6	7	8	9
2	1000							
3	319	1000						
4	−435	−643	1000					
5	368	607	−927	1000				
6	88	−3	−18	−358	1000			
7	558	785	−544	578	−179	1000		
8	−590	−560	371	−399	138	−918	1000	
9	−300	−863	626	−656	177	−762	444	1000

Table 2. Four factor matrix, principal axis model

	Factor loadings				Communality
	I	II	III	IV	estimates used
2	−0.546	0.310	0.404	0.105	0.590
3	−0.848	−0.045	−0.145	−0.382	0.836
4	0.804	0.423	−0.369	−0.007	0.927
5	−0.820	−0.465	0.046	0.315	0.927
6	0.158	0.119	0.422	−0.439	0.358
7	−0.914	0.342	−0.139	0.010	0.918
8	0.750	−0.578	0.005	−0.184	0.918
9	0.834	0.164	0.281	0.257	0.863
Eigen-values	4.456	0.986	0.598	0.549	

Varimax rotation can be employed to improve the interpretability of the model, being aimed at generating a large number of high and low weightings. The weightings produce high eigenvalues regardless of sign, and in turn the higher the eigenvalue, the more important is the factor for the structure as a whole. The present four-factor solution yielded a relatively clear interpretable set of four basic parameters once Varimax rotation had been applied (Table 3).

Varimax rotation causes the explanation of variance for the first factor to drop to 22.8 per cent, with high weightings remaining on the variables denoting the proportions of the two coniferous species in the growing stock, X_4 and X_5, although obviously with reversed signs. Thus this first factor may be referred to as the 'coniferous forest factor'.

The explanation of variance achieved by the second factor doubles upon Varimax rotation, now obtaining high weightings on variables X_2, proportion of forest land, X_7, proportion of sawlogs, and X_8, proportion of pulpwood. The last two again have opposing signs. This second factor can be termed the 'forest and timber reserves' factor.

Table 3. Four factor matrix, Varimax-rotation

	Factor loadings				Communalities
	I	II	III	IV	
2	−0.311	0.664	0.165	−0.063	0.568
3	−0.310	0.293	0.039	−0.839	0.888
4	0.881	−0.225	−0.070	0.361	0.962
5	−0.831	0.207	−0.375	−0.341	0.990
6	0.058	−0.014	0.635	0.043	0.409
7	−0.192	0.728	−0.188	−0.609	0.972
8	0.061	−0.895	0.162	0.317	0.931
9	0.342	−0.197	0.156	0.829	0.867
Eigen-values	1.820	1.989	0.664	2.115	6.588

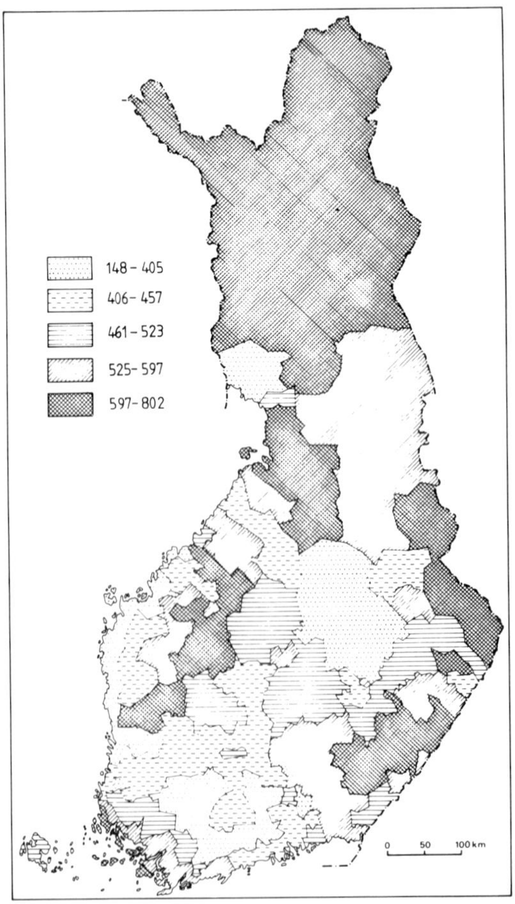

Figure 12. Scores on Factor I, coniferous forest factor.

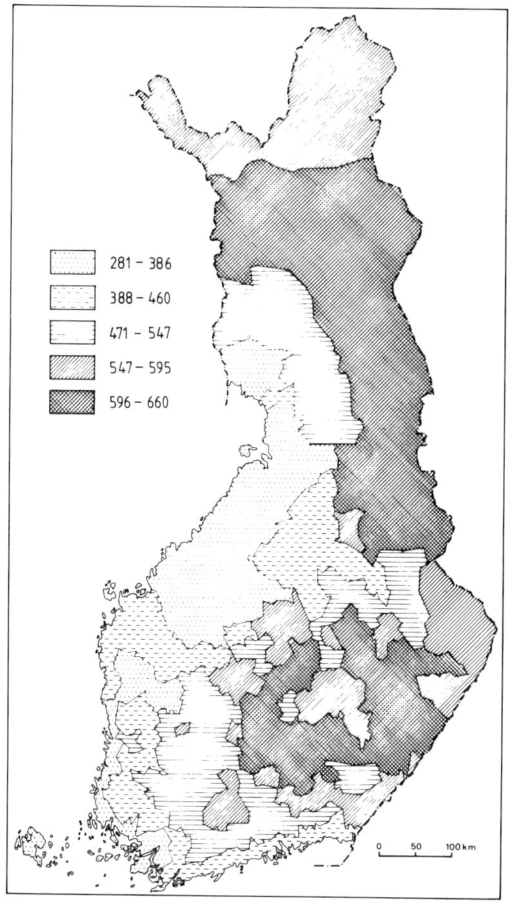

Figure 13. Scores on Factor II, forest and timber reserves.

The third factor still explains less than 10 per cent of the total variance even after Varimax rotation, but its character as a 'deciduous forest' factor emerges more clearly. It does not gain any high weightings other than on variable X_6. The fourth factor, on the other hand, achieves almost four times its previous explanatory power as a result of Varimax rotation, now accounting for 26.4 per cent of total variance, and takes the form of a 'timber yield' factor, with high weightings on variables X_3, mean increment, X_7, proportion of sawlogs, and X_9, proportion of logging waste. This factor is indicative both of the volume of the yield and also its nature. The four factors together explain 82.4 per cent of the total variance.

A useful approach in areal studies is to define estimates for the factors with respect of each areal unit. The theoretical values, or factor scores, for each factor with respect of the various areal units are calculated here by the 'Lederman' method (Harman 1960). The weighted factor scores for the areal units are standardized so that the mean of the scores is 500 and their standard deviation 100. The results are presented on quintile maps in Figures 12 to 15.

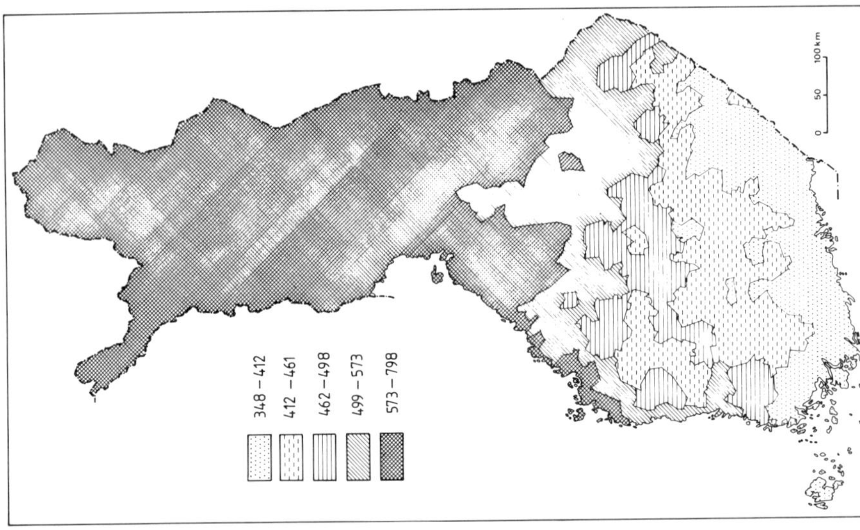

Figure 15. Scores on Factor IV, timber yield.

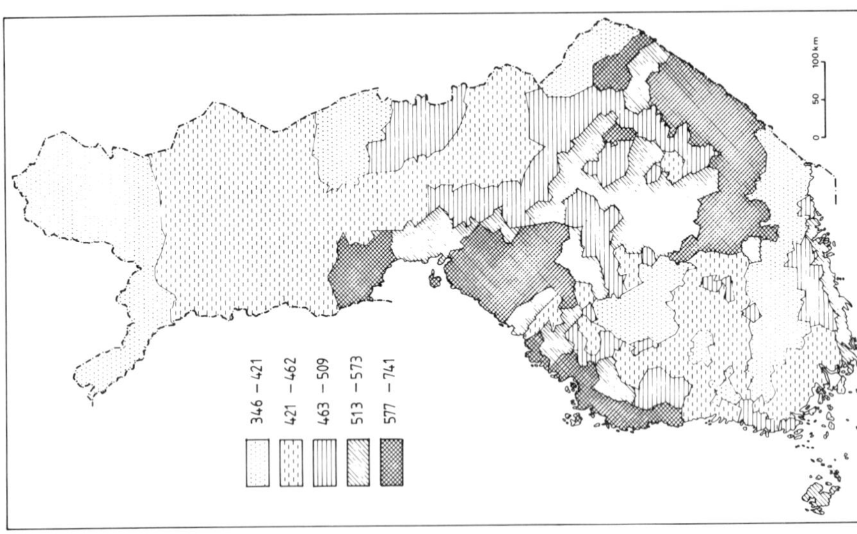

Figure 14. Scores on Factor III, deciduous forest factor.

The map depicting the factor scores for the first factor that associated with coniferous forests (Figure 12) may be interpreted as implying a high proportion of pine in the forests where the score is high, and correspondingly a low proportion of spruce. The map presents a similar areal aspect as that depicting the species composition of the forests.

In the case of the second factor, forest and timber reserves (Figure 13) higher factor scores imply greater forest reserves. This map serves well to bring out the variations existing in a west-east direction. The distribution of scores for the deciduous forest factor (Figure 14) is naturally the same as that observed in the map depicting the proportion of hardwood in the cut conforming to the annual increment (Figure 6), with higher scores representing greater proportions of these species.

In the last of the factor score maps (Figure 15), which depicts the areal distribution of scores for the 'timber yield' factor, the lighter shadings indicate more favourable yields.

Conditions are shown to worsen in a relatively strictly zonational manner from the south of the country towards the north.

The factors contributing to timber yields have thus been compressed into four basic dimensions, enabling a set of regional types to be defined by applying a grouping analysis to the factor scores for them. The grouping analysis procedure (HYLPS/GA) operates on the principle of dividing the data units into a preselected number of groups in such a way as to minimize the variance of the variables within each group. The measure of homogeneity within a group is its internal scaled sum of squares. As the grouping process advances, each areal unit is assigned to that group in which it deviates least from the mean values for the variables, after which the program calculates new mean values for that group. The internal homogeneity of the groups thus depends on the number of groups which the operator chooses to form (Rikkinen 1971; Talman 1980; Sänkiaho 1974).

The present grouping analysis was performed for prearranged numbers of groups ranging from three to eight and a map produced for each solution. Even the three-group solution serves to bring out the basic distinctions found to be present in almost every analysis of forests or forestry in Finland. The northern part of the country, i.e. Lapland and Kainuu, emerges as one region and the southern part falls into a coastal and an inland region. Although the areal picture becomes more precise as the grouping procedure advances, these basic outlines are still clearly visible even in the seven-group solution, chosen as representing the eventual areal synthesis (Figure 16). This latter solution divides the best forestry regions in the country among three types, types 2, 3 and 6, while the coastal zone is chiefly divided between two types, types 1 and 5, and the northern part of the country in a very logical fashion also into two zones, types 4 and 7. The numbers of areal units and means and standard deviations for the factor scores are given for each of the types in Table 4.

Table 4. Mean factor scores (M) and their standard deviations (SD)

Type	No. of unit areas		Factors I	II	III	IV
1	18	M	462	376	611	611
		SD	100	35	47	57
2	15	M	608	610	634	404
		SD	65	40	56	31
3	33	M	478	580	510	459
		SD	49	35	59	45
4	5	M	633	624	413	774
		SD	92	36	16	28
5	28	M	587	379	471	460
		SD	63	46	51	70
6	44	M	404	496	442	477
		SD	49	61	54	53
7	10	M	589	590	448	658
		SD	28	26	23	57

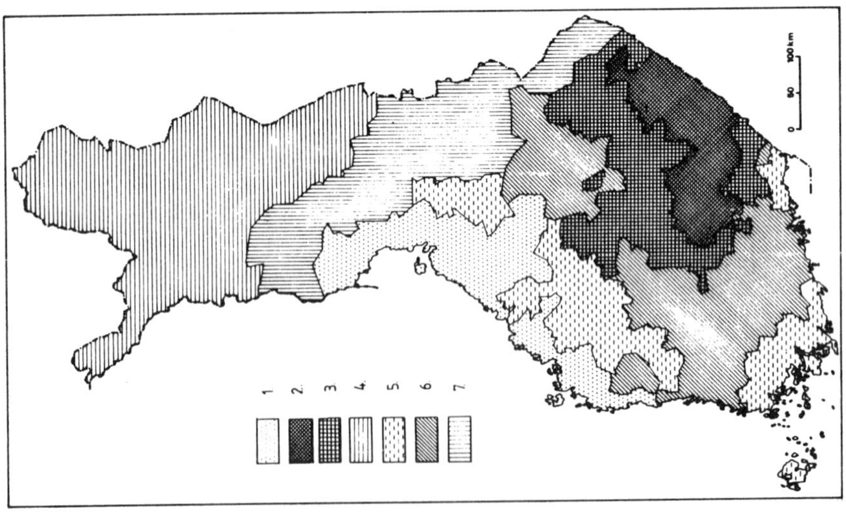

Figure 16. Distribution of timber yield types.

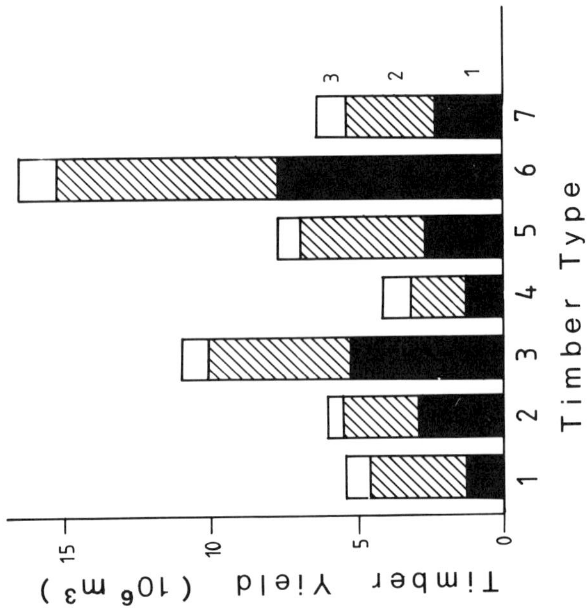

Figure 17. Total yield in different types and its distribution by timber types.

The properties and significance for forestry of the various types are shown in Figure 17, which indicates the total annual increment in the forests of each of the type areas and the distribution by timber type. The diagram illustrates well the importance of the inland areas of southern Finland for forestry, since types 2, 3 and 6 together account for almost 60 per cent of the total annual increment of 57.4 million m^3, solid measure, and have almost half of their timber in the largest diameter class, thus representing the most valuable assortment, sawlogs.

The types represented in the coastal area are of far less significance, the proportion of sawlogs in type 1, located in Ostrobothnia, is particularly low, since even types 4 and 7, which apply to areas in the extreme north of the country, represent very much better forests in terms of tree diameter. Although forest growth is slow in the north, this part of the country possesses vast areas of forest and the mean diameter is relatively high, especially in the more remote districts, so that northern Finland remains of substantial economic importance for its forest resources.

Literature Cited

Alho, P. 1968. Pohjois-Pohjanmaan metsien käytön kehitys ja sen vaikutus metsien tilaan. (Summary: Utilization of Forests in North Ostrobothnia and its Effect on their Condition.) Acta Forestalia Fennica, 89: 1-216.

Alho, P. 1972. Pohjois-Pohjanmaan metsien tuotantobiologiset edellytykset. (Summary: Qualifications of the North Ostrobotnian forests from the viewpoint of production biology.) Nordia Tiedonantoja, 2: 1-100.

Alho, P. 1975. Metsien tuoton alueellisista eroista Suomessa. (Summary: Regional differences in forest returns within Finland.) Acta Forestalia Fennica, 148: 1-25.

Atlas of Finland. 1976. Forestry. National Board of Survey and the Geographical Society of Finland. Helsinki, Finland.

Häkkilä, M. 1977. Geographical aspects of forest returns on Finnish farms. Fennia, 152: 1-86.

Häkkilä, M. 1980. Areal concentrations in certain aspects of Finnish forestry. Fennia, 158: 15-26.

Häkkilä, M. 1982. Background factors affecting areal differences in stumpage prices for spruce pulpwood in Finland. Nordia, 16: 185-226.

Harman, H. 1960. Modern factor analysis. University of Chicago Press. Chicago, Illinois.

Hiekurainen, L. and K. Seppälä. 1973. Ojitusalueiden puuston kasvun jatkumisesta ja alueellisuudesta. (Regionality and continuity of stand growth in old forest drainage areas.) Acta Forestalia Fennica, 132: 1-36.

Kalliola, R. 1973. Suomen kasvimaaniede. Porvoo.

Koivisto, P. 1970. Regionality of forest growth in Finland. Communicationes Instituti Forestalis Fenniae, 71: 2-76.

Kuusela, K. 1976. Metsävaranto. *In:* Atlas of Finland, Folio, 234: 2-12.

Kuusela, K. 1977. Suomen metsien kasvu ja puutavaralajirakenne sekä niiden alueellisuus vuosina 1970-1976. (Increment and timber assortment structure and their regionality of the forests of Finland in 1970-1976.) Folia Forestalia, 320.

Rikkinen, K. 1971. Typology of farms in Central Finland. Fennia, 106: 1-44.

Salminen, S. 1981. Vuosien 1971 – 1975 valtakunnallisia metsä-varatietdja karttamuodossa. (A cartographical presentation of forest resources in Finland 1971 – 1975.) Folia Forestalia, 483.

Sankinho, R. 1974. Temput ja kuinka ne tehdään. Monimuuttujamenetelmät kansan palvelijoina. (Tricks and how to do them. Multivariate methods for the people.) Reports from the Institute for Educational Research, University of Jyväskylä.

Talman, P. 1979. Population and settlement in Finland. *In:* The Nordic Countries: Man and Environment. A seminar in Finland May 27–June 3 1979. Geographical Society of Northern Finland.

Talman, P. 1980. Karjatalouden alueellinen rakenne Suomen maatilataloudessa 1970–luvulla. (Areal structure of livestock rearing on Finnish farms in the 1970's.) Nordia Tiedonantoja, 7: 1-51.

Yearbook of Forest Statistics 1980. Official statistics of Finland XVII A:12. Folia Forestalia, 460.

FIBRE PRODUCTION FROM BOREAL FOREST ECOSYSTEMS

G.F. Weetman
University of British Columbia

Abstract: The productivity of the boreal forest is considered in terms of cubic metres of stemwood production for commercial use. Both managed stand yields from the boreal forest in Scandinavia and net primary productivity studies suggest maximum mean annual increment values of about 8 m^3/ha/year on good sites. Unmanaged natural stands have empirical yields that are much lower. The projected yields of untreated cutovers are even lower and often involve a switch from valuable conifers to less valuable hardwoods. The opportunities and threats for yield improvement are noted.

The allowable cuts for forest management units in boreal forests are largely dictated by age class structures of the forests and are chosen based on computerized, whole forest yield projections. Institutional impediments are the primary reasons for poor forest management performance, not biological understanding. These impediments are: a workable landlord/tenant relationship to pay for silviculture work on corporate licence areas and ways to funnel more federal and provincial forest revenues back into the forest.

The boreal forest of Canada is a band which crosses the nation from Newfoundland to British Columbia and the Yukon. For over 50 years it has been the raw material base for the pulp and paper industry in eastern Canada. During this period most of the corporate licence cutovers have been left to regenerate naturally following horse logging up to the 1960s, and rubber-tire skidder logging since then. The last 10 years has seen a careful reappraisal of the productivity and drain on the forest resources. Rennie (1978) has already prepared a detailed appraisal of the potential of soils of the boreal for forest production, including the forest productivity base. His conclusions are as follows:

> 'Both nationally and internationally, the Canadian forest is large and important, forming a belt 1,000 km wide across Canada from the Yukon in the northwest to Newfoundland in the east. The forest is made up of over 130 tree species summating to 79% coniferous and 21% broadleaved. Not all the forest is Boreal, but the greater part or over 80% is. Some land is more Wildland than Production Forest, the areas being 519 $\times$ 10^6 ha and 323 $\times$ 10^6 ha respectively, compared with 67 $\times$ 10^6 for agriculture. On the 323 $\times$ 10^6 ha the standing wood volume is 19.5 $\times$ 10^9 m^3, but on only 161 $\times$ 10^6 ha is it currently economic to harvest tree crops.
>
> Over 90% of forest is Crown land and of that vested in the Provinces most is held by British Columbia, Ontario and Quebec. For just the Boreal, however, most is held by the latter two. About 0.65 $\times$ 10^6 ha are cut annually yielding 131.5 $\times$ 10^6 m^3 of wood volume valued at \$5.5 $\times$ 10^9 after manufacture. Intangible forest values are habitat for wildlife, visual aesthetics, the stabilization of terrain and preservation of water quality.
>
> For all Territories and Provinces, an overall picture emerges of a broad arc of

Boreal-Tundra transition on the Precambrian Shield of very low forest productivity — appreciably less than 1 m^3/ha/year. Underlying soils are Podzols, except in the northwest portions. Beyond this is the main band of Production forest, which is not as wide as the traditional Forest Regions map might suggest. Productivity is low to moderate west of the Manitoba-Ontario boundary – 0.6 to 1.5 m^3/ha/year, and greatest for the more southerly Boreal Forest Sections of Ontario and especially Quebec –1.2 to 3.0 m^3/ha/year. Soils of the west are primarily Gray Wooded Luvisols and those of the east Humo-Ferric Podzols. Climate clearly affects productivity across the transition from the more adverse western Boreal to the less adverse eastern Boreal.

Dangers threatening productivity include fire – 0.9×10^6 ha burnt annually, insect and diseases – 88.4×10^6 m^3 of merchantable timber lost annually, poor management practice – 17.8×10^6 ha of prime forest gone out of production, and airborne pollution – 2.5×10^6 ha affected causing annual growth-depletions of \$1.2 to \$$2.8 \times 10^6$. For more northern areas indiscriminate development practices wreak incalculable damage on heritage values. Opportunities to enhance productivity within the general constraints of managing predominantly natural even-aged forest include fire suppression, pest and disease control, genetically improved planting stock, fertilization and site-preparation techniques, and utilizing more of the tree crop. A particular need is better geographic information on forest productivity and soil characteristics so that the value of forest lands can be better assessed. In this way decisions on land-use developments, including the location of mills, in the context of increasing pressure on land, might be more rationally reached.'

Bickerstaff *et al.* (1981) have more recently presented mean annual increment data for all forest regions in Canada (Table 1). Nutrient cycling in boreal ecosystems was reviewed by Krause *et al.* (1978).

The purpose of this paper is to highlight and update some of these findings and present some ideas and views on problems in maintaining and increasing boreal forest productivity.

Table 1. Productivity potential of Canada's boreal continuous forests (Derived from basic data of Bickerstaff, *et al.* 1981)

	Area	Potential productivity	
	(10^6 ha)	(m^3/ha/year)	(ft^3/ac/year)
Actual growth	148	1.5	24
All lands fully stocked		3.0	48
Site improvement	14.8	3.0	48
Stand conversion	48.8	3.2	50
Genetic selection	16.3	3.2	51
Thinning	48.8	3.4	54

Measurement of Productivity

For conventional pulpwood and sawlog products productivity is measured by foresters in cubic metres of stemwood volume in trees which are merchantable i.e., to a top diameter of 7 cm. There must also be enough merchantable volume per hectare (conventionally about 50 m^3/ha or 7 cunits/ac) and trees of large enough size (11 trees/m^3 or 30 trees/cunit).

Minimal exploitable volumes are about 28 m^3/ha or 4 cunits/ac with 14 trees/m^3 or 40 trees/cunit. These minimal volumes must be in spruce, pine or fir. Poplar and birch utilization is still very constrained.

Ecologists measure forest productivity in Net Primary Productivity (NPP). For the boreal forests, the average value is about 800 g/m^2/year, which if we assume man can use about 50 per cent of the biomass roughly equals about 9 m^3/ha/year (Whittaker and Woodwell 1971; Spurr 1976). Table 2 compares some crude estimates of coverage increases in mean annual increment (MAI) attainable in 200 million hectares of temperate and 148 million hectares of boreal forest in North America.

Yield Tables

Most boreal forests are of fire origin. Their productivity is presented in empirical yield tables, based on plots in virgin stands. These give accumulated net merchantable volumes, assuming no recovery of mortality. Two of the better known ones are Plonski (1974) for Ontario and Boudoux (1978) for Quebec. Total accumulated volumes divided by the number of the years gives the mean annual increment (MAI). Stands are considered technically mature when current and mean annual increment curves cross; before this time they should not be cut. Site productivity, usually measured by height of dominant trees at age 50 (SI_{50}), influences the time it takes to grow to maturity. Technical maturity for black spruce $SI_{50}11$ is 70 years, for $SI_{50}8.2$ is 85 years, and $SI_{50}5.1$ is 105 years, associated MAIs are 3.2, 2.4 and 1.6 m^3/ha/year (Plonski 1974) (see Table 3). These MAI values are much less than the potential 8.6 m^3/ha/year from NPP.

What could these same black spruce do if they were planted on sites, rich in nitrogen and free of competition with controlled spacing? Ontario black spruce have been grown under such conditions in northern New Brunswick. Based on 15 years of growth and a stand simulation model, $SI_{50}22$ would generate an MAI of 8 m^3/ha/year in 50 years with 4 stems/m^3 (Erdle 1980). This is much closer to the average 50 per cent of NPP of 8.6 m^3/ha/year. We really do not know for sure because we do not have old plantations in the boreal forest grown under controlled spacings on rich sites. This type of data suggests the potential is in the trees if competition, spacing and nutrition can be controlled. The Canada Land Inventory (CLI) maps for potential forestry production are available, but they are based on empirical natural yields.

In Finland and Norway, boreal forests of Scots pine and Norway pine are under very intensive management. The climate and soils are very similar. They do have gross and net yield tables for managed stands. Recovery of mortality is generally expected to increase yield by 20 to 30 per cent. Norway spruce in Finland, fully stocked and thinned produces 4 to 10.5 m^3/ha/year of MAI ranging from poor to good sites on 100 to 70 year rotations (Vuokila and Valiaho 1980). Equivalent stand yields are probably attainable in the boreal forest with intensive management.

Table 2. Comparison of very crude estimates of increased productivity of temperate and boreal forests

U.S. TEMPERATE (Spurr 1976)		BOREAL	
NPP 1300 g/m^2/year (Whittaker and Woodwell 1971) ½ used by man = 5,800 lb/ac/year divide 14.29	≡ 11,600 lb/ac/year ≡ 200 ft^3/ac/year = 14 m^3/ha/year	800 g/m^2/year 3568 lb/ac/year	≡ 7136 lb/ac/year ≡ 123 ft^3/ac/year = 8.6 m^3/ha/year
Actual average growth, commercial forest: (1970 Outlook Study)	= 38 ft^3/ac/year or 2.6 m^3/ha/year	assume form (Bickerstaff *et al.* 1981)	= 14.3 ft^3/ac/year = 1.0 m^3/ha/year
All stands fully stocked to normal yields: giving to 100%	= 5.2 m^3/ha/year		= 2.0 m^3/ha/yr
Site improvement on 10% of 200 million ha to increase stand growth +15% = average improvement +1.5%	= 5.3 m^3/ha/year	10% of 148 million ha to increase growth 15%	= 2.03 m^3/ha/year
Hardwood conversion to faster growing conifer types on 12 million ha increasing stand yield +50%; average net gain +3%	= 5.4 m^3/ha/year	hardwood to conifer conversion on 20 million ha to increase stand yield +50% average net gain +7%	= 2.18 m^3/ha/year
Genetic improvement on 61 million ha of plantations at 1.2 million ha/year giving +1%/year growth gain in 50 years, average net gain 7.5%	= 5.8 m^3/ha/year	4 million ha of plantations at 50,000 ha/year giving 0.6%/year, growth gain in 50 years, average net gain +0.6%	= 2.19 m^3/ha/year
Recovery of mortality of thinning, +25%	= 7.2 m^3/ha/year	thin all stands	= 2.74 m^3/ha/year
Total production: 200 million ha × 7.2 m^3/ha/year	= 1440 million m^3/year	total production 148 million ha × 274 m^3/ha/year	= 401 million m^3/ha

Table 3. Rotations and mean annual increment for fully stock natural stands in Ontario (Plonski 1974)

Species	Site Class	Site SI_{50}(m)	Technical rotation (years)	MAI m^3/ha
Spruce (mixed)	1a	14.3	55	5.4
Black spruce	1	11.0	75	3.2
	2	8.2	85	2.4
	3	5.1	105	1.6
Jack pine	1	18.3	30	5.2
	2	15.3	40	3.8
	3	12.4	55	2.3
Aspen	1	23.5	55	6.3
	2	19.7	55	5.3
	3	16.0	60	3.7
White birch	1	18.1	50	3.8
	2	15.5	50	3.0
	3	13.1	50	2.5

In contrast to these impressive yield figures must be placed the dismal reality of cutover yield forecasts for the Canadian boreal forest. Unfortunately, commercial clearcutting does not usually duplicate a fire disturbance. Seeds in serotinuous cones are not released, seedbeds are not prepared and vegetative competition is not destroyed since birch and poplar are often left. Vast areas are cut annually, approximately 400,000 ha in the boreal forest; very little of this area is treated. We must take what comes naturally. There is ample evidence to suggest that it will often be much less than the original fire-origin stands. How much less we really do not know, it depends on site, chance, logging method and ecological region. Both our data base and our forecasting abilities are poor. A study based on a 10-year monitoring of 36 study areas cut in the late 1960s, plus the use of a stand simulation model suggested the following MAI values to 50 years (Frisque *et al.* 1978): lowland black spruce, 1.4 m^3/ha/year; upland black spruce, 3.5 m^3/ha/year; balsam fir, 4.1 m^3/ha/year; mixedwoods (all trees), 5.6 m^3/ha/year; and jack pine, 2.8 m^3/ha/year.

These are not very impressive figures and they are very crude estimates. Better data are needed for planning. The poor natural regeneration of boreal forests is the cause for much alarm and concern (Canadian Forestry Association 1977). Boreal mixedwoods with spruce, pine, poplar, birch and fir often convert to hardwoods following cutting. Since this type represents up to 90 per cent of the boreal (Weingartner and Basham 1979) loss of future conifer productivity is serious.

Table 2 assumes some stand level yield increases due to treatment and, by applying these increases to the whole boreal or temperate forest as indicated, it presents the average increase in MAI. The actions which really change whole boreal MAI are the achievement of full or normal stocking and recovery of mortality.

Improvement of Productivity at the Stand Level

There are several ways that stand productivity can be increased. Stands can be converted from hardwoods to conifers. This has very high priority on upland sites. Plantations could produce MAI of up to 6 m^3/ha/year. There can be better regeneration on jack pine sand plains, lowland black spruce and shallow sites on rocky ridges. This could match normal yield MAI values for these sites, 4–5 m^3/ha/year. Thinning to recover mortality can yield an additional 20–30 per cent.

No feasible solution to the unfavourable economics of the large-scale application of thinning boreal pulpwood stands is in sight. Pre-commercial thinning of dense young stands can decrease the number of trees per cubic metre and reduce rotation age. This is expensive, but as will be noted later, it can have a powerful effect on allowable cut when age classes are imbalanced.

Most of the boreal forests of the world are short of available nitrogen (N), net accumulated N in biomass and soil organic matter. Nitrogen supply is a factor limiting tree growth in the boreal forest, it is bound as organic N in the raw humus layer, largely unavailable to trees. Nitrogen fertilization increases the amount of needles and the leaf area index (LAI), resulting in more photosynthesis. Table 4 indicates the average five-year volume increases of 2 to 8 m^3/ha from single doses of 224 kg nitrogen per hectare as urea. The efficiency of conversions of fertilizer N to cubic metres of extra wood range from 10 to 40 kg N/m^3. The problem is that N application costs approach \$1.00/kg N. This produces wood at a cost greater than the provincial royalty sale price. There is evidence to indicate that high growth rates can be sustained by repeated N applications. Large scale N fertilization is a feature of boreal forests in Scandinavia, but stumpage (royalty) prices are much higher.

The large areas of boreal swamps could be drained, as is done in Finland on a vast scale. Nitrogen, phosphorus and potassium fertilizers would be required. However, huge areas of more fertile mineral soil sites are supporting poor tree growth where silviculture treatment is of higher priority.

Table 4. Average five-year total volume growth, over control, of various species fertilized with 224 kg N/ha as urea. Data are summarized from Interprovincial, CFS in-house and other trials (from Morrison 1981)

Species	Five-year growth over control ($m^3 \cdot ha^{-1}$)		
	Maximum	Minimum	Average
Douglas-fir	25.6	7.6	15.6
Jack pine	19.1	−2.7	8.5
Balsam fir	15.7	−3.8	7.1
Red spruce	10.9	−1.9	4.6
White spruce	14.5	−1.9	4.5
Black spruce	8.4	−7.8	2.2

Threats to Boreal Forest Productivity

There are threats to soil fertility and to forest growth. Acid rain as a threat to soil fertility is still to be proven, although it is suspect in long term. Full-tree logging, which removes more of the nutrients in biomass then does conventional tree-length logging, has received a lot of recent attention. The prevailing view appears to be that inputs of nutrients from dust, precipitation and soil weathering on long rotations on moderate to rich sites are probably enough to make up for full-tree export losses. All researchers qualify their findings carefully since so little is known about the nutrient cycles. Clearly full-tree harvesting and forest fires should be avoided on poor sites with low nutrient capitals (Freedman 1981). Physical removal of the humus layer by bulldozing is also to be avoided. In parts of the boreal forest, as in Newfoundland, replacement of forest by heathland vegetation leads to a tie-up of nutrients and loss of fertility, a situation which is very difficult to reverse. The long winters and nutrient poor soils of boreal forests result in 'within-plant' nutrient circulation strategies, recycling of nutrients by litterfall is very slow. Nutrient inputs in dust and precipitation are very important in maintaining productivity on poor sites. Capture of incoming nutrients by feather mosses, which decompose rapidly, cuts short the conventional litterfall recycling mechanism characteristic of more even climates and nutrient rich sites (Staaf and Berg 1981).

Much more important are the threats to the forest growth, i.e. from insects, disease and fire. Fire history dominates much of the boreal forest. Its role is well understood. The control level required depends on the economic values involved. The main problem is upsetting age class structures which govern allowable cuts. Insect losses, particularly the spruce budworm, are also a natural phenomenon. Man competes with fire and budworm to harvest timber. The losses of accumulated increment due to these two natural factors are enormous. Our understanding of these agencies has been aided by whole forest computer models which can grow forests over long periods. Strategies of protection and harvesting and their implications for allowable cut are thus possible in a dynamic way.

This dynamic analysis of forests has been particularly revealing of the severe implications on allowable cut of inadequate and poor regeneration. Yield curves are reduced or shifted to the right or both. When these new yield functions are multiplied by hectares and stored in the computer for the next 100 years of growth, the resulting harvestable growing stock scenarios can be radically changed. These new models of the forest have resulted in some profound changes in policy and administration of Canada's Crown forests.

Whole Forests and Allowable Cuts

Realistically, productivity of a forest is not area × average increment, but rather is the product of areas by age class and forest type × the yield of each combination accumulated over time. The age class structure of the forest, which is predominantly even-aged in the boreal forest, largely determines the rate of harvest or allowable cut within a sustained yield unit. All provinces are re-examining the supply side of the wood industry. An example, is given from Quebec (Gagne and Godbout 1978).

A province is divided into sustained yield management units (unités de gestion, Figure 1), and socio economic zones (Figure 1). For each unit, the area of harvestable forest is analyzed by a whole forest model for its sustainable cut and plotted on a map (Figure 2) in cunits/mile^2 (100 ft^3/mile^2 = 1.09 m^3/km^2). This is the allowable harvest (possibilité) for fir (SAB), spruce (EPI) and jack pine (PIG) with extensive management. These values are

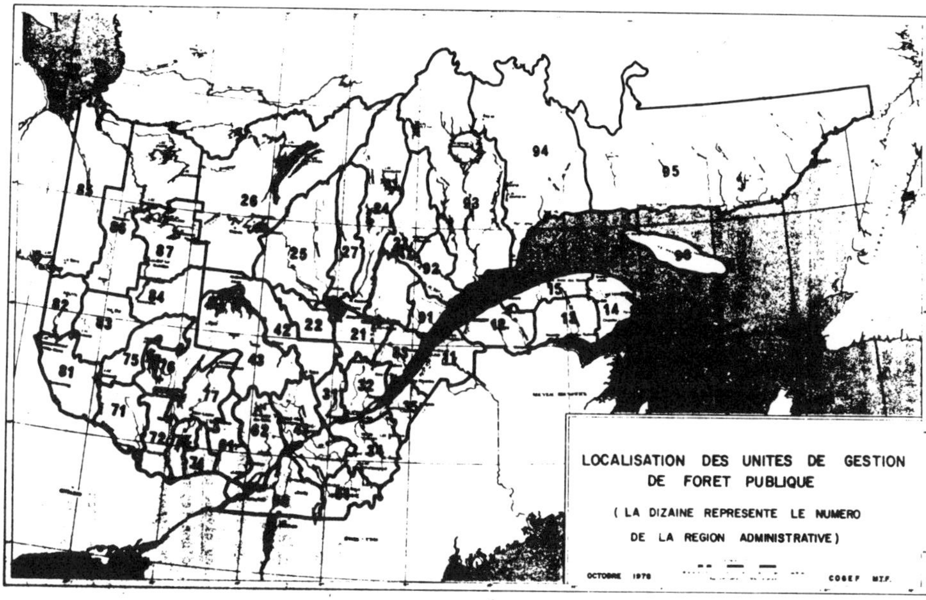

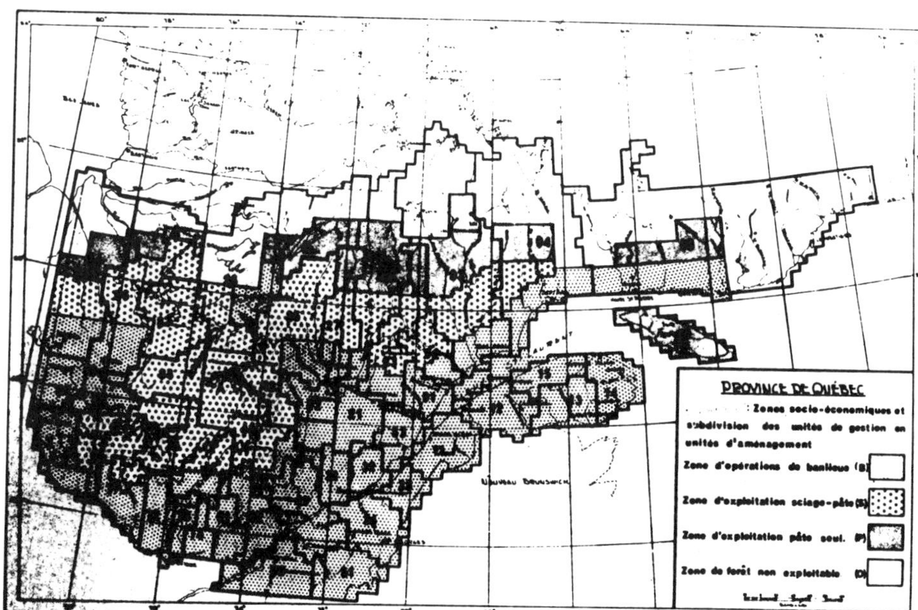

Figure 1. Forest management units (upper) and socio-economic zones (lower) in Quebec (Gagne and Godbout 1978).

reduced for the sawtimber dimensions, based on current practice (Figure 2 and Figure 4). By comparison, the increase in potential is also given according to Canada Land Inventory (CLI) mapping (Figure 3). Identification of surpluses and deficits for sawtimber and pulp become possible.

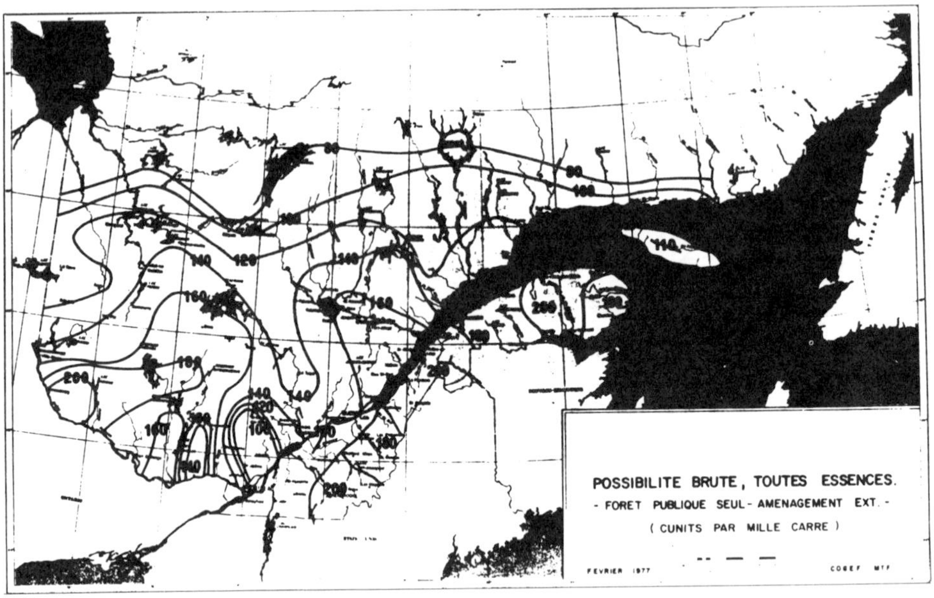

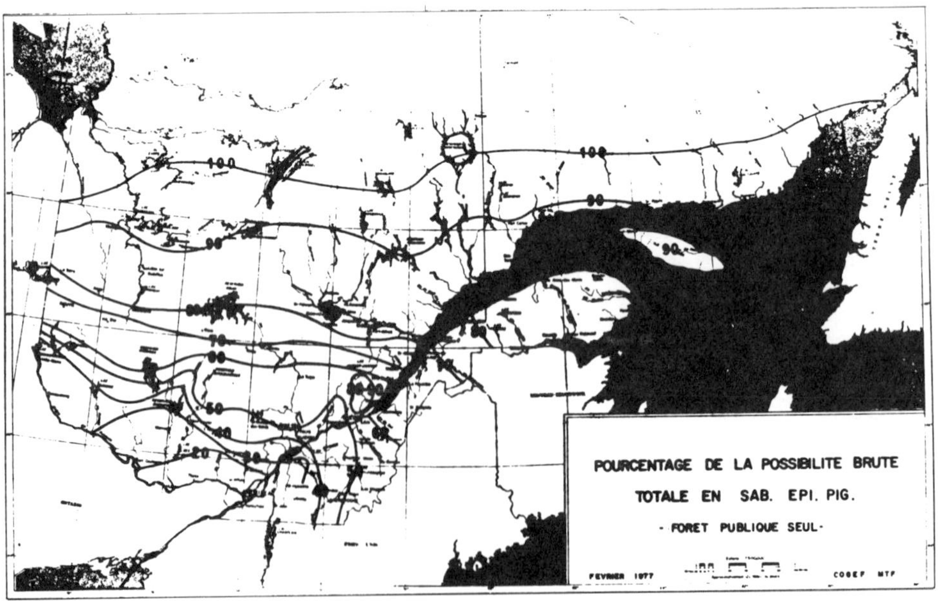

Figure 2. Allowable cuts in Quebec in 100 ft^3 per square mile ($=1.09\ m^3/Km^2$), based on a dynamic analysis of inventory and yield functions. Gross allowable cut all species (above). Percentage in balsam fir, black spruce and jack pine (below) (Gagne Godbout 1978).

These types of dynamic strategic analyses, have been carried out by government, and more recently by industry, all over Canada. They relate productivity, as expressed in yield functions, to forest inventory and allowable cut. They identify the problem areas for

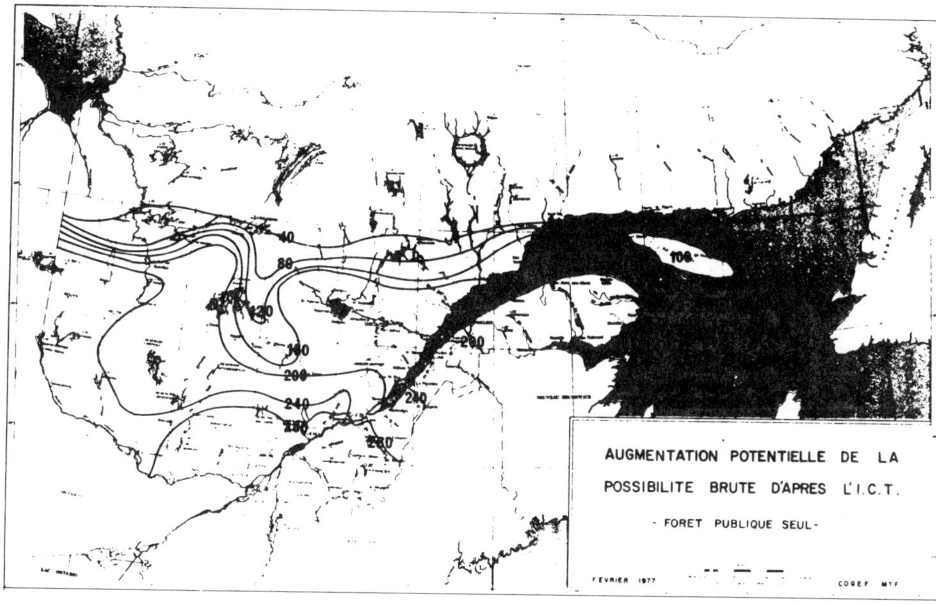

Figure 3. Potential increase in allowable cut based on Canada Land Inventory.

Date: 9/03/77

CALCUL DE LA POSSIBILITE DES RESINEUX EXPLOITABLES SUR FORET PUBLIQUE

- Plan de mise en valeur et d'allocation des ressources forestières du Québec -

Unité d'aménag.	Superficie productive brute (mi. ca.) Totale	Retenue	Facteurs de réduction de la superficie brute (%) Relief 50% +	Relief 30% à 50%	Zonage Bord. eau	Zonage Autres terr.	Superficie productive nette retenue (mi. ca.)	Possibilité brute (cunits/mi.ca.) R & F	R(1)	Facteurs de réduction, poss. brute (%) Tordeuse	Carie	Util.	Possibilité nette -Résineux- (cunits/mi.²)	Possibilité nette -Résineux- (M cunits)	% sciage	Possibilité nette par produit -Résineux- (M cunits) Pâte	Sciage
41B	1,970	1,970	1	2	6	-	1,795	135	70	5	2	5	62	111	70	33	78
42B	1,560	1,250	-	2	4	-	1,175	145	60	15	2	5	61	72	65	25	47
42S	2,285	2,270	-	2	6	-	2,090	145	120	-	1	5	113	236	70	71	165
43B	985	835	-	2	4	-	785	145	75	15	2	5	59	46	60	18	28
43S	4,855	4,325	-	2	8	-	3,895	170	120	-	1	5	113	440	65	154	286
Total rén	11,655	10,650	-	2	6	-	9,740	153	107	10	2	5	'93	905	67	301	604

(1) SAB - EP. - PIG seulement

Figure 4. Illustration of the reduction in allowable cuts due to relief, zones, budworm, rot and utilization in a management unit (Gagne et Godbout 1978).

supply and identify the most rewarding opportunities for more intensive management. Unfortunately for the boreal forest, these opportunities are usually greater south of the boreal. The consequence is that low productivity northern boreal forests are assigned lower socio-economic status and forest management money tends to be spent in the south. The boreal forests must still be given basic protection from fire and insects, but the forests are often left to regenerate without human intervention. Adjustments for the reduced productivity are then made by reduced allowable cuts. That is how the province-wide management systems work. There is of course only one owner, the province, to make these decisions.

Policy, Tenure and Implementation Problems in Productivity

It is clear that declines in stand productivity of usable conifer species and in allowable cut of these species, are occurring because of lack of adequate regeneration. This problem dominates boreal forestry. Achieving adequate regeneration on 300-400,000 hectares cut per year is a huge problem. It is not a problem so much in ecological understanding, which is good for the boreal forest, but in institutional arrangements. Reed (1978) identified these as the number 1 forestry problem in the country. Basically there continues to be a search for the following three agreements:

- a workable landlord/tenant relationship which will provide incentives and pay for the silviculture work which needs to be done by a corporation on licence areas (Table 5 gives the area of boreal forest under licence);

- a way to convince Provincial Treasury Boards to allow more money to be funnelled into silviculture;

- a way to funnel more of the federal revenues from the forest economy back into the forest.

These are political problems associated with public ownership of the resource. The money side is complicated by the historically low price set on Crown timber. The issue of maintaining boreal forest productivity on public lands is related to the degree of public concern. Corporate interest has historically been very weak. New legal forest management agreements which spell out obligations and financing and allowable cuts are awakening corporate involvement. One hoped-for product of this symposium will be aroused concern about the declining state of boreal forest productivity in both stand quality and in allowable cut of commercial species.

Table 5. Area (10^3ha) of boreal forest under licence to major forest corporations (Weetman 1982)

Province	Freehold	Leasehold	Total	Annual cutover	Annual cutover %
Alberta	0	3,300	3,300	8.8	0.27
Manitoba	0	6,800	6,800	8.8	0.13
Ontario	552	17,600	18,152	80.0	0.44
Quebec	759	15,800	15,959	175.0	1.10
Newfoundland	757	3,000	3,757	12.0	0.32

Literature Cited

Bickerstaff, A., W.L. Wallace and F. Evert. 1981. Growth of forests in Canada. Part 2. A quantitative description of the land base and the mean increment. Environ. Canada, C.F.S., Inf. Rep. P1-X-1.

Boudoux, M. 1978. Tables de rendement empiriques pour l'épinette noire, le sapin baumier et le pin gris au Québec. Le Ministère des Terres et Forêts, Québec, Québec.

Canadian Forestry Association. 1977. Proceedings National Forest Regeneration Conference. Ottawa, Ontario.

Erdle, T. 1980. A plantation yield forecasting system for black and white spruce in the Black Brook District. J.D. Irvings Woodlands. St. John, N.B. (Unpublished).

Freedman, B. 1981. Intensive budget harvest. A review of nutrient budget considerations. Environ. Canada, C.F.S., Inf. Rep. M-X-121.

Frisque, G., G.F. Weetman and E. Clemer, 1978. Reproduction and trial projected yields 10 years after cutting 36 pulpwood stands in Eastern Canada. Forest Engineering Research Inst. Tech. Rep. TR-23. Pointe Claire, Quebec.

Gagné, G. and C. Godbout. 1978. La matière ligneuse; sa disponibilité et la demande actuelle et future dans l'industrie forestière face à son approvisionnement. L'ordre des ingénieurs forestiers du Québec. 58ième Congrès annuel. pp. 14-42.

Krause, H.H., G.F. Weetman and P. Arp. 1978. Nutrient cycling in boreal forest ecosystems of North America. *In:* Forest Soils and Land Use. C.T. Youngberg (Ed.). Colorado State University. Fort Collins, Colorado. pp. 337-340.

Morrison, I. 1981. Assessment of the current state of forest fertilization in Canada. *In:* An industrial assessment of forestry research in Canada. Vol. II. K.M. Thompson (Ed.) Pulp Pap. Res. Inst. Can. Pointe Claire, Quebec. pp. 115-142.

Plonski, W.L. 1974. Normal yield tables (metric) for major forest tree species of Ontario. Ontario Ministry of Natural Resources. Toronto, Ontario.

Reed, F.L.C. 1978. Forest Management in Canada. Environ. Canada, C.F.S., Inf. Rep. FMR-X-102.

Rennie, P.J. 1978. Utilization of soils of the boreal for forest production. Proc. 11th Congress of International Soil Science 3: 208-331.

Spurr, S.H. 1976. American forest policy in development. University of Washington Press Seattle, Washington.

Staaf, H. and B. Berg. 1981. Plant litter input. Ecol. Bull. (Stockholm) 33: 147-162.

Vuokila, Y. and H. Valiaho. 1980. Growth and yield models for conifer cultures in Finland. Commen. Inst. Forestalis Fenn. 99.

Weetman, G.F. 1982. The Evolution and Status of Canadian Silviculture Practice. For. Chron. 58: 74-78.

Weingartner, D.H. and J.T. Basham. 1979. (Eds.) Forest management and research needs in the boreal mixedwood forest of Ontario. Spruce-Fir Aspen Research Committee. Ont. Min. Natural Resources. Toronto, Ontario.

Whittaker, R.H. and G.M. Woodwell (Eds.). 1971. Measurement of net primary production of forests. Proc. Brussels Symposium. UNESCO, Paris. pp. 154-175.

GENETIC RESOURCES OF THE NORTH AMERICAN BOREAL FOREST

Robert E. Farmer, Peggy Knowles and William H. Parker
Lakehead University

Abstract: The North American boreal forest contains a relatively small number of widely distributed tree species each of which is adapted to a broad range of environmental conditions. Existing knowledge of genetic variation is reviewed for major boreal tree species. Their genetic systems, which include wind pollination and other mechanisms promoting outcrossing, have resulted in maintenance of high levels of genetic variability both within and between populations. Much of the variation between populations is expressed along environmental gradients as clinal or ecotypic differentiation; but some does not correlate well with the present environment, perhaps reflecting adaptations acquired in glacial or preglacial times. Although little is yet known of variation in boreal shrubs and herbs, their patterns of variation appear less generalized than those of trees due to greater diversity in their genetic systems. High levels of genetic variability in boreal tree species have important implications for the management of boreal forest lands. Current efforts in regeneration have recognized the need to generally match seed source with planting area. Efforts are underway to genetically improve growth potential and other economically important characteristics. However, a thorough assessment of the genetic resource has not usually been made in conjunction with management and breeding, although these activities may presently be altering the nature of the resource. To ensure conservation of the genetic resources of the boreal forest, tree breeding programs require broadening to better delineate patterns of natural variation and describe genetic systems of all tree species.

The North American boreal forest is dominated by relatively few, genetically variable tree species with ranges extending across a long, complex environmental gradient. This gradient has changed cyclically during their recent evolutionary history. The life history and genetic characteristics of boreal forest components therefore, are as much the products of change as of present environmental gradients and conditions unique to boreal areas. The forest's genetic variation is a resource that represents the raw material for continued adaptation to changing environments as well as for genetic improvement. Today, the boreal forest is being subjected to anthropogenic forces that may have both ecological and evolutionary impacts and may effect genetic changes even more rapidly than past natural selection processes. Therefore, it is important that, as custodians, we understand and act upon our knowledge of current influences and the forest's biological history.

In this paper, we review major considerations related to the genetic structure of boreal tree species including the present genetic characteristics as a function of evolutionary change during Quaternary time, the current patterns of genetic variation in tree species, and the nature and genetic impact of forest management and 'tree improvement' programs.

Evolutionary Considerations

Although fossil evidence is scanty, it seems likely that most of our modern boreal tree species evolved well before the beginning of the Quaternary period several million years ago (Rudolph and Yeatman 1981; Liu 1971; Nienstaedt and Teich 1972; Wright 1955; Daubenmire 1974). Thus, modern boreal flora consists of relatively few species that have persisted over a fairly long period.

The histories of boreal plant species over the past two million years (the Pleistocene epoch) help account for their existing life history characteristics, patterns of natural variation and breeding systems. The key element in the success of modern boreal forest species has been the ability to accommodate to extremes of environmental change caused by glaciation. Most boreal plant species can be described as 'glacial transients', since their survival has depended upon following a cycle of southward migration during periods of glacial encroachment followed by northward migration during periods of glacial recession. Stratigraphic analyses indicate that boreal species have undergone four such cycles of migration during the Pleistocene epoch (Ericson and Wollin 1968).

To survive, boreal plant species must have migrated at a pace that matched the average rate (200 km/1000 years) of the encroaching ice sheets (Matsch 1976). This need for rapid migration likely accounts for the high incidence of seed dispersal by wind in boreal tree species. Wind dissemination guarantees that a certain percentage of seeds will be deposited regularly beyond the existing range of those species, and a small percentage of seeds may travel great distances under favourable circumstances (Roe 1967).

Equally important, however, is their breeding system, that allows the storage of the high levels of genetic variability required for rapid adaptation to changing climate. Wind pollination coupled with unisexual strobili or much reduced unisexual flowers is the prevalent breeding system of most woody boreal species. The rarity of this breeding system in plants from some other more stable vegetation zones (e.g. the tropics) suggests that it may be an adaptation to the boreal environment. While unisexual flowers effectively promote outbreeding, a price is paid in terms of energy, particularly for dioecious plants (Baker 1959). In addition, effective pollination by wind requires large numbers of gametes, and monoecious plants following this strategy generally possess mechanisms to reduce either the levels or effects of self-pollination. For example, in boreal coniferous species, this mechanism may consist of high levels of recessive lethal mutations (genetic load) that eliminate the bulk of embryos produced from self-pollinations (Wright 1976). For most boreal woody angiosperm species, this mechanism may take the form of an out-of-phase development sequence of male and female flowers (dichogamy).

Apparently, the advantages of wind-pollination and unisexual structures are worth the trade-off in terms of energy consumption for boreal plants. The resultant high level of outbreeding is almost certainly the chief advantage. As is the case for all outbreeding populations, variability is maintained in outcrossing boreal species as a reservoir of heterozygous recessive genes (Davis and Heywood 1973). Most of this variation need not be adaptive at any one time; but its storage has made possible adaptation shifts in response to rapidly changing Pleistocene climates. It is also possible that much of the apparently random, non-adaptive variation discernable in our modern boreal conifer species may well have been adaptive in nature during past stages of the Pleistocene (O'Reilly *et al.* 1982). The long-term genetic plasticity conferred by outbreeding suggests that future rapid adaptations by boreal plant species may be anticipated should environments change.

A few boreal tree species have abandoned outbreeding as a reproductive strategy. For example, certain boreal species of *Salix* are thought to be completely agamospermous (Cronquist 1964), thus achieving high immediate fitness, but with genetic rigidity.

Perhaps the main reason that successful boreal plants have maintained outbreeding via the energy-wasteful process of wind pollination is that this mechanism is not dependent upon the availability of insect or other animal vectors. Given the requirement for rapid migration during glacial advance and recession, the additional requirement for a dual synchronized migration of plant and pollinator species may have represented too fragile a link in the long-term survival strategies of these northern plants, particularly if simultaneous adaptations to a changing environment were required. However, wind pollination is not universal, and many boreal herbs, as well as certain shrubs, are insect pollinated. Generally, these species are not the dominant elements of boreal vegetation, and it is probable that they are unable to recolonize an area until the necessary buffering niches have been established for plants and pollinators by the dominant wind-pollinated elements of the community.

Recent paleobotanical studies indicate that the boreal vegetation zone was shifted about 1000 km southward during the maximum of the Wisconsin glaciation and early periods of the present interglacial (Whitehead 1973). However, this boreal forest was by all indications very different in species composition from modern boreal communities, being dominated by spruce (Ritchie 1976) with minor components of oak (Wright 1972), and with pine not becoming important until a later date. These differences in composition of the boreal forest in the early recolonization stages of the present interglacial likely reflect differences in the migration rate of the species. In addition, they probably also reflect the existence of an ancient climate differing from the modern and recent past boreal climates. The different plant associations shown by the paleobotanical reconstructions, therefore, suggest somewhat different adaptive norms. Thus, although modern boreal species generally are lineal descendents of boreal species existing for millions of years, these modern-day descendents are probably not adapted to precisely the same climates as their recent ancestors.

Interglacial periods have made possible interspecific gene flow west-east between closely related pairs of species. Evidence has accumulated that this interchange occurs when the species temporarily become sympatric in western Canada during periods of maximum northward recolonization. Species pairs following this pattern are *Abies lasiocarpa* and *Abies balsamea* (Parker *et al.* 1981), *Picea engelmannii* and *P. glauca* (Daubenmire 1974), and *Pinus contorta* and *P. banksiana* (Scotter 1974). Analogous species pairs of hardwood trees such as *Populus trichocarpa* and *P. balsamifera* are also apparently evolving in the same fashion. Thus, both east and west elements of the species pairs gain an enriched gene pool during interglacial periods with a resulting longitudinally zoned pattern of genetic variation that may not be entirely adaptive, but which enhances the potential for future genetic change.

Finally, since the last glaciation at least, the evolutionary impact of fire in the boreal forest is reflected in life history characteristics and the genetic structure of plant species. In the two dominant conifers, *Picea mariana* and *Pinus banksiana,* and to some extent in *Larix laricina,* cone serotiny with its consequent large seed banks is an important adaptation to the prevalence of fire. In *Populus tremuloides* and *P. balsamifera* outbreeding has been combined with a capability for surviving fire through vegetative propagation via root suckering. This has resulted in species with the capability for rapid genetic change and high current fitness with respect to a dominant environmental influence.

Current Variation Pattern by Species

Patterns of genetic variation in boreal plant species have only recently been investigated, mostly in conjunction with forest tree improvement research. A variety of methods most adopted from quantitative genetics, is being used.

A provenance test is a 'common-garden' experiment in which propagules are collected from geographically widespread natural stands and grown together under similar conditions at one or more sites. Variation among the source groups at a single site is assumed to have a genetic basis because the environment is relatively uniform.

The progeny test is an evaluation of parents in terms of their offsprings' performance in a common environment. It is used in numerous forms to evaluate within-population genetic variance and the breeding potential of parents.

Isozyme analysis infers the genetic characteristics of organisms through analysis of enzymes transcribed directly from nucleic acids. These single gene products (isozymes) can be resolved electrophoretically and offer the advantages of being directly quantifiable and free from environmental influences. However, at present, relationships between enzyme data and morphological traits in forest trees have yet to be established; this remains the major disadvantage of isozyme analysis technique.

A review of isozyme literature of forest tree species reveals several general trends. First, compared to other groups of plant species such as annuals, biennials and herbaceous perennials, forest trees are by far the most variable (Hamrick 1979). Second, genetic variation in enzyme traits in forest tree populations tends to be organized such that most of it occurs within populations, with only a small proportion of the variation actually differentiating populations (Conkle 1979). Third, characteristics associated with the most variable plant species included large ranges, high fecundities, an outcrossing mode of reproduction, wind pollination and a long generation time (Hamrick *et al.* 1979). Although few boreal species have been subjected to isozyme analysis, they would be expected to be highly variable, given these trends.

Other techniques to assess genetic variation include karotype analysis, assay of secondary chemical compounds, and measurement of DNA content and DNA hybridization ability.

Jack Pine

Jack pine ranges from the Canadian Maritimes (excluding Newfoundland and Labrador) westward to the foothills of the Rocky Mountains in Alberta, extending south into the Great Lakes–St. Lawrence forest and almost to tree line in the north. It is to date the most thoroughly investigated boreal tree species and is the subject of a recent literature review by Rudolph and Yeatman (1981) upon which this summary is based.

Modern jack pine migrated from refugia that survived the Wisconsin glaciation in the Appalachian and western Ozark mountains and possibly in the west. Analysis of present day cone serotiny and monoterpene characteristics indicates a very steep cline or partial discontinuity in Minnesota which has been interpreted as evidence of the interbreeding of two distinct lineages.

Variation patterns of morphological characters (bark thickness, needle length and width, cone serotiny) in natural populations of jack pine have shown clinal trends and some association with environmental factors such as precipitation, latitude and temperature (Schoenike 1976). These results suggest that 'environmental factors are the dominant selection forces responsible for the natural variation over the present day jack pine range' (Rudolph and Yeatman 1981). However, in examination of natural jack pine populations *in situ* one cannot distinguish between the environmental and genetic components of variation, a problem solved by the provenance test technique. Numerous studies have demonstrated the existence of provenance variation in growth, wood properties, and cold hardiness that could be related to environmental gradients. In contrast, there are substantial provenance differences in tree form, pathogen resistance and seed yield that are unrelated to environmental gradients. One noteworthy exception is variation in vulnerability to the North American strain of the fungal disease, scleroderris canker, which increases when southern material is moved northward in eastern Canada (Yeatman 1976a, b).

Progeny tests including either full or half-siblings have not been as extensively reported as provenance trials, but data for young trees (two to three years) indicate that there is substantial within-population genetic variance in growth and form.

White Spruce

White spruce is distributed throughout the boreal forest from Newfoundland to Alaska. It is believed to have survived the Pleistocene in several refugia in the Black Hills, the maritime region (Wright 1955) and possibly the Yukon. After the most recent glaciation, white spruce probably migrated outwards from two widely separated refugia, one in the northwest of the continent and one in the east (Neinstaedt and Teich 1972). Genetic evidence based on monoterpene analysis (Wilkinson *et al.* 1971) and DNA content (Miksche 1968) support the concept of two original populations.

Provenance trials have indicated that the following characteristics vary for genetic reasons over portions of the white spruce's range: height growth, wood density, branch angle, response to soil calcium ion concentration, germination temperature thresholds, and characteristics of needles, branches and buds (for review see Nienstaedt and Teich 1972). Patterns of variation have generally suggested clinal trends and an adaptive response to selective pressures imposed by variation in precipitation, photoperiod and temperature regimes. However, provenance tests have concentrated on southeastern populations planted mostly in the Great Lakes–St. Lawrence region, and data on the boreal forest portion of the species range are limited. Particularly noteworthy is the observation that materials from certain areas in the Ottawa Valley have proven superior in growth to other provenances at test sites throughout the southern portion of white spruce's range.

There have been only a few evaluations of genetic variance within populations, mostly conducted with relatively small samples of material from south of the boreal forest. Nevertheless, they do suggest that this component of variation is substantial. For example, date of budbreak, which has been thoroughly studied in spruce, varies widely in local populations, is under strong genetic control and closely related to growth and frost injury.

Black Spruce

Black spruce, the most common species in many parts of the boreal forest, has a distribution across North America almost entirely coincident with white spruce. Black spruce is usually considered to be a tree of northern interior lowlands, since it grows best on moist wet clay loams on gentle slopes and lowlands. However, in the northern part of its range it is also found on uplands, sometimes in association with jack pine.

Morgenstern (1978) initiated a range-wide provenance study of black spruce and has reported on variation in seedling height and timing of growth. The results to date indicate that black spruce in the boreal region is differentiated (in terms of these characters) from spruce in the Great Lakes area and the Maritime group. Plants in the boreal group flushed first, formed buds earliest, and were shortest at a central Ontario site. Association with climatic and geographic variables indicate that photoperiod and temperature are the major selective factors and that the variation is clinal in a north–south direction. Furthermore, significant variation exists between stands within regions. Data from previous studies have both supported the concept of a clinal variation pattern (Morgenstern 1969; Dietrichson 1969) and also suggested a more irregular, discontinuous (ecotypic) type of variation (Morgenstern 1972; Khalil 1973). Because of its occurrence on both upland and lowland sites some attention has been given to the possibility of edaphic ecotypes, but to date experimental evidence has not supported this notion.

While there are not yet published data on within-population variance from large progeny tests, the work of Morgenstern (1973, 1974) with young open-pollinated families and full-sibling progeny points to relatively wide variation in, and moderate genetic control over, juvenile characteristics. Of special significance is his observation that 'for traits related to reproductive fitness, such as germination and survival, large genes acting in dominance-recessiveness-fashion play a role' (Morgenstern 1979).

Quaking Aspen

Quaking aspen is the most widely distributed tree species in North America. In the north its range generally coincides with black and white spruce, but it also extends south of the boreal forest region along the Rocky Mountains and throughout New England. As with many of the other boreal tree species, aspen grows well on a variety of soils and occurs in mixed woods or small pure stands. Its capacity for vigorous vegetative regeneration by root suckers results in naturally occurring clones (Kemperman 1977; Schier 1981).

Neither large scale provenance nor progeny tests have been conducted for aspen. However, interclonal variation in numerous characteristics has been examined in natural stands, mostly in the southern and western parts of its range (Einspahr and Winton 1977). If boreal populations follow the same pattern, they are likely to be highly variable.

Barnes' (1969) study of intra- and interclonal variation in Michigan revealed marked interclonal variation in size and density of ramets, suckering ability, growth rate, and morphology of leaves and bark. Ramets of a clone were typically uniform in stem form, timing of leaf flushing, autumnal leaf coloration, and timing of leaf fall. These characteristics have been used to distinguish different clones in the field. Clonal variation in wood properties has also been reported (Einspahr and Winton 1977). A single study of natural clones in Ontario's boreal forest revealed variation in growth and susceptibility to stem defects (Kemperman *et al.* 1978).

A study of morphological variation in the western portion of aspen's range indicated a clinal trend along a north–south gradient in several characteristics, including leaf shape, size and tooth number. Aspen of the southern Rocky Mountains seems to be more closely related to central Asian aspen, while northern and southeastern aspen appear to be similar and perhaps evolutionarily younger (Barnes 1975).

Other Species

Because other major tree species in the boreal forest *(Abies balsamea, Betula papyrifera, Larix laricina, Populus balsamifera, Thuja occidentalis)* have had limited commercial use, there have been essentially no systematic efforts to assess their genetic variability. However, existing taxonomic literature about these species does contain numerous references to phenotypic variation in morphology (Parker *et al.* 1981), and there are some accounts of ecotypic differentiation in physiological characteristics (Bevington 1982; Musselman *et al.* 1975; Pauley and Perry 1954). Recent interest in *Betula, Larix* and *Populus* as either subjects of or contributors to breeding programs suggests that more coordinated effects may be soon mounted to study their genetic variation. The genetics of numerous ecologically important boreal shrubs and herbs is even less well known.

Conclusions

Data from all investigated species give evidence of geographic variation that has a genetic basis and is a product of adaptation to major environmental gradients. Genetic variance within populations also appears to be high; but, with the exception of jack pine, we have only sketchy information on the genetic features of populations. Moreover, these conclusions are based mostly on studies of highly heritable morphological traits or characters important in breeding, e.g. growth and form. These characteristics are not necessarily those of most adaptive value. Thus a better understanding of these particular characteristics would not necessarily reveal evolutionary trends and the way in which species are currently fitted to their environment. In short, a broader investigative approach than that required in breeding research would be useful. In particular, variation in reproductive biology merits closer examination within the context of evolutionary ecology. Questions of adaptive strategy must be addressed experimentally if we are to ultimately understand the region's genetic resources. This will not only broaden our knowledge of the genetic structure of populations, but should give greater depth to our understanding of life history characteristics essential to effective silviculture. Finally, while it is unlikely that mission-oriented research programs similar to tree breeding will ever develop around shrub and herb species, some (e.g. the nitrogen fixer *Alnus crispa*) are critically important to ecosystem function and should receive more coordinated research attention.

Impact of Management on Genetic Resources

The genetic constitution of the boreal forest is subject to anthropogenic change resulting from forest management, particularly silviculture, and from tree improvement (i.e. breeding) programs. We have few data of the effects of these activities, but one can surmise the manner and direction of their potential impact from information on species' genetics, ecology and silviculture

Silvicultural Impact

Under unmanaged conditions major coniferous species (principally black spruce and jack pine) normally regenerate after fire, and have characteristics (e.g., serotinous cones) that promote successful establishment of new stands on burned areas. A common post-fire scene is a stand of dead trees disseminating viable seed from their crowns to a fire-prepared seedbed beneath them. Under these conditions the genetic constitution of regeneration is similar to the parent stands. Imposition of forest management alters this scene by largely excluding fire and by removing trees. In some cases the best trees (in terms of growth and form) are removed leaving poorer trees as parents for the next stand. Unless special silvicultural procedures are followed to provide a good seed source and an appropriately prepared site for coniferous establishment, the forest will change floristically to species remaining on the site after cutting and/or those whose environmental requirements are met at the post-cutting site. Commonly, these species grow rapidly and have a capability for vegetative regeneration that allows them to quickly dominate cutover sites (e.g. *Betula, Populus, Alnus, Salix*). Unfortunately, these species are currently of low economic value. The silvicultural response to this potential transition is to simulate natural post-fire seedbed conditions by soil disturbance (i.e. scarification), to reduce vegetation competing with economically valuable conifers, and to either leave seed trees for natural restocking, plant seedlings, or artificially seed ('direct seed') the area. Establishment of large clearcuts and the associated use of large mechanized harvesting and site preparation equipment have tended to narrow silvicultural options and have increasingly necessitated planting seedlings or direct seeding. High harvesting pressure and low intensity management by a limited staff have accentuated this trend in northern Ontario.

This scenario has several genetic implications. The first of these is that planting seedlings entails the domestication of species and a possible consequent reduction of genetic variance. The history of domestication of other plant species provides ample evidence of drastic reductions (Harlan 1976). This pattern must not be repeated for boreal forest trees. At present, seed for planting stock (seedlings) are obtained from natural populations in the site region where seedlings will be planted. This measure avoids the disastrous consequence of planting material unadapted to local environmental conditions. However, nursery-produced material will have a much narrower genetic base than the surrounding natural population if only a small number of parents are represented. Thus, nursery production of seedlings combined with massive planting instead of natural regeneration could reduce within-region variation unless seed are collected from a broad sample of parents. In addition, competition and other conditions in the nursery and subsequent plantings are substantially different from those under natural conditions. As a result the genetic structure of populations may be altered in a number of ways. In situations where poorer trees are left for parents there may be a decrease in the genetic potential of the next generation, though there is at yet no experimental evidence of this effect.

On the other hand, currently undesirable species may be genetically altered by silvicultural efforts to reduce their impact on conifers or their abundance. A current field assessment will probably suggest that these efforts have been largely ineffective in terms of genetic alteration. Aspen and birch, for example, are among the most abundant species in northwestern Ontario, partly because of existing silviculture. However, ineffective silviculture notwithstanding, the genetic resource that these species represent should be considered in forest management policy. Some of our most productive species are considered 'weeds' only because of current utilization problems and product requirements; these economic circumstances can change within decades. The recent exploitation of such species as alternate energy sources in more densely populated areas of the world is a case in point.

Exclusion or reduction of fire over large areas of boreal forest has already had ecological effects and could eventually influence genetic resources since it involves elimination of a major natural selection force. For example, trees with relatively low levels of cone serotiny might be favored in a fire-free environment. Natural seed-bed conditions and early stand density, both of which influence population structure, are dramatically different in stands from which fire is excluded. Genetic changes resulting from these conditions will be difficult to evaluate in the short-term and may not be negative from the standpoint of long-term resource management.

Tree Improvement Programs

The second way in which the boreal forest may change genetically is through 'tree improvement' programs, which are closely associated with artificial regeneration. These breeding programs take advantage of wide genetic variation to increase genetic potential of planting stock for growth and other economically important characteristics. Conducted in Canada and the United States by government and industrial organizations, they are based on the assumptions that artificial regeneration will be extensively used and that genetic gains can be made relatively quickly (a single generation) through mass selection in natural populations and relatively simple breeding designs. To date these assumptions have usually proven valid in North America, and several economic studies have concluded that tree breeding is a good research and development investment (Carlisle and Teich 1971; Carlisle and Rauter 1979; Porterfield *et al.* 1975). Improvements in growth rate of 10 to 20 per cent over that of natural populations have been shown to be possible. Unfortunately, applied forest genetics has been frequently given a 'super tree' characterization in the popular press, where product expectations are often exaggerated.

Typically, tree improvement programs combine evaluating genetic variance, using it to breed better stock, and producing genetically improved seed or vegetative propagules. At present, there are breeding programs with the following boreal species: *Picea glauca, Picea mariana, Pinus banksiana,* and *Larix laricina* (see Proceedings, 17th Meeting Canadian Tree Improvement Assoc., Canadian Forest Service 1980).

As noted above, we now have numerous immediately useful data on geographic variation patterns and increasingly good data on within-population variance. We know the modes of inheritance for a few important characters, and we have learned a great deal about the silvical and physiological features of subject species. In the boreal forest region, however, we do not yet have forest tree material which is certifiably 'improved' on the basis of test data. It will be several decades before stock that is certified 'improved' for a 70-year rotation (the life span of a tree) is available.

In the meantime, there will be a strong tendency to use products with 'certification pending' on the assumption that they are at least as good as material now being used in artificial regeneration. This assumption is probably true. It is also possible that these interim products will have a narrower genetic base than currently used planting stock, though tree improvement does not necessarily reduce genetic variance. From the standpoint of conserving a broad base of natural variation, tree improvement programs may, therefore, present some hazards unless measures are taken to maintain a high level of genetic variance.

Recommendations

Loss of genetic variation via either silvicultural practices or tree improvement in major boreal forest species can be easily avoided. We have the following recommendations for the boreal forest.

Silvicultural and tree improvement programs should contain gene conservation efforts as a formal component. At present, this may seem superfluous in a ubiquitous species such as black spruce. But, until we know more about the genetic structure and operation of populations and the degree and nature of genetic change, such efforts are necessary. They can be simple and inexpensive. For example, it may be sufficient to simply designate some areas as 'gene pool reservations in situ' (Yeatman 1973), where only natural regeneration will be used. On the other hand, limited boreal extensions of more southern species such as *Pinus strobus, Pinus resinosa* and *Acer saccharum* may represent a unique genetic resource which required more extensive attention.

Natural regeneration in the boreal forest should be used whenever possible to ensure continuation of local populations. Planting should be considered as a tool for rectifying silvicultural failure, for obtaining regeneration in inherently difficult situations, and for changing species composition. Planting stock should have as broad a genetic base as the natural stand it replaces.

Tree improvement programs should include a large component of fundamental genetics research designed to more completely characterize patterns of natural variation. In the long run, we must have a much better understanding of the boreal species before either improvement can be made or adequate gene conservation programs instituted. In particular, we need more data on the degree to which silviculture and breeding are changing the nature of genetic variation.

Together these actions should dampen the silvicultural tendency (if it exists) to narrow the genetic base in major timber species and alter tree improvement programs to provide for a broader assessment of genetic change.

Literature Cited

Baker, H.G. 1959. The contribution of autecological and genecological studies to our knowledge of the past migrations of plants. Amer. Nat. 93: 255-272.

Barnes, B.V. 1969. Natural variation and delineation of clones of *Populus tremuloides* and *P. grandidentata* in northern lower Michigan. Silvae Genetica, 18: 130-142.

Barnes, B.V. 1975. Phenotypic variation of trembling aspen in western North America. For. Sci. 21: 319-328.

Bevington, J.M. 1982. Variation in the germination responses of seed of *Betula papyrifera* from different geographic sources. Abstract in Bull. Ecol. Soc. Amer. 63: 149.

Carlisle, A. and R.M. Rauter. 1979. The economics of tree improvement in Ontario. Environ. Canada, C.F.S., Proc. 0-P-7. pp. 215-229.

Carlisle, A. and A.H. Teich. 1971. The costs and benefits of tree improvement programs. C.F.S., Publ. No. 1302.

Conkle, M.T. 1979. Amount and distribution of isozyme variation in various conifer species. Proceedings Seventeenth Meeting of the Canadian Tree Improvement Association, Gander, Newfoundland. pp. 109-117.

Cronquist, A. 1964. *Salix* L. *In:* C.L. Hitchcock, A. Cronquist, M. Ownbey and J.W. Thompson. Vascular plants of the Pacific Northwest. Part 2 (Salicaceae to Saxifragaceae). University of Washington Press, Seattle. Washington. pp. 37-70.

Daubenmire, R. 1974. Taxonomic and ecologic relationships between *Picea glauca* and *Picea engelmannii*. Can. J. Bot. 52: 1545-1560.

Davis, P.H. and V.H. Heywood. 1973. Principles of angiosperm taxonomy. Krieger Publishing Co. New York.

Dietrichson, J. 1969. Genetic variation of cold damage, growth rhythm, and height growth in 4-year-old black spruce. (*Picea mariana* (Mill.) B.S.P.) Medd. Nor. Skogforsoeksves. No. 104, Bind 28. pp. 212-243.

Einspahr, D.W., and L.L. Winton. 1977. Genetics of quaking aspen. U.S.D.A., For. Serv., Res. Pap. WO-25.

Ericson, D.B. and G. Wollin. 1968. Pleistocene climates and chronology in deep-sea sediments. Science, 162: 1227-1234.

Hamrick, J.L. 1979. Genetic variation and longevity. *In:* Topics in Plant Population Biology. O.T. Solbrig, S. Jain, G.B. Johnson and R.H. Raven (Eds.). Columbia University Press, New York. pp. 85-113.

Hamrick, J.L., Y.B. Linhart, and J.B. Mitton. 1979. Relationships between life history characteristics and electrophoretically detectable genetic variation in plants. Ann. Rev. Ecol. Syst. 10: 173-200.

Harlan, J.R. 1976. Genetic resources in wild relatives of crops. Crop Sci. 16: 329-333.

Kemperman, J.A. 1977. Aspen clones: development, variability, and identification. Ontario Ministry of Natural Resources, Div. of Forests. Forest Research Information Paper No. 101.

Kemperman, J.A., S. Navratil, and J.T. Basham. 1978. Preliminary assessment of defect variation among aspen clones in northern Ontario. Ontario Ministry of Natural Resources. Forest Research Centre. Forest Research Report No. 104.

Khalil, M.A.K. 1973. Results of a four-year old nursery provenance experiment on black spruce in Newfoundland. Environ. Can., C.F.S., Inf. Rep. N-X-99.

Liu, T.S. 1971. A monograph of the genus *Abies*. Dept. For., Coll. Agric., National Taiwan University, Taipei, Taiwan.

Matsch, C.L. 1976. North America and the great ice age. McGraw-Hill, New York.

Miksche, J.P. 1968. Quantitative study of intraspecific variation of DNA per cell in *Picea glauca* and *Pinus banksiana*. Can. J. Genet. Cytol. 10: 590-600.

Morgenstern, E.K. 1969. Genetic variation in seedlings of *Picea mariana* (Mill.) BSP. Silvae Genet. 18: 161-167.

Morgenstern, E.K. 1972. Preliminary estimates of inbreeding in natural populations of black spruce, *Picea mariana*. Can. J. Genet. Cytol. 14: 443-446.

Morgenstern, E.K. 1973. Genetic research with black and red spruce at Petawawa Forest Experiment Station. Environ. Can., Inf. Rep. PS-X-45.

Morgenstern, E.K. 1974. A diallel cross in black spruce, *Picea mariana* (Mill.) B.S.P. Silvae Genetica, 23: 67-70.

Morgenstern, E.K. 1978. Range-wide genetic variation of black spruce. Can. J. For. Res. 8: 463-473.

Morgenstern, E.K. 1979. Status of black spruce tree improvement. *In:* Environ. Canada, C.F.S., Proc. 0-P-7. pp. 113-120.

Musselman, R.C., D.T. Lester, and M.S. Adams. 1975. Localized ecotypes of *Thuja occidentalis* L. in Wisconsin. Ecology, 56: 647-655.

Nienstaedt, H. and A. Teich. 1972. Genetics of white spruce. U.S.D.A., For. Serv., Res. Pap. W0-15.

O'Reilly, G.J., W.H. Parker, and J. Maze. 1982. Correlation of major environmental effects with phenotypic variation in *Abies balsamea* from northwestern Ontario. Botanical Society of America. Misc. Pub. No. 162: 40 (Abstract).

Parker, W.H., J. Maze, and G.E. Bradfield. 1981. Implications of morphological and

anatomical variation in *Abies balsamea* and *A. lasiocarpa* (Pinaceae) from western Canada. Amer. J. Bot. 68: 843-854.
Pauley, S.S. and T.O. Perry. 1954. Ecotypic variation of the photoperiodic response in *Populus*. J. Arnold Arboretum 35: 167-188.
Porterfield, R.H., B.J. Zobel, and F.T. Ledig. 1975. Evaluating the efficiency of tree improvement programs. Silvae Genetica, 24: 33-44.
Ritchie, J.C. 1976. The late-Quaternary vegetational history of the Western Interior of Canada. Can. J. Bot. 54: 1793-1818.
Roe, A.L. 1967. Seed dispersal in a bumper spruce seed year. U.S.D.A., For. Serv., Res. Pap. INT-39.
Rudolph, T.D. and C.W. Yeatman. 1981. Genetics of jack pine. U.S.D.A., For. Serv, Res. Pap. W0-38.
Schier, G.A. 1981. Physiological research on adventitious shoot development in aspen roots. U.S.D.A., For. Serv., Gen. Tech. Rep. INT-107.
Schoenike, R.E. 1976. Geographical variations in jack pine *(Pinus banksiana)*. (University of Minnesota, Stn.) Agric. Exp. Tech. Bull. 304.
Scotter, G.W. 1974. Distribution of pine *(Pinus* spp.*)* in the south Nahanni and Flat Rivers Region, Northwest Territories. Can. J. For. Res. 4: 555-557.
Whitehead, D.R. 1973. Late-Wisconsin vegetational changes in unglaciated eastern North America. Quat. Res. 3: 621-631.
Wilkinson, R.C., J.W. Hanover, J.W. Wright, and R.H. Flake. 1971. Genetic variation in monoterpene composition of white spruce. For. Sci. 17: 83-90.
Wright, H.E. 1972. Interglacial and postglacial climates: The pollen record. Quat. Res. 2: 274-282.
Wright, J.W. 1955. Species crossability in spruce in relation to distribution and taxonomy. For. Sci. 1: 319-349.
Wright, J.W. 1976. Introduction to Forest Genetics. Academic Press, New York.
Yeatman, C.W. 1973. Gene conservation in relation to forestry practice. Proc. 13th Meeting of the Committee on Forest Tree Breeding in Canada: Part 2. Prince George, British Columbia. pp. 19-24.
Yeatman, C.W. 1976a. Seed origin-first, last and always. Environ. Can., C.F.S., Infor. Rep. PS-X-64.
Yeatman, C.W. 1976b. A Canadian example of government-industry collaboration in tree improvement. For. Chron. 52: 283-288.

BIOCLIMATICAL REGIONS AS A FRAMEWORK FOR THE STUDY OF BOREAL FOREST ECOSYSTEMS

Vincent Gerardin and Jean-Pierre Ducruc
Environment Quebec

Abstract: We have attempted to put some order into the loosely defined approaches to bioclimatical zonation methods for northern territories. Our approach can be summarized as follows: a hierarchical bioclimatical subdivision is established on the basis of vegetation physiognomy and composition growing under similar mesic soil conditions (texture and drainage) and belonging to the same dynamic stage (final or quasi-final stage of evolution). The hierarchy comprises four levels. From the general to the specific, we classify them as follows: biological zones are a physiognomic subdivision (forests, wooded lands, tundra); biological sub-zones are a structural and cover composition subdivision (closed balsam fir forests, open black spruce forest, very open black spruce forests); domains are based on a physiognomic, structural and cover composition correlated with altitudinal or maritime climatical modifications (cloud vegetation, high altitude tundra) and land regions are a phytosociological, dynamic and productivity-based subdivision (high-productivity balsam fir forests; medium-productivity black spruce-mosses forests; low-productivity black spruce-lichens forests). Forest productivity is used to demonstrate the importance of bioclimatical zonation in the domain of primary production evaluation.

Natural production, primary as well as secondary, is first of all determined by climatic conditions. Human activities are also deeply controlled by climate. Hence no land management can be established outside a climatic framework. In this paper we intend to present a simple approach to bioclimatical zonation of large territories.

We know that it is impossible to define climatic thresholds for plants solely on the basis of climatic data, for two major reasons. One reason is the poor meteorological network (stations are too recent, there are not enough of them, and the existing ones are very poorly distributed). The second reason is more fundamental; plants do not react to independent climatic parameters, but are sensitive to a global climatical balance. There is no significant climatological threshold which can explain plant distribution. Climatic limits are not generally plant distribution limits. Climatic maps and charts help the ecologist understand the major climatic gradients but do not allow him to define bioclimatical regions. A bioclimatical region is an area of land within which a particular pattern of vegetation is in equilibrium with a particular climatic balance.

The Land Region

The most widely accepted definition in Canada of a land region or ecoregion is given by Lacate (1969): '...an area of land characterized by a distinctive regional climate as expressed by vegetation'.

This definition is so broad that everybody is pleased with it, since it can be given any meaning at all, without deviating too far from the original. For instance, it has recently been written: 'Ecoregions are loosely defined as areas of land that possess a recognized common identity from a regional and ecological perspective' (Wiken *et al.* 1981). Unfortunately, the term 'regional climate' has never been defined; nor have the elements of vegetation which are an 'expression' of regional climates. In this paper we have attempted to answer the second part of the question on the basis of our experience.

In Canada, few authors, if any, have expressed their methodological approach to land region mapping and classification. The taxonomic versus the mapping dimension of the question is still equivocal, if not muddled. The numerous studies of Emberger (1930, 1942, 1955) on bioclimatical regions by a phytosociology-climatology integrated approach (pluviothermic coefficients) are better known, particularly in Mediterranean regions. The works of Brisse and Grandjouan (1974, 1975, 1977) are also of interest. In the northlands, where the few meteorological stations are dispersed and recent, there seems to be no clearly defined methodological approach (except for Krajina's (1959, 1965, etc.) and Hills' (1961) studies). This has not been any more of a concern in Quebec than elsewhere. For a decade, however, many ecologists have tried to consistently define a hierarchy in the bioclimatic zonation of land regions across Quebec, particularly in boreal and subarctic territory (Gerardin *et al.* 1975; Zarnovican *et al.* 1976; Ducruc *et al.* 1976; Payette 1976; Gerardin 1980), and there are less recent papers on the macrozonation of the Quebec-Labrador peninsula (Hare 1950, 1959; Hare and Taylor 1956; Rousseau 1949, 1952; Hustich 1949, 1950). In the present paper we shall attempt to summarize the approach of our team towards defining and delineating the different levels of bioclimatical zonation.

Field Sampling

Soils and vegetation: Field sampling of soils and vegetation is basic to the establishment of a bioclimatical zonation. Data collected include origin and texture of the soils (parent material), soil drainage, vegetation physiognomy and structure, and stand floristic composition.

The combination parent material – soil drainage defines what we call 'preliminary land type' (L.T.). We have given a 'vegetation physiognomic type' (V.P.T.) (Figure 1), expressed through four parameters to each L.T. identified in the field. The four parameters defining the V.P.T. are the physiognomy (forest, shrub stand, meadow), the cover type (specific composition of the highest stratum of the stand), the stand structure (estimate crown density and average height), and the 'life-forms' of the ground layer (lichens, mosses, herbs, grasses, forbs). Although less frequently, we also add phytosociological *relevés* which will be later classified in 'preliminary vegetation stands' (V.S.). The L.T., the V.P.T. and the V.S. are thus the basic data used to define land regions.

Vegetation through all its expressions is the result of the combined effect of five major ecological variables. This is generally shown in the Major (1951) formula, as revised by Jenny (1958): V=f(C,P,R,B,T), where V=vegetation, C=climate, P=parent material, R=relief, B=biota, and T=time. Therefore, vegetation is an expression of the regional climate once the four other parameters of the equation are constant. This briefly, is the basic principle of our approach.

The vegetation physiognomic type describes four parameters:

(1) the physiognomy, (2) the cover type, (3) the structure, (4) the ground layer "life-forms".

(1) PHYSIOGNOMY

The following physiognomic classes are recognized:

AB: low shrub stand (shrubs <1 m)
AH: high shrub stand (shrubs >1 m)
DE: barren
FO: forest
FR: stunted forest
KR: krummholz
LB: wooded land
MU: moss stand
PB: low meadow (<0.5 m)
PH: high meadow (>0.5 m)

(2) COVER TYPE

Up to three species characterizing the physiognomy can be chosen from this category.

(3) STRUCTURE

The structure is described by a height-density grid (bottom of table).

(4) GROUND LAYER "LIFE-FORMS"

In addition to the species characterizing the physiognomy, up to three groups of plants which dominate in the ground layer are listed.

Some examples of vegetation physiognomic type:

FO/EN/D4/E,S: open (D) black spruce (EN) forest (FO), very low (4), with ericacees (E) and sphagnum (S)
KR/EN/B5/M,E: black spruce (EN) krummholz (KR), closed (B), high (5) with mosses (M) and ericacees (E)
MU/LI/B9/R: closed (B) lichens (LI) formation with creeping shrubs (R)

FORMATION	density	classes	TREES				SHRUBS		HERBS		MOSSES	BARREN
		height m	21	15-20	9-14	3-8	1-3	1	0,5	0,5	0,1	0
		classes	1	2	3	4	5	6	7	8	9	0
VERY CLOSED	80	A	A1	A2	A3	A4	A5	A6	A7	A8	A9	
CLOSED	60-79	B	B1	B2	B3	B4	B5	B6	B7	B8	B9	
MOD. OPEN	40-59	C	C1	C2	C3	C4	C5	C6	C7	C8	C9	
OPEN	25-39	D	D1	D2	D3	D4	D5	D6	D7	D8	D9	
VERY OPEN	5-24	E	E1	E2	E3	E4						
BARREN	5	F										F0

Figure 1. Parameters defining the vegetation physiognomic type.

In practical terms, this is achieved by using for analysis samples which are characterized by an identical substratum: origin and texture (factor P); the same type of drainage, reflecting a particular topographical position (factor R) and a vegetation cover that has attained a quasi-final or a final stage of evolution (factor T). The stet biota (factor B) which is too theoretical may be considered constant throughout large territories.

We attach a great deal of importance to the notion of 'mesic' conditions in defining land regions for the obvious reason that climate–vegetation relationships will be more clearly demonstrated if the other ecological factors do not have too strong an effect (positively or negatively) on vegetation itself. For instance, if the drainage is excessively rapid or very slow the vegetation growing on those sites will be the result of the combined effect of regional climate and soil drainage. And it will be difficult, if not impossible, to determine the respective influence of each factor. In Quebec northlands, we are using well to moderately well drained tills. Tills are the most important parent materials of Quebec northlands and are of an 'average' texture. From the well drained tills population we have eliminated every L.T. – V.P.T. combination where the V.P.T. represents a pioneer stage after fire or logging operations. For a certain proportion of the samples, we have phytosociological *relevés*. For these data we established a list of species for which the relative frequency within the *relevés* is greater than or equal to five per cent.

Climatic indicators: In general, climatic gradients can be correlated with the three geographical variables of latitude, longitude and altitude. Therefore, we intend to study the relationships between those geographic variables and the distribution of E (species), V.P.T. (vegetation physiognomic type) and V.S. (vegetation stands). The distance from the sea is also an important climatological factor and therefore should be taken into consideration. However, in the two major studies for which we have defined land region, this parameter was redundant either with longitude (James Bay territory) or latitude (St. Lawrence middle and lower north shore).

Biological zones and sub-zones

Biological zones are the first order of bioclimatical subdivision. Distribution of a species E within the classes of a variable L is what we call the ecological profile of species E towards variable L and reflects, more or less adequately, the ecological behavior of the species towards the given variable (Gerardin 1977).

Thus we compute, for the species we use, the ecological profiles for latitude, longitude and altitude (Figure 2). The information obtained is analysed through two complementary paths. The ecological profiles are classified through a hierarchical classification algorithm (Fortin 1976) which makes it possible to set up ecoclimatic groups of species having, within a given group, a similar behavior (distribution) with respect to the geographical variables (Figure 3). Maps of synoptic distribution of species or ecoclimatic groups of species are also set up in order to visualize the spatial relationships between species and geographical variables (Figure 4).

From the conjoined interpretation of the ecoclimatic groups and the maps, we formulate the first hypotheses concerning the bioclimatical limits of the territory. These hypotheses are compared with the findings of published studies and make it possible to define and map the biological zones (Figure 5).

Biological sub-zones are the second order bioclimatical subdivisions. All the V.P.T.

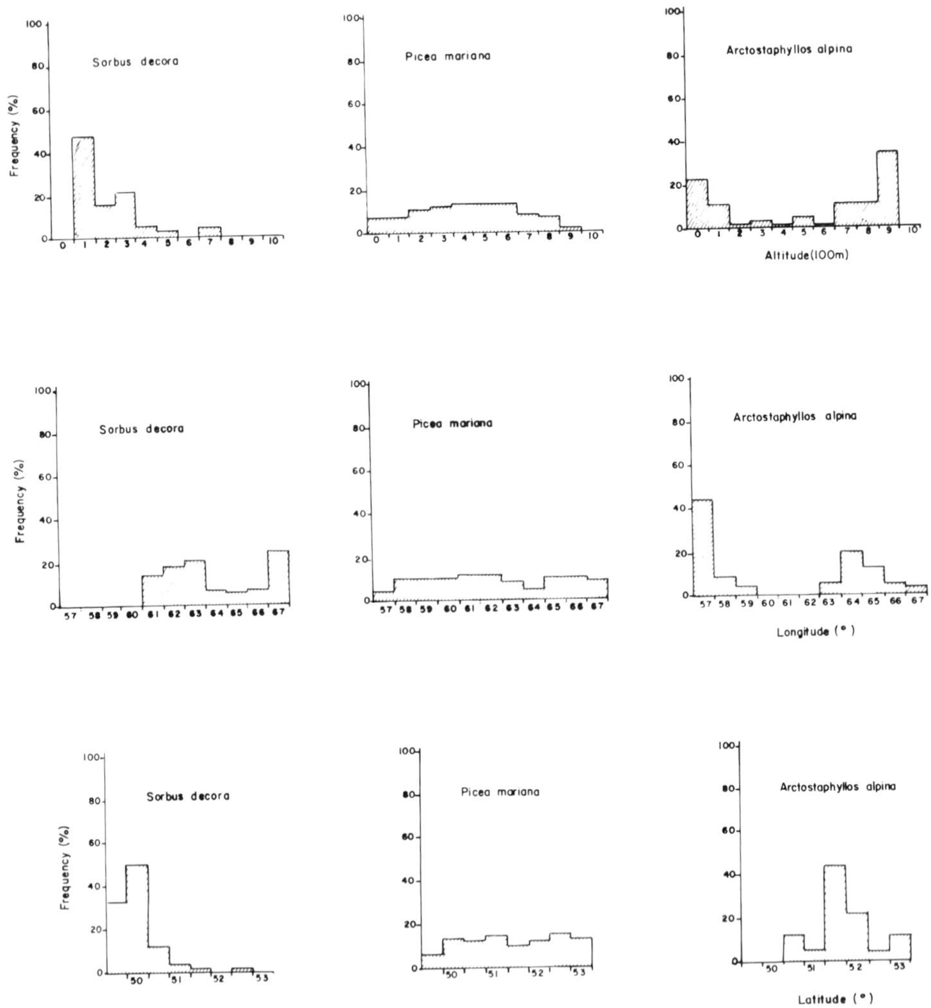

Figure 2. Ecological profiles of three species with respect to altitude, longitude and latitude on well drained tills on the St. Lawrence middle and lower north shore.

retained (as noted above) are then plotted on a small scale (1:500 000) map of the territory.

The study of the spatial distribution of stand density makes it possible to determine some large structural units. Their limits confirm the biological zones or subdivide them, thus defining the biological sub-zones. For example, the Quebec middle and lower north shore boreal zone has been subdivided into two sub-zones (Figure 5): a lower boreal sub-zone characterized by moderately open forests (40 per cent to 60 per cent crown density) and an upper boreal sub-zone dominated by open forest (25 per cent to 40 per cent crown density).

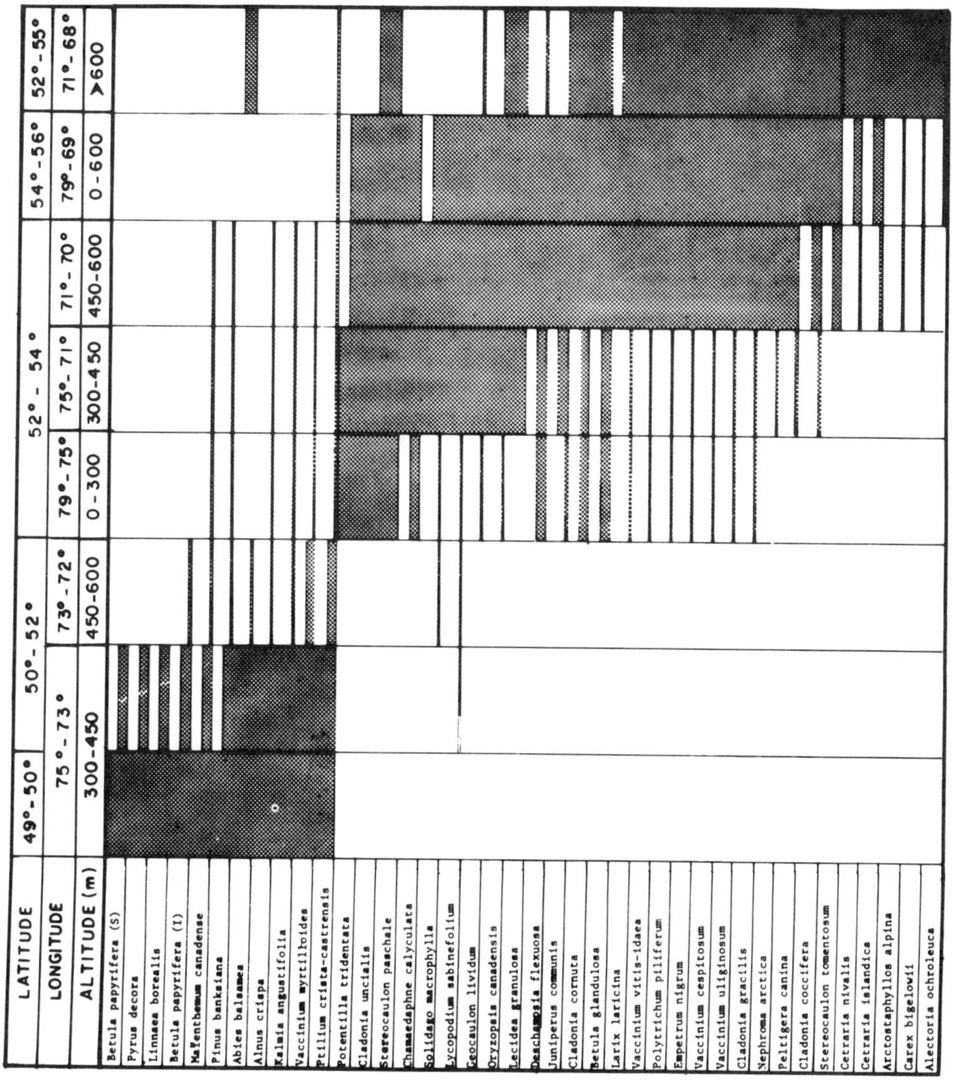

Figure 3. Ecoclimatical group of species with respect to latitude, longitude and altitude on well drained tills in James Bay territory (from Gerardin *et al.* 1975).

Domains

Climatic gradients defined according to latitude and longitude, disregarding the maritime and altitudinal influences, determine what we call the continental domains. Rey (1960) gives a more restrictive definition to continentality since it is defined solely on the basis of the fact that the territory is not subject to maritime influence. In our concept, continentality describes, in a sense, the "normal" bioclimatical conditions.

Climatic modifications due to altitude generate two other categories: the mountainous domain and the alpine domain. The mountainous domain comprises the average altitudinal conditions generating differences in vegetation cover, forest density and forest productivity. The upper limit of that domain corresponds to the upper forest limit. The

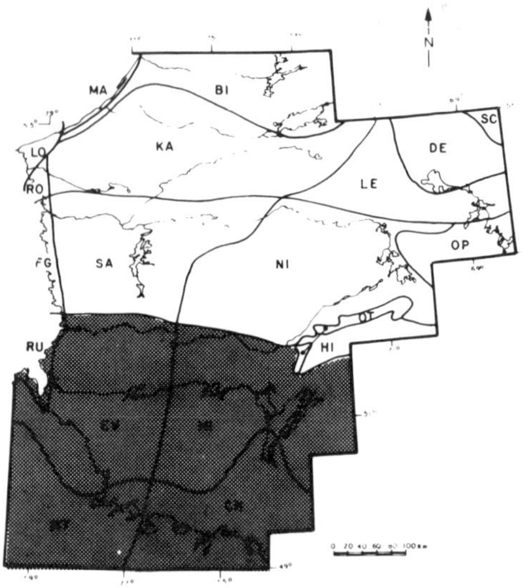

Distribution préférentielle de: *Sorbus decora, Alnus rugosa, Prunus pensylvanica, Mitella nuda, Cornus stolonifera, Amelanchier stolonifera, Cinna latifolia, Rhytidiadelphus triquetrus, Actaea rubra, Prunus pensylvanica, Dicranum drummondii*

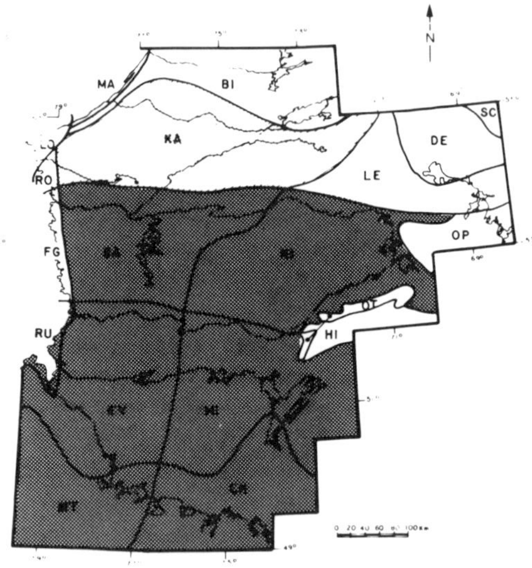

Distribution préférentielle de: *Pinus banksiana(s), Pinus banksiana(i), Kalmia angustifolia*

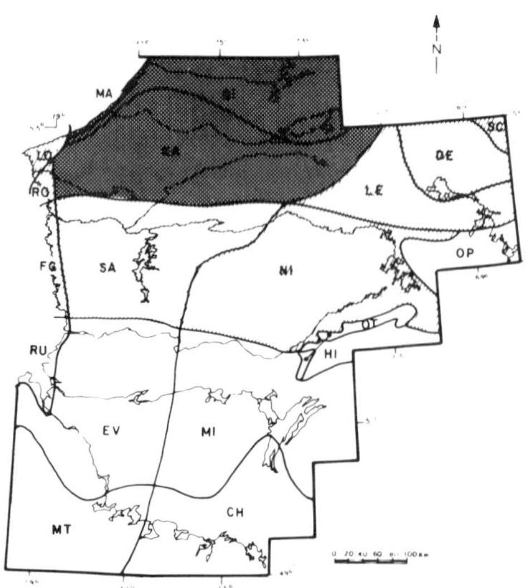

Distribution préférentielle de: *Potentilla tridentata, Lycopodium sabinaefolium*

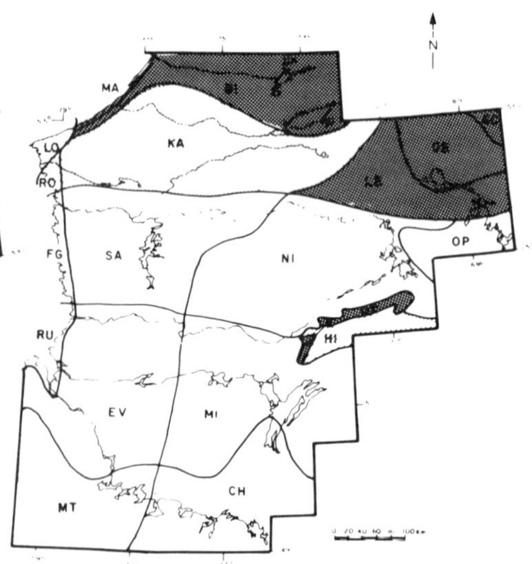

Distribution préférentielle de: *Loiseleuria procumbens, Cetraria nigricans*

Figure 4. Preferential distribution of some species on well drained tills in James Bay territory (from Gerardin 1980).

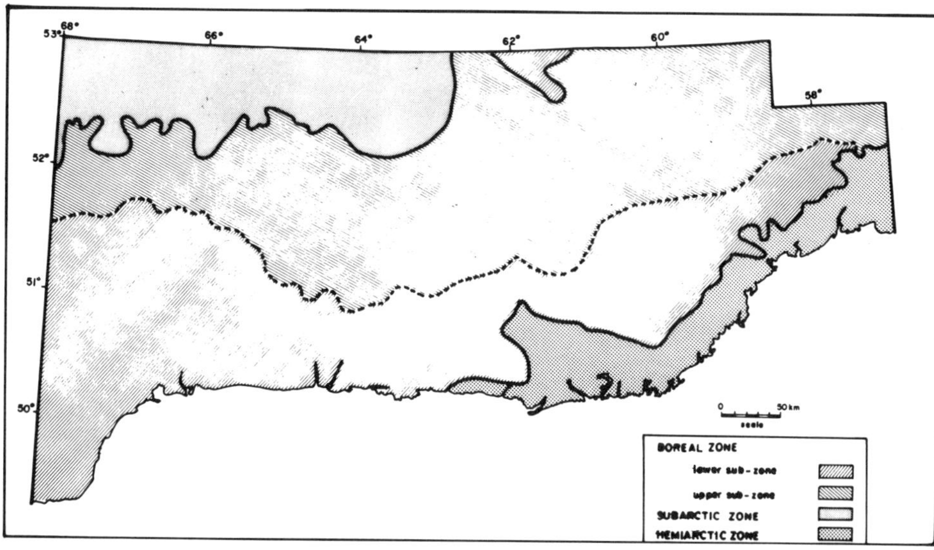

Figure 5. Biological zones and sub-zones of the St. Lawrence middle and lower north shore.

alpine domain encompasses what is generally accepted as alpine vegetation: krummholz and tundra.

Finally, proximity of the sea also introduces deviations with respect to continental conditions and therefore distinguishes a final category, the maritime domain. As opposed to mountainous and alpine domains, the limits of which are easy to define, it is often difficult in many coastal areas to pinpoint disruptions or discontinuities between continental and maritime domains. In some places, such as the hemiarctic zone of the St. Lawrence north shore, the structure and physiognomy of the vegetation are so different from those of the continental conditions (tundra and krummholz versus forest) that the limits can readily be seen on aerial photographs. In other places, only the presence, or simply a relatively greater abundance, of certain species, like white spruce (Payette and Filion 1975; Ducruc *et al.* 1976; Gerardin 1980) can be relied on — this makes the limits very delicate. Despite these factors, the maritime, mountainous and alpine domains are modifiers of the biological sub-zones.

Successive Approximation to Land Region Definition

Within the framework of zones, sub-zones and domains we analyse the distribution of the other components of the V.P.T.: the cover composition and the ground layer 'life-forms'. The geographical distribution of these parameters generates new subdivisions: the provisional land regions. From preliminary vegetation tables, a certain number of stands (V.S.) are defined where only the dominant floristic composition intervenes. For example, eleven V.S. have been set up for the St. Lawrence middle and lower north shore territory (Table 1). As already mentioned, that preliminary classification is valid for mesic sites only. From the *relevés,* then, the relative importance of V.S. is computed for each provisional land region (Figures 6, 7 and 8). This approach makes it possible to confirm and/or modify the contents and/or the limits of the provisional land regions.

Table 1. List of "vegetation stands" defined for the St. Lawrence middle and lower north shore territory.

Code	Vegetation stands
BB	white birch stands
SB.m	balsam fir — mosses stands
SB.eb	balsam fir — white spruce stands
SB.en.m	balsam fir — black spruce — mosses stands
EN.sb.m	black spruce — balsam fir — mosses stands
EN.m	black spruce — mosses stands
EN.m.lg	black spruce — mosses — ledum stands
EN.ce	black spruce — lichens stands
FR	stunted forests of balsam fir, white spruce and black spruce. Altitude stands
KR	krummholz formation of either balsam fir white spruce or black spruce
ALP	tundra; Arctic or alpine vegetation

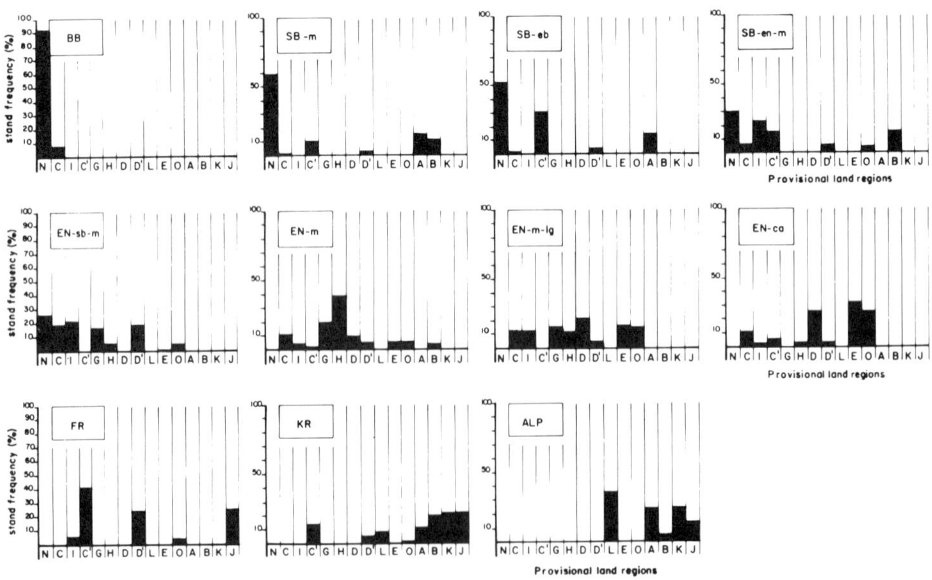

Figure 6. Frequencies of "vegetation stands" across the provisional land regions of the St. Lawrence middle and lower north shore (from Gerardin 1981).

The last step consists in evaluating the homogeneity of the land regions compared with vegetation chronosequences and natural primary productivity. This last point is of uttermost importance considering that the land region should be the first discriminant parameter in evaluating land productivity. Setting up chronosequences necessitates a final classification of land types and floristic data. When the floristic data are classified, the

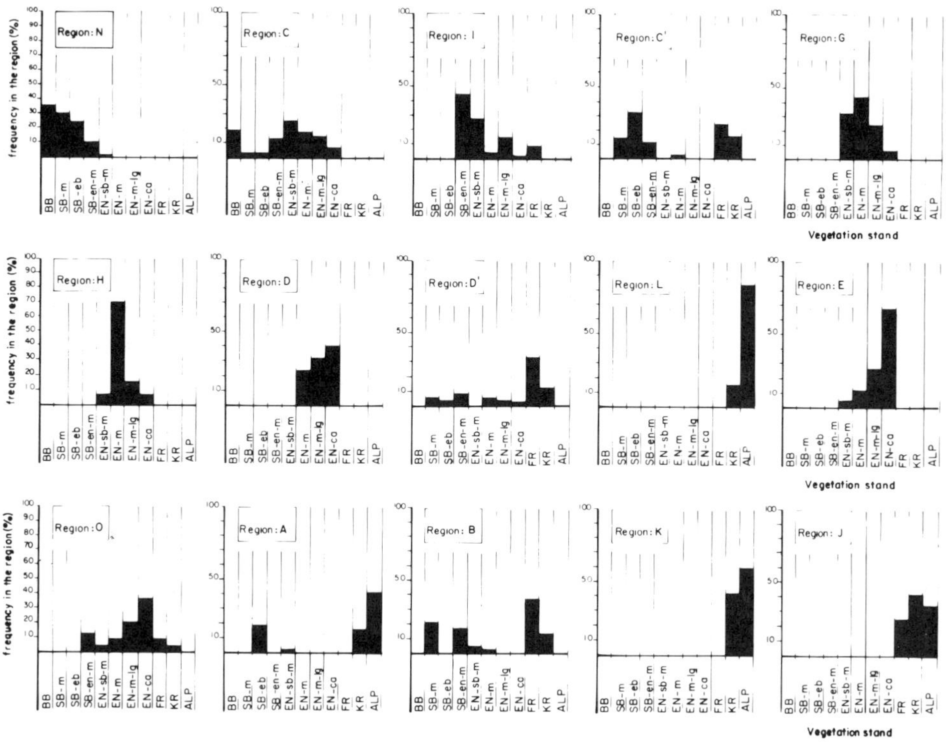

Figure 7. Relative frequency of "vegetation stands" for each provisional land region in the St. Lawrence middle and lower north shore (from Gerardin 1981).

vegetation stands thus defined are arranged chronosequentially by each land type to which they belong. The chronosequences are then compared region with region thereby generating possible adjustment of the land regions.

The analysis of forest productivity may also generate modifications in land region limits and definitions. This generally occurs when a forest productivity gradient appears (latitudinal, longitudinal or altitudinal) within a homogeneous vegetation pattern.

Land Region Mapping

As we understand it, the land region is mainly a taxonomical unit rather than a mapping unit in the sense that the mapping of land regions consists essentially in drawing a concept on a map. The concept here means an abstracted integration of phytoclimatological relationships. In other words, a land region mapping unit is a spatial unit encompassing a characteristic pattern of soil-vegetation (structure, composition, dynamics) which is visible neither in the field nor on aerial photographs.

However, under particular conditions land region limits can be drawn by interpreting satellite images and aerial photographs, with a certain amount of precision. This occurs when the climatic conditions change so drastically that the structure of vegetation is completely modified over short distances. The most pertinent example is given by the

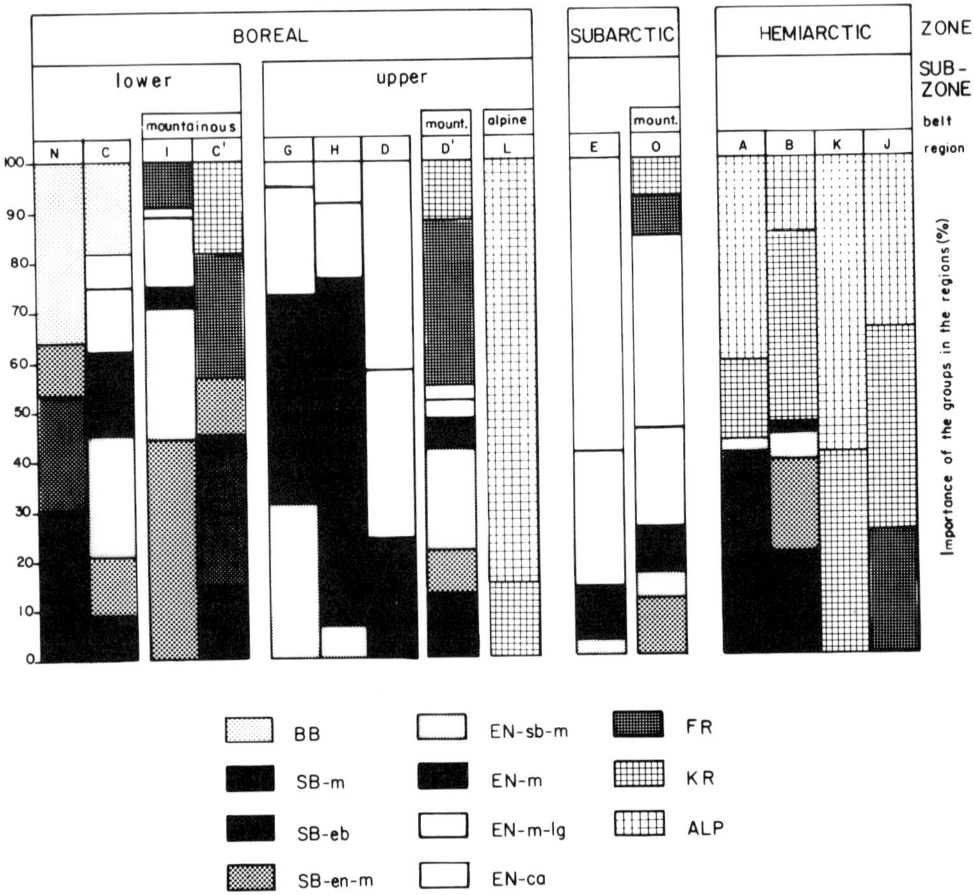

Figure 8. Pattern of "vegetation stands" for the provisional land regions of the St. Lawrence middle and lower north shore (from Gerardin 1981).

forest – tundra limits. In that case photo-interpretation makes it possible to draw limits more accurately than can be done between land regions dominated in one part by fir forests and in the other part by spruce forests.

The St. Lawrence Middle and Lower North Shore Land Regions

With the approach defined in this paper we have described 19 land regions (Table 2 and Figure 9). These belong to three biological zones. The boreal zone is dominated on mesic sites by black spruce and balsam fir forest (average cover density 45 per cent). The subarctic zone is dominated on mesic sites by black spruce wooded land to very open forest (average cover density 30 per cent). The hemiarctic zone is a pattern of tundra and krummholz on the interfluves and forests in the valleys.

The boreal biological zone is divided into two sub-zones. The lower boreal zone corresponds to the southern part of the zone and is characterized by moderately open forests of black spruce, balsam fir and mosses. The upper boreal zone is characterized by

Table 2. Land region classification of the St. Lawrence middle and lower north shore.

ZONES	SUB-ZONES	DOMAINS	LAND REGIONS
BOREAL	LOWER Moderately open forests	MARITIME Balsam fir, black spruce, white spruce, mosses.	TR HA RC
		CONTINENTAL Black spruce, balsam fir, mosses.	TO OL
		MOUNTAINOUS Balsam fir, white spruce, black spruce, mosses.	MM[1] PO
	UPPER Open forests	CONTINENTAL Black spruce, mosses.	FM FO AI SL
		MOUNTAINOUS Black spruce, balsam fir, white spruce, mosses.	CA[1]
		ALPINE Krummholz, tundra.	MG
SUBARCTIC	LOWER Wooded land	CONTINENTAL Black spruce, lichens.	AT
		MOUNTAINOUS Black spruce, balsam fir, lichens, mosses.	OP
HEMIARCTIC Tundra and forest		MARITIME Summits: tundra and krummholz Valleys: balsam fir, white spruce forests.	BS MU BR
		MOUNTAINOUS Krummholz	BU

1. Not mapped because these land regions are dispersed across their sub-zone.

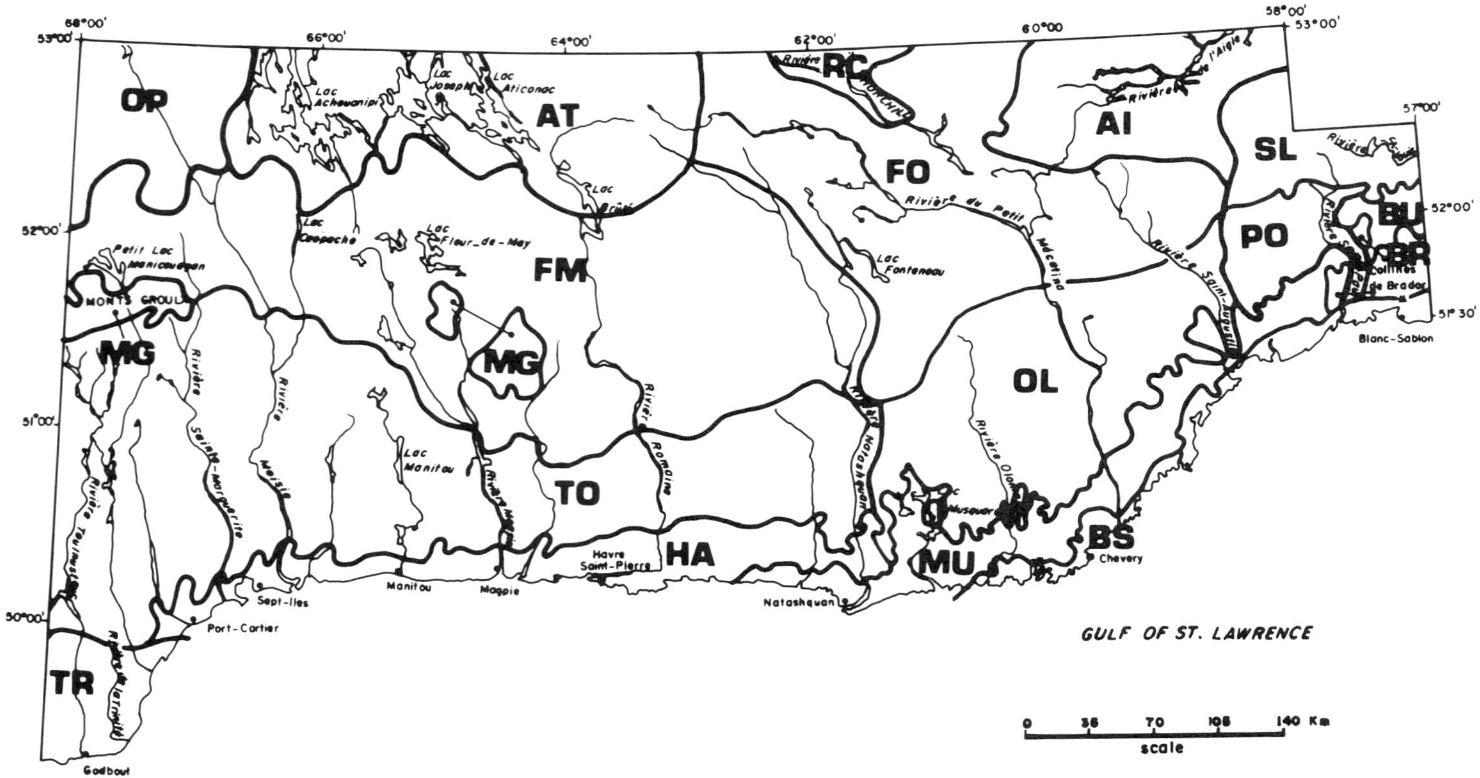

Figure 9. Land regions of the St. Lawrence middle and lower north shore (for the legend see Table 2).

open forests dominated only by black spruce. Since there is no upper subarctic zone in this territory, the lower subarctic zone has been defined according to the subarctic sub-zones of the adjacent James Bay territory (Gerardin 1980). The biological zones and sub-zones are then subdivided into two or three domains. A very succinct description of these domains is given in Table 2.

The Interpretative Potential of the Land Regions: An Example of Forest Productivity

In principle, a particular forest tree composition and a distinct tree growing rhythm must correspond to each land region. First, the forest cover composition has an important influence on the total wood production of a forest. Figure 10 shows this kind of relationship. Six different graphs representing six land regions display: the average stand crown density (D), the average stand height (H) and the product of these two terms (D × H) for four stands composition on well drained soils. It can be seen that an increase of balsam fir in the stand results in an increase of stand density and stand height. Since the crown density is strongly correlated with the basal area (Gerardin 1982), the product D × H is a good indicator of total wood production. Therefore, land regions dominated by balsam fir are already more productive than land regions dominated by black spruce, everything else being equal.

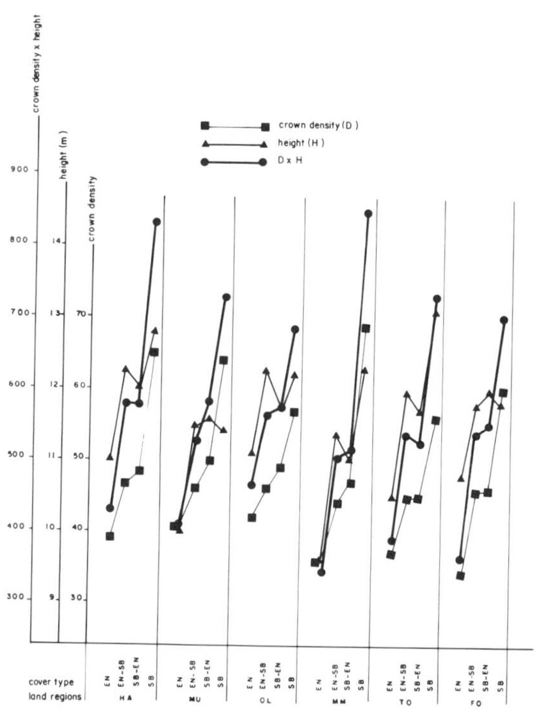

Figure 10. Relation between cover type and crown density and stand height for some land regions on the St. Lawrence middle and lower north shore, on mesic sites (EN = pure black spruce forest; SB = pure balsam fir forest, EN-SB and SB-EN = mixed black spruce – balsam fir forest).

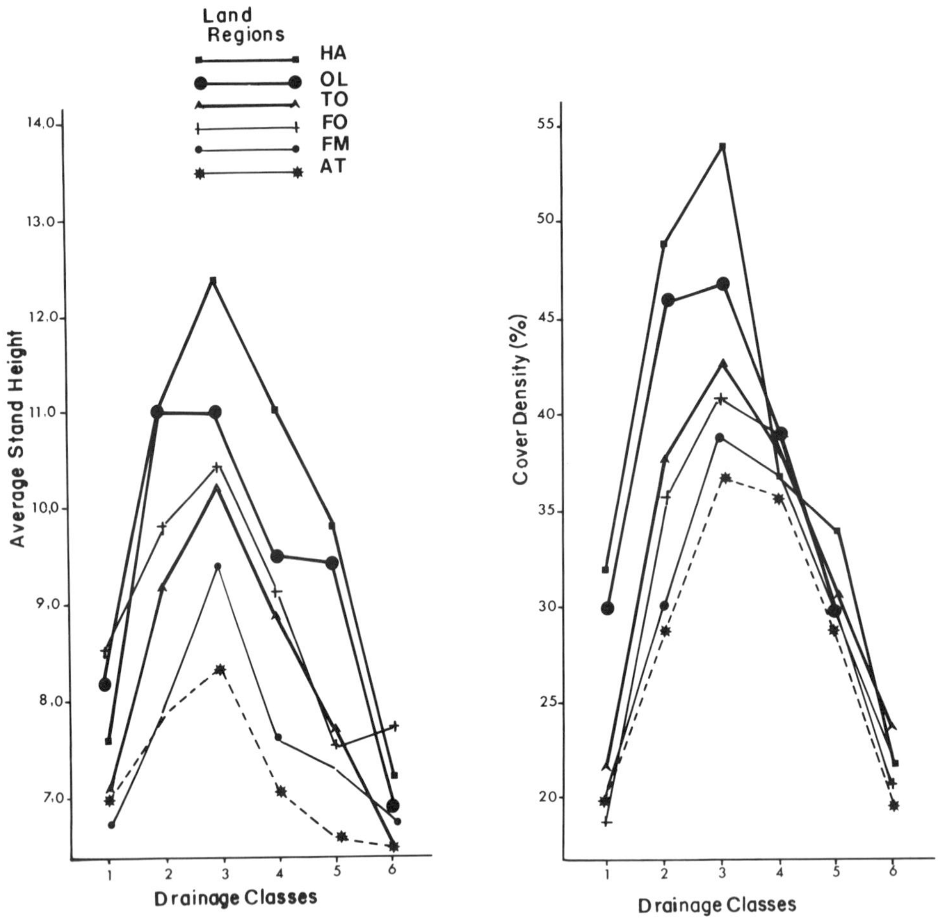

Figure 11. Average stand height (left) and average cover density (right) for coniferous stands with respect to drainage classes for some land regions of the St. Lawrence middle and lower north shore.

In Figure 11 the discriminant value of land regions is again demonstrated with regard to average stand height and crown density with respect to the six drainage classes for the principal land regions. These two figures clearly show that the lower boreal zone (land regions HA, OL, TO) is more productive than the upper boreal zone (FO, FM) which in turn is more productive than the subarctic zone (AT). Moreover, within each sub-zone, land regions are clearly distinguished. However, it can be seen that differences in productivity are greater for mesic drainage (drainage classes 2 and 3) than for poor drainage, especially with regard to forest density.

Similar results were obtained on the basis of dendrometric studies. However, there are not enough field data to permit an accurate evaluation of the forest productivity potential of the land regions. The influence of the major climatic gradients of the territory (latitudinal and altitudinal) on forest growth, however, is clearly shown in Table 3. Black spruce site-index curves have been computed for the territory (Gerardin 1982) and Table 3 shows

Table 3. Relationships between the black spruce site-index at 100 years by latitude and altitude on a well drained site

	Latitude		
	<51°30′N	51°30′N – 52°30′N	>52°30′N
Without seepage	11,6 m	10,5 m	9,4 m
With seepage	14,1 m	12,2 m	9,7 m
	Altitude[1]		
	<200 m	200 to 400 m	>400 m
Without seepage	11,9 m	11,1 m	9,8 m

[1] Within the boreal zone.

its average value on mesic sites with and without seepage for three latitudinal classes. The first class (<51°30′N) corresponds more or less to the lower boreal zone, the second class (51°31′-52°30′N) to the upper boreal zone, while the last class (>51°30′N) corresponds more or less to the subarctic zone. The interpretation of that table is similar to that of Figure 11: the lower boreal zone is more productive than the upper boreal zone, which in turn is more productive than the subarctic zone. We have compared, for well drained sites, the average black spruce height at 100 years for the boreal zone with respect to altitude (Table 3); the relationship is evident.

Conclusions

The main advantage of this hierarchical approach is its great simplicity since it is founded overall on the analysis of the physiognomic vegetation type which is excessively simple and rapid to use in the field and which is rich in information since with four parameters, the specific cover composition, the density, the average height and the dominant 'life-forms' of the ground layer are defined. Coupled with land types, the vegetation physiognomic type is therefore a very sensitive tool for evaluating macroclimatic influences.

Moreover, by its inherent simplicity, the vegetation physiognomic type seems advantageous for establishing the lower threshold in distinguishing the land region, thereby preventing the ecologist from concerning himself solely with eco-climatic relationships which could be of a lower level of perception than should be the land regions.

But overall any discussion about the inherent quality of the different approaches to the bioclimatical zonation of a country, it should be totally accepted that no land management which encompasses biological aspects can be realistic and efficient if it is not integrated in a bioclimatical framework.

Literature Cited

Brisse, H. and G. Grandjouan. 1974. Classification climatique des plantes. Oecol. Plant. 9: 51-80.

Brisse, H. and G. Grandjouan. 1975. Etalonnage climatique des plantes dans un territoire. Protocole des opérations pratiques. CNRS, Laboratoire d'écologie végétale. Strasbourg, France.

Brisse, H. and G. Grandjouan. 1977. Etalonnage et classification de 450 plantes en France. Berichte der Jut. Symp. der Jutern. Vereinigung für Vegetations kund Gerand gegeben von Reinhold Tüxen. Vegetation und Klima: 535-607.

Ducruc, J.P., R. Zarnovican, V. Gerardin and M. Jurdant. 1976. Les régions écologiques de la Baie-James: caractéristiques codominantes de leur couvert végétal. Cahiers de géographie de Québec 20: 265-392.

Ducruc, J.P. and G. Audet. 1980. Les régions écologiques provisoires (1ère approximation). Inventaire du Capital-Nature de la Côte-Nord. Rapport no 7.

Emberger, L. 1930. Sur une formule climatique applicable en géographie botanique. C.R. Acad. Sci. 191: 389-391.

Emberger, L. 1942. Un projet de classification des climats du point de vue phytogéographique. Bull. Soc. Hist. nat. Toulouse. 77: 97-124.

Emberger, L. 1955. Une classification biogéographique des climats. Recueil des trav. Lab., Géol. et Zool., Fac. Sci. Montpellier. Bot. 7: 3-43.

Fortin, M. 1976. Sut un algorithme pour l'analyse des données et la reconnaissance des formes. Rev. Stat. Appliquée, 23: 37-45.

Gerardin, V. 1977. An integrated approach to the determination of ecological groups in vegetation studies. Ph. D. thesis, unpublished.

Gerardin, V. 1980. L'inventaire du Capital-Nature du territoire de la Baie-James: les régions écologiques et la végétation des sols minéraux.

Gerardin, V. 1981. Inventaire du Capital-Nature de la Côte-Nord: les régions écologiques provisoires (2ième approximation). Service des inventaires écologiques, Environnement Québec.

Gerardin, V. 1983. Analyse de quelques facteurs contrôlant la production forestière sur le territoire de la moyenne et basse Côte-Nord. Service des inventaires écologiques, Environnement Québec. Série de l'inventaire du Capital-nature, no 1.

Gerardin, V., J.P. Ducruc, R. Zarnovican and M. Jurdant. 1975. Recherches de régions écologiques dans le Territoire de la Baie-James: définition d'une méthodologie. Compte-rendu de la Conférence circumpolaire sur l'écologie du Nord, 15-18 sept., 1975: National Research Council of Canada, I: 113-130.

Gerardin, V., J.P. Ducruc and R. Zarnovican. 1977. Analyse de la végétation dans le cadre d'un inventaire écologique intégré. *In:* "L'Inventaire du Capital-Nature", Jurdant *et al.,* 1977. Environnement Canada, Direction des Terres. Série de la classification écologique du territoire, no 2, pp. 57-64.

Hare, F.K. 1950. Climate and zonal divisions of the boreal forest formation in Eastern Canada. Geogr. Rev. 40: 615-635.

Hare, F.K. 1959. A photo reconnaissance survey of Labrador-Ungava. Geogr. Br. Mines and Tech. Survey, Memoir 6.

Hare, F.K., and G. Taylor. 1956. The position of certain forest boundaries in Southern Labrador-Ungava. Geogr. Bull. 8: 51-73.

Hills, G.A. 1961. The Ecological Basis for Land Use Planning. Ontario Dept. Lands and Forest, Res. rep. no 42.

Hustich, I. 1949. Phytogeographical regions of Labrador Arctic, 2: 36-42.

Hustich, I. 1950. Notes on the forests on the East coast of Hudson Bay and James Bay. Acta Geograph. 11.

Jenny, H. 1958. Role of the plant factor in the pedogenic functions. Ecology 39: 5-16.

Krajina, V.J. 1959. Bioclimatic zones of British Columbia. University of British Columbia, Botanical series No 1.

Krajina, V.J. 1965. Biogeoclimatic zones and classification of British Columbia. Ecol. Western N.A. 1: 1-7.

Lacate, D.S. 1969. Guidelines for Bio-physical Land Classification. Canada Dept. of Forestry, Canadian Forestry Service, Publ. No 4.

Major, J. 1951. A functional factorial approach to plant ecology. Ecology 32: 392-412.

Payette, S. 1976. Les limites écologiques de la zone hémi-arctique entre la mer d'Hudson et la Baie d'Ungava, Nouveau-Québec. Cahiers de Géographie de Québec 20: 347-364.

Payette, S., and L. Filion. 1975. Ecologie de la limite septentrionale des forêts maritimes, Baie d'Hudson, Nouveau-Québec. Nat. Can. 102: 783-802.

Rey, P. 1960. Essai de phytocinétique biogéographique. Centre National de la Recherche Scientifique. France.

Rousseau, J. 1949. La zone hémiarctique. Proc. Trans. Roy Soc. Can., 43 (Série 3) (app.F).

Rousseau, J. 1952. Les zones biologiques de la péninsule Québec-Labrador. Ecole pratique des hautes études, Centre d'Etudes Arctiques en Antarctiques. Sorbonne, France.

S.I.E. 1981. Vade-mecum des relevés écologiques. Service des inventaires écologiques, Ministère de l'Environnement. Québec, Québec.

Wiken, E.B., D.M. Welch, G.R. Ironside and D.G. Taylor. 1981. The northern Yukon: an ecological land survey. Ecological Land Classification Series, No 6. Ottawa, Ontario.

Zarnovican, R., V. Gerardin, J.P. Ducruc, M. Jurdant and G. Audet. 1976. Les régions écologiques du Territoire de la Baie-James. Rapport ETBJ no 28, Environnement Canada – Société de Développement de la Baie-James.

ECOLOGICAL LAND CLASSIFICATION AND EVALUATION IN SOUTHERN YUKON: AN AID IN IDENTIFYING RESEARCH AND MANAGEMENT REQUIREMENTS

John P. Senyk
Environment Canada

Abstract: Broad scale Ecological Land Classification and evaluation studies in the southern Yukon Territory have identified a host of poorly researched and understood phenomena related to high latitude ecosystems of the boreal forest region. Due to the exploratory and reconnaissance nature of studies to date, information on ecosystem dynamics and processes has had to be largely extrapolated from published data which are generally sparse, single subject matter-oriented and site specific in nature. To best avoid further impacts from present development and potential impacts from future development and exploitation it is suggested that Ecological Land Classification and research studies be carried out in priorized areas, initially focussing on high resource capability lands and those lands most sensitive to disturbance.

In the Yukon Territory the boreal forest region occurs mainly to the south of the Ogilvie Mountains or approximately south of 64°N. latitude. Mineral and petroleum exploration and development, forest harvesting, tourism and recreation have all had recognizable impacts on the biological and physical resources of this area. Historically, resource use and exploitation have taken place with little concern for, or awareness of, potentially negative effects on the environment. It has only been within the past decade that studies, initiated largely as a result of impending pipeline development, detailed major environmental concerns peculiar to these latitudes. These studies pointed to the urgent need for baseline resource information, preferably integrated within an ecological framework.

In 1975, an exploratory ecological land survey was undertaken for the entire Yukon Territory (Oswald and Senyk 1977). It identified 22 ecologically significant regions based largely on the modification of macroclimate by physiography and the visible results as displayed by distinct biological and physical assemblages or large scale ecosystems, (Figure 1). Subsequent to this, more detailed ecological land classification and evaluation studies undertaken in various parts of the southern Yukon Territory (Senyk *et al.* 1981; Oswald 1979) have revealed the need for a much better understanding of the genesis, evolution and reaction to disturbance of a variety of these high latitude ecosystems if sound resource management judgements are to be made.

Area Description

The southern Yukon Territory consists of rolling to hilly plateau lands enclosed by high, rugged mountains on the southwest, north and east (Bostock 1965) (Figure 2).

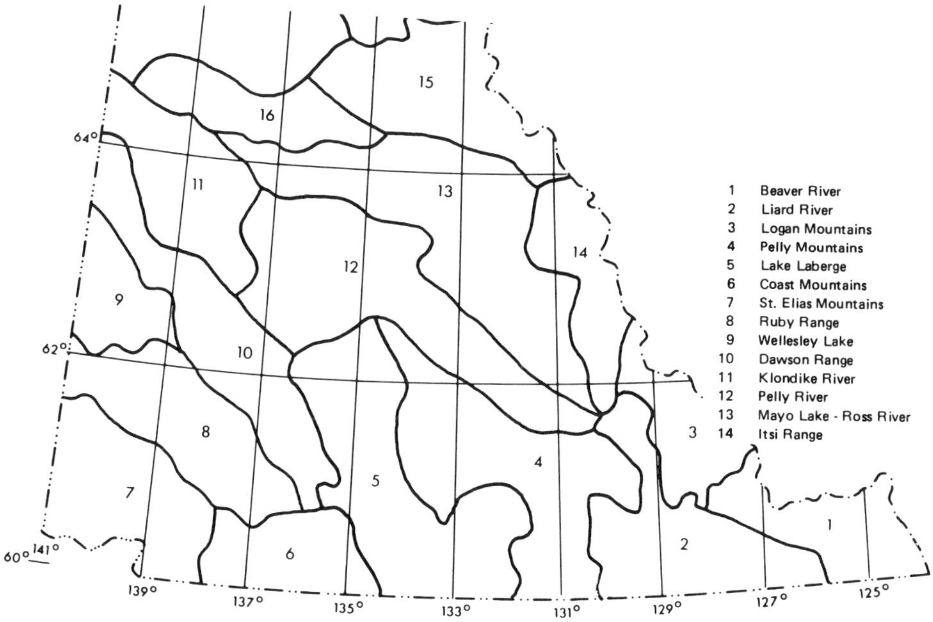

Figure 1. Ecoregions of the southern Yukon Territory.

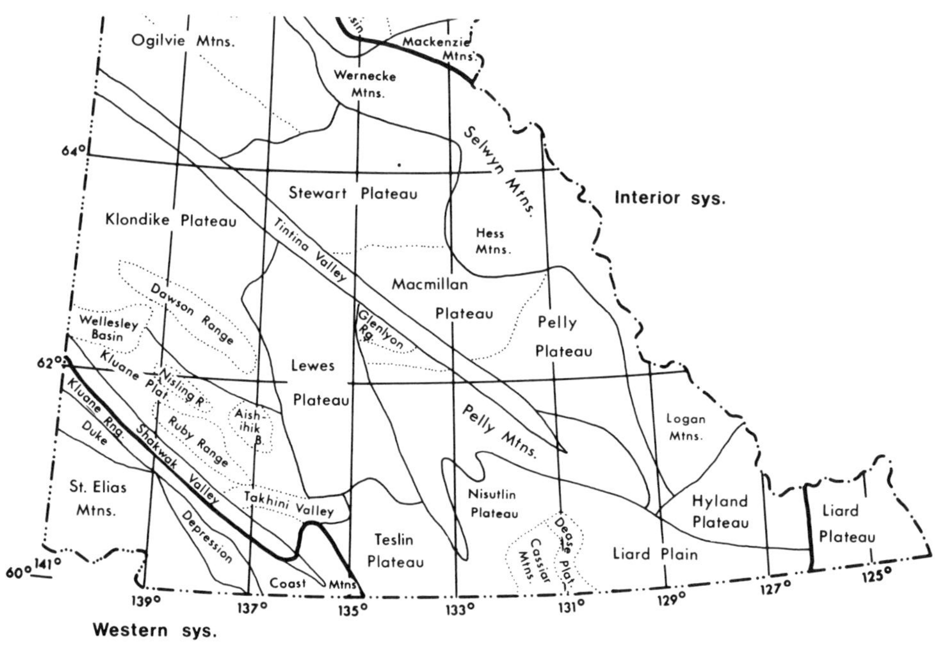

Figure 2. Physiographic subdivisions of the southern Yukon Territory.

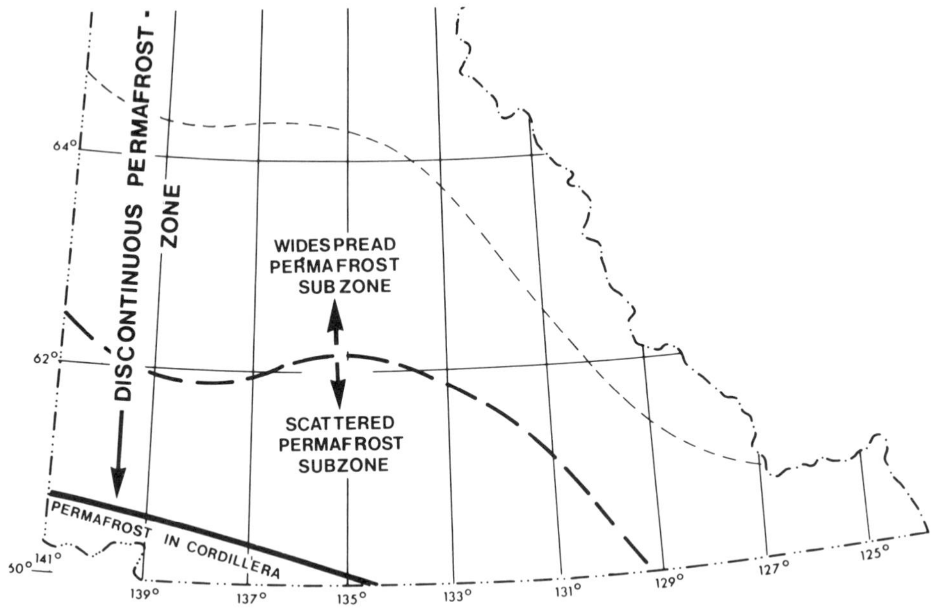

Figure 3. Permafrost subzones of the southern Yukon Territory.

The climate, described as subarctic continental, is influenced by the proximity of the Pacific Ocean, specifically the Gulf of Alaska and the relatively frequent intrusion of mild air (H. Wahl, personal communication, 1982). Air flow is predominantly from the south and west. As a result, areas located to the south and west of mountain ranges receive considerably more cloud and precipitation than areas on the north and east sides. Annual precipitation ranges from about 220 mm per year, immediately in the lee of the St. Elias and Coast Mountains, in a northerly and easterly direction, to 400 mm in the north and 750 mm in the Mackenzie Mountains in the east. Average annual temperatures decrease from about −1°C in the south centre to −7°C in the north and east. Temperature extremes increase in a northerly and easterly direction. These temperatures vary considerably with increases in elevation.

The area lies within the scattered and widespread permafrost subzones of the discontinuous permafrost zone, with the snow and ice covered St. Elias Mountains falling within the Cordilleran permafrost subzone (Brown. In press). (Figure 3).

A large area in the west was not glaciated during the Pleistocene. Volcanic ash, which decreases in depth and particle size from the southwest to the east and north, blankets the area. (Hughes *et al*. 1972).

Vegetation is typical of the northern sections of the boreal forest region, (Rowe 1972, Figure 4) with alpine tundra occurring on all the higher elevation terrain. White spruce *(Picea glauca)* lodgepole pine *(Pinus contorta)*, alpine fir *(Abies lasiocarpa)*, black spruce *(Picea mariana)*, trembling aspen *(Populus tremuloides)*, balsam poplar *(Populus balsamifera)*, and white birch *(Betula papyrifera)* are the common tree species with tamarack *(Larix laricina)* having limited distribution. White spruce occurs almost

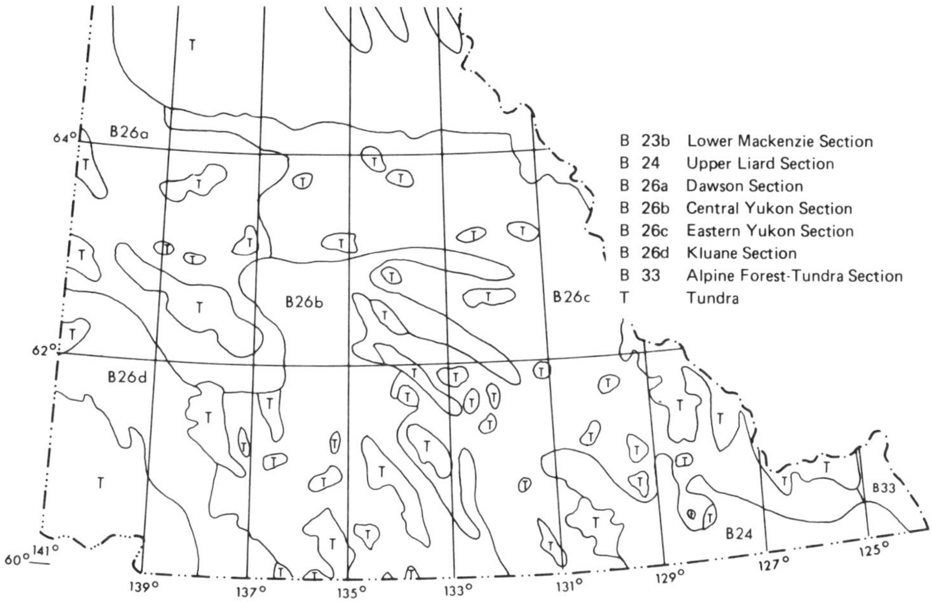

Figure 4. Forest sections of the southern Yukon Territory.

universally throughout while black spruce is scarce to absent in the south-central portion of the area. Tamarack is specific to the southeast and alpine fir is present at the higher elevations throughout the eastern half. Lodgepole pine has a range similar to that of fir. Forest productivity is highest in the southeast, lower in the south centre and lowest in the west and across the north.

Surficial materials are generally very shallow within mountainous areas and adjacent foothills and consist largely of rubbly colluvium with only minor occurrences of morainal, organic and fluvial deposits. Deposits are generally considerably deeper in the plateau lands and plain where morainal and fluvial materials dominate with glaciolacustrine colluvial and organic deposits occurring sporadically throughout.

Soils are mostly Eutric Brunisols in the dry, relatively warm south central part and grade to Dystric Brunisols, Luvisols and Podzols in the cooler more moist areas to the east and north. Cryosolic soils occur most commonly at high elevations and in imperfectly to poorly drained soils in valley bottoms throughout the area. They become more abundant with increasing latitude, occurring in a variety of materials, landscape positions and drainage conditions.

Ecological Land Classification

Ecological Land Classification (E.L.C.) is an integrated, ecologically based approach to gathering and displaying land resource data (ECS Task Force 1981). It is based on the concept that the earth's surface is spacially and ecologically differentiated into distinct natural systems. In E.L.C., those areas of land having ecologically significant and recognizable similarities, are delineated and described in terms of their more stable biological and physical resources. They can vary considerably as to size, components

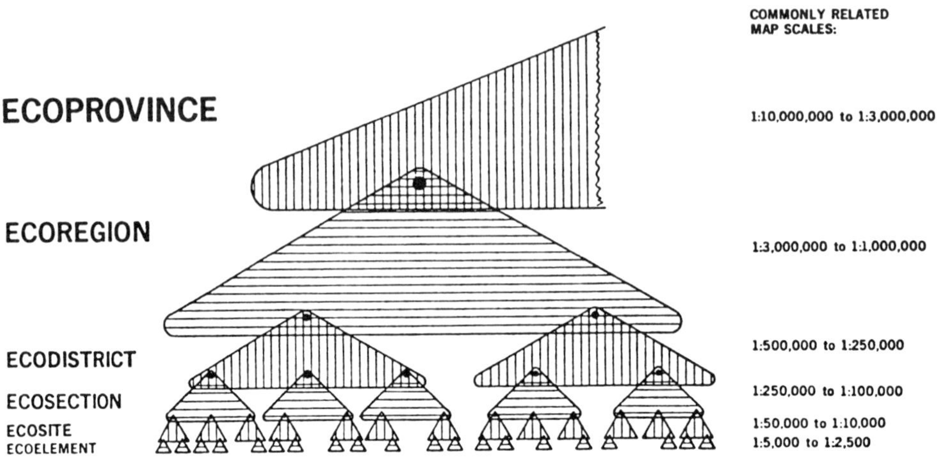

Figure 5. Ecological land classification hierarchy.

represented and variety of characteristics shared. To accommodate this variability, the E.L.C. system is structured within a scale-related hierarchy, comprised of six levels of generalization (Figure 5). Each level is nested within the one above and forms its constituent parts. These levels correspond favourably with various levels of resource management and land use planning, (Wiken *et al.* 1981).

The recognition and descriptive criteria for each level in the E.L.C. hierarchy parallels the detail associated with it. For example, the large order ecosystems (Ecoprovince, Ecoregion) are generalized, possessing fewer numbers of shared biological and physical characteristics than the small order ecosystems (Ecosection, Ecosite) which embody far less variation.

One of the basic assumptions in E.L.C. is that all ecosystem characteristics can be significant throughout the various levels of the hierarchy. Based on this premise, no singular component (i.e., terrain, climate) plays the dominant role in shaping ecosystems at any level, but rather ecosystems can be differentiated or biased towards terrain, vegetation, climate, etc., depending on the setting. The more stable components, (i.e., terrain, soils, water and climate to some extent) are stressed, however.

E.L.C. provides a baseline resource information package comprised of maps and descriptive material which gives a spacially defined overview of biological and physical resources and easily identified relationships. E.L.C. strives to characterize ecosystems in terms of their stable form or at a point in time where they have attained a state of equilibrium.

Historically, fires have had a profound and nearly universal effect on this part of the Yukon Territory, creating a mosaic of biological and physical patterns coincident with fire periodicity, which in some cases may be as short as 50 years or less. Under these circumstances, the classifier is faced with the problem of attempting to predict future biological, and in many cases physical conditions, that are likely to exist over a certain area given the present day disturbed ecosystem.

Although E.L.C. goes a long way towards supplying many of the immediate and long-term baseline resource information needs of land managers, it was not designed to undertake the study of detailed ecosystem dynamics and processes. For this information, which is a vital part of the evaluation process (the final phase of E.L.C.), it relies almost entirely on available background material. For much of the high latitude boreal region, this information is either not available or the research studies that have been carried out have largely been single subject matter-oriented, scattered and site specific in nature.

Research and Management Requirements

Resource managers must have the necessary information which allows them to make predictions on how particular landscapes will react to a variety of land uses and management prescriptions, in what geographic area these reactions will be nearly the same, and what can be done to avoid or ameliorate any negative effects due to land use and development.

The following examples depict a few specific ecosystems identified during ecological land survey at the ecodistrict and ecosection levels (Senyk *et al.* 1981). They are chosen because of their sensitivity to disturbance and their common occurrence throughout the southern Yukon Territory. The examples point out the need for intensive research and monitoring programs in order to obtain the necessary understanding of processes and dynamics associated with these and other ecosystems to allow realistic cause and effect analysis predictions to be made.

The Lake Laberge ecoregion (Oswald and Senyk 1977) has a large area of glaciolacustrine deposits situated within the Takhini Valley (Lat 60°45′N, Long 137°35′W). Although the largest portion of these deposits contains permafrost, the ice content and form varies considerably from the eastern to the western part of the basin. Thermokarst features occur throughout, but are most common in the west. They occur in undisturbed terrain, in areas that were burned about 25 years ago and adjacent to the Alaska Highway where presumably road construction disturbed the thermal equilibrium of the underlying permafrost and initiated thaw.

Thermokarst processes are most active in the burned over areas where present vegetation cover consists of scattered aspen, willow and grass and least active, in undisturbed terrain supporting closed stands of white spruce with a feather moss understory, (Figures 6 and 7). Adjacent to the Alaska Highway, the thaw ponds have become stabilized and white spruce and lodgepole pine are regenerating under aspen, willow and grass cover (Figure 8). Although vegetation plays a large part in insulating the underlaying permafrost (thus maintaining the thermal equilibrium and controlling to some extent the rate of ice degradation) the ice content of the permafrost plays the largest role in determining the rate of thaw once the collapse process is initiated.

Permanently frozen organic deposits, mainly peat plateaux and palsas and the collapse scars of these features, are particularly common in low-lying terrain in the eastern and northern parts of the area (Figure 9).

Some peat plateaux have mineral soils beneath one metre or more of peat, others have exaggerated hummocky surfaces, while still others have surface ice wedge polygon features. Some are nearly devoid of trees before collapse is initiated while others undergo active collapse under relatively dense black spruce tree cover, (Figures 10 and 11).

Figure 6. Thermokarst in 25-year-old burn.

Figure 7. Thermokarst in undisturbed spruce – feather moss type vegetation.

Figure 8. Stabilized thermokarst depression.

Figure 9. Peat plateau and collapse scars.

Figure 10. Collapsing peat plateau devoid of trees.

Figure 11. Tree-covered collapsing peat plateau.

Figure 12. North aspect permafrost, black spruce, ground birch; *Sphagnum*.

Vegetation sequences during and after collapse do not appear to be universal either, though a few species appear to be nearly always present. The variety of forms can be related to broad climatic regimes and local landscape settings, but these relationships are not well documented.

Although they are relatively easy to identify from remotely sensed imagery and to characterize superficially, an understanding of the genesis of these forms, their effects on the local and regional hydrology and events leading to collapse, is needed. It would appear that peat plateau growth and decay processes are cyclic; fire, decadence of overstory vegetation, changes in hydrology or various forms of man-made disturbance all have the potential to initiate thawing. Regrowth likely follows reestablishment of a vegetation mat on the collapsed surface of sufficient thickness to insulate frozen subsoil conditions.

The results of the interaction of climate and physiography are clearly visible throughout this part of the Yukon Territory although the biological and physical manifestations of these interactions are most prominent within and adjacent to mountainous landscapes. Besides the well-documented windward and leeward differences and the cool, moist north-facing and dry, warm south-facing aspect differences (Figures 12 and 13), there exists the phenomenon of temperature inversion which is clearly evident particularly in the eastern and northern parts of the area. This reversal of the lapse rate due mainly to strong radiational cooling during winter months, results in very cold temperatures in valley bottoms whereas the higher terrain remains relatively mild. The 'visible' ecosystem displays characteristics similar to subalpine, alpine and Arctic tundra on the valley floor, while as little as 50 meters above the valley floor closed spruce or spruce fir forests can occur, which continue upslope to the treeline, (Figures 14 and 15).

Figure 13. South facing grassland — aspen, dry warm slopes.

Figure 14. Valley bottom of perennially frozen hummocky terrain.

Figure 15. Spruce forest above critical cold air pool limits.

Much greater knowledge of the relationships of climate and landform is needed. This knowledge needs strong documentation through climate research, initially to determine microclimatic characteristics, (i.e., temperature, moisture), their ranges and extremes. These data should then be cross referenced with species tolerance information to try and give a more precise answer for the occurrence of these specific ecosystems in these situations.

The foregoing examples depict a few of the ecosystems common to northern latitudes about which definite knowledge is lacking.

Conclusions

Ecological Land Classification is one method of presenting baseline resource information within an ecological framework, in which resource data are integrated with available ecological knowledge. To improve the predictive capability of the E.L.C. data base and its application in resource management, rcsearch efforts need to be directed towards developing an improved understanding of ecosystem dynamics and processes, with initial priorities given to high capability and most sensitive lands.

Resource managers need to be shown the advantages of structuring all resource data within an ecologically based format. They must be persuaded that they are ultimately responsible for all resource components on lands within their jurisdiction.

Literature Cited

Brown, R.J.E. (In press). Permafrost in Canada. Map compiled for Hydrological Atlas of Canada.

Bostock, H.S. 1965. Physiography of the Canadian Cordillera, with special reference to the area north of the fifty-fifth parallel. Dept. Energy Mines and Resources, Geol. Surv. Can., Mem. 247.

Environmental Conservation Service Task Force. 1981. Ecological land survey guidelines for environmental impact analysis. Ecol. Land. Class. Series., No. 13, Federal Environmental Assessment Review Office, Environment Canada. Ottawa, Ontario.

Hughes, O.L., V.N. Rampton and N.W. Rutter. 1972. Quaternary geology and geomorphology, southern and central Yukon (northern Canada). Field Excursion Guidebook, Internat. Geol. Cong. 24th Session.

Oswald, E.T. and J.P. Senyk. 1977. Ecoregions of Yukon. Environ. Canada, C.F.S., BC-Y-164.

Oswald, E.T. (compiler) 1979. Forest Resource Assessment of the Nistulin Test Area, Yukon. Environ. Canada, C.F.S., P.F.R.C. Misc. Publ.

Rowe, J.S. 1972. Forest Regions of Canada Environ. Canada, C.F.S., Publ. No. 1300.

Senyk, J.P., E.T. Oswald, B.N. Brown and R.K. King. 1981. Ecological Land Classification and Evaluation of the Kusawa Lake Area, Yukon Territory. Environ. Canada, C.F.S., Misc. Rept.

Wiken, E., J.P. Senyk and E.T. Oswald. 1981. Inventorying land resources through an ecologically based approach. *In:* Resource Inventories, Principles and Practices. University of Maine. Orono, Maine.

A CLASSIFICATION AND ORDINATION OF FOREST ECOSYSTEMS IN THE GREAT CLAY-BELT OF NORTHEASTERN ONTARIO

R.K. Jones
Agriculture Canada
(Land Resource Research Institute contribution number 82-77)

G. Pierpoint
Ontario Ministry of Natural Resources

G.M. Wickware
Environment Canada, Burlington, and

J.K. Jeglum
Canadian Forestry Service

Abstract: A classification integrating 23 vegetation types and 14 soil types was developed using data collected from two hundred and fifty forest stands throughout the Great Claybelt of northeastern Ontario. Stands were first classified using TWINSPAN, a polythetic divisive classification technique, according to their vegetational attributes (species, per cent cover), and then according to various field recognizable soil attributes. Ordination using DECORANA, a detrended correspondence analyses technique, revealed vegetation types trending along complex environmental gradients related to moisture and nutrients. Soil types were related primarily to texture and moisture. A synthesis of vegetation and soil types resulted in the discrimination of 14 'operational groups'. These groups provide a satisfactory framework for future forest management activities as well as further research in the region.

Forest management is becoming more intensive and the large wood-using industries are conducting more of the silviculture on Crown lands, under the Forest Management Agreements. Forestry researchers have recognized the need for a well-founded ecosystem-based classification to serve as a framework for organizing and exchanging silvicultural knowledge, and for planning and conducting silvicultural activities such as harvesting, site preparing, regenerating and tending. In the spring of 1979, we were requested to consider the earlier classification work, both vegetation and soils-oriented, conducted by Hills and Boissonneau (1960), MacLean and Bedell (1955), Jeglum *et al.* (1974), Scott and Virgo (1977), and Gemmell (1979), and to develop a practical management level ecological classification for application in the commercial uncut forest stands throughout the Great Claybelt, site district 3 of site region 3E (Hills 1961).

After preliminary work throughout 1979, we conducted our main data collection program in the 1980 field season. This was followed through 1981 to the present, by data processing and analysis, the development of the classification, the testing of its application in the field in both uncut and cut-over conditions, the development of mapping techniques, the design of a field guide and a teaching program for the management staff, and the collating of over 160 man years of management experience to provide initial interpretations for various silvicultural activities. This last part of the program is not discussed in this present paper.

Methodology

During the 1980 field season, 250 plots, 10 metres × 10 metres, were established in Timmins, Cochrane, Kapuskasing and Hearst districts. These provided a sample of the range of soil parent materials, landforms, drainage classes and forest cover types occurring throughout the relatively mature uncut forest lands of the Claybelt.

All species of trees, shrubs, herbs, mosses and lichens were listed with estimate of percent cover by species by layer. Detailed profile descriptions of organic and mineral horizons were made and the soils were sampled. Representative heights, ages and diameters of trees were taken. Stocking was recorded by height or diameter classes in 20, 2 metre × 2 metre quadrats adjacent to the plot. Other site features including topography, position on slope, aspect and ground surface conditions were also noted.

After the field season, all data were compiled in computer data files at the University of Guelph. The soil samples were analysed at Guelph for a variety of physical and chemical properties and these data were also entered onto computer files.

Data Analysis Strategy

Our goal was to develop a practical classification system which could be applied operationally with relative ease, efficiency and repeatability. For the following reasons we decided to use vegetation information as the initial discriminator of ecosystem types and refine this, where necessary, with the subsequent evaluation of field recognizable soil properties. Our initial work showed that the vegetation could be used readily to predict a number of important soil properties. Second, topography, drainage pattern and moisture movement are very subtle in the Claybelt. This subtlety plus the preponderance of relatively deep organic mantles makes the differentiation of significant physiographic units much more difficult than is the case in upland areas of mineral soils, greater relief and more obvious drainage. Third, the present stratification of forest sites used by most field staff is based on vegetation.

It would be equally valid to analyse our sites using the soil data first and then evaluate the range of vegetation conditions within a soil type, and in fact, we have begun to express the types both ways, particularly for use in cut-over areas and with soil surveys. The vegetation types can be discerned from one another readily in the field and are reasonably well correlated with important soil attributes useful in forest management. Our data analysis, therefore, proceeded by developing a vegetation classification to give vegetation types, discriminating field recognizable soil properties by these vegetation types, developing a soil classification to give soil types based on the discriminated soil properties, then by identifying ecosystem types as combinations of a vegetation type and a soil type, and finally, by forming operational groups by grouping and/or subdividing the vegetation types based on their distribution of soil properties plus the consideration of forest management needs. Details are given below.

Vegetation Analysis

A number of vegetation analysis packages were reviewed. A recently-developed technique called TWINSPAN two way indicator species analysis — which is part of the Cornell Ecology Program package developed by Hill (1979a), was chosen.

TWINSPAN is a polythetic divisive classification strategy, which simply means that it hierarchically subdivides a population on the basis of all attributes. It reconciles the practical need to form vegetation classes or 'types' for forest management with the reality of vegetation continua, by using ordination to classify both the samples (first) and the species (second). A rearranged species by sample table is created which resembles the tabular analysis of the Braun-Blanquet approach (Braun-Blanquet 1928). TWINSPAN has the advantage of allowing for some degree of quantification of the input data, for example, classes of species percentage cover. Further, TWINSPAN provides a number of control features that allowed us to tailor the analysis to our specific objective.

The interpreted results of the TWINSPAN analysis are shown in Figures 1a and 1b in the form of a dendrogram. We use the term 'interpreted results' because the actual hierarchy subdivided the plots into 26 groups. In evaluating the finer level splits, we decided to regroup a few of the divisions when the species separation was considered too subtle and likely meaningless for the purpose of our classification. Twenty-three vegetation types were ultimately identified as being sufficiently different from one another and useful for the purpose of our program. We define a 'vegetation type' as a synthetic grouping of vegetation sample plots which demonstrates some level of internal homogeneity, but which has no assumed taxonomic rank. We might emphasize that at this stage of the analysis the ecological judgement relative to the objectives of the study, plays an essential role, using the classification analysis as a tool for decision making. Of great value to us is the feature of TWINSPAN which identifies objectively and efficiently the 'indicator species' for each division of the hierarchy. We used these to form a key (Figures 1a, 1b) for allocating new sites into the existing classification. Figures 1a, 1b show the indicator species and the maximum and minimum scores for deciding which way to go at each division point. Figure 2 gives a full explanation. To our knowledge, this TWINSPAN feature is unique and our use of it in the key is a vast improvement over the use of 'modal descriptions' only, when allocating new sites in the field.

For construction of the key, no more than seven indicator species at any one division were allowed to be designated and only 'presence/absence' of the key species was considered, not percentage cover class. Certain species that are difficult to identify in the field were not allowed to be chosen as indicators (although they were still used in the actual analysis).

The TWINSPAN dendrogram and table revealed many useful discontinuities and continuities in our vegetation data set, since the analysis employs principles of both classification and ordination. Some examples help to illustrate the application of TWINSPAN to our boreal forest vegetation. From the viewpoint of classification, Division 1 (see Figure 1a) subdivides the population into those plots (on the negative side) with a reasonably well represented herb species diversity from those plots (on the positive side) which have a poor to non-existent herb species diversity, i.e., stands with essentially a tree layer and a continuous moss forest floor. Division 5 subdivides the herb rich *mixedwood* forests from the herb rich *softwood* forests. Vegetation types 13 and 14 are quite similar from the viewpoint of ordination, yet at the first division they are positioned

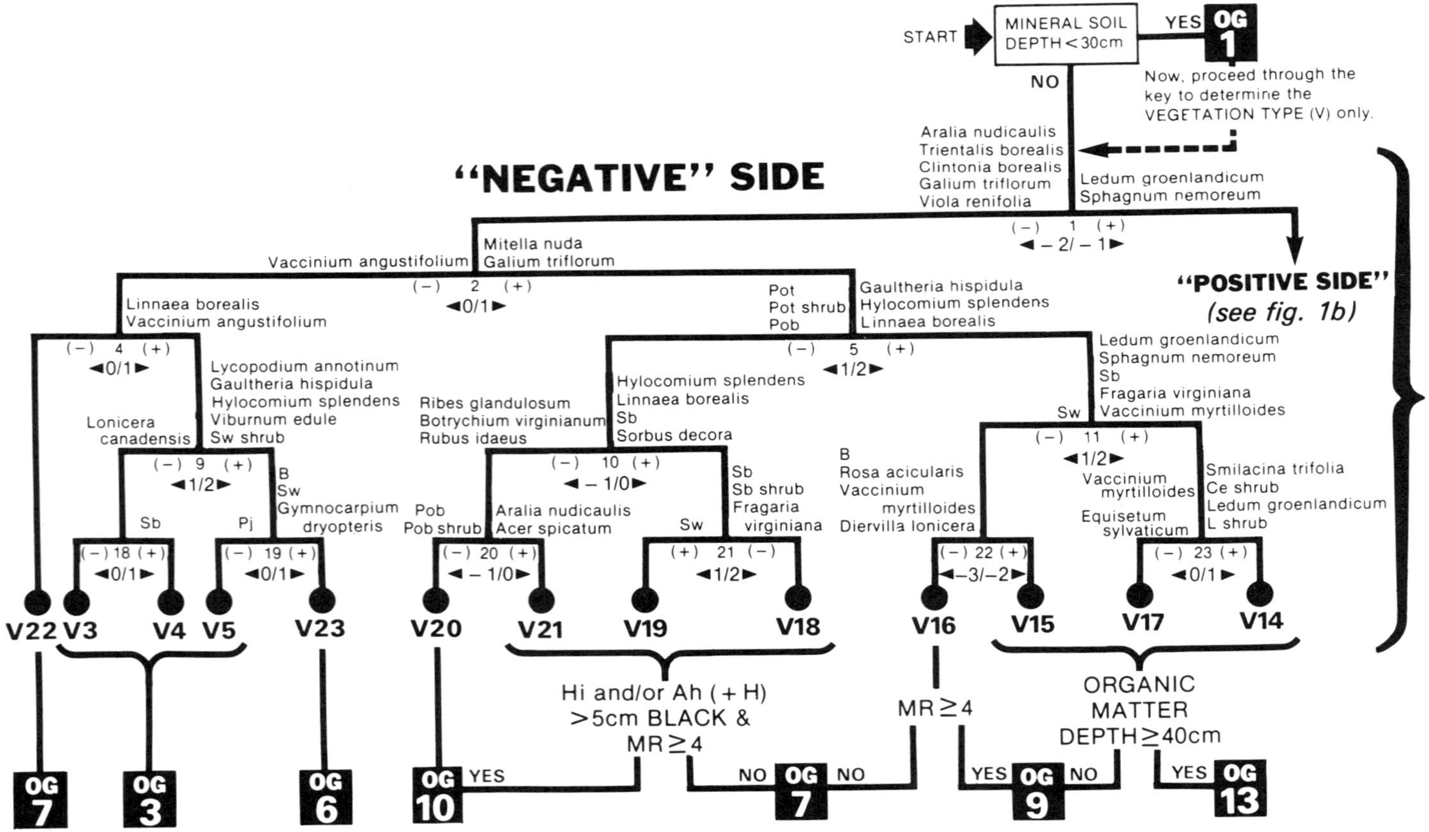

TREE & SHRUB ABBREVIATIONS

Pj = Pinus banksiana	Sb = Picea mariana
Pot = Populus tremuloides	L = Larix laricina
Pob = Populus balsamifera	B = Abies balsamea
Sw = Picea glauca	Ce = Thuja occidentalis

Figure 1a. TWINSPAN classification analysis dendrogram 'negative' (left) side, including the indicator species used to key to the vegetation types (V), plus the subsequent soil properties used to key further to the operational groups (OG). Refer to Figure 2 for explanation of format and how to key.

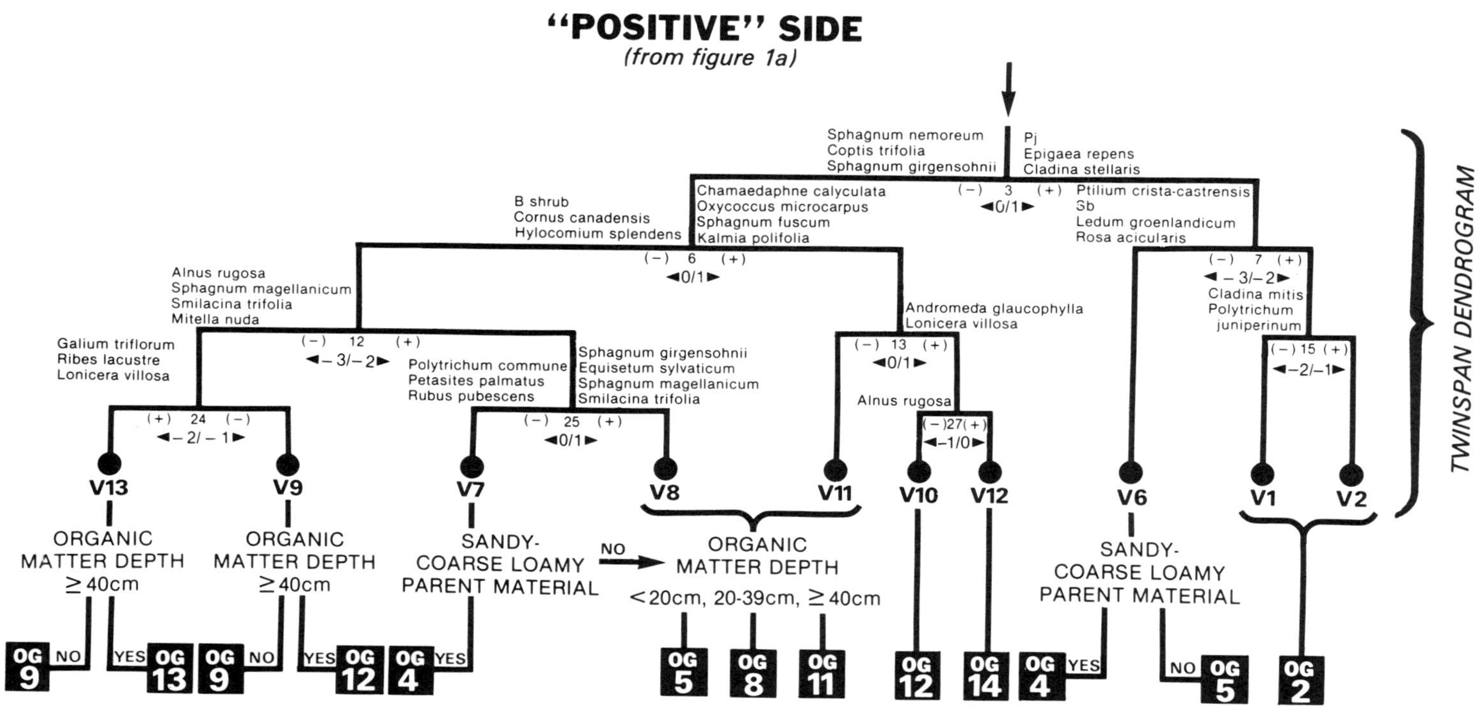

Figure 1b. Continuation of TWINSPAN classification analysis dendrogram: the 'positive' (right) side.

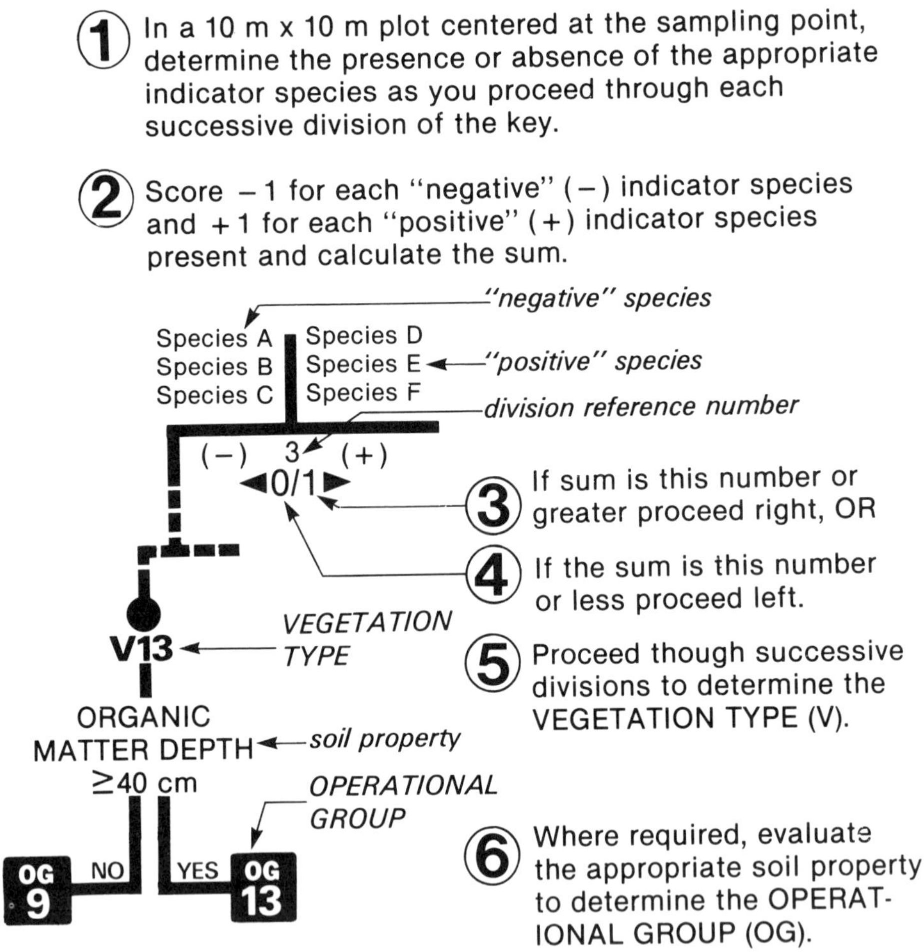

Figure 2. How to use the key to operational groups and vegetation types including explanation of format shown in Figures 1a, 1b.

on opposite sides of the dendrogram due to some differences in their herb diversity and sphagnum moss abundance. Both the distinction and similarity of these two types is conveniently shown by the species-by-sample tables.

Although we identified 23 vegetation types using TWINSPAN, we found that it was also useful in determining the relative similarity and dissimilarity of the vegetation types. As noted above, the dendrogram structure and sample/species order of the table help to reveal adjacent similarities in the otherwise divisive hierarchy, but they cannot particularly well demonstrate the relationship between types that are positioned farther apart, perhaps even on different sides of the primary split. In order to evaluate and display inter-type relationships, we created a modal or vegetation type data set. This data set was then analysed using the Cornell Ecology Program DECORANA − detrended correspondence analysis (Hill 1979b; Hill and Gauch 1980). Figure 3 shows the results of the DECORANA analysis of the 23 vegetation types plotting axis one against axis two. As

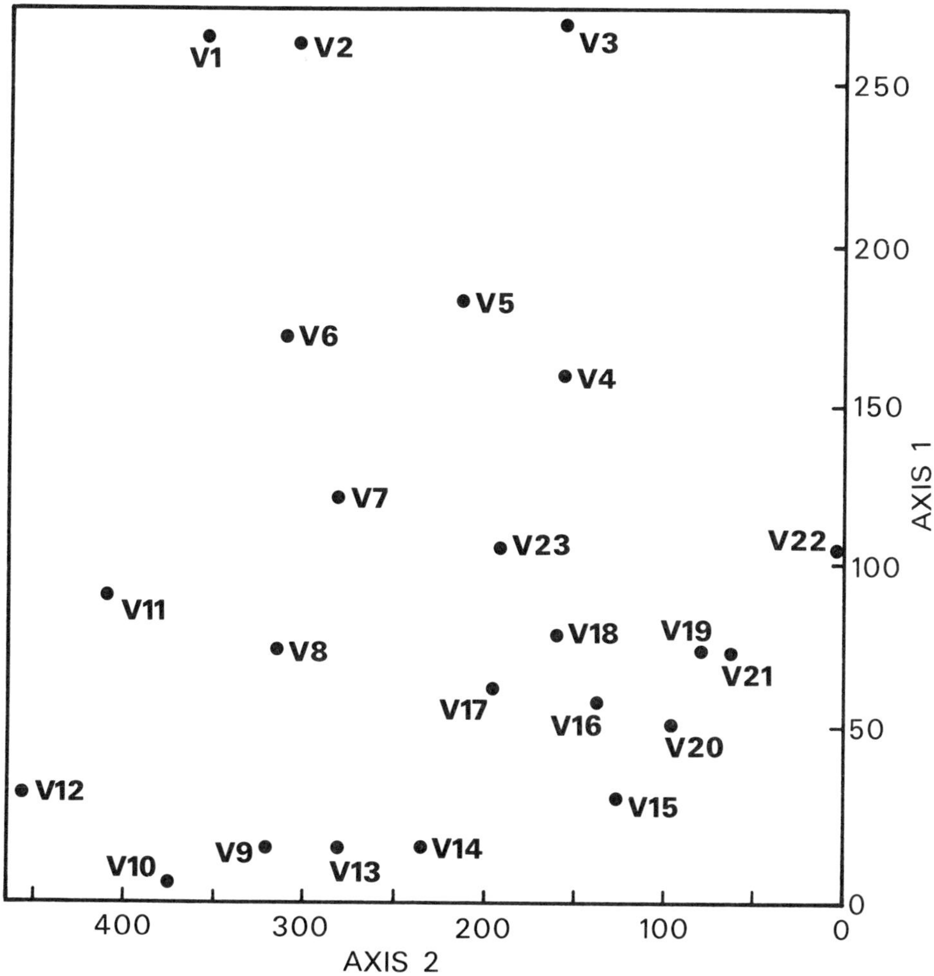

Figure 3. DECORANA ordination of the modal Vegetation Type data set derived from the interpretation of TWINSPAN. ''V1'' = Vegetation Type 1. A mirror image of the axis 1 × axis 2 plot is presented here so that it is similar in its orientation to traditional edatopic grid presentations (Pogrevnyak 1930; Krajina 1969) i.e., the top left representing dry, nutritionally ''poor'' sites and the bottom right representing moist-wet, nutritionally ''rich'' sites.

discussed earlier, vegetation types 13 and 14 are close to one another; similarly types 2 and 3 are in the same general area on the DECORANA ordination although they are positioned far apart on opposite sides of the TWINSPAN dendrogram.

Figures 4a and 4b display species groups and species diversity. First, in Figure 4a, moss-dominated systems are distinguished from those with herbaceous groundcover. Within this broad left-right split, there are upper-lower splits based on species composition and diversity. Overlaps of these conditions can be seen. Second, in Figure 4b tree species dominance is readily visualized. It illustrates how the DECORANA

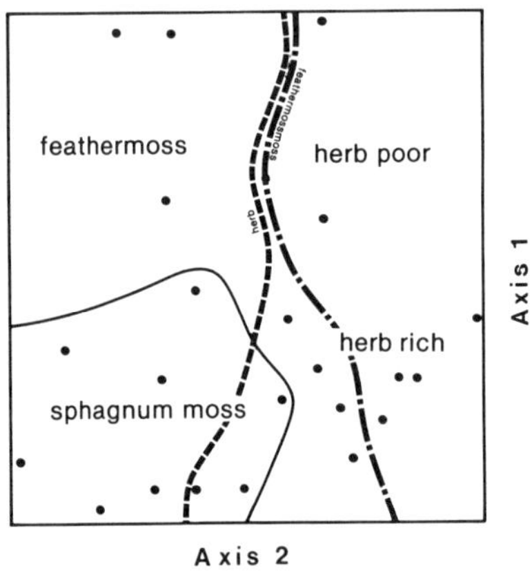

Figure 4a. Understory Vegetation

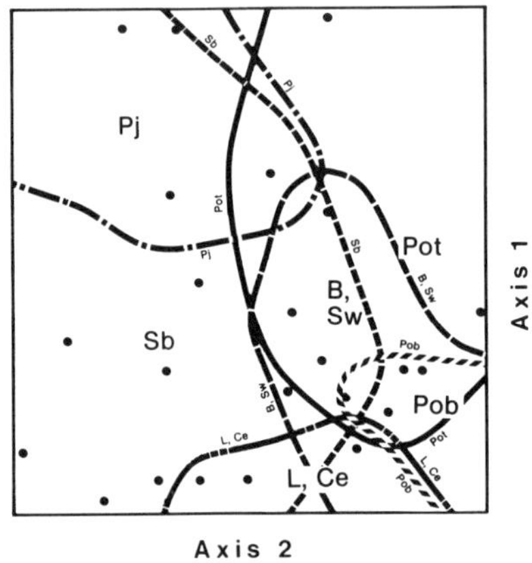

Figure 4b. Tree Species

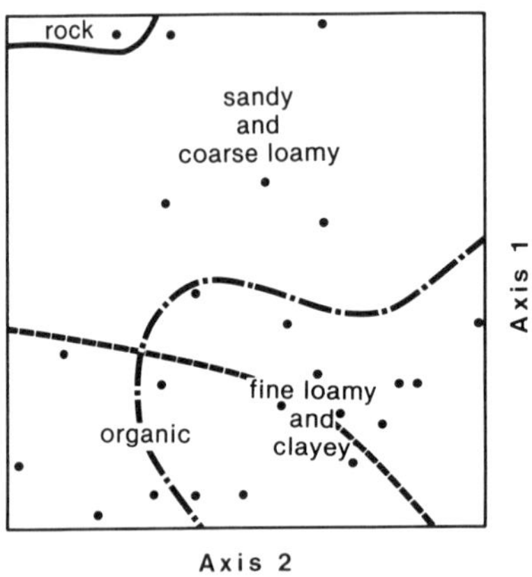

Figure 4c. Soil Texture

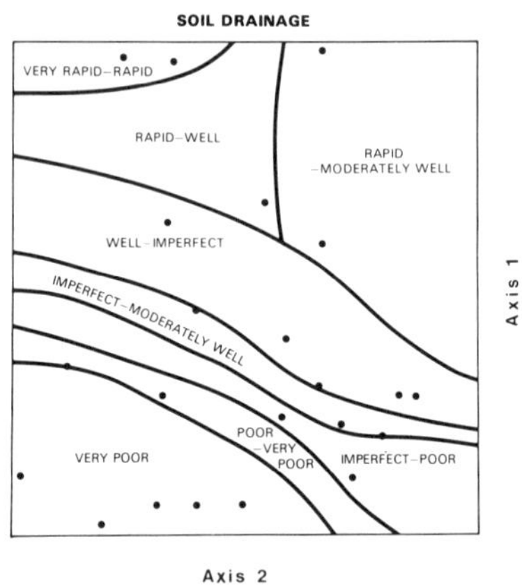

Figure 4d. Soil Drainage

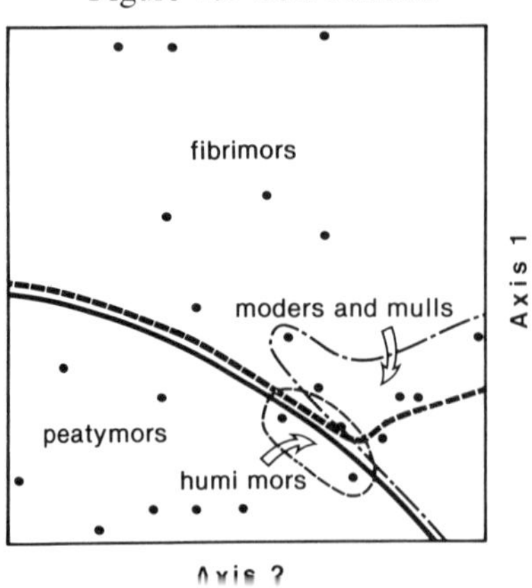

Figure 4e. Forest Humus Form
(Bernier 1968)

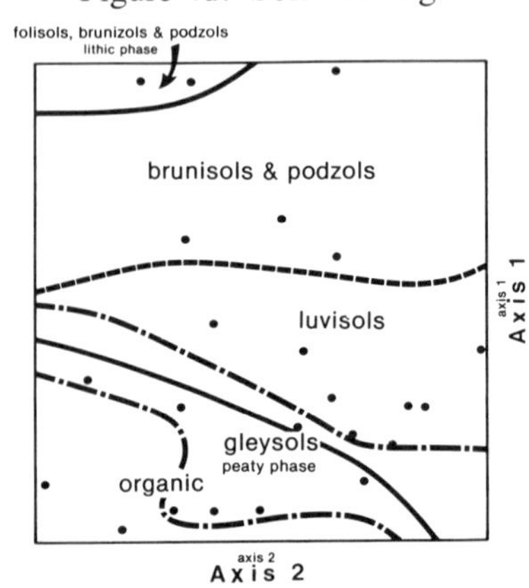

Figure 4f. Soil Taxonomy — Order Level
(CSSC, 1978)

Figure 4. Vegetation (Figures 4a & b) and soil (Figures 4c,d,e, & f) trends overlayed on the DECORANA ordination shown in Figure 3. The position of the lines is schematic. Symbols used in Figure 4b are given in Figure 1a.

ordination complements the TWINSPAN total plot analysis to reveal inter-type affinities and broad trends or 'core' features of the vegetation. We will now discuss our soil data analysis and later show how useful this DECORANA ordination of vegetation types is in evaluating and displaying correlated soil properties and operational groups.

Soil Data Analysis

Soil data analysis, in relation to vegetation types, began with 29 soil properties. Discriminant analyses were run to evaluate what soil properties best discriminated the vegetation types and how well. The vegetation type was discriminated by the soil data correctly on some 70 per cent of the plots. The most discriminating soil properties were sub-surface texture, thicknesses of forest humus horizons, moisture regime/drainage class, depth to carbonates and rooting zone thickness. Further properties obviously refined the discrimination.

Encouraged by these results, we used both polythetic agglomerative cluster and polythetic divisive analyses to develop a soil classification based on the most discriminating soil properties. Again we used DECORANA ordination to help us understand inter-class relationships during this work. Ultimately, we settled for a practical classification with 14 soil types (see Figure 5) which are useful management groupings of soil conditions in the Claybelt and are complementary to the vegetation types. These 14 soil types range from very shallow soils over bedrock, through dry and fresh sandy soils, fresh and moist loamy and clayey soils, to wet organic soils with various degrees of decomposition.

Vegetation – Soil Correlation and the Formulation of Operational Groups

We made a tabular sort of the soil data set according to the 23 vegetation types to observe what specific ranges of the discriminating soil properties were characteristic of each vegetation type.

Using the DECORANA ordination of the vegetation types as a graphical base we overlayed and observed obvious trends in soil characteristics such as soil texture, soil horizonation (soil taxonomy), soil depth, depth of organic matter, depth to carbonates, forest humus form and soil moisture regime/soil drainage. Figures 4c, 4d, 4e and 4f show some of the trends that we observed. Here we see the coincidence of soil texture, drainage, humus form and profile type with the vegetation type. We also studied the distribution pattern of our 14 soil types within the 23 vegetation types. This was done both graphically and analytically, using the DECORANA ordination and cluster analysis. From this, we recognized what the most characteristic combinations of vegetation and soil types were. These define in considerable detail the commonly occurring mature forest ecosystem types of the Claybelt.

At this stage of our analysis we reminded ourselves that the forest managers had asked for about 15 classes only. Against our inherent 'splitter' (versus 'lumper') natures, we began the task of integrating our knowledge of the vegetation and soil, and producing a management level classification which retained most of the integrity of our detailed analyses.

The management experience collated by Kent Virgo identified operational constraints

Soil Type Names and Description

S1 VERY SHALLOW to BEDROCK
Less than 30 cm mineral soil over bedrock. Usually dry and fresh (0-2) soils with sand, loamy sand or sandy loam parent material. Fibrimor forest humus form. Organic matter (5-10)^{7} (>10)^{2} (<5)^{1} cm.

S2 DRY-FRESH NON-CALCAREOUS SANDY
Dry and fresh (0-2) soils with non-calcareous sand or loamy sand parent material. Fibrimor forest humus form. Organic matter (2-10)^{9} (>10)^{1} cm.

S3 DRY-FRESH CALCAREOUS SANDY
Dry and fresh (0-2) soils with calcareous sand or loamy sand parent material. Fibrimor forest humus form. Organic matter (5-10)^{8} (<5)^{1} (>10)^{1} cm.

S4 FRESH-MOIST SANDY
Fresh and moist (3-6) soils with sand or loamy sand parent material, usually non-calcareous. Fibrimor forest humus form. Organic matter (5-15)^{8} (<5)^{1} (>15)^{1} cm.

S5 FRESH-MOIST NON-CALCAREOUS COARSE LOAMY
Fresh and moist (2-5) soils with non-calcareous very fine sand, loamy very fine sand or sandy loam parent material.Fibrimor forest humus form. Organic matter (5-10)^{8} (<5)^{1} (>10)^{1} cm.

S6 FRESH-MOIST CALCAREOUS COARSE LOAMY
Fresh and moist (2-5) soils with calcareous very fine sand, loamy very fine sand or sandy loam parent material. Generally fibrimor forest humus form. Organic matter (5-15)^{9} (>15)^{1} cm.

S7 FRESH NON-CALCAREOUS FINE LOAMY - CLAYEY
Fresh (2-3) soils with non-calcareous silt loam, loam and finer textured parent material. Generally fibrimor forest humus form. Organic matter (5-10)^{7} (<5)^{2} (>10)^{1} cm.

S8 FRESH CALCAREOUS FINE LOAMY - CLAYEY
Fresh (2-3) soils with calcareous silt loam, loam and finer textured parent material. Generally fibrimor forest humus form. Organic matter (5-10)^{8} (<5)^{1} (>10)^{5} cm.

S9 MOIST NON-CALCAREOUS FINE LOAMY - CLAYEY
Moist (4-6) soils with non-calcareous silt loam, loam and finer textured parent material. Generally fibrimor, occasionally peatymor forest humus form. Organic matter (5-10)^{7} (>10)^{2} (<5)^{1} cm.

S10 MOIST CALCAREOUS FINE LOAMY - CLAYEY
Moist (4-6) soils with calcareous silt loam, loam and finer textured parent material. Generally fibrimor, occasionally humimor or peatymor forest humus form. Organic matter (5-15)^{9} (>15)^{1} cm.

S11 MOIST BLACK HIAH CALCAREOUS FINE LOAMY - CLAYEY
Moist (4-6) soils with calcareous silt loam, loam and finer textured parent material. Generally moder, occasionally mull forest humus form. Organic matter (5-10)^{6} (11-15)^{3} (<5)^{2} cm.

S12 MOIST PEATY PHASE - MINERAL
Moist (5-6) soils with (20-29)^{6} (30-39)^{4} cm organic F, H or Of, Om, Oh over mineral layer of any texture. Peatymor, fibrimor and humimor forest humus form.

S13 WET THIN Of - ORGANIC
Wet (7-8) organic soils with less than 10 cm depth of surface organic fibric material (Of). Profile mesic to humic organic, generally moderately to strongly decomposed. Any texture of mineral soil may be found at depth. Organic matter (40-100)^{6} (>100)^{4}.

S14 WET THICK Of - ORGANIC
Wet (7-8) organic soils with greater than 10 cm depth of surface organic fibric material (Of). Profile fibric to mesic organic, generally very weakly to moderately decomposed. Any texture of mineral soil may be found at depth. Organic matter (40-100)^{5} (>100)^{5}.

Figure 5. Description of 14 soil types.

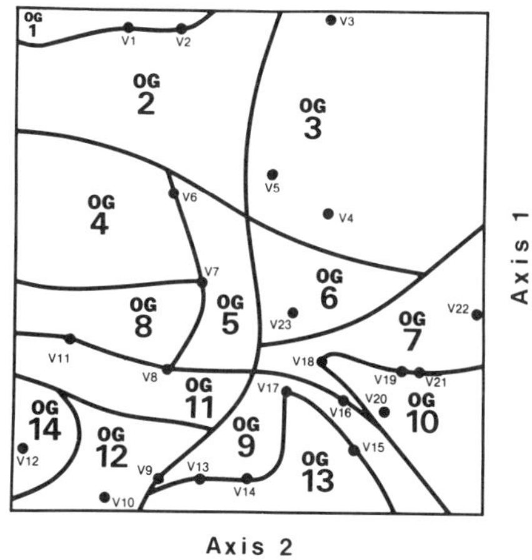

Operational Group Names

OG 1 **VERY SHALLOW SOIL OVER BEDROCK**
Sb &/or Pj - Feathermoss - Lichen on
Very Shallow Soil over Bedrock

OG 2 **VACCINIUM**
Pj (& Sb) - Vaccinium - Feathermoss - Lichen on
Dry or Fresh Sandy Soil

OG 3 **DIERVILLA**
Hardwood/Mixedwood - Vaccinium - Diervilla - Herb Poor on
Fresh Sandy or Loamy Soil

OG 4 **FEATHERMOSS - COARSE SOIL**
Sb (&/or Pj) - Feathermoss on
Fresh-Moist Non-Calcareous/Calcareous
Sandy or Coarse Loamy Soil

OG 5 **FEATHERMOSS - FINE SOIL**
Sb (&/or Pj) - Feathermoss on
Fresh-Moist Calcareous Fine Loamy-Clayey Soil

OG 6 **LYCOPODIUM**
Lycopodium - Herb Poor on
Fresh-Moist Non-calcareous/Calcareous Loamy Soil

OG 7 **MIXEDWOOD - HERB RICH**
Hardwood/Mixedwood - Herb Rich on
Fresh-Moist Calcareous Fine Loamy-Clayey Soil

OG 8 **FEATHERMOSS - SPHAGNUM**
Sb - Feathermoss - Sphagnum on
Moist Calcareous Fine Loamy-Clayey Soil with
20-39 cm. Organic Matter

OG 9 **CONIFER - HERB/MOSS RICH**
Conifer Mixed - Herb Rich on
Moist Calcareous Fine Loamy-Clayey Soil

OG 10 **HARDWOOD - ALNUS**
Hardwood - Alnus - Herb Rich on
Moist Calcareous Fine Loamy-Clayey Soils
with Thick Black Organic-Mineral Forest Humus Form

OG 11 **LEDUM**
Sb - Ledum on
Wet Moderately Decomposed Organic Soil with
Thick Surface Fibric Horizon

OG 12 **ALNUS - HERB POOR**
Sb - Alnus - Herb Poor on
Wet Moderately Decomposed Organic Soil with
Thick Surface Fibric Horizon

OG 13 **ALNUS - HERB RICH**
Sb (&/or Ce/L) - Alnus - Herb Rich on
Wet Well Decomposed Organic Soil with
Thin Surface Fibric Horizon

OG 14 **CHAMAEDAPHNE**
Sb - Chamaedaphne on
Wet Poorly Decomposed Organic Soil with
Thick Surface Fibric Horizon

Figure 6. The 14 Operational Groups overlayed on the DECORANA ordination shown in Figure 3, and legend. The position of the lines is schematic.

Ledum

OG 11

Sb - Ledum on Wet Moderately Decomposed Organic Soil with Thick Surface Fibric Horizon

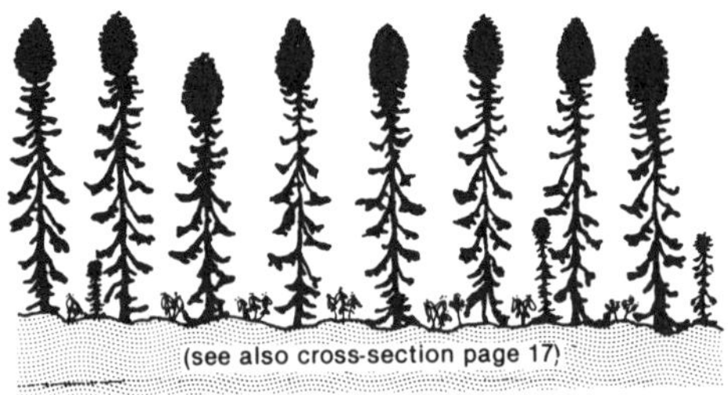

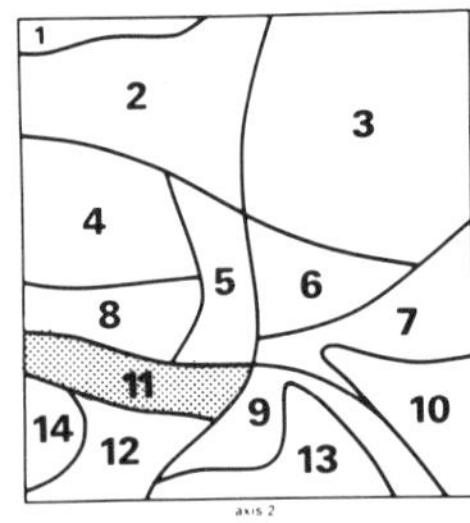

DESCRIPTION

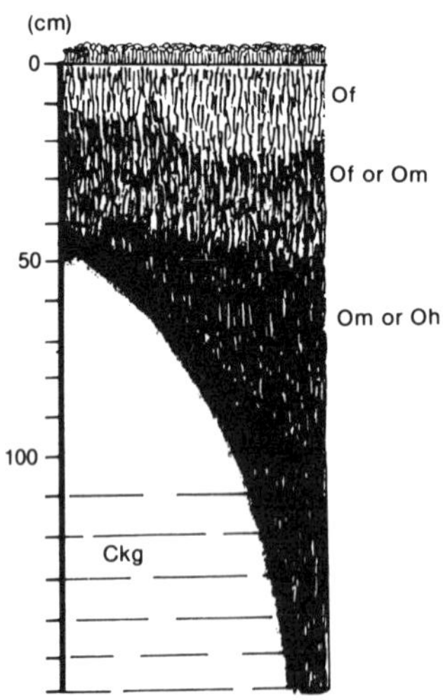

VEGETATION

COMMON FOREST COVER TYPES:
Sb

VEGETATION TYPES:
V8^{6} V11^{4}

TREE LAYER:
Black Spruce, occasionally Tamarack. Black Spruce also occurs in all layers.

SHRUB LAYER:
Ledum groenlandicum, Vaccinium mytrilloides, Chamaedaphne calyculata

HERB LAYER:
Few species. Gaultheria hispidula, Cornus canadensis, Coptis trifolia, Oxycoccus microcarpus, Smilacina trifolia, Carex trisperma.

MOSSES:
Sphagnum nemoreum, S. fuscum, S. magellanicum, S. girgensohnii, Pleurozium schreberi, Ptilium crista-castrensis, Dicranum polysetum.

LICHENS:
Cladina rangiferina

GROUND SURFACE:
Moderately hummocky with some "windowing." Sphagnum fuscum is the major mound forming species. S. magellanicum is associated with the windows.

SOIL

SOIL TYPE:
S14^{6} S13^{4}

FOREST HUMUS FORM:
Peatymor

DEPTH OF ORGANIC MATTER:
40-160+ cm

TEXTURE:
Organic soil.

MOISTURE REGIME/DRAINAGE:
(7-8)^{9} 6^{1} /VP

Figure 7. "Fact Sheet" for Operational Group 11.

imposed by various ecosystem features. This experience now aided substantially in the final grouping or subdividing of vegetation and soil types to form our management level ecological classification. This is expressed as 14 *operational groups* which are shown in Figure 6, overlaid on the DECORANA ordination of vegetation types (as shown originally in Figure 3). We define an operational group as a landscape segment with mature forest, which has an identified range of vegetation and soil conditions, and a probable response to specific management prescriptions.

In the field, we allocate any site to its operational group by using the key (Figures 1a, 1b). This operational group key is an extension of the vegetation type key taking into account certain soil conditions. For further description of operational groups, we have prepared fact sheets, one of which is shown in Figure 7. Information is given about soil and vegetation with both words and diagrams.

Our field testing of the operational group keys demonstrated that the system can be applied operationally with relative ease, efficiency and repeatability. Already we are very encouraged by the recent use of our classification by the field management staff in pre-cut site assessment and planning. We maintain that our classification, with the choice of detailed or operational level of differentiation, will provide a substantial ecosystem framework for both future forest management and research in the Claybelt.The study and classification of forest ecosystems in the boreal forest of Ontario will be extended beyond the Claybelt. The general approach will continue to consider and elucidate the important combinations of vegetation and soil. However, we will let the scientific data suggest the most useful way of expressing the ecosystem types and operational groups. Future allocation keys for application outside the Claybelt may well use soil and topographic features ahead of plant species, but that remains to be seen. We look forward to conducting and reporting on this work in the coming years.

Acknowledgements

We wish to acknowledge the professional support and technical help provided by a substantial number of people within our various organizations. We have already mentioned in the text the important roles of Kent Virgo and Rob Arnup. We would also like to name specifically, Jane Bowles, Doug Aspinall, Alison Munson and Chris Powter; but our thanks, of course, go to all of our colleagues. Our paper is Land Resource Research Institute contribution number 82-77.

Literature Cited

Braun-Blanquet, J. 1928. Pflanzensoziologie, Grundzüge der Vegetationskunde. Springer-Verlag. Berlin.

Bernier, B. 1968. Descriptive outline of forest humus form classification. *In:* Manual for Describing Soils in the Field (Revised 1978). J. Dumanski (Ed.). Land Resource Research Institute, Agriculture Canada. Ottawa, Ontario.

Canadian System of Soil Classification. 1978. Canada Soil Survey Committee, Subcommittee on Soil Classification.

Gemmell, J.R. 1979. The identification of sites in the Claybelt. *In:* Proceedings of the Second Ontario Conference on Forest Regeneration. Ontario Ministry of Natural Resources.

Hills, G.A. and A.N. Boissonneau. 1960. A Multiple Land-Use Plan for the Glackmeyer Development Area, Research Supplement. Ont. Dept. of Lands and Forests.

Hills, G.A. 1961. The ecological basis for land-use planning. Ont. Dept. Lands and Forests, Research Report No. 46.

Hill, M.O. 1979a. TWINSPAN — A FORTRAN Program for arranging multi-variate data in an ordered two-way table by classification of the individuals and attributes. Cornell University, Ithaca, New York.

Hill, M.O. 1979b. DECORANA — A FORTRAN Program for detrended correspondence analysis and reciprocal averaging. Cornell University. Ithaca, New York.

Hill, M.O. and H.G. Gauch, 1980. Detrended Correspondence Analysis, an improved ordination technique. Vegetatio 42: 47-58.

Jeglum, J.K., A.N. Boissonneau and V.F. Haavisto. 1974. Toward a wetland classification for Ontario. Environ. Canada, C.F.S., Info. Rep. 0-X-215.

Krajina, V.J. 1969. Ecology of forest trees in British Columbia. Ecology of Western North America 2: 1-146.

MacLean, D.W. and G.H.D. Bedell, 1955. Northern Clay Belt growth and yield survey. Can. Dep. North. Aff. Nat. Resour., For. Res. Div. Tech. Note No. 20.

Pogrebnayak, P.S. 1930. Über die Methodik von Standortuntersuchungen in Verbindung mit Waldtypen. Ver. II Int. Congr. Forstle. Versouchsanstalten, 1929. Stockholm, Sweden. pp. 455-471.

Scott, R.D. and K.D. Virgo. 1977. A Site Classification system for clearcut areas in the Claybelt. Ontario Ministry of Natural Resources, Kapuskasing District, Internal Report.

NUTRIENT CYCLING DYNAMICS IN DIFFERING SPRUCE AND MIXEDWOOD ECOSYSTEMS IN ONTARIO AND THE EFFECTS OF NUTRIENT REMOVALS THROUGH HARVESTING

Alan G. Gordon

Ontario Tree Improvement and Forest Biomass Institute

Ontario Ministry of Natural Resources

Abstract: Spruce forests are among the most important ecologically, geographically and economically in the North American boreal forest. Nutrient cycling dynamics and productivity of fully-stocked black spruce on two principal landtypes, peat and moist outwash sand, are elucidated and compared with those of two boreal mixedwood ecosystems consisting of white spruce, balsam fir, white birch and trembling aspen of moderate and high productivity. Comparisons are also made with north temperate red spruce ecosystems.

Mass budgets indicate that almost half of the total organic pools above- and below-ground are in the standing crops. When these are compared to their respective total organic and mineral soil mass reserves, only that of black spruce on peat is still nearly as great as reserves. Such sites could become vulnerable under full-tree harvesting. The effects of full-tree harvesting on the nutrient pools of these sites are shown. While simple budgets do not predict replacement times, data from nutrient cycles will. Differences in residence times for nitrogen in red and black spruce ecosystems demonstrate the importance of fire in accelerating rates in black spruce ecosystems. Replacement times provide estimates of the real tolerances to perturbations such as harvesting. Calculated at steady state input, they are relatively short, 20 years or less for the richer upland mixedwood sites and longer for black and red spruce sites. However on acidic tills, outwash sands and peat, replacement times for potassium can extend from 30 to 45 years. Upland mixedwood sites have greater resilience than black spruce stands on outwash or peat.

The areas in the boreal forest now being harvested by full-tree and whole-tree harvesting methods are massive. To conserve nutrients, harvesters should be utilized which strip the branches from the bole and leave them on the growing-site. How and what we harvest will have substantial effects on subsequent rotation ages.

Introduction

Spruce forests are among the most important ecologically, geographically and economically in the North American boreal forest. Spruce is actually more valuable, ton for ton, than any other boreal species. Recent concerns with respect to projected demands for wood relative to a diminishing supply, and current proposals to utilize forest biomass for energy production pose a number of questions important to the management of spruce forests.

Can the productivity of the forest and the forest land in all its variation be maintained with the type of management necessary to fulfill these objectives? Is it technologically and financially feasible? . . . and so on. The latter questions are relatively easy. This paper considers another question: can the productivity of the forest as it relates to the nutrient cycle, supply and availability, accumulation and loss of nutrients be maintained? Since this paper also deals with harvesting, full-tree logging refers to the removal of all above-ground portions of trees and whole-tree refers to all above and below-ground parts of the tree.

Forest ecosystems under natural controls are well regulated functional systems which evolve toward stability and persistence. Stability may be loosely defined as the integrity of composition and productivity. Many boreal forest communities operate in a form of cyclic stability. In some cases, for example the boreal mixedwood, all the species components may not be at the same stage of maturity in any given time frame. To maintain stability, ecosystems under perturbations such as harvesting regimes must be able to return to their original equilibrium following disturbance. In terms of the nutrient cycle this means sufficient site capital and atmospheric inputs to sustain the demands of the stand that may follow. The rate at which this occurs is known as resilience (DeAngelis 1980).

Study Area Description

In Ontario, black spruce *(Picea mariana)* characteristically grows on peats, moist outwash sand plains and on thin tills over bedrock. White spruce *(P. glauca)* grows predominantly in mixedwoods in association with balsam fir *(Abies balsamea)*, white birch *(Betula papyrifera)* and trembling aspen *(Populus tremuloides)*. Boreal mixedwoods characteristically occur on upland tills, till over bedrock, lacustrine fine sands, silts and clays and on some alluvial soils. The soils may have low to high lime content but are generally richer than those which support black spruce. In the mixedwoods, white birch occupies the stonier profile variants or thin till over bedrock. Red spruce *(P. rubens)* is not a boreal species but rather, a temperate conifer. It occurs on granitic tills of varying depth over acidic bedrock, and on limited areas of peat. It commonly grows in pure stands or in mixtures with balsam fir, white birch or red maple *(Acer rubrum)*, or with other later successional species such as yellow birch *(B. alleghaniensis)*, sugar maple *(A. saccharum)* and hemlock *(Tsuga canadensis)*.

Two study areas were utilized for boreal upland mixedwood studies; the first was east of Lake Nipigon approximately 12 km SE of Beardmore in the Blackwater River Watershed at 49°33′N latitude, 87°30′W longitude, elevation 442 m, Ontario site region 3W (Hills 1954). Supplemental data were also obtained from additional plots in the Black River Watershed, east of Heron Bay, Lake Superior. The climate is classed as la, the second coldest regime in Ontario (Ouellet and Sherk 1967). Annual precipitation is 78.6 cm of which 58.8 cm falls as rain. Mean annual maximum, mean and minimum temperatures are 6.1°C, −0.4°C, −7.0°C, respectively. This stand was considered as only moderately productive and had a rather low basal area of 20.3 m^2ha^{-1}, possibly in part a reflection of stand history. It was growing on fresh till.

The mean stand age of boreal mixedwoods is notoriously difficult to determine because of the multiple stages of stand development of all four principal species at any given time. Stand age was determined largely from frequencies of dominants and co-dominants of principal species in the overstory. Mean stand age was 100 years. Mean stand height was 21.0 m. The stand was considered mature.

The second mixedwood stand was located 97 km NW of Thunder Bay in the Dog River Watershed at 49°12′N latitude, 89°57′W longitude, elevation 472 m, Ontario site region 3W (Hills 1954). The area is classed as a 2a climate (Ouellet and Sherk 1967), with an annual precipitation of 78.7 cm of which 55.2 cm falls as rain. Mean annual maximum, average and minimum temperatures are 6.8°C, 0.6°C, and −5.6°C. This stand which was considered highly productive had a basal area of 28.8 m^2ha^{-1} and was growing on deep silt and fine sand. The mean stand age of principal species in the overstory (largely white spruce and trembling aspen) was 105 years and mean stand height was 23.0 m. The stand was considered mature. Both of these areas are located on broad massifs and the topography is gently undulating. There is no sharp topography in the immediatc vicinities and no differences are considered to be related to slope.

As noted in the foregoing, white spruce in Ontario scarcely ever grows in pure stands. In a previous paper, Gordon (1981a) showed that in boreal mixedwoods, differences in above-ground biomass and bioelement stores were small in relation to stand composition shifts from 25 per cent softwoods and 75 per cent hardwoods to the converse. Productivity did decrease, however. Plots for this study were selected to represent 50 per cent conifer and hardwood mixtures.

Two main areas were utilized for black spruce studies. The first was east of Lake Superior, north of Sault Ste. Marie in a boreal forest outlier at 46°42′N latitude, 84°21′W longitude, elevation 213 m, Ontario site region 4E (Hills 1954). Supplemental data were also obtained from additional plots 32 km west of Swastika at 48°15′N latitude, 80°27′W longitude, elevation 320 m, Ontario site region 3E (Hills 1954). The annual precipitation is 93.5 cm of which 66.2 cm falls as rain. Mean annual maximum, average and minimum temperatures are 8.4°C, 3.1°C and −2.2°C. This stand was developed on outwash sand and had a basal area of 32.6 m^2ha^{-1}. Mean stand age and height were 48 years and 15.5 m. The stand was immature.

The second black spruce stand was located east of Lake Nipigon about 32 km south of Beardmore at 49°15′N latitude and 87°45′W longitude, elevation 275 m, Ontario site region 3W (Hills 1954). The annual precipitation is 78.6 cm of which 58.8 cm falls as rain. Mean annual maximum, average and minimum temperatures are 6.1°C, −0.4°C and −7.0°C. This stand was developed on moist peat and had a basal area of 39.7 m^2ha^{-1}. The mean stand age and height were 108 years and 18.6 m. The stand was mature.

The red spruce stand was located in southern Algonquin Park at 45°32′N latitude, 78°49′W longitude, elevation 495 m, Ontario site region 5E (Hills 1954). The annual precipitation is 76.9 cm of which 59.1 cm falls as rain. Mean annual maximum, average and minimum temperatures are 10.8°C, 4.1°C and −2.1°C. This stand was developed on fresh till and had a basal area of 46.8m^2ha^{-1}. The mean stand age and height were 138 years and 18.3 m. The stand was mature.

Methods

Data were derived from a series of 50 plots (0.4 ha in size) distributed in red, black and white spruce-mixedwood types. All standard mensurational data were collected. Data on biomass were derived from equations developed for these species from destructive sampling of arrays of all diameter classes on several sites. Root data were collected by excavation and coring trees of large, medium and small diameter classes across site arrays. Apart from spruce, supplemental data for other species in the mixedwoods only, were utilized to augment our own data for balsam fir (Baskerville 1965; Honer and Heger 1971), white birch (Van Cleve and Noonan 1975; Troth *et al.* 1976; Jokela *et al.* 1979), and trembling aspen (Bray and Dudkiewicz 1963; Gorham and Bray 1964; Peterson *et al.* 1970; Pollard 1971, 1972; Van Cleve and Noonan 1975; Johnston and Bartos 1977; Alban *et al.* 1978).

Productivity data were derived from 10-year plot remeasurement, increment core data on more than 4000 trees, stem analysis on 152 trees and additional root sampling. Subordinate vegetation was surveyed and sampled by clipping by age class on ten 1.0 m^2 quadrats per plot. Soil and plant chemistry consisted of determining six macroelements: N, P, K, Na, Ca and Mg for all vegetative components and soil horizons as well as pH, loss-on-ignition and bulk densities for soils. Chemical determinations were in the order of several thousand.

Fluxes: precipitation (rain and snowfall), stemflow, throughfall, soil horizon moisture and litterfall were sampled by standard methods. Soil horizon moisture fluxes are still under study by the use of lysimeters and soil porous cups.

Analytical chemistry was accomplished by semi-micro Kjeldahl procedures for total N, in soil and plant tissue; exchangeable P for soils and total P for plant tissue, by the molybdate blue method on a spectrophotometer; exchangeable cations for soils and total cations for plant tissue, by atomic absorption procedures.

Results and Discussion

Biomass and reserves of fully stocked stands of spruce and boreal mixedwoods are shown in Table 1. Mass of shrubs and lesser vegetation, which are of greater consequence in the mixedwood sites than in the black spruce, are omitted from the table since it is expected they will be excluded from harvesting. Data on flowers and fruits were combined with foliage, while bark mass was included with bole. Since standing dead gradually falls down and is all knocked down during harvesting, it was included with soil reserves along with roots (Newbould 1981).

Soil mineral mass is defined by the maximum depth of rooting. Organic soil mass on the moist peat site is defined by the depth of the rooting zone usually delineated by the water table.

Foliar and subsequently branch biomass become stabilized shortly after crown closure and may be readily compared. It may be seen that foliage biomass declined from red spruce through black to the boreal mixedwood. This is, at least in part, accounted for by differing growth strategies of different spruce species. Red spruce is a very shade tolerant species. Black spruce is the least tolerant and white spruce is intermediate. It should be noted that spruce species strategies for crown characteristics are also important. White spruce has the longest conical crown of all three species. It appears that this species, being only of intermediate tolerance, has adapted to somewhat wider spacing to maximize crown surface and at the same time avoiding premature needle loss. Crown closure even in pure white spruce stands in Alaska is seldom over 75 per cent (Farr 1967). This in conjunction with the inclusion of aspen crowns could result in greater wood production for the stand as a whole. In the mixedwood stands aspen, predominantly, and white birch make up the remaining stocking. Aspen foliar mass is exceedingly light relative to that of spruce. This is in accord with data of Alban *et al.* (1978). These inferences could account for the lower

Table 1. Standing crop (10^3 kg ha^{-1}) of dry matter in spruce and mixedwood forests on typical land types

	Red spruce	Black spruce		Upland mixedwood	
		peat	outwash sand	fresh till	silt, fine sand
	fresh till				
Above-ground					
Foliage	20.6	17.7	15.9	6.1	11.3
Branches	36.4	22.7	19.9	19.5	23.0
Bole	206.9	120.3	104.8	76.5	101.6
Total	263.9	160.7	140.6	102.1	135.9
Below-ground					
Standing dead	8.5	6.7	6.6	3.7	30.4
Roots	62.0	36.6	26.8	30.7	40.7
Organic soil	187.3	127.3	107.0	110.7	68.5
Mineral soil	1235.8	0.0	1233.9	2417.8	4867.4
Total	1493.6	170.6	1374.3	2562.9	5007.0

foliar mass for the mixedwood sites than in the more densely stocked red and black spruce sites.

O'Neill and DeAngelis (1981), working with woodland data sets of the I.B.P., point out that wood increment is positively correlated with leaf biomass in conifers but relatively independent in broad-leaved species. It may be argued that within any species, leaf mass is also related to site, heavier leaf mass being carried on better sites. Gordon (1979, 1981b) provides evidence of this for red spruce across a moisture regime catena of dry, fresh, moist and wet sites. The greater the photosynthate available for growth, the greater will be the amount of wood tissue produced. There are, however, many strategies exploited by trees to accomplish this; that for trembling aspen with its light foliar biomass and trembling leaves is one. Spruce species strategies also go beyond simple biomass differences.

Age differences do not invite absolute comparisons of bole or total biomass. Age differences are, however, not great between any of the stands except the black spruce on outwash sand. It is, nevertheless, surprising that both black spruce stands had the same or greater bole and total biomass as did the upland mixedwoods.

It should be recalled that since understory shrubs and lesser vegetation are not included in harvesting, they are not included in total above-ground standing crop biomass in Table 1. Their mass (369 kg ha^{-1}), while greater on the mixedwood fresh site of lower basal area, is not considered sufficient to increase the total values beyond that of the black spruce stands. These components, particularly shrubs and herbs, are of minor consequence in the dark, fully closed stands of red and black spruce.

Biomass differences in the mixedwood sites were due primarily to site, as indicated by rooting depth mineral soil mass, and root mass. Stand history, possibly resulting in lower stocking density and basal area on the fresh site, may also have contributed to the differences. It should be noted that not only is the above-ground biomass lower on this mediocre mixedwood stand on the fresh site, but also the mass of unincorporated organic matter is greater than on the more productive moist site.

The relatively large mass of unincorporated organic matter is somewhat surprising for a mixedwood site. However, it is interesting to note that a mixed trembling aspen and white birch stand in the same general area (see Timmer *et al*. 1983), also had a substantial Ah horizon. This stand was previously a typical mixedwood and was growing on a sombric humo-ferric podzol.

Differences in biomass between the black spruce sites are due to site and age. Since the stand on the outwash is only half the age of that on the peat, it is expected that its final biomass will exceed that of the stand on the peat site.

The large bole and total biomass noted for the red spruce fresh site, actually more than 1½ times greater than that of site class 1 black spruce on moist peat, is in part due to its somewhat greater age and basal area. It is also due to the fact that it represents a fully occupied site by a temperate conifer at a lower latitude. Litterfall data indicate this is in accord with data presented by O'Neill and DeAngelis (1981).

Of greatest importance in Table 1 is to consider the proportion of harvestable biomass to that of soil reserves. This may be shown as a simple total reserves / above-ground standing crop ratio. For the upland mixedwoods the ratios are 37 for the deeper silt and fine sand site and 25 for the fresh till. For red spruce, fresh till the ratio is 6 and for black spruce outwash and peat sites, 10 and 1, respectively. These values rank in similar order as those of a related ratio: forest floor to tree biomass, compiled by Krause *et al*. (1978) for several boreal forest types.

More customarily, the total standing crop is compared as a percentage of the organic or total reserves. It may be seen from Table 1 that there is a striking pattern. In all cases except the mixedwood fresh site with its somewhat low above-ground biomass and relatively large amount of soil organic matter, the weight of the standing crop available for full-tree harvesting exceeds that of the organic soil reserves, exclusive of standing dead and roots, by more than 100 per cent. If, however, we include all the standing dead and the roots in the total organic reserves, it may be seen that in all cases, just about half of the total organic pool, above and below-ground, is in the standing crop. If total reserves are considered which include the mineral soil mass, only black spruce sites on peat support a crop in which the biomass is still nearly as great as in the reserves. This could make these sites particularly vulnerable to full-tree harvesting. It should be remembered that reserves in peat are only applicable to forest growth down to the water table.

Biomass and Basal Area

Over time, the principal biomass increase occurs on boles with increasing height and diameter growth. Height in turn is dependent on age for any given site (site index). Continued diameter growth is accommodated by negative shifts in stocking density at gradually increasing basal area.

It has been indicated that in deciduous trees similar biomass distributions may be produced at similar basal areas through a considerable range in stocking density (Gardner

Table 2. Pattern of basal area (m^2ha^{-1}) and basal area increment ($m^2ha^{-1}year^{-1}$) decline with age (years) in mature spruce stands

Species	Locality	Age	Basal area	Basal area increment	Reference
Picea abies	Karelia, USSR	109	38.9	0.5	Kazimirov and Morozova (1981)
Picea abies	Karelia, USSR	126		0.3	Kazimirov and Morozova (1981)
Picea abies	Karelia, USSR	138	38.0	0.2	Kazimirov and Morozova (1981)
Picea abies	Solling Project, F.R. Germany	87	44.8	0.88	Ellenberg (1981)
Picea abies	Solling Project, F.R. Germany	115	37.7	0.56	Ellenberg (1981)
Picea mariana (Site Class 1)	Northern Ontario	110	40.0	0.3	Plonski (1956)
Picea mariana (Site Class 1)	Northern Ontario	115		0.2	Plonski (1956)
Picea mariana (Site Class 2)	Northern Ontario	120	37.8	0.3	Plonski (1956)
Picea mariana (Site Class 2)	Northern Ontario	125		0.2	Plonski (1956)
Picea rubens (fresh site)	South central Ontario	138	46.8	0.25	Gordon (1981b)

and Mankin 1981). The latter is a function of age. In spruce, basal area gradually increases with age until the stand starts to open up or fall apart, often at considerable age. In Alaska, long-lived white spruce stands will remain well stocked (crown cover 50-75 per cent) for 180 years or more (Farr 1967). Basal area increment (BAI) however, starts to diminish far sooner.

Table 2 provides data on the pattern of decline, with age, of basal area increment in relatively mature stands of several species of spruce. Younger stands demonstrate higher basal area increments, often greater than 1.0 $m^2ha^{-1}year^{-1}$, while in old stands, BAI falls to 0.1 $m^2ha^{-1}year^{-1}$ or less. The pattern of declining productivity with age is doubtless related to that of basal area increment.

Productivity and Stand Composition

It has long been known that productivity is related to age. Aging effects begin to appear in conifer plantations between 50 and 100 years and in natural conifer stands between 150 and 200 years (O'Neill and DeAngelis 1981).

Many tree species are thought of as even-aged, maximizing their performance under uniform conditions, and with all individuals developing aging symptoms about the same time. Some forests, however, composed of these very species and also thought of as even-aged, follow a somewhat different strategy. The boreal mixedwood is one of these. Aging effects develop at different times for each of the species in the mixture. As one species begins to die out, another species experiences release and increment increase. In this way, productivity of the forest is actually sustained beyond the life of some of its component species. Related sequences have been elucidated for willow *(Salix)*, balsam poplar *(Populus balsamifera)*, white spruce, white birch and black spruce in Alaska (Van Cleve *et al*. 1980; Juday and Zasada, In press).

Similarly, in the case of temperate forests, mixtures of shade tolerant species or species of varying tolerance, such as red spruce, eastern hemlock and northern tolerant hardwoods, the canopy and immediate understory may contain species and individuals of varying ages. Equilibrium between mortality and replacement may develop and be prolonged well beyond the age of some of the initial species. Coniferous mixtures of relatively tolerant conifers of differing longevity such as white spruce or red spruce and balsam fir; Engelmann spruce *(P. engelmannii)* and subalpine fir *(Abies lasiocarpa)* are further examples.

Red, white and black spruce are species of differing growth strategies (Gordon 1976). Red spruce, characteristically has a rather even and relatively constant increment which is maintained with little reduction to considerable age. The pattern of increment reflects environments in which changes occur gradually. Red spruce tends to become established over a considerable period and is long-lived. It commonly reaches ages of 255 years and is known to reach 324 years (author's data). Incremental decline to mortality is gradual.

The increment of white spruce is noticeably responsive, strongly reflecting environments which change relatively suddenly. Foliar mass can be increased rather quickly following release and is reduced quickly following budworm attack. Increment decline to mortality is rather steep. White spruce seedlings tend to become established over a relatively short time. While this species is the longest-lived in the boreal mixedwood, it is not exceptionally long-lived in eastern Canada. It commonly reaches ages of 180 years and may live to 225 years. In Alaska on flood plains, the oldest recorded is 329 years (Farr 1967), but in the uplands it scarcely surpasses 200 years as a result of recurring fire patterns (Van Cleve and Viereck 1981). Davidson and Redmond (1957) found that decay in eastern white spruce rises from 2.7 per cent at 100 years to 9.4 per cent at 200 years, whereas in red spruce, decay does not rise above 1.1 per cent even at 230 years. The decay is customarily located in the butt and is usually enough to bring the tree down in a wind. In their study, no white spruce trees were encountered over 200 years.

The increment of black spruce tends to be rather uniform from tree to tree reflecting the free-to-grow strategy of an even-aged species. Increment is evenly maintained among dominants as the stand closes, and decline to mortality is gradual. While black spruce is a successful even-aged species in closed crown boreal forests, Rencz and Auclair (1980) point out that in the subarctic open lichen black spruce woodlands, where the spruce can maintain its crown canopy to the ground, it assumes a reverse J-shaped frequency/diameter distribution typical of all-aged stands, despite the fact the trees are all the same age. This is similar to the circumstances apparent when black spruce colonizes open bog or outwash sand plains in the boreal forest.

It is evident from the foregoing that at least for red spruce and white spruce growing in intolerant mixtures, stand growth is sufficiently prolonged in equilibrium to enable measurement and comparisons of productivity at greater ages than would apply for truly

Table 3. Above-ground productivity (litterfall plus increment) in kg ha^{-1}year^{-1} for five spruce and mixedwood stands

Lat.	Site	
45°30′N	Red spruce mature — fresh till	8703.7
46°42′N	Black spruce immature — fresh outwash sand	5742.3
49°15′N	Black spruce mature — moist peat	4568.3
49°12′N	Mixedwood mature — fresh-moist silt, fine sand	5755.0
49°33′N	Mixedwood mature — fresh till	4742.4

even-aged stands where aging effects would be expected to appear earlier.

Productivity

Above-ground productivity is presented in Table 3. Of the boreal mixedwood sites it may be seen that productivity on the deeper silt fine sand site is substantially greater than that on the fresh till. While the productivity of both mixedwood sites exceeds that of site class 1 black spruce on peat, the best mixedwood site far exceeds it despite the efficiency of black spruce on good moist peat soils.

The productivity of the black spruce on the outwash also exceeds that of black spruce on peat despite the fact that the basal area of the former (32.6 m^{2}ha^{-1}) was less than that of the latter (39.7 m^{2}ha^{-1}). This probably reflects both the site (the reserves are greater although not in organic matter) and age. This stand on outwash is 48 years old and is classed as immature. It may be noted that in an age sequence study of Norway spruce *(Picea abies)* in southern Karelia, USSR, the greatest productivity (6210 kg ha^{-1}year^{-1}) occurred at age 54 (Kazimirov and Morozova 1981).

The productivity of red spruce on a fresh site is substantially greater than that of all the boreal sites. While not the best site on which red spruce grows (Gordon 1981b), this is a very good site. The productivity is again an example of a temperate conifer at a lower latitude. It is interesting that a relatively high productivity is being maintained although the stand is somewhat older than the boreal stands.

It is of interest for purposes relating to energy efficiency, etc. to make comparisons of productivity of different ecosystems (Rodin *et al.* 1975), for example, boreal and tundra, as well as agricultural ecosystems. Comparisons of boreal forest productivity with best expected agricultural productivity at similar latitudes are of importance.

Annual yields of dry matter production of grasses on agricultural lands at Kapuskasing, Ontario, 49°23′N latitude, 82°30′W longitude, Ontario site region 3E (Hills 1954) ranged from 2534 kg ha^{-1} with no fertilizer, to 8026 kg ha^{-1} maximally fertilized with 404 kg ha^{-1} of N, as well as P and K (Guillemette 1982). To attain the same productivity with grass forage production as that of upland mixedwoods on similar sites, about 130 kg ha^{-1} of N, plus ameliorating amounts of P and K are required annually. To attain the same productivity for forage grasses as that of black spruce, about 90 kg ha^{-1} of N plus P and K are required. Similarly, barley yields of most varieties and best cultural methods including annual fertilizing with N, P and K, reach 4857 kg ha^{-1}year^{-1} (Guillemette 1982).

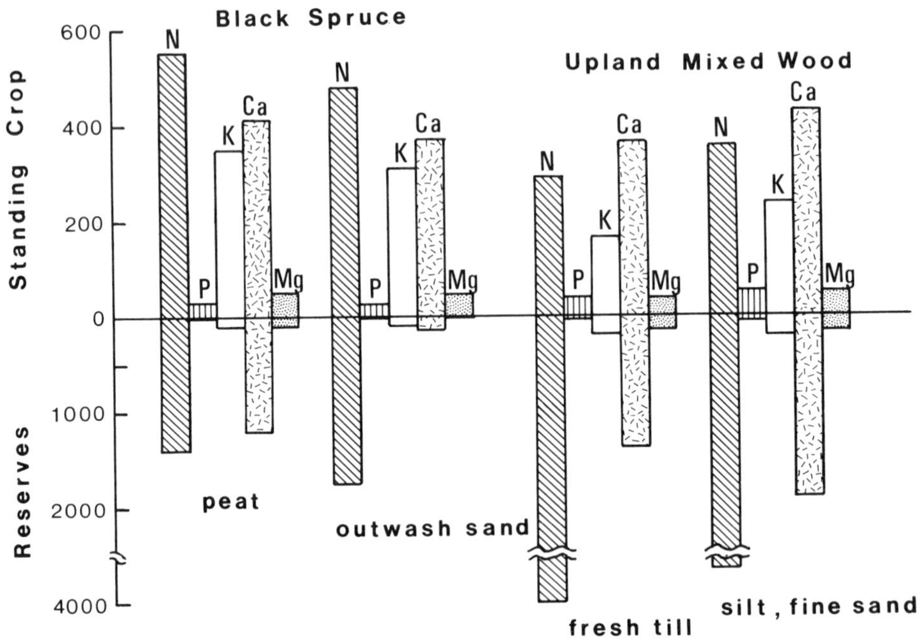

Figure 1. Bioelement stores in kg ha^{-1} for black spruce and upland mixedwoods of the boreal forest on typical land types. (See text for further details.)

Bioelement Stores

Bioelement stores are the accumulated nutrients in the biomass of the above-ground standing crop and the nutrient stores in the reserves. In Figure 1 above-ground tree components are grouped. The small amounts of these five macroelements in the subordinate vegetation are omitted since it is expected they will not be removed in harvesting. The amounts are insignificant in the pools of these sites. Standing dead as well as roots are included in the organic reserves and these are grouped with soil mass reserves in the rooting zone.

It may be seen that the nitrogen content of the standing crop on all four sites was substantially less than that in the reserves. Except on the black spruce outwash site, calcium also was in excess in the reserves over that of the standing crop. Surprisingly, it was in excess in the black spruce peat site. This seems unusual but there may be an additional source in ground water. Magnesium in the reserves was also in excess of standing crop on all sites but the black spruce outwash sand.

On the other hand, phosphorus on both black spruce sites and potassium in all sites but one of the mixedwood sites, was greater in the standing crop than in the soil reserves. It is in these sites, with lower nutrient reserves in the soil than in the standing crop, that we might expect future nutrient stress following harvesting by full-tree logging.

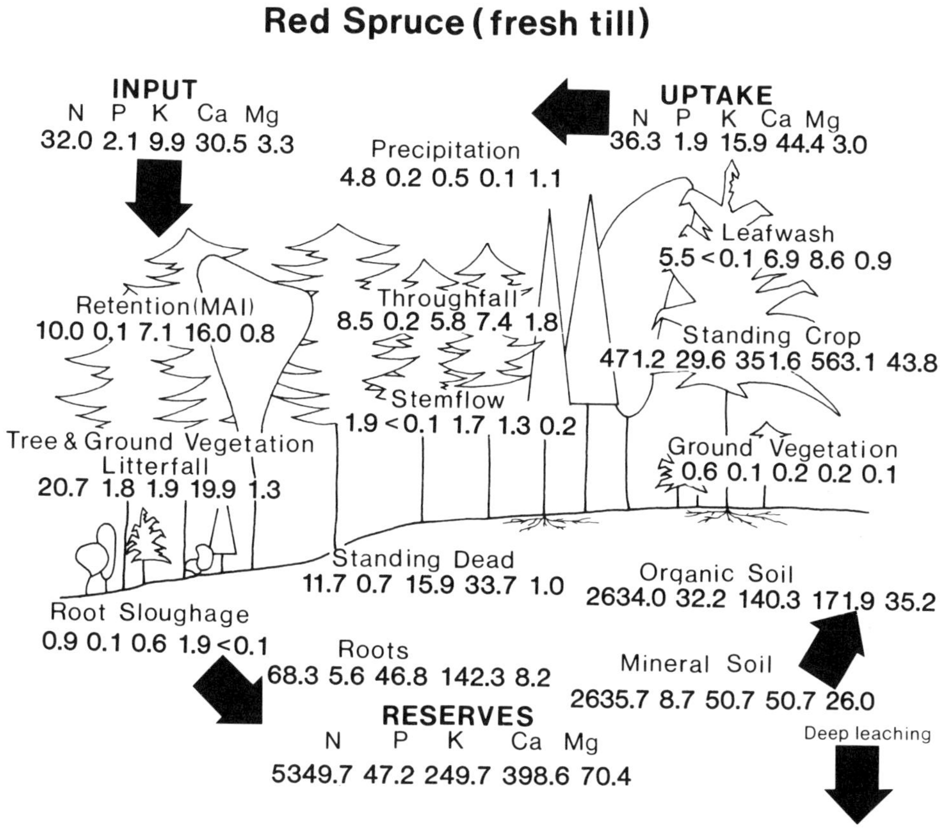

Figure 2. Stylized nutrient cycle of a north temperate red spruce forest on fresh granitic till. Species represented are red spruce, balsam fir, eastern hemlock, yellow birch and sugar maple.

Nutrient Cycling

Figures 2 to 6 present stylized nutrient cycles for a number of spruce forest types. These closed-crown stands are in steady state. Inputs essentially balance output and internal cycling within the stand is maximal. Small net gains and losses occur but generally the cycles are tight. Such measurements of input/output are generally accomplished by a synthesis of several interactive processes. Input, for example, is a summation of throughfall and stemflow + total litterfall + root sloughage. Uptake includes nutrients in litterfall and those retained in the individual increments of all components of the trees and lesser vegetation, as well as fluxes of water in leafwash, etc. For black spruce on peat, the reserves were limited to organic soil and organic components.

In the upland mixedwood fresh till site, the mean stand height was 21 m, while that on the mixedwood silt and fine sand was 23 m. Because these were all largely coniferous ecosystems having trees of the same general life form, large differences in input and uptake were not expected. However, minor differences occasioned by locality, site and

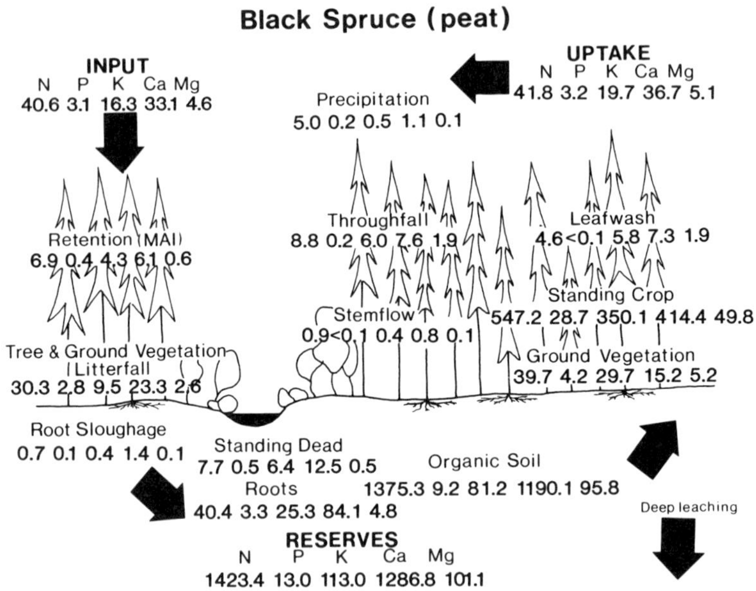

Figure 3. Stylized nutrient cycle of a boreal black spruce forest on peat. Species represented are black spruce, *Ledum groenlandicum, Alnus rugosa* and *Chamaedaphne calyculata*.

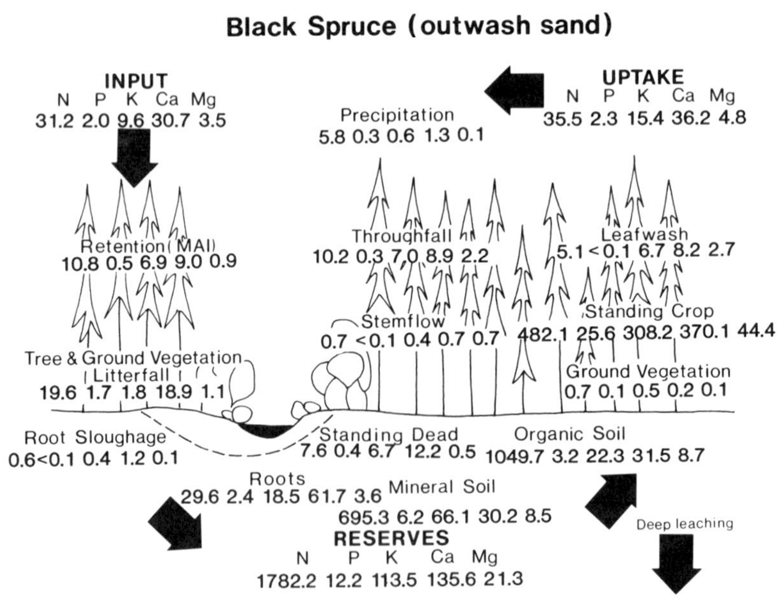

Figure 4. Stylized nutrient cycle of a boreal black spruce forest on outwash sand. Species represented are black spruce, *Nemopanthus mucronata, Alnus rugosa* and *Ledum groenlandicum*.

Upland Mixedwood (fresh till)

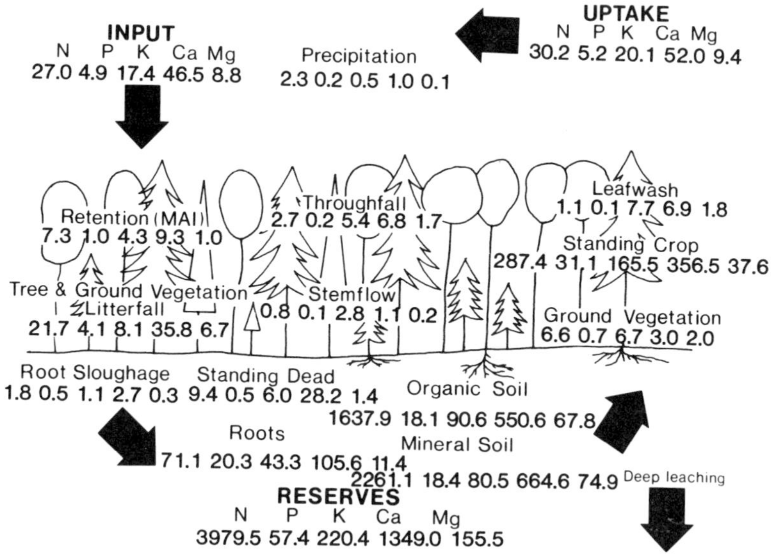

Figure 5. Stylized nutrient cycle of a boreal upland mixedwood on fresh till. Species represented are white spruce, balsam fir, trembling aspen and white birch. The mean stand height is 21.0 m.

Mixedwood (silt, fine sand)

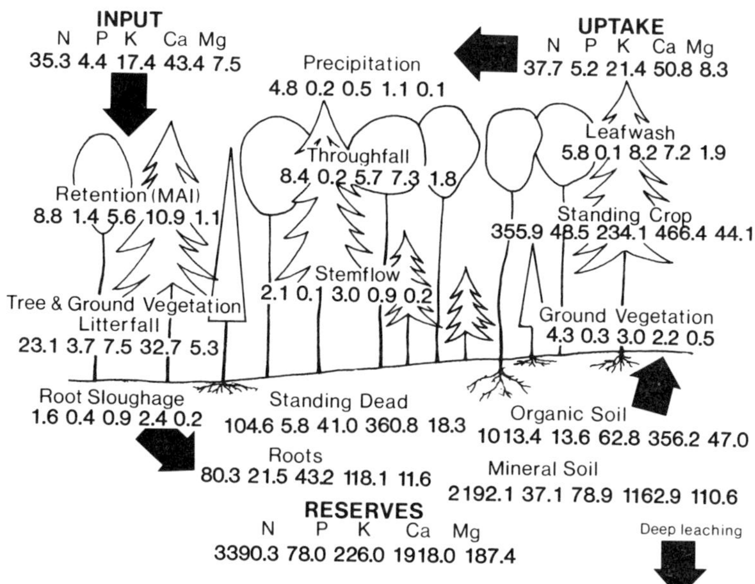

Figure 6. Stylized nutrient cycle of a boreal upland mixedwood on silt, fine sand. Species represented are white spruce, balsam fir, trembling aspen and white birch. The mean stand height is 23.0 m.

stand characteristics were evident. For example, both input and uptake values for all elements were greater on the black spruce peat than on the black spruce outwash as shown in Figs. 3 and 4. Those for the mixedwood stands were somewhat similar and apart from nitrogen, were greater than those for both black spruce stands.

Residence Time

In a nutrient cycle, nutrients are returned to the forest floor by litterfall, throughfall and stemflow. Some remain in a labile form and are available for uptake by trees or ground vegetation, or are lost by leaching. Other forms become locked up in the organic matter of the forest floor for many years. They may gradually be released and re-enter uptake or output flux systems. In steady state where input balances output, the rate of nutrient input to the forest floor may be compared to the total accumulated in the forest floor. Residence time is the amount of a nutrient in the mass of the soil organic matter divided by the amount of that nutrient in the input (litterfall, net stemflow + throughfall, etc.). The reciprocal of residence time is the turnover rate ($year^{-1}$) or the fractional turnover. Residence time could be considered an index of constraint to nutrient availability and loss.

Table 4 presents residence times for the four boreal and one temperate stands. Nitrogen has the longest residence time because it is associated with the carbon cycle and may be bound up in organic complexes for long periods. The silt fine sand mixedwood site had among the shortest times for most elements. In general, mixedwood sites had the most active forest floor horizons and shorter residence times were expected. Resilience, therefore, will be greatest in upland mixedwoods. Gordon (1981a) showed that as hardwood composition of mixedwood stands increased, residence time decreased.

Residence times on the black spruce outwash sand site were remarkably short and much shorter than those for red spruce. This reflects both the very low nutrient status of the granitic outwash sands supporting black spruce, relative to the granitic tills supporting red spruce; and also the difference in stand origins of these two species. Red spruce stands seldom originate following fire but rather repeat themselves in a form of gap-phase replacement in grouped multi-aged or all-aged stands. Organic matter may build up over very long periods. Black spruce stands very commonly originate following fire. Fire cycles are sufficiently frequent to affect origins of black spruce at least every other rotation (Heinselman 1971; Yarie 1981). The stand in this study originated following a very hot fire which burned the organic matter down to the mineral soil.

Table 4. Residence time* for mass and five nutrients in five spruce and mixedwood sites

	Mass	Total N	P	K	Ca	Mg
Red spruce — fresh till	65	83	16	15	7	12
Black spruce — peat	35	35	4	6	37	22
Black spruce — outwash sand	39	35	2	3	2	3
Mixedwood — fresh till	33	62	5	6	13	9
Mixedwood —silt, fine sand	19	30	4	5	9	7

*Residence time = 1/k (years).

$$K = \text{turnover rate} = \frac{\text{Input}}{\text{Input + Organic Soil.}}$$

Table 5. Mass and nutrient loss in standing crop removal (trees only) as a percentage* of organic reserves and organic and mineral reserves

	Organic reserves						Organic and mineral reserves					
	Mass	Total N	P	K	Ca	Mg	Mass	Total N	P	K	Ca	Mg
Red spruce — fresh till	102	17	77	177	162	99	18	9	63	141	141	62
Black spruce — peat	94	38	221	310	32	49	94	38	221	310	32	49
Black spruce — outwash sand	100	44	420	649	351	349	10	27	208	271	273	209
Mixedwood — fresh till	70	17	80	118	52	47	4	7	54	75	26	24
Mixedwood — silt, fine sand	97	30	119	159	54	57	3	10	62	104	22	24

$$* \frac{\text{Standing Crop (Trees Only)} \times 100}{\text{Reserves}}$$

Crop Removal: Full-tree Harvesting

Comparative data for different spruce types are presented in Table 5 for mass and nutrients of the standing crop which will be exported, as a percentage of the soil reserves. Values over 100 indicate by what percentage crop removal exceeds the reserves.

Since a high percentage of the fine feeding roots of spruce trees are located in the organic layers of forest soil profiles, it is interesting to note that potassium export in all cases and phosphorus in almost as many, were in excess of that in the organic reserves. The values for all elements exported except nitrogen were substantially in excess of organic reserves on the outwash sand site. When the total reserves in both the organic and mineral soil down to maximum rooting depth were considered, the condition applying on the outwash sand still remained. All elements exported except nitrogen were two-fold or more greater than what was left in the site. Potassium exports on all sites but one, and phosphorus on both black spruce sites — peat and outwash, were greater than those remaining in the reserves.

It should be noted that while with full-tree logging the total nutrient pool was massively disrupted, the reserves in the soil were still intact and available for seedling growth. Early nutrient deficiency might be expected for potassium and phosphorus, whose export from some sites was excessive relative to reserves, but growth may more certainly be limited as the stands close and demand becomes maximal for a time. After the stand closes, internal cycling also becomes maximal. Calculations based on data from these cycles (Figures 2-6) indicate that spruce and spruce-mixedwood forests retranslocate quite moderate amounts of nutrient elements. In addition, Gordon (1975) showed that in three species of spruce, nutrient demands of developing current foliage were accompanied at least by small and proportionate nutrient losses in one, two and three-year-old foliage. Cole and Rapp (1981) indicated that in conifers little, if any, nutrient elements were translocated from old to new tissue; demand almost entirely being supplied by uptake. Nevertheless, with the onset of tight cycling, demand on the actual soil reserves may decline for some time (Miller, *in* Heal *et al*. 1982).

It is clear from the foregoing data that total nitrogen should be sufficient in the soil reserves along with atmospheric input to sustain growth. However, long residence times on some sites (Table 4) indicated that much of the nitrogen is locked up in unavailable soil organic form. Assuming that a portion of this becomes available through N mineralization following harvesting, there should be sufficient for seedling growth at least until stand closure. Weetman and Webber (1972) however, cautioned that most of the available nitrogen for the next tree crop comes from decomposition of logging slash. This is particularly true of sites occupied by black spruce. With full-tree logging there is often very little remaining debris.

Replacement Times

The time required to return the organic and nutrient pools to former levels and condition following harvesting or other perturbation is known as replacement time. Estimates of replacement times following full-tree harvesting of mature stands at full rotation are shown in Table 6. These values are somewhat liberal since they were calculated at steady state input. Internal cycling was maximal and losses were considered to be small.

Replacement times were relatively short, mostly 20 years and under, for the richer, upland mixedwood soils and somewhat longer for sites supporting black and red spruce types. For most elements, times were not excessive but in the case of potassium it can take from 30 to 45 years for the system to return to preharvesting levels. These stands consisted mostly of site class 1 spruce. It will be understood that if site quality is lower, some nutrient elements (e.g., potassium) will already be at limiting levels. Replacement times may prove to be very long.

The values shown in Table 6 represent harvesting by full-tree logging. Stump axes and roots were presumed left on site. If whole-tree logging was practised where stumps and roots are harvested or biomass harvesting where all subordinate vegetation is also taken, replacement times would be substantially longer.

The foregoing estimates do not include data for losses due to deep leaching. Such data are being collected in these studies but evaluation for a short weather cycle is not yet complete.

Table 6. An estimate of the number of years required to replace, through input, nutrients lost in a single crop removal (left) and single crop removal plus two years of leaching (right)

	Single crop removal					Crop removal plus leaching	
	Total N	P	K	Ca	Mg	Total N	Ca
Red spruce — fresh till	23	23	45	28	20	31	34
Black spruce — peat	22	18	31	22	20	28	27
Black spruce — outwash sand	24	21	42	22	22	31	27
Mixedwood — fresh till	20	16	19	17	14	28	21
Mixedwood — silt, fine sand	19	21	23	20	16	25	23

It has been shown (Likens *et al.* 1978) that losses from leaching were greatest in the first two years following harvesting. From leachate data recorded for forest land elsewhere (Likens *et al.* 1978) and from the foregoing cycling and crop removal data, estimates are shown in Table 6 of increased replacement times for nitrogen and calcium. Increases were relatively small, 7 years on the average for nitrogen and 5 years for calcium. Revegetation following harvesting tightens up the cycle and reduces losses rather quickly but losses may still continue to exceed inputs for several more years. This circumstance will have the effect of extending replacement times. The extent and type of site preparation and the level of herbicide use may also extend this period.

Weathering inputs from bedrock and glacial parent material should be greatest on the deep mixedwood soils because of particle size and petrography. In general, however, inputs from weathering in cool northern climates are expected to be slight in soils which are poorly drained, or shallow soils over bedrock where water residence time may be short, and soils of low clay content, or outwash (Gordon 1981a, and more particularly Gorham *et al.* 1979; Clayton 1979).

Finally, the disposal of slash at harvesting has been briefly mentioned. Modern harvesting methods put far more stress on the nutrient pools and the structure and function of cycling pathways of forest ecosystems than any previous type of so-called 'conventional' clearcutting, where stumps and slash were left on the site. Full-tree logging is probably the most widespread and common form of harvesting in the Canadian boreal forest. Whole-tree logging is becoming more widespread in eastern North America, both in the boreal forests and southward in the forests of the eastern seaboard. It is practised in Scandinavia particularly in Finland (Hakkila 1976a, 1976b; Mälkönen 1976, 1979). It is gaining in acceptance since absolute yield, at least in the current rotation, is increased on any site. Biomass harvesting, where the whole stand including all shrubs and all other subordinate vegetation, as well as all the so-called 'waste' is taken, is advocated by many who regard it as a means of solving interim or long-term energy shortage problems, and by others as a means of converting silvicultural slums into constant yield fibre forests.

For the present, the practices of full and whole-tree logging are still very limited on areas of stony and thin-soiled shield soils of rough or steep topography. They are most frequently practised on stone-free soils: outwash sands and gravels, silts and clay plains of flat to moderate topography and peat soils in the frozen condition. There are many attractions to these forms of harvesting, both economic and silvicultural. Immediate yields are increased and site preparation costs reduced, while at the same time planting chances are improved through slash reduction or removal. Competition may be more easily controlled.

Vitousek and Melillo (1979) and Vitousek (1981) have outlined processes in the nitrogen cycle which are important to nitrogen loss and conservation. Principal losses occur following harvesting through mineralization of relatively stable and previously unavailable forms of organic nitrogen in the reserves. When a forest crop is removed, soil surface temperatures increase, evapotranspiration is reduced and increased mineralization follows. Since uptake has also been drastically reduced by crop removal, excessive losses through leaching of the mineralized nitrogen (nitrate and ammonium) may occur until predisturbance uptake levels are re-established by incoming pioneer vegetation.

The importance of slash on harvested land should be emphasized (Likens *et al.* 1978; Heal *et al.* 1982). Slash acts both as an environment where decomposers immobilize large quantities of nitrogen and prevent loss from the ecosystem, and also, as part of the reserves, a nutrient sink from which nutrients will gradually become available as the slash decomposes. These effects are particularly consequential in cool northern ecosystems and in the generally acid soils of the Canadian Shield.

Harvesters, therefore, which strip the branches from the boles before extraction, and leave them on the growing site, are much to be preferred to those which as in full-tree logging, draw the tops to a landing with the boles. On sites of moderate to low fertility, two advantages are apparent: 1) substantial amounts of the nutrient capital are left on site, and 2) increased residues will assist in lowering initial levels of N mineralization, and increase immobilization, thereby making less nitrogen available for leaching. Similarly, in the case of whole-tree logging, stump axes and major roots left on-site will both assist in retaining residual reserves, and in providing immobilizing substrates. Wiklander (1981) showed that the uprooting of spruce and pine stumps resulted in increased losses of nitrate even on low quality sites.

Conclusions

1. Foliar and branch biomass was greatest on pure red spruce stands and least in boreal mixedwoods.

2. Bole and total biomass were the same or heavier in boreal black spruce stands than in boreal mixedwoods. Temperate red spruce on fresh till carried one-and-a-half times the bole and total biomass of site class 1 black spruce in the boreal forest.

3. Organic and mineral soil mass reserves were least on the black spruce sites, outwash sands and peat, particularly, and greatest on the upland mixedwood sites.

4. In all cases almost half of the total organic pool above and below-ground was in the standing crop. When total reserves, organic plus mineral soil mass, were considered in relation to standing crop, only black spruce on peat supported a crop still nearly as great as the reserves. This could make these sites vulnerable under full-tree harvesting.

5. Productivity in red spruce stands and in white spruce growing in intolerant mixtures was prolonged to greater ages than that of species in even-aged stands. Productivity is related to basal area increment.

6. In the boreal forest, productivity was greatest on the best upland mixedwood sites. The productivity of black spruce on outwash was nearly as good. It was better than that on peat. The productivity of the mixedwood sites compared favourably with the best agricultural yields at similar latitudes. The productivity of red spruce, a temperate conifer, was substantially greater than that of the boreal sites and species.

7. Bioelement stores of phosphorus on black spruce outwash and peat sites and potassium in all sites but one of the mixedwoods were greater in the standing crop than in the reserves. It is on these sites that nutrient stress might be expected following full-tree harvesting.

8. Residence times were influenced by site quality and nutrient mobility. Nitrogen had the longest residence times. In general, upland mixedwoods had the shortest times. In

this context, it may be noted that upland mixedwoods had the greatest resilience in the boreal forest. Sites of very low nutrient capital such as outwash with low organic matter and nutrient content, resulting from an intensive fire history, also had low residence times.

9. Crop removal by full-tree harvesting: on the outwash site all elements exported except nitrogen were more than two-fold greater than what remained on the site in both organic and mineral reserves. Potassium exports on all sites but one, and phosphorus on both peat and outwash soils were greater than the remainder in the reserves. Upland mixedwood soils appeared to be the least vulnerable.

10. Replacement times which take into account the foregoing reserves, nutrient inputs and cycling, were relatively short, mostly 20 years or less. Potassium, however, may take up to 45 years to return to preharvesting levels for site class 1 spruce. Extended times to account for losses due to initial leaching were not great, only increasing replacement times by 5 to 7 years. There are many limitations to predicting exact rotation ages, however long replacement times can lead to increasing rotation ages.

11. Since most harvesting in the boreal forest is now full-tree logging, with whole-tree harvesting becoming much more widespread, it is advocated that harvesters be used which strip the branches from the boles before extraction and leave them on site. This will retain substantial nutrient capital on site and, particularly if stumps and roots are also left, will provide immobilizing substrates and lessen losses due to leaching. This will be particularly important on outwash sands and gravels and peat sites.

Acknowledgements

I should like to acknowledge the unstinting support of my assistants, T. Norbert Lussier, Nancy Jondreau and Maria Famiglietti. I have been privileged to have had an excellent group of energetic and enthusiastic students from various disciplines, forestry, biology, chemistry, etc., working on our projects over a number of years. They know who they are. I also wish to acknowledge the encouragement of Jeanne Gordon throughout the study and for help in the editing.

This paper is a contribution to the UNESCO Program Man and the Biosphere (MAB).

Literature Cited

Alban, D.H., D.A. Perala and B.E. Schlaegel. 1978. Biomass and nutrient distribution in aspen, pine, and spruce stands on the same soil type in Minnesota. Can. J. For. Res. 8: 290-299.

Baskerville, G.L. 1965. Dry matter production in immature balsam fir stands. For. Sci. Monogr. No. 9.

Bray, J.R. and L.A. Dudkiewicz. 1963. The composition, biomass and productivity of two *Populus* forests. Bull. Torrey Bot. Club 90: 298-308.

Clayton, J.L. 1979. Nutrient supply to soil by rock weathering. *In:* Proc. Impact of Intensive Harvesting on Forest Nutrient Cycling. Syracuse, New York. pp. 75-96.

Cole, D.W. and M. Rapp. 1981. Elemental cycling in forest ecosystems. *In:* Dynamic properties of forest ecosystems. D.E. Reichle (Ed.). International Biological Programme 23, Cambridge University Press. Cambridge, England. pp. 341-409.

Davidson, A.G. and D.R. Redmond. 1957. Decay of spruce in the Maritime provinces. For. Chron. 33: 373-380.

DeAngelis, D.L. 1980. Energy flow, nutrient cycling, and ecosystem resilience. Ecology 61: 764-771.

Ellenberg, H. 1981. Woodlands data set. *In:* Dynamic properties of forest ecosystems. D.E. Reichle (Ed.). International Biological Programme 23, Cambridge University Press. Cambridge, England. pp. 666-671.

Farr, W.A. 1967. Growth and yield of well-stocked white spruce stands in Alaska. U.S.D.A., For. Serv., Res. Pap. PNW-53.

Gardner, R.H. and J.B. Mankin. 1981. Analysis of biomass allocation in forest ecosystems of the IBP. *In:* Dynamic properties of forest ecosystems. D.E. Reichle (Ed.). International Biological Programme 23, Cambridge University Press. Cambridge, England. pp. 451-497.

Gordon, A.G. 1975. Productivity and nutrient cycling by site in spruce forest ecosystems. *In:* Energy flow — its biological dimensions. Cameron and Billingsley (Eds.). A summary of the I.B.P. in Canada 1964-1974, Royal Society of Canada. Ottawa, Ontario. pp. 119-126.

Gordon, A.G. 1976. The taxonomy and genetics of *Picea rubens* and its relationship to *Picea mariana*. Can. J. Bot. 54: 781-813.

Gordon, A.G. 1979. Productivity and nutrient cycling in natural forests. *In:* Canada/MAB Report No. 12. Biomass Strategy Consultation, Ottawa, Ontario. Co-sponsored by The Canadian Committee for the Unesco Program on Man and Biosphere (MAB) and The Science Council of Canada. pp. 34-49.

Gordon, A.G. 1981a. Impacts of harvesting on nutrient cycling in the boreal mixedwood forest. Environ. Canada, C.F.S., Proc. O-P-9. pp. 121-140.

Gordon, A.G. 1981b. Woodlands data set. *In:* Dynamic properties of forest ecosystems. D.E. Reichle (Ed.). International Biological Programme 23, Cambridge University Press. Cambridge, England. pp. 576-579.

Gorham, E. and J.R. Bray. 1964. Litter production in forests of the world. Adv. Ecol. Res. 2: 101-157.

Gorham, E., P.M. Vitousek and W.A. Reiners. 1979. The regulation of chemical budgets over the course of terrestrial ecosystem succession. Ann. Rev. Ecol. Syst. 10: 53-84.

Guillemette, L. 1982. An overview of agriculture in the Great Clay Belt. Presented at the Northeastern Forest Soils Conference Annual Meeting, Cochrane, Ontario.

Hakkila, P. 1976a. Complete-tree logging and utilization potentials of softwoods. Applied Polymer Symposium 28: 383-392.

Hakkila, P. 1976b. Stumpwood as industrial raw material. (English summary) Folia Forestalia 292: 1-39.

Heal, O.W., M.J. Swift and J.M. Anderson. 1982. Nitrogen cycling in United Kingdom forests: the relevance of basic ecological research. Phil. Trans. R. Soc. Lond. B 296: 427-444.

Heinselman, M.L. 1971. The natural role of fire in northern conifer forests. *In:* Fire in the northern environment. C.W. Slaughter, R.J. Barney and G.M. Hansen (Eds.). Pacific Northwest For. and Range Exp. Stn. Portland, Oregon. pp. 61-72.

Hills, G.A. 1954. Field methods for investigating site. Site Res. Manual No. 4. Ont. Dept. Lands Forests, Res. Div. Maple, Ontario.

Honer, T.G. and L. Heger. 1971. Stem weight and volume distribution in six balsam fir trees. Bi-monthly Res. Notes 27: 35-36.

Johnston, R.S. and D.L. Bartos. 1977. Summary of nutrient and biomass data from two aspen sites in western United States. U.S.D.A., For. Serv., Res. Note INT-227.

Jokela, E.J., C.A. Shannon and E.H. White. 1979. Biomass and nutrient equations for 66-year-old paper birch. *In:* Proc. Impact of Intensive Harvesting on Forest Nutrient Cycling. Syracuse, New York.

Juday, G.P. and J.C. Zasada (In press). The structure and development of an old-growth white spruce (*Picea glauca* (Moench) Voss) forest on an interior Alaska floodplain site, Willow Island.

Kazimirov, N.I. and R.M. Morozova. 1981. Woodlands data set. *In:* Dynamic properties of forest ecosystems. D.E. Reichle (Ed.). International Biological Programme 23, Cambridge University Press. Cambridge, England. pp. 629-645.

Krause, H.H., G.F. Weetman and P.A. Arp. 1978. Nutrient cycling in boreal forest ecosystems of North America. *In:* Forest soils and land use. C.T. Youngberg (Ed.). Proc. 5th North American Forest Soils Conference, Colorado State University. Ft. Collins, Colorado. pp. 287-319.

Likens, G.E., R.H. Bormann, R.S. Pierce and W.A. Reiners. 1978. Recovery of a deforested ecosystem. Science 199: 492-496.

Mälkönen, E. 1976. Effect of whole-tree harvesting on soil fertility. Silva Fennica 10: 157-164.

Mälkönen, E. 1979. Biomass and nutrient removal in whole-tree harvesting in some coniferous stands thinned for the first time. *In:* Proc. Impact of Intensive Harvesting on Forest Nutrient Cycling. Syracuse, New York.

Miller, H.G. 1982. In discussion: O.W. Heal, M.J. Swift and J.M. Anderson. 1982. Nitrogen cycling in United Kingdom forests: the relevance of basic ecological research. Phil. Trans. R. Soc. Lond. B 296: 427-444.

Moore, T.R. and E. Verspoor. 1973. Aboveground biomass of black spruce stands in subarctic Quebec. Can. J. For. Res. 3: 596-598.

Newbould, P. 1981. Terrestrial nitrogen cycles: problems, present knowledge and future research needs. Ecol. Bull. (Stockholm) 33: 671-691.

O'Neill, R.V. and D.L. DeAngelis. 1981. Comparative productivity and biomass relations of forest ecosystems. *In:* Dynamic properties of forest ecosystems. D.E. Reichle (Ed.). International Biological Programme 23, Cambridge University Press, Cambridge, England. pp. 411-449.

Ouellet, C.E. and L.C. Sherk. 1967. Map of plant hardiness zones in Canada. Can., Dept. of Agric. Ottawa, Ontario.

Peterson, E.B., Y.H. Chan and J.B. Cragg. 1970. Aboveground standing crop, leaf area, and caloric value in an aspen clone near Calgary, Alberta. Can. J. Bot. 48: 1459-1469.

Pollard, D.F.W. 1971. Mortality and annual changes in distribution of above-ground biomass in an aspen sucker stand. Can. J. For. Res. 1: 262-266.

Pollard, D.F.W. 1972. Above-ground dry matter production in three stands of trembling aspen. Can. J. For. Res. 2: 27-33.

Plonski, W.L. 1956. Normal yield tables for black spruce, jack pine, aspen and white birch in northern Ontario. Ont. Dept. Lands Forests, Rep. No. 24.

Rencz, A.N. and A.N. Auclair. 1980. Dimension analysis of various components of black spruce in subarctic lichen woodlands. Can. J. For. Res. 10: 491-497.

Rodin, L.E., N.I. Bazilevich and N.N. Rozov. 1975. Productivity of the world's main ecosystems. *In:* Productivity of world ecosystems. D.E. Reichle, J.F. Franklin and D.W. Goodall (Eds.). National Academy of Sciences, Washington, D.C. pp. 13-26.

Timmer, V.R., H.M. Savinsky and G.T. Marek. 1983. Impact of intensive harvesting on the nutrient budgets of boreal forest stands. *In:* Resources and dynamics of the boreal zone. R.W. Wein, R.R. Riewe and I.R. Methven (Eds.). Association of Canadian Universities for Northern Studies.

Troth, J.L., F.J. Deneke and L.M. Brown. 1976. Upland aspen/birch and black spruce stands and their litter and soil properties in interior Alaska. For. Sci. 22: 33-44.

Van Cleve, K. and L.L. Noonan. 1975. Litterfall and nutrient cycling in the forest floor of birch and aspen stands in interior Alaska. Can. J. For. Res. 5: 626-639.

Van Cleve, K., T. Dyrness and L.A. Viereck. 1980. Nutrient cycling in interior Alaska floodplains and its relationship to regeneration and subsequent forest development. *In*: Murray and Van Veldhuizan (Eds.). Forest Regeneration of High Latitudes. U.S.D.A., For. Serv., Gen. Tech. Rep. PNW-107. pp. 11-18.

Van Cleve, K. and L.A. Viereck. 1981. Forest succession in relation to nutrient cycling in the boreal forest of Alaska. *In:* Forest succession concepts and application. D.C. West, H.H. Shugart and D.B. Botkin (Eds.). Springer-Verlag, New York. pp. 185-211.

Vitousek, P.M. and J.M. Melillo. 1979. Nitrate losses from disturbed forests: patterns and mechanisms. For. Sci. 25: 605-619.

Vitousek, P.M. 1981. Clear-cutting and the nitrogen cycle. Ecol. Bull. (Stockholm) 33: 631-642.

Weetman, G.F. and B. Webber. 1972. The influence of wood harvesting on the nutrient status of two spruce stands. Can. J. For. Res. 2: 352-369.

Wiklander, G. 1981. Clear-cutting and the nitrogen cycle. Ecol. Bull. (Stockholm) 33: 642-647.

Yarie, J. 1981. Forest fire cycles and life tables: a case study from interior Alaska. Can. J. For. Res. 11: 554-562.

SEASONAL PATTERNS OF NITROGEN MINERALIZATION FOLLOWING HARVESTING IN THE WHITE SPRUCE FORESTS OF INTERIOR ALASKA

Andrew M. Gordon and Keith Van Cleve
University of Alaska

Abstract: The effects of commercial timber harvesting upon nitrogen transformations were evaluated for the forest floor and mineral soils of mature white spruce *(Picea glauca)* forest sites in interior Alaska. Analyses of forest floor incubated *in situ* in mature forest and two recently harvested areas of different ages indicated an ammonium-dominated soil system for the unharvested area. Maximum NH_4-N mineralization rates (300 μg N / 100 g dry soil / day) were found in mid-summer and generally declined with the onset of fall. In the harvested areas, rates of NH_4-N accumulation were almost invariably lower than in the uncut area, a consequence of elevated levels of nitrification. Shortly after harvesting, NO_3-N concentrations were extremely high. Thereafter, they declined to levels slightly higher than in the mature forest. Nitrification was strongly enhanced by harvesting, and regular patterns within season were evident. For the youngest clearcut, the combined processes of ammonification and nitrification only occasionally supplied more nitrogen to the site on a daily basis than was supplied to the mature forest site. For the oldest clearcut, the supply from the combined mechanisms was variable and depended on time since clearcutting.

Until recently forest ecologists have considered one of the most important consequences of forest harvesting to be the removal of a large portion of the site nutrient capital in the harvested trees. Many past studies were concerned with the effects of clear-cutting upon soil nutrient reserves. For example, the effects of clear-cutting on chemical changes in forest floor organic matter (Covington 1981), woody litter decomposition (Abbott and Crossley 1982), soil microbiological properties (Smith *et al.* 1968; Theodorou and Bowen 1981) and soil solution chemistry (Hart *et al.* 1980) have been documented. With the advent of whole-tree logging, and consequently, the removal of branch material and potential needle litter, a substantial fraction of the nutrient pool formerly present on the site will be lost to subsequent generations of trees. Recent studies at the ecosystem level in a variety of forest site types have evaluated nutrient losses and changes in site productivity resulting from whole-tree harvesting (Adams and Boyle 1982; Mann and West 1981; Silkworth and Grigal 1982, and many others). With any type of logging, replacement of these losses will largely arise from the remaining stock of nutrients. Under whole-tree harvesting, it is likely that this reserve will be taxed severely. The rate at which nutrients, especially mineral nitrogen, are supplied from this reserve will undoubtedly have a substantial effect, in conjunction with other stand parameters, upon not only the re-establishment of the stand but also upon the growth rate of young trees.

The biochemical aspects of nitrogen mineralization and nitrification have been well understood for some time (Robertson 1982). Increased importance has been placed upon nitrogen mineralization as a major factor in the maintenance of site productivity as indicated by a number of recent studies evaluating the effects of certain forest management practices upon nitrogen mineralization and nitrification. Urea fertilization of forest soils has been found to stimulate net nitrogen mineralization (Johnson *et al.* 1980) and nitrification (Popovic 1977). Similarly, mineral nitrogen production was increased by afforestation (Williams and Cooper 1979) and by ploughing cultivation (Powlson 1980). The latter could readily be likened to forest site scarification. In Greece, Nakos (1980) reported that only one herbicide (asulam) considerably reduced soil nitrification. Unfortunately, while the importance of soil nitrogen mineralization and nitrification to forest production processes has been recognized (Likens *et al.* 1969), few studies have dealt with the effects of commercial timber harvesting upon nitrogen mineralization. Glavac and Koenies (1978a,b) in Europe, indicated that clear-felling increased net nitrogen mineralization by a factor of three. In North America, net nitrogen mineralization is almost always increased by clear-felling (Matson and Vitousek 1981) but this is strongly dependent upon the physical and biochemical conditions of the soil. Cold, dry, or anaerobic soils may prevent substantial nitrification even in the presence of substantial ammonium substrate (Vitousek 1981).

Soil temperature and moisture content are the important factors controlling the mineralization of nitrogen. It follows that, in cold-dominated soils such as those found in portions of interior Alaska, the processes of organic matter decomposition and nitrogen mineralization occur at slower rates than in warmer, temperate-latitude soils. The underlying question to be asked is this: if physical variables controlling nitrogen mineralization are adversely affected by the removal of forest cover, what can the mineralization and nitrification rates be expected to do? Rapid short-term increases in the rate due to increased surface temperature and available moisture may be offset by excessive leaching. Over a longer period of time, the mineralization rate may decline as the forest floor dries up. Competition for mineral nitrogen reserves by weed species may further limit the nitrogen available for uptake by desired species such as white spruce. As well, upon decomposition, the remaining mass of primary root tissues with its typically high C/N ratio will likely become a carbon source for microbial activity. Thus, soil nitrogen reserves may become temporarily unavailable while supplying these microbial demands.

Increasing development of Alaska's interior forests for a variety of wood products dictates the need for an understanding of changes in productivity of the land base with use and time. Nutrient cycling studies in interior Alaska have, to date, documented many of the important nutrient pool sizes and fluxes for the predominant covertypes but there is a lack of information on the effects of forest management activities upon these nutrient reserves and processes.

There are four principal objectives in our present program. They are:

- to quantitatively assess and model the effect of timber harvesting disturbance upon the seasonal breakdown and transfer of organically-bound nitrogen to available form within the forest floor and mineral soil of mature white spruce stands;

- to determine, using these mineralization rates and other estimates of N-pool sizes for system components, the short-term and potential long-term effects of fluctuations in these rates due to disturbance upon the site and system nitrogen balance;

- to provide a predictive mechanism for the potential replenishment and/or depletion of nitrogen supplies with disturbance; and
- to elucidate and test, in conjunction with the above, methodology associated with the use of polyethylene bags in field incubation experiments, through a variety of laboratory and field experiments.

The present paper presents an overview of aspects of nitrogen mineralization in the forest floor, within the context of the first objective.

Materials and Methods

Study Area

The study site is located in the Bonanza Creek Experimental Forest, 32.2 km west of Fairbanks, Alaska (64°45′ N, 148°15′ W). Three sites were selected for study. The control area is a mature, 110-year old stand of white spruce, with a minor component of white birch *(Betula papyrifera)* and alder *(Alnus crispa)*. The moss layer in this stand is continuous, consisting principally of the feathermosses *Pleurozium schreberi*, and *Hylocomium splendens*. A harvested area resulting from commercial logging in 1977 is located immediately adjacent to the control on the east; a 1978 cut lies immediately to the west. Clear-cuts were chosen in order that a sequence of 'months after clear-cutting' could be portrayed. For the most part, the integrity of the forest floor has been destroyed within the clear-cut areas, resulting in a layer of mixed organic material, logging slash and fine roots and twigs to a maximum depth of approximately 15 cm. Pioneer ground vegetation within these areas is comprised principally of *Mertensia paniculata, Rosa acicularis, Epilobium angustifolium, Galium boreale, Equisetum arvense,* and *Calamagrostis canadensis*.

This area of Alaska was unglaciated; the mineral soil underlying all three sites is a micaceous loess silt (Péwé 1955) over precambrian schist and gneiss bedrock (Johnson and Hartman 1971). The study sites are in the zone of discontinuous permafrost at about 200 m elevation, with gently-sloping southern aspects. Some climatic data for the Bonanza Creek Experimental Forest are given by Barney and Berglund (1973). Generally, the climate of the region is continental, with a growing season limited to about 100 days by late-spring and early-fall frosts. Mean annual air temperature is −3.5°C and average annual precipitation is 28 cm, the bulk of which falls as rain.

Field Methods

In situ field incubations using buried polyethylene bags (Eno 1960) were used to estimate nitrogen mineralization. This method prevents leaching and plant uptake of mineral nitrogen (NH_4-N and NO_3-N) while at the same time maintaining aerobic conditions and preventing undesirable moisture loss.

Within each of the disturbed sites a preliminary vegetation survey by line-point revealed the basic ground cover pattern. This allowed for the stratification of sample points within the vegetatively heterogeneous clear-cut areas. Thirty permanent sample points were established randomly within each site. Sampling was on a monthly basis for August and September, 1979 and from June to September, 1980. In addition, the overwinter periods 1979 to 1980, and 1980 to 1981 were also monitored. At the start of each sampling period a standard soil core (15 cm diameter) was removed from the vicinity of each sampling point. The mineral soil and forest floor were separated and each was thoroughly mixed. Major roots and woody debris were removed. A subsample of each was taken for initial mineral nitrogen (NH_4^+, NO_3^-) content. The remaining mixed material was molded into a thin, 1 cm disk, enclosed in a 0.8 mil polyethylene bag (Presto Products, Appleton, WI) and re-inserted into the core hole at the appropriate depth (8-10 cm for forest floor; 20-22 cm for mineral soils). The core-holes were re-covered and the samples left to incubate under a natural temperature regime for the period of interest. Subsamples were removed to the lab and frozen at −10°C, pending analysis for NH_4-N and NO_3-N. At the end of the incubation period, the field samples were removed, transported to the lab, and kept frozen until analyzed.

Laboratory Methods

Ten grams of all samples were thawed and immediately extracted for NH_4-N and NO_3+NO_2-N using 2 *N* KCl (Bremner 1965), in a 7.5:1 ratio (soil:KCl). Duplicate extracts were shaken for 1 hour, suction-filtered and analyzed colorimetrically using a modified dual channel Technicon II Autoanalyzer system with a sensitivity of 0.01 ppm. NH_4-N was measured on 1 channel by a variation of the Technicon industrial method No. 98-70W. On the second channel, the extracts were passed over a copperized cadmium column, reducing the nitrate to nitrite. Total nitrite was analyzed using a variation of the Griess reaction and Technicon industrial method No. 100-70W. Results are expressed on an oven-dry weight concentration basis. Moisture content was estimated by oven-drying subsamples at 65°C (forest floor) and 100°C (mineral) for 48 hours.

Results and Discussion

Results of available nitrogen analysis for the organic horizon (sub- and incubated samples) are shown in Figure 1. The blocks of bars on the far left illustrate NH_4-N and NO_3-N levels from mid 1979 through May, 1981, 22 continuous months of incubation, for the control area of 110-year old white spruce. Similarly, the blocks in the center and on the right express the same factors over the same period for clear-cuts resulting from harvesting in 1977 and 1978.

In the uncut situation, the concept of an ammonium-dominated climax ecosystem as reported by Rice and Pancholy (1972) can be seen. In such systems, ammonium is probably the preferred nitrogen species by plants since less energy must be expended to incorporate nitrogen in NH_4 as opposed to nitrogen in NO_3, into proteinic structures (Haynes and Goh 1978). Little or no nitrification and associated nitrate are present. Unincubated ammonium levels (instantaneous net pool size) in the soil varied from approximately 5,100 μg N/100 g dry soil to over 11,000 μg N/100 g dry soil, indicating a wide variation in background ammonium levels seasonally. Despite this, seasonal patterns of nitrogen mineralization were detectable and are listed in Table 1. August of each year shows the highest values, with subsequent decreases in fall and overwinter. This is likely in response to favorable post-summer microclimatic and site conditions. Earlier in the season the soil temperature regime may have favored large amounts of heterotrophic decomposition, resulting in the creation of organic residue with wide C/N ratios. Available nitrogen may have been at a premium while supplying these microbial demands.

In the disturbed areas, background levels of NH_4-N do not differ drastically from the

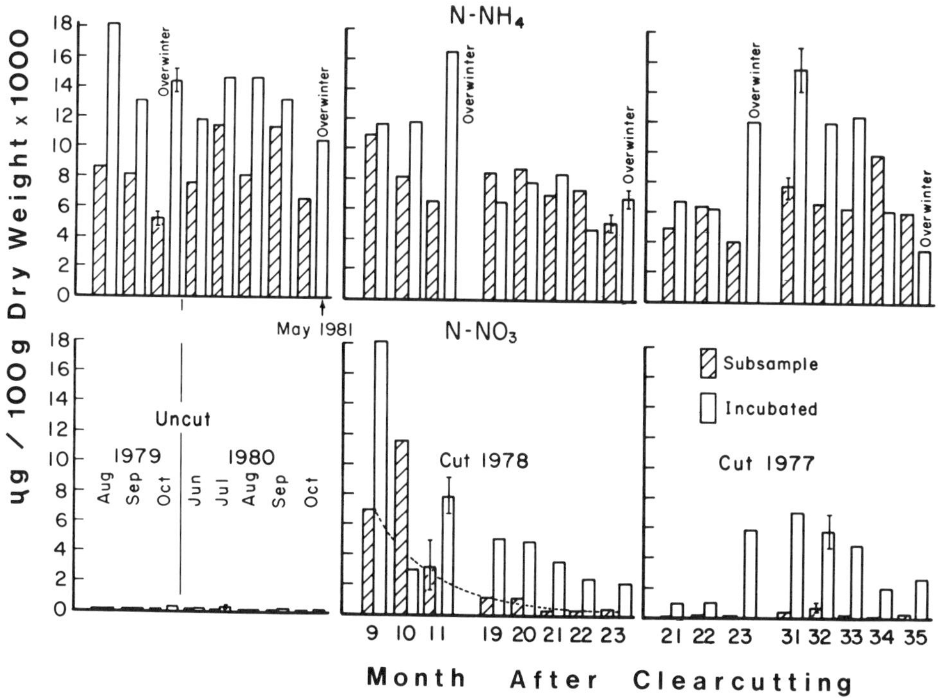

Figure 1. Distribution of available nitrogen (N-NH_4 upper; N-NO_3 lower) by month in the forest floor of cut and uncut stands of 110-year old white spruce. Each bar represents 30 field points plus 2 lab reps. Some typical standard error bars are shown. Incubation period was approximately 30 days in all cases, except where overwinter *in situ* incubations are indicated.

Table 1. Seasonal values of soil nitrogen mineralization (ammonification) in the forest floor of a 110-year-old stand of white spruce[a]

Period	Mineralization (ammonification) rate (μg N/100 g dry weight/day)
August 1979	311.40
September 1979	166.40
Overwinter 1979-1980	37.97
June 1980	154.22
July 1980	86.51
August 1980	223.01
September 1980	60.44
Overwinter 1980-1981	16.93

[a]The values reported here and in subsequent tables have as of yet not been corrected by weight-volume estimates for expression on an area basis. Each value is based on differences between averages from sets of 30 points.

Table 2. Seasonal values of soil nitrogen mineralization in the forest floor following harvesting in 110-year-old white spruce

Time after harvesting (months)	Mineralization (ammonification) rate (μg N/100 g dry weight/day)	Mineralization rate from uncut area (μg N/100 g dry weight/day)
9	21.19	311.40
10	128.44	166.40
11-18	42.40	37.97
19	net loss	154.22
20	net loss	86.51
21	50.87[a]-55.77[b](53.32)[c]	223.01-311.40[d]
22	net loss	60.44-166.40
23-30	7.16[a]-33.51[b](20.34)[c]	16.93- 37.97
31	281.10	154.22
32	149.37	86.51
33	214.45	223.01
34	net loss	60.44
35-42	net loss	16.93

[a]Area harvested in 1978.
[b]Area harvested in 1977.
[c]Average.
[d]Where two values appear in this column, compare the first to the "a" value in the previous column, and the second to the "b" value.

uncut area, although variation in levels from the clearcut areas is reduced. However, net rates of NH_4-N accumulation, as indicated by accumulation of NH_4-N above background levels were lower than rates of accumulation for the uncut area with the exception of two mid-summer periods nearly three years after harvesting (Table 2). In many cases, a net loss of NH_4-N can be noted in the cleared areas. Several years of decomposition may be initially required before NH_4-N is finally released to the site. However, since harvesting might be generally expected to enhance conditions of temperature and moisture favorable to mineralization, it is more likely that nitrification has been stimulated by favorable conditions, one of which might be a readily-available source of NH_4-N. This is verified by the data for the distribution of NO_3-N (Figure 1). Net NH_4-mineralization rates, averaged over all sampling periods by cut and uncut area, are approximately twice as large for the control situation. This is in part due to the number of net losses recorded for the harvested sites. It is important to note that the gross mineralization (ammonification) rate in the harvested areas is at least equivalent to that of the uncut situation. Net NH_4-mineralization is less in the cleared areas due to bacterial oxidation of NH_4-N into NO_3-N.

Nitrate is much more prevalent in the disturbed areas and it is possible that it is preferentially taken up by many of the plant species growing there. Plants with a large photosynthetic capacity, such as those found in disturbed situations, might be expected to have a net energy balance which would allow the expenditure of energy necessary to push the oxidation state of nitrogen from +5 to −3, as it occurs in ammonium, and many proteinic associations. Middleton and Smith (1979), working with grasses, indicate that the difference in total energy costs for nitrogen uptake is only eight per cent greater when the nitrate form is assimilated in favor of the ammoniacal form. The excess NO_3-N found in the harvested areas is likely rapidly leached from the site, but may also be assimilated as *Calamagrostis* becomes established on the site. On the other hand, early successional plants may follow a strategy of preferentially utilizing reduced nitrogen (NH_4) in favor of NO_3-N, thus saving energy which could be used for carbon fixation and growth. The form in which nitrogen is taken up is important when considering the ability of white spruce seedlings, for example, to compete effectively for nitrogen with other species.

Nine to 11 months after harvesting, NO_3-N levels are extremely high and varied. Thereafter, they decline to relatively low levels, although remaining higher than in the mature forest. The hypothetical decline of background levels of NO_3-N is indicated by the dotted line in Figure 1. In all cases except for one, nitrification substantially increased NO_3-N levels in incubated bags. After the 19th month following harvesting the levels of NO_3-N from incubated samples were always higher than sub-samples taken from the site (for the next incubation period) immediately following removal. This indicates that NO_3-N is being rapidly leached from this system or taken up by plants. (NO_3-N is excluded from root uptake and is unable to leach from the samples incubated in polyethylene bags and is essentially ''captured'' for analysis.) Rates of nitrification are given in Table 3. If these values are plotted against 'months after harvesting', as in Figure 2, seasonal patterns of nitrification can be readily seen. High rates of nitrification appear to cluster around mid-summer with subsequent decreases as the year progresses. Between nine and 40 months after clear-cutting, the maximum rate of nitrification does not appear to be related to time after harvesting, but rather to seasonal (climatic) phenomena. These patterns of fluctuating rates should be superimposed upon a general decline in absolute levels of soil NO_3-N (Figure 1) as nitrate produced is lost from the forest floor.

When the mineralization rate for total available nitrogen is calculated (Table 4) for the harvested areas, and compared to ammonification (the principal source of available nitrogen) rates from the uncut areas, some interesting patterns of nitrogen supply emerge. These are plotted in Figure 3. For the youngest clear-cut (cut 1978) the combined processes of ammonification and nitrification only occasionally supply more nitrogen to the site on a daily basis than is supplied to the mature forest site in the uncut situation. In fact, the patterns of supply from two essentially different strategies are very similar on a seasonal basis. This is an indication of the shift in microbial processes that might occur in a disturbed ecosystem. However, since NO_3-N may be rapidly leached from the system, the net amount of nitrogen made available consists to a large extent of ammonium. Thus, a net decline in total available nitrogen supplied to the site may occur in the cleared areas. For the older clear-cut (cut 1977), the rate of nitrogen supply to the site is much less than the uncut area for the latter part of the 1979 growing season, but much greater for most of the 1980 growing season. The depression/resurgence of the supply mechanisms for this site, as contrasted between 1979 and 1980, reflect seasonal fluctuations of both nitrification and ammonification. The patterns suggest that shortly after harvesting, soil nitrogen mineralization processes are governed more by substrate chemistry than by environmental influences. In 1979, the youngest (1978) clear-cut (see Figure 3) receives nitrogen at almost exactly the same rate as the harvested area. A year later, in 1980, when it is two years old, small deviations in the supply rate can be noted. In contrast the older (1977) clear-cut, shows moderate deviations two years after cutting (1979), but larger deviations from the uncut supply rate in 1980, 3 years after harvesting. After 30 months ammonification once again provides net gains as nitrification declines, and nitrogen may be supplied to the site at a rate greater than would occur in the uncut situation. This is exemplified for the rate of supply in 1980 for the older cut (cut 1977); the pattern of supply is very similar, but the rate, at least for the summer months, is at least twice that of

Table 3. Seasonal values of soil nitrification in the forest floor following harvesting in 110-year-old white spruce

Time after harvesting (months)	Month of year	Nitrification rate (μg N/100 g dry weight/day)
9	August	369.98
10	September	net loss
11-18	Over-winter	19.79
19	June	123.32
20	July	103.45
21	August	32.62[b]-92.66[a](62.64)[c]
22	September	30.17[b]-36.96[a](33.57)[c]
23-30	Over-winter	9.79[a]-24.06[b](16.93)[c]
31	June	233.53
32	July	142.49
33	August	168.48
34	September	30.03
35-42	Over-winter	10.51

[a]Area harvested in 1978.
[b]Area harvested in 1977.
[c]Average.

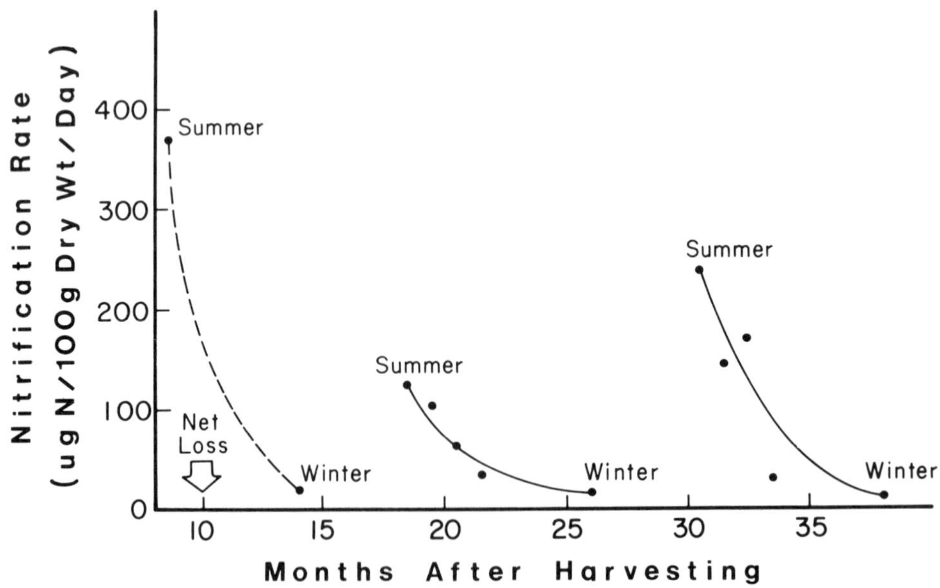

Figure 2. Seasonal patterns of soil nitrification. Lines represent best fit by eye. The dashed line reflects a hypothetical decrease in the rate of nitrification, given an unexplained net loss in month 10 (see Table 3). Based on differences between averages from sets of 30 points.

Table 4. Seasonal values of total soil nitrogen mineralization (ammonification + nitrification) in the forest floor after harvesting in a 110-year-old white spruce stand

Time after harvesting (months)	Corresponding month of year	Ammonification + nitrification rate (μg N/100 g dry weight/day)
9	August	391.97
10	September	128.44
11-18	Over-winter	62.19
19	June	123.32
20	July	103.45
21	August	88.39[b]-143.53[a](115.96)[c]
22	September	30.17[b]- 36.96[a] (33.57)[c]
23-30	Over-winter	16.95[a]- 57.57[b] (37.26)[c]
31	June	514.63
32	July	291.86
33	August	382.93
34	September	30.03
35-42	Over-winter	10.51

[a]Area harvested in 1978.
[b]Area harvested in 1977.
[c]Average.

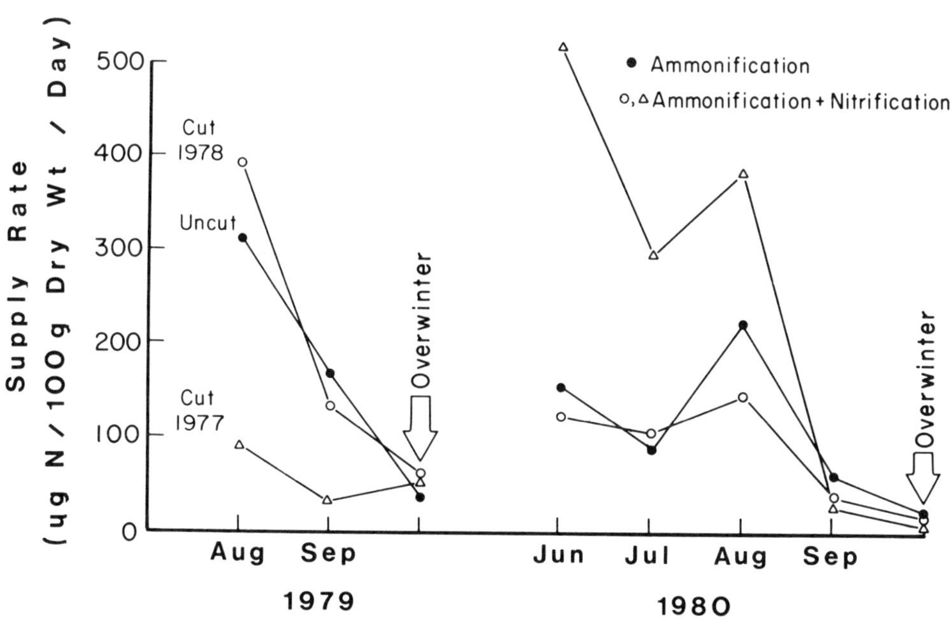

Figure 3. Comparison of soil nitrogen supply mechanisms for harvested and unharvested areas of 110-year-old white spruce.

Table 5. Net gains and variation in NH_4-N concentrations from cut and uncut stands of 110-year-old white spruce. Based on a composite sample of data sets from all sampling periods

	Clearcuts		Control
	Youngest	Oldest	110 year-old White Spruce
% of bags showing positive gains in NH_4	58	68	88
Average $(S/\bar{X})$ for subsamples	0.86	0.51	0.58
Average $(S/\bar{X})$ for incubated samples	0.92	0.50	0.52

the uncut area or the youngest clear-cut. Soil processes may be under strong climatic control at this point in time after harvesting, although decomposition of roots and other organic debris left on the site, will certainly have commenced, and variations in the C/N ratio of newly-deposited litter from pioneer plants may also be asserting an effect.

Table 5 shows net gains and variation in NH_4-N values from all three areas. Almost 90 per cent of the bags in the control area of mature spruce showed positive gains in NH_4-N concentration. Conversely, not quite 60 per cent of the bags incubating in the youngest clear-cut showed positive gains. The oldest clear-cut was intermediate. Variability in data, as illustrated by the coefficient of variation, is increased upon harvesting, but with time, may be reduced to its original level or below.

Matson and Vitousek (1981) also evaluated clearcutting effects upon nitrogen mineralization processes. They found similar increases in nitrification which they attributed in part to the initial population size of nitrifying bacteria. They also indicated that mineralization (total NH_4-N and NO_3-N) was increased in one clear-cut over the control after a period of only 1½ years. We have reported on a similar pattern, about 3 years after harvesting (Figure 3). We have also shown that net ammonification by itself was increased in one of the cleared areas, for this same point in time (Table 2).

It becomes very hard to predict what the nature of nitrogen supply from the site's soil resource to a growing forest might be 100 years from now, given the wide range of management options available and their effects on soil properties. For example, C/N ratio, a strong control of microbial and chemical activity, can be changed drastically simply by time and type of harvesting and post-harvesting site treatment such as chipping of slash. The nature of the competing vegetation will also be important. In the long-term, seeding, planting, mulching, herbicides, thinning, fertilization, no-treatment, etc., will all have differing effects upon soil nutrient supply. Ultimately, as with most aspects of forestry, the final question to be asked is: can the management strategy be continually repeated successfully on the variety of site qualities encountered in a given region every 100 years? As well, is it possible to have rotations, within which nitrogen availability is not an important limiting factor to growth, and where successful forestry could be realized, in a shorter period of time?

Certainly, management decisions based on previous knowledge of the impact of site preparation, seedfall, seedling recruitment, stand growth and soil fertility are invaluable. As far as the latter is concerned, for example, it would be desirable to know what the mineralization rates are of nitrogen in the soil immediately after harvesting. It is also important to know, for the most common management regimes, what levels of soil nitrogen supply might be expected at the end of rotation, just prior to the second cut, based on the uptake rate of the currently growing forest, and the mineralization rate of soil N in an aging ecosystem. We do know that within certain temperate coniferous forest ecosystems it may take 20 to 30 years to replace, through de-novo inputs to the system, the total nitrogen lost from the site under conventional logging (Gordon 1979). Under whole-tree logging, or in climatically-cool regions such as Alaska, this period may be much longer. Information on the rate of nitrogen mineralization in relation to other element recycling is also becoming available. Sulphur, for example, is mineralized at a much faster rate (Tabatabai and Al-Khafaji 1980). From the management point of view, the rate of supply of nitrogen and other nutrients to seedlings of a desirable species and the survival of these seedlings is of utmost importance. Thus N-mineralization should be evaluated as a strong component of a nutrient replenishment/depletion model, where the nitrogen available for seedling uptake reflects the net amount available after supplying competing vegetation and losses to other unavailable pools and pathways.

Literature Cited

Abbott, D.T. and D.A. Crossley, Jr. 1982. Woody litter decomposition following clearcutting. Ecology 62: 35-42.

Adams, P.W. and J.R. Boyle. 1982. Soil fertility changes following clearcut and whole-tree harvesting and burning in central Michigan. Soil Sci. Soc. Amer. 46: 638-640.

Barney, R.J. and E.R. Berglund. 1973. Summary of climatic data for the Bonanza Creek Experimental Forest, Interior Alaska. U.S.D.A., For. Serv., Res. Note PNW-201.

Bremner, J.M. 1965. Nitrogen availability indexes. pp. 1149-1341. *In:* Methods of soil analyses. Part II. C.A. Black, D.D. Evans, J.L. White, L.E. Ensminger and F.E. Clark (Eds.). Agronomy Monogr. 9, Amer. Soc. Angr., Inc. Madison, Wisconsin.

Covington, W.W. 1981. Changes in forest floor organic matter and nutrient content following clearcutting in northern hardwoods. Ecology 62: 41-48.

Eno, C.F. 1960. Nitrate production in the field by incubating the soil in polyethylene bags. Soil Sci. Soc. Proc. 24: 277-279.

Glavac, V. and H. Koenies. 1978a. Mineralstickstoff-Gehalte und N-Nettomineralisation im Boden eines Fichtenforstes und seines Kahlschlages wahrend der Vegetations periode 1977. Oecol. Plant. 13: 207-218.

Glavac, V. and H. Koenies. 1978b. Vergleiche der N-Nettomineralisation in einem Sauerhumus-Buchenwald (Luzulo-Fagetum) und einem benachbarten Fichtenforst am gleichen Standort vor und nach dem Kahlsag. Oecol. Plant. 13: 219-226.

Gordon, A.G. 1979. Productivity and nutrient cycling in natural forests. *In:* Canada/MAB Report No. 12. Report of the Biomass Strategy Consultation, February 26-28, 1979. Ottawa, Canada. pp. 34-49.

Hart, G.E., N.V. Debyle and R.W. Hennes. 1980. Soil solution chemistry and slash disposal. *In:* Symposium on watershed management, Vol. 1, Am. Soc. Civ. Eng. New York. pp. 318-325.

Haynes, R.J. and K.M. Goh. 1978. Ammonium and nitrate nutrition of plants. Biol. Rev. 53: 465-510.

Johnson, D.W., N.T. Edwards and D.E. Todd. 1980. Nitrogen mineralization,

immobilization and nitrification following urea fertilization of a forest soil under field and laboratory conditions. J. Soil Sci. Soc. Amer. 44: 610-616.
Johnson, P.R. and C.W. Hartman. 1971. Environmental atlas of Alaska. 2nd Ed. Inst. Arctic Environ. Engineer. and Inst. Water Res., University of Alaska. Fairbanks, Alaska.
Likens, G.E., F.H. Bormann and N.M. Johnson. 1969. Nitrification: importance to nutrient losses from a cutover forested ecosystem. Science 163: 1205-1206.
Mann, L.K., and D.C. West. 1981. Whole-tree harvesting: first year progress report — impacts on productivity and nutrient change. ONRL/TM-7873. Oak Ridge National Laboratory. Oak Ridge, Tennessee.
Matson, P.A. and P.M. Vitousek. 1981. Nitrogen mineralization and nitrification potentials following clearcutting in the Hoosier National Forest, Indiana. For. Sci. 27: 781-791.
Middleton, K.R. and G.S. Smith. 1979. A comparison of ammoniacal and nitrate nutrition of perennial ryegrass through a thermodynamic model. Plant and Soil 53: 487-504.
Nakos, G. 1980. Effects of herbicides used in forestry on soil nitrification. Soil Biol. Biochem. 12: 517-519.
Péwé, T.L. 1955. Origin of the upland silt near Fairbanks, Alaska. Geol. Soc. Amer. Bull. 66: 699-724.
Popovic, B. 1977. Effect of ammonium nitrate and urea fertilizers on nitrogen mineralization, especially nitrification, in a forest soil. Avdelningarna for skogsekologi och skoglig marklara. Rapporter Och Uppsatser, No. 30.
Powlson, D.S. 1980. Effect of cultivation on the mineralization of nitrogen in soil. Plant and Soil 57: 151-153.
Rice, E.L. and S.K. Pancholy. 1972. Inhibition of nitrification by climax ecosystems. Amer. J. Bot. 59: 1033-1040.
Robertson, G.P. 1982. Nitrification in forested ecosystems. Phil. Trans. R. Soc. Land. B296: 445-457.
Silkworth, D.R. and D.F. Grigal. 1982. Determining and evaluating nutrient losses following whole-tree harvesting of aspen. J. Soil Sci. Soc. Amer. 46: 626-631.
Smith, W.H., F.H. Bormann and G.E. Likens. 1968. Response of chemautotrophic nitrifiers to forest cutting. Soil Sci. 106: 471-473.
Tabatabai, M.A. and A.A. Al-Khafaji. 1980. Comparison of nitrogen and sulphur mineralization in soils. J. Soil Sci. Soc. Amer. 44: 1000-1006.
Theodorou, C. and G.D. Bowen. 1981. Microbiological studies of first and second rotation pine forest soils in south Australia. Aust. For. Res. 11: 129-139.
Vitousek, P.M. 1981. Clear-cutting and the nitrogen cycle. Ecol. Bull. (Stockholm) 33: 631-647.
Williams, B.L. and J.M. Cooper. 1979. Some effects of afforestation with lodgepole pine on rates of nitrogen mineralization in peat. Forestry 52: 151-160.

IMPACT OF INTENSIVE HARVESTING ON NUTRIENT BUDGETS OF BOREAL FOREST STANDS

V.R. Timmer and H.M. Savinsky
University of Toronto

G.T. Marek
Ontario Ministry of Natural Resources

Abstract: The effect of intensive logging systems on future forest productivity was assessed on four contrasting stand conditions in the Nipigon area: a young coniferous *(Abies balsamea)* stand, a young deciduous *(Populus tremuloides – Betula papyrifera)* stand, and two 120 year old *Picea mariana* stands, one established on deep mineral soils, the other on shallow soils restricted by bedrock. Parts of these stands were harvested by Domtar Forest Products using a whole-tree chipping process. The uncut portion of the stands were sampled to determine biomass and nutrient exports associated with intensive (whole-tree) logging and conventional (stem only) harvesting. Although biomass removal increased 56-116 per cent with full tree utilization corresponding nutrient removals from these sites increased 83-224 per cent. Higher nutrient losses reflected large differences in elemental concentrations between stems and foliage. Nutrient content of the stands varied with age and species.

Nutrient exports at harvest were compared with the existing nutrient reserves of the forest floor and mineral soil, and with the quantities of nutrients returned as logging residue. Assuming steady state nutrient cycling results indicated that nutrient losses from conventional harvesting were relatively modest with regard to soil nutrient supplies. However, on three out of the four sites studied intensive logging methods would lead to deficiencies in P, K or Ca in the succeeding rotation. Forests on deeper soils, managed at long rotations were less susceptible to nutrient depletion than short rotation stands established on shallow soils. Silvicultural strategies minimizing potential nutrient loss from intensified harvesting are discussed.

Increasingly, harvest operations in the boreal forests of Canada are changing from the conventional tree-length (stem only) method to a more mechanized, less labour-intensive full-tree logging system. The major advantage of this system is greater utilization of the standing crop. Not only are the entire above-ground parts of the tree virtually completely utilized, but trees formerly regarded as undersized, or unmerchantable are also processed. In addition, the technique also provides an economical means of cropping small timber, facilitates site preparation for planting, and to many is aesthetically more acceptable because of the lack of unsightly logging slash (Tufts 1976).

Increased yields associated with full-tree logging systems, however, are also associated with increased nutrient exports from the logged areas since much of the residual stand and logging debris, which contain nutrients and would remain under a conventional harvest, are removed for chipping. Such a loss of nutrients may impede the growth of future stands, especially on sites where nutrient supplies are limiting. Hence potential impacts of intensive harvesting on forest productivity have generated much interest in Canada (Foster and Morrison 1976; Kimmins 1977; Carlisle and Methven 1979; Gordon 1979, 1981; Morrison and Foster 1979; Rennie 1955; Krause *et al*. 1980; Freedman *et al*. 1981). One of the earliest studies in the boreal forest region found that full-tree harvesting of natural stands may adversely affect nutrient budgets on shallow soils of low fertility (Weetman and Webber 1972).

Recognizing the potential for declining site productivity, the objectives of the present study were to assess the impact of intensive harvesting on typical boreal forest sites in the Nipigon district, and to recommend silvicultural and harvesting strategies to reduce the possibility of site deterioration.

Methods

Study Area

In 1979–1980, Domtar Forest Products operated a full-tree chipper harvesting system in several stands representing various ages, species and sites. On a number of sites only a portion of a stand was cut. These locations offered an opportunity to document the pre-cut forest following harvesting. Intensive sampling of the soil, forest floor, vegetation and logging residue was carried out on four contrasting sites selected for this study. These stands were representative of the major boreal forest types in the Nipigon area.

The study area was located in the Superior (B.9) and Nipigon (B.10) sections of the boreal forest region (Rowe 1972) approximately 200 kilometers northeast of Thunder Bay, Ontario. Study sites were established near the eastern shore of Lake Nipigon (49°39′N, 88°05′W) (Figure 1). Plots were located at an elevation of approximately 400 m above sea level.

The glacial drift of the area is predominantly shallow, undulating ground moraine of fine sandy and silty texture, often with high stone and boulder content (Zoltai 1967). Moraines and drumlins are common as are glaciofluvial deposits such as eskers, kames and outwash plains. Soils of the Nipigon district include brunisols and luvisols, generally beneath mixedwood and hardwood forests; podzols generally beneath conifer forest; and gleysols and organic soils in wet areas (Canadian Soil Survey Committee 1978).

Mean annual precipitation for this area is 740 mm (Chapman and Thomas 1968). Winters are long and summers are short. The average frost-free period is 103 days. The growing season extends 155 days from mid-May to mid-September. During this period there is an accumulation of about 2000 growing degree days (°F). Westerly winds off Lakes Superior and Nipigon have a moderating effect on the local climate.

Data characterizing the four stands selected for this study are shown in Table 1 and Figures 2 and 3.

The balsam fir site was composed of predominantly balsam fir *(Abies balsamea)* admixed with black spruce, white spruce *(Picea glauca)* and trembling aspen (Figure 2). The stand was established on outwash sands, underlain by gravel, and originated from a clear cut 41 years ago. The ground vegetation was characterized by a carpet of feather mosses of

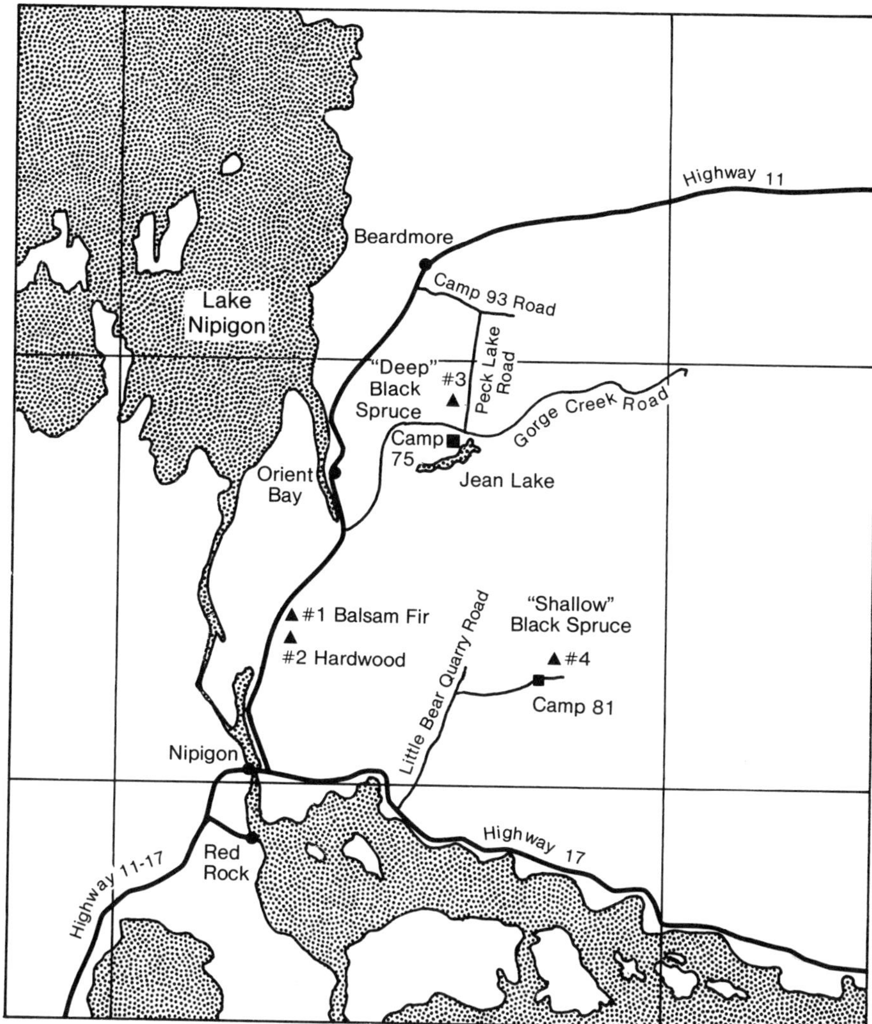

Figure 1. Location of balsam fir (#1), hardwood (#2), "deep" black spruce (#3) and "shallow" black spruce (#4) sites in the Lake Nipigon area.

Pleurozium schreberi, Dicranum spp., *Hylocomium splendens,* and *Ptilium crista-castrensis.* The soil was a deep, well-drained medium to coarse sand, with an orthic humo-ferric podzol profile overlain by a thin mor humus layer.

The hardwood site was a fire-origin mixed stand of trembling aspen and white birch *(Betula papryifera,* Figure 2) which was situated north of Gretel Lake. The trees were on a deep, fine sand which had a seasonally fluctuating water table between the B and C horizons. The soil was a sombric humo-ferric podzol. The understory of shrubs and herbs was dominated by *Galium triflorum, Epilobium angustifolium, Carex trisperma, Athyrium filix-femina, Viola incognita, Rubus pubescens, Rubus strigosus, Equisetum sylvaticum, Osmunda claytoniana,* and *Cinna latifolia.*

Table 1. Stand characteristics of the four study sites

Site	Effective rooting depth (cm)	Age (years)	Height (m)	Average dbh (cm)	Basal area (m^2/ha)	Stand density (#/ha)[1]
Balsam fir	32	41	13.3	7.2	35.7	3380
Hardwood	50	39	20.4	11.1	32.5	2740
Black spruce ''deep''	23	126	20.4	17.1	49.2	1960
Black spruce ''shallow''	12	126	17.3	12.9	32.0	2140

[1]Number of trees/hectare.

The black spruce 'deep' site was composed of even-aged, fire origin black spruce, (Figure 3) established on a deep (75 – 120 cm), moderately well-drained, medium to fine sandy loam (gleyed humo-ferric podzol). The thick fibrous, mor humus layer supported a moss groundcover characterized by *Pleurozium schreberi, Gaultheria hispidula, Cornus canadensis, Ledum groenlandicum, Vaccinium myrtilloides* and *Maianthemum canadense*.

The black spruce ''shallow'' site was an even-aged, black spruce stand of fire origin, (Figure 3) established on a shallow glacial till overlying bedrock. The well-drained, fine sandy loam varied in depth from 0 – 100 cm, and was classified as an orthic humo-ferric podzol. Bedrock and mineral soil were capped by a thick layer of fibrous mor humus. Ground vegetation consisted of a carpet of *Pleurozium schreberi,* with *Cornus canadensis, Vaccinium myrtilloides, Dicranum* spp., *Ptilium crista-castrensis* and *Mitchella repens*.

Harvesting Operations

Stands were harvested by a Drott 40 track feller-buncher. The operating procedure was described by Hilliker (1980). Full-trees were skidded to a Morbark Chipharvestor 22 XL equipped with a sliding boom loader. The entire tree was chipped and blown into large tri-axle chipper vans. Light material, consisting largely of bark and leaves, was separated from the chips during the chipping operation by means of a sorting mechanism. This waste material remained at the central landing and was not redistributed over the cutover. The ''deep'' black spruce, hardwood, and balsam fir sites were harvested in January, February, and March respectively. These operations were conducted in deep snow over frozen ground and consequently, caused little disturbance of the humus and mineral soil. The shallow-soil black spruce site, however, was harvested during mid-summer. The more severe impact of heavy equipment in the summer was obvious from a disturbance of the mineral soil and exposure of bedrock. A rotary flail, mounted on a skidder was used at the balsam fir site to reduce the foliage component of trees prior to chipping.

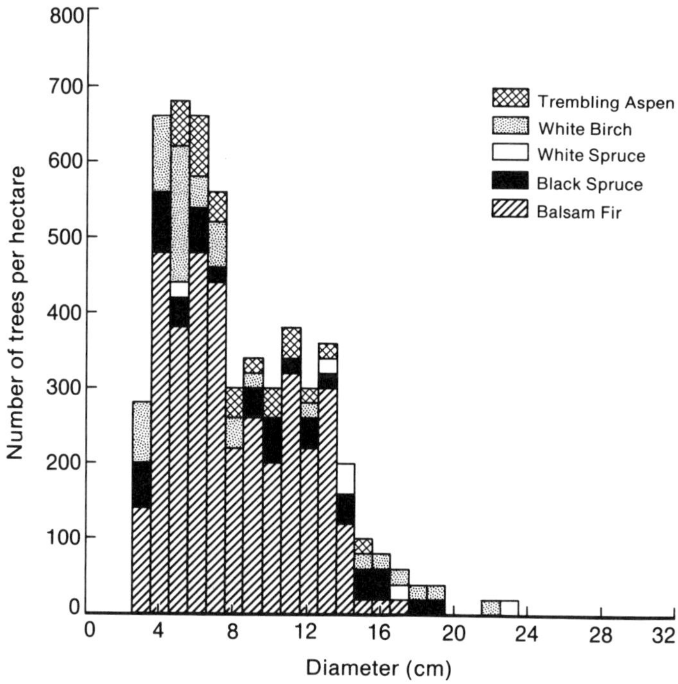

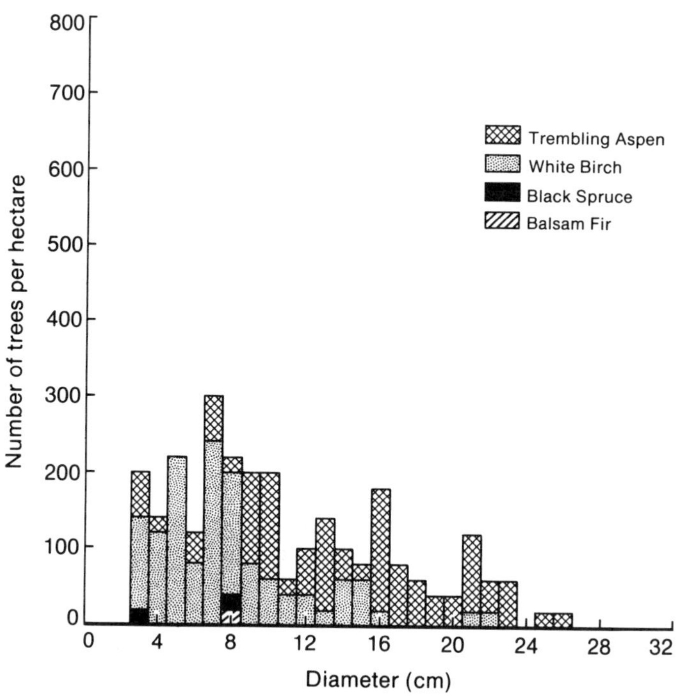

Figure 2. Distribution of trees by species and diameter class — balsam fir site (above) and hardwood site (below).

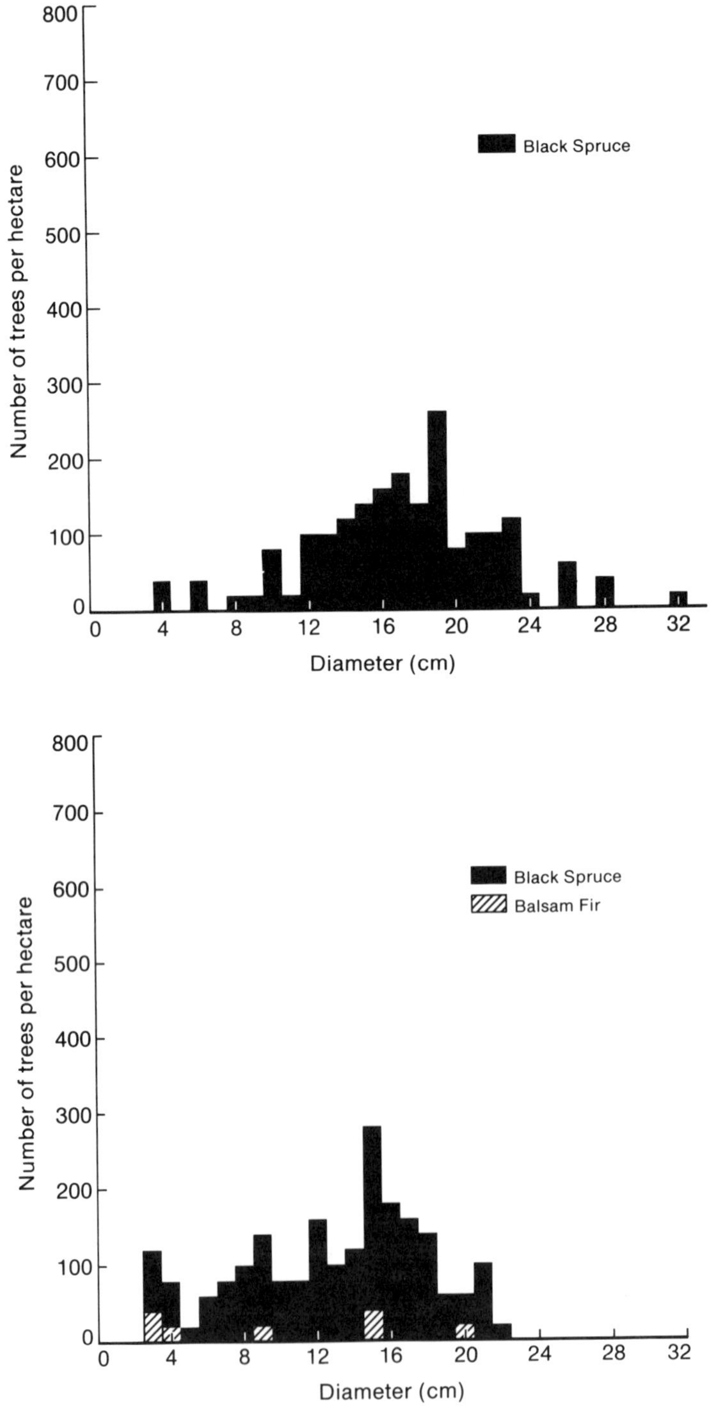

Figure 3. Distribution of trees by species and diameter class — deep-soil black spruce site (above) and shallow-soil black spruce site (below).

Field Sampling Procedures

Trees: Five 0.01 ha plots were established in the stands of each study area. Trees greater than 10 cm DBH located within the plots were tallied by species and diameter for the compilation of stand tables (Figures 2 and 3).

Within each stand, sample trees were selected to represent the range of diameters present. Twenty white spruce, 20 balsam fir, 20 trembling aspen, 12 white birch and 12 black spruce were sampled. Trees were felled and measured for height, age and diameter at breast height. Each tree was divided into components of bole, branches, foliated twigs, and dead branches. Cones were recorded as twigs. Fresh tree components were weighed in the field, using a portable hanging scale. Subsamples of each component were placed in plastic bags and transported to the laboratory for weighing and oven-drying at 70°C in a forced-air oven. Foliated twig samples were separated into foliage and twig components. All components were then reweighed oven-dry. Total oven-dry weights of merchantable bole, unmerchantable bole, branches, twigs, and foliage were calculated for each sample tree by using subsample dry-to-fresh-weight ratios. Five subsamples from each component were selected randomly and retained for chemical analysis.

Logging residue: Ten 1-m quadrats were randomly located in each clear-cut area. Logging residue within each quadrat was oven-dried, and sorted into components — bolewood, dead branches, live branches, foliage. Where slash samples were considered too large to be sorted, only a small fraction was processed and proportions were applied to the whole.

Mineral horizons: A and B horizons were sampled as cores concurrently with the forest floor. Soil samples from the lower mineral horizons were taken from soil pits located randomly on each site. Two samples were taken from opposite ends of each of three soil pits. Mean depth, rooting abundance and bulk density were also determined for each horizon. A-horizon weights were determined using an areal conversion factor, while lower soil horizon weights were calculated as the product of bulk density and horizon volume.

Forest floor: Dry weight and nutrient content of the forest floor were based on 20 randomly collected core samples (5 cm dia.) in the undisturbed stand. Forest floor materials were oven-dried at 70°C and weighed. A per-hectare weight was obtained by using a conversion factor corresponding to the surface area of the core samples. Ten composite samples, each representing two sampling points, were taken for chemical analysis.

Laboratory Analysis

Estimates of soil nutrient reserves were based on analyses of total N, P, K, Ca, and Mg in the forest floor, and the total N, available P, and exchangeable K, Ca and Mg in the mineral horizons. It was reasoned that the faster rate of decomposition of organic matter relative to weathering rates for soil minerals warranted total elemental analysis for the forest floor and for N in the mineral soil, and that the extractable portion of nutrients would be more representative of the plant-available nutrients in the mineral profile. Gordon (1979) has pointed out, that estimates of reserves based on total analysis, may overestimate nutrients available to trees during a rotation. Pritchett (1979) has suggested that an integration of the two approaches may prove to be the best method for reporting soil nutrient reserves until a more accurate system for estimating nutrient supply can be determined. The choice of analytical methods was based on these considerations.

Mineral soil: Soil samples were stored in open paper bags and allowed to dry at room temperature. Dry samples which contained aggregates were crushed using a mortar and pestle. Analysis of the fine earth fraction was as follows: total N was estimated by the Kjeldahl procedure (Bremner 1965), available P was determined using an acid extracting solution of ammonium fluoride (Bray and Kurtz 1945), and exchangeable K, Ca, and Mg were determined by atomic absorption after extraction with 1*N* ammonium acetate buffered at pH 7 (Jackson 1958).

Plant and humus material: Dried samples were ground using a Wiley cutting mill. Prior to weighing for analysis, samples were redried at 70°C in a forced draft oven and allowed to cool in a desiccator. Samples were dry ashed at 500°C in a muffle furnace, and dissolved in hydrochloric acid (Piper 1945). Nitrogen concentrations were determined by the Kjeldahl digestion and distillation method (Bremner 1965). Phosphorus was determined by the molybdenum blue method using stannous chloride as a reducing agent (Allen 1974), and K, Ca, and Mg were determined by atomic absorption.

Calculations

Height-diameter curves were obtained by regressing the logarithm of height on that of diameter. Both quadratic and logarithmic functions were tested, however the logarithmic equation yielded higher coefficients of determination. Logarithmic regressions of component weights (stem, live branches, foliage) on diameter, and diameter with height were also performed. Generally, the inclusion of height in the equation explained variation by more than one per cent. Similar regressions for merchantable bole were also obtained for poplar, white birch and black spruce. Unmerchantable bole weights of white spruce and balsam fir were not found to vary with diameter, thus merchantable boles were calculated as stem weights less average top (unmerchantable bole) weight. In order to arrive at a bole nutrient content figure it was necessary to distinguish between bark and wood since their respective nutrient concentrations vary greatly; therefore, the proportions of bolebark to bolewood constituting the stem was obtained from the literature (Baskerville 1966; Weetman and Webber 1972; Alban *et al*. 1978).

Results and Discussion

Biomass Removals

As shown by comparative harvest yields in Table 2, full-tree chipping operations resulted in substantially greater yields per unit area than conventional tree-length logging. While the greatest gain (74.9 t/ha) was made on the deeper black spruce site, the largest relative increase (116 per cent) occurred on the balsam fir site. The large proportion of the total tree represented by the crown (Table 3) and the high occurrence of undersized trees, unmerchantable by conventional harvesting standards, account for this difference (Figure 3). On the other hand, the additional harvest of the crown provided the smallest relative biomass increases on the hardwood site (Table 2).

Not all above-ground components of the tree were removed from the site during full-tree harvesting. Losses often occurred during the felling and forwarding operations and would accumulate as slash and logging debris (Table 4). Bolewood losses were attributed to inefficiencies of the harvesting system. In the conifer stands, winter harvesting resulted in more logging residue than summer harvesting, most likely because of increased breakage of crown components at low temperatures. Mechanical flailing of cut trees also contributed to deposits of foliage, twigs and branches on the site (Table 4).

Table 2. Biomass (t/ha) removed from the study sites as harvest material

Harvesting system	Balsam fir	Hardwood	Black Spruce "Deep"	Black Spruce "Shallow"
		– t/ha –		
Tree length	57.3	112.1	100.1	86.3
Full tree	123.7	160.0*	225.0	138.3
Difference	66.4	47.9	74.9	52.0
Difference (%)	116	43	50	60

* Winter harvest – excluding leaves.

Table 3. Biomass (t/ha) of above ground stand components by site

Component	Balsam fir	Hardwood	Black spruce "Deep"	Black spruce "Shallow"
		– t/ha –		
Stem	94.6	147.0	185.9	115.6
Live branches	13.4	13.7	19.6	10.9
Dead branches	5.4	3.7	6.9	4.2
Foliage	10.2	4.5*	12.6	7.6
Crown/stem	0.30	0.15*	0.21	0.20

* Sampled in mid-summer (crown/stem reduced to 0.11 after leaf fall).

Table 4. Logging residue (t/ha) on site after full-tree harvesting

Component	Balsam fir*†	Hardwood*	Black spruce "Deep"*	Black spruce "Shallow"**
		– t/ha –		
Stem	8.8	0	6.8	0
Live branches	3.8	2.8	4.4	1.3
Dead branches	3.4	0.8	6.5	0.3
Foliage	2.7	0	1.4	1.1
Loose bark	0.1	0	0.4	0
Total	18.8	2.8	19.0	2.7
		– Per cent –		
% of harvest	15.2	1.7	8.6	1.9

* Winter harvest.
** Summer harvest.
† Rotary flail employed.

Nutrient Removals

Changing utilization standards markedly affected the export of nutrients from a site. Full-tree logging increased nutrient losses from 59 to 238 per cent over conventional harvest (Table 5), while corresponding biomass exports (Table 2) were substantially smaller (56 to 116 per cent). The difference between proportional increases of biomass and nutrients was attributed to the elevated elemental concentrations within the crown (which constituted the added harvest) relative to the bole.

Generally, the nutrient losses due to full-tree chipping were twice those of conventional harvests (Table 5). Balsam fir incurred a three-fold increase in nutrient drain, and seems to be an exception. A high crown/stem ratio (Table 3) and generally higher concentrations of foliar nutrients in the younger balsam fir may account for this relatively greater loss.

Within the ecosystem, leaf litter is a major source of decomposable nutrients for future plant uptake. On the hardwood site, which contained a large component of white birch and trembling aspen, foliage represented up to 20 per cent of the total elemental content of the stand, comprising a major potential loss of nutrients under full tree harvesting. Under such circumstances, harvesting during the leafless period may be an appropriate conservation measure.

Nutrient Reserves

Nutrient supply to effective rooting depth and 1 m depth is presented in Table 6. Except for phosphorus, the hardwood site was the most abundantly supplied with nutrients. High calcium levels in this soil were attributed to a strongly calcareous C horizon (pH 7.8), while comparatively higher nitrogen levels were related to high organic matter content in a thick Ah horizon. The major source of nitrogen in the three conifer stands was derived from the humus layers rather than the mineral soil. The soil of the black spruce growing on the shallow site contained the lowest nutrient capital, presumably because of restricted rooting depth and soil volume.

When soil horizons that are presently not exploited by roots are considered in stand nutrient budget, nutrient reserves are much larger (Table 6). These supplies at lower depths are potentially available to trees capable of downward root extension, except where root penetration is inhibited by soil compaction, excess water, or physical barriers such as bedrock.

Impact Assessment

The primary objective of impact studies which examine forest nutrient budgets is to assess and predict the consequences of specific management practices on site productivity. Researchers dealing with this topic are reluctant to reach definitive conclusions and recommendations because of the problems associated with quantifying the dynamic components of the nutrient cycles in forest ecosystems. Accurate values of the rates, fluxes, and pools of nutrients involved in the critical processes of mineralization, immobilization, leaching, weathering, and fixation are difficult and costly to measure precisely, especially in the short term (Kimmins 1977; Stone 1979). Although standard chemical analysis techniques to assay forest soils have been well established, these are based on agronomic methods and experiences; consequently they may not reflect nutrient

Table 5. Nutrient exports (kg/ha) associated with tree-length (TL) and full-tree (FT) harvesting operations

Nutrient	Balsam fir		Hardwood		Black spruce "Deep"		Black spruce "Shallow"	
	TL	FT	TL	FT	TL	FT	TL	FT
				– kg/ha –				
N	119	398	306	602	433	897	244	536
P	14	45	29	54	59	108	33	65
K	65	210	103	207	84	220	53	141
Ca	236	637	333	628	1066	1696	602	1024
Mg	28	77	44	84	56	106	32	66
				– Per cent –				
Increase*								
N	234		97		107		120	
P	221		86		83		97	
K	223		101		162		166	
Ca	170		89		59		70	
Mg	175		91		89		106	
Mean	205		93		100		112	

* Over conventional tree length harvesting.

Table 6. Nutrient reserves (t/ha) in the forest floor and mineral soil to effective rooting depth, and to root exploitable depth (maximum 1 m)

Nutrient	Balsam fir	Hardwood	Black spruce "Deep"	Black spruce "Shallow"
		– t/ha –		
Effective rooting depth				
N	3626	6371	3676	2284
P	131	25	129	111
K	166	211	241	178
Ca	1173	11027	2324	178
Mg	296	1148	673	148
Exploitable depth				
N	4655	7370	4541	2284
P	275	51	199	111
K	237	416	642	178
Ca	2884	11456	8878	178
Mg	514	1832	1333	148

* Soil depth less than 1 m.

Table 7. Nutrient loss from sites as a percentage of soil nutrient reserves

					Black spruce			
	Balsam fir		Hardwood		"Deep"		"Shallow"	
Nutrient	TL	FT	TL	FT	TL	FT	TL	FT
				– Per cent –				
Effective rooting depth								
N	3	11	4	9	10	24	9	23
P	1	3	49	165	31	75	22	54
K	19	102	30	87	21	84	19	73
Ca	14	50	3	6	34	68	85	359
Mg	8	25	4	7	7	16	17	43
Exploitable soil depth*								
N	2	8	4	8	8	19	9	23
P	4	16	33	92	23	50	22	54
K	15	76	19	47	10	33	19	73
Ca	7	21	3	5	11	19	85	359
Mg	5	15	2	5	4	8	17	43

(TL) Tree-length logging; (FT) full-tree logging.
* 1 m maximum.

Table 8. Sufficiency period of soil nutrient reserves following intensified harvesting

					Black spruce			
	Balsam fir		Hardwood		"Deep"		"Shallow"	
Nutrient	TL	FT	TL	FT	TL	FT	TL	FT
				– years –				
Effective rooting depth								
N	1293	361	903	440	1212	521	1336	541
P	3765	1133	84	25	402	165	562	229
K	210	38	136	47	592	148	654	169
Ca	276	78	1403	724	364	183	146	35
Mg	495	158	1129	561	1703	797	737	291
Exploitable soil depth*								
N	1629	462	1037	508	1459	641	1336	541
P	872	242	121	45	549	246	562	229
K	252	51	217	88	1186	374	654	169
Ca	559	183	1457	752	1126	662	146	35
Mg	795	268	771	891	3222	1563	737	291

(TL) tree-length logging; (FT) full-tree logging.
* 1 m maximum.

supply over the life of the stand (Pritchett 1979; Stone 1979). Other uncertainties arise in estimating nutrient losses incurred after logging by accelerated decomposition, soil erosion and lack of regeneration, as well as in characterizing and quantifying the important site recovery processes that take place by rapid regrowth and invasion of pioneer vegetation (Marks and Bormann 1972; Kimmins 1977). Despite these constraints, it is often necessary and mandatory to include potential harvesting impacts in the planning stages of an intensive forest management plan. Impact assessment must thus rely on the best available short-term data, which may not account fully for the dynamic character of the forest nutrient cycle. They do, however, provide generalized and approximate estimates of the nutrient pools and fluxes involved. This form of approach was adopted in the present study, therefore the assessment and conclusions which follow must be viewed in the light of these limitations.

Nutrient losses relative to site nutrient capital are given in Table 7 for tree-length, and for full-tree logging. Under a steady-state model of ecosystem dynamics (Marks and Bormann 1977; Alban *et al*. 1978), inputs are balanced by outputs, thus values exceeding 100 per cent indicate that insufficient nutrients remain on the site to support a second crop of equal size.

The results show that conventional tree-length logging makes only modest demands on site nutrient capital; sufficient nutrients remain on the site to sustain a second generation of trees. On the other hand, shortages of Ca, P, and K will occur on the shallow spruce, hardwood, and balsam fir sites respectively, when harvested by full-tree methods. Calcium losses on the shallow-soil black spruce site were the most severe, exceeding soil reserves by a magnitude of 2.5. Weetman and Webber (1972) regarded such losses on a black spruce site in Northern Quebec as serious since calcium indirectly stimulates the mineralization of nitrogen in the humus layer (Dahl *et al*. 1967). The deep-soil black spruce site appears to be the most suitable for full-tree harvesting, supporting the largest nutrient reserves.

An estimate of the period under which current management practices can be maintained is provided by the sufficiency period (Table 8). This time period is derived by dividing the soil nutrient supply by the net annual rate of nutrient accumulation (content) in the trees. Values lower than present stand age (in years) predict the onset of deficiency in the next rotation. Thus, phosphorus may become exhausted in the hardwood site in 25 years, potassium in the balsam fir site in 38 years, and calcium in the shallow-soil black spruce in 35 years. Sufficiency periods are longer of course, when root exploitable horizons are considered.

Predictions like those suppose that future crops are grown at rotation lengths equivalent to the previous crop. This may not be realistic, however, considering current trends towards more intensive, fast-rotation forest management. Sufficiency periods will be drastically reduced if rotation ages are shortened by modern silvicultural means without using adequate fertilization. For instance, if an equivalent volume of black spruce on the shallow site could be grown in half the rotation age (63 versus 126 years), soil Ca will probably become limiting in 17.5 instead of 35 years. It seems reasonable therefore, to foresee that the replacement of nutrients by forest fertilization will play a more important role in the future management of fragile, shallow sites, if such stands are harvested by full-tree methods.

Table 9. Summary of comparative impacts of full-tree harvesting operations on differing stand and site conditions

Harvesting effect on	Quantitative impact	Refer to
Nutrient removal in crop		
Young vs old stands	nutrient loss (205 vs 100%) higher in younger stands	Bf and Sb "deep" vs 126 yr old stands (Table 5)
Conifers vs hardwoods	nutrient loss (205 vs 93) higher in conifer stands	Bf vs Hwd (Table 5)
Leaf vs leafless periods	winter (leafless) harvests reduce nutrient loss 9% in hardwoods	Hwd stand, + and − leaf components and other data* (Tables 3 and 5)
Winter vs summer	winter harvests reduce nutrient loss (2 vs 9) 7% in conifers by breakage	logging residue and other data* (Table 4)
Mechanical flailing	on site flailing devices reduce nutrient loss (15 vs 9) 6%	logging residue and other data* (Table 4)
Soil nutrient supply		
Deep vs shallow soils	nutrient sufficiency period (246 vs 35) 211 years shorter on shallow sites	Sb stands, "deep" vs shallow sites (Table 8)
Long vs short rotations	nutrient sufficiency period (246 vs 51) 195 years shorter under short rotation management	Sb "deep" and Bf stands, 126 vs 41 year rotation (Table 8)

* Nutrient concentration data of individual components, available on request.

Conclusions

An evaluation of the macro nutrient budgets of four forest stands representative of the major sites in the Nipigon area indicate that the nutrient losses associated with conventional (stem only) harvesting methods are relatively modest with regard to nutrient supplies in the soil. These findings agree with other impact studies carried out on boreal forest sites (Weetman and Webber 1972; Boyle and Ek 1972; Foster and Morrison 1976; Gordon 1979).

The potential consequences of whole-tree harvesting on forest nutrient budgets are, however, more severe, and may well cause nutrient deficiencies in the next generation of trees (as in three out of the four sites samples here). Stands supported on deeper soils and on longer rotations are most able to withstand intensive harvesting. Areas of shallow-soil black spruce which comprise a large part of the Nipigon area (Marek 1975; Jeglum 1980) may be the most susceptible to nutrient depletion by increased crop utilization. This fact was also noted by Weetman and Webber (1972) working on similar sites in Quebec.

The impacts of whole-tree logging on different stand and site conditions vary in degree and magnitude and are summarized in Table 9. Generally, the proportions in which nutrients are taken up vary with species. It appears that hardwood crops place a greater demand on site nutrient reserves than do conifers. Short rotations and intensive logging practices directly affect nutrient drain. Their impact on forest yields is expected to be greatest on poor (shallow) sites. Winter harvests and mechanical flailing will markedly lengthen the sufficiency period of soil nutrient reserves.

This study was conducted on only four stands in the Nipigon area. Each stand was representative of a major forest type in the district. However, because of financial and time constraints replication in each forest type was not possible, hence the recommendations that can be drawn from this impact assessment are generalized and applicable only to the site and forest conditions of this region. The following guidelines for minimizing potential nutrient losses by forest harvesting appear warranted.

It is recommended that:

- conventional logging methods exclusively be utilized on fragile, shallow-till sites;
- full-tree and complete-tree chipping operations be restricted to stands supported by relatively deep mineral soils;
- marginal sites which are sensitive to full-tree or complete-tree logging be harvested in the winter with snow present;
- hardwood sites be harvested by full-tree methods in the dormant (leafless) season;
- mechanical flailing or delimbing devices be operated on sites which are sensitive to intensive harvesting;
- rapid regeneration or active revegetation be ensured after whole-tree chipping to accelerate nutrient capture and site recovery processes; and
- intensive, short-rotation forest management be restricted to the deeper, more productive sites in the Lake Nipigon-Beardmore area.

Acknowledgements

This study was funded by the Ontario Ministry of Natural Resources as a research contract to V. R. Timmer at the University of Toronto. The authors are grateful to Dr. I. K. Morrison, Canadian Forestry Service, Great Lakes Forest Research Centre, Sault Ste. Marie, for assistance, advice, and review of this study.

Literature Cited

Alban, D.H., D.A. Perala and D.E. Schlaegel. 1978. Biomass and nutrient distribution in aspen, pine and spruce stands on the same soil type in Minnesota. Can. J. For. Res. 8: 290-299.

Allen, S.E. (Ed.). 1974. Chemical analysis of ecological materials. John Wiley and Sons, New York.

Baskerville, G.L. 1966. Dry matter production in immature balsam fir stands. For. Sci. Monogr. 9.

Boyle, J.R., and A.R. Ek. 1972. An evaluation of some effects of bole and branch pulpwood harvesting on site macronutrients. Can. J. For. Res. 2: 407-412.

Bray, R.H. and L.T. Kurtz. 1945. Determination of total organic and available forms of phosphorus in soils. Soil Sci. 59: 39-45.

Bremner, J.M. 1965. Total nitrogen. *In:* Methods of soil analysis. C.A. Black (Ed.). Amer. Soc. Agron. Monograph No. 9. pp. 1149-1178.

Canadian Soil Survey Committee, Subcommittee on Soil Classification. 1978. The Canadian system of soil classification. Can. Dep. Agric. Publ. 1646. Ottawa, Ontario.

Carlisle, A. and I.R. Methven. 1979. The impact of whole tree harvesting and short rotations. Biomass Strategy Consultation Proc., Canada/MAB Rep. No. 12. pp.70-76.

Chapman, L.J. and M.K. Thomas. 1968. The climate of Northern Ontario. Climatological Studies No. 6. Can. Dept. Trans. Toronto, Ontario.

Dahl, E., O. Gjems and J. Kielland-Lund. 1967. On the vegetation of Norwegian conifer forest in relation to the chemical properties of the humus layer. Medd. Nor. Skogforkswesen. 23: 503-531.

Foster, N.W. and I.K. Morrison. 1976. Distribution and cycling of nutrients in a natural *Pinus banksiana* ecosystem. Ecology 57: 110-120.

Freedman, B., R. Movash and A.J. Hanson. 1981. Biomass and nutrient removals by conventional and whole-tree clear-cutting of a red spruce-balsam fir stand in Central Nova Scotia. Can. J. For. Res. 11: 249-257.

Gordon, A.G. 1979. Productivity and nutrient cycling in natural forests. Biomass Strategy Consultation Proc., Canada/MAB Rep. No. 12.

Gordon, A.G. 1981. Impacts of harvesting on nutrient cycling in the boreal mixedwood forest. Environ. Canada, C.F.S., Proc. O-P-9. pp. 121-140.

Hilliker, R.G. 1980. Whole tree chipping. *In:* Proc. 61st Annual Meeting. Woodlands Sec., Can. Pulp and Pap. Assoc., Montreal, Quebec. pp. 67-70.

Jackson, M.L. 1958. Soil chemical analysis. Prentice-Hall, Inc.

Jeglum, J.K. 1980. Strip cutting in shallow-soil upland black spruce near Nipigon, Ontario. I. Study establishment and site conditions. Environ. Canada, C.F.S., Inf. Rep. O-X-315.

Krause, H.H., G.F. Weetman and P.A. Arp. 1980. Nutrient cycling in boreal forest ecosystems of North America. *In:* Forest soils and land use. C.T. Youngber (Ed.). Colorado State University. Ft. Collins, Colorado. pp. 287-319.

Kimmins, J.P. 1977. Evaluation of the consequences for future tree productivity of the loss of nutrients in whole-tree harvesting. For. Ecol. Manag. 1: 169-183.

Marek, G.T. 1975. Ecosystem management on shallow sites in Lake Nipigon-Beardmore area. Black spruce symposium. Environ. Canada, C.F.S., O-P-4. pp. 195-200.

Marks, P.L. and F.L. Bormann. 1972. Revegetation following forest cutting: mechanism for return to steady-state nutrient cycling. Science, 176: 914-915.

Morrison, I.K. and N.W. Foster. 1979. Biomass and element removal by complete-tree harvesting of medium rotation forest stands. *In:* Proc. Impact of intensive harvesting on forest nutrient cycling. State University of New York, Syracuse, New York. pp. 111-129.

Piper, C.S. 1945. Soil and plant analysis. Interscience Pub. Inc., New York.

Pritchett, W.L. 1979. Properties and management of forest soils. John Wiley and Sons, New York.

Rennie, P.J. 1955. The uptake of nutrients by mature forest growth. Plant Soil, 7: 49-95.

Rowe, J.S. 1972. Forest regions of Canada. Environ. Canada, C.F.S., Pbl. No. 1300.

Stone, E.L. 1979. Nutrient removals by intensive harvest — some research gaps and

opportunities. *In:* Proc. Impact of intensive harvesting on forest nutrient cycling. State University of New York, Syracuse, New York. pp. 366-386.
Tufts, E.M. 1976. Whole tree chipping. Tappi, 59: 60-62.
Weetman, G.F. and B. Webber. 1972. The influence of wood harvesting on the nutrient status of two spruce stands. Can. J. For. Res. 2: 351-369.
Zoltai, S.C. 1967. Glacial features of the north-central Lake Superior Region, Ontario. Can. J. Earth Sci. 4: 515-528.

DYNAMICS OF A BLACK SPRUCE ECOSYSTEM IN COMPARISON TO OTHER FOREST TYPES: A MULTI-DISCIPLINARY STUDY IN INTERIOR ALASKA

Keith Van Cleve
University of Alaska
and
Leslie A. Viereck and C.T. Dyrness
Institute of Northern Forestry,
USDA Forest Service

Abstract: For the past several years, the University of Alaska and the Institute of Northern Forestry have conducted a multi-disciplinary study of a black spruce ecosystem in interior Alaska. The black spruce type, widespread in interior Alaska, is the most fire-prone forest type. Black spruce ecosystems are also the most nutrient limited and least productive forest type, especially in late stages of succession. The central hypothesis of this study is that ecosystem differences in productivity and degree of nutrient limitation are controlled mainly by soil and forest floor temperatures. A corollary hypothesis is that in the black spruce ecosystem, which is characterized by low productivity and slow nutrient cycling, fire acts as a rapid decomposer and is essential to replenish depleted available nutrient pools. In order to test these and other hypotheses a number of semi-intensive sites were studied for comparison with the black spruce permafrost-dominated intensive site. These semi-intensive sites represent a complete spectrum from the coldest sites to the warmest and driest sites that support tree growth in interior Alaska. The sites were also selected to represent both successional and mature stands.

The taiga of interior Alaska includes a gross area of approximately 136,799,600 hectares. Commercial forest land comprises seven per cent of this total or 9,000,000 hectares. Intermingled with the commercial forest land and extending beyond to western and northern limits of tree growth are 33,200,000 hectares of sparse or open woodlands considered non-commercial forest land. Within this region, on the extensive poorly drained sites, sites underlain by permafrost and cool sites on north facing slopes, the dominant tree species is black spruce *(Picea mariana)*.

Environmental factors unique to the northern parts of the boreal forest, including Alaska, affect a generally lower primary production and rate of nutrient cycling compared with temperate latitude forest ecosystems. These factors include drastic seasonal fluctuations in daylength and temperature, a short growing season, low soil temperature, and the occurrence of permafrost. It is primarily the forest floor and mineral soil temperature and moisture, in conjunction with forest floor chemistry (organic matter quality), which interact through a series of feedback mechanisms to control organic matter decomposition, nutrient availability and tree productivity (Figure 1). The black spruce ecosystem is

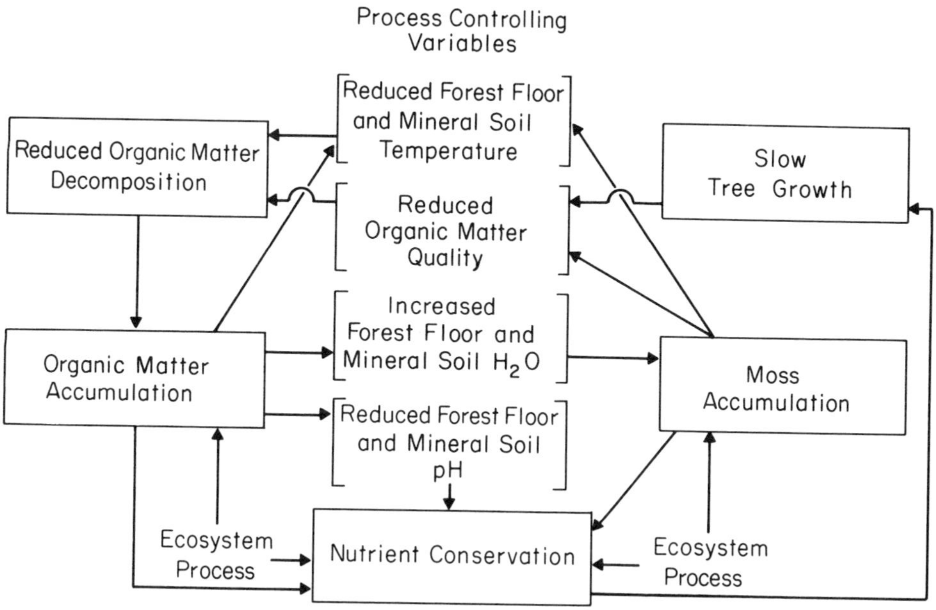

Figure 1. Hypothesized synergistic action of ecosystem processes and process controlling variables in taiga forest ecosystems (Van Cleve and Viereck, 1981).

the extreme example of low productivity and nutrient limitation in interior Alaskan forests. Another hypothesis is that in the black spruce ecosystem, fire acts as a rapid decomposer and is essential to replenish depleted available nutrient pools.

The objective of the research reported in this paper was to examine these hypotheses within the framework of a multidisciplinary, integrated research program dealing with taiga forest ecosystem structure and function. The program was conducted near Fairbanks (64°51′N latitude, Figure 2) from 1975 to 1980 with support from the National Science Foundation, U.S. Forest Service and the University of Alaska.

One additional reason for studying the taiga now is that large-scale development is just beginning. The trans-Alaska oil pipeline extends through a broad band of taiga from the south slope of the Brooks Range to the north slope of the Chugach Mountains. Along with the pipeline have come transportation corridors and associated human developments. More development is certainly on the horizon with the plans for a gas pipeline and with increased mineral, coal and oil exploration and extraction. Forestry and agriculture are in the initial phases of development. At this juncture, we are obligated to know more about the ecosystems in which much of the future development will take place.

Most of the Alaskan taiga lies in the intermontane plateau between the Brooks Range on the north and the Alaska Range on the south (Figure 2). Highest peaks reach 3,000 and 6,400 metres respectively in the two ranges. The region is characterized by small isolated mountain ranges, large areas of gently sloping uplands and broad rounded ridges, meandering rivers with broad plains and extensive flat lowland plains dotted with thaw lakes and occasional pingos. Permafrost underlies as much as 75-80 per cent of the terrain except for south facing slopes and floodplains adjacent to major rivers. Black spruce

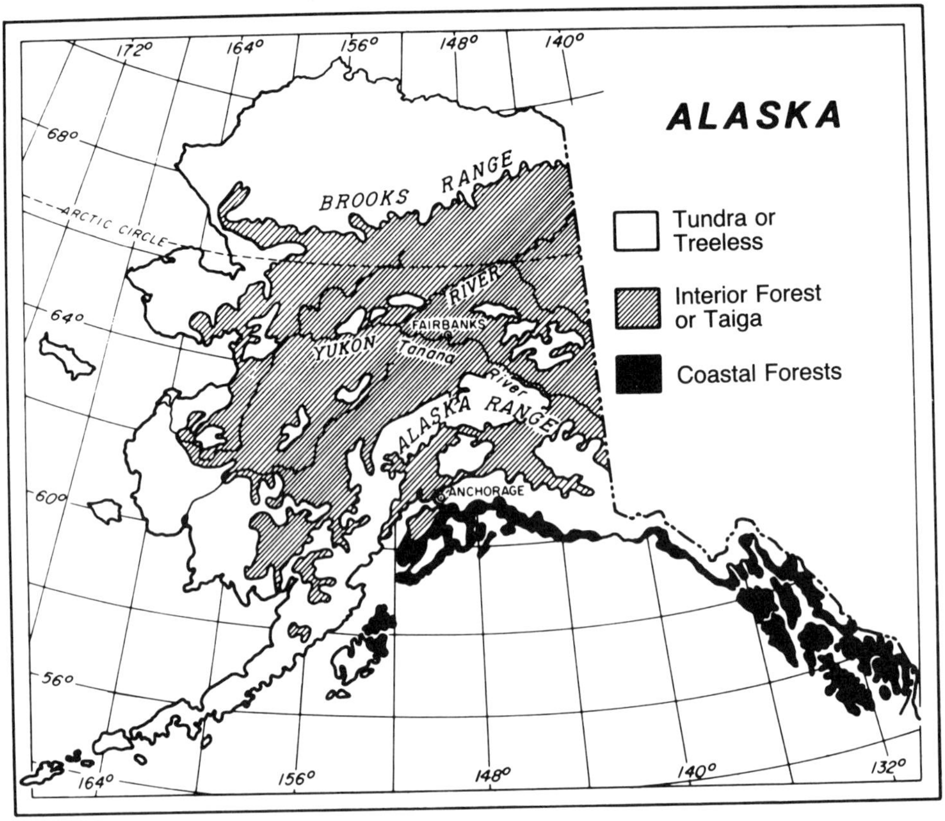

Figure 2. Distribution of taiga in Alaska (Van Cleve *et al.* 1983).

muskeg and forest are the dominant vegetation types below treeline covering 44 per cent of the area. They are closely associated with permafrost. Seventeen per cent of the area lies above the treeline (Selkregg 1975) at 900 metres elevation and is covered by subalpine shrub stands, tundra sedge meadows, heaths and fell fields. The treeline forest is characterized by open stands of black and white spruce. In the environmental mosaic below the treeline, mature forest types change with slope, aspect, and drainage. All these site factors apparently exert their influence primarily through soil and air temperature and permafrost.

Upland taiga forest types range from moderate to highly productive coniferous and deciduous tree stands on south-facing, well-drained slopes, to permafrost and moss-dominated black spruce forest of low productivity on gentle north-facing slopes (Van Cleve and Viereck 1981). Forest succession on the floodplains (Viereck 1966, 1970) is a primary succession with a sequence of growth forms, productivity and nutrient accumulation more typical of middle latitude forest succession (Van Cleve and Viereck 1981).

Fire recurs in some forest types as often as every 50-60 years. Tree stands older than 170 years are rare except on river flood plain islands or within moist drainages (Quirk and Sykes 1971). Between 1940 and 1970 an average of almost 400,000 ha (one per cent of

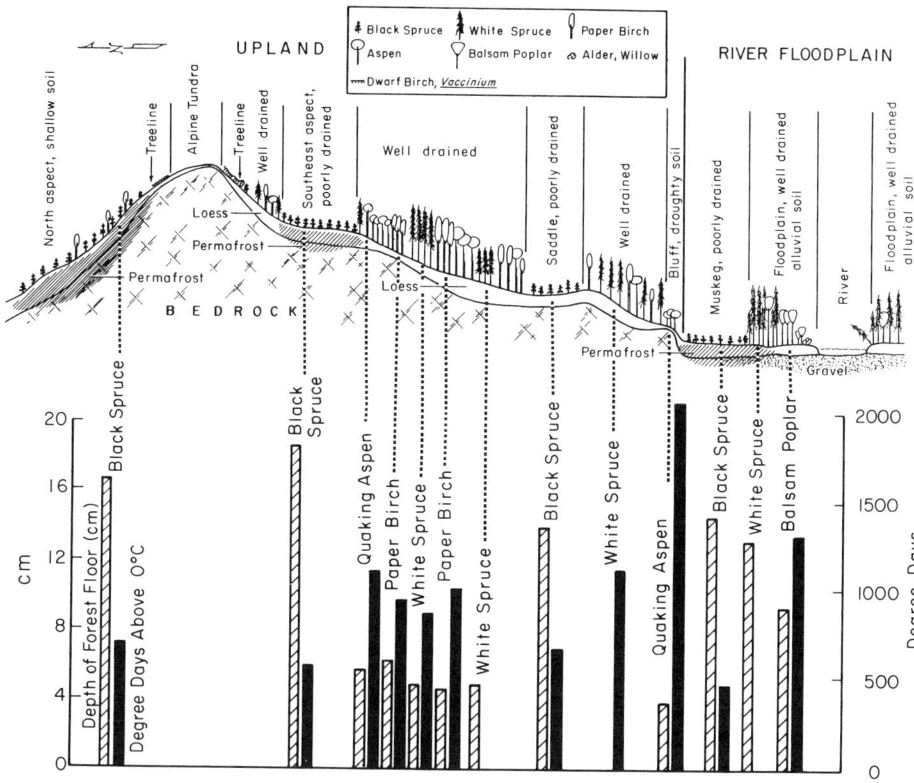

Figure 3. Composite cross section of Yukon Tanana Uplands showing forest type distribution and associated forest floor depth and soil heat sum (Viereck and Van Cleve, in press).

interior Alaskan forested land) were burned annually in interior Alaska (Barney 1971). Since 1970 slightly more than 240,000 ha (0.6 per cent of interior Alaskan forested land) has burned annually (Viereck and Schandelmeier 1980; United States Department of the Interior 1978).

Because of recurrent, patchy fires, taiga forest vegetation is a mosaic of stands of different ages. In the uplands this variation caused by fire and the periodicity of recruitment is superimposed upon variations due to slope and aspect, creating a mosaic of black and white spruce *(Picea glauca)*, birch *(Betula papyrifera)* and aspen *(Populus tremuloides)* patches best developed on warmer, more southerly aspects.

Soils of the region show relatively slight morphological development. This reflects the cold and semi-arid subarctic climate, which reduces the intensity of chemical weathering. Thus for interior Alaska, Inceptisols, Entisols and Histosols occupy approximately 78, 12 and 7 per cent respectively, of the land area, a total of 97 per cent or approximately 33 million hectares (Rieger *et al*. 1979). Soils showing more extensive profile development, including Mollisols and Spodosols encompass only 3 per cent or approximately 840,000 hectares of interior Alaska.

Description of Study Area and Methods

The intensive study site for the black spruce ecosystem was located in the Washington Creek Fire Ecology Experiment Area, about 45 km north of Fairbanks. This site is representative of a large area of interior Alaskan taiga, with the dominant vegetation type being black spruce/feathermoss growing on soils with shallow permafrost (Viereck and Dyrness 1980).

The terrain is gently rolling (average slope about 10 per cent), loess-mantled, and underlain by Precambrian schist. Elevation of the intensive site is approximately 400 m and the dominant aspect is south-east. Soil is the Saulich silt loam, a Histic Pergelic Cryaquept.

In addition to the intensive site, 19 semi-intensive sites were also selected for study. These sites, all within 50 km of Fairbanks, represent the complete spectrum from the coldest sites to the warmest and driest sites that support tree growth. Their position along a cross section of the floodplains and upland is shown in Figure 3.

On the floodplain the stands under study represent a successional series developing in newly exposed silt on the actively meandering Tanana River. Younger forest stages include 50- and 80-year-old balsam poplar/alder/*Equisetum* stands and two 120-year-old white spruce-feathermoss stands. Older stages on terraces underlain by permafrost include a mixed white spruce-black spruce/*Sphagnum* stand and an open black spruce/feathermoss stand. This successional sequence is described in detail by Van Cleve and Viereck (1981).

The upland semi-intensive sites represent a wide variety of environmental conditions. The warmest driest site is occupied by an aspen/grass/*Artemisia* stand adjacent to an open grassland on a south-facing bluff above the Tanana River. A 50-year-old aspen/*Viburnum*/*Linnaea borealis* stand and two paper birch/alder/*Calamagrostis* stands represent the deciduous forest stage following fire in productive permafrost-free upland sites. The final stage of this successional sequence is represented by a 140-year-old closed white spruce/feathermoss type. On the least productive black spruce site, annual moss production may equal above-ground tree production of about 100 g/m^2.

The colder sites in the uplands, most of which are underlain by permafrost, are represented by stands of closed black spruce/feathermoss, black spruce/feathermoss/*Cladonia* and an open black spruce/*Sphagnum* stand. The coldest site is an open treeline stand of the black spruce-white spruce/dwarf birch/feathermoss type at an elevation of 750 metres.

The cycling and compartmental distribution of biomass and nutrients were studied at all sites. Phenology, vegetation, soils and soil and air miroclimate were also characterized. For graphical comparisons, average values were calculated for the respective ecosystem parameters based on information collected from the replicate stands examined in each of the five major forest types: floodplain balsam poplar, white spruce (upland and floodplain), black spruce, birch and aspen.

The forest floor was a focal point in our studies because soil temperature and moisture regimes, organic matter accumulation, decomposition and soil chemical status appeared to be correlated with striking differences in ecosystem structure and function among taiga forest types.

Table 1. Above-ground tree biomass and annual tree production and vascular plant litterfall for the principal forest types in interior Alaska (g/m^2)

Vegetation type	Number of stands	Forest floor biomass		Tree biomass		Annual tree production		Litterfall	
		Range	$\bar{x} \pm SE$	Range	$\bar{x} \pm SE$	Range	$\bar{x} \pm SE$	Range	$\bar{x} \pm SE$
Black spruce	4	4469-9215	7646±602	2589-10984	5094±2107	72-148	113±17	15-142	43±11
White spruce	4	4707-10533	7426±1496	6152-24577	17442±4279	238-540	366±63	95-316	155±78
Quaking aspen	2	2102-5328	4750±1223	4653-17490	11021±6419	363-760	565±199	27-222	156±65
Paper birch	3	4374-6838	5797±3347	9192-14713	11156±1782	343-572	470±61	233-265	251±8
Balsam poplar	3	1720-2641	2195±266	4067-18027	12097±4165	264-952	552±206	268-653	389±49

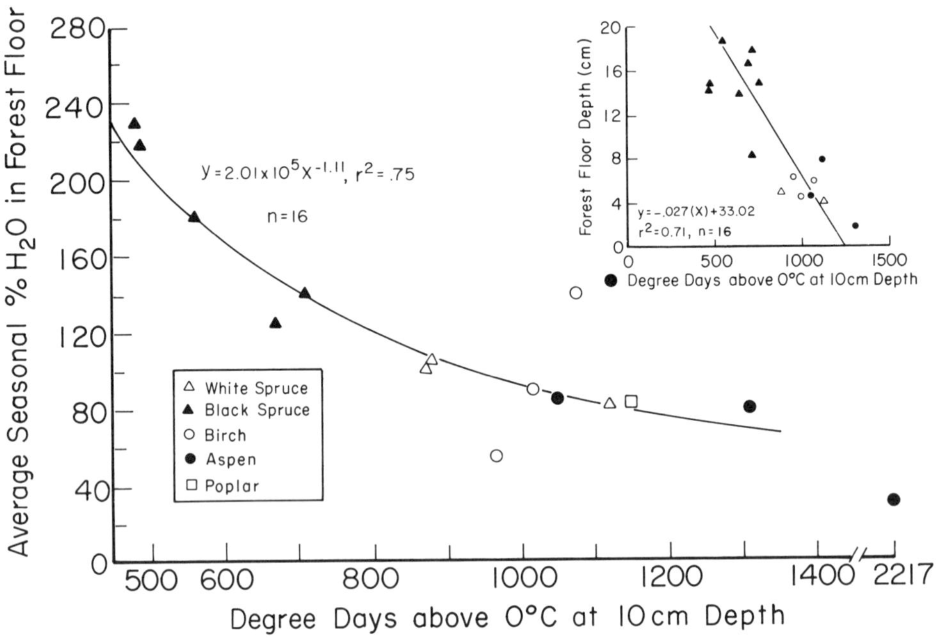

Figure 4. Relationship between soil heat sum and forest floor depth and average seasonal moisture content (Van Cleve *et al*. 1983).

Methods for vegetation classification are outlined in Viereck and Dyrness (1980). Vegetation and soil sampling procedures and methods for chemical analysis of soil and plant tissue samples are found in Dyrness and Grigal (1979), Van Cleve *et al*. (1981), Viereck and Dyrness (1979), Viereck *et al*. (1979) and Skre and Oechel (1981).

Results and Discussion

On the average, black spruce forest floors are significantly thicker (average thickness of 14.9 cm, α=0.01) and they contain substantially more biomass (7646 g/m^2, α=0.05, Table 1) compared with the average for the other forest types (4.4 cm thickness, 5191 g/m^2 biomass). These conditions are reflected in a strong correlation between soil temperature and forest floor depth (r^2=0.71) across the array of vegetation types considered in this study (Figure 4). A yearly heat sum (degree days, (DD), 0°C at the 10 cm depth) showed a significantly lower average temperature of 640 DD (n=8, α=0.01) for black spruce compared with average of 1132 DD (n=8) for all other forest types (Figures 3 and 4). In addition, average seasonal forest floor and mineral soil moisture content are closely related to soil temperature. Values for black spruce range between 120 and 240 per cent for the forest floor and 60 to 160 per cent for mineral soil respectively, while the other forest types generally are encountered on warmer, drier sites (Figure 4).

Differences in soil heat sums exist primarily because northern, western, and eastern aspects or level ground sites receive less radiation compared with south-facing aspects. Moreover, the thick forest floor and moss layer of a black spruce forest have low bulk density and thermal conductivity and therefore, strongly insulate the mineral soil.

Table 2. Residence times (years) for selected elements in forest floor.

	n	Biomass	N	P	K	Ca	Mg
Black spruce (taiga)	5	50 ± 5	61 ± 10	99 ± 23	74 ± 13	39 ± 3	73 ± 12
All others (taiga)	11	15 ± 3	38 ± 6	28 ± 8**	28 ± 7**	21 ± 4**	30 ± 7*
Taiga coniferous	8	33 ± 11	56 ± 6	84 ± 17	65 ± 12	36 ± 3	66 ± 10
Taiga deciduous	8	13 ± 3*	33 ± 8*	17 ± 6**	23 ± 5**	18 ± 3**	24 ± 5**
Temperate coniferous[1]	13	17	18	15	2	6	13
Temperate deciduous[1]	14	4	6	6	1	3	3

1. Temperate coniferous and temperate deciduous estimates from Cole and Rapp (1981).
**,* Difference between means for black spruce and others significant at $\alpha = 0.01, 0.05$, black spruce vs. all other taiga; taiga coniferous vs. taiga deciduous.

Considerable heat flux is required to seasonally thaw and heat the active layer of the soil profile. Consequently the densest plant rooting occurs at a shallow depth of about 14 to 20 cm (021 layer of the forest floor). Transpiration losses of mineral soil water are thus reduced.

Chemical characteristics of the forest floor also show substantial ranges among forest types examined in this study. Forest floor nitrogen and lignin concentration display correlations with temperature (r^2=0.69 and 0.45, respectively). Black spruce ecosystems display the lowest N but highest lignin concentrations compared with the other forest types. In addition, the ratios of C/N and C/P in the forest floor are significantly higher (α=0.01) in black spruce compared with the other vegetation types. Black spruce shows an average forest floor C/N of 44 and a C/P of 404. For white spruce and the hardwood types the C/N ratio ranges between 28 and 23 while the C/P ranges between 250 and 240. These ratios indicate potentially a greater demand for resources (N and P) by competing microbial and plant flora in black spruce ecosystems.

Of all the forest types strongest forest floor acidity is encountered in black spruce. The average pH ranges from 4.5 in the 01 (L) to 4.1 in the 022 (H) horizons. This compares with a range of 5.5 to nearly 6.0 in the other ecosystems. This condition is associated with two to four-fold lower concentrations of Ca and Mg respectively in black spruce forest floors. As a result base saturation is about 12 per cent of these systems compared with 44, 61 and 69 per cent in white spruce, birch and aspen, respectively.

The physical and chemical environment of the forest floor acts to modify the rate of organic matter mineralization and nutrient recycling (Figure 1; Fogel and Cromack 1977; Aber and Melillo 1980; Sharpe *et al.* 1980; Moore 1981; Melillo *et al.* 1982). For example, forest floor decomposition constants (k), calculated according to Jenny *et al.* (1949), increase with increasing N and decreasing lignin concentrations (r^2=66 for k vs. per cent N and 0.55 for k vs. per cent lignin). In addition k increases with increasing base saturation of the forest floor to a maximum of about 0.06 in birch at 60 per cent saturation, then declines to 0.04 in aspen at base saturation above 60 per cent. Moisture in addition to temperature and organic matter quality may be limiting on aspen sites. In black spruce, these environmental controls are the least favorable to support rapid mineralization of organic matter and recycling of chemical elements for tree growth. This situation is reflected in substantially longer residence times for biomass and chemical elements in the forest floor of black spruce compared with other taiga ecosystems. Values range from

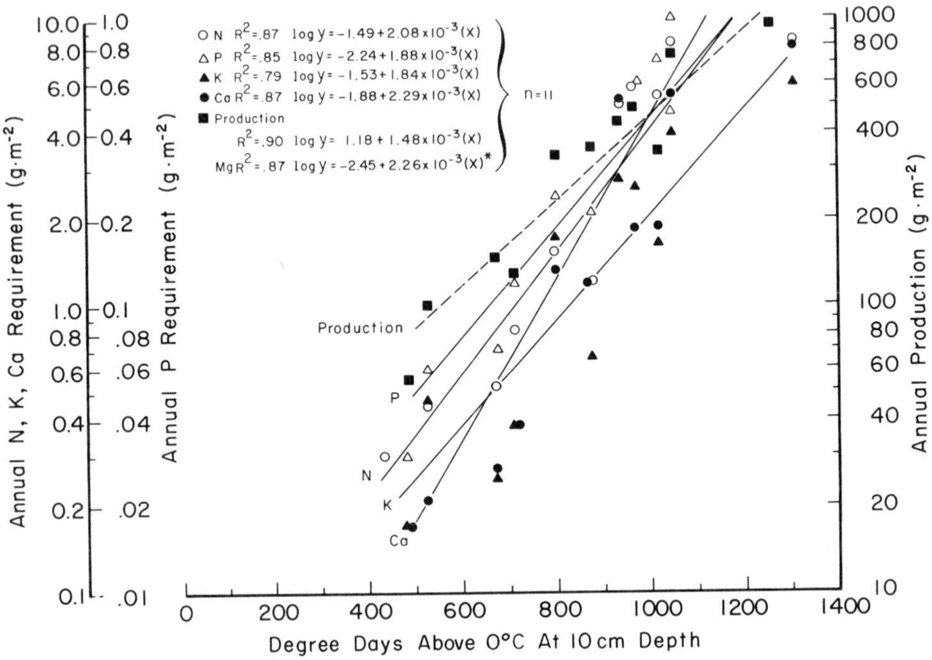

Figure 5. Relationship between soil heat sum and annual requirement by aboveground tree components for N, P, K, Ca and net annual aboveground tree production (Van Cleve *et al*. in press).

three-fold longer for organic matter (50 years vs. 15 years) to between two- and three-fold for N, P, K, Ca and Mg (Table 2). The net result of this degree of physical and chemical control on forest floor decomposition processes is the accumulation of the thick, slowly decaying organic debris (Figures 1 and 4). This further translates to reduced annual tree element requirement and production which is encountered in black spruce. Strong correlations between soil temperature, indices of forest floor organic matter quality (C/N, C/P) and annual element requirement and production support the hypothesis that black spruce is the least productive most nutrient limited taiga ecosystem (Figures 5, 6 and 7). In each case, black spruce is encountered at the lower end of the respective curves. This forest type, developing on nutrient poor substrates and under the influence of low soil temperature, has the lowest element requirement and annual production. This condition is caused by the interaction of temperature, moisture and substrate quality which restricts soil biological processes associated with element supply and tree growth. That the controls act in the synergistic manner outlined in Figure 1 is illustrated in Figures 5 through 9. In Figures 8 and 9 the ratio of biomass to N and P for litterfall and the forest floor was employed as an analog of the C/N and C/P ratios because carbon analysis was not available for litterfall samples. These relationships demonstrate that forest types developing on substrates of low quality and in which the recycling of N and P potentially is limited (such as the case of black spruce) will recycle organic matter which also is of low nutrient content. This condition reinforces the development of low tree productivity. Organic matter from which N and P may more rapidly be recycled generally is produced by the hardwood forest types. This rate of nutrient cycling supports greater forest tree productivity.

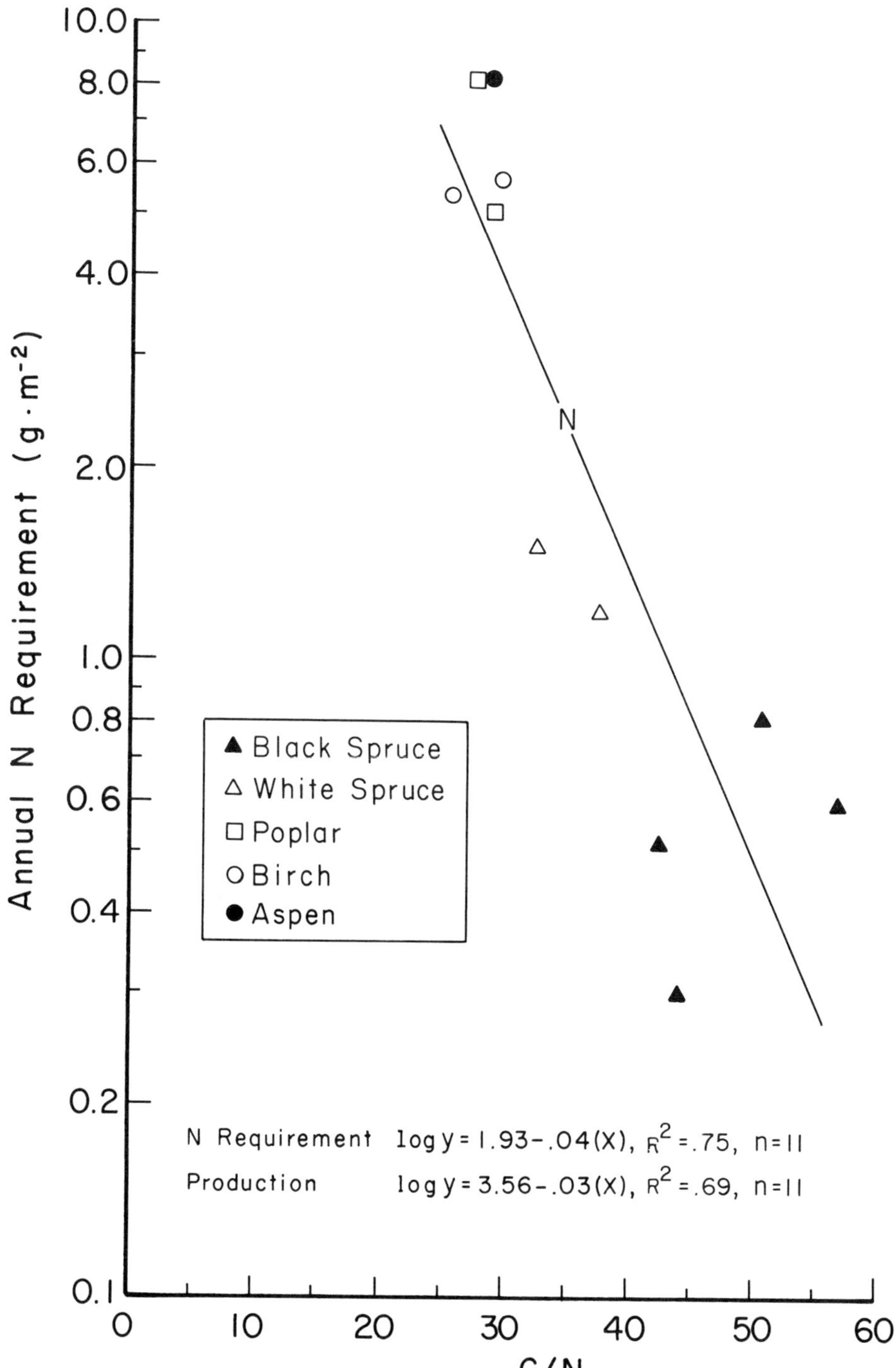

Figure 6. Annual above-ground tree N requirement compared with forest floor C/N ratio (Van Cleve *et al*. in press).

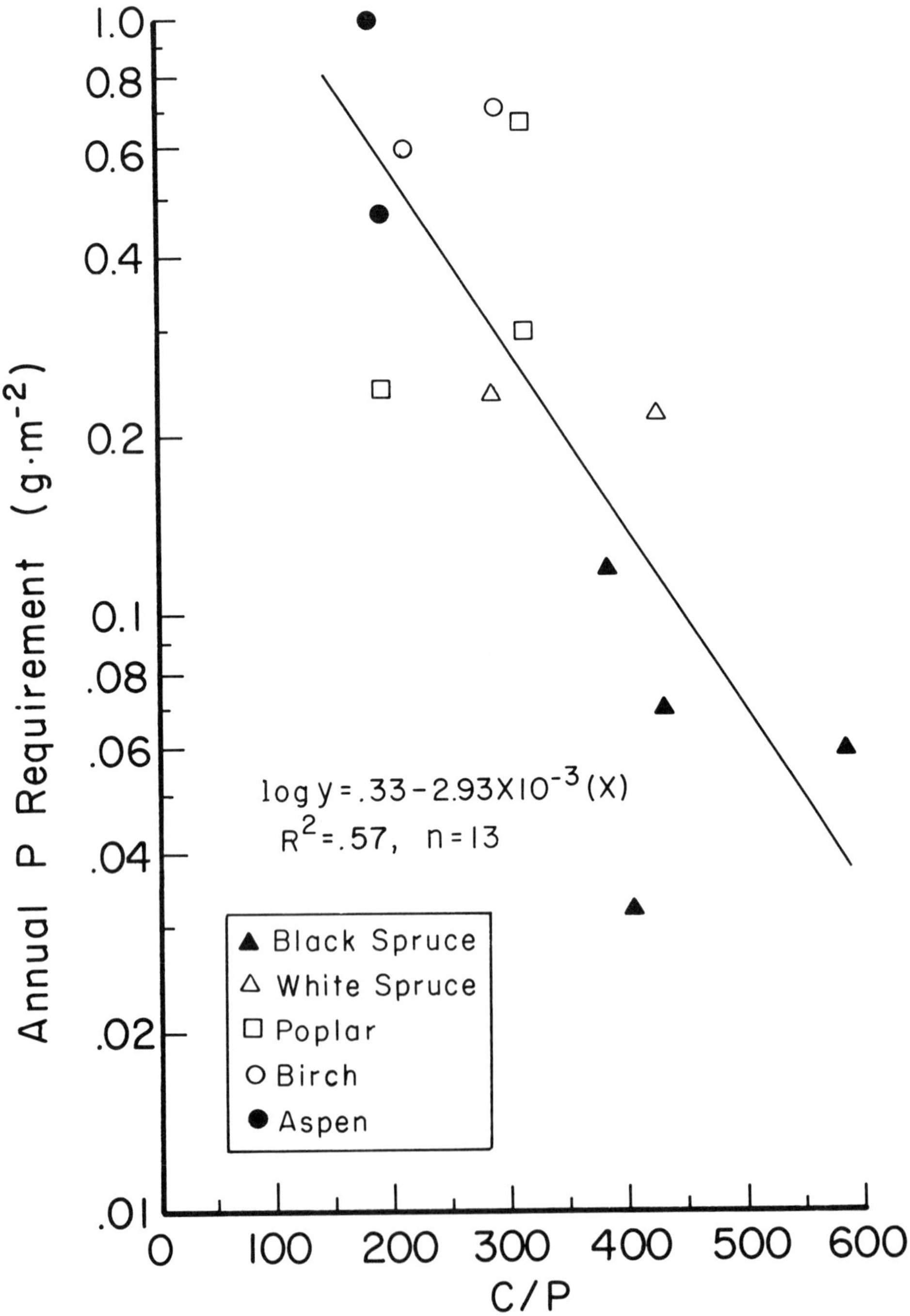

Figure 7. Annual aboveground tree P requirement compared with forest floor C/P ratio (Van Cleve *et al*. in press).

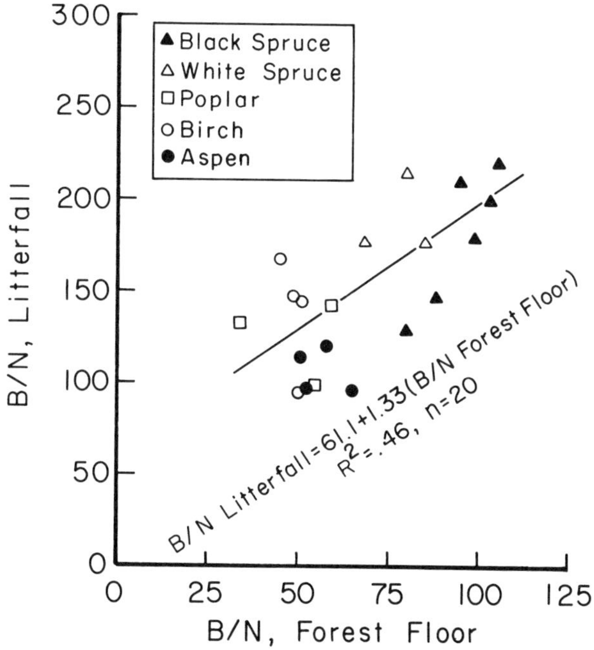

Figure 8. Biomass/N ratio of litterfall compared with biomass/N ratio of forest floor (Van Cleve *et al.* in press).

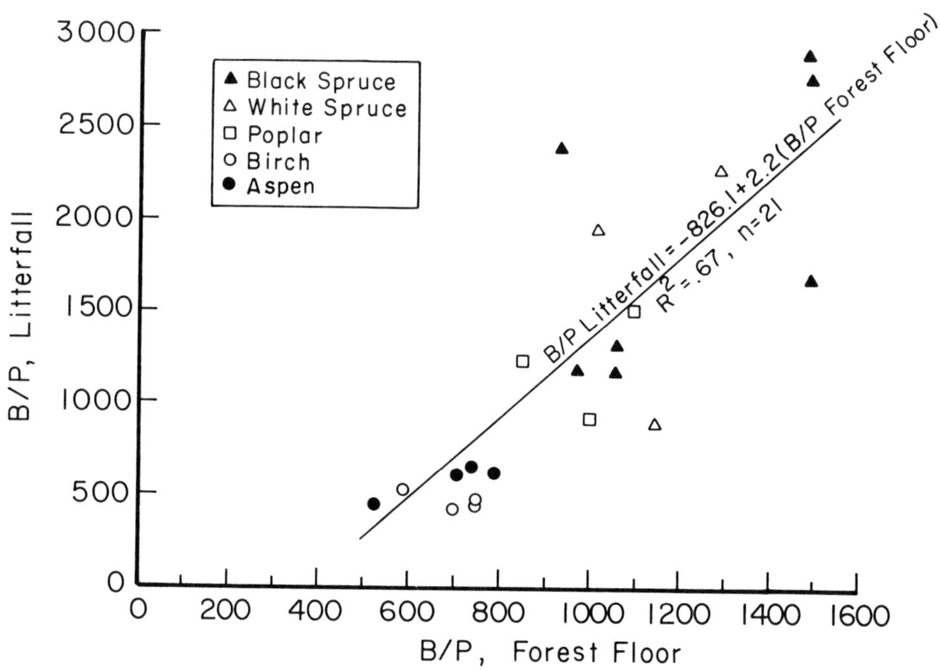

Figure 9. Biomass/P ratio of litterfall compared with biomass/P ratio of forest floor (Van Cleve *et al.* in press).

The question of dominant environmental control or controls with regard to ecosystem structure and function is difficult to establish. Correlation analysis of system processes on the basis of field data may not clearly separate interaction among supported independent variables. An example is the difficulty in separation of the temperature-moisture control from forest floor chemistry (quality) control with regard to decomposition processes, nutrient cycling and tree growth. Two experiments were conducted which separately examined the influence of soil temperature and forest floor organic matter quality on organic matter decomposition and nutrient supply for plant growth. In the first experiment, heating of the black spruce forest floor to approximately 9°C above ambient temperature for three summers resulted in a 20 per cent reduction in forest floor biomass (8360 to 6488 g/m^2). During this period the 022 horizon (H-layer) showed a significant decline in weight of 742 g/m^2. This weight change was associated with significant increases in exchangeable NH_4 and available P in the forest floor and in Kjeldahl-N in soil solution. As a result of the more favorable nutrient and temperature regimes, spruce foliage showed increased rates of photosynthesis and significantly higher concentrations of N, P and K compared with untreated trees.

In the second experiment forest floors from representative stands of six of the principal forest types were tested for ability to supply N for paper birch seedling growth. The forest types included alder (*Alnus incana*), birch, aspen, poplar, white spruce and black spruce. Organic material was uniformly mixed with sand. The potted mixtures were planted to birch seedlings and regularly supplied with a complete nutrient solution minus nitrogen. Temperature (20°C) and lighting regime (21 hour photoperiod) were constant for the experiment. This bio-assay showed strong decreasing exponential relationships between initial forest floor C/N ratios and average total seedling yield (R^2=0.96). Similar strong relationships also existed between the C/N ratios and average seedling mass of total N and mass of seedling root N (R^2=0.96 and 0.99, respectively). A further comparison between average percentage N content in seedling leaf, stem and root tissue and forest floor C/N ratios showed the same general relationship (r^2=0.80, 0.94, and 0.75, respectively).

The reduced tree productivity of black spruce ecosystems is illustrated by the fact that the standing crop of tree biomass is less than half (5 kg/m^2 vs. 10 to 12 kg/m^2) that encountered in birch, aspen and poplar ecosystems (Table 1). Net annual tree production may be three to five times greater (100 g/m^2 vs. 360 to 560 g/m^2) in white spruce, birch, aspen and poplar than in black spruce (Table 1). Moreover, an order of magnitude greater annual requirement is encountered for the respective elements in the more productive forest types compared with black spruce (Figure 10). In excess of 10 times more of N, P and K may be returned to the forest floor in these ecosystems compared with black spruce.

It would be hard to overestimate the importance of fire in shaping the vegetation on the upland of interior Alaska. The distribution of aspen and paper birch can often be traced to patterns of past fires. In the mosaic of burned conditions generally the hardwoods invade areas where the entire forest floor was consumed, whereas black spruce replaces itself in areas where a less intense fire left the forest floor intact. This mosaic of highly diverse conditions in the wake of a fire causes difficulties in our ability to generalize the effects of fire on the ecosystem. However, due to increased soil temperatures and rates of nutrient cycling the net effect of periodic fire in black spruce ecosystems appears to be a simulation of warmer, more productive sites at least for a period of 10 to 20 years after a fire.

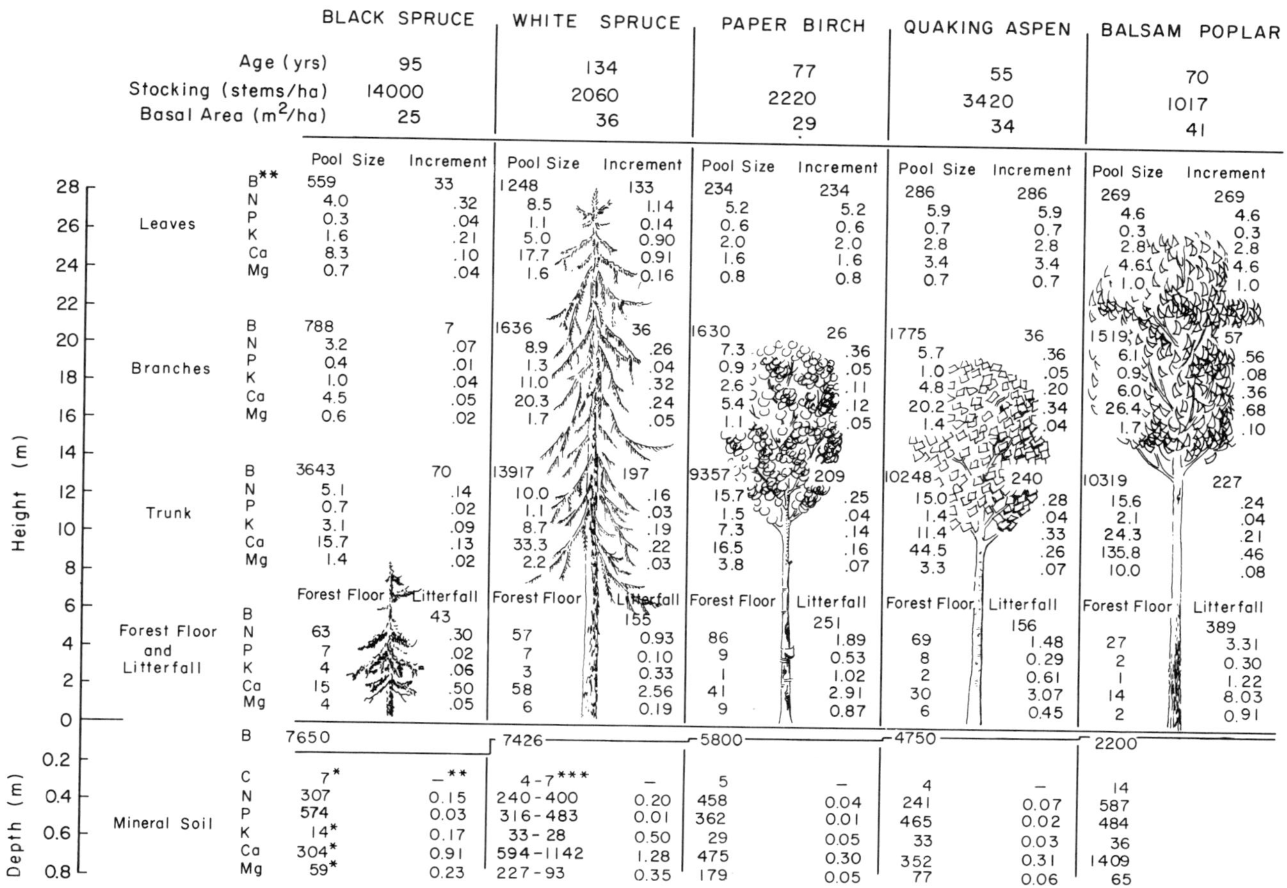

Figure 10. Distribution and flux of selected nutrients in principal taiga forest types of interior Alaska. *Carbon in kg·m^{-2}, base elements estimated from ion exchange analysis, N and P are total chemical analysis. **Left-hand figures are for uplands, right hand figures are for floodplains.

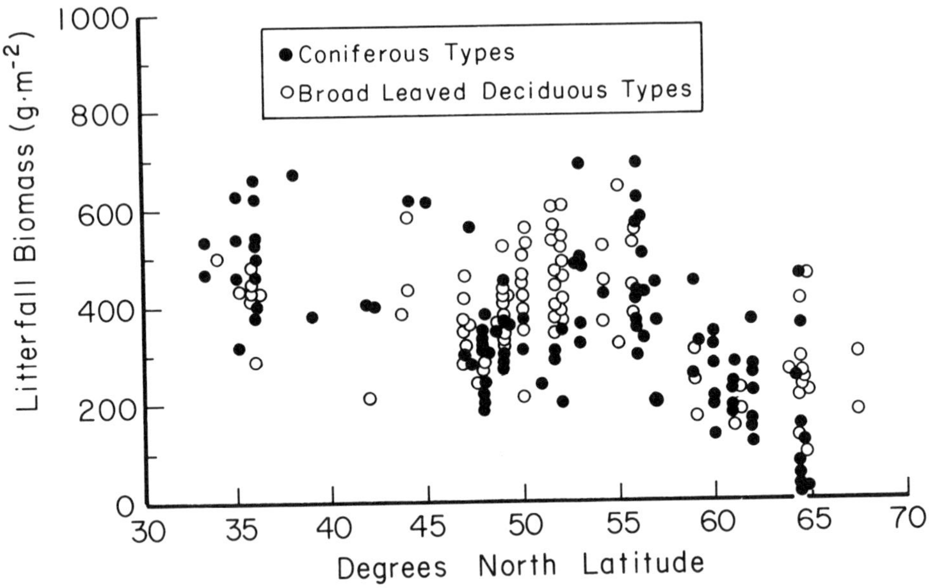

Figure 11. Forest litterfall biomass in relation to latitude. Data were abstracted from Bringmark (1977), Cragg *et al*. (1977), Lousier and Parkinson (1976), Malkonen (1977), Nihlgård and Lindgren (1977), Rodin *et al*. (1965) and Van Cleve and Noonan (1975).

Comparison With Other Temperate Forest Regions

Several indices of ecosystem function were compared among geographic regions. Residence time for forest floor organic matter and selected elements generally is substantially greater for taiga compared with temperate latitude forest types (Table 2). The values are from two to three times longer for biomass in coniferous and deciduous types respectively. Between 20- and 30-fold longer residence times are encountered for K with other elements showing from three- to eight-fold longer residence times for the taiga forest floor.

Annual litterfall biomass generally appears to increase with decreasing latitude to an average of 400 to 500 g/m^2 at 35°N (Figure 11). O'Neill and DeAngelis (1981) show a generally increasing trend for litterfall from 0° to 70°N. An average of over 1100 g/m^2/year was estimated at the equator. However, the majority of their data comes from forests north of 35°N. Closer evaluation of the trend might be obtained by adjusting for age and basal area. Average taiga litterfall biomass appears to be about 100 to 200 g/m^2 annually with a minimum of less than 20 g/m^2 and a maximum approaching 500 g/m^2.

The mass of nitrogen returned in litterfall first increased with decreasing latitude to a maximum between 8 and 9 g/m^2 around 55°N, then declines to between 3 and 4 g/m^2 at 35°N (Figure 12). The average maximum in temperate forests appears to fall between 5 and 6 g/m^2. It would be well to keep in mind, however, the limited data base employed in the figure. Average annual return of N in taiga forest litterfall appears to lie between 1 and 2 g/m^2, 16 per cent to 40 per cent of the average maximum (5 to 6 g/m^2) encountered in temperate latitude forests.

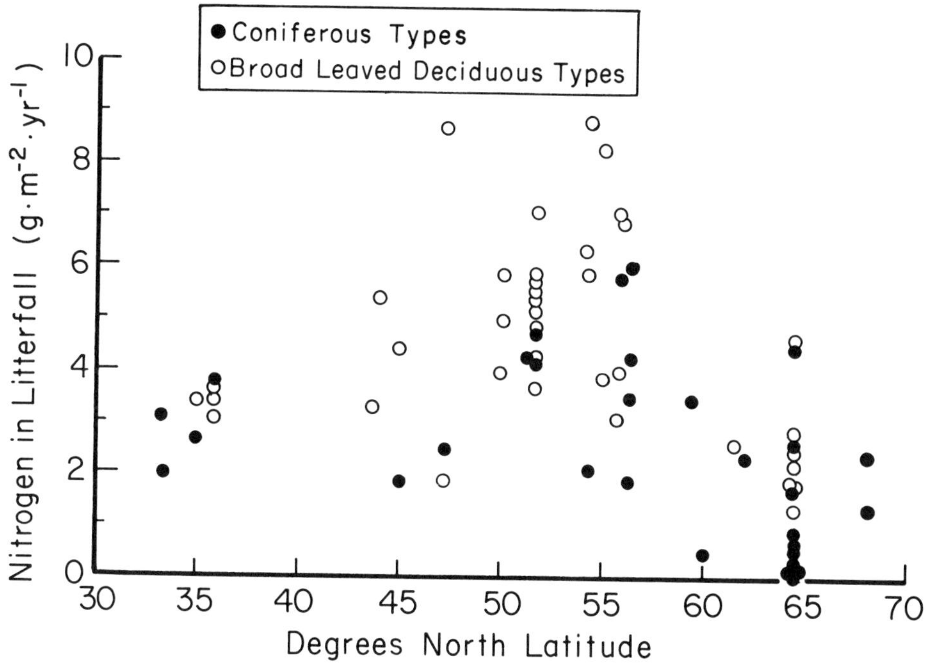

Figure 12. Mass of nitrogen in forest litterfall in relation to latitude. Data were extracted from Bringmark (1977), Cragg *et al.* (1977), Lousier and Parkinson (1976), Malkonen (1977), Nihlgård and Lindgren (1977), Rodin *et al.* (1965) and Van Cleve and Noonan (1975).

Compared with the most productive temperate latitude forests, annual tree dry matter production in taiga forests appears to range from about three per cent in low productivity black spruce to 50 per cent in productive balsam poplar. However, the most productive white spruce stands, located along the Yukon River north of Fairbanks, may annually produce aboveground 2500 g/m^2 of dry matter (Yarie and Van Cleve ms.). This exceeds total net tree production (tops and roots) of some temperate latitude pine plantations (2000 g/m^2, Cole and Rapp 1981).

Conclusions

Results of our research indicate that taiga forest productivity and nutrient cycling are directly and principally controlled by forest floor and mineral soil temperature. An important secondary control, the chemical composition of forest floor organic matter, through control of decomposition rates, modifies the flow of nutrient elements important to plant growth. Primarily in response to these controls, black spruce ecosystems display the lowest net annual tree production and smallest flux of nutrient elements of any taiga forest type.

White spruce, quaking aspen, paper birch, and balsam poplar show marked contrasts with black spruce. Annual tree biomass production in these stands is over three to five times greater than in black spruce. This large increase in tree growth rate correlates well with increases in soil temperature and consequent higher rates of organic matter and nutrient cycling.

Although we have demonstrated the general validity of our central hypothesis that forest productivity is primarily controlled by soil temperature, we have also discovered exceptions. The warmest site we studied, for example, had sharply reduced tree productivity as a result of inadequate supplies of available soil water. Such sites, however, are rare in the interior Alaskan taiga as most soils contain available water throughout much of the growing season due to frequent summer rainfall.

Our research has concentrated on the structure and function of the low productivity, cold soil black spruce type. However, we have examined ecosystem processes over a wide range of forest types and have learned much about system function in relation to site productivity and to successional sequences as well as mature vegetation types. It is important that future research be directed toward examining the structure and function of the most productive ecosystems, especially those developing on the flood plains of the major rivers.

The current research program has resulted in a substantial increase in our knowledge and development of models of forest primary production and nutrient cycling in upland secondary and river floodplain primary successional sequences (Van Cleve and Viereck 1981). The successional models provide a new framework within which future research will deal with hypothesized contrasting controls of structure and function in upland and floodplain taiga forest ecosystems.

Literature Cited

Aber, J.D. and J.M. Melillo. 1980. Litter decomposition: measuring relative contributions of organic matter and nitrogen to forest soils. Can. J. Bot. 58: 416-421.

Barney, R.J. 1971. Wildfires in Alaska — some historical and projected effects and aspects. *In*: Fires in the northern environment. C.W. Slaughter, R.J. Barney and G.M. Hansen (Eds.). Pacific Northwest For. and Range Exp. Stn. Portland, Oregon. pp. 51-59.

Bray, J.R. and E. Gorham. 1964. Litter production in forests of the world. *In*: Advances in ecological research II. J.B. Cragg (Ed.). Academic Press. New York. pp. 101-157.

Bringmark, L. 1977. A bioelement budget of an old scots pine forest in central Sweden. Silva Fennica, 11: 201-209.

Cole, D.W. and M. Rapp. 1981. Elemental cycling in forest ecosystems. *In*: Dynamic properties of forest ecosystems. D.E. Reichle (Ed.). Cambridge University Press. Cambridge, England. pp. 341-409.

Cragg, J.B., A. Carter, C. Leischner, E.B. Peterson and G.N. Sykes. 1977. Litterfall and chemical cycling in an aspen *(Populus tremuloides)* woodland ecosystem in the Canadian Rockies. Pedobiologia, 17: 428-443.

Dyrness, C.T. and D.F. Grigal. 1979. Vegetation-soil relationships along a spruce forest transect in interior Alaska. Can. J. Bot. 57: 2644-2656.

Fogel, R. and K. Cromack, Jr. 1977. Effect of habitat and substrate quality on Douglas-fir litter decomposition in western Oregon. Can. J. Bot. 55: 1632-1640.

Jenny, H., S.P. Gessel and F.T. Bingham. 1949. Comparative study of decomposition rates of organic matter in temperate and tropical regions. Soil Sci. 68: 319-432.

Lousier, J.E. and D. Parkinson. 1976. Litter decomposition in a cool temperate deciduous

forest. Can. J. For. Res. 54: 419-436.
Malonen, E. 1977. Annual primary production and nutrient cycle in a birch stand. Comm. Instit. Forest Fenniae.
Melillo, J.M., J.D. Aber and J.F. Murtore. 1982. Nitrogen and lignin control of hardwood leaf litter decomposition dynamics. Ecology, 63: 621-626.
Moore, T.R. 1981. Controls on the decomposition of organic matter in subarctic spruce-lichen woodland soils. Soil Sci. 131: 107-113.
Nihlgård, B. and L. Lindgren. 1977. Plant biomass, primary production and bioelements of three mature beech forests in south Sweden. Oikos, 28: 95-104.
O'Neill, R.V. and D.L. DeAngelis. 1981. Comparative productivity and biomass relations of forest ecosystems. *In*: Dynamic properties of forest ecosystems. D.E. Reichle (Ed.). Int. Biol. Pro. 23. Cambridge University Press. Cambridge, England. pp. 411-449.
Quirk, W.A. and D.J. Sykes. 1971. White spruce stringers in a fire-patterned landscape in interior Alaska. *In*: Fires in the northern environment. C.W. Slaughter, R.J. Barney and G.M. Hansen (Eds.). Pacific Northwest For. and Range Exp. Stn., Portland, Oregon. pp. 179-197.
Rieger, S., D.B. Schoephorster and C.E. Furbush. 1979. Exploratory soil survey of Alaska. U.S.D.A. Soil Conservation Service.
Rodin, L.E. and N.I. Bazilevich. 1965. Production and Mineral Cycling in Terrestrial Vegetation. English Translation, G.E. Frogg (Ed.). Oliver and Boyd, Edinburgh and London. 1967.
Rodin, L.E., N.I. Bazilevich and N.N. Rozov. 1975. Productivity of the world's main ecosystems. *In*: Productivity of world ecosystems. D.E. Reichle, J.F. Franklin and D.W. Goodall (Eds.). National Academy of Sciences. Washington, D.C. pp. 13-26.
Selkregg, L.L. 1975. Alaska regional proviles — Yukon region. University of Alaska Arctic Environmental Information and Data Center.
Sharpe, D.M., K. Cromack, Jr., W.C. Johnson and B.S. Ausmus. 1980. A regional approach to litter dynamics in southern Appalachian forests. Can. J. For. Res. 10: 395-404.
Skre, O. and W.C. Oechel. 1981. Moss functioning in different taiga ecosystems in interior Alaska. I. Seasonal, phenotypic and drought effects on photosynthesis and response patterns. Oecologia, 48: 50-59.
United States Department of the Interior, Bureau of Land. 1978. Fire statistics. The Land, 6. 24 pp.
Van Cleve, K., R. Barney and R. Schlentner. 1981. Evidence of temperature control of production and nutrient cycling in two interior Alaska black spruce ecosystems. Can. J. For. Res. 11: 258-273.
Van Cleve, K., C.T. Dyrness, L.A. Viereck, J. Fox, F.S. Chapin, III and W. Oechel. 1983. Taiga ecosystems in interior Alaska. BioScience, 33: 39-44.
Van Cleve, K. and L.L. Noonan. 1975. Litterfall and nutrient cycling in the forest floor of birch and aspen stands in interior Alaska. Can. J. For. Res. 5: 626-639.
Van Cleve, Keith, Lola Oliver, Robert Schlentner, Leslie A. Viereck and C.T. Dyrness. 1983. Productivity and nutrient cycling in taiga forest ecosystems. Can. J. For. Res. Vol. 13.
Van Cleve, K. and L.A. Viereck. 1981. Forest succession in relation to nutrient cycling in the boreal forest of Alaska. *In*: Forest succession, concepts and application. D.C. West, H.H. Shugart and D.B. Botkin (Eds.). Springer Verlag. New York. pp. 185-210.
Viereck, L.A. and C.T. Dyrness. 1980. A preliminary classification system for vegetation of Alaska. U.S.D.A., For. Serv., Gen. Tech. Rept. PNW-106.

Viereck, L.A., M.J. Foote, C.T. Dyrness, K. Van Cleve, D. Kane and R. Seifert. 1979. Preliminary results of experimental fires in the black spruce type of interior Alaska. U.S.D.A. For. Serv., Pac. Northwest For. and Range Exp. Stn., Portland, Oregon. Res. Note, PNW-332. 28 p.

Viereck, L.A. and L.A. Schandelmeier. 1980. Effects of fire in Alaska and adjacent Canada — a literature review. U.S. Dept. Interior, Bureau of Land Management, BLM — Alaska Technical Report 6. 124 p.

Viereck, L.A. and Keith Van Cleve. In press. Some aspects of vegetation and temperature interactions in Alaska forests. *In*: The potential effects of carbon dioxide-induced climatic changes in Alaska. The Proceedings of a Conference. Fairbanks, Alaska, April 7, 8, 1982. School of Agriculture and Land Resources Management, University of Alaska, Fairbanks. Miscellaneous Publications.

PARKS CANADA IN THE BOREAL FOREST ECOSYSTEM (A PILGRIM'S PROGRESS)

N. Lopoukhine
Natural Resources Division, National Parks Branch
Parks Canada, Ottawa, Ontario, Canada

"Every species and every society represents a column of pilgrims not only travelling in space, but also in time."

J. Rousseau (1966)

Abstract: Canada's national parks are presented in the guise of an itinerant 'pilgrim' on a perpetual journey. Changes witnessed during his travels, reflect society's outlook on the role of national parks in the conservation of natural resources. Within the ten boreal forest national parks, individual management objectives vary but the ten parks share a common Act, Policy and finally management guidelines. Each of these are designed to steer the national park pilgrim along the true path, conservation. In this context, future challenges facing the national park pilgrim are explored. The opportunities for research and the potential contributions to the understanding and management of the boreal forest ecosystem within these ten parks are detailed while also presenting the framework for carrying out such research. A common thread in the opportunities for research is the role of fire. The steps taken by Parks Canada towards the management of fire are presented. In conclusion, the responsibility of Canadians and in particular researchers who are in a position to affect the future journey of the national park pilgrim is stated.

Columns of species moving and changing over time are perhaps easier to visualize than their alloted space. It is not, of course, the finiteness of space which is in question here but rather the quality of the space which collectively is definable as a species habitat. The difficulty in perception arises principally when one considers that this space, the habitats, has been altered by man at such a rate that in historic times, man is considered as the principal cause of species extinctions (Frankel and Soulé 1981).

Yet, the society which is causing these alterations to take place has also set aside areas designated as national parks. This, plus the changes in the statutes and policies governing such areas reflect the changes in our society insofar as resource conservation is concerned. National parks are a barometer of these attitudes or, in the context of the Rousseau quote, are, as a societal institution, also a pilgrim on a journey.

From the perspective of resource conservation, the journey, just begun, is for perpetuity since according to governing legislation a park's purpose includes the protection of values for future generations of Canadians (Canada 1974). This alone provides incentive to ensure that the journey is not interrupted or that the pilgrim is not tempted, coerced or forced from the 'narrow and righteous' path of conservation.

Through both hindsight and foresight, mechanisms to prevent straying from this path, have been developed. There is added urgency caused by the continued space alterations by man outside of national parks. As a result, national parks are rapidly becoming the sole purveyors of unaltered space where the columns of species pilgrims have a hope of continuing on their journeys unabated. The added incentive then is to reduce the rate at which columns of species have been truncated (made extinct) in historic time. The importance as well as the adequacy of national parks and other protected areas in this role and in particular genetic pool conservation is well treated by Roche (1974), Soulé and Wilcox (1980), and Frankel and Soulé (1981), among others.

This paper, focusing on the boreal forest national parks traces the formative years of the national park pilgrim and the development of the mechanisms to keep the pilgrim on the right path. The importance of research in this context and the framework for research in national parks is detailed. Finally, one active resource management initiative, fire management, is discussed because it is particularly relevant to the boreal forest national parks.

National Parks in the Boreal Forest

There are ten parks situated in the boreal forest region as defined by Rowe (1972). Table 1 identifies these parks by name and size while Figure 1 situates them within this region. These ten parks are obviously different in individual size and management objectives and the species mix they contain and thus protect. The total area contained within these ten parks approximates the size of such European countries as Portugal and Hungary.

The boreal forest ecosystem, a relatively species-poor biome, is characterized by a north-south and east-west continua. The representation of this ecosystem within Canada's parks, though impressive area-wise, falls short in capturing all of its variations. One measure of this short fall is that the current ten parks represent only 15 of the 48 Boreal Forest Sections delineated within the boreal forest region by Rowe (1972). Another such measure, of more pertinence to this paper, is the representation within the Natural Regions defined by Parks Canada. There are 39 such terrestrial regions defined principally according to physiographic and ecological, geographical and geological considerations (Parks Canada 1972). Of these regions, 19 occur in the boreal forest region and the ten boreal forest national parks represent but eight (Figure 2). Two additional areas (in natural regions 9 and 17) set aside by Orders-in-Council seem likely to become national parks once negotiations are completed.

Table 1. National Parks in the boreal forest

National Parks	Location	Year Established	Square km.	Visitation 1980-1981 (F.Y.)
Elk Island	Alberta	1913	194.3	422,637
Wood Buffalo	Alta., N.W.T.	1922	44,807.0	———
Prince Albert	Saskatchewan	1927	3,874.6	233,734
Riding Mountain	Manitoba	1929	2,975.9	839,329
Terra Nova	Newfoundland	1957	496.5	467,141
Forillon	Quebec	1970	240.4	727,179
Gros Morne	Newfoundland	1970	1,942.5	192,903
Kluane	Yukon	1972	22,015.0	28,914
Nahanni	N.W.T.	1972	4,765.6	———
Pukaskwa	Ontario	1978	1,877.8	———
Total			83,089.6	

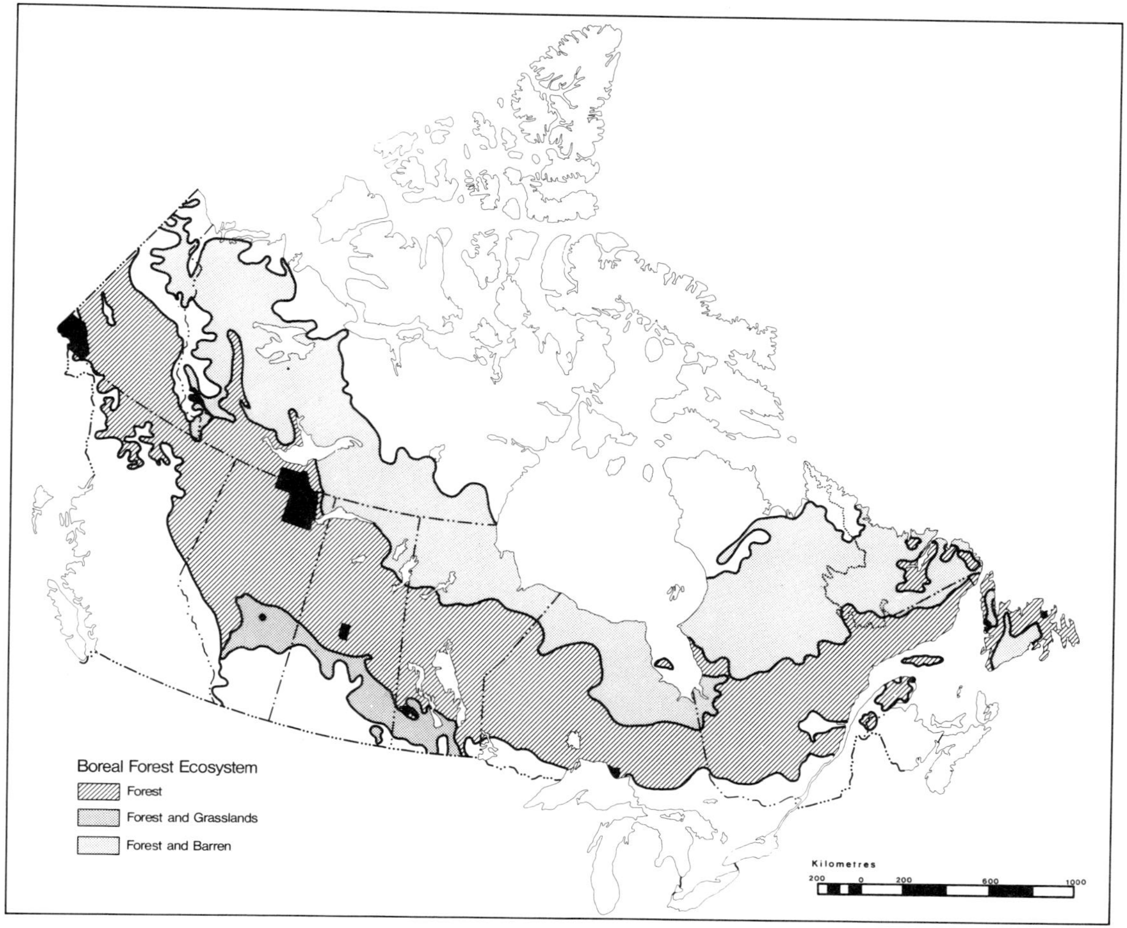

Figure 1. Location of National Parks within the boreal forest region defined by Rowe (1972).

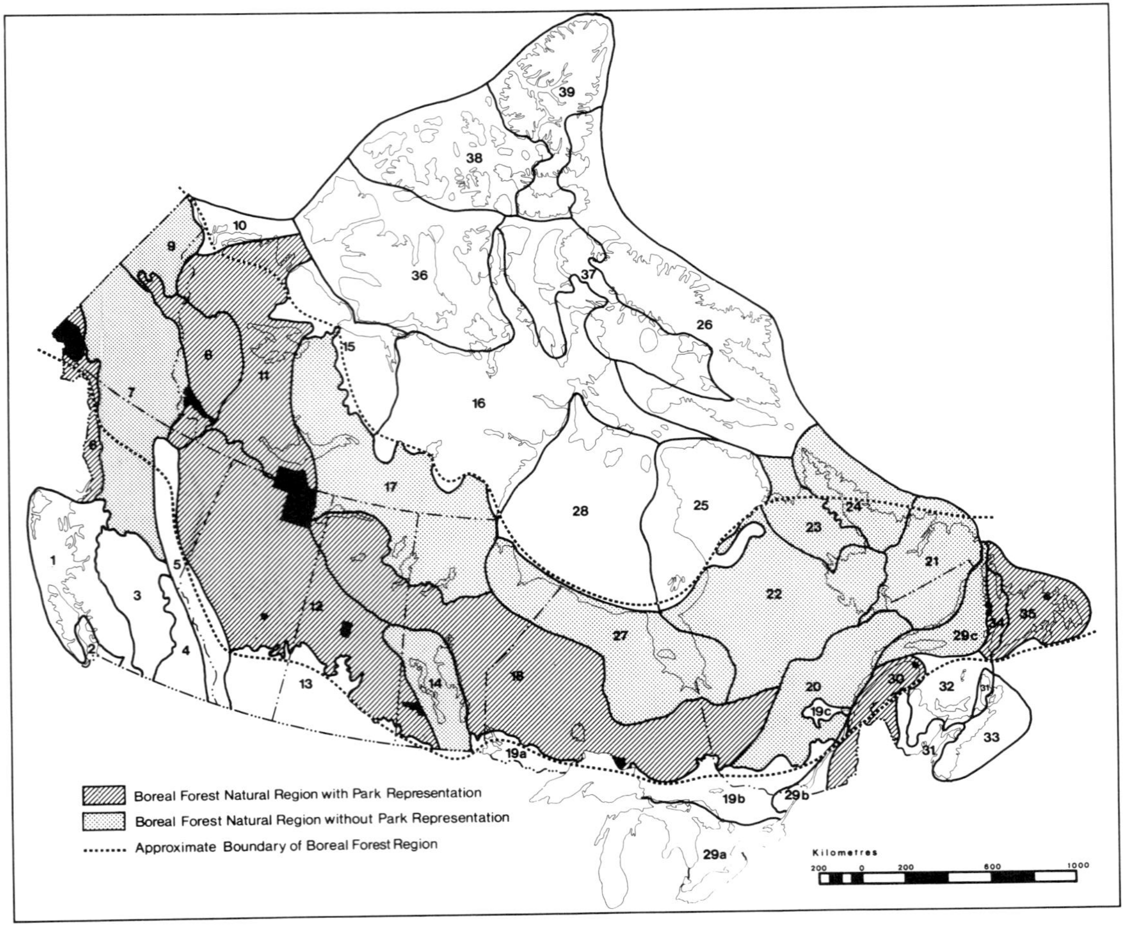

Figure 2. Boreal Forest natural regions with or without National Park representation.

To date, genetic variability within an ecosystem such as the boreal forest has not been taken into consideration in park system planning. However, the role of parks in achieving genetic pool conservation objectives is growing in recognition. Thus, as the genetic variability and values become better known, this parameter will undoubtedly gain greater importance in park system planning.

Historical Notes (The Pilgrim's Initial Steps)

National Parks will be celebrating their first centennial in 1985. When the Banff Springs were set aside in 1885 by an Order-in-Council the reasoning for this action was principally economic. In fact, the declaration of this area as a park two years later was supported with the following:

> ". . . if carefully managed, it will more than many times recuperate or recoup the Government of any expenditures" (Canada 1887).

During the next 31 years a total of 14 parks were established. Economic determinism led by tourism development and some resource exploitation dominated the purpose of these parks. Resource conservation was such that systematic predator elimination programs were common and fires were suppressed.

Changes were forthcoming as the Commissioner of Dominion Parks signaled in his 1929 annual report when he wrote

> ". . . (when the) National Park movement was in its infancy, it was perhaps necessary to emphasize this aspect (economic) of their value. Today, when this is so widely recognized it is permissible to point out that there are other values, equally if not more important in the long run.".

In the next year, the National Park Act was passed which introduced the use without abuse purpose of national parks. In the subsequent 30 years, the formative years of the pilgrim, exploitive activities such as mineral extraction, grazing and commercial forestry were prohibited in parks but entertainment, recreation and other essentially consumptive uses were condoned.

Over the course of time, transgressions did occur. However, these were useful to point out the error of the pilgrim's ways and to provide the impetus by which such temptations could be resisted. These include minor National Park Act revisions in the '50s, and in the mid '60s, the adoption by Cabinet of a policy on National Parks which called for master planning within parks accompanied by public hearings. It is this latter item perhaps which gave rise to the currently strong conservation ethic advocated by national parks. During one such public hearing a lack of knowledge regarding the natural resources of individual parks becomes evident. In response, inventories of national parks' natural resources using an integrated approach (now called Ecological Land Survey) were launched (East *et al.* 1978). Concurrently, development of processes by which planning and management of these park resources could be undertaken were initiated. The resulting two processes, Natural Resources Management Process and the Park Planning Process, are illustrated in Figure 3. It is not within the scope of this paper to expand on these processes. Suffice it to say that they transcend benign neglect and are based upon the essential principles of gathering data first, analysing them, then formulating action plans and implementing these while allowing an appropriate series of checks for adjustment throughout the process (Parks Canada 1978, 1980).

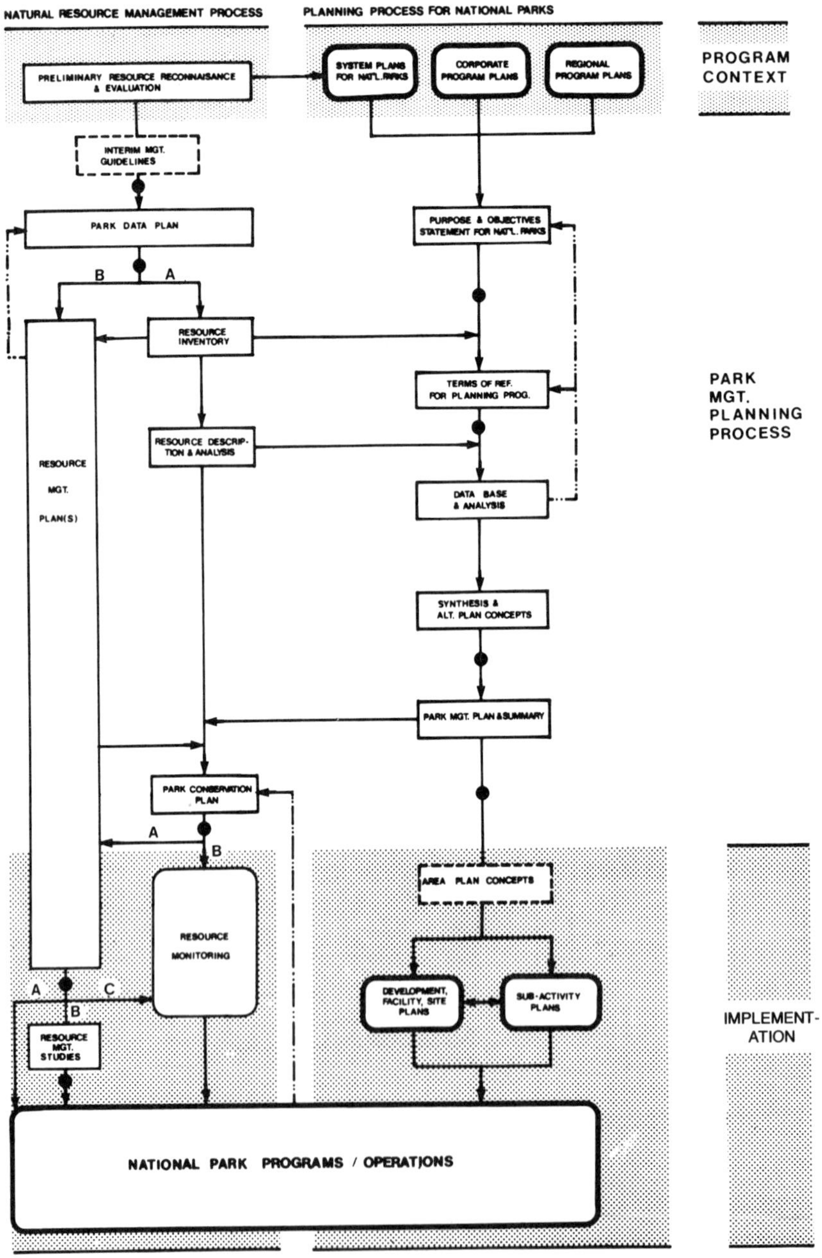

Figure 3. Natural resource management process and planning process for National Parks. Rounded rectangles represent programs. Rectangles represent products or optional products (dashed lines). Solid circles represent approval requirements. Solid lines and arrows represent flows of process; dashed lines represent feedback pathways; letters A, B & C represent pathway options. Evaluation and monitoring occur at all stages of Park Management Planning Process and Implementation. Natural Resource Management Process is one of the many inputs into the Planning Process for National Parks.

Attention is nevertheless focused here on these processes because they provide the current framework for planning and resource management in national parks. As such they are essentially the driving mechanisms for actions within national parks. They implicate everyone from the senior managers who approved these processes and now participate at various decision points along the processes (dots between steps in Figure 3) to the individual park staff member who will implement approved plans. It is this commitment to natural resource management which bodes well for the future progress of the pilgrim along his journey.

The approval of the Parks Canada Policy by Cabinet in 1979 marked another major step in the journey of the national park pilgrim. This policy states clearly that, within national parks, preservation of park values for future generations is paramount. Furthermore, natural processes such as fire, insect infestations and disease outbreaks are given recognition in that whenever possible, they should be permitted to continue unhindered. These natural processes are now rightfully recognized as a national heritage worthy of protection. They have been included as one of the columns of pilgrims.

These "historical notes" provide a rough sketch of the road the national park pilgrim has travelled. These initial 100 years are important as they have defined the pilgrim's shape and function. To have achieved this current stature is perhaps more impressive if one considers the following resource levels. The 1981-1982 main estimates Parks Canada program consists of $245 million and 5,054 person years. The National Park Activity comprises $153 million and 2,566 person years. Of these resources, the Conservation Subactivity totals $15 million and 390 person years which includes 190 full time park wardens and a maximum of 174 seasonal wardens.

The point of bringing these numbers forward is to show that those in national parks who are ensuring conservation or use without abuse are not as numerous nor are they on an equal economic basis to those promoting use without abuse. This is not to be construed as criticism but simply to point out that, while the national park pilgrim has grown in the first 100 years of the journey and achieved placing the necessary mechanism required to ensure a righteous journey along the man-dated path of conservation, the actual work force supporting this endeavour and associated funding is low and will undoubtedly require bolstering.

National Parks: A Place for Research and Researchers

Increases in human and fiscal resources alone will not keep national parks on the 'true' path of conservation. First of all, information gaps will need to be filled in order that strategies with a reasonable certainty of success can be formulated. To do this, obviously, requires research and theory testing which in their own right are resource drains. Such resource expenditures will need to become commonplace particularly when management of natural resources increasingly abandons the laissez-faire or benign neglect approach and initiates an holistic active resource management approach for purposes of achieving specific objectives such as restoring suppressed processes, protecting endangered species or fulfilling a vegetation plan.

The active approach refers to the manipulation of natural resources to achieve an approved objective. Fire suppression, a form of active resource management is not included in this definition whenever the objective is simply to extinguish the fire. Currently, in the boreal forest national parks, except in Elk Island National Park where fire is used to achieve resource objectives, resource management is not yet in the active mode. It is one of the next challenges facing the national park pilgrim.

Another challenge is the integration of research into park management practices. Research within national parks is recognized as a key to proper park management. The Parks Canada policy clearly states, "Parks Canada will conduct and encourage research for, the identification, protection, understanding and use of Canada's heritage resources". Furthermore the Policy eloquently encourages baseline research within national parks; "Parks Canada will encourage basic research which may expand man's knowledge of his world and which requires study of natural or cultural resources protected in the Parks Canada system" (Parks Canada 1979).

For the purpose of this paper, managers of boreal forest national parks were canvassed for natural resource management issues requiring research. A summary of the responses received plus others are given below under appropriate headings.

Fauna

— cycles, role of fire, logging and trapping, habitat requirements (J. Barlow, personal communication, 1982)
— the "neglected" fauna (invertebrates, amphibians, reptiles) usually in the lower trophic levels need attention (F.A. Bamber, personal communication, 1982)
— decomposers are also an unknown

Flora

— successional issues
— role of fire and logging (Barlow, personal communication, 1982)
— fire's role in checking *Kalmia angustifolia* invading spruce stands in Terra Nova National Park (K.H. Deichmann, personal communication, 1982).

Processes

— fire as a management tool (D.A. Allan and F.A. Bamber, personal communication, 1982)
— comparative fire histories (J. Barlow, personal communication, 1982)
— role in recycling nutrients (J. Barlow, personal communication, 1982)
— correlation of infestations with fire histories (J. Barlow, personal communication, 1982)
— impacts of fluctuations on ecosystem components (J. Barlow, personal communication, 1982).

Management techniques

— defining easily perceived benchmarks useful to gauge air, water and site quality changes (F.A. Bamber, personal communication, 1982).
— remote sensing techniques useful for monitoring and updating existing data bases (F.A. Bamber, personal communication, 1982).

Decision taking

- — development and integration of ecosystem models (preferably based on existing ecological land classification data) with the natural resource management process thus facilitating decision making. (F.A. Bamber, personal communication, 1982).
- — quantification of resource inventory data into a format useful for managing park uses. (P.A. Benson, personal communication, 1982).

This list identifies a number of crucial knowledge gaps which hamper sound natural resource management decision making in the boreal forest parks. Moreover, it clearly identifies that park managers are cognizant of the importance of research. This latter point is perhaps the most important as it, along with the Parks Canada Policy regarding research, indicates that researchers are welcome within Parks Canada.

The intrinsic advantages of national parks for study sites coupled with the extensive data collected on individual parks should provide added incentive for a researcher to conduct research within a national park. These data are on individual ecosystem components as well as ecosystems, the latter in an ecological land classification format. It is unlikely that a comparable area in Canada's boreal forest is as extensively documented. As a final incentive, cooperative research can be an obvious economic benefit to both national parks and the researcher.

The invitation set out to researchers above must be placed in a proper framework. It has been suggested that researchers impose a tax on an already busy field staff and usually with a minimal return (P.A. Benson, personal communication, 1982). This important point was not expressed to hinder research but rather to encourage, if not pragmatic research *per se*, then the transmittal of results and in a form useful to park management.

To complete the framework within which research is conducted in national parks a repeated reference to the Natural Resource Management Process is in order. Specifically, there are two critical steps in the Process which define the priority for research within individual parks: the resource description and analysis step, which synthesizes existent natural resource data and identifies information gaps; and the conservation plan step which assigns priority to studies required to complete specific management objectives. Any one contemplating research within a park must consult the two documents which emanate from these two steps.

Finally, research in national parks is currently monitored through the issuance of a collection permit. A management directive on the issuance of research and collection permits in national parks is in preparation.

The Future Journey

Returning for a moment to the allegory of a pilgrim's progress it is appropriate to dwell on the future journey of the national park pilgrim. Above we reviewed the formative stages of the pilgrim where mechanisms to keep the pilgrim on the proper (conservation) route were developed. It was also noted how, in the boreal forest in particular, the national park pilgrim expanded representation. This is in keeping with the historical phase of reserve establishment out of which society appears to be easing (Frankel and Soulé 1981). Indeed in Canada one-third of the National Park System was established in the 1970s and the boreal forest complement was doubled (see Table 1). Accordingly, the next phase will most likely concentrate on ensuring the protection of the resources already in reserves. Our pilgrim's future journey will consist of meeting this challenge principally by utilizing the mechanisms developed in the formative years.

These mechanisms will require periodic review and adjustment as the pilgrim is forced out of a reactionary mode of resource management and into an active mode. The inevitability of this is evident and perhaps can be illustrated as follows. Roche (1974) wrote ". . . the boreal forests of Canada are no more immune to man's intervention than the tropical rain forests of Africa, and the day is rapidly approaching when natural forest ecosystems, undisturbed by man, will be found only as remnants in national parks and other protected areas". In 1978, the Forestry Branch of the Saskatchewan Department of Tourism and Renewable Resources launched a tree improvement program and established additional nurseries for the purpose of increasing forest renewal. For a white spruce *(Picea glauca)* seed source they turned to Prince Albert National Park, since within the province only this protected area contained the tall and straight phenotypes sought. Apparently, elsewhere in the province evident sources exhibiting these traits were eliminated through repetitive saw timber cuts since the 1800s (Anonymous 1978). Thus, in one part of Canada's boreal forest, Roche's predicted day came to pass in only four years.

The above example can be construed as a benefit of passive conservation biology. National Parks, by protecting, (from the forester) the forester's prototype of a best tree, were able to provide assistance to the forester at a later date. In this instance, the protection was of phenotypic traits as the genotypic traits of those specific individuals and of other boreal species are only now beginning to be investigated (albeit economically spurred). Also, it is important to note that there was no specific objective of protecting this phenotype by Parks Canada. It happened, because of the longevity and successional status of white spruce. Should a need arise for a genetic source for a rarer species, or one which is a pioneering species, we may not be as fortunate.

It is feasible that specific management plans designed to propagate a boreal plant species as a measure of conservation may be adopted by a park. However, the most likely approach is safeguarding the occurrence of natural processes and the reintroduction of processes where they have been excluded. Through these means the complement of species common to the park area will develop and change over time and, most importantly, in concert with their normal complement of pests and pathogens. The added benefit of course is the retention of the high genetic variability common to wild stock relative to the cultivars used in forest plantations.

In Canada's boreal forest, plantations are still uncommon. What is common are extensive clear cuts left for natural regeneration to restock. Subsequent plant successions are often manipulated on the economic and usually fertile sites. The implications of these forestry practices in the long term on the genetic resources of the boreal forest, at this point in time, can only be speculative. Nevertheless, since a difference exists, no matter how comparable clear cuts and fire are made out to be, an impact on an otherwise fire evolved ecosystem must be expected.

Partly, to ensure that fire continues its role in developing the genetic variability of Canada's boreal forest, Parks Canada is committed to implementing its fire policy. In summary, it is to have fires burn whenever feasible and reintroduce fire where man has altered the natural balance through fire's exclusion. The activities in this regard within Canada's boreal forest national parks has been reviewed by Alexander and Dubé (1983).

Briefly, some of these activities are described below. A guiding philosophy on fire in national parks was published by Parks Canada (Van Wagner and Methven 1980). Prescription burns have been carried out in some parks and most boreal parks have commissioned fire history studies. These, and the appointment of the author as the national fire policy coordinator, are preparatory steps to a full fledged implementation of the fire policy.

There are three other co-dependent initiatives designed to serve the overall objective of utilizing fire in national parks as a tool for achieving natural resource management objectives. These initiatives consist of the following:

Education

— Training in the ecological principles of fire, the business of fire management with particular emphasis on fire suppression capability, and the management of fire effects are priorities for Parks Canada staff. Hopefully most of the opportunities to acquire such education will be made available in Canada.

— External audiences, mostly laymen, will be targeted for promotion of the idea that fires can be beneficial from the perspective of safety as well as ecology. The onus of this lies not only with Parks Canada but all those versed in fire ecology as well as those in the business of fire management.

Research

— Parks Canada has designated Kluane and Nahanni as pilot parks within which fire experiences will be accumulated and later transferred to the system as a whole.

Operational considerations

— The quality of fire data recording across the system is highly variable. A national standard complementary to national and provincial needs will be introduced in the near future. Weather data pertinent to fires is a particular need that requires attention.

— The development of vegetation plans for individual parks is perhaps the most difficult consideration. It must include objectives for the whole park system within an ecosystem the size of the boreal forest in Canada as well as for an individual park. The former must determine which boreal forest parks will represent what part of the boreal forest vegetation continuum. Per park, the objectives must include vegetation mixes complementary to faunal, visitor services and other park objectives as well as for vegetation (a resource unto itself) objectives. Within these objectives, fire's role requires definition.

— Finally, and to bring all these initiatives together a formalized strategic plan for the implementation of the fire policy which would define resource requirements, dates and specific objectives is envisioned as an eventual requirement.

Conclusions

It is once again *à propos* to bring forward the societal institution of national parks in the guise of a pilgrim. John Bunyan's original pilgrim culminated his journey when trumpets sounded marking the arrival into heaven. The national park pilgrim differs in that his journey is for perpetuity and thus no trumpets will sound to mark the journey's end. Nevertheless, it is plausible to imagine a real achievement of sorts if the pilgrim does not waver from the true path (conservation) and continues to provide that needed space for the column of species and processes to proceed on their journey relatively unabated.

Failing this, the pilgrim's journey will become marked by transgressions committed while away from the true path. The historic evidence of this is quite clear. The future implications are graver since it will mark the conscious decision by society that the existence of ecosystems, both impoverished and vulnerable to disaster is acceptable for future generations. What happens to and within national parks is the responsibility of all Canadians and in particular the researcher who is in a position to affect the future of national parks.

Acknowledgements

Assistance in developing the list of priority research topics came from D.A. Allan, Prince Albert National Park, F.A. Bamber, Elk Island National Park, J. Barlow, Wood Buffalo National Park, P.A. Benson, Western Regional Office, Parks Canada, Calgary and K.H. Deichmann, Terra Nova National Park. A critical review of an earlier draft was provided by staff of the Natural Resources Division, Hull and of the Resource Conservation Division, Ontario Regional Office, Parks Canada, Cornwall.

Literature Cited

Alexander, M.E. and D.E. Dubé. 1983. Fire management in wilderness areas, parks and other nature reserves. *In*: The role of fire in northern circumpolar ecosystems. Ross W. Wein and David A. MacLean (Eds.). SCOPE 18, John Wiley and Sons, Chichester, England. pp. 273-297.

Annual Report of the Commissioner of Dominion Parks. 1912-1934. Department of the Interior. Ottawa, Ontario.

Anonymous. 1978. Press Release, Information Directorate, Parks Canada. Winnipeg, Manitoba.

Canada. 1887. Parliament, House of Commons Debates. Vol. 2.

Canada. 1974. National Parks Act, R.S., C 189, S.I.

East, K.M., D.L. Day, D. LeSauteur, W.M. Stephenson and L. Charron. 1978. Parks Canada application of Biophysical Land Classification for Resources Management. *In*: Application of ecological (biophysical) land classification in Canada. Proceedings of 2nd Meeting of CCELC, Lands Directorate, Environment Canada. Ottawa, Ontario. pp. 209-219.

Frankel, O.H. and M.E. Soulé. 1981. Conservation and Evolution. Cambridge University Press, Cambridge, England. 379 pp.

Parks Canada. 1972. National Parks System Planning Manual. Ottawa, Ontario.

Parks Canada. 1978. Planning Process for National Parks. Ottawa, Ontario.

Parks Canada. 1979. Parks Canada Policy. Environment Canada, R 62-109/1979. Ottawa, Ontario. 69 pp.

Parks Canada. 1980. Natural Resources Management Process manual. Ottawa, Ontario.

Roche, L. 1974. Preface. *In:* Ecological Studies: Analysis and Synthesis. J. Jacobs, O.L. Lange, J.S. Olson, and W. Wieser (Eds.). Vol. 6: K. Stern and L. Roche, Genetics of Forest Ecosystems. Springer-Verlag, Berlin. 330 pp.

Rousseau, J. 1966. Movement of Plants under the influence of man. *In*: The evolution of

Canada's flora. R.L. Taylor and R.A. Ludwig (Eds.). University of Toronto Press, Toronto, Ontario. pp. 81-99.
Rowe, J.S. 1972. The Forest Regions of Canada. Environ. Canada, C.F.S., Publ. No. 1300. Ottawa. 172 pp.
Soulé, M.E. and B.A. Wilcox. (Eds.). 1980. Conservation Biology: An Evolutionary — Ecological Perspective. Sinauer Associates, Sunderland, Massachusetts. 395 pp.
Van Wagner, C.E. and I.R. Methven. 1980. Fire in the management of Canada's national parks; philosophy and strategy. National Parks Occas. Pap. No. 1. Ottawa. 18 pp.

CHANGES IN TREE SPECIES COMPOSITION IN NATURALLY REGENERATING STRIP CLEARCUTS IN SHALLOW-SOIL UPLAND BLACK SPRUCE

J.K. Jeglum
Canadian Forestry Service
Great Lakes Forest Research Centre,
Sault Ste. Marie, Ontario

Abstract: A long-term study of regeneration in alternate strip clearcuts in shallow-soil upland black spruce *(Picea mariana)* near Nipigon, Ontario, has revealed shifts in tree species composition. The original forest was dominated by black spruce and had minor tree size components of jack pine *(Pinus banksiana)*, balsam fir *(Abies balsamea)*, trembling aspen *(Populus tremuloides)* and white birch *(Betula papyrifera)*. The regenerating first cut strips are a mix of black spruce-white birch-trembling aspen, and the proportion of hardwood to conifer is about 50:50. Scarification increased the level of regeneration in the upland site positions for all species except balsam fir. In lower slopes and peaty drainageways, black spruce regenerated slightly better in nonscarified strips. Regeneration of black spruce increased from 51 to 65 per cent between two and four years of natural seeding. Comparisons of these results with regeneration in clearcuts on the limits of American Can Canada Inc. indicate that strip cutting provides much more regeneration of both conifers and hardwoods. Shallow-soil sites should be classified and managed on the basis of at least two subtypes — fragile nonplantable, and plantable. The manager should make explicit plans as to how he intends to manage hardwoods in cutover upland black spruce forests.

I have been involved, since 1974, in a long-term study of regeneration in alternate strip clearcuts in shallow-soil upland black spruce (Figure 1) near Nipigon, Ontario. This study was undertaken as a cooperative project among three parties, Domtar Forest Products (Red Rock, Ontario), the Ontario Ministry of Natural Resources (OMNR), and the Canadian Forestry Service, to determine if strip clearcutting is a biologically and economically acceptable system for harvesting and naturally regenerating black spruce.

Several reports and papers have been published on both biological and economic studies (Jeglum 1980, 1982; Ketcheson 1979, 1982). As well, a number of other related studies, such as blowdown associated with strip clearcutting (Fleming and Crossfield 1983) and watershed impacts (water quality and quantity) are in progress. This paper will deal with one specific theme, namely, species composition changes in the regenerating strips. The results are from the first of three study areas, which has been analysed for all assessments up to five years after cutting (Jeglum 1982). These results are compared with regeneration survey data from clearcuts on the limits of American Can Canada Inc., Marathon, located relatively close to the Nipigon strip cuts.

Figure 1 Alternate strip clearcuts in shallow-soil upland black spruce, near Nipigon, Ontario. Strip widths shown are 20 m and 40 m. Note scattered individuals and clumps of trembling aspen and white birch.

Table 1. Stocking percentage (S) and mean density (D) in the preharvest forest — Nipigon strip cut experiment, study area I. Based on 1400, 2 × 2 m quadrats

	Trees		Saplings		Seedlings	
	S	D	S	D	S	D
Black spruce	19	0.24	2	0.03	54	3.94
Balsam fir	4	0.05	7	0.10	70	7.81
Jack pine	5	0.05	0.1	0.00+	0.4	0.01
Trembling aspen	2	0.02	0.1	0.00+	9	0.26
White birch	1	0.01	0.1	0.00+	32	2.18
All species	28	0.38	10	0.14	86	14.34

+ = <0.005.

Study Areas

The preharvest forest was predominantly a black spruce dominated forest (Figure 1, Table 1). Considerably smaller proportions of tree size balsam fir, jack pine, trembling aspen, and white birch were present. At the time of this study, about 115 years after its fire origin, the stand was overmature and opening up in spots; blowdown was common, and regeneration was particularly abundant in the throwmound scars. The main advance growth was balsam fir and black spruce. Seedlings of white birch were also abundant (Table 1).

Regeneration survey results from clearcuts were provided by American Can, Marathon. Their limits are located approximately 150 km ESE of the Domtar strip cut area. Both areas are in the central plateau section (B.8) of the boreal forest (Rowe 1972). Yang and Fry (1981) have published data on natural regeneration following clearcutting in boreal mixedwoods and hardwoods, but in their report they did not include any data for the softwood cover types. Such data were requested and generously provided for comparisons with the regeneration data from clearcut strips at Nipigon.

The black spruce cover type at Marathon is composed of more than 75 per cent by volume of black spruce. It occurs on a wide variety of sites ranging from deep and wet organic lowlands to very shallow soil over bedrock. The predominant type is lowland.

Methods

A number of main factors which could have an important influence on regeneration such as strip width, leave time, scarification, topographic position, seedbed conditions, and vegetative regrowth were included in the study (Jeglum 1980). Strip widths were 20 m, 40 m, and 80 m. Leave periods for the residual strips were two and four years. The study was replicated in three areas, each of which had 1400 permanent quadrats sampled before the cut, and half this number (the first cut strips) sampled at years one, three and five after the cut.

The first cut strips were scarified soon after cutting, in July 1975, by rubber-tired skidders pulling flanged barrels. Some 40-m strips were not scarified, and these served as controls. American Can regeneration data consisted of assessments for 5 year old cutovers, and assessments for 15 to 30 year old cutovers. Assessments were based on single examination plots for each of the ages rather than on permanent sample plots. Surveys of natural regeneration in the black spruce cover type five years after cutting were analyzed by moisture regime and terrain (relief) type. The dry plus fresh, moisture regimes, and the medium relief sites, were interpreted to be, predominantly, the shallow-soil black spruce uplands. The other categories which were used (moist and wet moisture regimes, and gentle and flat relief types) are probably wetlands, including moist upland sites transitional to wetlands. The 15 to 30 year old assessments were not subdivided by these. moisture and relief categories.

Results

Domtar Strip Cuts

In the postharvest regeneration of the first cut strips, fifth year assessment, the absolute stocking of black spruce had the same level as in the preharvest conditions, whereas that of balsam fir was much lower (Table 2, *cf.* Table 1). All other species increased dramatically from their preharvest levels.

The level of stocking of black spruce increased from 51 per cent when the leave strips were harvested after two years, to 65 per cent when the leave strips were harvested after four years (Table 2). Most of the increase in black spruce stocking occurred during the third fall and spring. This was most likely owing to favorable moisture supply during this

Table 2. Stocking percentage (S) and mean density (D), five years after harvesting — Nipigon strip cut experiment, study area I, scarified strips.

Species	Postharvest seedlings			
	Seeded 2 yrs		Seeded 4 yrs	
	S	D	S	D
Number of quadrats	300		265	
Black spruce	51	2.74	65	4.12
Balsam fir	29	0.61	32	1.45
Jack pine	15	0.85	19	0.56
Trembling aspen	40	1.68	38	1.66
White birch	56	5.46	54	6.36
Conifers	65	4.21	77	6.17
Hardwoods	71	7.41	68	8.23
All species	86	11.61	87	14.40

Table 3. Composition of the original forest (trees >10 cm DBH) and of the postharvest regeneration (seedlings <2.5 cm DBH) five years after harvesting — Nipigon strip cut experiment, study area I, scarified strips. Stocking percentage (S) and density (D) (from Tables 1 and 2) are expressed as relative percentages.

Species	Pre-harvest (trees)		Postharvest (seedlings)			
			Seeded 2 yrs		Seeded 4 yrs	
	S	D	S	D	S	D
Number of quadrats	1400		300		265	
Black spruce	61	65	27	24	31	29
Balsam fir	13	14	15	5	15	10
Jack pine	16	14	8	8	9	4
Trembling aspen	6	5	21	15	18	12
White birch	3	3	29	48	26	45
Conifers	89	89	48	36	53	43
Hardwoods	11	11	52	64	47	57

period, and possibly to a good cone crop brought on as a stress crop three years after the first strips were cut (Jeglum 1982). These results suggest that residual strips should be left three or four years to boost the level of stocking of black spruce to fully acceptable levels. It must be cautioned, however, that the additional leave cannot be too long, owing to continuing blowdown of the trees in the residual strips on these shallow-soil sites. The results of a one year survey of blowdown have provided a first approximation of how much loss there is for each additional leave year (Fleming and Crossfield 1983).

The relative stocking percentages of the main species in the original forest and in the postharvest regeneration are compared in Table 3. The original forest was dominated by conifers (89 per cent). The postharvest regeneration, however, is a mixture of about equal quantities of conifer and hardwood. The relative proportions of black spruce and jack pine decreased, while those of trembling aspen and white birch went up. The same trends were observed for relative density, except that white birch was even higher because of the relatively high number of seedlings. In terms of density, the proportion of hardwood to conifer was about 60:40.

There was an increase in hardwood composition relative to conifers in all topographic site types (Table 4). However, there was slightly lower increase of hardwoods in the wet drainageways than in the other three topographic site positions. Hence, regardless of site position in this irregular, bedrock-controlled topography, there is an increase in the relative proportion of the hardwood component.

There are some differences within sites between scarified and nonscarified treatments (Table 4). In drainageways, black spruce has higher quantities in nonscarified strips, whereas on upper slope and crest it is higher in the scarified. This suggests that for the moister site types (drainageways and lower slopes) scarification may be detrimental to regeneration, whereas on the upper slopes and crests it seems to be helpful. For balsam fir there were usually higher values in nonscarified strips. The reduction of balsam from the precut condition (Table 1), and further reduction by scarification, suggest that existing advance growth is destroyed by harvesting and scarification and, that little new regeneration has occurred as with the other species. For jack pine there were higher stockings (relative and absolute) for all site types in the scarified strips. Jack pine had very low values or nil in the drainageways, showing a preference for the three upland sites. White birch had higher values in the scarified strips for all sites. Trembling aspen obtains similar values among site types and does not appear to be influenced by scarification.

Among site types which were scarified, all species showed lower absolute stockings in drainageways, higher on the upland site positions (Table 4). On the nonscarified strips, there were higher relative and absolute stockings in drainageways for black spruce. Jack pine still showed preference for upland sites. The other species tended to have slightly higher values in drainageways, such that there were generally similar values across all site types.

American Can Clearcuts

The American Can surveys of clearcuts provide interesting comparisons with the Nipigon results. For the fifth-year regeneration, the dry plus fresh and medium sites showed comparable relative stocking levels, 30 to 36 per cent, for black spruce and trembling aspen (Table 5). Aspen appears to have increased in the regeneration and become as relatively important as black spruce, similar to the Nipigon strip cuts (Table 4). Balsam fir is almost as important as spruce and aspen. However, white birch is relatively much less important on American Can clearcuts than Domtar strip cuts. It is significant that the absolute stocking levels for spruce and aspen, between 22 and 23 per cent are considerably less than those for the regenerating strip cuts (Table 2).

Table 4. Composition of natural regeneration (seedlings <2.5 cm DBH) five years after harvesting, in four topographic site types, and in scarified and non-scarified first cut strips (all widths and leave times combined) — Nipigon strip cut experiment, study area I.

	Topographic site types									
	Drainage-way		Lower slope		Upper slope		Crest		All	
	S[a]	D[a]	S	D	S	D	S	D	S	D
Black spruce										
Scarified[b]	39	1.6	57	3.4	62	3.5	62	4.8	58	3.4
Non-scarified[c]	73	6.9	61	5.5	41	2.2	54	1.7	59	4.6
Balsam fir										
Scarified[b]	20	0.4	22	0.6	41	1.5	32	1.1	30	1.0
Non-scarified[c]	35	1.1	35	1.9	41	1.6	23	1.5	35	1.6
Jack pine										
Scarified[b]	3	0.3	19	0.9	20	0.7	13	0.4	17	0.7
Non-scarified[c]	0	0	4	0.1	4	<.05	5	0.2	3	0.1
Trembling aspen										
Scarified[b]	30	1.2	37	1.9	43	1.6	43	1.7	39	1.7
Non-scarified[c]	37	2.8	41	1.8	47	1.5	46	1.4	42	1.9
White birch										
Scarified[b]	45	3.5	50	5.0	60	6.2	59	10.0	55	5.9
Non-scarified[c]	45	2.3	47	4.7	41	1.6	41	1.5	44	3.0
All conifer										
Scarified[b]	47	2.3	68	5.0	80	5.8	70	6.4	71	5.1
Non-scarified[c]	83	10.1	70	7.9	69	3.9	72	3.3	73	6.9
All hardwood										
Scarified[b]	61	5.0	67	7.1	74	8.0	73	12.3	70	7.8
Non-scarified[c]	60	5.1	65	6.6	69	3.3	62	3.0	64	5.1
All species										
Scarified[b]	72	7.3	86	12.0	91	13.8	87	18.7	86	12.9
Non-scarified[c]	92	15.3	85	14.5	78	7.2	85	6.3	85	12.0

[a] S = stocking percentage; D = density, mean number per 2-m × 2-m quadrat.
[b] Numbers of quadrats for values in this row are 64, 218, 220, 63, and 565.
[c] Numbers of quadrats for values in this row are 60, 105, 51, 39, and 255.

For the wet and flat site types, the lowland black spruce, spruce and balsam fir achieved highest relative stockings (Table 5). Again, the levels of stocking to spruce (ca. 40 per cent) were lower than those at Domtar (ca. 65 per cent). Aspen was less important in these lowland sites, than in upland sites, and white birch was again relatively unimportant.

Data for 15 through 30-year old cutovers in black spruce show highest relative densities for spruce and balsam fir (Table 6). Values for spruce ranged between 31 and 62 per cent stocking. None of the hardwoods achieved importance in these assessments.

The lower absolute stockings for most species on American Can limits is explained mainly by the method and season of harvest, and the low degree of ground disturbance caused by the logging. The black spruce cover type was usually cut in the winter, and bundles of cut logs were yarded by high leads. With this method there was a low degree of disturbance to

Table 5. Composition of natural regeneration 5 years after logging in the black spruce cover type — American Can Canada Inc. limits. Stocking percentages, based on milacre (2 m × 2 m) quadrats, are expressed as relative percentages. Absolute values are in parentheses.

	Moisture regime			Terrain type		
Species or species group	Dry and fresh	Moist	Wet	Medium	Gentle	Flat
Number of stands	51	78	367	10	310	175
Spruce	31 (23)	25 (14)	42 (21)	36 (23)	38 (21)	37 (18)
Balsam fir	28 (21)	35 (20)	34 (17)	20(13)	31 (17)	39 (19)
Jack pine	—	—	—	—	—	—
Trembling aspen	30 (22)	35 (20)	18(9)	36 (23)	24 (13)	18 (9)
White birch	11 (8)	5 (3)	6 (3)	8 (5)	7 (4)	6 (3)
Total	100 (74)[1]	100 (57)	100 (50)	100 (64)	100 (55)	100 (49)

Source: Unpublished data provided by R.D. Fry, Chief Forester, American Can Canada Inc., Marathon, Ontario P0T 2E0.

[1] Totals of stocking percentages for all species are absolute totals.

Table 6. Composition of the second growth forest 15, 20, 25 and 30 years after logging in the black spruce cover type — American Can Canada Inc. limits. Densities are expressed as relative percentages. Absolute values (mean number per milacre, ca. 2 m × 2 m, quadrat) are in parentheses. All size classes of seedlings, saplings and trees are included.

	Years since logging			
Species	15	20	25	30
Black spruce	44 (0.98)	62 (1.51)	31 (1.01)	42 (1.19)
White spruce	6 (0.13)	4 (0.10)	1 (0.02)	+ (0.01)
Balsam fir	46 (1.02)	25 (0.60)	61 (2.03)	52 (1.46)
Jack pine	1 (0.03)	1 (0.02)	+ (+)	− (−)
Trembling aspen	2 (0.04)	− (−)	− (−)	+ (0.01)
Balsam poplar	1 (0.02)	5 (0.12)	5 (0.18)	4 (0.12)
White birch	+ (0.01)	3 (0.07)	2 (0.07)	1 (0.04)
Total	100 (2.23)	100 (2.42)	100 (3.31)	99 (2.83)

Source: Unpublished data provided by R.D. Fry, Chief Forester, American Can Canada Inc., Marathon, Ontario P0T 2E0.

+ =<0.005.

the ground surface and hence little receptive seedbed (e.g., mineral soil, lower humus layers adjacent to mineral). Black spruce seed requires such receptive surfaces for germination and establishment. Most of the large black spruce would have been removed in the cut. Hence, no or little seed would have been available, except from uncut unmerchantable trees or uncut adjacent timber. Most of the black spruce regeneration must have been from advance growth existing prior to cutting.

Aspen and birch would not have been felled in the clearcut operation. Birch seeds also require receptive seedbed which would not be present, hence explaining its low stocking. Aspen regeneration would undoubtedly have been mainly of sucker origin (Heeney *et al.* 1975). The lower levels of aspen in the 15 to 30-year assessments is probably owing to the fact that these areas were mostly lowlands, initially containing little hardwood. In the case of fifth year assessments, the mix of site conditions may have been more similar to those in the Domtar area. This is because the precut density for aspen trees at Domtar was about the same as that at American Can, five to six trees per ha. It could be that the summer skidding and scarification operations at Domtar acted to sever or injure the aspen roots, thus stimulating more sucker production than at American Can where little disturbance took place.

Discussion

Compositional Changes

Results to date suggest that the alternate strip clearcut system, when combined with the appropriate coverage of scarification, yields fully acceptable levels of black spruce regeneration in the first cut strips (Jeglum 1982). However, the results also indicate a shift towards a higher proportion of hardwood in the regeneration than was present in the original stand. If the proportions remain similar as the stand matures, and if no tending operations are carried out, it will become a mixedwood dominated by black spruce, white birch, and trembling aspen. Of course, it is not known if the hardwoods will drop out rapidly, or if some sizeable proportion of them will still be present at the time of the next harvest, which may occur after a much shorter rotation for the second cut, say 60 to 80 years instead of 115.

The shift in composition toward hardwood is explained by the original forest cover conditions, by the method of harvest, and by site preparation. Scattered throughout the study area were small clumps and individuals of trembling aspen and individuals of white birch. In the harvesting of these conifer stands, usually the hardwoods that are present were left standing. These residuals may produce a stress crop of fruits and seeds, and the aspen suckers prolifically. Scarification provides a receptive seedbed for all of the species with the possible exception of balsam fir. (Apparently fir does not produce abundant seed to exploit these disturbed surfaces.) Hence, by leaving the hardwoods standing, the manager is inadvertently favoring a compositional shift toward more hardwood.

The Influence of Different Logging Methods

How widespread is this compositional shift from softwood to hardwood, and how is it related to the method of logging and its impact on the site? There is little question that with modern mechanized harvesting there is a widespread shift in cutovers from softwood to hardwood. In the horse-logging era, during which logging was usually conducted in the winter, there was relatively little destruction of advance growth and little ground surface disturbance. Hence, balsam fir advance growth tended to be preserved on upland sites (MacLean 1960; Hughes 1967; Ellis and Mattice 1974; Sleep 1975). Without site disturbance, species like black spruce, jack pine and white birch, which require exposure of the lower humus layers and mineral soil for seed germination, are not favored (Weetman *et al.* 1973; Ellis and Mattice 1974).

As machine logging and mechanical site preparation have come on the scene, there has been more destruction of advance growth, and more ground scarification. This has caused a shift in composition of natural regeneration from balsam fir to higher proportions of hardwoods. Frisque *et al.* (1978) documented a shift in composition of uplands from a ratio of 88:12 softwood to hardwood before logging, to 51:49 after logging. Several authors have noted that site preparation, although effective in increasing softwood regeneration (spruces and pine), also stimulates hardwood species which grow faster and offer vigorous competition to the conifers (Hughes 1967; Ellis and Mattice 1974; Fry 1975).

The cutover assessment of Clemmer and Atkins (1980) on the same forest limits as the Nipigon data gives a general picture of significant conversions from spruce and jack pine working groups to balsam fir, mixedwood and hardwood working groups. They note that 39 per cent of cutovers are not satisfactorily restocked, and that two-thirds of these NSR lands were originally spruce or jack pine dominated forests.

Clemmer and Atkins (1980) noted differences in the proportions of different working groups regenerating after cutover, depending on the type of disturbance during logging. During the horse-logging era, there was a low level of damage to the advance growth. Balsam fir typically developed to dominate upland sites; spruce developed successfully only on wetter lowlands; and jack pine was almost completely excluded. With the shift to machine-logging and mechanical site preparation within the last 15 years, conversions have been toward more hardwoods on all sites. Spruce and jack pine now comprise 39 per cent of the working groups, where previously, they amounted to 65 per cent. At the same time balsam fir, mixedwood and hardwood working groups combined have shifted from 35 per cent before to 61 per cent after harvesting.

It would appear that the mechanized logging followed by mechanical site preparation and/or fire has had the effect of stimulating shrub species as well as the hardwood trees. Both of these can act as strong competition to softwood regeneration. The present strip cut study showed a shift toward more aspen and birch; however, the fir decreased. Possibly, the clear-cutting in combination with site preparation disturbance caused considerable fir mortality. Fir appears to be particularly abundant as advance growth on upland sites with deeper, rich tills (Eiber 1982). It is probably not as vigorous on the shallow-soil sites, and here it may not contribute as much to the advance growth and subsquent stand composition.

The above trends suggest that the governmental regeneration program has not been too successful. As Clemmer and Atkins (1980) note, ". . . the only consistent result of the government's reforestation program has been to accelerate the rate of stand conversion from softwood working groups to mixedwood and hardwood working groups. Silvicultural treatment has done little to improve the rate of cutover success/failure and future softwood yields. Further, unless steps are taken to better control the incursion of hardwoods onto treated acreage the situation is unlikely to improve at all."

Clemmer and Atkins (1980) projected rather poor yields of the second growth forests, particularly for the spruce and jack pine working groups. However, they cautioned that since most (87 per cent) of the cutover plots were less than 35 years old, the projections should be viewed cautiously. They recommended the development of, or use of existing, computer simulation models of forest growth. By utilizing whatever long-term growth information is available, and extending this through simulation models, the manager will be able to judge how various species mixes may produce over time, and what the effects will be of favoring certain species over others. For example, Botkin (1982) used the JABOWA forest growth model to indicate the growth over time (to 210 years) of various forests, given different species mixes and removals.

The understanding of compositional shifts in cutover boreal softwoods requires thorough documentation including precut stand conditions, understory vegetation and site conditions, the season and method of cutting, the frequency of residuals and distance to nearest forest edge as a seed source, and subsequent treatments such as site preparation, burning and chemical weed control. In addition, biotic factors such as the production of stress crops of seeds, distance of the area from sources of seed, and weather and shading influences require documentation.

It is clear that the kind of harvesting, its accompanying disturbances, residual uncut trees, and size of cutover patch and subsequent site treatments will influence greatly the mix of species which naturally regenerate. Composition may also be influenced by management efforts, via artificial regeneration and active tending to favor some species over others. In this way direction is given to processes of forest development, in order to obtain the type of forest product that is desired and also that optimally utilizes the productive capacities of any given site.

Management Decisions for Shallow-soil Sites

One of the conflicts in managing shallow-soil sites is the need to remove or minimize slash in order to facilitate regeneration, vis-a-vis the need to conserve nutrients and prevent soil erosion (see papers by Timmer *et al*. (1983), Gordon (1983) and Gordon and Van Cleve (1983)). Regeneration, whether by seed or planting (sometimes the latter is difficult or impossible on the shallowest sites), is favored or facilitated by low amounts of slash, which can be accomplished by full tree logging (taking full trees to the landings before they are delimbed and topped). However, a large proportion of the nutrients in a tree are concentrated in the needles, and to preserve the small nutrient stock in these sites it is argued that perhaps needles and branches should be left on site.

Another conflict is that scarification is currently recommended to provide a receptive seedbed for seeding, or to provide rows for the planters to follow. Yet by scarifying, the thin organic mor mat is disturbed and the mineral soil exposed to potential erosion, at least for a few years until vegetation regrowth restabilizes the surface. To reduce the disturbance, it has been suggested that the scarification be done 'delicately', with just enough disturbance to provide adequate levels of regeneration (Marek 1975). This report presents some evidence that the drainageway and lower slope sites are moist enough to obtain adequate levels of regeneration without any further scarification (Jeglum 1982). However, this finding demands further verification before it can be accepted as an operational guideline.

A good deal of care and innovative management is required in handling these shallow-soil sites to obtain regeneration and yet minimize nutrient and soil loss. One of the important elements in this management is to classify these sites more finely to reflect differences in fragility. The most fragile sites are the rock crests or plateaus with very shallow to no soil over bedrock, steep cliffs, boulder outwash, and patterned sites consisting of rocky ridges and peaty drainageways. These sites are difficult to plant and should be distinguished from less fragile 'plantable' shallow-soil sites. A definition of plantable shallow-soil sites might be a mapping unit (polygon) with 75 per cent or more of the soil depths from a depth survey being 10 cm or more. If the soil depth averages 30 cm or more deep the site is no longer a shallow-soil site (see Jones *et al*. 1983).

In the rockiest and most rugged sites, the manager may decide not to cut at all, or to only cut with narrow strip cuts. This would leave behind residual strips to hold the soil and to regenerate the openings.

In plantable shallow-soil sites, the manager may still decide to use the alternate strip clearcut method, combined with scarification if not enough seedling stock or resources are available. In scarifying he should avoid the lowland portions since this tends to reduce regeneration by 'mucking-up' and development of graminoid competition (Jeglum *et al*. 1983). The manager may decide to accept conversions to mixedwood, in which case the sites will produce more aspen and birch, along with black spruce in the next rotation. This option may be forced on him if resources are limited and he decides to devote the greater part of his regeneration effort to richer and/or deeper soil sites.

The most appropriate season for harvesting these sites depends on how fragile they are. In the most fragile sites (steep, little soil, sizeable proportions of peatland) they should be harvested in winter to minimize damage. However, in others that have some soil and are not too steep, it may be preferable to harvest in summer in order to obtain some site preparation (disturbance) from the logging machines. There is a possibility that with full-tree logging to reduce slash, the site preparation by logging may be enough to obtain adequate black spruce regeneration, without further site preparation (Jeglum 1982). However, this possibility must be documented further before being accepted.

A basic decision that the manager has to make is whether or not to attempt to control the hardwoods. He may decide not to control them in these shallow soil sites, possibly because resources are limited, and there may be other richer, more productive sites that he decides are higher priority for tending. Indeed, it is not well-known just how much of a problem these hardwoods will be in the shallow soil sites. It may be that competition here is not great, relative to other sites, and that treatment costs do not warrant the returns in better volume production by the conifers. In fact it may be that the hardwoods may provide some positive benefits such that they do not have to be removed. Gardiner (1968) conducted a literature review on birch as a soil improver, and concludes that it does have a number of silvicultural benefits. It protects in terms of reducing climatic extremes and reducing insect attacks, it improves soil through the decomposition of leaves and effective exclusion of weeds and grasses, and it increases total volume production.

Whether or not birch and aspen may have this effect in shallow soil sites with the species we are concerned with is not known. It is known, however, that the forest floor litter layer beneath many hardwoods, including large-toothed aspen, have a higher percentage ash content than pines (Loomis 1982). Even if the hardwoods do have a positive effect early in the development of the stand, owing to nutrient enrichment or shelter, they may have to be removed later (15 to 20 years) to remove competition effects.

The manager also may decide to leave the hardwoods and harvest them along with the conifer the next time. This of course depends on the development of a capacity to utilize the hardwoods. Aspen can be utilized in the mills, although the fiber it provides is not as long or as favored as that of spruce. Large quality logs of aspen and birch are used for veneer, and birch is used locally as firewood. However, there are other products that could be developed for both species. It should be noted however, that the shallow soil sites are poor for both aspen and birch growth compared to deeper, richer till soils. Consequently, the manager may decide to discourage the hardwoods.

If the manager decides to discourage hardwoods, the problem is how to do it most efficiently. By cutting the parent trees along with the conifers, the regeneration from seed is prevented, and subsequent silvicultural operations, such as scarification and chemical tending, are thereby greatly facilitated. Also, regeneration by seed from the hardwoods will be prevented. However, this does not prevent suckering by aspen. It is possibly worse to cut the large parent aspen trees at the time of harvest, owing to the high production of suckers.

A field study in Manitoba by Steneker (1974) suggested that the apical dominance effect primarily controls suckering. Once this effect is interrupted, increased soil temperatures will promote suckering. Another study in Utah by Schier and Smith (1979) showed that clearcutting resulted in the greatest number of suckers. Partial cuts (removal of 67 per cent of the basal area) yielded less than 50 per cent as much regeneration as clearcuts, although there were still high numbers of suckers. Scarification had no apparent effect on sucker production. Girdling of mature trees initially produced high numbers of suckers, although less than either clearcuts or partial cuts. However, mortality was high on girdled plots, and by the 12th year after treatment few suckers had survived. Girdling is a potential treatment to discourage aspen suckering if aspen are not cut in the initial harvesting, and if manpower is available.

Recommendations

1. If trembling aspen and white birch are present in a proposed area of cutover of upland black spruce, management plans should contain explicit prescriptions of how the natural hardwood regeneration is to be managed. In the precut inspection the potential for regeneration by hardwoods should be judged and plans made for how to deal with it. It should be stated whether hardwood regeneration is to be accepted, or discouraged by felling the existing hardwoods at the time of cutting and subsequent hand cleaning or chemical control.

2. More site work needs to be done in the shallow-soil upland units. The unit should be subdivided into two or more units to reflect more specific forest management problems related to forest growth, regeneration, and potential nutrient loss and erosion.

3. More research needs to be done to determine the relative advantages and disadvantages of allowing black spruce on shallow soil sites to convert to mixedwoods. This might include such studies as simulation models for forest growth with various mixes of species or pure stands, potential site improvement by hardwoods, competition effects, rates of hardwood mortality as the stand develops, and alternative methods of discouraging and favoring various species.

4. More documentation is required of the development of cutovers of all boreal forest types. This should include thorough definition of preharvest conditions (stand conditions, vegetation, soils, and site); the season and method of harvest; the number and spacing of residuals; the size of clearcuts; the location of uncut forest edges or residuals; disturbance during logging; slash abundance; seedbeds; vegetative regrowth and competition influences; and subsequent site treatments such as scarification,

burning, and chemical control. Considerable permanent sample plot information exists within the forest industry and within the OMNR that has not been thoroughly analyzed and evaluated. As well, more long-term studies using permanent sample and growth plots are required (Jeglum 1980). Such studies should always be orientated in the context of a good site classification, such as the Forest Ecosystem Classification (Jones *et al*. 1983).

5. Studies of later stand development of both natural and regenerated cutovers should assess volume changes with mortality (Wannamaker and Carleton 1983). These would shed light on such management questions as spacing and thinning, and rates of wood loss by various species in mature and overmature forests.

Acknowledgements

I gratefully acknowledge the use of regeneration survey data from American Can Canada Inc., Marathon, Ontario, provided by R.D. Fry, Chief Forester.

Literature Cited

Botkin, D.B. 1982. Causality and succession. *In*: Forest succession: concepts and application. D.C. West, H.H. Shugart and D.B. Botkin (Eds.). Springer-Verlag, New York, Chapt. 5, pp. 36-55.

Clemmer, E. and T. Atkins. 1980. St. Lawrence licence cutover assessment final report. Ont. Min. Nat. Resour., Nipigon District, Nipigon, Ontario. Unpubl. Rept. Mimeo.

Eiber, T.G. 1982. Site, budworm, and forest management. Proc. Central Internatl. For. Insect and Disease Conf., Thunder Bay, Ont., Sept. 27-29, 1982.

Ellis, R.C. and C.R. Mattice. 1974. Stand development following harvesting at the Experimental Lakes Area in northwestern Ontario. Environ. Canada, C.F.S., Inf. Rep. O-X-207.

Fleming, R.L. and R.M. Crossfield, 1983. Strip cutting in shallow-soil upland black spruce near Nipigon, Ontario. III. Windfall and mortality in the leave strips, preliminary results. Environ. Canada, C.F.S., Infl Rep. (In press).

Frisque, G.J., G.F. Weetman and E. Clemmer 1978. Reproduction and trial projected yields 10 years after cutting 36 pulpwood stands in eastern Canada. Tech. Rep. No. TR-23, Forest Engineering Res. Inst. Can.

Fry, R.D. 1975. Tree planting and stock quality — a view from the forest. Environ. Canada, C.F.S. Proc. O-P-4. pp. 80-85.

Gardiner, A.S. 1968. The reputation of birch for soil improvement: A literature review. For. Comm. Res. and Devel. Pap. No. 67. London, England.

Gordon, A.G. 1983. Nutrient cycling dynamics in differing spruce and mixedwood ecosystems in Ontario and the effects of nutrient removals through harvesting. *In*: Resources and dynamics of the boreal zone. R.W. Wein, R.R. Riewe and I.R. Methven (Eds.). Association of Canadian Universities for Northern Studies, Ottawa, Ontario.

Gordon, A.M. and K. Van Cleve. 1983. Seasonal patterns of nitrogen mineralization following harvesting in the white spruce forests of interior Alaska. *In*: Resources and dynamics of the boreal zone. R.W. Wein, R.R. Riewe and I.R. Methven (Eds.). Association of Canadian Universities for Northern Studies, Ottawa, Ontario.

Heeney, C.J., J.A. Kemperman and G. Brown, 1975. A silvicultural guide to the aspen working group in Ontario. Ont. Min. Nat. Resour., Toronto, Ontario.

Hughes, E.L. 1967. Studies in stands and seedbed treatment to obtain spruce and fir reproduction on the mixedwood slope type in northwestern Ontario. Can. Dep. For. Rural Devel., C.F.S., For. Br. Publ. 1189.

Jeglum, J.K. 1980. Strip cutting in shallow-soil upland black spruce near Nipigon, Ontario. I. Study establishment and site conditions. Environ. Canada, C.F.S. Inf. Rep. O-X-315.

Jeglum, J.K. 1982. Strip cutting in shallow-soil upland black spruce near Nipigon, Ontario. II. Regeneration in the first study area. Environ. Canada, C.F.S., Inf. Rep. O-X-337.

Jeglum, J.K., V.F. Haavisto and A. Groot. 1983. Peatland forestry in Ontario: An overview. *In*: Peat symposium '82, Sept. 13-15, Shippagan, N.B. Sponsored by Govts. of Canada and New Brunswick, and the Can. Natl. Comm. of Internatl. Peat Soc. (In press).

Jones, R.K., G. Pierpoint, G.M. Wichware and J.K. Jeglum. 1983. A classification and ordination of forest ecosystems in the Great Claybelt of northeastern Ontario. *In*: Resources and dynamics of the boreal zone. R.W. Wein, R.R. Riewe and I.R. Methven (Eds.). Association of Canadian Universities for Northern Studies, Ottawa, Ontario.

Ketcheson, D.E. 1979. A study of the cost of strip cutting black spruce stands in northern Ontario. Environ. Canada, C.F.S., Inf. Rep. O-X-301.

Ketcheson, D.E. 1982. The impact of strip cutting on logging costs: Additional road construction and maintenance are the key factors which incur extra costs. Pulp and Pap. Can. 83: 29-34.

Loomis, R.M. 1982. Seasonal variations in ash content of some Michigan forest floor fuels. U.S.D.A. For. Serv., Res. Note NC-279.

MacLean, D.W. 1960. Some aspects of the aspen-birch-spruce-fir type in Ontario. Can. Dep. For., For. Res. Div. Tech. Note 94.

Rowe, J.S. 1972. Forest regions of Canada. Environ. Canada, C.F.S. Publ. No. 1300.

Schier, G.A. and A.D. Smith, 1979. Sucker regeneration in a Utah aspen clone after clearcutting, partial cutting, scarification, and girdling. U.S.D.A., For. Serv., Res. Note INT-253.

Steneker, G.A. 1974. Factors affecting the suckering of trembling aspen. For. Chron. 50: 32-34.

Timmer, V.R., H.M. Savinsky and G.T. Marek. 1983. Impact of intensive harvesting on nutrient budgets of boreal forest stands. *In*: Resources and dynamics of the boreal zone. R.W. Wein, R.R. Riewe and I.R. Methven (Eds.). Association of Canadian Universities for Northern Studies, Ottawa, Ontario.

Wannamaker, B.A. and T.J. Carleton. 1983. Self-thinning and growth of natural black spruce stands in the Clay Belt of northern Ontario. *In*: Resources and dynamics of the boreal zone. R.W. Wein, R.R. Riewe and I.R. Methven (Eds.). Association of Canadian Universities for Northern Studies, Ottawa, Ontario.

Weetman, G.F., Grapes, W.W. and Frisque, G.J. 1973. Reproduction and ground conditions 5 years after pulpwood harvesting: Results from 37 study areas in eastern Canada. Logging Res. Rep. LRR/51, Pulp and Pap. Inst. Can.

Yang, R.C. and R.D. Fry. 1981. Natural succession following harvesting in the boreal mixedwood forest. Environ. Canada, C.F.S., Proc. O-P. 9. pp. 65-77.

SELF-THINNING AND GROWTH OF NATURAL BLACK SPRUCE STANDS IN THE CLAY BELT OF NORTHERN ONTARIO

B.A. Wannamaker and T.J. Carleton
University of Toronto

Abstract: Even-aged black spruce *(Picea mariana)* stands in the Clay Belt of northern Ontario were sampled in order to investigate the −3/2 self-thinning power law. This phase of stand growth is characterized by a period of substantial density-dependent mortality due to intraspecific competition, the relationship between density and mean tree volume on a logarithmic scale having a slope of −1.5. Plots were established in five age classes, covering a span of 40 to 85 years, with a density range of 2000 to 9800 trees per hectare. Six trees per plot were felled and sectioned for stem analysis. All stems (living and dead) were aged from basal cores or discs. Stand density reconstruction and mean tree volume estimation for one stand resulted in the volume-density relationship

$$\log V = 11.92 - 3.75 \log p$$

Possible reasons for deviation from the idealized −1.5 are considered, among these being spatial distribution, allometric relationships and the shade-tolerant behaviour of black spruce. The universal applicability of the −3/2 law is thus questioned as a means of stand density management of shade-tolerant species such as black spruce.

The biologically-based −3/2 self-thinning power law and its applicability as a simplistic stand growth model in black spruce forests in northern Ontario's Clay Belt is the subject of this paper.

Significant studies of self-thinning in plant species populations were initiated by Japanese scientists, Koyama, Kira and Yoda in the 1950s. For a variety of herbaceous species, these researchers found a consistent relationship between mean above-ground weight and density in monospecific populations where density dependent mortality was occurring. This relationship is expressed mathematically as:

$$w = k_1 \cdot p^{-1.5}$$

or, $\log w = \log k_1 - 1.5 \log p$

where, k_1 = species − specific constant
w = mean plant weight
p = density.

Yoda *et al*. (1963) have provided a geometrical derivation of the law, with the following assumptions:

(1) that weight (w) is directly proportional to the cube of a linear plant dimension (1)

$$w \propto 1^3$$

(2) mean plant area occupied (s) is directly proportional to p

$$s \propto l^2$$
$$s \propto p^{-1}$$
$$\text{therefore} \quad s \propto l^2 \propto (l^3)^{2/3} \propto w^{2/3}$$
$$\frac{1}{p} \propto w^{2/3},$$
$$\text{therefore} \quad w \propto p^{-3/2}$$

Applications of the self-thinning rule as a model for stand density management are few. Tadaki (1964), Ando (1968) and Aiba (1975) have utilized yield-density concepts as a basis for predicting yield of managed forest lands under various thinning regimes. Drew and Flewelling (1977, 1979) have applied these Japanese theories to stand density management of coastal douglas fir plantations, utilizing the $-3/2$ rule as a growth model from which they can relate tree size, stand yield and stand density.

Methods

The Clay Belt of northern Ontario has provided an excellent location for the investigation of self-thinning in natural stands, as the area supports extensive even-aged black spruce forests of fire origin. This study is concentrated in the black spruce-feather moss *(Pleurozium schreberi)* site type, with consistent soil texture (silty clay loam), organic matter (fibrimor, about 10 centimeters in depth) and a fresh to moist moisture regime. Plots were established in five areas of known fire origin, each of differing ages and with two extremes of density. The resulting sample has covered an age range of 45-85 years, and a density range expressed on a per hectare basis from 2000 – 9800 trees, with mortality ranging from 38 to 55 per cent.

Mensurational data were collected; DBH, total height, crown radius and depth; stem mapping was done. Living stems were cored at the base and dead stems were cored or disced. Once heights were obtained, a height class distribution was drawn up in the field, and a stratified random sample of six trees across the range of height classes was taken for stem analysis. These trees were cut at the estimated root collar and were sectioned at one meter intervals.

Results and Discussion

Once all trees were aged in the laboratory and stem analysis was completed, the procedure in establishing a density-mean tree volume relationship was begun. This involved the study of individual plot age structure based on at least 80 living trees per plot (Figure 1). The stand density was then reconstructed to show change with time by assuming a postfire peak year of establishment (Figure 1) for all dead stems. Finally, the cumulative annual volume estimates, and mean tree volume for survivor growth were calculated.

The resultant curve for one plot illustrates the volume-density relationship in the self-thinning phase, where density is plotted against mean tree volume at ten-year intervals (Figure 2). The relationship in this particular example is expressed as:

$$\log V = 11.92 - 3.75 \log D$$

Further refinement of the data to improve mean tree calculations was done. A model relating volume to age and DBH, derived from stem analysis data, was applied to increment cores of living trees. As well, an estimate of mortality volume was made. A weighted mean tree volume was then calculated, i.e., a mean tree volume based on trees felled for stem analysis, all other living cored trees and trees destined to die over a ten-year period. This recalculation resulted in increasing the negative value of the slope (i.e., in the above case the slope increased negatively to −3.90).

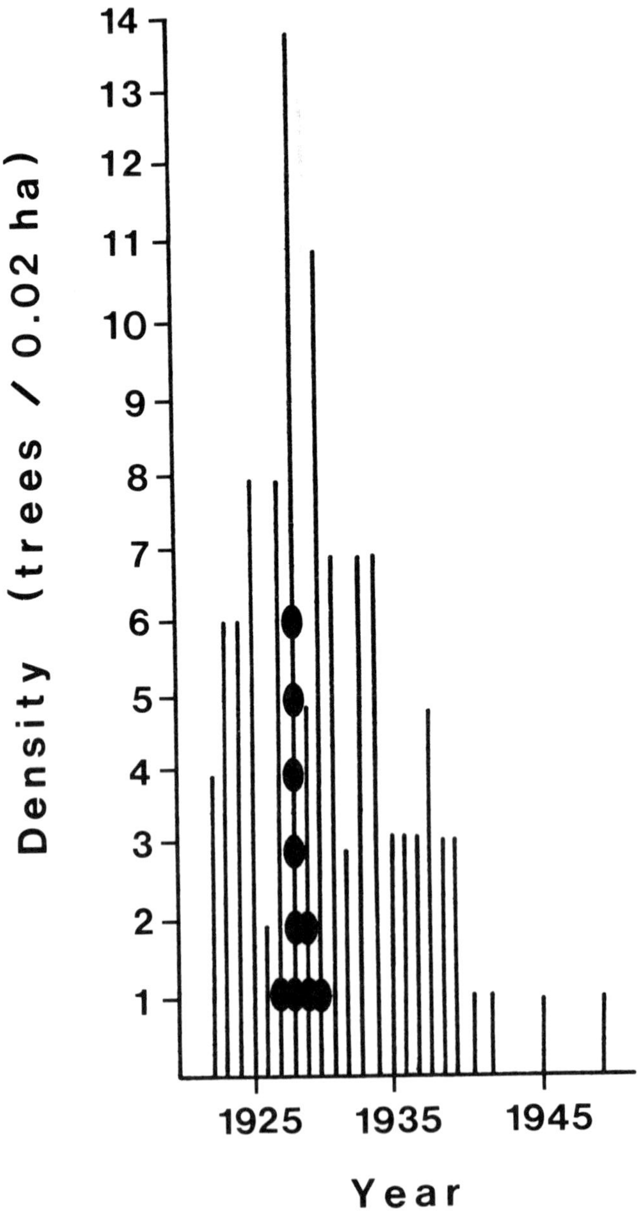

Figure 1. Frequency histogram illustrating the age structure of a postfire black spruce stand, and the apparent postfire peak year of seedling establishment. The dots represent trees felled for stem analysis, aged from basal discs.

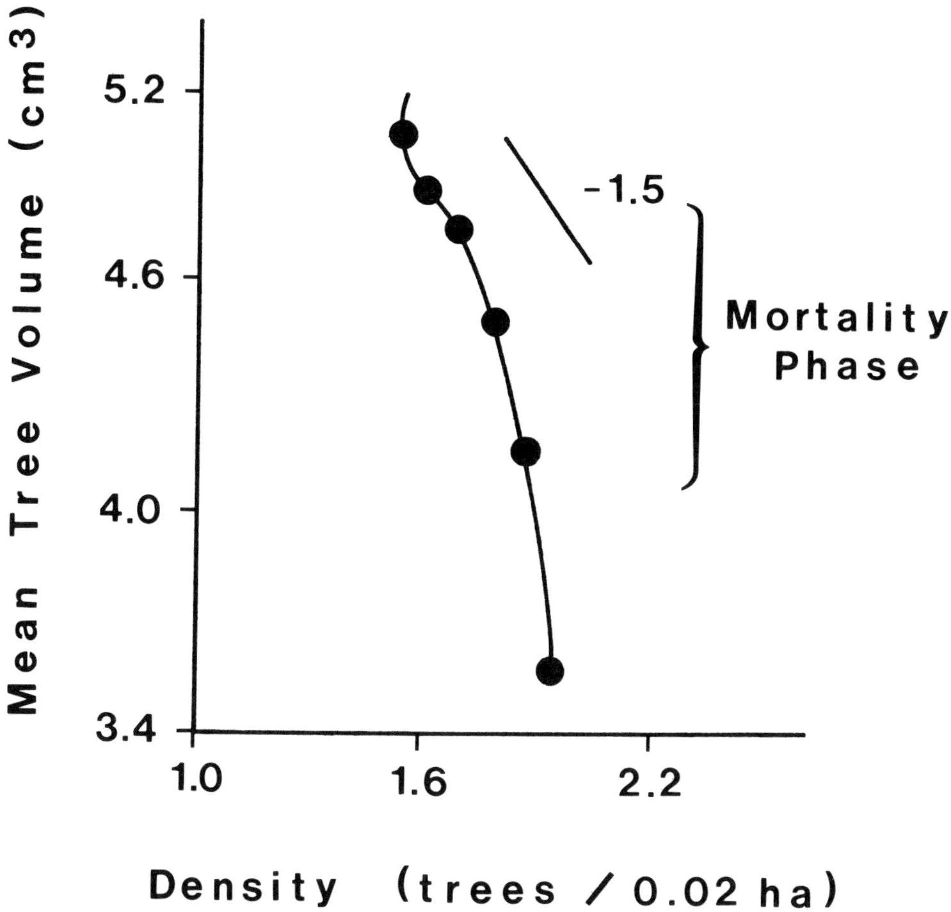

Figure 2. The relationship between volume and density on a logarithmic scale (log V = 11.92−3.75 log D) in a self-thinning black spruce stand, plotted at 10-year intervals. Note the steepness of the slope as compared to the idealized −1.5. (The year of burn is 1906.)

Possible reasons for deviation from the idealized −1.5 power function have to be considered. When one examines the spatial distribution of trees in these stands, one finds a definite irregular, often clumped stem distribution, often with large canopy gaps. A regression between mean inter-tree distance and the self-thinning power function revealed that with increasing inter-tree distance there is an increasing negative deviation from the ideal. It appears that stands only at very high densities, spaced such that intensive inter-tree competition inherent in the −3/2 law is ongoing, are subject to the −3/2 law.

Certainly, spatial distribution and allometric relationships are areas to be investigated in the development of a self-thinning model for natural black spruce forests. The shade-tolerant behaviour of black spruce may also be a factor in the lack of conformance of the species to the −3/2 self-thinning law. In a given stand, those trees that are destined to die (up to 55 per cent of the total density) remain in a suppressed, subcanopy position for many years, before mortality finally occurs. These individuals may never exert a severe depression on the survivor growth of the stand. The $-3/2$ power law implies a maximum size-density relationship where vigorous competition restricts mean tree survivor growth. It appears that survivor growth in the stands investigated in this study is not restricted in this manner. Thus stand density management of these forests based on a simplistic growth model such as the $-3/2$ power law, becomes a difficult task.

Further research is required in natural stands older than those addressed in this study, to examine the possibility of a gradual approach of the self-thinning curve to the -1.5 slope. As well, examination of volume-density relationships in black spruce plantations of varying age, density and site type is essential, should stand density management become a management priority in the future.

This research will hopefully bring to light the problems in developing simplistic stand growth models such as the $-3/2$ power law, especially as it applies to natural forest stands of shade-tolerant black spruce.

Acknowledgements

The authors are grateful for assistance provided by the Ontario Ministry of Natural Resources and the Spruce Falls Power and Paper Co. staff in Kapuskasing throughout the duration of the study. The research was supported by a DREE grant.

Literature Cited

Aiba, Y. 1975. Effects of cultural systems on the stand growth of Sugi-plantations *(Cryptomeria japonica)* III. Estimate of the stem volume yield under actual stand density (H-D-p-v diagram) J. Jap. For. Soc. 57: 67-73.

Ando, T. 1968. Ecological studies on the stand density control in even-aged pure stands. Gov. For. Exp. Stn. (Tokyo) Bull. 210: 1-153.

Drew, J.T. and J.W. Flewelling. 1977. Some recent Japanese theories of yield-density relationships and their applications to Monterey pine plantations. For. Sci. 23: 517-534.

Drew, J.T. and J.W. Flewelling. 1979. Stand density management: an alternative approach and its application to douglas fir plantations. For. Sci. 25: 518-532.

Tadaki, Y. 1964. Effects of thinning on stem volume yield studies with competition-density effects. Gov. For. Exp. Stn. (Tokyo) Bull. 154: 1-19.

White, J. 1981. The allometric interpretation of the self-thinning rule. J. Theor. Biol. 89: 475-500.

Yoda, K., T. Kira, H. Ogawa and K. Hozumi. 1963. Intraspecific competition among higher plants. XI. Self-thinning in overcrowded pure stands under cultivated and neutral conditions. J. Biol. (Osaka City Univ.) 14: 107-129.

POPULATION DENSITIES AND HABITATS OF SPRUCE GROUSE IN ONTARIO

K.J. Szuba and J.F. Bendell
University of Toronto

Abstract: We studied the association between population density and habitat of spruce grouse *(Canachites canadensis)* in 13 even-aged forests of jack pine *(Pinus banksiana)* and eight of black spruce *(Picea mariana)* in the boreal forest region of north-central Ontario. The number of grouse observed per hour of search with dogs (grouse/h) was used to estimate grouse populations since it was strongly related ($R^2 = 0.968$) to total counts of banded grouse in six intensively worked forests. The highest densities of spruce grouse (0.50 to 0.80 grouse/ha), greater than any recorded elsewhere in North America, occurred in young (11 to 21 years) pine plantations. Similar forests have not been studied outside Ontario. Older plantations (28 to 62 years) were not as productive. However, one (46 years) contained a high population. Populations were generally lower in spruce forests, possibly because of differences in the herb layer and in microclimate. In jack pine, mean stand height was correlated with population density ($r = -0.745$, $P < 0.05$). However, canopy height alone was not a good indicator of the level of spruce grouse populations, since four forests of comparable height differed in grouse/h. These differences appeared to be related to the extent of the coniferous understory. In some cases differences in trees/ha were also important. A new variable, foliage area index (FAI), was used to represent the total amount of forest cover provided by the canopy and understory. FAI was significantly correlated with population density in jack pine forests ($r = 0.867$, $P < 0.01$). Spruce grouse may be good indicators of the effect of forestry practices on wildlife, since they occur in a wide range of population densities and appear to respond to differences in habitat. Implications of the results to forest management are discussed.

If we are to successfully manage forests for wildlife, or to predict the consequences of forest management practices on wildlife populations, we must know how wildlife responds to habitat. Many studies consider selection of habitat by wildlife within one large heterogeneous tract. Thus, the general applicability of the results is often unclear. Moreover, little is known about how large almost monospecific stands, common in some regions of the boreal forest, affect wildlife populations.

We studied the impact of modern forestry practices on wildlife by comparing population densities of spruce grouse *(Canachites canadensis)* in a range of ages of even-aged monocultures of jack pine *(Pinus banksiana)* and black spruce *(Picea mariana)* in north-central Ontario. Spruce grouse are good study animals for several reasons. They are common members of boreal communities, and are important in food chains as prey for raptors and fur-bearers. As well as being important small game, they are relatively easy to study. The results we present demonstrate that spruce grouse populations occur in a wide range of densities and appear to respond to differences in habitat. Thus, spruce grouse may be good indicators of the effect of forestry practices on wildlife. Since their range is continent-wide and is roughly coincident with the boreal forest region (Aldrich 1963), results from Ontario may be applicable to populations elsewhere in North America.

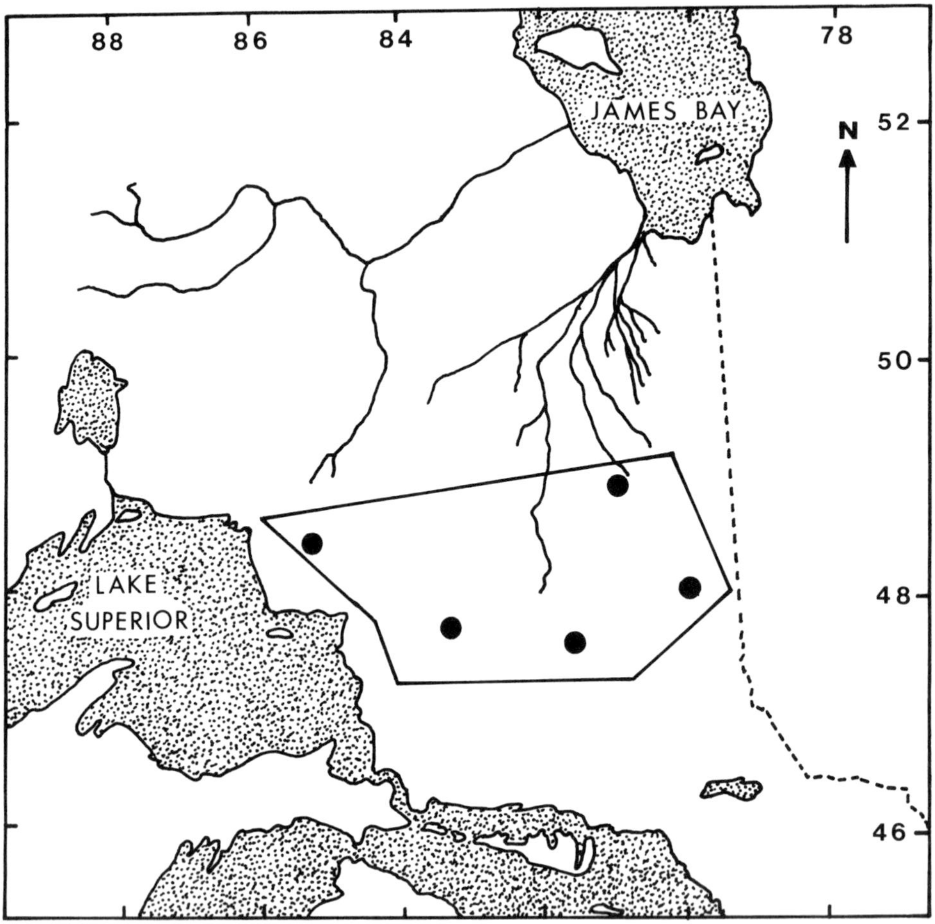

Figure 1. Region of study in north-central Ontario and location of the towns of White River, Chapleau, Gogama, Kirkland Lake and Cochrane (left to right). Numbers indicate degrees west longitude and north latitude.

Within-forest selection of habitats by spruce grouse has been studied in Alaska (Ellison 1971), Alberta (McCourt 1969; McLachlin 1970), Montana (Herman 1980), Minnesota (Haas 1974), Michigan (Robinson 1969) and Maine (Hedberg 1980). However, no studies have been conducted in Ontario, and the range of densities of spruce grouse populations is poorly understood. Our main objective was to determine the density of spruce grouse populations and associated habitat relationships in monocultures in boreal Ontario.

Methods

Study areas

The region of study was bounded by 47° and 49°N latitude and 79° and 86°W longitude. The towns nearest the study areas were White River, Chapleau, Gogama, Kirkland Lake and Cochrane, Ontario (Figure 1). Carleton and Maycock (1978) describe the vegetation of the region in detail, concluding that the forests are predominantly boreal and a large proportion are monospecific.

An extensive survey of forests was conducted from May 1 to August 31 in 1979 and 1980. Additional intensive work was carried out on two study areas in 1981. Thirteen plantations of jack pine, seven natural forests of black spruce, and one mixed black and white spruce *(Picea glauca)* plantation were sampled to determine the level of spruce grouse populations and characteristics of the tree and shrub layers.

Potential study areas, at least 40 ha in size, were identified on Forest Resource Inventory maps, then checked in the field. Those selected for study had these features in common: regular stand shape, canopy relatively uniform in species composition and distribution, and relatively homogeneous topography and soil throughout. We purposely included forests differing in age (young, medium, old) and trees/ha (open, dense). All study areas were at least five kilometres apart.

Estimation of grouse populations

We examined the association between population density of grouse and features of habitat by comparing relative levels of spruce grouse populations as reflected by the number of grouse observed per hour (grouse/h) of search with dogs. The census procedure and its accuracy are documented elsewhere (Szuba 1982), but will be briefly described here. The census was conducted in a similar manner in all forests by the same two observers, each accompanied by a trained pointing dog. The two observers searched different portions of a forest at the same time, attempting to cover it evenly by walking at a relatively constant speed along suitable compass bearings. One day of search in a forest constituted one sample. To ensure that a grouse was counted only once on a given day, grouse were captured with extendable snare-poles (Zwickel and Bendell 1967) and were banded with unique combinations of coloured leg bands. Each observer recorded effective search time (excluding time spent handling birds, making notes) to the nearest minute. Data for both observers were pooled to allow for possible differences in the efficiency of the observers and dogs. Grouse/hour was simply the number of individual grouse observed by two observers in one day divided by the effective search time in that day. Daily estimates were weighted by the total effective hours of search in that day for the calculation of overall mean grouse/h for a forest. Weighted variances were calculated in standard fashion (Sokal and Rohlf 1969).

We think this is an acceptable census method. Linear regression analysis of grouse/hour and known population densities showed that the two were strongly related (Szuba 1982). Known densities were determined by total counts of banded and resighted grouse in six study areas. The high R^2 value ($R^2 = 0.968$) and highly significant slope ($P<0.001$) indicated a good fit of the data to the regression line.

We use the regression equation developed by Szuba (1982) to estimate absolute population density (grouse/h = 0.379 + 2.119 grouse/ha) by inverse prediction. Predicted and known population densities are compared with known densities recorded elsewhere in North America to determine the generality of results from Ontario.

In all cases means are compared by t-test or Mann-Whitney U-test and are considered significantly different at $P < 0.05$. Means are presented ± 1 SD. Throughout the text "adult" refers to adults and yearlings; chicks are excluded.

Grouse populations may cycle or fluctuate annually (Watson and Moss 1979). There is no good evidence that spruce grouse populations cycle, however, annual fluctuations may occur (Boag *et al*. 1979). Thus, we view our results, obtained in two years of fieldwork, conservatively, and emphasize only large differences in density when comparing populations over a wide geographic area.

Vegetation sampling

Mean canopy height, trees/ha, stand age, and species composition of the tree layer (trees are ⩾ 3 m in height) were quantified using a variable plot cruise (Grosenbaugh 1952) of 20 points in each forest. At each point we also estimated the percentage cover of all species of trees and shrubs (shrubs are woody plants <3 m, ⩾ 1 m in height) in a circular plot of radius 15 m. This represented the available habitat. A similar circular plot was described at each grouse location to examine the selection of habitats within a stand. The structure of the vegetation from ground level to the top of the canopy was described in at least 100 m of line intercept, where the species, height and position of all trees and shrubs crossing the line were recorded.

Emphasis is placed on jack pine forests. Spruce forests are used for comparative purposes and are not treated in detail.

Results

Spruce grouse occupy a variety of forests over the region of study and occur in a range of population densities (Table 1). Grouse populations reached higher levels in jack pine than in black spruce. Population density appeared greatest in forests of young dense jack pine, with some exceptions.

Attributes of canopy and understory

In jack pine, high populations of grouse occurred in forests ranging from 4.3 to 14.7 m in height, 11 to 46 years of age, and 2295 to 4863 trees/ha. Medium and low density populations were in pine forests from 8.5 to 19 m in height, 20 to 62 years of age, and 827 to 3105 trees/ha. All lowland black spruce forests had sparse populations. A medium density population occurred in one mixed upland and lowland spruce forest 103 years of age (Table 1).

Neither trees/ha nor stand age were significantly correlated with grouse populations as measured by grouse/hour in jack pine or black spruce forests. However, spruce grouse populations tended to decrease with increasing stand height in jack pine forests (Figure 2).

Although the correlation between grouse/hour and stand height was significant, major exceptions weaken the utility of canopy height alone as an indicator of population density. For example, two young stands of similar height differed significantly in grouse/hour (Table 1: CH-EGH, CH-OGH). A difference in trees/ha may be important in this case. However, even when trees/ha and stand height were comparable (as in KL-CF, CO-P29 and CO-P25), grouse populations were different (Table 1). These three forests were characterized by a relatively continuous canopy beginning 8 to 9 m above ground level and a virtual absence of tall deciduous shrubs such as *Alnus* spp. and *Salix* spp. In terms of the canopy and understory, the three forests differed primarily in quantity of understory black spruce. Increasing cover of black spruce appeared to be associated with increasing

Table 1. Predicted and known population densities of spruce grouse in 21 forests sampled in north-central Ontario from 1980 to 1981

Forest type	Forest** code	Canopy height (m)	Age (years)	Trees/ha (canopy)	Grouse/h* (x ± 1 SD)	Population density (♂♂+♀♀/ha) Known	Predicted	95% confidence limits of predicted density
Jack pine	CH-EGH	8.2	21	3458	2.13 ± 0.22^a	0.80	0.83	0.62 - 1.08
	GO-VA	4.9	13	4446	2.01 ± 0.19^a	—	0.77	0.56 - 1.03
	GO-I	7.0	15	2295	2.00 ± 0.05^a	—	0.76	0.56 - 1.02
	CO-P25	14.7	46	2890	1.67 ± 0.30^a	—	0.61	0.40 - 0.86
	GO-VB	4.3	11	4863	1.52 ± 0.77^a	0.50	0.53	0.33 - 0.79
	KL-BE	12.5	28	1437	0.80 ± 0.19^b	—	0.20	0.00 - 0.45
	CO-P29	14.0	32	2794	0.69 ± 0.17^b	—	0.15	0.00 - 0.39
	CH-PP	15.0	29	3105	0.63 ± 0.21^b	—	0.12	0.00 - 0.37
	GO-G	15.2	32	1297	0.59 ± 0.26^b	0.17	0.10	0.00 - 0.35
	CH-OGH	8.5	26	981	0.59 ± 0.31^b	—	0.10	0.00 - 0.35
	WR-CL	9.7	20	988	0.56 ± 0.14^b	0.12	0.08	0.00 - 0.33
	KL-CF	14.0	32	2826	trace	—	—	—
	CH-VTP	19.0	62	536	trace	—	—	—
Black spruce	GO-SB	17.7	103	1909	1.02 ± 0.06^b	—	0.30	0.09 - 0.55
	WR-SW	6.6	19	3537	0.53 ± 0.08^b	—	0.07	0.00 - 0.32
	CO-PS	12.2	120	2584	0.53 ± 0.03^b	—	0.07	0.00 - 0.32
	KL-SB	11.6	86	2969	0.42 ± 0.10^b	—	0.02	0.00 - 0.27
	CH-VTSB	14.3	62	2902	0.36 ± 0.05^b	—	0.00	0.00 - 0.24
	CO-SNA	10.3	73	6032	0.26 ± 0.30^b	—	0.00	0.00 - 0.19
	CO-SA	10.0	90	18952	0.25 ± 0.10^b	—	0.00	0.00 - 0.19
	KL-SL	6.4	25	3290	trace	—	—	—

** Prefix designates location by district: CH=Chapleau, GO=Gogama, CO=Cochrane, KL=Kirkland Lake, WR=White River.
* Means sharing a common letter are not different ($P > 0.05$); Trace = only droppings found. Densities too low to measure with grouse/hour.

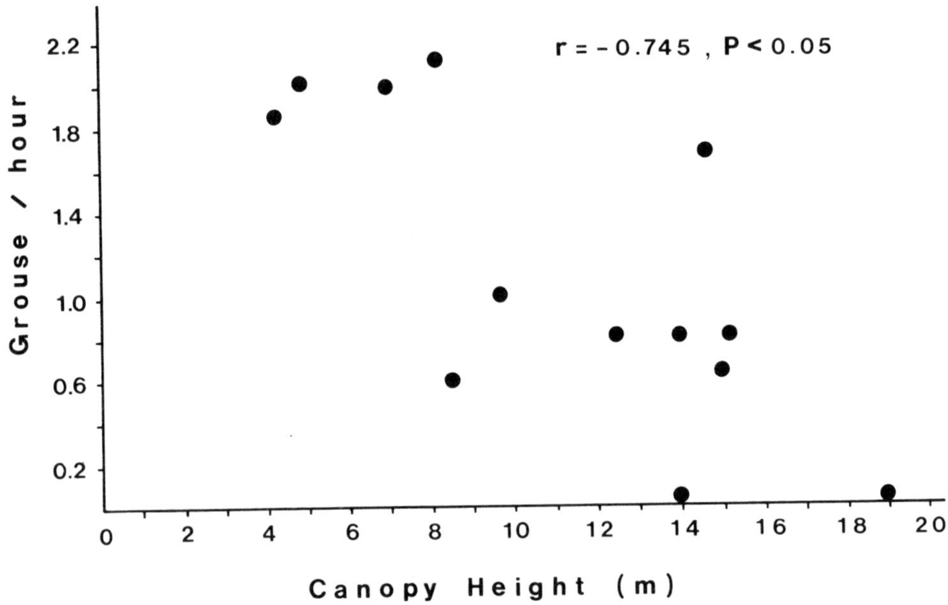

Figure 2. Relationship between canopy height and relative densities of spruce grouse populations in jack pine forests.

Table 2. Canopy and understory characteristics of four jack pine forests differing in relative population densities of spruce grouse. (Means sharing a common letter vertically are significantly different, P<0.05).

Forest code	Grouse/h* ($\bar{x}$ ± 1 SD)	Canopy Height (m)	Canopy Trees/ha	Understory % Cover black spruce	Understory %Cover ** deciduous
KL-CF	trace	14.0	2826	0.3 [c]	trace
CH-PP	0.63 ± 0.21[a]	15.0	3105	2.7 [e]	18.8
CO-P29	0.69 ± 0.17[b]	14.0	2794	5.4 [cd]	trace
CO-P25	1.67 ± 0.30[ab]	14.7	2890	13.3 [cde]	trace

* Trace = only droppings found; populations too low to measure in grouse/hour.
** Trace = < 1%.

spruce grouse populations (Table 2). Furthermore, within these forests spruce grouse appeared to select for black spruce. There was significantly more black spruce at grouse locations than was available in the stands as a whole (CO-P25: $U_s = 168.5$, $U_{0.01\ (16,12)} = 151$; CO-P29: $U_s = 200.5$, $U_{0.01\ (12,20)} = 186$).

To determine the generality of the above results, all 13 jack pine forests were plotted against canopy height, trees/ha, and percentage cover of understory black spruce (Figure 3). The amount of understory spruce separated forests with high, medium and low density grouse populations only in the case of the three forests just discussed (KL-CF, CO-P29, CO-P25; c.f. Table 2, Figure 3). In young jack pine forests understory black spruce was

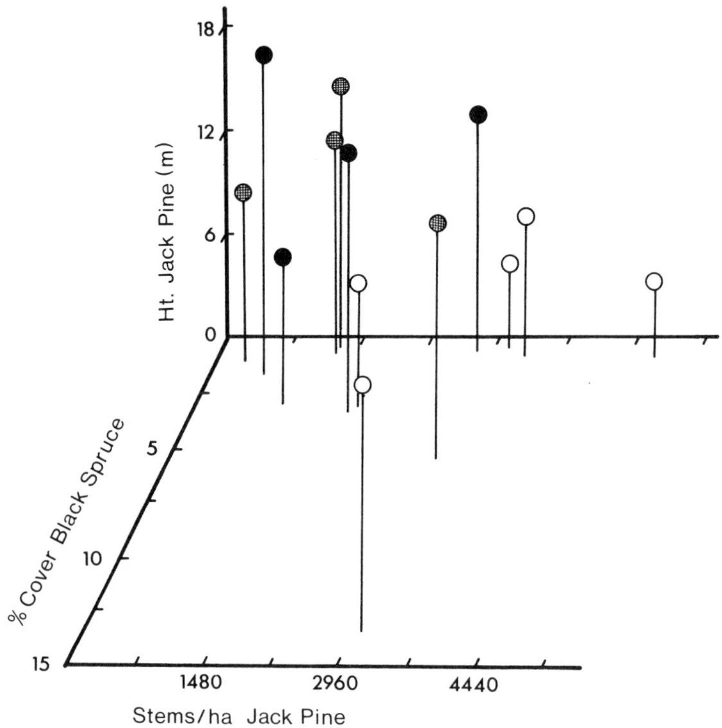

Figure 3. Association between relative densities of spruce grouse populations and three features of forest cover in jack pine forests. (Grouse populations: ○ high, ◍ medium, ● low).

unnecessary for high populations of grouse to occur. Some high populations existed where there was less than one per cent black spruce cover (Figure 3).

If understory cover is essential for high densities of grouse to occur in tall (old) jack pine forests, then an extensive deciduous component may serve as well as black spruce. Forest CH-PP was similar in height and trees/ha to forests KL-CF, CO-P29 and CO-P25 (Table 2). However, unlike the latter forests the understory consisted primarily of deciduous species such as *Acer spicatum, Alnus crispa, Betula papyrifera, Corylus cornuta,* and *Sorbus* spp. In terms of the extent of the understory (all species combined) CH-PP was similar to CO-P25. However, grouse/hour was different (Table 2). In terms of the percentage cover of black spruce, CH-PP was similar to CO-P29, and grouse populations were comparable. Thus, in older stands, the type of understory (deciduous or black spruce) appeared as important as its extent.

Combined influence of canopy and understory

No single attribute of the tree layer accounted for differences in grouse populations across the entire spectrum of jack pine forests (Figure 3). Canopy height, trees/ha and the percentage cover of understory spruce appeared to be important in specific cases. We developed a new variable, foliage area index (FAI), to quantify their combined effect on amount of available cover.

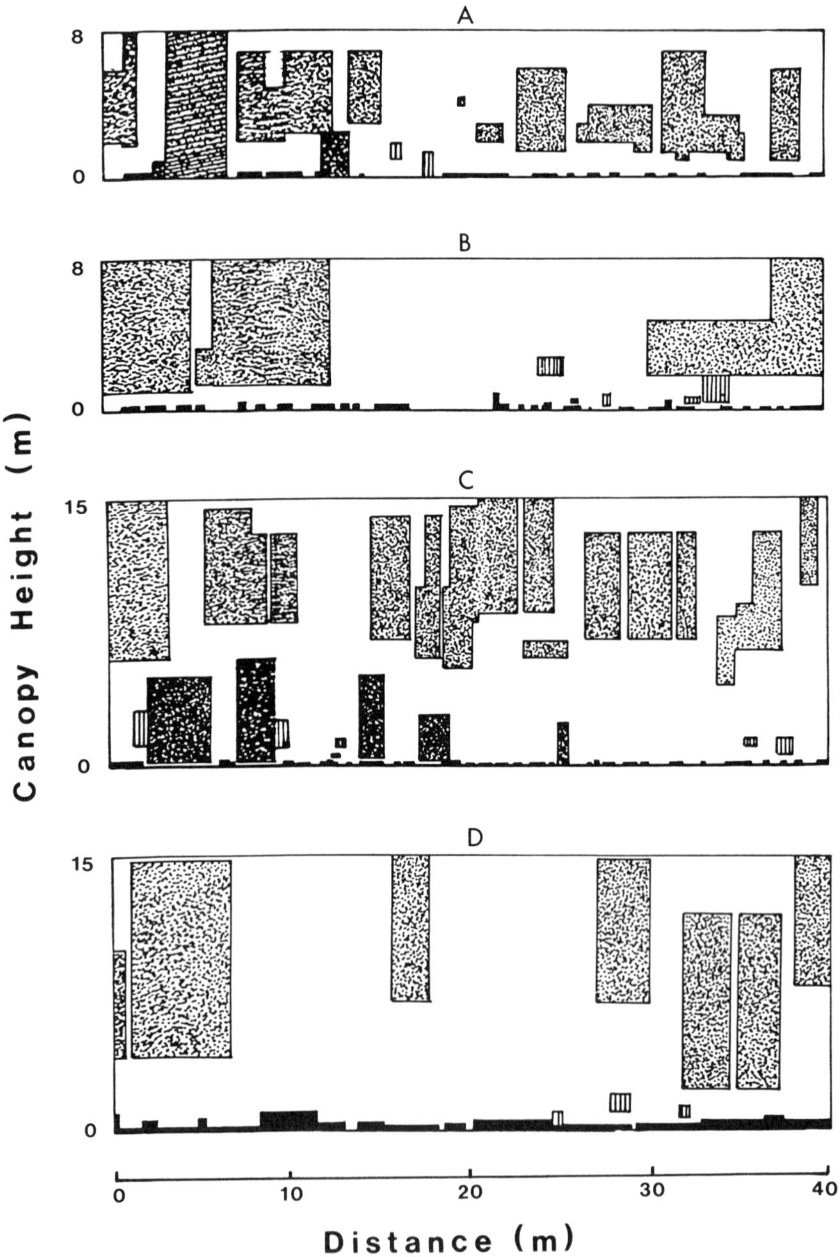

Figure 4. Illustration of the contributions of vertical and horizontal cover to foliage area index (FAI). The profiles represent actual 40 m sections of line intercept in two short (approx. 8 m) and two tall (approx. 15 m) jack pine forests. (A = stand CH-EGH, FAI = 53 per cent; B = stand CH-OGH, FAI = 39 per cent; C = stand CO-P25, FAI = 47 per cent; D = stand GO-G, FAI = 34 per cent). (Legend: lightly stippled — jack pine, stippled and slanted bars — balsam fir, heavily stippled — black spruce, solid black — low shrubs and herbs, vertical bars — deciduous shrubs).

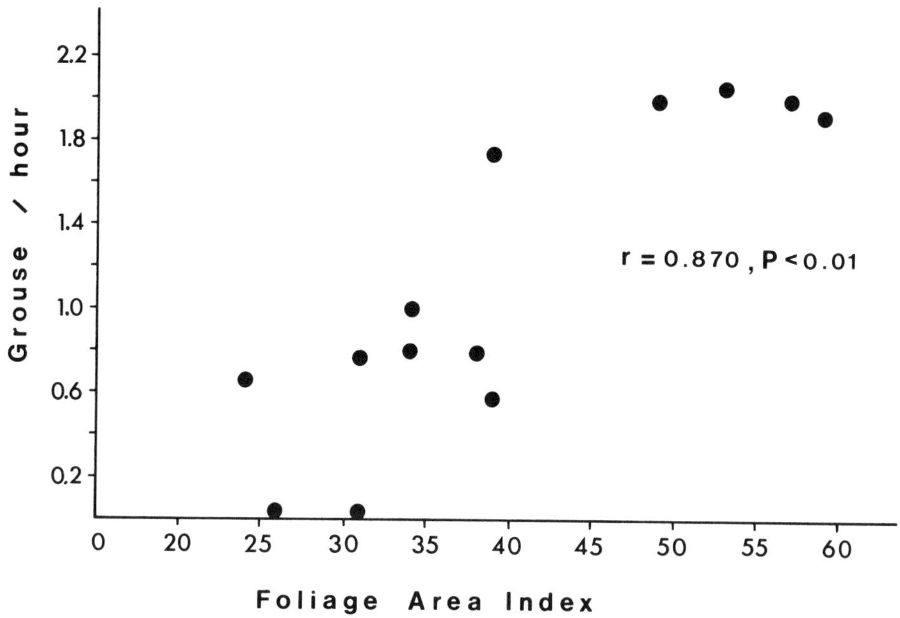

Figure 5. Association between relative densities of spruce grouse populations and foliage area index in jack pine forests.

Line intercept data were used to construct two-dimensional profiles of the 13 jack pine forests. The area of all portions of trees and shrubs intercepted by the line from ground level to the top of the canopy was summed. The accumulated area of foliage was expressed as a percentage of the area available. Thus, FAI was:

$$100 \times \frac{\text{area of foliage}}{\text{length of line intercept} \times \text{mean canopy height}}$$

Differences in canopy height, trees/ha, amount of understory vegetation, and distance from ground level to the lower living canopy are all accounted for in FAI. Thus, two forests equal in all but one of these variables have different values of FAI. Short and tall forests with relatively high and low FAI are shown in Figure 4.

FAI was highly correlated with grouse/hour (Figure 5). Grouse populations tended to increase with increasing FAI.

Comparison of spruce grouse populations in North America

The highest population density of spruce grouse in Ontario, determined by a total count of individuals (approx. 80 adult males and females/100 ha, Table 1), was considerably greater than the highest density recorded elsewhere in North America (19.3 adult males and females/100 ha, Table 3). The highest elsewhere was sparse relative to most others sampled in Ontario (Tables 1 and 3).

When forests with similar features of the canopy are compared across North America, spruce grouse populations appear to be comparable, with a few exceptions. The lodgepole pine forest studied in Alberta (Table 3) is probably most like forests CO-P25, CO-P29, CH-PP and KL-CF (Table 1). Spruce grouse populations in forests of this age and trees / ha appear to be comparable, except for CO-P25 and KL-CF. The black spruce cover in the understory of CO-P25 may account for its unusually high population. However, the low population in KL-CF cannot be explained on the basis of the canopy. Differences in the herb and low shrub layer may be responsible. The forest studied in Michigan can be compared in canopy features and grouse population levels to forests CH-OGH and WR-CL in Ontario (Tables 1 and 3). Ellison (1974) worked in mature spruce and found densities as high as the lower limits of the predicted density for stand GO-SB in Ontario. Both of these were upland forests. The study area in New York State (Table 3) appears to have a denser population than most lowland black spruce forests in Ontario.

Discussion and Conclusions

Population density

Spruce grouse populations exist in a wide range of densities in jack pine and black spruce forests in north-central Ontario. They appear to respond to differences in the canopy and sub-canopy, at least in jack pine. Thus, spruce grouse offer good material to show how forests and forestry affect wildlife. Populations ranged from almost undetectable low levels to a high of approximately 80 adult males and females / 100 ha, or four times the highest density recorded elsewhere in North America. The highest population was in a jack pine plantation.

No other studies of spruce grouse have been conducted in young dense jack pine forests 11 to 21 years old. Hence, it is difficult to speculate on the generality of the high populations recorded in these forests in Ontario. The region near Chapleau and Gogama, Ontario (Figure 1), where the densest spruce grouse populations were found, is intensively managed for jack pine. However, it is doubtful that these extreme populations are a local phenomenon, since one population in Cochrane (Table 1, CO-P25) was also high. Therefore, dense spruce grouse populations may exist in other regions. More studies of density in other parts of spruce grouse range are needed.

Comparison of population density in similar forests across the continent is difficult because data are available for only a few areas, and most are at the edges of spruce grouse distribution. Studies of spruce grouse generally present population density or habitat information, but not both. Only those providing both were considered here. The limited data suggest that populations of spruce grouse throughout North America are likely similar in jack pine and black spruce forests of equal age, canopy height and trees / ha, with some exceptions.

Habitat relationships

Stand structure is considered important to selection of habitats by grouse within (McCourt 1969; Robinson 1969; McLachlin 1970; Hedberg 1980) and among forests (Frandsen 1980). In jack pine forests spruce grouse populations tend to decrease with increasing canopy height. However, the association is weak since there were significant differences

Table 3. Densities of spruce grouse populations studied in coniferous forests in North America in spring. (Hatched lines indicate that data were unavailable)

Location	Forest type	Canopy height (m)	Age†	Trees/ha	Maximum recorded grouse population (♂♂+♀♀/100 ha)	Source
Southwestern Alberta	lodgepole pine*	7-15	35-48	3835	19.3	Herzog and Boag (1978), Boag *et al.* (1979)
Northern Michigan	jack pine	8	(16)	447	9.6	Robinson (1969, 1980)
Southern Alaska	black and white spruce	—	mature	—	10.0	Ellison (1974)
New York State	black spruce	—	—	—	16.0	Chambers (1980)

† Age in brackets is inferred from information on stand origin.
* *Pinus contorta*

in grouse/hour among forests of comparable height. High populations of grouse occurred in all young dense jack pine forests sampled, but were not restricted to these forests alone.

In jack pine forests approximately 14 m in height, where there was typically a space of 8 to 9 m from the ground to the lower living canopy, spruce grouse populations were relatively sparse. In one forest the grouse population was higher than expected, a condition we relate to the abundance of understory black spruce.

Robinson (1980) suggested that mixed jack pine and black spruce are preferred by spruce grouse in Michigan. In Ontario, this appeared to be true only in taller forests. Black spruce was unnecessary for high populations to occur in young forests.

A new variable, foliage area index (FAI), was used to express the total amount of cover provided by the canopy and understory combined in jack pine forests. FAI was positively correlated with grouse/hour, and was the best single habitat feature indicating the level of grouse populations. However, there was considerable variation in grouse/hour for given values of FAI. Thus, the relationship between habitat and spruce grouse populations may be more complex than this simple correlation suggests.

While there are good associations between population density and forest cover in jack pine, more research concerning habitat requirements of spruce grouse is needed. The correlations we report do not explain how habitat acts on grouse to produce population density.

Habitat requirements

Gullion (1970) stated that ruffed grouse *(Bonasa umbellus)* prefer deciduous forests with abundant short vertical cover. He suggested less predation results in higher populations in such habitats. Thus, young dense jack pine may produce high densities of spruce grouse primarily because its low canopy decreases mortality to predators. Similarly, in tall forests understory black spruce provides a structural link to the lower canopy, and may be important for the additional cover it affords.

Although the amount of understory cover was important to population density in tall forests, a large quantity of deciduous shrubs did not appear to have the same effect as black spruce. Furthermore, while black spruce was beneficial as a minor component in jack pine, pure black spruce was not as good a habitat. The highest population we found in spruce forests was only one-third the level in jack pine. Perhaps differences in the herb layer, and possibly in microclimate, also influence the density of spruce grouse populations, since spruce grouse feed on a variety of plants (Pendergast and Boag 1970; Lattner 1982) and are mostly terrestrial during the breeding season. Information concerning the herb layer is required to help understand the characteristics of dense and sparsely populated forests.

Relationships between population density and the amount and quality of food have been documented for many species of birds (Lack 1966; Gullion 1967; Miller *et al*. 1970; Armleder 1980; Frandsen 1980; King and Bendell 1982). Lattner (1982) suggested that the abundance of invertebrates in the ground and herb layer, almost the only food of young spruce grouse chicks, is related to population density of grouse in jack pine forests. However, preliminary results (unpub. data) from two years of intensive work in three jack pine plantations suggest that invertebrate food available for chicks is not a crucial factor. Older forests with sparse grouse populations may have fewer invertebrates, but chick growth is comparable and their survival is better than in young forests.

Bendell and Zwickel (1978) suggested that nesting requirements, including food for incubating females, may be a limiting factor in the ecology of North American forest grouse. Comparative studies of demography and habitat-use in jack pine forests with high and low grouse populations are needed to determine how differences in habitat affect reproduction and predation.

Implications to forest management

The results suggest a number of management strategies for sandy upland sites. High populations of spruce grouse occurred in essentially pure even-aged plantations of jack pine from 43 to 766 ha in size. Thus, present plantation design may require a minimum of alteration to produce habitat capable of supporting dense spruce grouse populations. We have demonstrated that jack pine plantations are by no means sterile wildlife habitats. A similar conclusion was reached by Naylor and Bendell (1983) for small mammal populations.

The simplest strategy is to ensure that young (less than 25 years) jack pine plantations are dense (at least 2300 trees/ha). This density can be achieved by planting at 1.83 × 1.83 m (6 × 6 ft.) spacing, providing that survival of planted stock is almost complete. A more realistic approach may be to space stands (to no more than approximately 4500 trees/ha, all species) after aerial seeding. With either of these two options high spruce grouse populations may occur for relatively few years (from stand age less than 11 years to approximately 25-30 years). The potential capability of a jack pine forest to support high grouse populations over a long period may be enhanced if black spruce forms a sizable portion of the understory. This may be achieved by mixed plantings of jack pine and black spruce (approximately 10-15 per cent evenly spaced black spruce), or by retention of advanced spruce regeneration after logging preceding planting to jack pine. Short rotations (less than 60 years) may be beneficial to spruce grouse populations.

In this study we have a somewhat truncated view of jack pine forests, considering only those 11 to 62 years of age with 536 to 4863 trees/ha. Thus, the management strategies we outline may be appropriate only for forests similar to those we have examined. It must be recognized that these suggestions are derived from work in relatively few forests, and are based on correlations where some exceptions were evident.

A most important qualification is that we do not know how jack pine plantations produce the population densities we observed. The mere presence of dense forest cover may not ensure high grouse populations. We have not included aspects of the ground vegetation or possible effects of forest openings. It is our impression that dense jack pine plantations, where survival of stock is high throughout, may not be as desirable to spruce grouse as situations where unplanned death of stock has created small openings in the canopy. Research into these areas is needed before realistic predictions concerning the probability of success of the suggested strategies can be made.

Acknowledgements

The following were fundamental to the successful completion of the project: Julie E. Murphy, Brian J. Naylor, Lucille Szuba, Ted Szuba, Tom Szuba and Yvonne Bendell.

We thank Ontario Ministry of Natural Resources personnel in Chapleau, Cochrane, Gogama, Kirkland Lake and White River for their help in locating study forests. Special thanks are extended to R.E. Jerrard, Norman W. Quinn and H.G. Lumsden. We appreciate the financial support of the Natural Sciences and Engineering Council, the Ontario Ministry of Natural Resources, and Faculty of Forestry grants to J.F. Bendell. We thank General Foods Ltd (Gravy Train), and S.C. Johnson and Son Ltd (Deep Woods Off!) for their contributions.

Literature Cited

Aldrich, J.W. 1963. Geographic orientation of North American Tetraonidae. J. Wildl. Manage. 27: 529-545.

Armleder, H.M. 1980. Habitat use and movements of hens and broods in relation to demography of blue grouse. M.Sc.F. Thesis. University of Toronto, Toronto, Ontario.

Bendell, J.F. and F.C. Zwickel. 1978. Problems in the abundance and distribution of blue, spruce and ruffed grouse in North America. *In*: The ecology of woodland grouse. T. Lovel (Ed.). World Pheasant Assoc., Andover, England. pp. 48-63.

Boag, D.A., K.H. McCourt, P.W. Herzog and J.H. Alway. 1979. Population regulation in spruce grouse: a working hypothesis. Can. J. Zool. 57: 2275-2284.

Carleton, T.J. and P.F. Maycock. 1978. Dynamics of the boreal forest south of James Bay. Can. J. Bot. 56: 1157-1173.

Chambers, R.E. 1980. Distribution and abundance of spruce grouse in the Adirondack region of New York. College of Environmental Science and Forestry, State University of New York, Syracuse, New York. Unpubl. Rep.

Ellison, L.N. 1971. Territoriality in Alaskan spruce grouse. Auk 88: 652-664.

Ellison, L.N. 1974. Population characteristics of Alaskan spruce grouse. J. Wildl. Manag. 38: 383-395.

Frandsen, D.H. 1980. Density, habitat and behaviour of blue grouse. M.Sc.F. Thesis. University of Toronto, Toronto, Ontario.

Grosenbaugh, L.R. 1952. Plotless timber estimates — new, fast, easy. J. For. 50: 32-37.

Gullion, G.W. 1967. Selection and use of drumming sites by male ruffed grouse. Auk 83: 87-112.

Gullion, G.W. 1970. Factors influencing ruffed grouse populations. Trans. N. American Wildl. Nat. Res. Conf. 35: 93-105.

Haas, G.H. 1974. Habitat selection, reproduction, and movements in female spruce grouse. Ph.D. Thesis. University of Minnesota, Minneapolis, Minnesota.

Hedberg, J. 1980. Habitat selection by spruce grouse in eastern Maine. M.Sc. Thesis. University of Maine, Orono, Maine.

Herman, M.F. 1980. Spruce grouse habitat requirements in western Montana. Ph.D. Thesis. University of Montana, Missoula, Montana.

Herzog, P.W. and D.A. Boag. 1978. Dispersion and mobility in a local population of spruce grouse. J. Wildl. Manag. 42: 853-865.

King, R.D. and J.F. Bendell. 1982. Foods selected by blue grouse *(Dendragapus obscurus fuliginosus)*. Can. J. Zool. 60: 3268-3281.

Lack, D. 1966. Population studies of birds. Clarendon Press, Oxford, England.

Lattner, M.T. 1982. Foods, mobility, habitat selection and density of spruce grouse. M.Sc.F. Thesis. University of Toronto, Toronto, Ontario.

McCourt, K.H. 1969. Dispersion and dispersal of female and juvenile Franklin's grouse in lodgepole pine forest in southwestern Alberta. M.Sc. Thesis. University of Alberta, Edmonton, Alberta.

McLachlin, R.A. 1970. The spring and summer dispersion of male Franklin's grouse in lodgepole pine forest in southwestern Alberta. M.Sc. Thesis. University of Alberta, Edmonton, Alberta.

Miller, G.R., A. Watson and D. Jenkins. 1970. Responses of red grouse populations to experimental improvement of their food. *In*: Animal populations in relation to their food resources. A. Watson (Ed.). Blackwell, Oxford, England. pp. 323-355.

Naylor, B.J. and J.F. Bendell. 1983. Influence of habitat diversity on the abundance and diversity of small mammals in jack pine forests in Ontario. *In*: Resources and dynamics of the boreal zone. R.W. Wein, R.R. Riewe and I.R. Methven (Eds.). Association of Canadian Universities for Northern Studies, Ottawa, Ontario.

Pendergast, B.A. and D.A. Boag. 1970. Seasonal changes in diet of spruce grouse in central Alberta. J. Wildl. Manag. 34: 605-611.

Robinson, W.L. 1969. Habitat selection by spruce grouse in northern Michigan. J. Wildl. Manag. 33: 113-120.

Robinson, W.L. 1980. Fool hen. University of Wisconsin Press, Madison, Wisconsin.

Sokal, R.R. and F.J. Rohlf. 1969. Biometry. W.H. Freeman and Co., San Francisco, California.

Szuba, K.J. 1982. Population densities and weights of spruce grouse in Ontario. M.Sc.F. Thesis. University of Toronto, Toronto, Ontario.

Watson, A. and R. Moss. 1979. Population cycles in the Tetraonidae. Ornis Fennica, 56: 87-109.

Zwickel, F.C. and J.F. Bendell. 1967. A snare for capturing blue grouse. J. Wildl. Manag. 31: 202-204.

LIMITATIONS TO PREDICTABILITY OF PLANT SUCCESSION IN NORTHERN ECOSYSTEMS

Ross W. Wein and Mohamed A. El-Bayoumi
University of New Brunswick

Abstract: Concepts of community succession have been viewed from many perspectives because it is such an important aspect of ecology. This paper discusses some of the limitations of the classical concepts of succession when they are applied to northern ecosystems. We attempt to illustrate the range of community changes that are recognized by providing examples of random fluctuations, cyclic fluctuations and succession (*sensu* Odum 1969) under the heading of the temporal dimension. We also recognize that there is a spatial dimension to community succession which has received less attention in the literature and which, unfortunately, cannot be studied readily through traditional techniques. We highlight two models that have successfully captured the dynamics of ecosystems and we suggest that other simple models should be constructed to test management strategies. The results of alternate management strategies can be simulated and the output will determine if the strategy should be adopted or rejected.

Successional replacement sequences and their cause have been of prime scientific interest to plant ecologists since the late 1800s. Different ecologists formulated different hypotheses about this phenomenon. Since Warming (1896) and Cowles (1901) described the stages of sand dune succession, the Clements (1904, 1916) classical concept about succession influenced many ecologists' understanding of the phenomenon. The model case of that concept is that of secondary succession in the deciduous forest, in which there is a 'disturbance' followed by a series of predictable changes that are initiated by the invasion of 'pioneer' species, sustained by subsequent invasions and ended with an equilibrium climax state.

General dissatisfaction with the classical description of succession as presented by the above model and as derived from the Clements (1916, 1936) concept has existed for a long period (Gleason 1926; Cooper 1926). Events that qualify as 'disturbance' in one ecosystem may not be considered so in another and accordingly, they may produce little, if any, compositional change. For instance, in boreal forest ecosystems, fire has been widely recognized as an induced disturbing factor, or a built-in natural random factor (Heinselman 1981). Furthermore, the classical concept, viewing succession as a deterministic, mechanistic, regeneration of the disturbed community, has limited applicability to many ecosystems and allows only a deterministic 'single pathway' prediction of succession (Cattelino *et al*. 1979). The assumption that the path is independent of the nature of the disturbance and vegetation at the time of disturbance is no longer acceptable.

The quantitative identification of different seral stages in the progression toward the equilibrium state became the subject of many studies (Field 1947; Kershaw 1957, Lawrencc 1953, Odum 1969). Prediction of succession based on simulating the characteristics of the seral stages regardless of the species attributes (Noble and Slatyer 1980) or the dynamics of the concerned system (Botkin *et al*. 1972) have been used to a limited extent in northern ecosystems. It is only recently that alternative descriptions of vegetation replacement sequences (plant succession) and suggestions for modelling them have been proposed (see West *et al*. 1981).

The objectives of this paper are to discuss the limitations of the classical concepts of succession, to illustrate the range of community changes that have been recognized as succession, and to suggest approaches that may be used in predicting successional patterns in northern ecosystems.

Limitations to Succession Concept

Secondary succession is usually defined as an orderly replacement of species of a community over time, which leads to a stable (climax) condition or mature stages (*sensu* Odum 1969). This definition may hold for temperate region ecosystems where organisms strongly modify the physical and chemical environmental factors and where one species, while playing out its life cycle, prepares the way for other species. At some point a community tends to perpetuate itself. In the Arctic and boreal ecosystems, this definition does not appear to hold particularly well because the physical environment appears to dominate. For instance, under conditions of extreme environmental stress (drought, cold soils, low nutrient levels) such as in the High Arctic or Arctic-alpine areas, there appears to be only a random replacement of individuals so that community species number and abundance change extremely slowly, if at all. Even in the boreal forest, where the biota modifies the environmental factors to a greater extent, it has long been recognized that succession is not easily predicted because the ecosystem is influenced by repeated disturbance (Rowe 1961). The physical factors, particularly fire, permit community variability to be expressed while preventing the expression of a self-perpetuating community.

We wish to discuss the northern examples of succession under two main dimensions: the temporal and spatial. When discussing the temporal dimension we have recognized that the time of recovery to predisturbance conditions is related positively to the severity of the disturbance. When discussing the spatial dimension, we recognize the scale of the disturbance and that disturbance severity can be spatially-specific.

Temporal dimension

The study of community changes ranges from primary succession over thousands, or hundreds of years, to secondary succession over hundreds of years to decades. The focus for this paper is secondary succession and the examples have been chosen primarily from forested ecosystems of the Mackenzie River Basin in northwestern Canada. Since the maximum ages reached by southern boreal tree species ranges up to hundreds of years (Fowells 1965) this is the time span focus if no individuals establish during succession. For the purpose of discussing the views of secondary succession in northern ecosystems, we are using three divisions: namely, random fluctuations, cyclic fluctuations and succession (*sensu* Odum 1969). These divisions can be viewed to be increasing in disturbance severity and require increasing time for the community to make a full recovery.

Random fluctuations: We are defining random fluctuations as those plant community changes caused by disturbances which result in no species change, although there may be a change in the relative abundance of species. There are few quantified studies on vegetation recovery after this type of disturbance. Essentially this would be unpredictable, short-term (year-to-year) variability. Examples would be due to the somewhat random collection of food by animals, such as caribou cratering in snow to find lichens, hares clipping shoots of shrubs and small trees or ptarmigan utilizing buds of shrubs not covered by snow. Lightning strikes killing individual or small groups of trees could also be considered as a random fluctuation. Human activities such as the use of snowmobiles and even larger machinery (Hernandez 1973) may also cause this type of disturbance if sufficient snow is present at the time of the activity. Environmental factors are changed little by these low severity disturbances and, therefore most plant species that have the ability to re-sprout from adventitious stem buds or buds at the root collar will recover within one or two years.

Cyclic fluctuations: We are defining these changes as those that show cyclic change in the relative abundance of species but where there are no changes in species over time. This means that there are feedback mechanisms operating that reinforce or respond to environmental cycles (Horn 1976). The time period involved may be in terms of decades. A number of examples can be found in the Mackenzie River Basin. For instance, bogs do not develop uniformly but often show a hummock-hollow cycle over time. Gill (1972) has drawn attention to the fact that beavers dam small channels near the *Populus balsamifera* habitat of the Delta where they harvest the trees. This harvesting is somewhat cyclic since the habitat is usually over-utilized and must be abandoned until the *Populus* recovers. Another example that is found in physically-dominated ecosystems is that of the vegetation regeneration wave. Sprugel (1976) in the northeastern United States recognized that in balsam fir forests, trees mature and die, then as the tree canopy opens, seedlings in the understory grow quickly to establish the forest cover again. This distinct wave action moves across the plant community in the direction of the prevailing winds. This specific example does not have a northern counterpart but Pruitt (1958) described a similar type of wave action that radiates from a central point. He described the accumulation of snow (qali) on trees during windless periods to the point that trees will break, particularly if they are already leaning. As the loaded or breaking tree lodges against similarly snow-loaded trees, the breaking trees create an opening in the forest. The opening grows until wind circulation prevents the accumulation of the snow on the trees. Apparently this phenomenon has not been quantitatively documented which is not surprising since few plant ecologists work under winter conditions when trees are dormant.

Another physical factor that can perpetuate cycles is freezing and thawing of soils. In many upland areas the freeze-thaw cycle obviously creates unstable soil conditions which, in turn, create plant community pattern changes on a micro-scale around mudboils. Often it is quite difficult to identify changes from year-to-year but in frost boil areas of the northern forest, the post-fire 'drunken forests' described by Zoltai and Pettapiece (1974) and Zoltai (1975) are the ultimate expression of this phenomenon.

It is generally recognized that on these wet permafrost soils fire does not usually burn to mineral soil so protected underground plant parts lead to fast recovery. While studying post-fire plant community recovery near the treeline, Black and Bliss (1978) noted that there were few vascular species number changes because most of these plants re-sprouted. Many of the tundra fires recorded near Inuvik (Cody 1965; Wein 1975, 1976) also showed this phenomenon.

In these examples, plant community re-establishment is predictable over a short period of the cycle. It should also be noted that the post-fire cyclic pattern of community re-establishment may be delayed, especially under harsh environmental conditions. For example, regrowth of trees may be delayed because of poor seed availability so that the former understory shrubs increase in abundance (Hawkes 1983).

Succession (sensu Odum *1969):* Rowe (1961), Methven *et al.* (1975) and others have recognized that succession in the traditional sense does not occur in the boreal forest. Other authors have found difficulty in using the term 'succession' when viewing post-fire vegetation and have sought other terminology. Johnson and Rowe (1977) suggested 'vegetation composition change' to describe post-fire black spruce-lichen stands while Black and Bliss (1978) described a 'vegetation recovery sequence' in black spruce communities.

As with any developing science that has been largely descriptive, the perspective of the researcher will determine how boreal succession is viewed. If the objective of the research was primarily to study trees, then there may be no species change and therefore no succession after a disturbance (Johnson and Rowe 1977). If, however, the objective of the research was to study overstory and understory postdisturbance changes, then it could be argued that secondary succession was described because there was an orderly replacement of species on the site. Without denying the value of the available studies, it should be recognized that the state of the science is such that more succession examples are based on experienced opinion and hypotheses than on quantitative data. The few studies that are available are scattered over large geographic areas and have such varying detail that generalizations are difficult to identify.

In examining the range of studies termed succession studies, it is readily apparent that authors recognize that succession is disturbance-specific. Within each type of disturbance there is a range of severity that will dictate the recovery period. In general, recovery periods may be decades to several hundred years.

Postfire succession studies received a good deal of attention in the boreal ecology literature since fire is one of the all-pervasive disturbing agents. A few of the recent reviews include Auclair (1983), Cayford and McRae (1983), Furyaev *et al.* (1983), Heinselman (1981), and Viereck (1983). As recently reviewed by Alexander (1982), the influence of fire on the ecosystem depends very much on the behavior of fire. In general, with increasing removal of organic matter during the fire, there is a post-fire release of nutrients and a change in the soil thermal balance so that nutrients cycle at a faster rate and in greater quantities. As the vegetation cover re-establishes and shades the soil surface, decomposition of the organic matter decreases and the soils remain cooler in summer. The organic matter continues to accumulate, immobilizing nutrients until the original low nutrient status of the system is reached (Heilman 1966, 1968). Strang (1973) has suggested that if fire does not occur in fire-prone areas within the life span of the trees, the area may become treeless.

To give the state of the descriptive science, the reader is referred to the two succession diagrams presented in Viereck (1983) and Damman (1964). These qualitative descriptions show the relationship of boreal forest types that occur because of a range of fire frequencies interacting with moisture conditions of a specific site. These diagrams represent a high probability of vegetation changes and are based on years of field observations.

In post-fire succession studies, vascular plant community changes may show species replacement to a certain extent but in many cases, species are found within the original community. Even the so-called invasion of species such as *Epilobium angustifolium* or *Calamagrostis canadensis* is really a change in abundance since the individuals, of low vigour, are present in the predisturbance community. In contrast, for cryptogamic flora, Black and Bliss (1978) have shown that there is an orderly replacement of species over time following fire. Cryptogams are usually oxidized or killed by fire so invasion must take place from outside the burned area.

Post-insect attack succession has received some attention in the literature from the point of view of tree composition changes (Baskerville 1975; Baskerville and MacLean 1979) but there are some short-term vegetation change studies on the understory species (Ghent *et al*. 1957). In general, insect attack succession studies focus primarily on the insect feeding on specific tree species. When these are removed from the community, this releases understory species and gives the surviving species a competitive advantage.

Industrial disturbances of forested areas, such as through-tree harvesting, has received attention since it is recognized that this particular disturbance can cause significant changes in forest composition. In general, there are no species changes but it is widely recognized that broad-leaved species rapidly invade areas originally dominated by conifers (Jeglum 1983; Yang and Fry 1981). Since harvesting has been focussed largely on the southern boreal forest, most succession studies are relatively short-term or may not apply directly to the more northern portions of the boreal forest.

A very severe disturbance that can strongly influence plant succession occurs in ice-rich permafrost soils in the northern portions of the boreal forest. Wave action or moving water can remove over-burden so that the thermal conditions of the soil changes. Soil or earth flows can occur which lead to very distinct revegetation patterns (Lambert 1968). Mechanical disturbance caused by road building, or seismic activity related to oil exploration can change terrestrial systems to a series of ponds within the seismic trail when the ice in the soil melts. For this degree of disturbance, species that are adapted to the new conditions must colonize from surrounding habitat (Hok 1969; Hernandez 1973).

In most of the above examples, the temporal dimension of succession is given in extremely general terms. Prediction is not quantified but this is understandable because of a number of sources of variability. As with all ecosystems that have organisms with life spans of decades or centuries, researchers are forced to reconstruct change from one point in time over a small portion of the total time period. To understand this time related change, researchers have examined spatially distinct areas that have been disturbed at a different time in the past. This results in the confounding of the severity of the disturbance factors since the severity probably was not the same in each case and the confounding of the relationship with the inherent biologically and substrate differences in spatially distinct areas.

Spatial dimension

The temporal dimension of community succession has occupied the greatest proportion of researchers time but one must recognize the spatial dimension because it can strongly influence the temporal dimension (Rowe 1983). In a review of ordination and multivariate analysis in identifying successional trends, Austin (1977) differentiates between a 'static' and a 'dynamic' approach to this problem. In the static approach, data are collected in one point of time (from supposedly different temporal phases), while in the dynamic approach, data from an area are collected at several points in time (re-measurements are made). The former approach is based on the grossly simplified assumption that the temporal phases are products of similar disturbances occurring under similar biological and environmental conditions. This assumption implies a 'single pathway' for community development.

There are two aspects to the spatial dimension; the first is that there is a gradient of scales, and the second is that the intensity of the disturbance may be spatially-specific.

On a small-scale, it is relatively easy to determine the changes that have occurred over time because one can examine communities that have been disturbed at different times. On a larger scale, there is no succession because all of the stages are present as pockets within the larger stable ecosystem boundary.

Succession can be confounded by spatial dimensions, especially if the severity of the disturbance is spatially-specific. Certain areas on the landscape where succession is taking place, should be recognized as being distinct from general succession patterns. These exceptions may include flood plains, transportation routes or areas of localized disturbance, such as pollution. Other disturbances, such as insect outbreaks or fire, affect larger areas of the landscape and yet the severity of these factors may be quite different within the landscape. For instance, if a fire is relatively small, the fire is likely burning under weather conditions that are not conducive to fire propagation and the proportion of organic matter removed is small. There may be little or no change to species diversity but because the area was a small concrete area, it can be seen as a distinct unit. On the other hand, if a fire burns over a large area of the landscape, species diversity may change in a localized area due to the fire burning more deeply into the convex portions of the landscape and less deeply on the concave portions. Thus for the same disturbance agent the effect on ecosystems can be quite different. It may be argued, that the smaller the area of disturbance, the more it will be influenced by the surrounding undisturbed area and therefore the faster the area should recover.

The traditional views of succession have tended to be static but it is well documented that the driving variable may influence succession so that a range of end products results. It is recognized, for instance, that when spruce-fir-pine conifer mixtures are burned, fir tends to decrease while spruce and pine increase in abundance. If a similar mixed conifer forest is clearcut spruce and especially fir increase in abundance. Unfortunately these are only a few end points. We need to recognize others but more importantly, it would be extremely useful to know the probability of achieving any end point, given certain uncontrollable or management treatment influences on a given stage of succession.

Prediction of Succession

One final aspect of plant succession that must be addressed is that of predictability. In much of the literature on succession the short-term predictions are better based because field measurements have been made within the life span of a research study period. However, the long-term predictions are really a poorly-quantified guess.

The most recent approach to studying plant succession is by modelling. Early concepts (Clements 1916) assumed that following a disturbance, a community regenerated a semblance of itself by an orderly and predictable series of species replacements. This assumption allows a deterministic 'single pathway' prediction of succession. Such a classical concept, viewing succession as a deterministic, mechanistic regeneration of the disturbed community, has limited applicability to, for example, dynamic fire-prone ecosystems (Cattelino *et al*. 1979). The assumption that the path is independent of the nature of fire and vegetation at the time of fire (fire is fire) is no longer acceptable. Holling (1978) has emphasized that systems cannot be understood simply by measuring successional changes over a short survey period. More recently, other researchers have developed different perspectives. The JABOWA model (Botkin *et al*. 1972) mimicked successional changes on an individual tree basis by considering competition, environmental variables, growth rate, survival characteristics and above all the interaction of the plants and environment by allowing the plants to alter the light environment. A different approach was taken by Horn (1974) who assigned an event probability to each individual tree in the canopy and the understory tree species that may replace it. The replacement probabilities were then used to predict successional changes.

Alternatively, Whittaker (1975) characterized succession by developmental trends; that is, characteristics of the ecosystem that change through succession. These include species diversity, productivity, biomass and nutrient stock. These characteristics are good measures of succession, nevertheless, they are not actual processes of change but rather the results of processes. Similarly, Odum (1969) viewed succession as a process of ecosystem evolution. The three models of Connell and Slatyer (1977) described succession in terms of species facilitation, tolerance and inhibitive properties.

These models dealt with general disturbances with emphasis on replacement sequences observed in the plant communities (either species populations or community characteristics). They addressed neither the adaptive traits of a species persisting through a disturbance (Gill 1977) nor the periodicity of the disturbance (Vogl 1977).

Shugart and West (1977) developed a modified version (FORET) of the JABOWA model (Botkin *et al*. 1972) and used it to simulate the impact of the chestnut blight on an Appalachian deciduous forest succession. Another version of FORET, (BRIND) was developed to mimic the long-term behavior of *Eucalyptus*-dominated forest ecosystems in upper-elevational Brindabella Range near Canberra, Australia by Shugart and Noble (1981). El-Bayoumi (1982) used FORET to develop the SMAFS model that mimicked the long-term dynamics of a stand in the Acadian Forest of eastern Canada. An important asset of these modified versions was the biologically based model structure and the ease with which the model parameters could be estimated.

A forest stand is a highly dynamic entity that can, at any given time, reflect the ecological response of tree populations to a long history of environmental conditions when such a stand is considered over relatively long intervals. The period of time that may be needed to measure changes in forest ecosystems (Emamuel *et al*. 1978) and the long recovery period of forest systems following perturbations combine to make it difficult to ascertain system behavior.

Models such as JABOWA and FORET are useful to augment the incomplete empirical data available for investigating forest dynamics when used as an intentional tool for exploring the possible implications of known (or hypothesized) interactions among trees. Models are also extremely useful in suggesting productive areas for further research. Missing data or inadequately explored relationships will be quickly identified.

Now we wish to question if it is possible to predict the vegetation dynamics in northern ecosystems for management purposes. For a particular ecosystem, predicting the vegetation dynamics as a whole would be either a naive approximation of the complexity of the natural system, or an inflated research project. In addition, it should be recognized that the degree of prediction based on autecological information alone may not be high (Rowe 1983). With respect to the temporal dimension we are suggesting that predictions of the dynamics should be based on simple, yet realistic, and specific simulation models. These models should be built with the least number of variables and processes that can simulate the ecosystem's behavior within the range expected by experienced field researchers and managers.

To further illustrate this point we will refer to two examples of models built specifically to simulate northern ecosystems, those of Walters *et al.* (1975) and El-Bayoumi (1982).

In the first example (Walters *et al.* 1975), a computer model was built to simulate Barren Ground caribou dynamics. The natural system was represented in terms of the values of a series of indices indicating the system's condition at any point in time. These indices included the population size classified by age and sex; mortality, reproduction and harvest rates of each class; food habitat classified by successional stage or vegetation classes and the net production of each class; lichen standing crop and browse in each habitat class; and indices to indicate weather severity.

This model proved to be successful in that it demonstrated the problem of over-exploitation of the barren ground caribou. A solution to the over-exploitation was to reduce the harvest. Another hypothesis regarding the effect of the destruction of the winter habitat by fire, on the caribou population, was tested. Model output suggested that this was not as serious a threat to the herd as had been previously suggested.

The second example (El-Bayoumi *et al.* 1984) is based on a sequence of models developed for North American forests (Botkin *et al.* 1972; Shugart and West 1977; Mielke *et al.* 1978). In this SMAFS model the long-term succession of a mixed species stand in the Acadian Forest is simulated. The SMAFS model simulates a 1/12 ha forest stand by computing the growth of each individual tree in the stand. It considers establishment and death of trees on a tree-by-tree basis using stochastic functions. The growth of each tree is a function of its current size, maximum size (indicated by its maximum height and diameter), competition for space and light, and the climate. This model also succeeded in capturing some of the important dynamics of the complex systems that could be found in this type of forest. It has been used to study the natural succession and to make inferences about the behavioral response of the system with respect to different experimental treatments. Recent runs (not published yet) validate the hypothesis that the successional development in northern ecosystems is largely influenced by changes of the physical conditions of a site rather than the biological characteristics of the species.

These two examples are among the few successful attempts to simulate different northern ecosystems with respect to management practices in these ecosystems.

The treatment of the spatial dimension in modelling practices has differed according to the objective of the modelling exercise and the system to be modelled. In the model of Walters *et al.* (1975) the spatial dimension was not emphasized. The simulated system was considered a homogeneous independent unit with no interaction with neighbouring units. Baskerville and Kleinschmidt (1981) took a different approach in simulating growth in defoliated fir stands. They divided the large area to be simulated into smaller units, then simulated these units independently. Although infestation among neighbouring stands was recognized in the fir-dominated forest, it was not simulated. In the SMAFS model (El-Bayoumi *et al.* 1984) as well as in models based on the JABOWA model (Botkin *et al.* 1972) another approach was taken. Since these models are stochastic in nature the inherent variability among the units to be simulated was accounted for. Each unit was simulated 100 times under the same set of conditions before a mean overall simulation result for this unit was produced. In this technique, these 100 replicates may correspond to the tally sheets for 100 stands found within the boundaries of simulated forest types.

Future Challenges

It seems that we should concentrate on the development of simpler yet realistic and specific models. These models would need specialized data dealing with species characteristics under a particular set of conditions. We should look at successional stages as identified by the specific species behavior and not by community characteristics as has been done in the past. Once the model has captured the dynamics of the system alternate management strategies can be simulated into the future to determine which strategies achieve particular goals.

We recognize the difficulty in including both temporal and spatial dimensions in prediction of successional changes and we hope more attention will be given to this problem.

Acknowledgements

We wish to acknowledge the interactions with our colleagues within U.N.B. Fire Science Centre and with H.H. Shugart of Oak Ridge National Laboratory, Oak Ridge, Tennessee as we prepared this paper. Funding for the studies supporting the paper was obtained from Operation Grants from the Natural Sciences and Engineering Research Council, Ottawa by the senior author.

Literature Cited

Alexander, M.E. 1982. Fire behavior in aspen slash fuels as related to the Canadian Fire Weather Index. Can. J. For. Res. 12: 1028-1029.

Auclair, A.N.D. 1983. The role of fire in lichen-dominated tundra and forest-tundra. *In*: The role of fire in northern circumpolar ecosystems. R.W. Wein and D.A. MacLean (Eds.). SCOPE 18, John Wiley and Sons, Chichester, England. pp. 235-256.

Austin, M.P. 1977. Use of ordination and other multivariate descriptive methods to study succession. Vegetatio 35: 165-175.

Baskerville, G.L. 1975. Spruce budworm: super silviculturist. For. Chron. 51: 138-140.

Baskerville, G.L. and S. Kleinschmidt. 1981. A dynamic model of growth in defoliated fir stands. Can. J. For. Res. 11: 206-214.

Baskerville, G.L. and D.A. MacLean. 1979. Budworm-caused mortality and 20-year recovery in immature balsam fir stands. Environ. Canada, C.F.S., Info. Rept. M-X-102.

Black, R.A. and L.C. Bliss. 1978. Recovery sequence of *Picea mariana - Vaccinium uliginosum* forests after burning near Inuvik, Northwest Territories, Canada. Can. J. Bot. 56: 2020-2030.

Botkin, D.B., J.F. Janak and J.R. Wallis. 1972. Some ecological consequences of a computer model of forest growth. J. Ecol. 60: 849-872.
Cattelino, P.J., I.R. Noble, R.O. Slatyer and S.R. Kessell. 1979. Predicting the multiple pathways of plant succession. Environ. Manage. 3: 41-50.
Cayford, J.H. and D.J. McRae. 1983. The ecological role of fire in jack pine forests. *In*: The role of fire in northern circumpolar ecosystems. R.W. Wein and D.A. MacLean (Eds.). SCOPE 18, John Wiley and Sons, Chichester, England. pp. 183-199.
Clements, F.E. 1904. Development and structure of vegetation. Rep. Bot. Surv., Nebraska 7.
Clements, F.E. 1916. Plant succession: an analysis of the development of vegetation. Carnegie Inst., Publ. No. 242. Washington, D.C.
Clements, F.E. 1936. Nature and structure of the climax. J. Ecol. 24: 252-284.
Cody, W.J. 1965. Plants of the Mackenzie River Delta and reindeer grazing preserve. Plant Research Institute, Canada Dept. Agriculture. Ottawa, Ontario.
Connell, J.H. and R.O. Slatyer. 1977. Mechanisms of succession in natural communities and their role in community stability and organization. Amer. Nat. 111: 1119-1144.
Cooper, W.S. 1926. The fundamentals of vegetation change. Ecology, 7: 391-413.
Cowles, H.C. 1901. The physiographic ecology of Chicago and vicinity: A study of the origin, development and classification of plant societies. Bot. Gaz. 31: 73.
Damman, A.W.H. 1964. Some forest types of central Newfoundland and their relation to environmental factors. For. Sci. Monogr. 8.
El-Bayoumi, M.A. 1982. Simulation model of Acadian Forest succession (SMAFS). M.Sc. Thesis, University of New Brunswick, Fredericton, New Brunswick.
El-Bayoumi, M.A., H.H. Shugart Jr. and R.W. Wein. 1984. Application of the modelling approach to simulate Acadian Forest succession. Ecol. Model. 10: In press.
Emamuel, W.R., D.C. West and H.H. Shugart. 1978. Spectral analysis of forest model time series. Ecol. Model. 4: 313-326.
Field, W.O. 1947. Glacier recession in Muir Inlet, Glacier Bay, Alaska. Geog. Rev. 37: 369-399.
Fowells, H.A. 1965. Silvics of forest trees of the United States. U.S.D.A., For. Serv., Agric. Handbook 27.
Furyaev, V.V., R.W. Wein and D.A. MacLean. 1983. Fire influences in *Abies*-dominated forests. *In*: The role of fire in northern circumpolar ecosystems. R.W. Wein and D.A. MacLean (Eds.). SCOPE 18, John Wiley and Sons, Chichester, England. pp. 221-234.
Ghent, A.W., D.A. Fraser and J.B. Thomas. 1957. Studies of regeneration in forest stands devastated by the spruce budworm. I. Evidence of trends in forest succession during the first decade following budworm devastation. For. Sci. 3: 184-208.
Gill, A.M. 1977. Plant traits adaptive to fires in Mediterranean land ecosystems. pp. 17-26. *In*: Proc. Symp. Environment consequences of fire and fuel management in Mediterranean ecosystems. U.S.D.A., For. Serv., Gen. Tech. Rept. WO-3.
Gill, D. 1972. The evolution of a discrete beaver habitat in the Mackenzie River Delta, Northwest Territories. Can. Field-Nat. 86: 233-239.
Gleason, H.A. 1926. The individualistic concept of the plant succession. Bull. Torrey Bot. Club 53: 7-26.
Hawkes, B.C. 1983. Fire history and ecology of forest ecosystems in Kluane National Park — Fire management implications. *In*: Resources and dynamics of the boreal zone. R.W. Wein, R.R. Riewe and I.R. Methven (Eds.). Association of Canadian Universities for Northern Studies. Ottawa, Ontario.
Heilman, P.E. 1966. Changes in distribution and availability of nitrogen with forest

succession on north slopes in interior Alaska. Ecology, 47: 826-831.

Heilman, P.E. 1968. Relationship of availability of phosphrous and cations to forest succession and bog formation in interior Alaska. Ecology, 49: 331-336.

Heinselman, M.L. 1981. Fire and succession in the conifer forests of northern North America. *In*: Forest succession: Concepts and application. D.C. West, H.H. Shugart and D.B. Botkin (Eds.). Springer-Verlag, New York. pp. 374-405.

Hernandez, H. 1973. Revegetation studies: Norman Wells, Inuvik, and Tuktoyaktuk, N.W.T. and Prudhoe Bay, Alaska: a report to the E.P.B. and ALUR. Interim Rept. No. 3. E.P.B., Winnipeg and ALUR.

Hok, J.R. 1969. A reconnaissance of tractor trials and related phenomena on the North Slope of Alaska. U.S.D.I., Bureau Land Management.

Holling, E.S. 1978. Adaptive environmental assessment and management. John Wiley and Sons, New York.

Horn, H.W. 1974. The ecology of secondary succession. Ann. Rev. Ecol. Syst. 5: 25-37.

Horn, H.W. 1976. Succession. pp. 187-204. *In*: Theoretical ecology: principles and application. Blackwell Sci. Publishers, Oxford, England.

Jeglum, J.K. 1983. Changes in tree species composition in naturally regenerating strip clearcuts in shallow-soil upland black spruce. *In*: Resources and dynamics of the boreal zone. Association of Canadian Universities for Northern Studies. Ottawa, Ontario.

Johnson, E.A. and J.S. Rowe. 1977. Fire and vegetation change in the western subarctic. Canada, Department of Indian Affairs and Northern Development, ALUR Report 75-76-61.

Kershaw, K.A. 1957. The use of cover and frequency in the detection of pattern in plant communities. Ecology, 38: 291-299.

Lambert, J.D.H. 1968. The ecology and successional trends of tundra plant communities in the Low Arctic Subalpine zone of the Richardson and British Mountains of the Canadian Western Arctic. Ph.D. Dissertation, University of British Columbia, Vancouver, British Columbia.

Lawrence, D.B. 1953. Development of vegetation and soil in south-eastern Alaska, with special reference to the accumulation of nitrogen. Final Report, DNR. pp. 160-183.

Methven, I.R., C.E. Van Wagner and B.J. Stocks. 1975. The vegetation on four burned areas in northwestern Ontario. Environ. Canada, C.F.S., Info. Rept. PS-X-60.

Mielke, D.L., H.H. Shugart and D.C. West. 1978. A stand model of upland forests of southern Arkansas. ORNL/TM-6225, Oak Ridge National Laboratory, Oak Ridge, Tennessee.

Noble, I.R. and R.O. Slatyer. 1980. The use of vital attributes to predict successional changes in plant communities subject to recurrent disturbances. Vegetatio, 43: 5-21.

Odum, E.P. 1969. The strategy of ecosystem development. Science, 164: 262-270.

Pruitt, W.O., Jr. 1958. Qali, a taiga snow formation of ecological importance. Ecology 39: 169-172.

Rowe, J.S. 1961. Critique of some vegetational concepts as applied to forests of northwestern Alberta. Can. J. Bot. 39: 1007-1017.

Rowe, J.S. 1983. Concepts of fire effects on plant individuals and species. *In*: The role of fire in northern circumpolar ecosystems. R.W. Wein and D.A. MacLean (Eds.). SCOPE 18, John Wiley and Sons, Chichester, England. pp. 134-154.

Shugart, H.H. and I.R. Noble. 1981. A computer model of succession and fire responses of the high-altitude *Eucalyptus* forests of the Brindabella Range, Australian Capital Territory. Aust. J. Ecol. 6: 149-164.

Shugart, H.H. and D.C. West. 1977. Development and application of an Appalachian deciduous forest succession model. Environ. Manag. 5: 161-179.

Sprugel, D.B. 1976. Dynamic structure of wave generated *Abies balsamea* forests in north-eastern United States. J. Ecol. 64: 889-911.
Strang, R.M. 1973. Succession in unburned subarctic woodlands. Can. J. For. Res. 140-143.
Viereck, L.A. 1983. The effects of fire in black spruce ecosystems of Alaska and northern Canada. *In*: The role of fire in northern circumpolar ecosystems. R.W. Wein and D.A. MacLean (Eds.). SCOPE 18, John Wiley and Sons, Chichester, England. pp. 201-220.
Vogl, R.J. 1977. Fire frequency and site degradation. *In*: Proc. Symp. Environ. Consequences of Fire and Fuel Management in Mediterranean Climate Ecosystems. U.S.D.A., For. Serv., Gen. Tech. Rept. WO-3. pp. 193-201.
Walters, C.J., R. Hilborn and R. Peterman. 1975. Computer simulation of barren-ground caribou dynamics. Ecol. Model. 1: 303-315.
Warming, E. 1896. Lehrbuch der Oekologischen. Pflanzengeographie.
Wein, R.W. 1975. Recovery of arctic vegetation after burning. ALUR Program Report, Ottawa, Ontario.
Wein, R.W. 1976. Frequency and characteristics of arctic tundra fires. Arctic, 29: 213-222.
West, D.C., H.H. Shugart and D.B. Botkin (Eds.). 1981. Forest succession: Concepts and application. Springer-Verlag, New York.
Whittaker, R.H. 1975. Functional aspects of succession in deciduous forests. *In*: Sukzessions-forschung Ber. Int., Symp. R. Tuxen (Ed.). Rinteln. 1973. pp. 377-405.
Yang, R.C. and R.D. Fry. 1981. Natural succession following harvesting in the boreal mixedwood forest. *In*: Environ. Canada, C.F.S., COJFRC Proc. O-P-9. pp. 65-77.
Zoltai, S.C. 1975. Tree ring record of soil movements on permafrost. Arctic and Alpine Research, 7: 331-340.
Zoltai, S.C. and W.W. Pettapiece. 1974. Tree distribution on perennially frozen earth hummocks. Arctic and Alpine Research, 6: 403-411.

METHODOLOGY TO DETERMINE LONG TERM FIRE HISTORIES: An Examination of Charcoal and Pollen from Mashagama Lake, Ontario

R.A. Battson
University of Western Ontario

and

K.B. Cawker
University of Toronto

Abstract: This study involved an evaluation of the various measures of fire occurrence as recorded in lake sediments. A short core (90 cm) was extracted from Mashagama Lake, Ontario. The basin was burned in 1948 and 1967, yielding two zones, one burned and one unburned in the sediment core. Charcoal percentage composition (based on the total number of fragments) best represented the burned zone (the top 26 cm) with a mean of 46.7 per cent compared to a mean of 23.1 per cent for the rest of the core. Charcoal concentration (based on the total number of fragments) and charcoal area were found not to represent the two zones as expected. Charcoal percentage was also significantly correlated with loss on ignition ($r = -0.680$) and the S/C ratio ($r = 0.734$) while charcoal concentration and area were not. Examination of the size class data indicated that small fragments (100-199 μm^2) were more prominent in burned samples than in unburned samples and that larger fragments (2000 μm^2+) were more numerous in unburned samples as compared to burned samples. This has serious implications for the use of the charcoal area as an indicator of fire in lake sediments. The large fragments contributed a great deal toward the total charcoal area of the sample. For example, one fragment of charcoal represented 32 per cent of the total area of the sample.

It has been well established that fire has had a significant impact on the formation of present forest communities (Maissurow 1941; Spurr 1954; Ahlgren and Ahlgren 1960; Mutch 1970; Frissell 1973; Kilgore 1973; Loope and Gruell 1973; Rowe and Scotter 1973; Wright and Heinselman 1973). In order to better understand the magnitude of this impact, research on the frequency of forest fires based on recent fire histories and long term fire histories has taken place. Recent fire history has been examined through the use of written records, stand origins and fire scar dendrochronology (Spurr 1954; Frissell 1973; Heinselman 1973; Loope and Gruell 1973; Taylor 1974; Arno 1976; Cwynar 1976; Kilgore and Taylor 1979; Tande 1979; Wein and Moore 1979).

These studies of recent fire histories are limited by the age of vegetation and therefore, only yield direct information on the fire regime for a period of two to three hundred years, much of which may fall within the era of European settlement. A fire record derived from the post settlement period may very well indicate a greater periodicity of fires due to anthropogenic factors. In addition the accuracy of the record decreases with time often yielding only spotty evidence of older fires (Heinselman 1973). The studies to date, however, do indicate that fire has been widespread in a variety of forest regions including the Great Lakes-St. Lawrence, Subalpine, Boreal, Acadian and Montane forest regions.

In order to extend the fire record back in time, other types of evidence must be used. The sequential deposition of material into a lake bottom provides such a record. Through the examination of changes in microfossils, fine charcoal abundance and clastic versus organic content of lake sediments, long term fire histories have been attempted (Waddington 1969; Swain 1973, 1978, 1980; Byrne *et al.* 1977; Cwynar 1978; Amundson and Wright 1979; Green 1981, 1982).

As a fire severely burns a watershed, the vegetation responsible for the pollen production is largely destroyed. Local pollen production, therefore, decreases causing an increased representation of pollen arriving from a greater distance. As regeneration occurs, a different mix of species emerges changing the pollen spectra. This shift in the relative abundances may be detected in lake sediments. This shift, however, would only indicate that some disturbance had occurred causing secondary succession to proceed.

Charcoal fragments are also deposited in lake sediments along with pollen. As a fire burns a basin, charcoal will be produced and deposited into the sediment through input from streams and from airborne deposition. Attempts have been made to quantify this fine charcoal component in fossil pollen preparations in order to provide an indication of occurrence of fire. The size fraction typically prevalent in samples prepared for pollen analysis ranges from very small (less than 100 μm^2) to fragments of approximately 10,000 μm^2 and very occasionally larger pieces. Samples derived from sediment which were deposited during a fire should have a greater amount of charcoal present relative to samples from adjacent strata. Charcoal, however, is found in moderate quantities throughout the sediment. This charcoal component may represent charcoal derived from long distance transport and charcoal washed in from the soil surface during erosional events.

Erosion is often increased following a fire as much of the stabilizing vegetation is removed (Connaughton 1935; Sartz 1953; Biswell and Schultz 1957; Rich 1962; Berndt 1971; Klock and Helvey 1974). This results in a greater input of clastic substances into the sediment matrix. Through analysis of the inorganic/organic ratio, this increase in clastic input may be detected and interpreted as an erosional event. A shift in the pollen spectra and an increase in the inorganic/organic ratio, therefore, provides evidence that a disturbance has occurred while an increase in charcoal levels suggests the cause.

The quantification of charcoal in lake sediments is a crucial first step in the determination of a long term fire history. Too little critical examination has occurred of the various charcoal measures available and the interpretation of the size fractions of the charcoal fragments. It was the objective of this research to address these problems and to provide some insight into the difficulties of interpreting these charcoal particles.

To date, only a charcoal area index has been used to quantify charcoal abundance. Inherent with this measure is the assumption that local fires produce a greater proportion of large fragments relative to a non-local fire, or relative to the background charcoal falling into the lake during periods devoid of fire. It was the purpose of this paper to discover if this measure was adequate or if alternate charcoal measures, avoiding an area component, would yield more satisfactory results. In order to explore these problems a site was chosen with a well-documented fire history. The results of the charcoal analysis could then be compared with what was already known of the fire history.

Study Area

In this study charcoal was analyzed from a short core taken from Mashagama Lake, located in the Algoma district of northern Ontario at approximately 46°55′N and 83°21′W, west of County Road 129, 80 km north of Thessalon (Figure 1). The study area falls into the Timagami section of the Great Lakes-St. Lawrence forest region.

Two major fires have occurred nearby since 1948 (Figure 1). Based on fire records and the large size of the remaining trees (a 350 year old red pine (*Pinus resinosa*) was found along the shore), it was judged that the area had been free from a major burn for a long period of time prior to 1948.

Methods

A 90 cm core was extracted using a modified surface Livingston piston corer in 4.2 m of water. The core was taken to the laboratory while contained in the coring device, frozen, extruded, and cut into 1 cm slices. Based on visual inspection there was no obvious stratigraphy, with the core consisting of a homogenous dark-brown gyttja.

In order to extract the pollen a procedure similar to that outlined by Faegri and Iversen (1975) was used. *Lycopodium* marker grain tablets were added to obtain concentration data. Five hundred or more grains were counted per sample. Loss on ignition was determined by incinerating a small subsample at 375°C for 12 hours.

Charcoal particles and pollen were counted. Charcoal was identified as being black to dark brown fragments usually planar in nature. Larger charcoal fragments with a distinct cell structure were used to determine the characteristics of smaller particles which lacked recognizable cell structure.

The area of each charcoal particle encountered was estimated with the use of a graticule grid with individual squares 30 μ each side. Sampling continued until a total of 200 fragments were measured. This upper limit was used as this procedure was extremely time consuming. Fragments less than 100 μm^2 were not counted to avoid confusion with pyrite crystals. The number of *Lycopodium* marker grains counted to that point were also noted so that the relative proportion and the concentration of charcoal in each sample could be determined.

Charcoal percentage represented the total number of charcoal fragments present in each sample divided by the pollen sum plus the number of charcoal particles. It is important to note that no distinction based on the size of particle has been made in this measure. Charcoal percentage is very similar to the charcoal-pollen (C/P) ratio used by Swain (1973) except that charcoal was treated outside the pollen sum in the C/P ratio calculation. Charcoal concentration again was based simply on the total number of fragments.

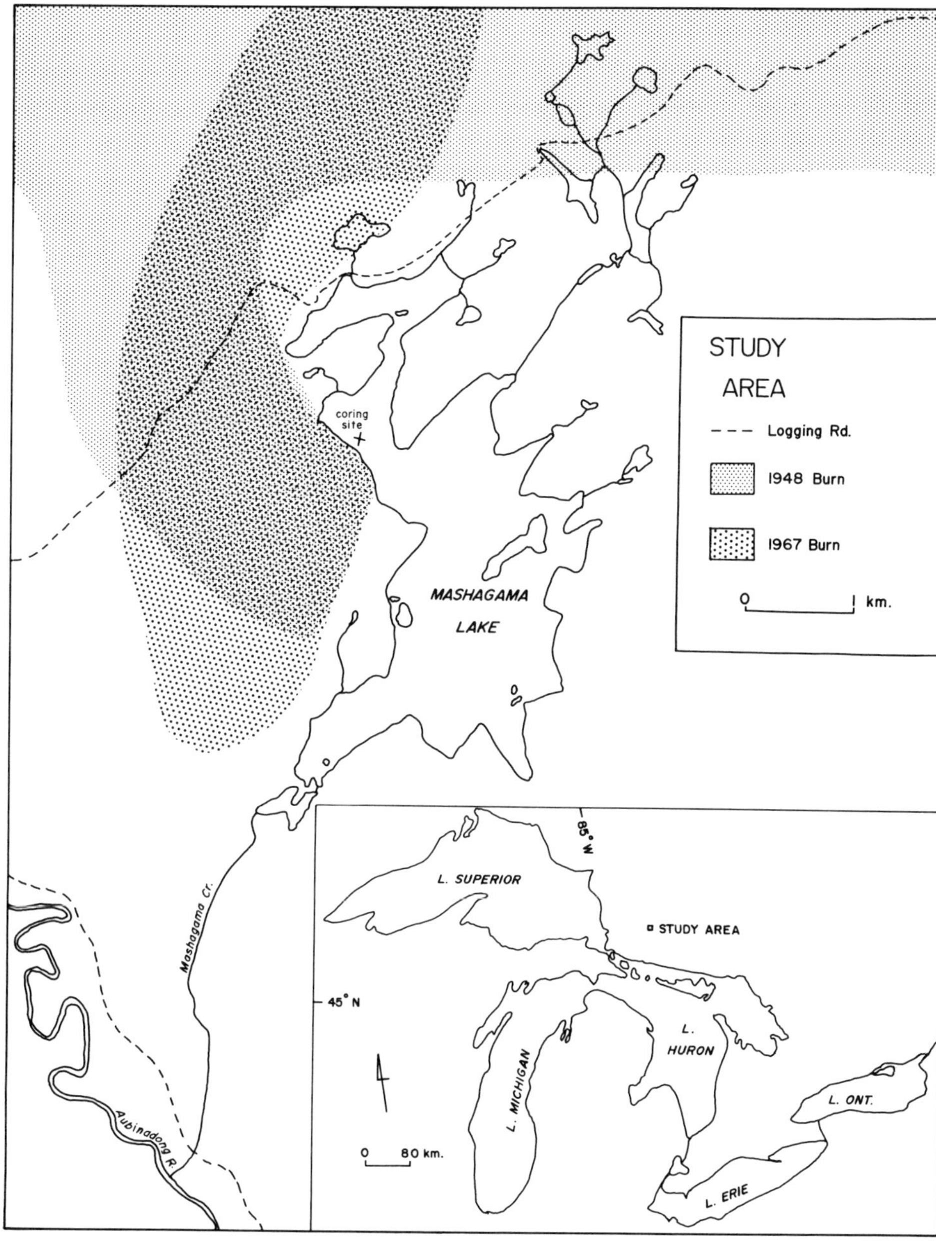

Figure 1. Coring site in Mashagama Lake, Ontario. Note the location of the coring site immediately adjacent to the 1948 and 1967 burns.

In order to produce results comparable with the literature, a charcoal area index was calculated using a method similar to that of Waddington (1969), Swain (1973, 1978, 1980) and Cwynar (1978). This method involved placing the charcoal fragments into eight size classes (in μm^2): 100-199, 200-299, 300-399, 400-499, 500-599, 600-899, 900-1199 and 1200-1999. Fragments 2000 μm^2 and larger were counted separately. The size classes were chosen to ensure that a reasonable sample size existed in each class. The total area per sample was calculated by taking the sum over all size classes of the number of particles in the size class multiplied by the midpoint of the class. Areas of fragments larger than 2000 μm^2 were added to the total. That value was then converted to the concentration in the form $\mu m^2/g$ dry weight. Since the previous authors had adequate time controls, their concentrations were then converted to influx values. The lack of suitable time control for the Mashagama Lake core made this procedure unacceptable.

The sprouter-conifer (S/C) ratio was used to show a combined vegetative response to disturbance. The taxa used to calculate the ratio were similar to those used by Swain (1973, 1980), and were sprouters (*Betula*, *Populus*, *Alnus*, Gramineae, *Ambrosia*, *Artemesia*, Compositae and Chenopodaceae) and conifers (*Pinus*, *Picea*, *Abies*, and Cupressaceae).

Results

Loss on ignition varied from 14.0 per cent to a high of 53.2 per cent (Figure 2). There was a general trend of decreasing loss on ignition (L.I.) values from the bottom of the core to the top. The most significant drop occurred from 38 cm (a L.I. value of 37.1 per cent) to 14 cm (a L.I. value of 14.0 per cent). Loss on ignition values then remained relatively constant to the surface of the core.

The pollen sequence has been divided into four zones based on the pollen percentage diagram (Figure 3). The zones were derived by visual inspection. Zone 1 (70-86 cm) was characterized by low but rising levels of *Pinus* and high but falling levels of *Betula*, *Picea* and *Alnus*. The high S/C ratio in this zone was really a reflection of the *Betula* curve. Loss on ignition maintained high levels during this period, indicating low erosion rates.

Zone 2 (38-69 cm) showed little variation and was clearly dominated by *Pinus* with values ranging from 79.1 per cent to 93.7 per cent. *Picea* was the next most abundant taxon with values ranging from 0.4 per cent to 8.9 per cent. All other taxa remained at very low levels throughout this zone. Zone 3 (24-37 cm) showed a decrease in *Pinus* which seemed to be associated with an increase in *Picea*. *Betula, Alnus, Salix* and the S/C ratio all rose slowly while loss on ignition decreased gradually.

Zone 4 (0-23 cm) was characterized by a decrease in *Pinus* from 60 to 90 per cent to levels ranging from 30 to 60 per cent as well as a less marked decrease in *Picea*. *Betula*, the most notable of the sprouters, increased drastically from levels below 20 per cent to levels above 30 per cent. Other indicators of disturbance including *Ambrosia, Artemesia,* Compositae and Chenopodaceae increased as well. This trend was reflected clearly in the S/C ratio. *Alnus* increased initially but then dropped to low levels. While there was no abrupt change in loss on ignition, generally lower values predominated in the top 23 cm of the core indicating increased erosion during this period.

The three charcoal measures are presented in Figure 4 with the pollen zones superimposed. Charcoal percentage values varied relatively little in Zones 1 and 2 dropping steadily, however, throughout the two zones to a low of 8.4 per cent at 45 cm. Zone 3 showed a remarkably steady increase from 11.9 per cent to 27.1 per cent. Charcoal percentage values then increased very dramatically in Zone 4 reaching peaks of 60.1, 53, and 64.4 per cent at 2, 7 and 14 cm, respectively.

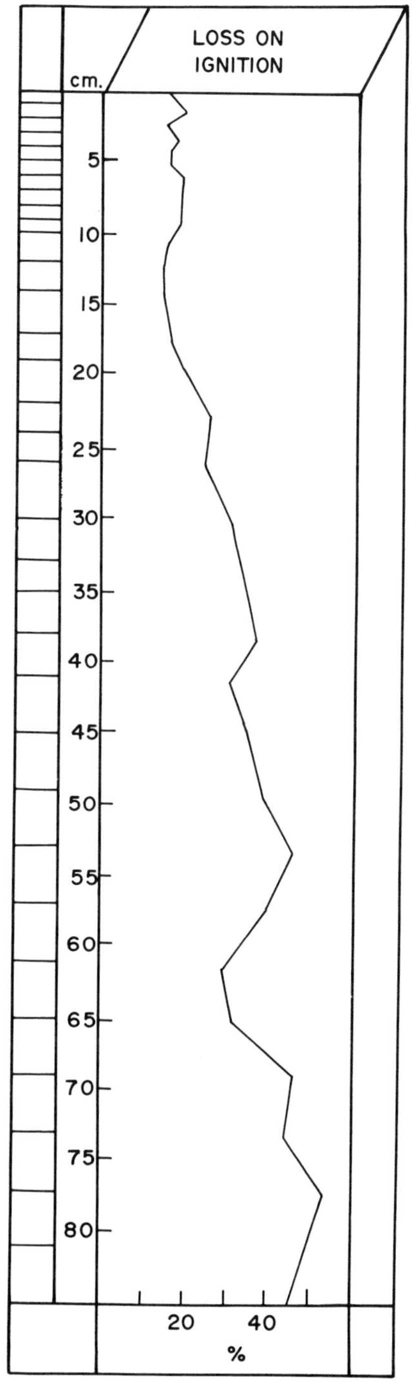

Figure 2. Loss on ignition from Mashagama Lake core. Loss on ignition was used as an indicator of changes in erosion rates. The curve indicates that the top 20 cm of sediment were derived in an environment of higher erosion rates.

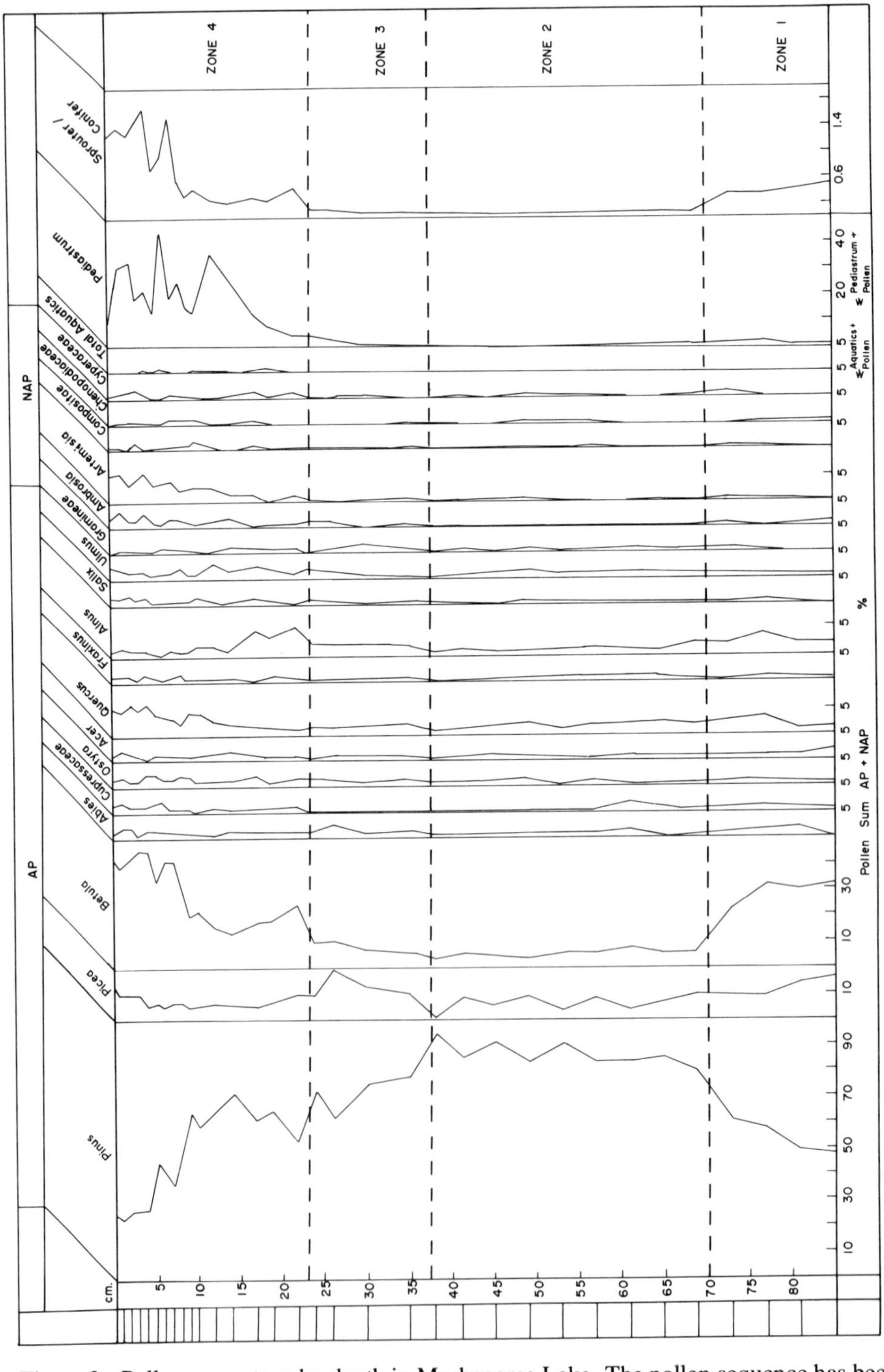

Figure 3. Pollen percentage by depth in Mashagama Lake. The pollen sequence has been divided into 4 zones. The left-hand column indicates the sampling interval.

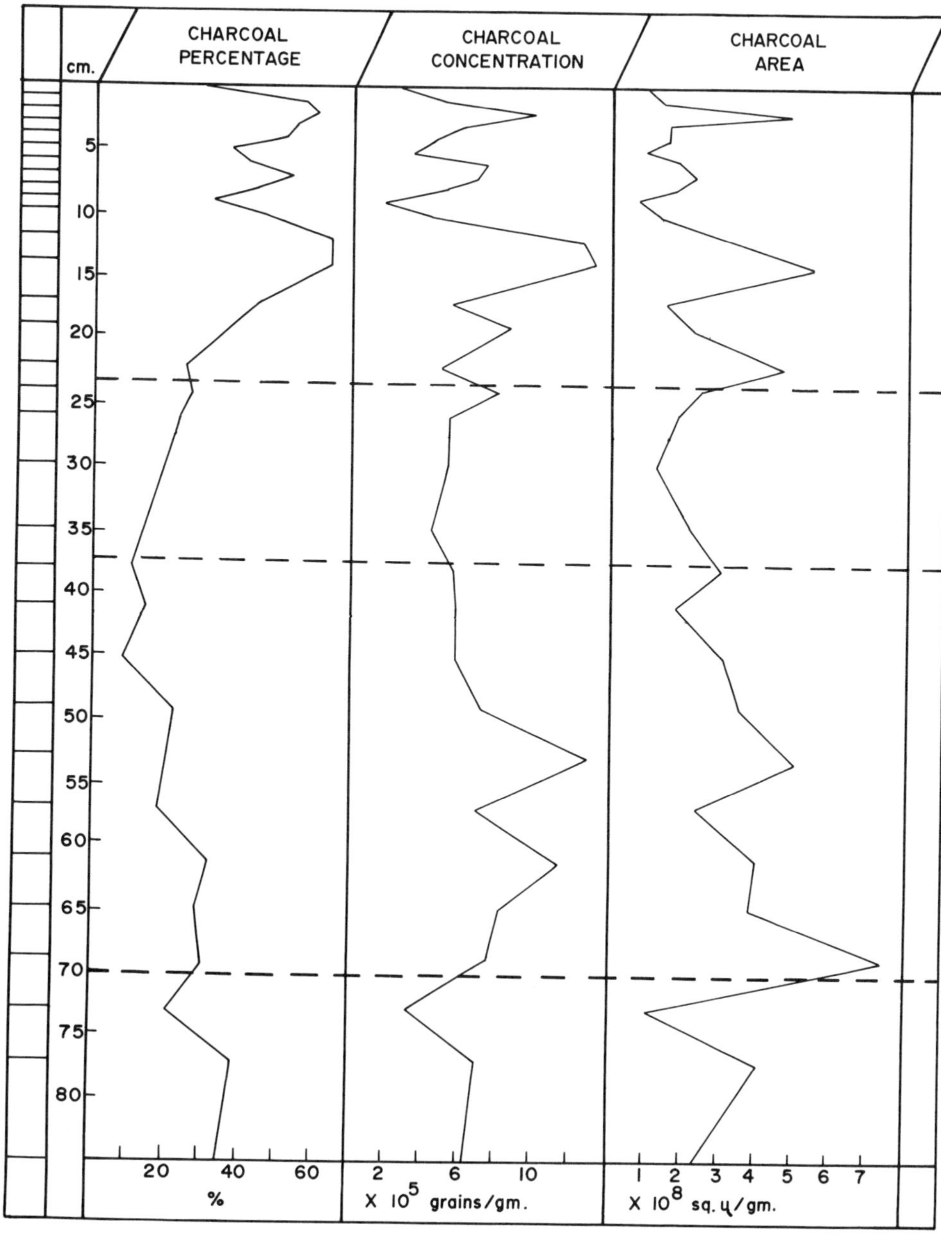

Figure 4. Charcoal measure profiles from Mashagama Lake on which the pollen zones have been superimposed. Note the strong relationship with charcoal concentration and charcoal area. This is due to the fact that area is a concentration measure. Charcoal percentage shows generally higher values in Zone 4 except for sample 81-82.

Charcoal concentration was much more variable with peaks of 13.28, 12.87, 12.50, 11.44, 9.88 and 7.49 fragments/g $\times$ 10^5 at 14, 53, 12, 61, 2 and 6 cm, respectively. Much of this variability may be due to variation in clastic input. While no pattern in terms of abundance occurred in any one zone, Zone 4 experienced the greatest variability in charcoal concentration.

Charcoal area exhibited a pattern similar to that of charcoal concentration, probably reflecting the fact that the area index is an area concentration measure. Two major anomalies do exist, however. The charcoal area peaked dramatically to 7.5×10^8 μm^2 while concentration values only rose moderately at 69 cm. At 22 cm, charcoal concentration decreased while charcoal area increased markedly. Similar variability was evident in Zone 4.

A total of 6,400 charcoal fragments were measured and placed into size classes. As was noted by Amundson and Wright (1979) the smaller charcoal fragments dominated the sample. In this case 48.97 per cent of all charcoal fragments belonged to the 100-199 μm^2 size class. This domination of the smaller size classes is graphically presented in Figure 5. Only 1.7 per cent of the particles were 2000 μm^2 and larger with the largest fragment counted being 62,100 μm^2.

Discussion

Pollen Zones

The pollen diagram has been divided into four zones. Zone 1 appeared to represent the later stages of a recovery from disturbance. Pollen percentages from sprouter species including *Betula* and *Alnus* were relatively high and declining while *Pinus* was increasing. Loss on ignition values varied very little and remained at high levels indicating low erosion rates. The soil surface may have become stabilized while *Pinus* was steadily out-competing the seral species to reach the high levels characteristic of Zone 2.

Pinus varied little in Zone 2 and remained at strikingly high levels (85 to 90 per cent). High pollen concentration and high loss on ignition values could indicate that this zone was very stable. At 39 cm, pine dropped rapidly, matched by an increase in the proportion of spruce. This could represent the onset of logging in the area in the late 1800s. It was common practice at that time to go through an area and take out the pine first, and then, if warranted, return for the spruce. This interpretation was supported by the increase in sprouter species (*Betula, Alnus,* Gramineae, *Artemesia,* and *Ambrosia*) in Zone 3. Loss on ignition decreased in the zone indicating increased erosion as one would expect in a logging operation. The higher erosion levels would have resulted in an increase in nutrient levels perhaps leading to the observed increase in *Pediastrum*, a colonial green algae. It is possible that Zone 3 could represent an environment frequented by fires. If this were the case, however, it would seem doubtful that spruce would increase relative to other taxa. The decline in pine and increase in sprouters (manifested in the S/C ratio) occurred very slowly, consistent with the gradual process of logging.

Zone 4 appeared to represent a fire influenced environment. Fire often results from the quantities of slash left following a logging operation. Perhaps a small burn occurred at 26 cm resulting in an increase in the S/C ratio and a decrease in loss on ignition. *Pediastrum* also increased slowly possibly representing continued high levels of nutrient input into the lake. The S/C ratio also increased and remained at a slightly higher level. A major burn must have occurred at 15 cm. At this point *Pediastrum* rose sharply and loss on ignition reached its lowest level. This is believed to represent the 1948 burn. The S/C ratio, however, did not increase markedly until 8 cm. This lag would be expected as

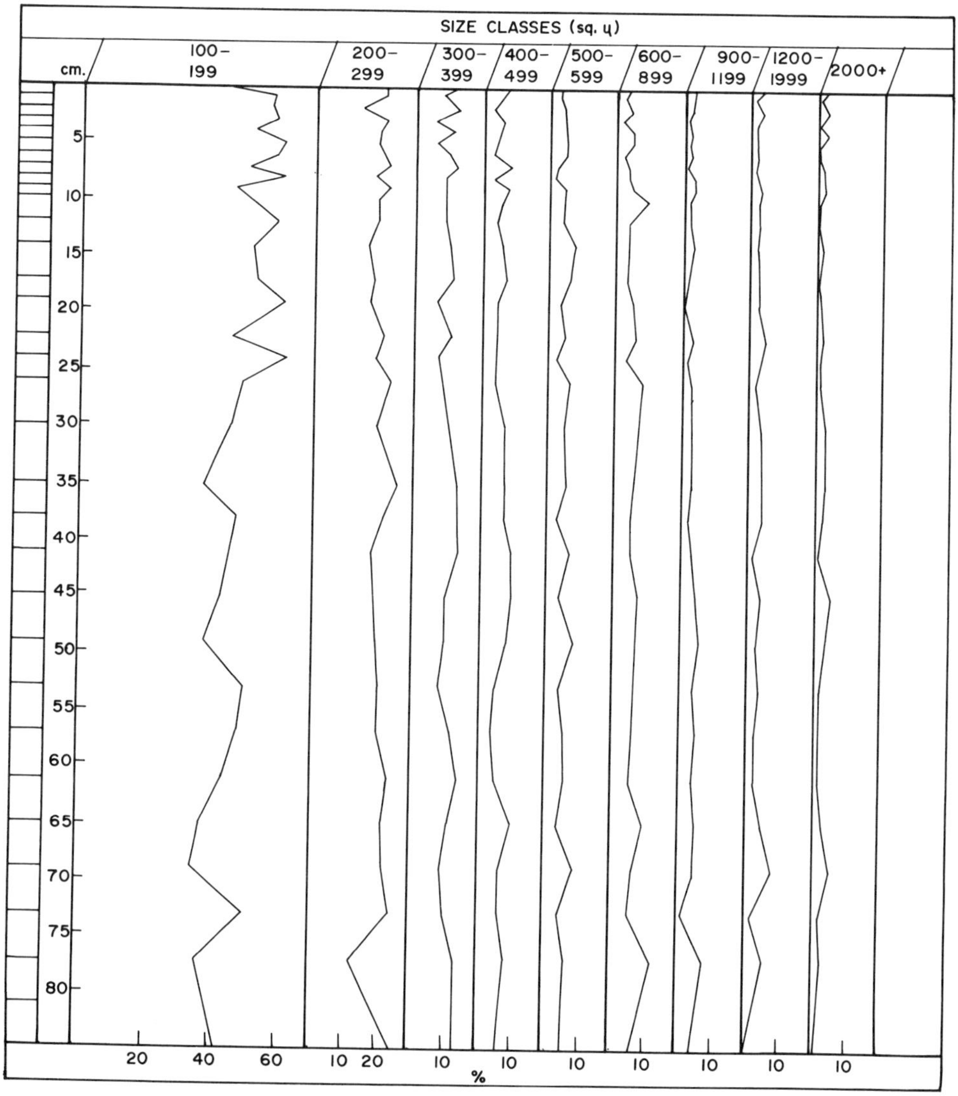

Figure 5. The percentage of charcoal particles allocated to each size class. Note the dominance of the smallest size classes. Forty-nine per cent of all fragments fall into this size class.

several years are needed for trees to reach sexual maturity and produce pollen. The 1967 fire was difficult to place because of the prior disturbances. The larger peak in *Pediastrum* at 7 cm could represent a nutrient flush from this event. Loss on ignition, however, varied little throughout this period.

It seems reasonable, based on the fossil and sedimentary evidence and what is known of the recent fire history of the lake basin, that Zone 4 represented a pollen and charcoal spectrum characteristic of a locally burned environment. While Zone 3 represented a partially disturbed environment, it did not appear that fire was principally responsible since an increase in *Picea* accompanied the *Pinus* decrease. Zone 2 clearly represented a stable unburned environment. Zone 1, while at the latter stages of recovery, represented a stabilizing environment at least. High levels of loss on ignition and low *Pediastrum* levels indicated that erosion was minimal. Since this zone is approaching a stabilized state, inputs of charcoal would be similar to those expected in an unburned environment. This zone, therefore, was considered to be uninfluenced by fire.

Discussion of Charcoal Measures

Of the three charcoal measures reported, (Figure 4), the charcoal percentage seemed to best represent the burned environment. The mean charcoal percentage for Zone 4 was 46.7 per cent while the mean for the remainder was only 23.1 per cent. Inspection of the charcoal percentage curve clearly indicated this trend. The highest charcoal peak of 64.4 per cent occurred at 15 cm and is believed to represent the 1948 fire. Charcoal percentage and *Pediastrum* behaved in a similar manner in Zone 4, further substantiating the assertion that local fires resulted in nutrient flushes into the lake causing an increase in levels of *Pediastrum*.

After a fire has burned a watershed there is usually an increase in erosion rates. In addition, the proportion of pollen from sprouter species after a lag period increases relative to conifer pollen. If this is the case, loss on ignition regressed against charcoal abundance should produce a negative relationship while the S/C ratio regressed against charcoal abundance should produce a positive relationship. The three measures of charcoal abundance studied have been regressed against both the Ln of loss on ignition and Ln of the S/C ratio. The results are presented in Figure 6.

Ln of loss on ignition versus Ln of percentage charcoal gave an $r = -0.680$, as would be expected. Likewise, a positive relationship ($r = 0.734$) was obtained when the Ln of the sprouter/conifer ratio was regressed against Ln of percentage charcoal. The r values were both significant at the 0.01 level of confidence. It must be noted, however, that the percentage charcoal is not independent of the S/C ratio and may not be independent of loss on ignition. High levels of *Pinus* as experienced in the Mashagama Lake core affect the S/C ratio in an obvious manner. Since percentage charcoal was determined as a percentage of the pollen, it was also affected by highly varying levels of *Pinus*. Swain (1973) suggested that the charcoal/pollen ratio (in effect percentage charcoal) would eliminate false charcoal peaks due to erosion as pollen and charcoal would be deposited together. This may not be the case, however, as pollen may degrade on the soil surface being exposed to aerobic degradation, while charcoal would not. In this case, an erosional event would decrease loss on ignition but increase charcoal percentage. This situation, however, would be uncommon and would probably not be a significant factor responsible for the positive relationship noted. The two variables, in most cases, would act independently of one another.

Charcoal concentration versus the natural log of loss on ignition (Figure 6), showed an almost totally random scattering of points. This could possibly be due to the response of the loss on ignition values to the increased levels of charcoal being added to the sediment, increasing the organic component of the matrix. This process would act to counter the increase in clastic input which would occur following a severe fire. The same pattern, however, held true for charcoal concentration versus the natural log of the S/C ratio (Figure 6). Again the relationship was negative, contrary to what would be expected. If

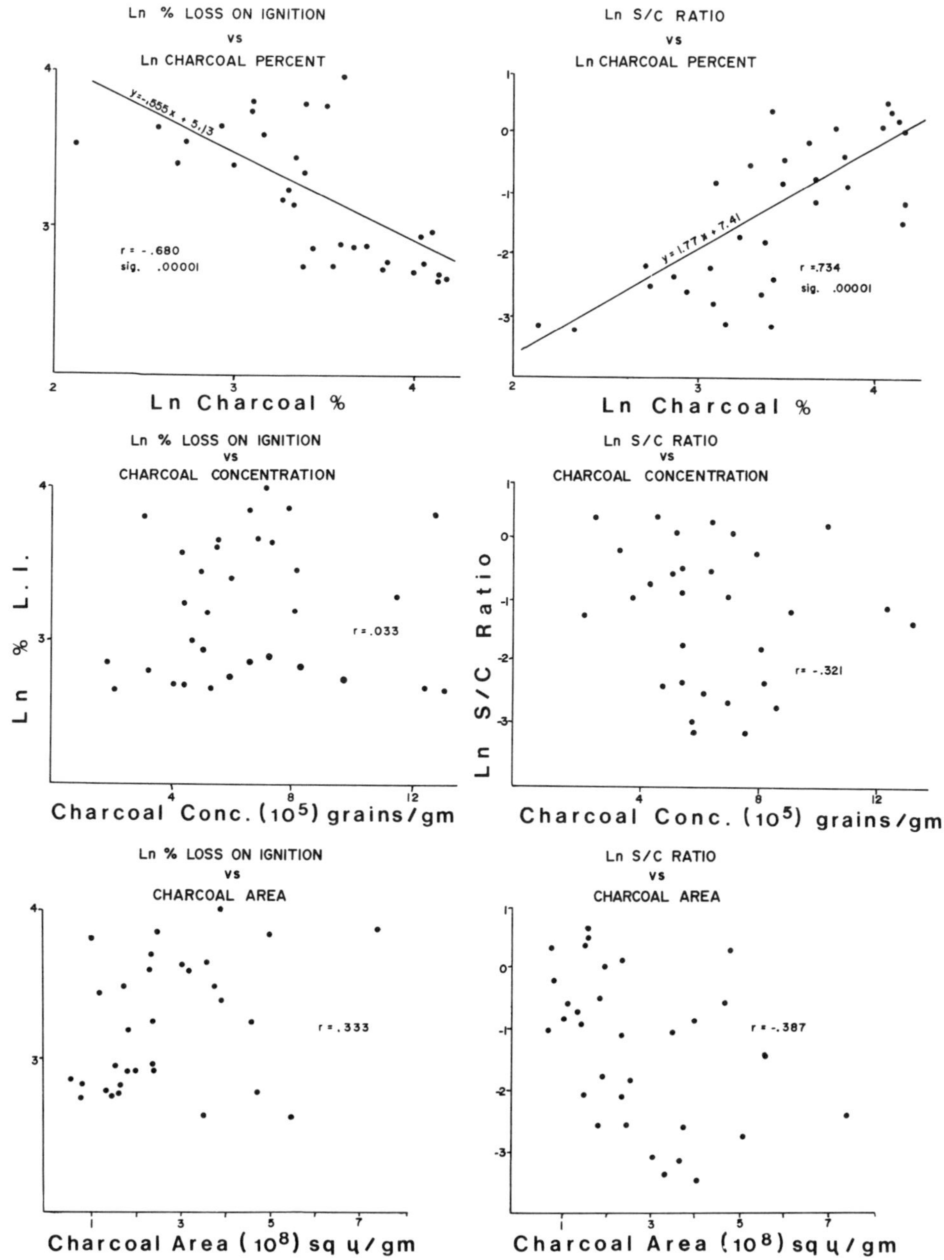

Figure 6. Regression diagrams for the 3 charcoal measures against Ln Loss on Ignition and Ln of the S/C ratio. Charcoal percentage shows the best relationship with Ln per cent L.I. and Ln S/C ratio with r^2 values of 0.463 and 0.539 respectively. That is to say that approximately one half of the variation in L.I. and the S/C ratio was explained by charcoal percentage. Charcoal concentration and area both show an almost totally random scattering of points. No more than 11 per cent of the variability of L.I. is explained by either charcoal concentration or area and no more than 14 per cent of the variability of the S/C ratio is explained by either charcoal measure. This seems to indicate that charcoal percentage is a better indicator of fire than the other two measures. The poor results experienced with charcoal area, however, may be due to its association with the concentration data.

charcoal concentrations were high, indicating a burned environment, then one would expect higher levels of sprouter species resulting in an elevated S/C ratio.

Charcoal area exhibited a pattern very similar to that of charcoal concentration. This was not at all surprising as the area measure was calculated as a concentration. The mean area concentration for Zones 1, 2 and 3 combined was 3.14 ($\mu m^2/g$) $\times$ 10^8 and was 2.34 ($\mu m^2/g$) $\times$ 10^8 for Zone 4. Note that since charcoal area was calculated using the midpoint of size classes, a mean was not a proper measure. It was used only to give an indication of trend. The procedure would be satisfactory if the data were normally distributed. The natural log of loss on ignition versus charcoal area (Figure 6) indicated a positive relationship (r = 0.333) again not to be expected. As with the charcoal concentration, charcoal area experienced a negative relationship (r = −0.387) with the Ln of the S/C ratio. It was difficult to isolate the area component from the area concentration measure. It was quite possible that the poor results experienced with the charcoal area-concentration were really a reflection of problems associated with the concentration data.

Relationship of Size Class Data to Fire Measures

The purpose of this section is to discover whether any differences exist in charcoal size class data between burned and unburned environments. The delineation of the two environments was determined using four methods so that each method could be evaluated against what is known of the recent fire history at Mashagama Lake. Three of the groups were stratified by charcoal percentage, charcoal concentration and charcoal area. The samples with the highest values for each measure constituted Group 1 (burned zone) and the lower values Group 2 (unburned zone). In addition, a stratification independent of charcoal measures was completed by clustering loss on ignition and the S/C ratio.

The charcoal size class distribution characteristics for each group were then compared in order to determine if any significant differences existed between the distribution of charcoal particles in a burned environment and those in an unburned environment. If a particular measure showed a strong difference in size class structure, then the implication was that the charcoal for each group were deposited in the lake in different manners. If there was no significant difference between groups then either there were no differences in the mechanism of transport of charcoal in burned and unburned environments, or differences in transport had no effect on the size class structure or the particular measure tested had incorrectly identified samples derived from a burn.

Stratification of Burned and Unburned Samples

Clustan, a package program designed to form statistical clusters was utilized to form two groups, independent of stratographic position by clustering Ln loss on ignition and Ln of the sprouter-conifer ratio (Figure 7). The Clustan package (Whishart 1978) divided the data by 'hierarchy' and then applied the 'Wards' method in which the two closest individuals were joined to form a cluster. The cluster then became an individual and the next two closest individuals were joined. The process continued until all individuals were in one cluster. Through examination of the fusion coefficient the researcher can decide where to stop the clustering procedure. This is usually chosen when continued clustering begins to lose its discriminating power. In this case the procedure was allowed to continue until two groups were formed so that the data could be stratified into disturbed and undisturbed groups.

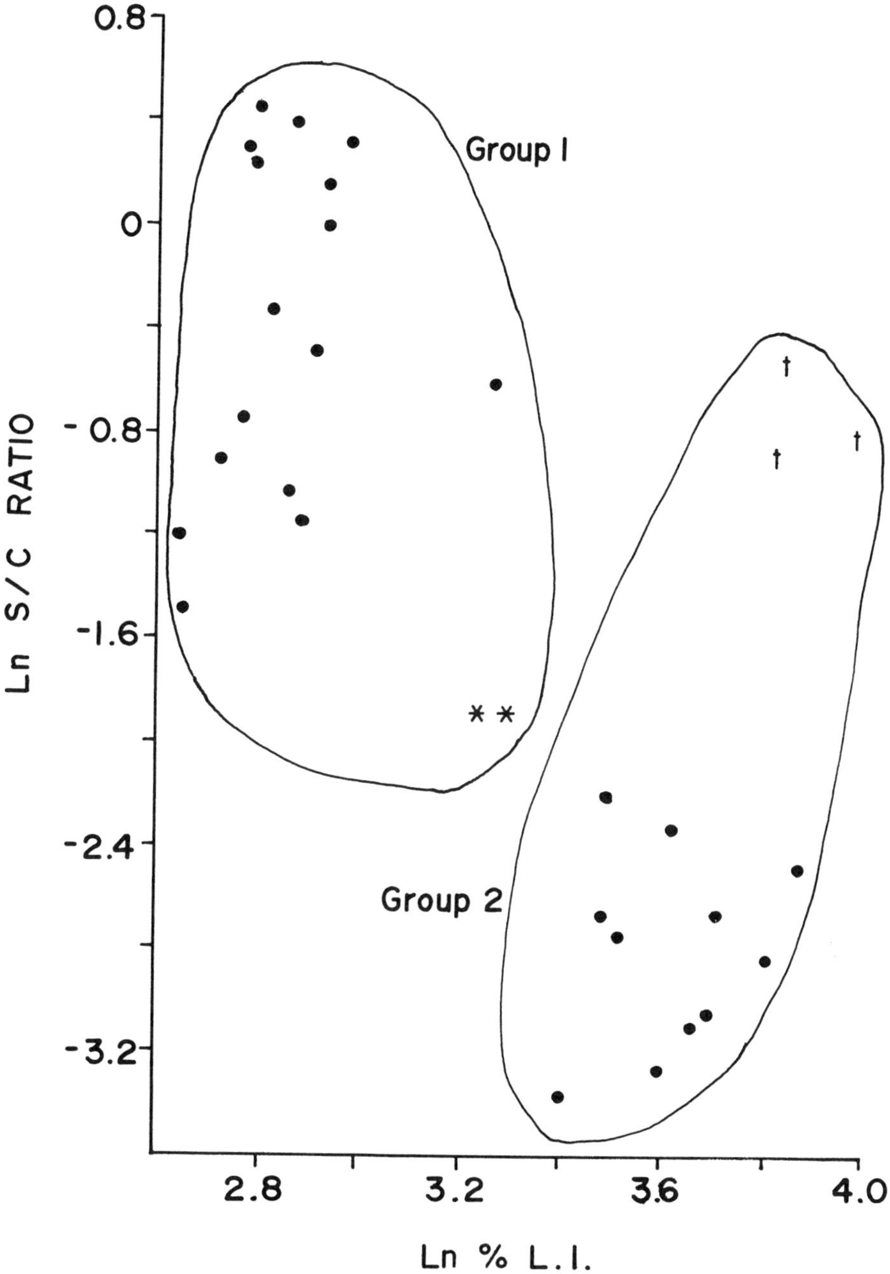

Figure 7. Ln Loss on Ignition vs. Ln of sprouter-conifer ratio. The two groups were determined through a cluster procedure. Group 1, representing samples with a high S/C ratio and low loss on ignition is composed entirely of samples from the top 27 cm of the core. Group 2, representing samples with a low S/C ratio and high loss on ignition includes the rest: * samples 24 – 25, 26 – 27;
† samples 73 – 74, 77 – 78, 85 – 86.

Group 1 (burned) represented the cases with a high S/C ratio and a low loss on ignition value, or samples representing a disturbed environment. Group 2 (unburned) represented samples with a low S/C ratio and high loss on ignition, probably a stable environment free from large scale disturbance. It is surprising to note that group 1 (burned) was composed of all samples from the top of the core to 27 cm exclusively (18 samples) and group 2 (unburned) contained the rest (14 samples).

Special note must be made of samples marked * and †. Samples with the asterisk were 24-25 and 26-27 or the transition samples between group 1 (burned) and group 2 (unburned). Their position between the two groups was therefore, not surprising. The samples marked with the † represented the lowest three samples (73-74, 77-78, 85-86). As discussed earlier, these samples probably represented the latter stages of recovery from a disturbance and were considered to be representative of an unburned environment. They have been included in the unburned group (group 2).

Stratification of the charcoal measures into the 2 groups posed a problem. It was difficult to determine at which level one should separate samples into burned and unburned groups. Since the cluster technique was independent of the charcoal measures, it was decided to stratify the samples in the same proportion as was done above. Therefore, the 18 highest ranking samples for each charcoal measure constituted group 1 (burned) and the lower 14, group 2 (unburned).

The Mann Whitney U statistic was used to determine if there were any significant differences in the number of charcoal fragments in each of the nine size classes between the burned (group 1) and unburned (group 2) samples, as stratified by the various measures discussed above. (i.e., whether the charcoal 'peaks' contained a higher proportion of larger fragments). Table 1 shows the results of this analysis. The medians of each group for each size class have been added so that the direction of the difference can be noted, as well as the significance levels.

Clearly the cluster procedure yielded the most significant differences between groups with four of the size classes being significant at the 0.05 level of confidence (300-399, 400-499, 900-1199, 2000+) and two being significant at 0.02 (100-199, 600-899). Charcoal percentage gave less striking results with only two size classes being significant (100-199 and 600-899) at the 0.05 level. Charcoal concentration showed the poorest results with no size class exhibiting significant differences.

Charcoal area requires additional discussion. The largest two size classes showed significant differences between groups (1200-1999 being significant at 0.02 level of confidence). This was not at all surprising as the area index was calculated by multiplying the midpoint of size class by the number of fragments in that class and, in the case of the 2000 μm^2+ category each fragment was added to the total. Samples with a high number of large fragments therefore, would have a high total area and thereby fall into group 1 (burned) and exhibit a distinct size class difference from other samples. In terms of the area index the results of the Mann Whitney U statistic were, therefore, a statistical artifact.

Table 1. Significant differences between burned (G-1) and unburned (G-2) samples based on the Mann Whitney U test

Size classes and groups*	Cluster of L.I. and S/C ratio Median	Sign.	Charcoal percentage Median	Sign.	Charcoal concentration Median	Sign.	Charcoal area Median	Sign.
100-199								
G-1	106.5	P 0.02	103.5	P 0.05	96.5		92.5	
G-2	85.5		92		98.5		102.5	
200-299								
G-1	37.5		37.5		38.5		39.5	
G-2	40.5		40.5		41		38.5	
300-399								
G-1	17	P 0.05	17.5		19.5		21	
G-2	21.5		20.8		19.5		17.5	
400-499								
G-1	9	P 0.05	10.8		11		11.5	
G-2	13		13		11.5		11.2	
500-599								
G-1	7.8		8.5		8		7.5	
G-2	9.8		7.5		8.5		8.3	
600-699								
G-1	7.5	P 0.02	8.5	P 0.05	9.5		10.5	
G-2	11.5		11.5		9		8.5	
900-1199								
G-1	3.5	0.05	4		6		5.8	
G-2	6.5		5.5		4.5		4	
1200-1999								
G-1	3.8		3.7		4.3		5.5	P 0.02
G-2	5.5		5.5		4.2		3.2	
2000 +								
G-1	2.2	P 0.05	2.5		2.9		3.5	P 0.05
G-2	4.5		4.2		3.8		1.5	

* Group 1 (G-1) cluster represents high S/C, low L.I., highest 18 cases for charcoal percentage, charcoal concentration and charcoal area.
Group 2 (G-2) cluster represents low S/C, high L.I., lowest 14 cases for charcoal percentage, charcoal concentration and charcoal area.

The surest independent measure of fire was the cluster procedure. The cluster procedure yielded results most similar to charcoal percentage. In both measures the smallest size class (100-199) showed significant differences with medians of 106.5 and 103.5 for group 1 (burned) samples from the cluster and charcoal percentage measures respectively and medians of 85.5 and 92 for group 2 (unburned) samples. The larger size classes were less similar. Charcoal area, on the other hand, had the largest three size classes being significant but not any of the smaller size classes.

Dispersion of Charcoal from a Fire

The assumption made by most researchers, implicit in the charcoal area index used, is that a local fire (burning within the basin) would produce a greater proportion of large charcoal fragments, depositing these into the lake via either the turbulent winds which accompany a fire or through a waterborne component washing from the surrounding surface. Little notice has been given to the background charcoal, present in all samples, and to its mode of deposition. Some insight into this problem can be achieved by comparing the size class distribution in group 1 (burned) and group 2 (unburned) (Figure 8).

The distributions were opposite to what was expected. It has generally been assumed that local fires would add a pulse of charcoal and that the charcoal would be composed of large fragments relative to the background charcoal. The use of the area index is evidence of this assumption. Group 1 (burned), however, had a greater proportion of smaller fragments (54 per cent) in size class 100-199 as compared to group 2 (unburned) which had only 42.5 per cent of the total number of charcoal fragments in the smallest size class. The largest size class (2000+) accounted for 2.4 per cent of the total number of particles as compared to only 1.2 per cent in group 1 (burned).

This same trend was evident statistically (see Table 1). Note that for groups 1 and 2 stratified by clustering of loss on ignition and the S/C ratio that the fire-influenced environment resulted in a significantly ($P<0.02$) greater proportion of charcoal fragments in the 100-199 size class with a median of 106.5 as compared to the non-fire-influenced samples with a median of 85.5. Group 1 (burned) had significantly ($P<0.05$) fewer larger fragments than group 2 (unburned) in size classes 300-399, 400-499, 600-899 ($P<0.02$), 900-1199 as well as the 2000+ size class. A similar trend was evident with charcoal percentage although not as pronounced. Charcoal area exhibited the reverse trend, however, as already mentioned, this was a statistical artifact.

These data suggest that the background charcoal (represented by the charcoal present in non-burned zones) was brought to the lake either from direct erosion of the soil surface or via streams. If the background charcoal were to represent a continual atmospheric rain, the size class distribution would be expected to exhibit fewer larger fragments and a greater proportion of small ones in comparison with a burned situation. There is no doubt that following a fire a great deal of charcoal is produced that remains on the soil surface and that increased erosion washes an amount of this charcoal into the lake. It would be expected that this charcoal component would have a similar size class structure as the background charcoal assuming the background charcoal was deposited from erosion off the soil surface. The fact still remains, however, that the charcoal incorporated into the sediment during a fire exhibited a size class distribution which had fewer large fragments and a greater proportion of smaller ones. This phenomenon may be due to the large scale dilution of the streamborne component by an airborne source. The deposition from the smoke produced by a fire must be of such a magnitude that it supplements the charcoal record with large quantities of fine grained charcoal masking the streamborne component.

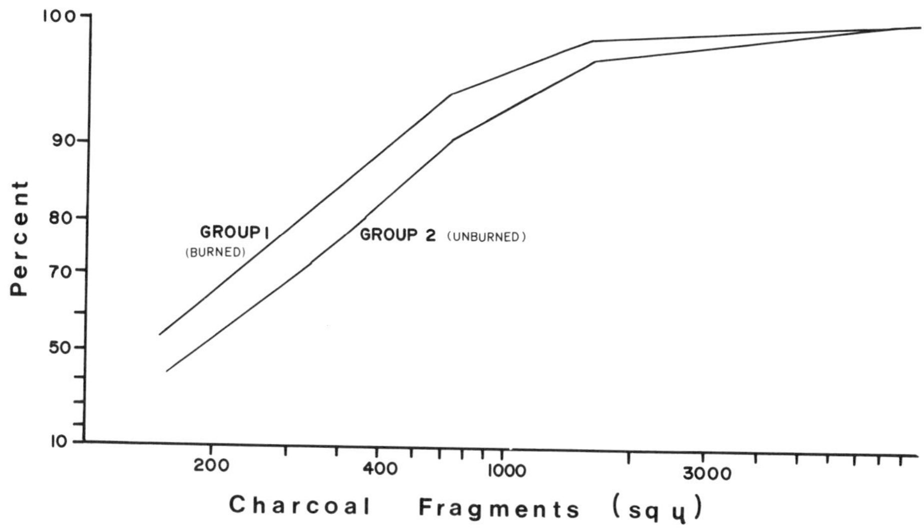

Figure 8. Cumulative percentage of the number of charcoal fragments plotted at the midpoints of size class intervals. Note that group 1, (burned), samples have a greater proportion of smaller fragments than group 2 (unburned). This is opposite to what one would expect given the assumption that larger fragments point to local fires.

One problem with this dispersion hypothesis is the episodic nature of fires. Samples exhibiting characteristics of a burned environment began at 27 cm and continued to the surface. If just two fires have occurred in this period of record, it is difficult to understand how the small fragments resulting from the airborne component would be prevalent in all samples included in group 1 (burned). Problems of sediment mixing and long distance charcoal deposition may contribute to the subtle shift in size class distribution. In addition, much of the streamborne charcoal would be fine as it would contain material from a number of sources. The fact still remains, however, that the assumptions that local fires produce a greater proportion of larger fragments does not appear to be valid and this creates severe problems with the validity of the charcoal area index as an indicator of fire.

Evaluation of Charcoal Measures

Based on the above discussion it appears clear that of the charcoal measures, charcoal percentage best indicated a burned environment. This measure, as well, was the easiest to calculate in that only the presence of the particle had to be noted. The main drawback to the measure involved a common problem to all percentage data in pollen analysis, the fact that it is not independent of abundances of other taxa. As has been demonstrated increased erosion often follows severe fires. The increased clastic input results in a decrease in pollen and charcoal concentrations. If increases in charcoal concentration are used as an indicator of fire, this process acts in reverse, negating the actual increase in charcoal which, of course, occurred. This problem certainly was evident with the Mashagama Lake core. Since percentage values are independent of the amount of clastic input, errors of this type were not encountered.

The charcoal area index is the measure used by most authors (Waddington 1969; Swain 1973, 1978, 1980; Cwynar 1978; Amundson and Wright 1979; Green 1981, 1982). The area index also had the most severe problems of the three measures examined. Charcoal area was calculated by multiplying the number of fragments in each size class by the midpoint of the class. Given the size class distribution presented, it is clear that the mean of each size class was not the midpoint but something less than that. This problem is not severe when a large number of size classes are used, but, when only few are utilized the use of the midpoint of the class may introduce significant error.

If one assumes that each particle of charcoal is a piece of information, each being equally important (a reasonable assumption considering virtually nothing is known of actual charcoal dispersion mechanisms), then the amount interpreted from each fragment should be equal. This is certainly not the case with the area index (Figure 9). The diagonal line across each figure is a line of equal information. The greater the slope of the plotted line the less information (total area) is being extracted from the number of particles. (A large number of small particles may be equal to one large fragment.) The lower the slope the greater the amount of information being extracted from the number of particles. In an extreme case, (samples 22-23) (Figure 9p) 46 per cent of the total number of charcoal particles were in the 100-199 size class. This represented only 7.1 per cent of the total area. Only four charcoal particles were in the 2000 + size class but they represented 65.6 per cent of the total area of the sample with one fragment representing 32.0 per cent of the total area. Inspection of Figure 9 shows that this was not an isolated case. In addition, there does not seem to be any trend between groups (based on cluster of L.I. and S/C). Groups 1 (burned) and 2 (unburned) exhibited both extremes (samples in which only a few fragments represented a large proportion of the total area and samples which were closer to the line of equal information).

One could argue that larger fragments by their presence are worth more information than smaller fragments. This assumption is inherent with the area index in that larger fragments represent local fires. The data presented, however, refute this claim and raise serious doubts about the validity of the charcoal area index. Even if larger fragments did represent local fires, it is unwise to obtain 32 per cent of the total charcoal area of a sample from one fragment of charcoal. The results in such a case would be statistically unstable. There is a great probability that one piece of charcoal may be missed or that one larger may be found, totally altering the final area index value obtained. The problem is compounded when the area index value is converted to a concentration value with all of the problems already discussed of the concentration measure. A detailed time control such as that provided by varved sediments alleviates this problem to a great degree.

Conclusions

Three methods of quantifying charcoal abundance in lake sediments were evaluated. The top 23 cm of sediment were considered to represent a burned environment based on evidence from the pollen spectra, loss on ignition and what was known of the recent fire history of the region. The lower 62 cm were considered to be uninfluenced by fire. Charcoal percentage exhibited the best relationship between the two zones with high levels dominating the burned zone and lower levels dominating the non-fire-influenced zones. Charcoal percentage also was significantly correlated with loss on ignition and the S/C ratio further indicating the success of this measure in indicating the occurrence of fire.

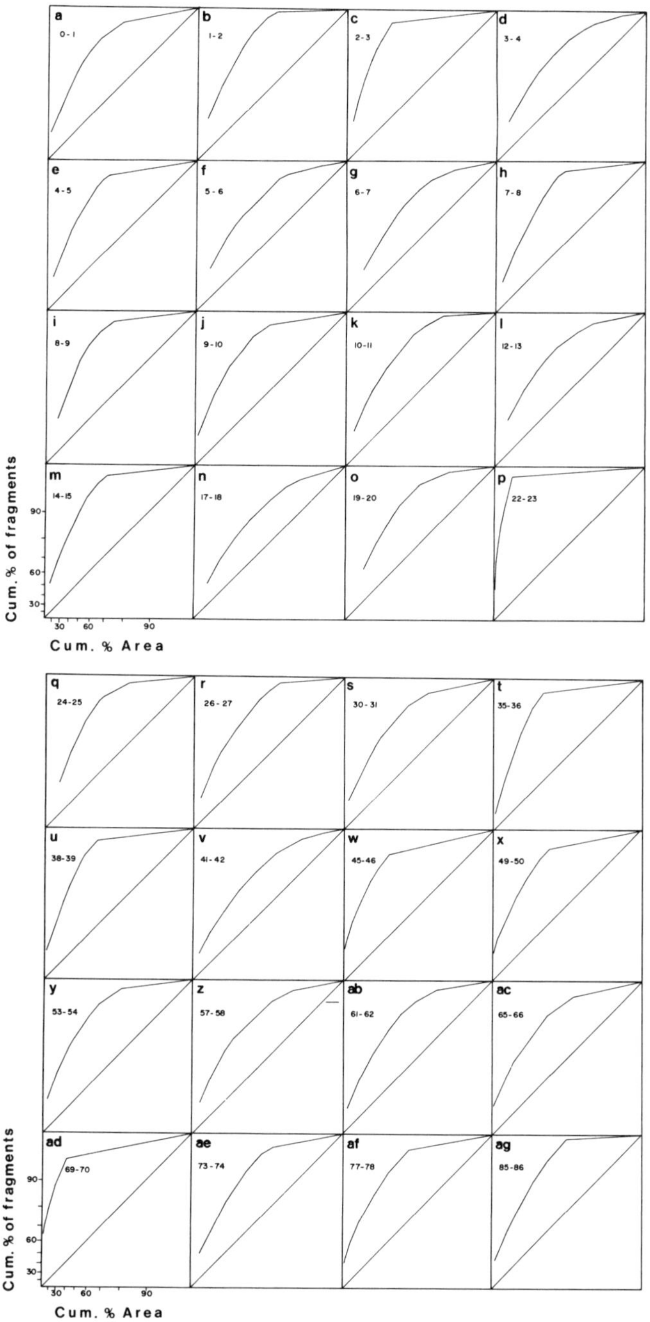

Figure 9. Cumulative percentage of the number of charcoal fragments — charcoal area. This figure shows the relationship between the number of charcoal fragments and charcoal area. The smaller size classes have the majority of fragments but only represent a very small portion of the total area. The largest size class, on the other hand, has very few fragments but those few particles contribute greatly to the total area of the sample.

Both charcoal concentration and charcoal area poorly indicated increased charcoal abundances in the burned zone. Problems associated with increased clastic input due to erosion, and dilution of the charcoal record may have been responsible for this trend. A detailed time control such as that obtained from varved sediments would be needed to alleviate this problem. Charcoal area shares the problems of charcoal concentration and in addition, problems of over-representation of a few large fragments in the total area index to render this measure misleading. Both charcoal area and charcoal concentration showed poor correlation with other indices of past fires. While one would expect a negative correlation with loss on ignition a positive one resulted. Charcoal concentration and area were negatively associated with the S/C ratio. Again, this relationship was the reverse of what would be expected.

Analysis of the size class data suggested that the background charcoal, that being continually deposited into the sediment, was streamborne in nature due to the relatively greater proportion of larger fragments as compared to smaller ones. During a burn, it appeared that the additional charcoal added to the sediment was done so via an airborne source. The size class distribution of charcoal showed that in such situations there was a greater proportion of smaller fragments and fewer large particles. This added airborne component diluted the charcoal brought to the lakes via a water source giving rise to the size class distribution found.

A great deal of research still needs to be completed. Studies of this type should be repeated to ensure that the results found from Mashagama Lake are not anomalous. Different basin configuration, vegetation cover, soil types, fire intensity and wind direction may each play a role in the amount and size class distribution of charcoal preserved in lake sediment. The effect of sample preparation on the size class frequency of charcoal needs to be documented. If the procedure itself alters the size class structure, little information would be obtainable from the record.

Of extreme importance is research undertaken to quantify actual dispersion mechanisms of charcoal following fires of various intensities. Data including the amounts and size class distribution of charcoal brought to the lake via airborne and waterborne routes before, during and following a fire would be invaluable. The behaviour of charcoal once in the lake would also be of interest. Factors such as settling velocities, effects of winds and currents may affect the actual incorporation of charcoal into the sediment. Empirical data such as these are essential in order that the dispersal mechanisms of charcoal be understood and interpretation of charcoal in lake sediment be more accurately undertaken.

The information required by forest managers in terms of fire history is the frequency at which particular forest types burn. It does not seem likely, given the present state of the art, that analysis of charcoal in lake sediments can provide such a record. Problems of sediment mixing, sampling intervals and variable sedimentation rates (particularly following disturbances such as occur in a fire) act to mask the charcoal record. Perhaps, with a great deal of additional calibration of charcoal inputs, an understanding of the relative occurrence of fire in different vegetation types may be established.

Acknowledgements

This research was partially funded by the Natural Sciences and Engineering Research Council of Canada. Travel expenses to the conference were funded in part by the Northern Research Committee, University of Western Ontario. We would also like to thank the Geography Department, University of Western Ontario for the use of laboratory facilities.

Literature Cited

Ahlgren, I.E. and C.E. Ahlgren, 1960. Ecological effects of forest fires. Bot. Rev. 26: 483-533.

Amundson, D.C. and H.E. Wright, Jr. 1979. Forest changes in Minnesota at the end of the Pleistocene. Ecol. Monogr. 49: 1-16.

Arno, S.F. 1976. The historical role of fire on the Bitteroot National forest. U.S.D.A. For. Serv. Res. Pap. INT-187.

Berndt, H.W. 1971. Early effects of forest fire on streamflow characteristics. U.S.D.A. For. Ser. Res. Note PNW-148.

Biswell, H.H. and A.M. Schultz, 1957. Surface runoff and erosion as related to prescribed burning. J. For. 55: 372-374.

Byrne, R., J. Michaelsen and A. Souter, 1977. Fossil charcoal as a measure of wildfire frequency in southern California: a preliminary analysis. U.S.D.A. For. Ser. Gen. Tech. Rep. WO-3, pp. 361-373.

Connaughton, C.A. 1935. Forest fires and accelerated erosion. J. For. 33: 751-752.

Cwynar, L.C. 1976. The recent fire history of Barron Township, Algonquin Park. Can. J. Bot. 55: 1524-1538.

Cwynar, L.C. 1978. Recent history of fire and vegetation from laminated sediment of Greenleaf Lake, Algonquin Park, Ont. Can. J. Bot. 56: 10-21.

Donnelly, R.E. and J.B. Harrington, 1978. Forest Fire History Maps of Ontario, Ont. Min. of Nat. Res., Aviation and Fire Management Centre, Sault Ste. Marie, Ont.

Faegri, K. and J. Iversen. 1975. Textbook of Pollen Analysis. Second ed. Hafner Publ. New York, NY.

Frissell, S.S., Jr. 1973. The importance of fire as a natural ecological factor in Itasca State Park, Minnesota. Quat. Res. 3: 397-407.

Green, D.G. 1981. Time series and postglacial forest ecology. Quat. Res. 15: 265-277.

Green, D.G. 1982. Fire and stability in the postglacial forests of southwest Nova Scotia. J. Biogeo. 9: 29-40.

Heinselman, M.L. 1973. Fire in the virgin forests of the Boundary Waters Canoe area, Minnesota. Quat. Res. 3: 329-382.

Kilgore, B.M. 1973. The ecological role of fire in Sierran conifer forests. Quat. Res. 3: 496-513.

Kilgore, B.M. and D. Taylor. 1979. Fire history of a sequoia-mixed conifer forest. Ecology 60: 129-142.

Klock, G.O. and J.D. Helvey. 1974. Soil-water trends following wildfire on the Entiat Exp. forest. Tall Timbers Fire Ecol. Conf. 15: 193-200.

Loope, L.L. and G.E. Gruell, 1973. The ecological role of fire in the Jackson Hole area, Northwestern Wyoming. Quat. Res. 3: 425-443.

Maissurow, D.K. 1941. The role of fire in the perpetuation of virgin forests of northern Wisconsin. J. For. 39: 201-207.

Mutch, R.W. 1970. Wildland fires and ecosystems: A hypothesis. Ecology 51: 1046-1051.

Rich, L.R. 1962. Erosion and sediment movement following a wild fire in a ponderosa pine forest of central Arizona. U.S.D.A., For. Ser. Res. Note RM-76.

Rowe, J.S. and G.W. Scotter, 1973. Fire in the boreal forest. Quat. Res. 3: 444-464.

Sartz, R.S. 1953. Soil erosion on a fire denuded forest area in the Douglas Fir region. J. Soil and Water Conservation 8: 279-281.

Spurr, S.H. 1954. The forests of Itasca in the nineteenth century as related to fire. Ecology 35: 21-25.

Swain, A.M. 1973. A history of fire and vegetation in northeastern Minnesota as recorded in lake sediments. Quat. Res. 3: 383-396.

Swain, A.M. 1978. Environmental changes during the past 2,000 years in north central Wisconsin: Analysis of pollen, charcoal, and seeds from varved lake sediments. Quat. Res. 10: 55-68.

Swain, A.M. 1980. Landscape patterns and forest history in the Boundary Waters Canoe area, Minnesota: a pollen study from Hug Lake. Ecology, 61: 747-754.

Tande, G.F. 1979. Fire history and vegetation pattern of coniferous forests in Jasper National Park, Alberta. Can. J. Bot. 57: 1912-1931.

Taylor, D.L. 1974. Forest fires in Yellowstone National Park. J. For. Res. 18: 68-77.

Waddington, J.C.B. 1969. A stratographic record of the pollen influx to a lake in the Big Woods of Minnesota. U.S. Contrib. Quat. Res., Geol. Soc. Amer., Spec. Pap. 123: 263-282.

Wein, R.W. and J.M. Moore, 1979. Fire history and recent fire rotation periods in the Nova Scotia Acadian Forest. Can. J. For. Res. 9: 166-178.

Whishart, D. 1978. Clustan Users Manual. Program Library Unit, Edinburgh University.

Wright, H.E., Jr. and H.E. Heinselman. 1973. The ecological role of fire in natural conifer forests of western and northern North America — Introduction. Quat. Res. 3: 319-328.

FOREST FIRE WEATHER AND WILDFIRE OCCURRENCE IN THE BOREAL FOREST OF NORTHWESTERN ONTARIO

B.J. Stocks
Canadian Forestry Service

and

R.B. Street
Atmospheric Environment Service

Abstract: During the period between 1974 and 1980 northwestern Ontario experienced an unprecedented number of severe forest fire years. Well-entrenched high pressure systems over central North America effectively blocked spring precipitation from reaching northwestern Ontario in 1974, 1976, 1977 and 1980, creating prolonged warm, dry periods and resulting in significant fire losses. An analysis of northwestern Ontario fires during this period, by size, cause, and distribution, shows that, with the exception of 1980, a large proportion of the fires that grow largest occur in areas of limited protection in the far north.

The overwhelming impact of wildfires on ecosystem development and forest composition in the boreal forest region of northwestern Ontario is readily apparent upon even casual observation. Large tracts of even-aged jack pine (*Pinus banksiana*) and black spruce (*Picea mariana*) dominate the landscape in an irregular patchwork mosaic, the result of periodic severe wildfire years. Charcoal discovered in lake sediments and peat bogs provides positive evidence that fire has always been strongly associated with the boreal forest. The advent of organized forest fire suppression in the early 1900s has had an impact on wildfire activity in this area, but it remains to this day a classic example of a fire dependent ecosystem, capable, during periods of extreme fire weather, of sustaining the very large, fast moving forest fires which are responsible for its existence.

The period between 1970 and 1980 produced extremes in meteorological conditions that impacted significantly on many social and economic activities throughout the world. In the west-central part of Canada, climatic anomalies produced a dramatic increase in forest fire activity. Nowhere was this more evident than in northwestern Ontario where in excess of two million hectares of forest was burned over between 1970 and 1980.

This paper will examine in detail the occurrence and distribution of wildfires in northwestern Ontario during the 1970-1980 period and will discuss in general the fire weather patterns that contributed to the particularly severe forest fire losses in specific years. The 1980 fire season was the most significant in recent memory in northwestern Ontario, as a number of large fires in commercially valuable areas resulted in large-scale evacuations and private property losses. Where possible, fire statistics and fire weather for the 1980 fire season will be compared to averages for the decade from 1970-1980. In this paper the term northwestern Ontario refers to the combined area of the Northwest and Northcentral Regions of the Ontario Ministry of Natural Resources.

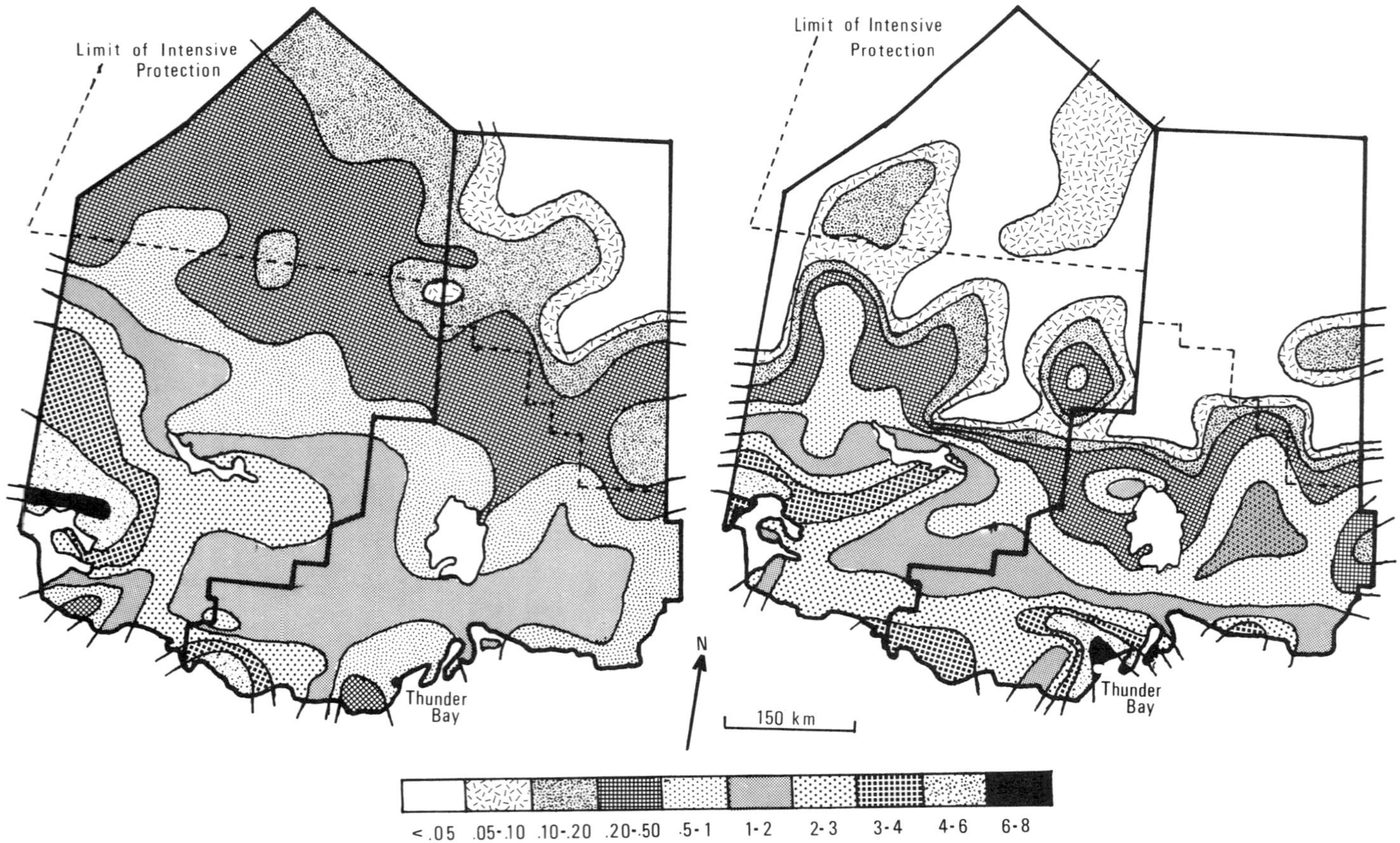

Figure 1. Lightning fire distribution (left) and man-caused fire distribution (right) in northwestern Ontario, 1970-1979.

Fire Occurrence and Distribution

During the 11-year period from 1970 up to and including 1980, northwestern Ontario experienced an annual average of more than 700 fires (46 per cent of the provincial total) which on average burned over an area greater than 190,000 hectares (95 per cent of the Ontario total) yearly. Tremendous fluctuations from year to year, in both fire occurrence and area burned, were common during this period. Fire numbers varied ten-fold (from slightly more than 200 fires in 1978 to a high of more than 2100 fires in 1976), while annual area burned varied 200-fold (from a low of approximately 2600 hectares in 1978 to a high of more than 550,000 hectares in 1980).

Lightning fires are much more common in this area than throughout the remainder of the province, accounting for 59 per cent of the fires and 93 per cent of the area burned in northwestern Ontario between 1970 and 1979. Figure 1 illustrates the distribution of lightning fires in northwestern Ontario during the 1970-1979 period. The heaviest concentration of lightning fires occurs in the Red Lake-Kenora-Quetico Park area where dry lightning storms are common and two to eight lightning fires per 1000 km^2 (including water) occur annually, commonly on shallow, well-drained upland soils. Lightning fire incidence tapers off as one moves north and east away from the lightning storm track and towards wetter lowland fuel types.

Man-caused fire distribution in northwestern Ontario is presented in Figure 1 and shows a much tighter isoline gradient than was the case with lightning fires. The impact of population levels is readily apparent as man-caused fire occurrence increases dramatically around Indian reservations, cities and towns, high use recreational areas, and along highways and railways. The highest incidence of man-caused fires occurs in the Kenora-Thunder Bay area with up to six fires per 1000 km^2 per year, while much of the sparsely populated area north of 51°N latitude experiences virtually no man-caused fires. As would be expected, given the ease of detection and close proximity of suppression forces, man-caused fires are generally kept to a much smaller size than lightning fires. In northwestern Ontario man-caused fires constituted 41 per cent of the total number of fires reported (as compared to 75 per cent provincially) but accounted for only seven per cent of the total land area burned during the 1970-1979 period.

Figure 2 shows fire occurrence and area burned by months for both 1980 and the 1970-1979 period in this part of Ontario. On average the 1970-1979 fire occurrence levels remained relatively constant from late May until early August, while in 1980 the fire season started and ended earlier. By comparison, the vast majority of area burned occurred in the months of May and June in the 1970-1979 period and somewhat earlier in 1980. In general, May and June appear to be the problem fire months in northwestern Ontario.

When evaluating forest fire statistics for northwestern Ontario it is essential to recognize the fact that two relatively distinct zones of fire protection exist in this region. The area classified as being Outside Intensive Protection (OIP), approximately 111,000 km^2 of land, receives a very limited form of fire protection. Fires are generally only observed in this zone, and are fought only when private property is threatened. Within the Intensive Protection Zone (IIP), 226,000 km^2 of land, all fires are aggressively attacked. For purposes of this analysis, the line separating the OIP and IIP zones has been, as it is for administrative purposes, considered static. Actually the line delineating active suppression may fluctuate depending on the magnitude of the fire load elsewhere in the

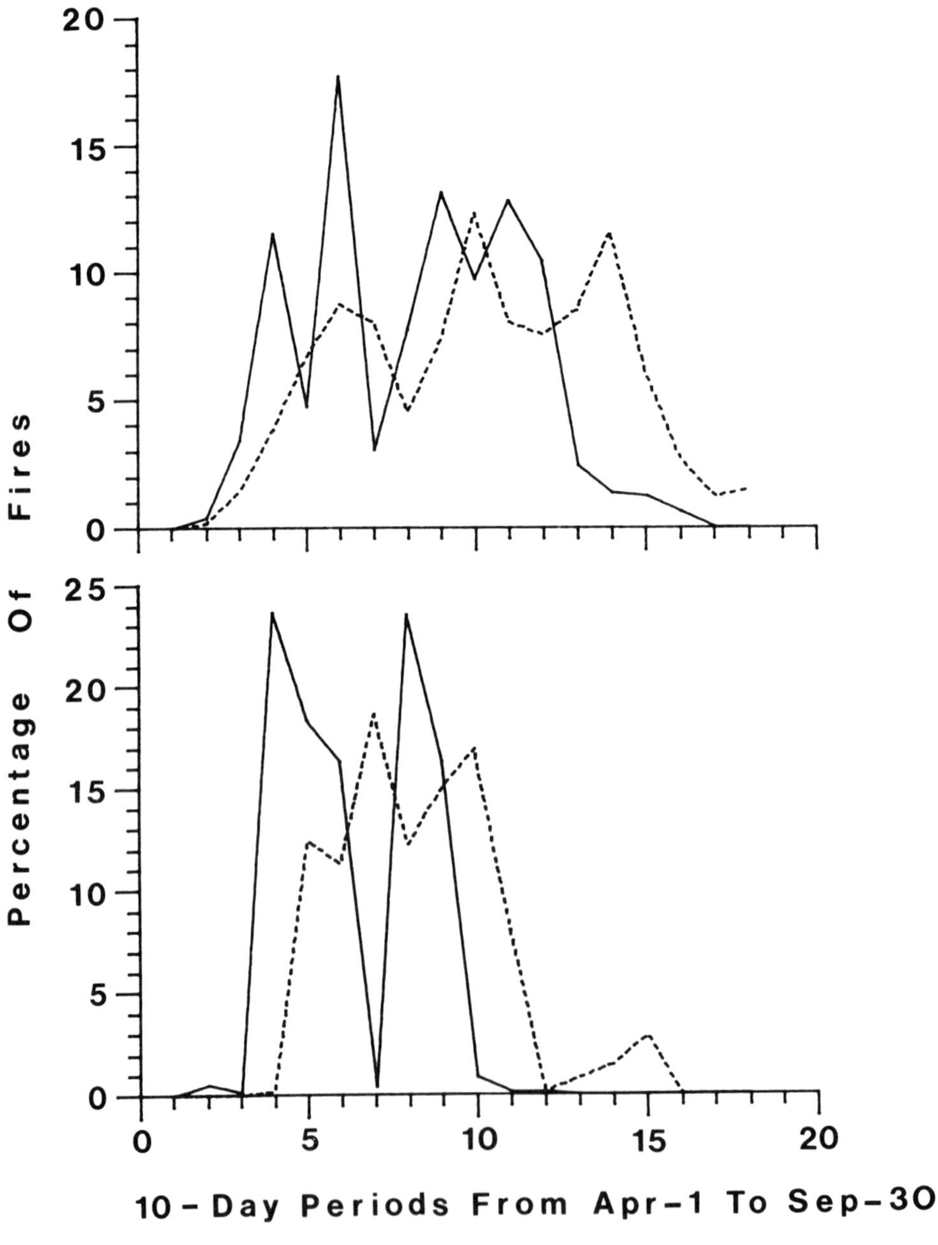

Figure 2. Percentage of fires (upper) and per cent area burned (lower) (lightning and man-caused) by 10-day periods for northwestern Ontario for 1970-1979 (dotted line) and 1980 (solid line).

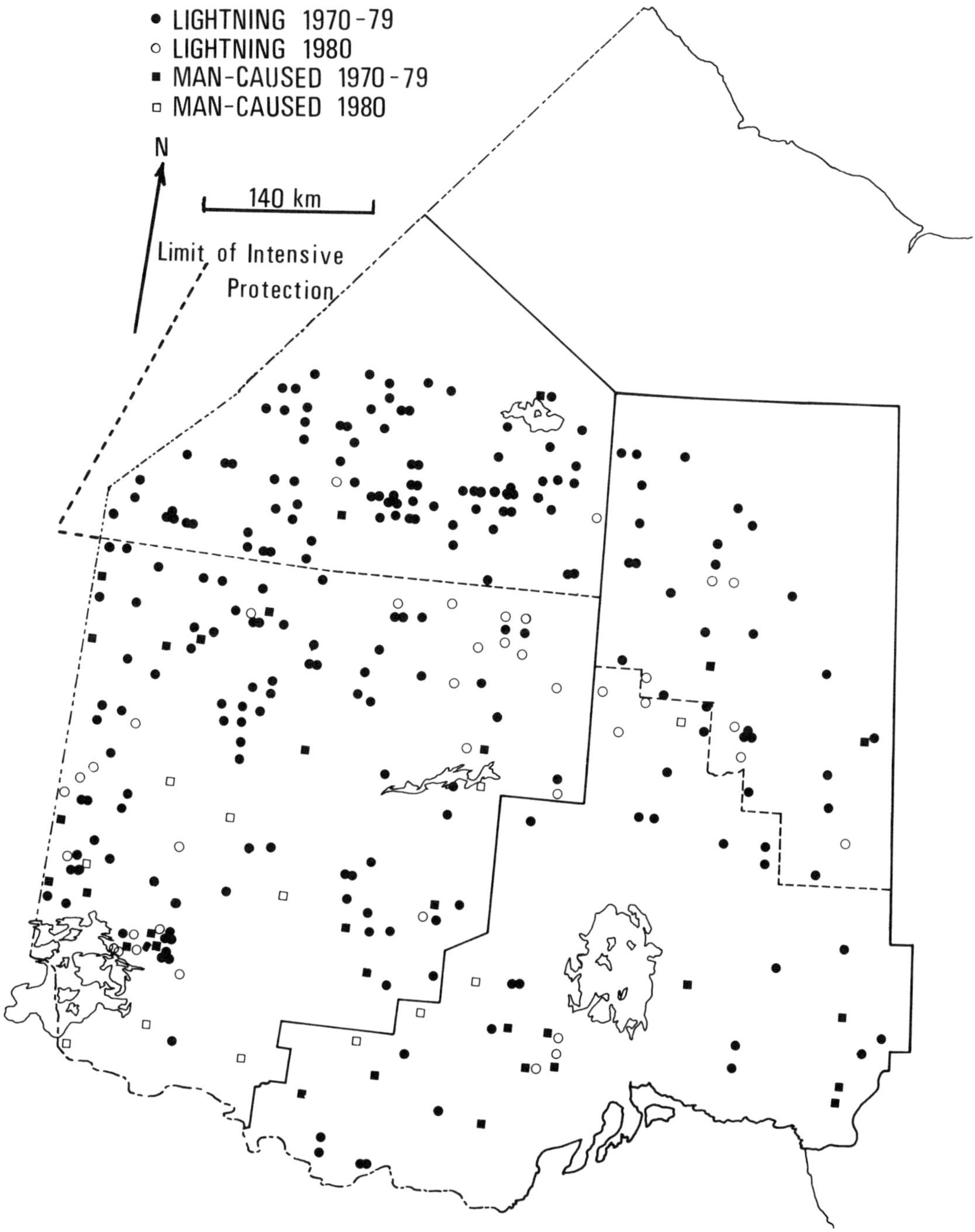

Figure 3. Distribution of northwestern Ontario fires ≥200 ha for 1970-1980 period.

province. Some fires below the line may be assigned a low priority in times of peak fire business and, conversely, during periods of low fire activity in the protected zone, certain fires north of the line may be actioned.

Figure 3 illustrates the distribution of all fires ≥200 ha in size that occurred in northwestern Ontario during the 1970-1980 period. When viewed alongside Table 1, it becomes readily apparent that the lack of fire control activity within the OIP zone profoundly affects area burned figures both within this zone and, consequently, within the whole of northwestern Ontario. Most striking is the fact that, during the 1970-1979 period, only 4.5 per cent of the fires in northwestern Ontario were outside the limit of intensive protection, and yet these fires accounted for 59.6 per cent of the total area burned in this region. Significantly, more than 17 times as many OIP fires reach 200 ha or greater in size than is the case with IIP fires, and the average OIP fire size during this period is 30 times that of IIP fires. The vast majority of OIP fires originate from lightning (86 per cent), and the few man-caused fires occur in close proximity to remote native settlements. In general forest cover north of the limit of intensive protection is similar to that south of the line, particularly in the most western portion of northwestern Ontario where decent quality stands of jack pine and black spruce are common. The eastern portion of this area generally supports low quality, marshy areas of black spruce, trailing off into the Hudson Bay Lowlands.

A further glance at Table 1 indicates that, in comparison to averages for the decade of the 1970s, 1980 was a particularly severe fire year in this area of Ontario, as was the case in much of west-central Canada. Fire occurrence (particularly lightning fire occurrence) was significantly higher in 1980 and area burned figures increased dramatically. Particularly noteworthy is the fact that the area burned within the intensively protected zone was almost ten times the average for the previous decade, much of this being attributable to man-caused fires. The resulting economic and social impact, while not easily quantified, was significant.

As a further means of illustrating the distribution of northwestern Ontario fire sizes, Figure 4 shows average fire size and annual percentage area burned for the 1970-1979 period. Fire data were grouped for areas approximately 15,000-16,000 km^2 in size and isolines drawn. Increasing fire size as one moves north and away from highly populated and intensively protected areas is quite evident. Annual percentage area burned figures are calculated using land area only, water areas being excluded by applying a 4:1 land/water ratio measured for northwestern Ontario as a whole. Annual area burned figures range from lows of 0.02 per cent in the Thunder Bay area to in excess of 1.00 per cent in the extreme northwestern portion of the region. Average annual percentage area burned figures are 0.75 per cent and 0.25 per cent for the OIP and IIP zones, respectively.

General Fire Weather in Recent Years

The frequent number of severe fire years in northwestern Ontario during the decade of the seventies can be directly attributed to prolonged periods of dry weather. The primary cause of these extended dry weather intervals was the development of longwave ridging of the atmospheric circulation pattern over central North America. The development of a stable high pressure system over the centre of the continent would steer moisture from the Pacific Ocean and the Gulf of Mexico northward across the southern Northwest Territories and eastward below the Great Lakes respectively. As a result, the flow of moist air from these traditional sources into northwestern Ontario would be effectively blocked for extended periods.

Table 1. A comparison of OIP* and IIP** fire statistics for the 1970-1980 period in northwestern Ontario[1]

		Annual No. of fires		Annual area burned (ha)		Average fire size (ha)		% Annual area burned		Percentage of fires ≥200 ha	
		1970-1979	1980	1970-1979	1980	1970-1979	1980	1970-1979	1980	1970-1979	1980
OIP	Lightning	26.7	43	81627.0	34004.8	3057.2	755.7	0.73	0.31	44.9	17.8
	Man-caused	4.5	5	1069.2	26.6	459.8	5.3	0.02	0.00	8.9	0.0
	Total	31.2	50	83696.2	34031.4	2682.6	680.6	0.75	0.31	39.7	16.0
IIP	Lightning	385.5	662	49089.3	227715.5	127.3	344.0	0.22	1.01	3.3	4.7
	Man-caused	284.4	332	7577.9	287516.7	26.7	866.0	0.03	1.27	0.9	3.6
	Total	669.9	994	56667.2	515232.2	84.6	518.3	0.25	2.28	2.3	4.3

* Outside intensive protection.

** Inside intensive protection.

[1] Figures do not exactly match OMNR statistics, as only fires occurring when OMNR weather stations were in operation (late April — late September) were used in this analysis.

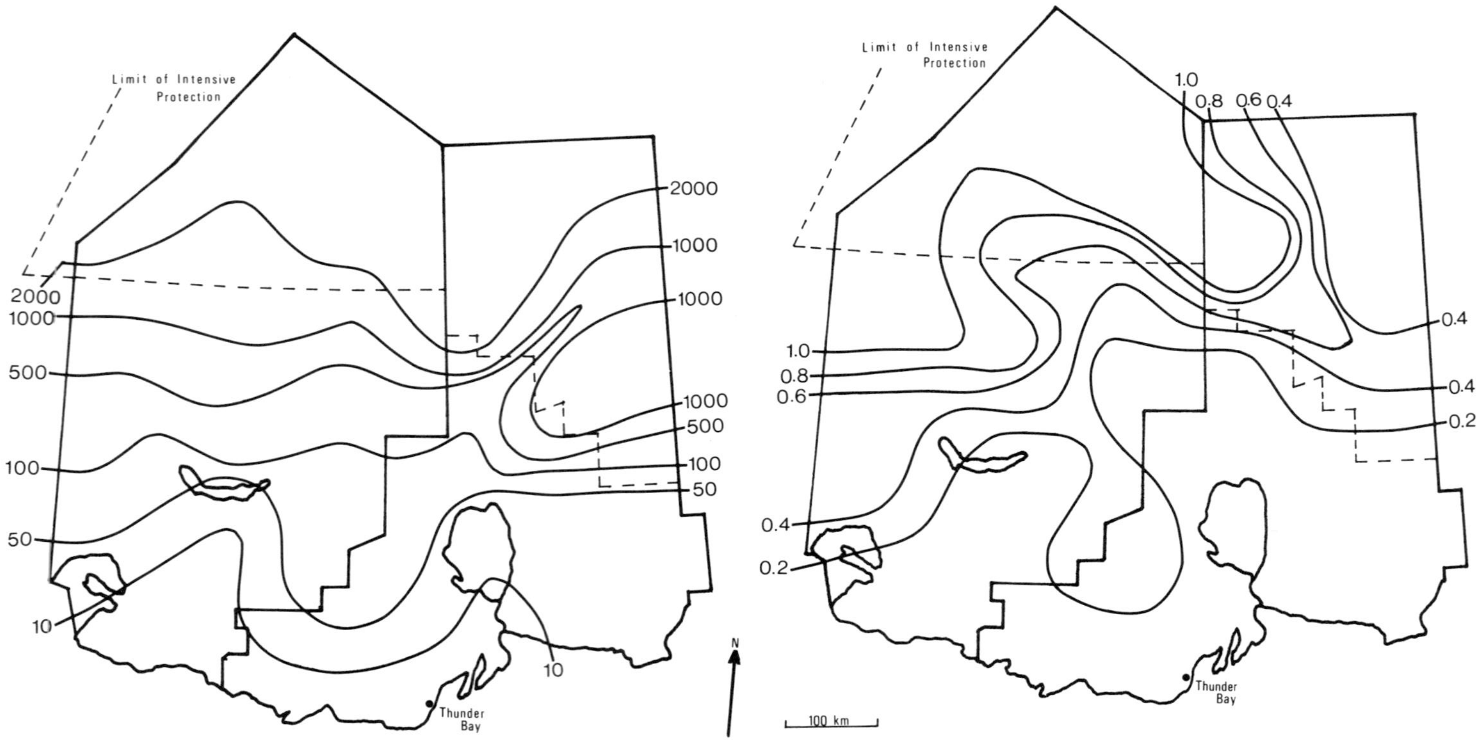

Figure 4. Average fire size, in hectares (left) and annual percentage area burned (right) for northwestern Ontario (1970-1979).

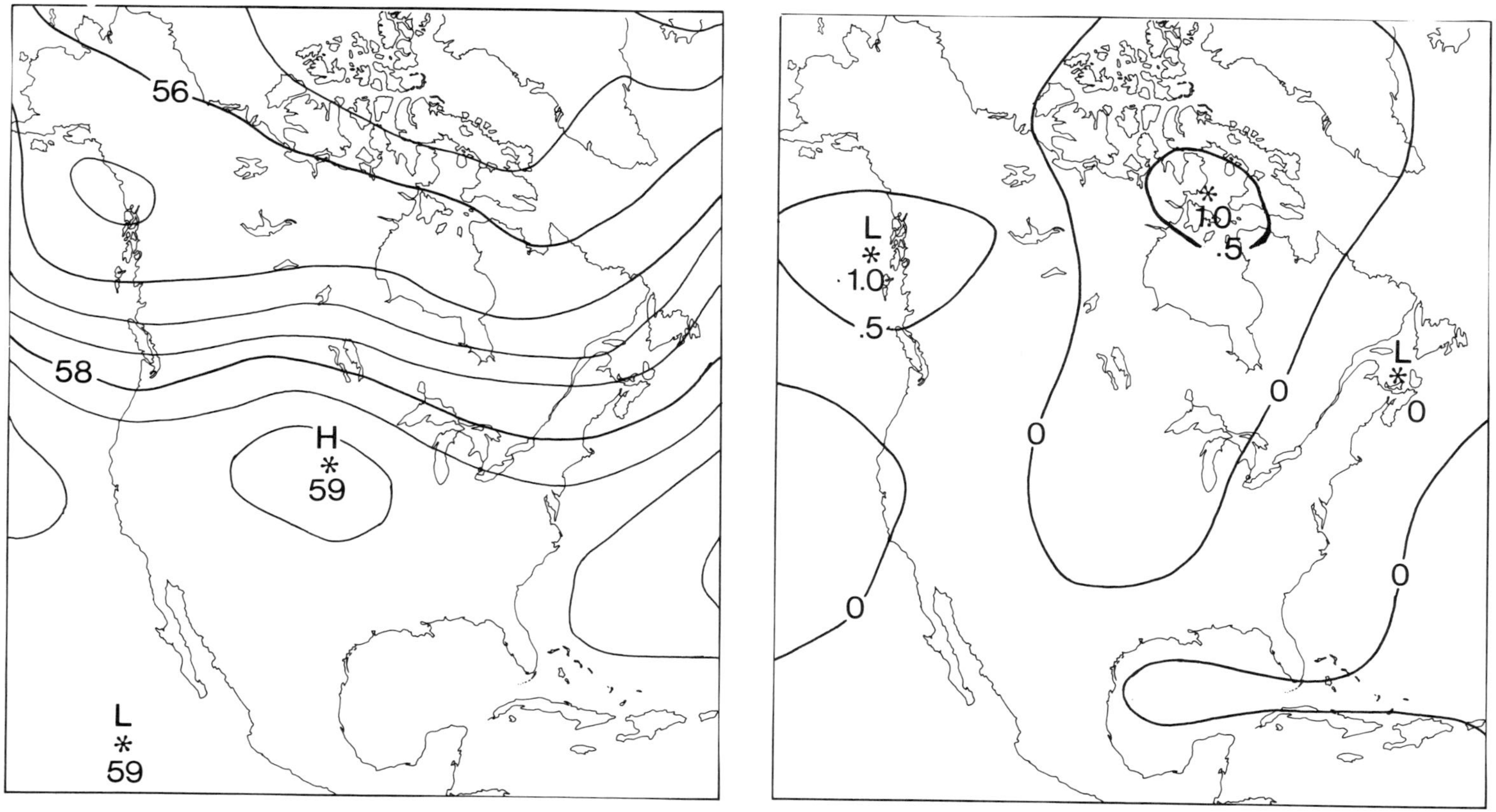

Figure 5. July 1974 mean 50 kPa height and anomalies using 30-year mean. Heights are shown in hundreds of metres.

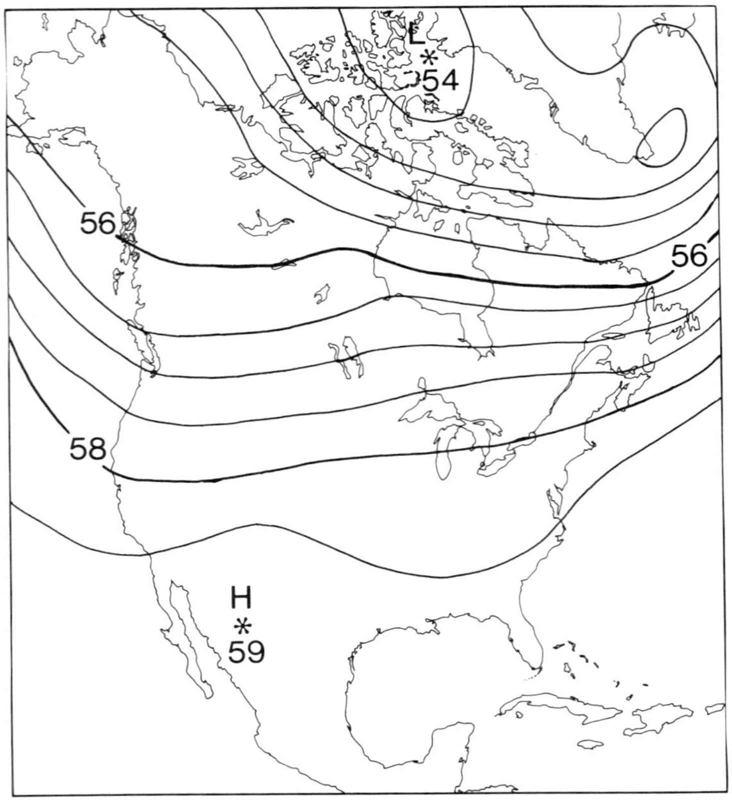

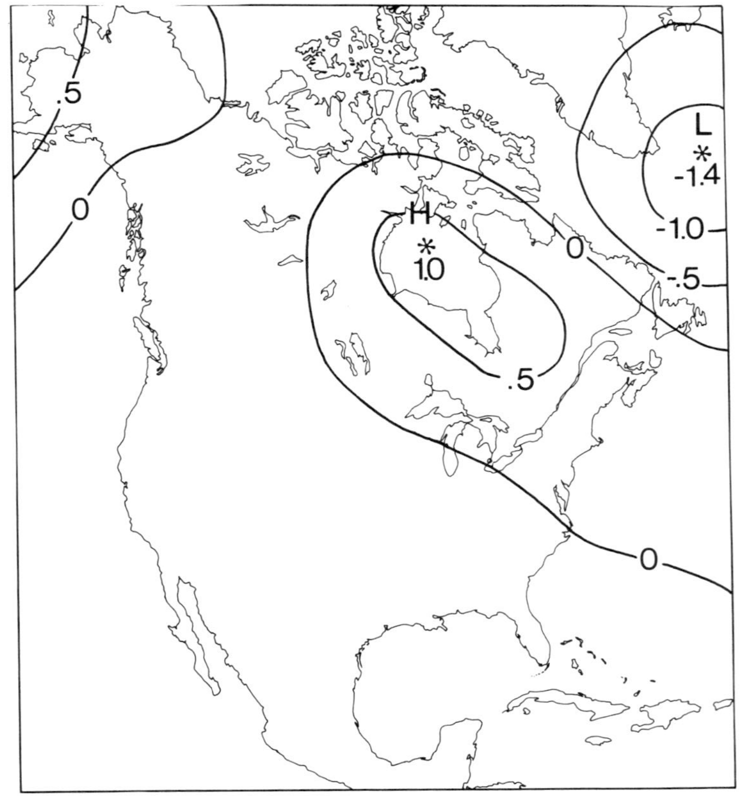

Figure 6. June 1976 mean 50 kPa height and anomalies using 30-year mean. Heights are shown in hundreds of metres.

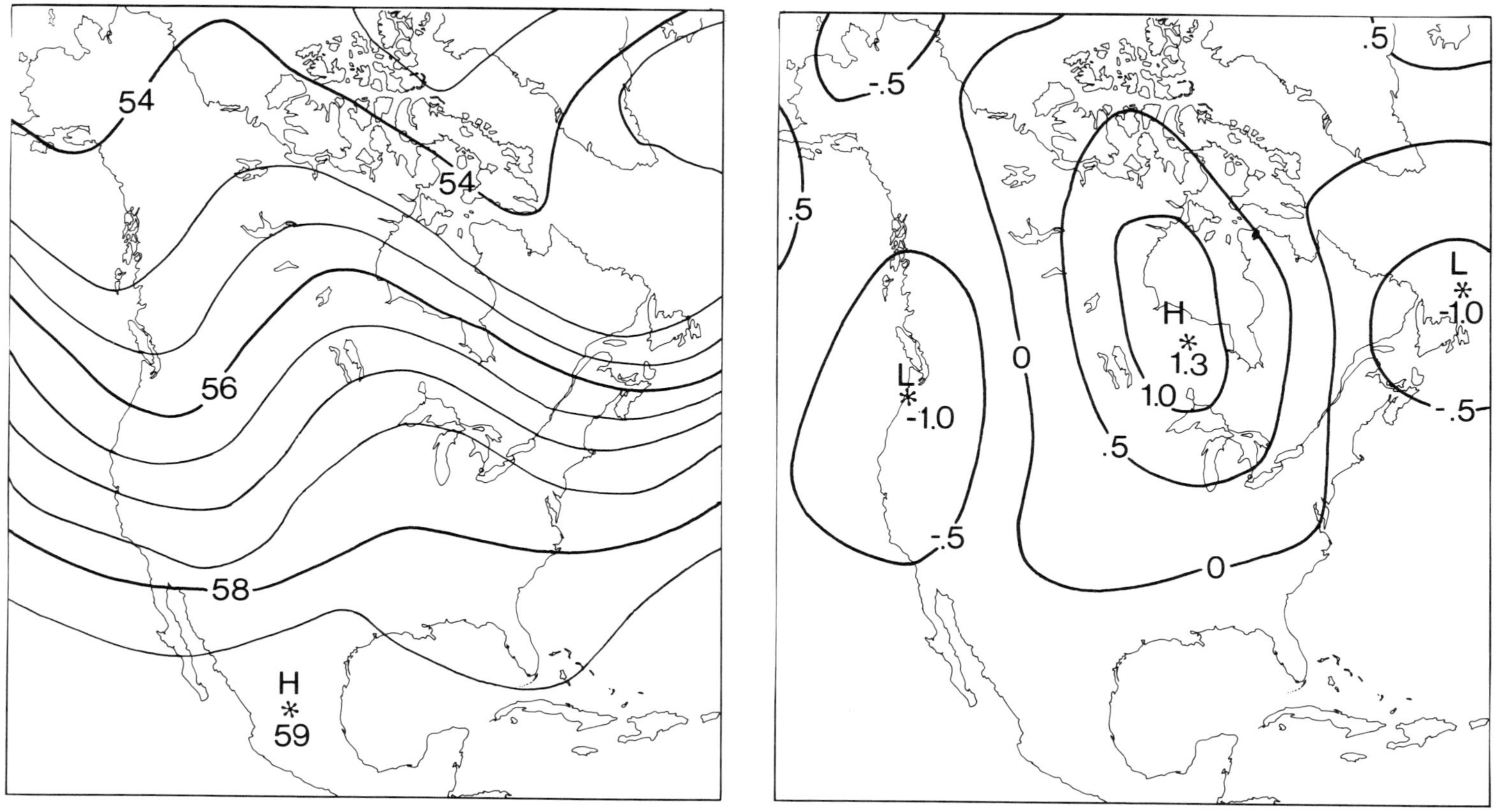

Figure 7. May 1977 mean 50 kPa height and anomalies using 30-year mean. Heights are shown in hundreds of metres.

A number of months during the 1970-1979 period, when the area burned in northwestern Ontario was significantly above average, were selected for comparison with monthly mean 50 kPa height and anomaly charts in order to determine if a correlation existed. Figures 5 to 7 depict mean heights and mean height anomalies for 50 kPa of atmospheric pressure for the months of July 1974, June 1976, and May 1977, respectively. The height analysis portion of each figure depicts the mean height of the 50 kPa pressure level above mean sea level (MSL) during the month (based on two observations daily), and heights shown are in units of hundreds of metres above MSL with an analysis interval of 50 metres. The anomaly analysis shows the deviation of the month's 50 kPa heights from the 30-year monthly normal heights and anomalies are in units of hundreds of metres difference from normal with analysis interval of 50 metres. Areas labelled 'H' or with positive values have 50 kPa mean monthly heights that are above the 30-year normal for the month. Conversely, 'L' or negative anomalies denote areas that have 50 kPa mean monthly heights that are below the 30-year normal. In each case depicted in Figures 5-7, prominent upper-air longwave ridges and associated above normal heights can be identified, even when averaged over a complete month, over the Prairie Provinces and northwestern Ontario. The type of upper air flow pattern depicted in these figures provides the persistent blocking necessary to prevent the influx of atmospheric moisture into northwestern Ontario, and also favours the flow of warm dry surface air into this region from the south-central United States.

For comparison, two months (June 1975 and July 1978) in which very little area in northwestern Ontario was burned over had upper air flow patterns which were near normal or exhibited long-wave troughs and below normal heights in the vicinity of the Great Lakes (Figures 8 and 9). This type of flow pattern would allow the influx of moist air from the south, resulting in a low probability of significant fire behavior.

The forest complex in the boreal forest of northwestern Ontario consists mainly of upland, shallow-soiled ridges, on which jack pine grows in pure even-aged stands or in a mixture with black spruce, interspersed with lowland marshy areas. Most sites are well-drained and any interruption in regular precipitation can cause rapid drying of organic fuels and subsequent escalation in forest fire danger levels. One to two weeks of dry weather can set the stage for catastrophic wildfire behavior. Overwinter precipitation amounts do not seem to have an appreciable effect on the subsequent fire season. Studies show that snowfall amounts and lake inflow data do not correlate with area burned figures for the following fire season in northwestern Ontario. A short period of dry weather, particularly during the spring, is all that is necessary. Long-term drought is important only in determining whether lowland areas will burn.

As was the case in previous severe fire seasons during the seventies, the most significant meteorological feature affecting fire weather in northwestern Ontario during the month of May 1980 was an upper-air high pressure system located, for an extended period, over central North America (Figure 10). Once again, the flow of atmospheric moisture into northwestern Ontario was effectively blocked, and warm, dry air was pumped into this region from the central United States. Air temperatures in northwest Ontario were dramatically above normal. Kenora, for example, averaged 4.6°C and 5.5°C above normal temperatures during the months of April and May, respectively. These warm temperatures, with attendant low relative humidities, resulted in the early and accelerated evaporation of the snow-pack and subsequent rapid drying of forest fuels. April and May precipitation amounts at Kenora were only 16 per cent and 41 per cent of normal, respectively. Figure 11 outlines the average Severity Index (a function of the Canadian

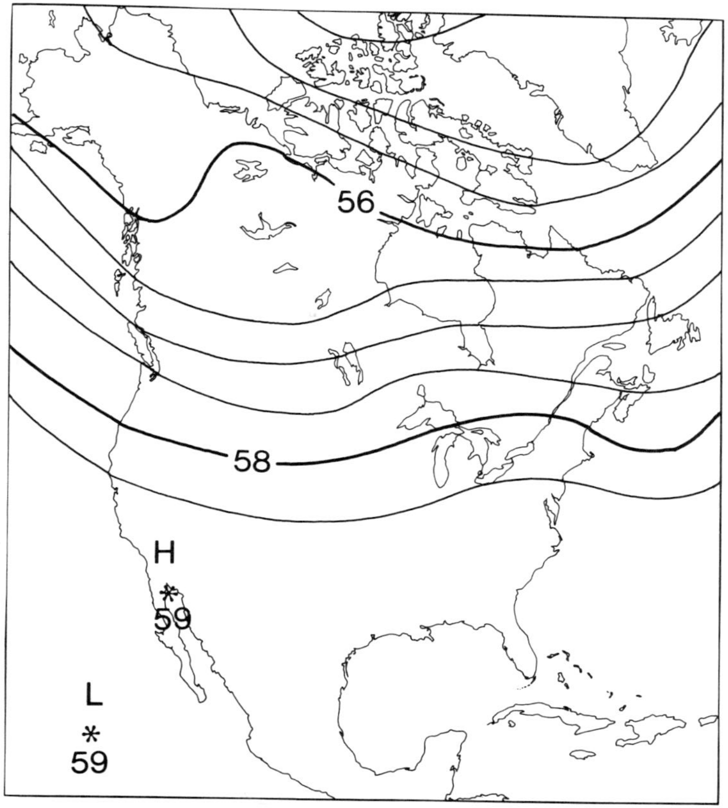

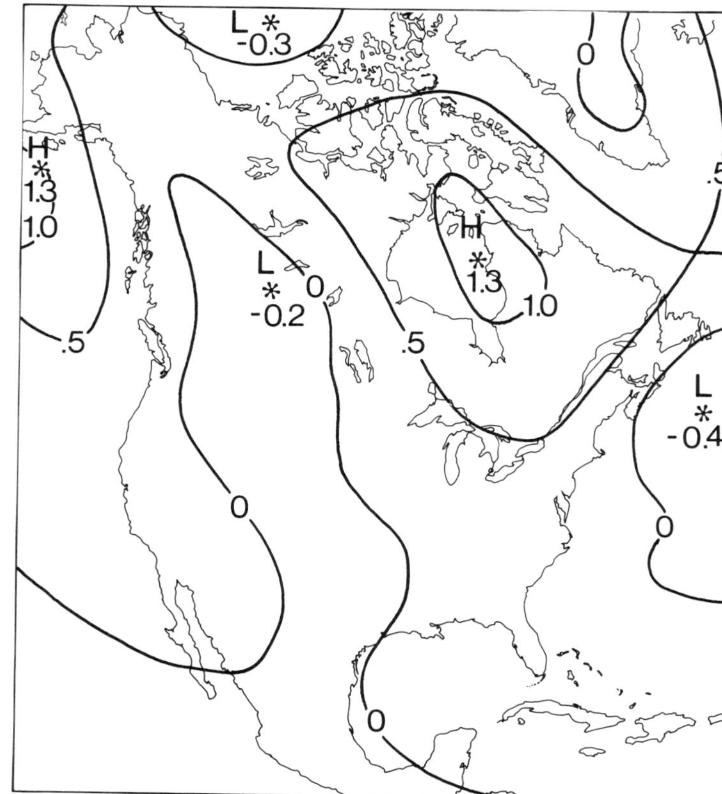

Figure 8. June 1975 mean 50 kPa height and anomalies using 30-year mean. Heights are shown in hundreds of metres.

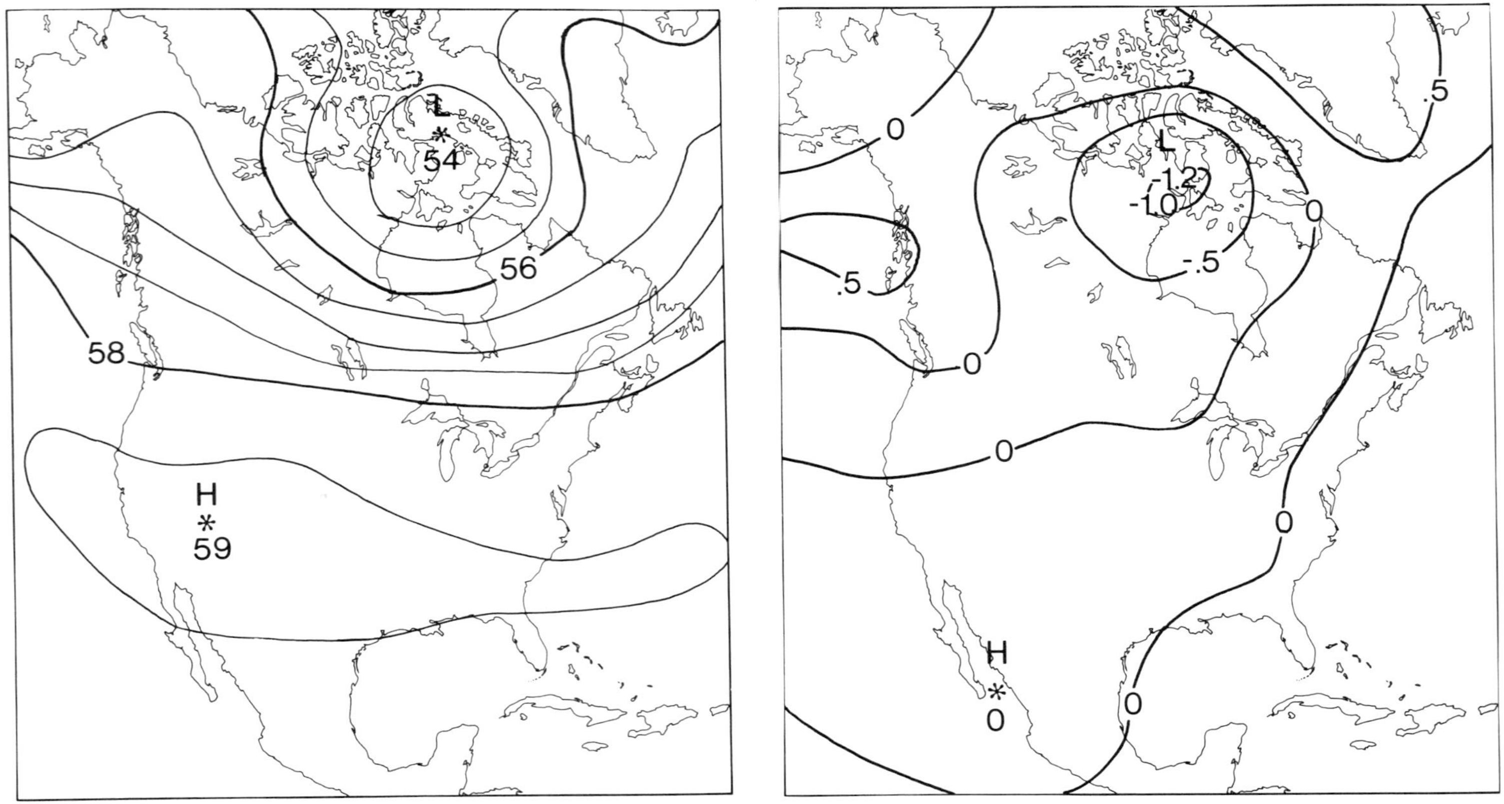

Figure 9. July 1978 mean 50 kPa height and anomalies using 30-year mean. Heights are shown in hundreds of metres.

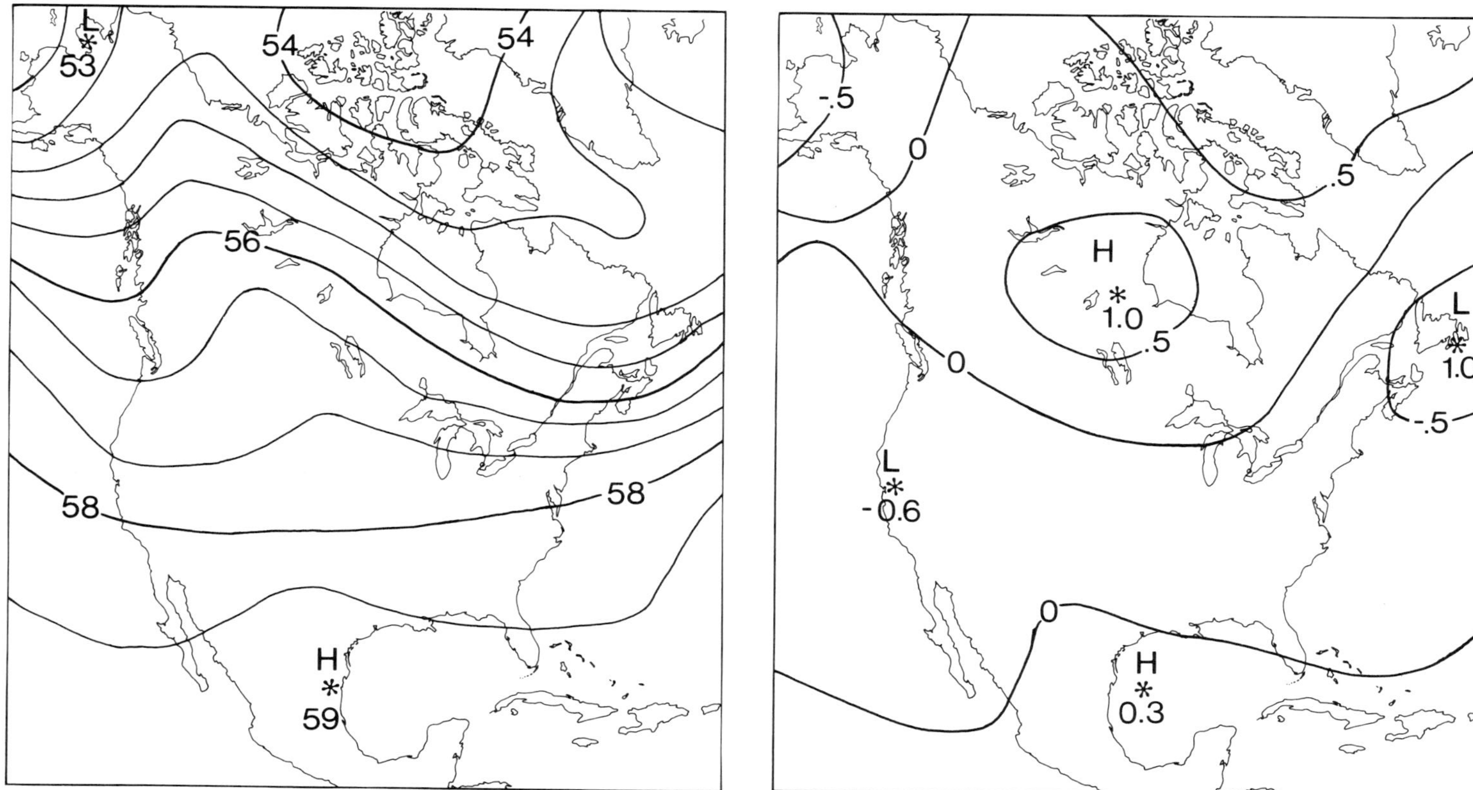

Figure 10. May 1980 mean 50 kPa height and anomalies using 30-year mean. Heights shown are in hundreds of metres.

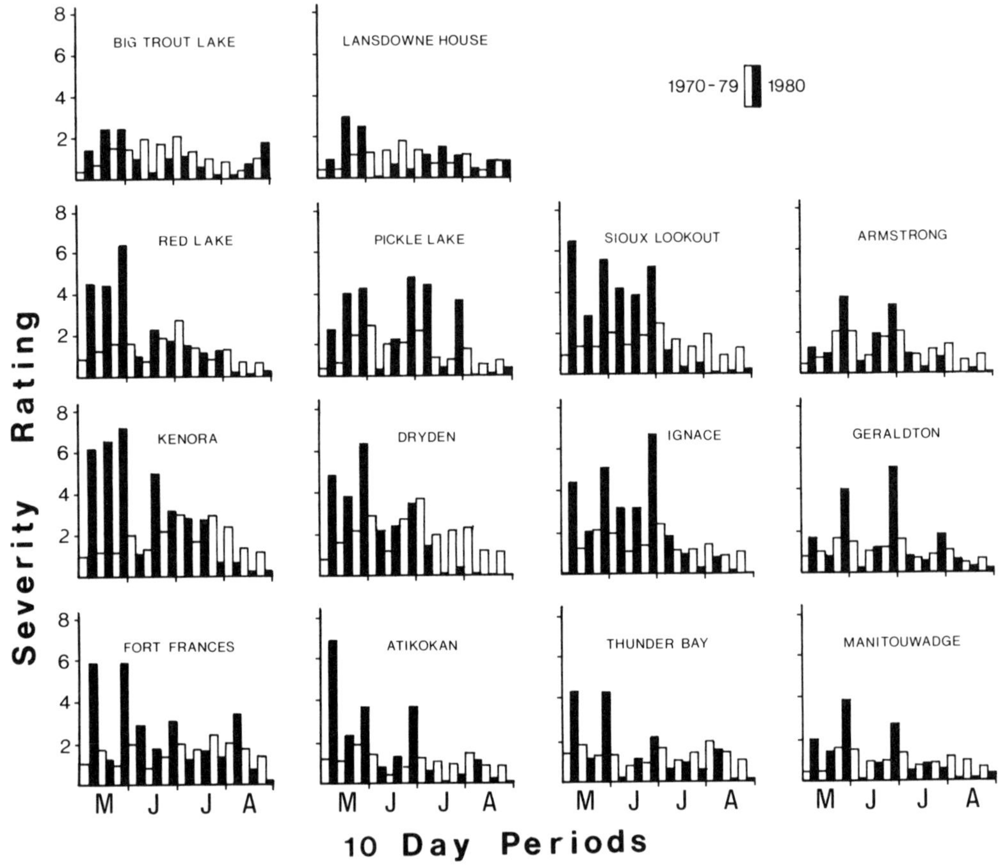

Figure 11. Severity indices by ten-day periods for northwestern Ontario fire weather stations.

Forest Fire Weather Index) for ten-day periods for a variety of northwestern Ontario OMNR weather stations, for both 1980 and the 1970-1979 period. The impact of the early spring in 1980 and the continued warm, dry weather during the month of May is quite evident from Figure 11 — Kenora, Dryden, Fort Frances, and Red Lake show dramatically high SI values compared to averages for the previous decade. The receptivity of forest fuels to fire ignitions during this period was reflected in the fact that, by the end of May, a total of 419 fires had burned over almost 325,000 ha in northwestern Ontario. Continued severe fire weather in the Sioux Lookout-Ignace-Armstrong area during June resulted in an additional 246 fires burning approximately 230,000 ha during this month.

Summary

The decade of the 1970s was the worst on record in terms of forest fire numbers and area burned in northwestern Ontario, with significant fire losses occurring in 1974, 1976 and 1977. Following this 1980 proved to be the most significant individual forest fire year since 1923 in northwestern Ontario.

During the 1970-1979 period many of the largest fires in northwestern Ontario burned in the far northern portion of the region, where little organized fire protection is undertaken and fires are usually allowed to burn freely. In 1980, however, the largest fires occurred in the intensively protected portion of northwestern Ontario, burning over high-value wood supply areas, resulting in widespread social and economic impacts.

The direct cause of the severe forest fire problems in northwestern Ontario in recent years can be traced to periods of prolonged dry spring weather, the result of stable high pressure systems becoming entrenched over central North America and deflecting the flow of moist air from the southern and western United States away from northwestern Ontario. The resulting absence of precipitation, in combination with extremely flammable, fast-drying forest fuels and multiple lightning fire starts, readily explains the severe forest fire problems in northwestern Ontario in recent years.

FIRE HISTORY AND ECOLOGY OF FOREST ECOSYSTEMS IN KLUANE NATIONAL PARK: Fire Management Implications

Brad C. Hawkes
Canadian Forestry Service

Abstract: To aid in the development of a fire management plan for Kluane National Park, a study of the fire history and ecology of the park's forest ecosystems was undertaken to determine the ecological role of fire in vegetation renewal and succession. Results of the study indicated that lightning is a very infrequent ignition source. Man-caused fires were important in vegetation renewal, especially since the late 1800s as indicated by the difference in fire frequency between remote and heavy human-use study areas within the park. The present vegetation mosaic is partially the result of man-caused fires (early Europeans and Indians) and supports a wide variety of wildlife. In addition to fire, glacial movements have exposed new material, caused lake formation and drainage, and resulted in vegetation renewal and succession. The fire management strategies developed by Parks Canada for Kluane National Park should consider these vegetation cycling mechanisms. Decisions will be made as to what vegetation mosaic will be perpetuated by Parks Canada as "the natural resources within the park". If only lightning fires were considered for re-cycling vegetation, the average age of forest stands would increase, stand composition would change, and vegetation mosaics would tend towards less landscape diversity.

Fire's role in national park ecosystems was not recognized in Parks Canada policy until 1979 (Parks Canada 1979). This policy statement recognizes wildfire as a natural phenomenon and expresses the need to incorporate fire as a force in ecosystem development.

Implementation of this policy will require development of fire management plans in Canadian national parks. This investigation was undertaken to assist in the preparation of a fire management plan for Kluane National Park, Yukon Territory. The present study was initiated in 1980 and field work was conducted during the summer of 1981. Information on fire history and ecology has been collected in other national parks such as Nahanni, Northwest Territories and Wood Buffalo, Alberta and Northwest Territories. Alexander and Dubé (1983) presented a review of past and current fire research and status of fire management plans and programs in Canadian parks and other nature reserves of northern circumpolar countries.

Early work by Johnson and Raup (1964), as part of their geobotanical and archaeological investigations in southwest Yukon Territory, examined the role of fire in vegetation succession in the Kluane area during the late 1940s. They found that fire and lake formation were major factors in vegetation renewal. Theberge's (1972) initial investigation into the fire history of Kluane National Park was limited in detail and left some unanswered questions on the ecological role of fire in vegetation renewal and succession. The two major questions were (1) what was the past fire regime, and (2) to what extent did early Europeans and natives influence this fire regime.

The objectives of the present paper are:

- to describe the ecological role of fire in vegetation renewal and succession in the vegetation types,
- to determine the fire history of selected areas to examine the importance of man-caused fires (i.e., early Europeans and natives) in the more recent fire history and the influence of climatic regions on mean fire intervals as described by Romme (1980), and
- to discuss the fire management implications of the fire history and ecological information.

Study Area Description

Kluane National Park encompasses approximately 22,100 km^2 in the southwestern corner of the Yukon Territory and borders the Wrangell-St. Elias reserve in Alaska on its western border (Figure 1).

Only 18 per cent of the park is covered with alpine, subalpine and montane forest vegetation. The rest of the park area consists of permanent snow and ice fields, water, rock and gravel floodplains. Montane forest vegetation encompasses only seven per cent of the park area mainly along the river valleys in the eastern and northern sections of the park. Much of the forest is concentrated along the eastern border which is formed by the Haines and Alaska highways.

The forested areas are classified by Rowe (1972) as the Kluane Section (B.26d) of the boreal region and by Oswald and Senyk (1977) as being in Ecoregions 6-Coast Mountains, 7-St. Elias Mountains, and 8-Ruby Range. The major forest species found are white spruce (*Picea glauca*), trembling aspen (*Populus tremuloides*) and balsam poplar (*Populus balsamifera*), while lodgepole pine (*Pinus contorta* var. *latifolia*) and paper birch (*Betula papyrifera*) are represented by only a few individuals (Douglas 1980). Three major vegetation zones have been described (Douglas 1980).

Most of the fire history and ecology work was concentrated in the montane zone where fire has been more prevalent than in the subalpine and alpine zones. The montane zone occurs on the lower valleys and slopes (up to 760 to 1080 m) of the Kluane Ranges and is dominated by essentially continuous white spruce forests with some mixtures of trembling aspen and balsam poplar (especially areas with recent disturbance by fire, snow or water). Within the montane zone there are also numerous marsh, fen, shrub and herbaceous vegetation types (Douglas 1980).

Methods

Selection of Study Areas

The vegetated area of Kluane National Park is much too large for a detailed fire history study of all areas in the time available for this study. The study approach used was similar to that developed by Barrett (1980) to determine the influence of Indian fires in the forests of western Montana prior to European settlement. The importance of man-caused fires in the more recent fire history of the park was to be determined. This fire history included those caused by early Europeans from the 1880s (start of Dalton Trail and placer mining activities — see Wright (1980)) to present and Indian fires before and after the 1880s.

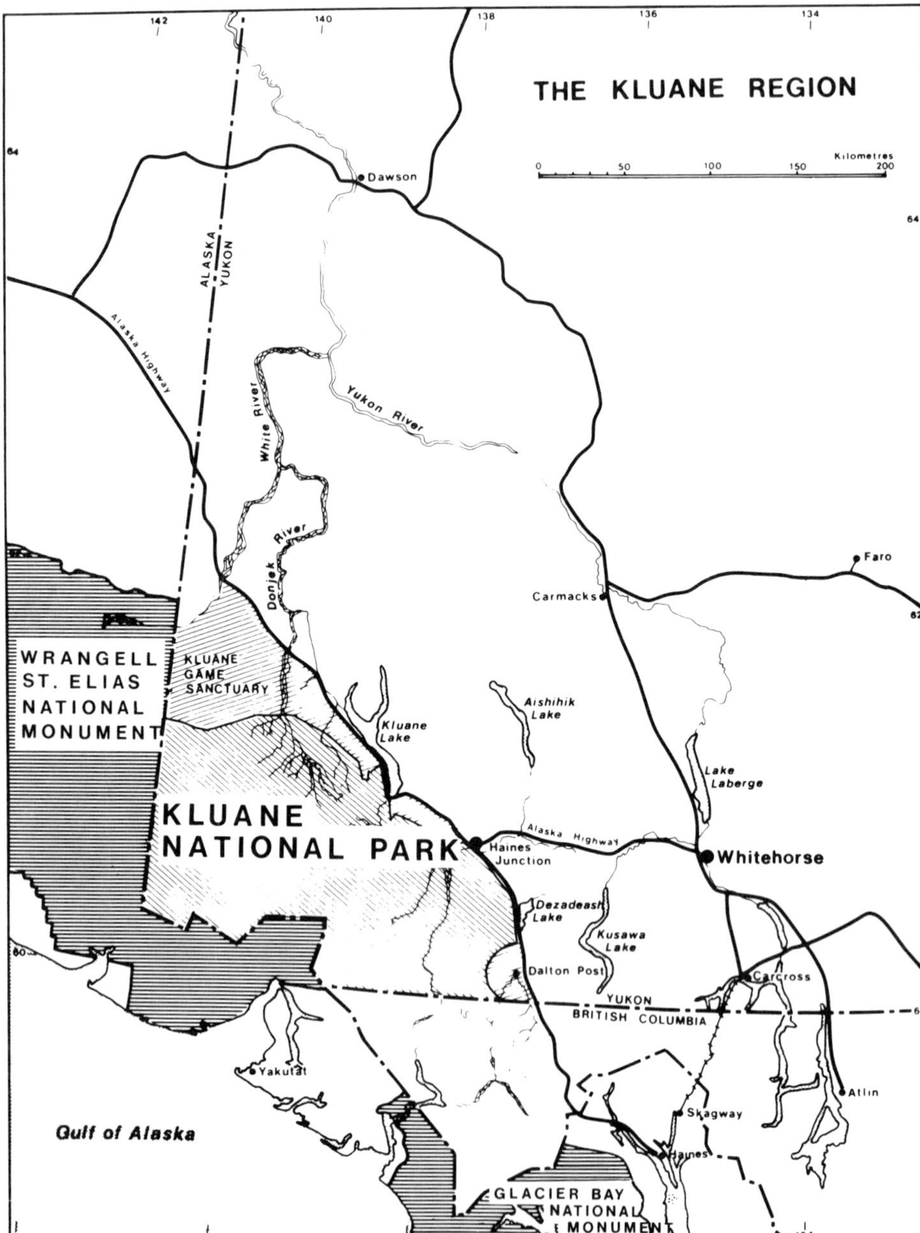

Figure 1. Location of Kluane National Park in southwestern Yukon Territory. Wrangell-St. Elias National Monument and Glacier Bay National Monument are also identified.

Barrett (1980) used a sampling method where a number of stands were examined for differences in fire frequency (here defined as the number of fires per unit time in some designated area (which may be as small as a single point) (Romme 1980)) due to differences in Indian use. Kluane National Park can also be stratified into areas which have had concentrated use by early Europeans for mining, trapping, hunting and surveying (Stevenson 1978; Wright 1980) and the Southern Tutchone Indians for fishing, hunting, trapping, and settlements (Workman 1978). Figure 2 identifies six areas which were chosen for the fire history and ecology portion of the study. These areas were selected on the basis of amount of human use (i.e., 'heavy human-use' or 'remote human-use' areas) and the climatic region as described by Douglas (1980). The areas that were selected are:

1. Northern climatic region:
   A. heavy human-use — Slims and Jarvis Rivers
   B. remote human-use — Upper Donjek River

2. Central climatic region:
   A. heavy human-use — Kathleen Lakes — Quill Creek
   B. remote human-use — Dusty and Disappointment Rivers and Trout Lake.

3. Southern climatic region:
   A. heavy human-use — Alder Creek, Fraser River and Dezadeash Lake
   B. remote human-use — Bates Lake and River and Onion Lake.

Data Collection and Analysis

Data were collected on the ecological role of fire in vegetation renewal, succession and fire history. Information from Douglas (1980) was used to document the community type. Ten to 15 fire history and ecology plots were established in each study area depending on the complexity of fire history and time available for sampling.

The fire history information collected was different from that of Barrett (1980) due to the difference in the type of fire history information that was available. Barrett (1980) was working in an area which had a high frequency of low-intensity surface fires which left multiple fire-scars on individual trees. The reconnaissance survey in September 1980 indicated that most fires were stand lethal and left few scarred trees except on fire boundaries. White spruce was the main conifer species and therefore had the most fire scars.

The shortage of fire scar information meant that one or more of the following three options was used to estimate the fire interval(s) (defined as the number of years between two successive fires in a designated area with the size of the area clearly specified (Romme 1980)) on each fire history plot (the plot size was a small stand and could be considered a point estimate).

The first option was used when two or more fire-scars were available to date multiple fires. The difference between the dates of any two successive fires gave an estimate of one fire interval. The most common calculation was the difference between the most recent fire date and the date of the previous fire which originated the stand. This option seldom occurred since white spruce is killed by even low intensity fires and the probability of a particular tree surviving two fires is low.

The second option was used when one or more fire scars was available to date the most recent fire but only the standing dead (snags) or nearby live stand was available to obtain

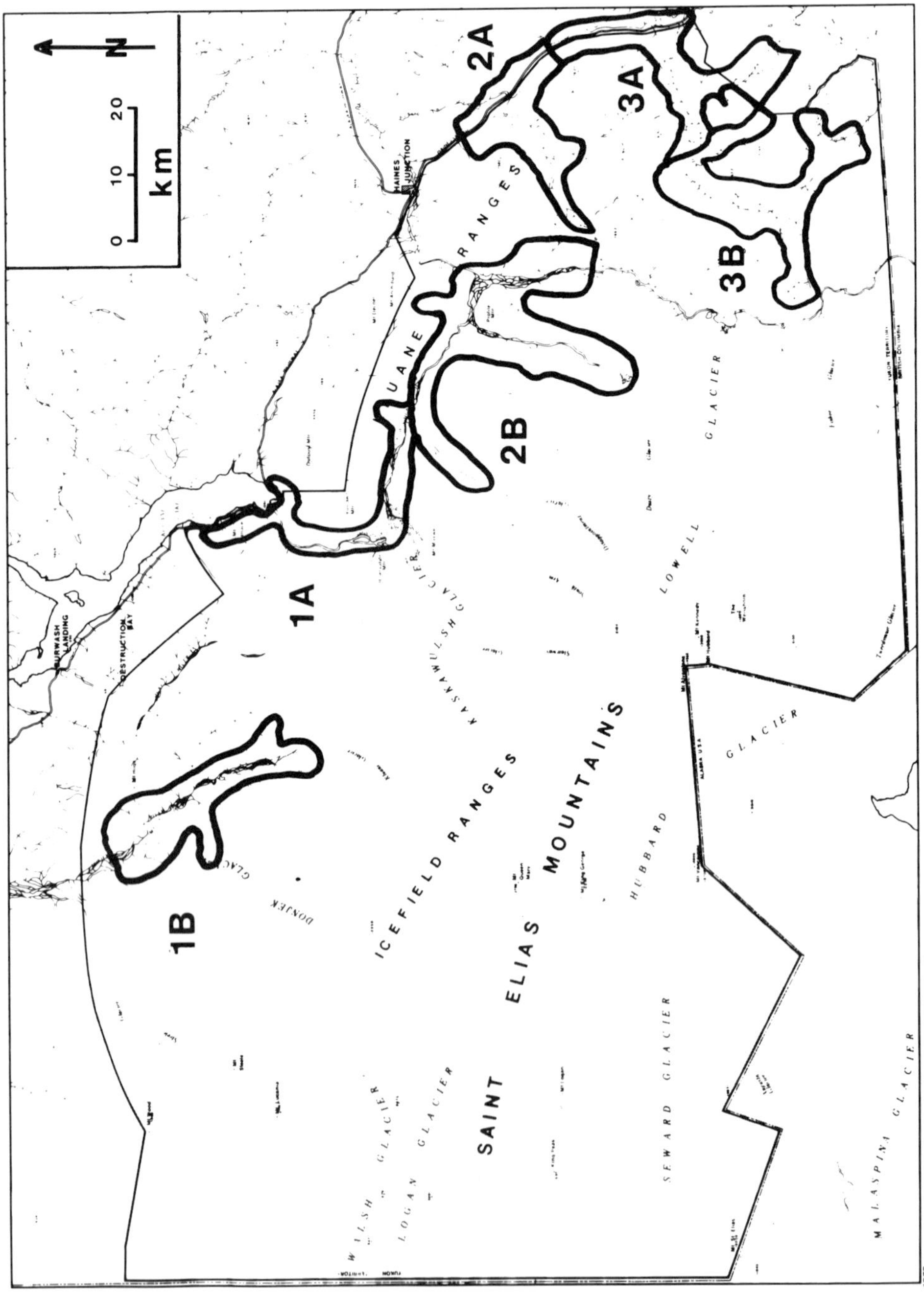

Figure 2. Fire history and ecology study areas located in Kluane National Park.

the approximate year of stand origin. Not all stands originated following fire. Fire origin was indicated by the presence of charcoal in the duff or soil or on a stump or snag. Stand origin (for fire or other causes such as glacial retreat, old lakebeds, floodplains or outwash fans) was determined by taking the age of the oldest tree found and adding a correction factor for regeneration delay after stand disturbance. This correction was determined by comparing the average age of regeneration to that of the fire (date from fire scar record). The regeneration delay ranged from 7 to 55 years with an average of 25 years. The fire interval for this option was calculated by subtracting the date of stand origin from the fire date.

The third option was used when no recent fire evidence, such as a fire scar, was found to date the present stand, but fire evidence such as the presence of charcoal and early fire successional species, such as trembling aspen, were still present. Stand origin was estimated by aging several of the dominant overstory trees and adding the correction factor for regeneration delay (25 years). Stand origin of stands with no evidence of past fires was estimated in the same manner as those with evidence of fire.

The first two options were used most frequently in heavy human-use and remote human-use areas, respectively. The third option gave the least reliable estimate of the fire interval of a stand because the stand has not burned again to document an actual fire interval and charcoal was difficult to find due to movement by wind or water where a stand may have originated from fire.

Fire interval estimates from fire history and ecology plots were averaged to give mean fire intervals (defined as the arithmetic average of all fire intervals determined in a designated area during a designated time period (Romme 1980)) for each of the six study areas and the three climatic regions. Mean fire intervals were calculated for each study area and climatic region from all plot fire intervals.

Fire dates were not corrected for false or missing rings with cross-dating as described in Stokes (1980). Fire dates were confirmed in many cases by exact agreement between the number of fire scars in the same area. Madany *et al.* (1982) felt this approach was accurate enough for most managerial recommendations and some ecological interpretations.

Fire boundaries were mapped from aerial photographs and observations made from a helicopter. These boundaries were used to estimate area burned during the pre-suppression period of 1880 to 1940.

Results and Discussion

Post-fire Vegetative Recovery

Recent reviews by Viereck and Schandelmeier (1980) and Kelsall *et al.* (1977) provide information on the effects of fire on vegetation, soils and wildlife in the boreal forest. This discussion will be directed to the immediate effects of fire on vegetation and post-fire vegetative recovery with some references to soils in terms of fire's effect on organic layer reduction, seedbed condition and nutrient availability. Vegetative recovery will be presented as a general picture of secondary succession after fire in the closed white spruce montane forest. Nomenclature and authorities of plants follow Douglas (1974).

Fire immediately kills most or all of the above-ground parts of plants, consumes part, but rarely all, of the organic layers, leaving variable amounts of below-ground portions of vegetation to sprout, and creates some exposed mineral soil seedbed for germination of seed from nearby plants. Wildfires have left a mosaic of vegetation on any individual burn which has developed from sprouts, seedlings and vegetation which survived the fire due to the variability of fire intensity and depth of burn. This variability is due to variations in moisture conditions of the surface fuel, organic layers and soil both spatially and over the duration of the burn.

Viereck and Schandelmeier (1980) concluded that the effects of fire on soil nutrients vary depending on fire intensity and depth of burn, site condition and original soil properties. They stated that nutrients which were tied up in the vegetative (tree) overstory and the moss and organic layers (such layers in Kluane National Park were over 20 cm deep in stands greater than 100 years old) are released after a wildfire. No specific studies were undertaken to determine the effect of fire on nutrient cycling, although some work in the boreal forest indicates that the shrub/herbaceous community that dominates the early years of post-fire vegetation probably utilizes a large portion of these available nutrients (Auclair 1982).

Low soil temperatures can also limit plant productivity in northern climates by slowing the activity of decomposing and nitrogen fixing organisms (Viereck and Schandelmeier 1980). The organic layer and closed tree overstory limits summer warming of the soil and, in northern areas of Kluane National Park, such as the Donjek River valley, keeps the level of permafrost close to the surface. Wildfire removes most or part of the overstory and organic layer and blackens the soil surface allowing the soil to absorb more of the sun's heat. An increase in juvenile diameter growth was observed in a fire-disturbed white spruce stand which showed no evidence of previous fire and was located in the northern area of the park where permafrost is common and organic layers are deep.

Depth of burn was difficult to estimate on most fires in Kluane National Park because all fires were older than 40 years and the immediate effects of the fires were no longer in evidence. Where depth of burn was estimated from the burned organic layer depth compared with a nearby remnant stand, it varied from 58-95 per cent reduction of the original 6-26 cm depth. This estimate has been influenced by decomposition and accumulation of litter since the fire. Less than five per cent mineral soil exposure could be found on any of the burns sampled in 1981. More might have been found immediately after some of the fires. This amount of duff reduction suggests that the organic layer moisture must have been low for most of these fires, which suggests early to late summer burns.

To become established after a burn, white spruce requires a mineral seedbed for germination and seedling survival, survivors nearby to provide seed, a good seed year (these can be ten or more years apart), and a lack of vegetative competition (Viereck and Schandelmeier 1980). The latter two factors are probably the most important in white spruce regeneration delay. White spruce was found to establish itself under a deciduous shrub community if mineral soil was still exposed or was being exposed by the previous fire-killed stand falling down. The seed source was usually the survivors present because of the mosaic nature of the burns.

Trembling aspen and balsam poplar are better adapted for revegetating burns because they have numerous light seeds and can also sprout from live roots in the mineral soil and unburned portions of organic layers. Agreement of stem ages of aspen with known fire dates indicated an immediate sprouting of some individuals. Seeding by live trees nearby was illustrated by stem ages less than the age indicated by the fire date.

Foote (1982) described six general developmental stages of forest succession in Alaska: newly burned, seedling/herb, tall shrub/sapling, and three tree dominated stages. These are used to describe a general picture of vegetative recovery after fire in the closed white spruce montane forest (Figures 3-8).

The newly burned stage probably lasts from a month to a year depending on the time of year the burn occurred, the depth of burn and the available seed source. Fires outside Kluane National Park had to be used to describe the newly burned and seedling/herb stage. The forest floor is dominated by charred mosses, shrub snags and mineral ash. No live vegetation is present in this stage. This is illustrated in Figure 3 which shows a white spruce stand one year after a fire in 1980 near Canyon Creek. The area in the photograph was dominated by exposed mineral soil.

The seedling/herb stage probably lasts from one to 15 years. If depth of burn is low to moderate, sprouts from live roots and rhizomes in the organic layer and soil start to appear within several weeks after the fire. Shrub and tree species such as trembling aspen, balsam poplar, willows (*Salix* spp.), alder (*Alnus* spp.), buffaloberry (*Shepherdia canadensis*), rose (*Rosa acicularis*) and shrub birch (*Betula glandulosa*) sprout quickly if the depth of burn is low to moderate. Herbs such as fireweed (*Epilobium angustifolium*) and grasses sprout or seed in from nearby areas. Pioneer mosses such as *Polytrichum juniperium* invade on mineral soil. Except for the feathermosses (*Pleurozium schreberi* and *Hylocomium splendens*) and lichens, most of the plant species, including the tree species, become established in this stage and early in the next stage. An example of this stage, shown in Figure 4, is the site of the Canyon Creek fire one year after a fire in 1980. This photograph shows a burned area with a lesser depth of burn than in Figure 3 and a trembling aspen pre-fire stand.

The next stage, the tall shrub/sapling stage lasts from five to 80 years. In this stage the tall shrubs (willows and alder) and sapling trees (white spruce, trembling aspen and balsam poplar) dominate the vegetation. The *Salix glauca* shrub community is the most common vegetation type described by Douglas (1980) in this stage. An extreme case was a *Salix glauca* shrub community mapped by Douglas (1980) that was present 120 years after fire due to a 55 year white spruce regeneration delay. This stage is illustrated in Figure 5 which shows the Kathleen Lake burn near Haines Road, which occurred around 1915.

The pole stage lasts from 50 to 100 years and is dominated by pole size trees of small diameter with the canopy starting to close. This stage can be dominated by deciduous, mixed deciduous-conifer, or conifer trees. The shrubs and herbs have been shaded out and feathermosses start to dominate the forest floor in the conifer dominated stands. The deciduous and mixed deciduous-conifer stands still have an extensive low shrub/herb understory. Figure 6 illustrates a trembling aspen pole stand which originated around 1889 with a well developed shrub and herbaceous understory that includes some white spruce saplings. White spruce pole stands are most common near fire boundaries where seed supply was adequate after the last disturbance.

The next stage, the mature tree stage, which lasts from 100 to 200 years, is dominated by mature deciduous, mixed deciduous-conifer, and conifer stands. The hardwood components, if present, are reaching their pathological age and are being replaced by the longer lived white spruce. The understory of shrubs and herbs is still extensive in the mature hardwood and mixed stands. The mature white spruce stands have mainly feathermosses in their understories. Figure 7 illustrates this stage with a mature white spruce stand which originated around 1835 with few shrubs and herbs remaining in the understory.

Figures 3-8. Six developmental stages of postfire forest succession in Kluane National Park: white spruce stand (3) and trembling aspen stand (4) near Canyon Creek burned one year previously; *Salix glauca* community (5) near Haines Road in 1915 Kathleen Lake burn; trembling aspen pole stand (6) between Kathleen and Dezadeash Lakes in 1889 fire; mature white spruce stand (7) near 1835 Kathleen Lake burn; and mature white spruce stand (8) located near Kathleen Lake campground in 1592 fire. See text for further details.

The final stage is a mature white spruce stand 150 to 400 years old where only minor changes in overstory and understory cover might occur. White spruce tends to develop stem rot resulting in decreasing stand density and openings in the stand. White spruce is also susceptible to the spruce bark beetle (*Dendroctonus rufipensis*) at this stage. An outbreak in the 1940s in Kluane National Park created many openings in the white spruce canopy. The forest floor is still dominated by feathermosses, but the shrub and herbaceous understory is moderately developed. Figure 8 illustrates a mature white spruce stand which originated around 1592 with much windfall and a shrub and herbaceous understory. Deciduous trees are usually no longer in the canopy.

In southern areas of the park the most common succession of shrub to forest community is from *Salix glauca* to *Picea glauca/Salix glauca* (Douglas 1974). More common in northern areas is succession from the *Salix glauca* community to the *Picea glauca/Thuidium abietinum* or *Picea glauca/Aulacomium palustre* forest communities. A deciduous forest community sometimes occurs between the shrub *Salix glauca* and conifer forest communities. The common deciduous forest communities in this role are *Picea glauca — Populus tremuloides/Shepherdia canadensis — Linnaea borealis, Populus tremuloides/Arctostaphylos uva-ursi,* and *Salix scouleriana/Shepherdia canadensis*. In the eastern portion of the park along the Haines Road on the drier sites, the succession indicated by Douglas (1980) is from *Shepherdia canadensis*, to a deciduous type, and then to the *Picea glauca/Shepherdia canadensis* closed or open forest community. In the Slims River Valley, succession would be from a dry shrub type like *Juniperus horizontalis* or *Shepherdia canadensis* to the *Picea glauca/Arctostaphylos* spp. or *Picea glauca/Hypnum revolutum*.

Early Miner and Indian Use of Fire

The early Europeans started trading and placer mining activities in the 1880s and continued until the 1930s. Hunting and guiding were important activities from the early 1900s to the establishment of the Territorial Game Sanctuary in 1942. The native Southern Tutchone Indians had settlements at Klukshu Lake and near Dalton Post at Nesketakeen.

Johnson and Raup (1964), in their investigations during the late 1940s of the Southwest Yukon for geobotanical and archaeological features, noted that "there can be no doubt that fire has been a major factor in the life of the valley from the beginning of occupance". They could trace charcoal in loess and silts back to the end of the last ice age, about 12 000 years ago. The Southern Tutchone have lived with fire as part of their environment for centuries, but also have used this powerful tool to manipulate their environment (Workman 1978). Lewis (1982) has recently reviewed the use of fire by North American and Australian hunter-gatherers and points out that the few existing studies of the use of fire by North American Indians show that fire technologies were employed to control the distribution, diversity, and relative abundance of plant and animal resources. Lewis (1982) also points out that the most important general feature of aboriginal applications of fire is that they differed significantly from natural fires in terms of seasonality, frequency, intensity and selectivity.

Early work by Catharine McClellan in the work *My Old People Say* reported in Workman (1978) shows that the Southern Tutchone in historic times used fire to create moose browse. Fires were also used in the early spring by women to burn off banks of streams and lakes to thaw the ground to dig and collect *Hedysarum alpinum* roots (McClellan as cited by Workman 1978). In times of starvation, the Indians would build fires to melt snow so they could collect old bearberries and crowberries in winter and spring. Wright (1980) reported that in early May, 1891, Dalton and Glave travelled up the main Shakwak Valley and their Indian guides would light up a spruce tree when they reached an exposed position to see if there was any answering column of smoke. Fire was, of course, used for cooking and smoking purposes.

The use of fire by placer miners is not well known, although they apparently burned off the vegetation and organic layers along creeks to help melt the permafrost to ease removal of placer deposits (Larry Tremblay, personal communication, 1982). These would have been early spring fires. Many of the placer miners that came to the Yukon Territory were also hard rock miners and prospectors and wanted to remove the forest to get a better view of the underlying rock. Also, the miners preferred fire killed and dried timber rather than green wood for shaft building (Larry Tremblay, personal communication, 1982).

Mean Fire Intervals by Climatic Region and Study Area

The mean fire interval seems to increase from the southern to the northern climatic region (133-234 years) (Table 1). Statistical testing of fire history data to determine if the mean fire interval is significantly different among climatic regions or heavy and remote human-use areas was not possible because the fire history plots were not randomly located within each study area. Most fire history plots were established near fire boundaries to help locate fire scar information and document fire effects on more recent fires. The mean fire intervals reported in this paper are probably shorter than if plots had been randomly located, because more older stands would have been included in the selection.

Mean fire intervals by study area ranged from 113 to 238 years with individual fire intervals ranging from 9 to 403 years (Table 1). Mean fire intervals for the heavy and remote human-use study areas were similar except in the southern area (113 years for 3A-Mush Lake and 162 years for 3B — Bates River). An extensive review of the history of Kluane National Park after completion of sampling, indicated that remote and heavy human-use areas in the central and especially in the northern area had similar early European and native use. The southern area (3A-Mush Lake) had the most concentrated and lengthy use by Europeans and natives.

The overall park mean fire interval was 179 years (Table 1) which might increase to 200-300 years if man-caused fires since the late 1800s were eliminated. This mean fire interval is longer than those reported for other parts of the boreal forest, such as the 40-100 years mean fire interval reported by Johnson (1980) for an area east of Great Slave Lake in the Northwest Territories of Canada.

Records of recent fire occurrence were obtained from Yukon Lands and Forest Service for the period 1963-1981. During this 19 year interval, eight recorded fires burned less than 0.5 ha in total. These fires were man-caused and, in common with observations in the rest of the Yukon Territory, had no distinct seasonal (May-October) peak (Kiil 1971). The main source of man-caused fire was recreational activity. The number of man-caused fires was low, probably because of recent fire prevention activities and the low level of human use in this area.

There were no recorded lightning-caused fires during this 19 year period. Unpublished data from the Bureau of Land Management, Alaska indicate that lightning fire occurrence

Table 1. Mean fire intervals and average fire size in Kluane National Park. Ranges are given in parentheses

	Plots (n)	Mean fire interval (Years)	Average fire size (ha)
By climatic region			
Northern	25	234 (80-403)	
Central	31	164 (9-373)	
Southern	27	133 (14-274)	
By study area*			
1A-Slims River	14	231 (80-403)	983 (92-2700)
1B-Donjek River	11	238 (135-301)	179 (2-819)
2A-Kathleen Lakes	15	152 (30-290)	200 (5-790)
2B-Dusty River	16	175 (9-373)	394 (5-1800)
3A-Mush Lake	16	113 (14-274)	1600 (1-5000)
3B-Bates River	11	162 (105-218)	140 (3-329)

* Overall mean fire interval is 179 years (113-238).

for the period 1957-1979 in the Wrangell Mountains of Alaska was one fire every 24 years for a forested area the size of Kluane National Park. The park is south of the main lightning fire occurrence belt which extends from Beaver Creek up to Dawson, and swings southeast across the Yukon Territory. The Wrangell Mountains and Kluane National Park fire occurrence and history data indicate that lightning-caused fires in the park occur once every 20-50 years.

Fire Size, Shape and Behavior

The 3A-Mush Lake study area had the largest average fire size (1600 ha) (Table 1), and the 3B-Bates River study area (140 ha) the lowest. These average fire sizes are for fires which occurred up to 200 years ago. After stands reach this age it is difficult or impossible to determine older fire boundaries from aerial photographs or observation from an aircraft, especially if there is little or no deciduous component. Except for one fire (approximately 1759 in 3A-Mush Lake) fire size and shape could not be determined prior to the 1850s. Only a systematic survey of stands older than 120 years would provide such information and this was beyond the scope of this study.

The range in fire size was the largest in 3A-Mush Lake (1-5000 ha) and reflects the variable fire weather, topographic and fuel conditions which can occur. There has not been a fire greater than 1.0 ha in size since 1940. This probably results from both better fire suppression capabilities and fewer man-caused ignition sources.

Three study areas had fires over 1000 ha in size (1A-Slims, 2B-Dusty River, 3A-Mush Lake). These large fires usually have elongated shapes which reflect the influence of the frequent strong down-valley winds. Smaller fires tend to have variable shapes and impacts on surface and aerial fuels, which probably indicates less severe and more variable fire weather conditions during the burn.

Fire behavior of past wildfires was difficult to determine because of their age (≥40 years). Estimates of past rate of spread and fire intensity are almost impossible to make, although visual evidence of consumption of crown foliage-bearing twigs and branches indicates crown fires have occurred. This would be especially true where slope and strong valley winds have played a part in fire behavior.

Fire Management Implications

The data and literature on fire history and plant succession indicate that fire has played a role in ecosystem renewal and development. Parks Canada policy (Parks Canada 1979) indicates that natural fire regimes will be re-introduced into Canadian park ecosystems wherever possible.

There are three key issues and challenges for Parks Canada in implementing this new fire policy in Kluane National Park.

1. There has been no distinction between lightning and man-caused fires in the discussion of fire management strategies. Van Wagner and Methven (1980) stated that what the park manager really desires is not the natural fire regime *per se*, but rather the vegetation complex that the natural fire regime would have created. The only worthwhile distinction then is between wanted and unwanted fires, the ignition agent being irrelevant. Van Wagner and Methven (1980) clearly point out that the vegetation plan is the major consideration for decisions regarding wanted vs unwanted fires. If this philosophy is adopted in Parks Canada policy, it will give some flexibility for developing fire management strategies. For example, prescribed fire may have to be used in high human-use areas where a fire exclusion policy is necessary because of public safety.

2. Parks Canada staff will have to make a decision on what vegetation complex to perpetuate before a fire management plan can be developed. A number of vegetation management options are open to Parks Canada depending on their decision as to what constitutes the "natural" vegetation complex. The decision may be to perpetuate the vegetation complex that would result from only a lightning fire regime. If this option were implemented, the average age of forest stands would increase, species changes would take place (i.e., trembling aspen to white spruce), and vegetation mosaics would tend toward less landscape diversity because of the influence of man-caused fires in the more recent fire history. These changes would take place slowly as indicated by the vegetation succession rate.

   Alternatively, Parks Canada may decide to perpetuate the vegetation complex that would result from a fire regime which included lightning and Indian fires. This would be the vegetation complex that probably existed before Europeans arrived around 1880. If this option were implemented, the previously stated results would also occur but to a lesser degree, especially in the highest Indian-use areas.

   Or again, Parks Canada may choose to perpetuate the vegetation complex that resulted from a fire regime which included lightning, Indian- and European-caused fires. Such a vegetation complex would be very similar to what is found today except there would probably be more stands in the younger age classes (<40 years) (i.e., there is a lack of area burned since 1940 (<0.5 ha)).

Finally, it is impossible to perpetuate the vegetation complex that exists today which has resulted from a combination of lightning and man-caused ignitions sources and reflects some influence of fire suppression activities, because the present age class distribution could not be maintained.

3. Whatever vegetation complex is chosen by Parks Canada to perpetuate, the fire management plan should include a long range goal of the average area burned per year which will create the age class distribution desired. Fluctuations in the number, location and seasonality of fires from year to year necessitate a long term approach in vegetation management. Flexibility in fire management strategies is needed to deal with constraints imposed by park boundaries and facilities, public safety concerns, scale problems due to park size and location, and designated special preservation areas.

Acknowledgements

Thanks are extended to Parks Canada, Prairie Region, Winnipeg, Manitoba for providing funding for this study and for making Allan Wright's unpublished report available and to Larry Tremblay, Chief Warden, Kluane National Park, for his helpful comments.

Literature Cited

Alexander, M.E. and D.E. Dubé. 1983. Fire management in wilderness areas, parks and other nature reserves. *In:* The role of fire in northern circumpolar ecosystems, Ross W. Wein and David A. McLean (Eds.). SCOPE 18, John Wiley and Sons, Ltd., Chichester, England. pp. 273-297.

Auclair, A.N.D. 1982. The role of shrubs in post-fire nutrient retention in subarctic lichen woodland. Abstract Bull. Ecol. Soc. Am. 63:182.

Barrett, Stephen W. 1980. Indian fires in the Pre-Settlement forests of Western Montana. U.S.D.A. For. Serv. Gen. Tech. Rep. RM-81. pp. 35-41.

Douglas, G.W. 1974. Montane zone vegetation of the Alsek River region, southwestern Yukon. Can. J. Bot. 52: 2505-2532.

Douglas, G.W. 1980. Biophysical inventory studies of Kluane National Park. Volume 1 Draft Copy, submitted to Parks Canada, Prairie Region, Winnipeg, Manitoba.

Foote, J. 1982. General patterns of forest succession in Interior Alaska. Paper presented at the Natural Resource Officer Training Course, April 12-16, 1982. Sponsored by the Bureau of Land Management, Fairbanks, Alaska.

Johnson, E.A. 1980. Fire recurrence and vegetation in the lichen woodlands of the Northwest Territories, Canada. U.S.D.A. For. Serv. Gen. Tech. Rep. RM-81. pp. 110-114.

Johnson, F. and H.M. Raup. 1964. Investigations in southwest Yukon: geobotanical and archaeological reconnaissance. Papers of the Robert S. Peabody Foundation for Archaeology 6: 1-198.

Kelsall, J.P., Telfer, E.S. and T.D. Wright. 1977. The effects of fire on the ecology of the boreal forest, with particular reference to the Canadian north: a review and selected bibliography. Can. Wildl. Serv. Occas. Pap. 32.

Kiil, A.D. 1971. A statistical summary of forest fires in the Yukon Territory 1966-1970. Environ. Canada, C.F.S. Inf. Rep. NOR-X-10.

Lewis, H.T. 1982. Fire technology and resource management in aboriginal North America and Australia. *In:* Resource managers: North American and Australian hunter-gathers. N.M. Williams and E.S. Hunn (Eds.), American Association for the Advancement of Science selected symposium. pp. 45-67.

Madany, M.H., Swetnam, T.W. and Neil E. West. 1982. Comparison of two approaches for determining fire dates from tree scars. For. Sci. 28: 856-861.

Oswald, E.T. and J.P. Senyk. Ecoregions of Yukon Territory. Environ. Canada, C.F.S. BC-X-164.

Parks Canada. 1979. Parks Canada policy. Published under authority of the Hon. J. Hugh Faulkner, Minister responsible for Parks Canada, Ottawa, QS-7079-000-EE-A1, Catalogue No. R62-109/1979, ISBN 0-662-10425-0, 69 p.

Romme, W.H. 1980. Fire history terminology: report of the ad hoc committee. U.S.D.A. For. Serv. Gen. Tech. Rep. RM-81. pp. 135-137.

Rowe, J.S. 1972. Forest Regions of Canada. Environ. Canada, C.F.S. Publ. No. 1300.

Stevenson, M. 1978. Inventory and assessment of the historic resources of Kluane National Park, Yukon Territory. Parks Canada, Environment Canada, National Historic Parks and Sites Branch, Publ. No. QS-7090-0310EE-A1. (Restricted Distribution).

Stokes, M.A. 1980. The dendrochronology of fire history. U.S.D.A., For. Serv., Gen. Tech. Rep. RM-81. pp. 1-3.

Theberge, J.B. 1972. Considerations for fire management in Kluane National Park. Contract Report submitted to Can. Wildl. Serv., Western Region, Edmonton, Alberta.

Van Wagner, C.E. 1978. Age-class distribution and the forest fire cycle. Can. J. For. Res. 8: 220-227.

Van Wagner, C.E. and I.R. Methven. 1980. Fire in the management of Canada's National Parks: Philosophy and Strategy. National Parks, Occasional Paper.

Van Wagner, C.E. and T.L. Pickett. 1975. Equations and Fortran IV program for the 1976 metric version of the forest fire weather index. Environ. Canada, C.F.S. Inf. Rep. PS-X-58.

Viereck, L.A. and L.A. Schandelmeier. 1980. Effects of fire in Alaska and adjacent Canada — a literature review. U.S.D.I. BLM-Alaska Tech. Rep. 6.

Workman, W.B. 1978. Prehistory of the Aishihik-Kluane area, Southwest Yukon Territory. Archaeological Survey of Canada. Paper No. 74. National Museums of Canada, Ottawa.

Wright, A. 1980. Human history of Kluane National Park. Unpublished Internal Draft Report Parks Canada, Winnipeg, Manitoba.

POSTFIRE ESTABLISHMENT OF JACK PINE AND BLACK SPRUCE FROM SEED ON ORGANIC SOIL SEEDBEDS

P.A. Thomas and Ross W. Wein
University of New Brunswick

Abstract: Studies of jack pine and black spruce germination and establishment requirements have shown that jack pine is more successful in establishing from seed on warmer and drier microsites. It is hypothesized that differential seedling establishment of these species accounts in part for the greater abundance of black spruce in the northern part of the boreal forest where there are fewer ''safe sites'' for jack pine. Direct seeding of the two species was discussed in relation to the above but also in relation to seed supply and shelter afforded to seedlings by postfire vegetative regeneration of plants.

For a variety of reasons, including poor drainage, much of the boreal forest has a slowly decomposing layer of organic matter overlying the mineral soil. If this layer is sufficiently dry, it may ignite, burn by glowing combustion and expose the mineral soil beneath. However, the long periods of low rainfall and high evapotranspiration necessary to dry the organic soil layers are relatively rare, so that many fires remove only the top dry layers of litter, leaving behind a layer of organic matter of variable thickness (Wein 1983). It is on this common postfire seedbed that we wish to concentrate.

It has long been appreciated that organic matter left after a fire is not a favourable seedbed, as compared to exposed mineral soil, for the regeneration of certain trees from seed (see reviews by Arnott 1973; McRae 1979). The blackened surface has a low albedo and thus can become very hot during the day. The top few centimetres of the soil (where most germination is taking place) dry rapidly (Isaac 1938; Geiger 1965; Uggla 1973) due to high rates of evaporation (Viro 1974). The low water holding capacity of charred organic matter is also attributed to the lower infiltration capacity (Rowe 1947) often produced by a water repellent layer (DeBano 1969).

Because of the unfavourable conditions on charred organic matter, not all seeds falling on the seedbed will be successful. Only those species with germination and establishment requirements suited to the conditions found on the seedbed will survive. This stage is therefore seen as a sieve (Harper 1977); species that can tolerate the conditions will survive to form a seedling community; other species intolerant of the conditions will not survive. With the addition of stump sprouts from hardwood species (see review by McRae 1979) this seedling community is the starting point of the postfire forest. It can therefore be seen that a sieve, acting at the seed and young seedling stage of different species, is an important factor in determining the ultimate composition of the forest. Mathematical models that have been produced to predict forest regeneration have tended to treat this establishment stage as an unknown ''black box'' governed by a sometimes simple set of formulated assumptions (El-Bayoumi 1982). To check the validity of these formulae and to be able to predict which seedling populations might appear after fire requires a sound knowledge of the interaction between the seed and/or young seedling and the seedbed conditions.

As seedbed conditions change, different species will be favoured. In the boreal forest, black spruce (*Picea mariana*) has a more northerly distribution than jack pine (*Pinus banksiana*) (Fowells 1965; Hosie 1969). It is suggested that this can be at least partially explained by the environmental conditions found in the north and the south of the boreal forest and their suitability to the requirements for germination and establishment of these two species.

Requirements for Germination and Establishment

Given that the conditions to overcome dormancy have been met, a seed can only germinate if environmental conditions around it are suitable. As the young seedling grows, appears above ground and becomes independent of seed food reserves (established), environmental conditions must not exceed the tolerance range for its survival at this delicate stage.

Much of the literature from field studies suggests that water supply and temperature are the most important factors limiting the germination and establishment of conifers (see review by Arnott 1973). There are, however, few data on the specific moisture and temperature requirements of jack pine and black spruce.

Moisture Requirements

Using solutions of known osmotic potential to provide conditions of constant water availability it has been shown that some conifer species are capable of germinating under a greater water deficit than others (e.g., Tesche 1975); jack pine and black spruce were not included in these studies.

Experiments that have been carried out using solutions of polyethylene glycol (Thomas 1983) have shown that the germination of jack pine and black spruce respond in a similar manner to an osmotic potential up to -20 bars (Figure 1). Neither species shows a significant reduction ($P \leq 0.001$) in germination until below -12 bars osmotic potential.

Under field conditions, constant water potentials rarely occur; periods of precipitation are followed by subsequent drying, producing repeated fluctuations in soil moisture availability. The effect of fluctuating water regimes on seedling emergence (germination of the seed and growth to appear above ground) and survival has been investigated under greenhouse conditions for a variety of species using beds of soil watered at different frequencies (Barr 1930; Alexander and Noble 1971; Shepperd and Noble 1976; Noble *et al.* 1978; Jeglum 1979). Most experiments using this technique have used single species; it is difficult to compare their data due to the different and largely unknown conditions of each study. Boerker (1916) did, however, use several species under the same experimental conditions and found that some species displayed greater emergence of seedlings on less frequently watered beds.

In an extension of the polyethylene glycol studies, jack pine and black spruce seed was sown onto separate beds of soil and watered at frequencies of 2, 8 and 12 days (Thomas 1983). From Table 1 it can be seen that, compared to jack pine, black spruce showed a greater reduction in emergence with less frequent watering. It was concluded that the ability to emerge well under a fluctuating water regime is not solely dependent upon the ability to germinate under low water potentials. There are several potential mechanisms for withstanding periods of low soil moisture between periods of soil wetting (Thomas

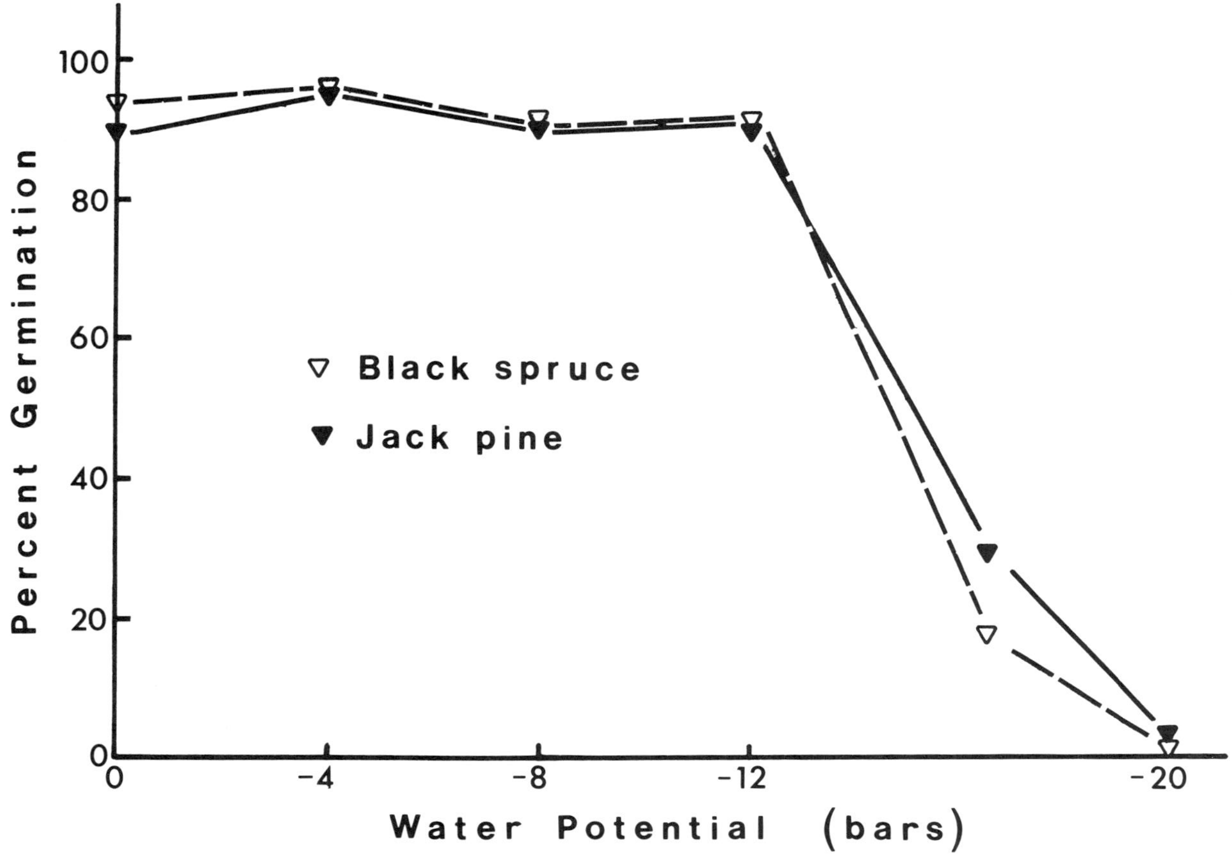

Figure 1. The germination response of jack pine and black spruce to constant water availability in solutions of polyethylene glycol 6000.

Table 1. The emergence of jack pine and black spruce from beds of soil watered at different frequencies. Data are percentage emergence ± standard error (n = 6 in all cases)

	Watering frequency (days)		
	2	8	12
Jack pine	94 ± 4.0	68 ± 5.7	43 ± 7.0
Black spruce	94 ± 4.2	42 ± 12.8	1 ± 1.0

1983). Preliminary investigations suggest that in the field the rate of radicle elongation is important to the survival of the germinated seed. After germination the emerging radicle of black spruce grows slowly and probably does not penetrate to more abundant moisture before moisture in the surface layer of the soil becomes limiting (Toumey and Neethling 1923). This may result in the early death of the seedling even before it has appeared above ground.

It is known that after one growing season the root penetration of black spruce is between 2.5 and 5.0 cm (Place 1955; Heinselman 1957) compared to at least 14 cm for jack pine (Stoeckeler and Limstrom 1942). It is suggested that although both black spruce and jack pine can germinate in relatively dry soils, black spruce, to be successful, needs soils that dry slowly to allow adequate time for root penetration.

Temperature Requirements

The range of temperatures over which these two species will germinate has been documented for a limited number of provenances. In studies by Fraser (1970a) and Winston (1973), jack pine did not germinate below 8°C nor above 44°C (Table 2). This was true for all three seed provenances used. In contrast, black spruce did not germinate below 4°C nor above 38°C; again no difference was found in the 7 seed provenances used (Fraser 1970b; Winston 1973). Thus the range of temperatures within which black spruce can germinate is lower than that of jack pine. Perhaps of more significance is the ability to germinate at sub-optimal temperatures. From Table 2 it can be seen that at 10°C black spruce is capable of higher germination than jack pine, almost reaching the percentage germination expected under optimal conditions at 20°C. Thus, not only can black spruce begin germination at a lower temperature than jack pine, it also shows high emergence potential at temperatures just above this minimum. This may be an important factor in determining the distribution of these two species.

It is interesting to note from Table 2 that the more northerly provenances of both species have a higher germination capacity at 10°C than more southerly provenances. Although the minimum temperature at which germination occurred was the same for all provenances tested, the ability to germinate at low temperatures increased with increasing latitude.

Table 2. The relationship between temperature and the germination of jack pine and black spruce for several provenances

	Seed source	Temperature (°C) at which germination will occur: Minimum	Maximum	Germination at 10°C as a percentage of the germination observed under optimal conditions
Jack pine	Southern Ontario (46° Lat.)	8	44	30
	Northern Ontario (52° Lat.)	8	44	40-45
Black spruce	Southern Ontario (46° Lat.)	4	38	65
	Northern Ontario (60° Lat.)	4	38	70-80

Seedbed Conditions in the North and South of the Boreal Forest

The seeds of conifers rarely survive in the soil for more than one year (Frank and Safford 1970), although this has been disputed (Riley 1980). Because of this, the seedbed can be said to extend only to the depth to which seeds can penetrate cracks and holes soon after falling. Generally, this germination zone is restricted to the top few centimetres of the soil (Dengler 1925; Rubner 1927; see review by Richardson 1969) and it is about this zone that quantitative data on environmental factors are needed.

During germination a seed responds to the environmental conditions surrounding it, which, for an individual seed, are on a very small scale. Within a small distance, possibly millimetres, conditions may change from suitable to unsuitable. Since it is not possible to identify and count individual "safe sites" (Harper 1977) for seed germination, an index of suitability of a particular area must be made. This is done by taking environmental measurements on a gross scale. To be at all useful, such gross measurements should be restricted to the relevant volume of soil, in this case, the top few centimetres. Most field measurements are averaged over deeper soil profiles, and while such data are useful in predicting growth and survival of established seedlings, they are of little value in determining the course of germination and early survival.

Local variations in environmental conditions, such as wet and dry areas, produce a mosaic of seedbeds suitable for each species. Searching the literature, however, it appears that on postfire organic seedbeds, there are superimposed on this mosaic gradients in soil moisture and temperature within the germination zone from the north to the south of the boreal forest.

Soil Moisture Gradients

Generalizations can be made concerning soil moisture gradients, but it should be stressed that there are few quantified data. Moisture throughout the soil profile is known to increase with increasing latitude (Figure 2), especially where there is ground ice which is likely to melt throughout the summer because of increased soil temperatures after fire (Black and Bliss 1980; Auclair 1983; Brown 1983). It is expected that the reduced amounts of solar radiation per unit time with increasing latitude (Frank and Lee 1966), aided by the abundant soil moisture (Pruitt 1978), will cause the soil surface to dry slower after wetting in the north than would be found further south. It is known that in central New Brunswick moisture in the top 1 cm of the soil may reach a water potential of −15 bars only 1-2 days after saturation to above field capacity (Thomas 1983).

It should be noted that due to reduced summer precipitation with increasing latitude (Larsen 1972; Pruitt 1978; Black and Bliss 1980; Wein 1983), soil moisture in the germination zone in northern regions may reach low levels for longer periods than found further south once spring melt water has evaporated. Soil moisture may therefore be less available for germination and early growth in the latter part of the growing season in more northerly regions.

Soil Temperature Gradients

Soil temperatures after a fire are somewhat better quantified. Measurements of the mean daily temperature during summer in the top 2 cm of the soil have been made, ranging from 25°C in New Brunswick to 10°C in areas of permafrost (Brown 1983). It should be noted that soil surface temperatures are generally inhospitable in the north and south, reaching 65°C even in areas overlaying permafrost in the Northwest Territories (Rouse 1976). These high surface temperatures produce an extremely steep temperature gradient in the germination zone.

The Success of Black Spruce and Jack Pine on Different Seedbeds

Southern Boreal Forest

In the south, high levels of solar radiation in spring rapidly increase soil temperatures above the minimum required for both black spruce and jack pine germination (Table 2). The mean summer temperature in the germination zone remains suitable for the germination of both species. However, the higher temperatures at which jack pine can germinate gives this species an advantage if the seed is near the higher temperature of the soil surface. It can therefore be said that in the south there are probably fewer "safe sites" with suitable temperatures for the germination of black spruce than for jack pine.

Probably of more importance in the south is the higher rate of drying of the upper soil layers (Figure 2); this will tend to favour jack pine because of its greater ability to avoid (and possibly withstand) long dry periods. Following germination, root penetration of black spruce appears to be less able to keep pace with the downward desiccation of the surface soil during periods of dry weather. It would seem from considerations of temperature and moisture requirements that jack pine is better adapted to start life in the south.

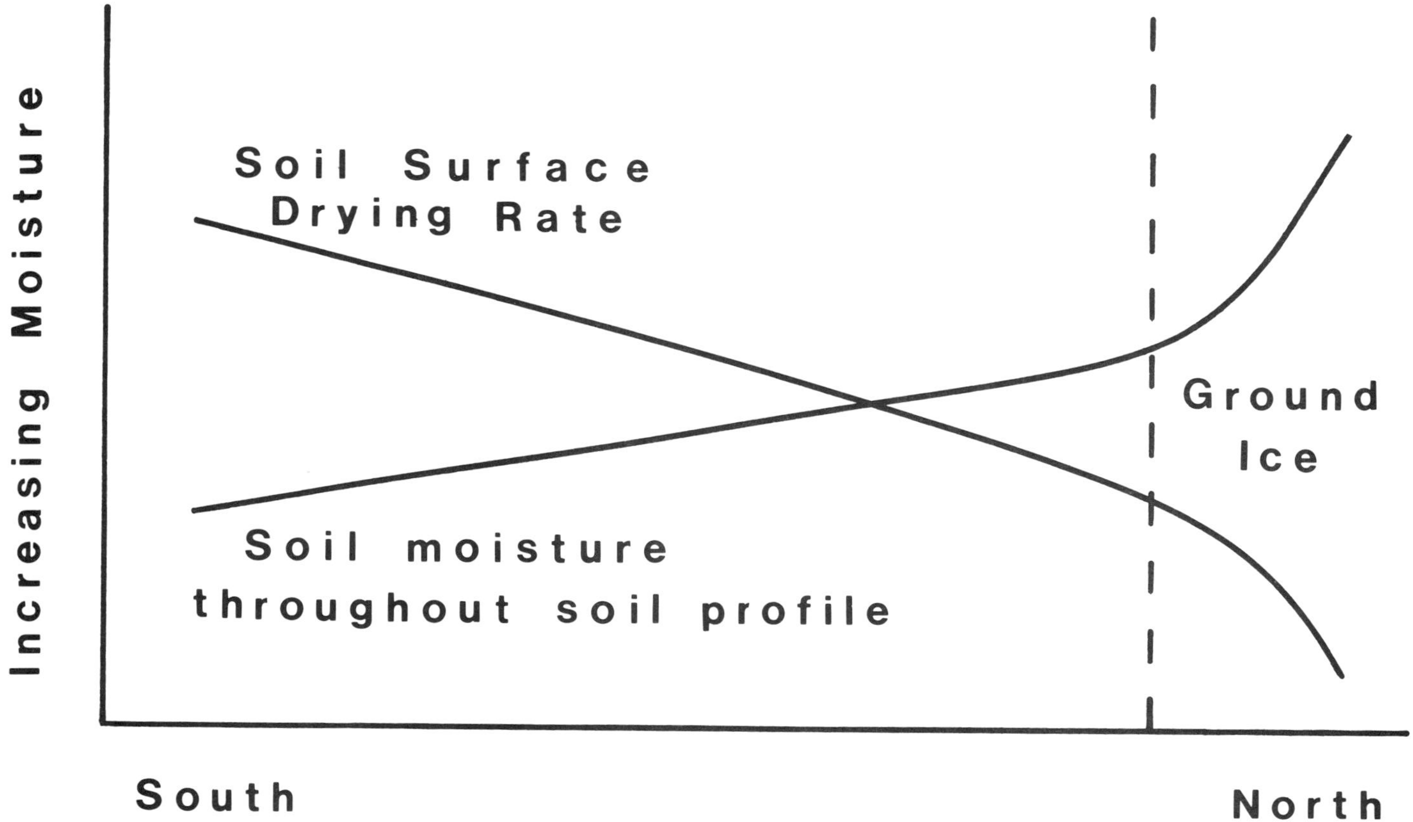

Figure 2. Gradients of soil moisture characteristics from the south to the north of the boreal forest.

Northern Boreal Forest

In the north, as the soil slowly warms up in the spring from below freezing, black spruce may begin germination before jack pine, due to a lower temperature limit on germination and greater germination at low temperatures (Table 2). This advantage may last for days or possibly weeks depending upon the rate of warming of the soil. It is suggested that the earlier germination of black spruce coupled with the slower drying of the upper soil layer will allow the young root to reach the more moist sub-surface layer before summer droughts begin. Conversely, when temperatures are more suitable for the germination of jack pine, soil moisture may well be lacking in the germination zone, due to the reduced summer precipitation, thus leading to lower establishment. In addition, the tap root of jack pine may quickly reach cold, even frozen, layers, giving the shallow-spreading root system of black spruce a further advantage in the north.

The distribution of jack pine and black spruce in the boreal forest obviously does not follow the north to south zonation as closely as suggested here. Local variations in environmental conditions, such as wet and dry areas, are superimposed on any north to south gradients, producing a mosaic of seedbeds suitable for each species. Also, for a species to enter the seedling community, is indicative but no guarantee of future success. It can be said, however, that the more northerly distribution of black spruce can be at least partially attributed to the germination and establishment requirements of this species.

Given known seedbed conditions, it should be possible to predict which species will form the new seedling community from the seed supply strewn on a charred organic matter surface. If prediction is possible the next step is one of active management, by direct seeding of the desired species and by choosing and manipulating the seedbed conditions to favour this species.

Direct Seeding in the Boreal Forest

Conditions and success: The success of direct seeding on postfire seedbeds in the boreal forest has been variable and unpredictable (Leslie 1953; Heinselman 1957; Smithers 1965; Chrosciewicz 1970; Johnston 1971; Waldron 1973; Johnson 1974). As a consequence, seedbed preparation by fire formed only ten per cent of the area seeded directly in Canada from 1900 to 1972 (Waldron 1973). It is usually concluded that mineral soil should be exposed by scarification for successful direct seeding. However, an analysis of the costs of site preparation by O'Donnell (1976 as given in McRae 1979) in Ontario showed scarification to cost 2 to 5 times as much as the initial prescribed burning. Presumably these costs would increase as accessibility decreased in higher latitudes and would be a major part of site preparation costs. It is therefore asked whether adequate establishment from seed can be obtained on charred organic matter without further site preparation. Insights into possible answers can be gained by considering what is known of the germination and establishment requirements of jack pine and black spruce.

In the southern boreal forest, jack pine has been the most successful species in direct seeding (Waldron 1973) and accounts for 90 per cent of the seed sown in the boreal region of Ontario (Brown 1973). This has included successful seeding onto charred organic matter in central Ontario (Chrosciewicz 1970, 1974; Riley 1980). The success of black spruce seeding has been found to be higher on wet peatlands (Heinselman 1957; Richardson 1970; Johnston 1971; Haavisto 1979) compared to drier areas (Richardson 1979). Such observations are consistent with jack pine being able to withstand drier and hotter conditions. Individual seeds of jack pine falling on such charred surfaces are more likely to find safe sites having the conditions required for germination than black spruce seed. The key to direct seeding, therefore is to ensure that seeds find safe sites for germination and establishment.

Seed supply: If it is assumed that the availability of safe sites is limiting the establishment on inhospitable charred surfaces then the first aim should be to ensure that most safe sites contain seed. Chrosciewicz (1970, 1974) observed that the first year establishment of jack pine in central Ontario increased with increasing duff removal by fire. Chrosciewicz (1976) also observed increased black spruce establishment with greater duff removal five years after burning in southeastern Manitoba lowland peatlands. In both cases it is expected that the removal of loose organic matter and surface litter resulted in a seedbed composed of more compact and decomposed duff, kept moist by the close proximity of the water table. Thus, as more organic matter was removed, a greater number of safe sites was available with a consequent increase in establishment. A corollary to these observations is that to produce a similar number of seedlings with decreasing seedbed quality requires that a larger number of seeds be sown. This has been demonstrated by Riley (1980). On favourable seedbeds with a high proportion of exposed mineral soil there was little increase in the high numbers of jack pine seedlings when the seed supply was increased from 50,000 to 75,000 seeds ha^{-1}. However, when the seedbed was mostly charred organic matter there were few seedlings at 50,000 seeds ha^{-1} but a significantly greater number when 75,000 seeds ha^{-1} were sown. Further evidence on the low probability of seed being successful on unfavourable seedbeds comes from a study by Zasada *et al.* (1979) who estimated that only 0.2 to 1.8 per cent of the seed that fell on charred organic matter in Fairbanks, Alaska, produced seedlings three years after the fire. Chrosciewicz (1970) using jack pine in central Ontario suggested that the seeding rate be from 70 $g \cdot ha^{-1}$ (when much organic matter is removed) to 700 $g \cdot ha^{-1}$ (when little is removed). This corresponds to approximately 25,000-250,000 seeds ha^{-1}. It should be noted that Waldron (1973) recommended from a review of direct seeding studies that 50,000-75,000 jack pine seeds ha^{-1} are adequate but recognized that a considerable number of seedlings may be from cone-bearing slash. The high densities of seeding in these studies are still below the natural seed fall. Jack pine may release up to 5 million seeds ha^{-1} and black spruce up to 1.25 million ha^{-1} after a fire (Day and Harvey 1981).

It is suggested that the probability of obtaining a high number of seedlings from direct seeding on charred organic matter can be increased by removing at least some of the organic matter (by fire severity) and scattering large numbers of seed. Vyse (1973) has estimated that most of the costs involved in direct seeding are from the cost of the seed supply. To increase the seeding rate would thus cause a significant increase in the cost of direct seeding but may still be cheaper than scarification. The results of scarification are probably more predictable than high sowing rates since the latter is more prone to the vagaries of seed predation and climate. Despite this, it is considered worthwhile to investigate high seeding rates, especially in remote northern areas (and thus probably with black spruce).

Shelter from inhospitable conditions: A second method of improving the results of direct seeding is to increase the number of safe sites by providing shelter from direct solar radiation, thus reducing soil surface temperature and evaporation. Shelter from strip cutting aids establishment of seedlings from seed on burned and unburned organic matter (Haavisto 1979; Jeglum 1983). From field studies that we are currently conducting in New Brunswick, it appears that understory vegetation, regenerating quickly after a fire, may provide such shelter. Black spruce showed a higher emergence when sheltered by understory vegetation, jack pine was not significantly affected. The regeneration of vegetation from roots and rhizomes surviving the fire is dependent upon the penetration of lethal temperatures into the soil (Flinn and Wein 1977). Thus it should be possible to manipulate the seedling community through altering fire severity.

Seeds are most likely to be in a safe site if they are buried; they are then less exposed to high temperatures and rapid drying. In large scale seeding operations it is impractical to mechanically bury seeds, although this is possible with "spot seeding" (see reviews by Arnott 1973; McRae 1979). If sown in the fall or winter there is likelihood that seeds will become buried in the surface of the seedbed by the actions of wind and frost. The disadvantage is that there is a longer period for seed loss to predators and pathogens. Immediately after a fire there is usually a layer of ash that does not consolidate until the first rain. If seed is sown during this period of loose ash it may become buried when the ash settles. This would require direct seeding within days of burning. Beaufait (1962) has shown that, after heating, the serotinous cones of jack pine open within several minutes. Thus, after a fire sufficiently intense to heat the cones on the trees, jack pine will drop seeds within the first few days after the fire. The semi-serotinous cones of black spruce may hold the seed for longer periods, especially in northern areas. It is suggested that by imitating natural seed fall through direct seeding immediately after a fire the likelihood of seedling establishment is increased. Such seeding after a fire in the autumn may have little advantage over winter seeding if there is insufficient time for germination and establishment before the end of the growing season.

Conclusions

To be able to predict the outcome of management requires a sound knowledge of the response of each species to environmental conditions. Much of the story given here for jack pine and black spruce is based on speculation. Data are still needed on the basic responses of seeds and seedlings of boreal forest trees to environmental factors such as temperature and moisture and how these influence establishment success. The effects of other environmental factors such as light and gases are even less quantified.

To date, the majority of field research on the regeneration of trees from seed has considered the establishment process as a "black box" — a prepared site is seeded and the outcome is observed. Such studies are potentially useful to the silviculturalist by giving empirical relationships between site treatment and seeding success. For example, data presented by Chrosciewicz (1970, 1974) and Riley (1980) from Ontario allow prediction of seeding success based on the degree of duff depth reduction and area of mineral soil exposed for different seeding densities. These desired seedbed conditions can be linked to the Canadian Forest Fire Weather Index (Van Wagner 1974) through empirical relationships given for Ontario, for example, by McRae (1980). Thus the forest manager has a means of prescribing under what conditions a fire should be set (or a wildfire allowed to burn) to produce the desired seedling community. As Riley (1980) admits, however, "These trials were undertaken in only one segment of [Ontario] . . . The results may therefore be considered directly applicable to that area only or to areas with similar soil, climatic and tree species characteristics.". Each set of empirical relationships is the outcome of a unique set of conditions. This approach is a good starting point to provide a foundation for future work but we are now in a position to continue to move on to find out why such observations occur. To be able to predict what will happen under any given combination of conditions requires a working model. Such a model can best be built based on sound understanding of the processes and responses of each species to given conditions.

Literature Cited

Alexander, R.R. and D.L. Noble. 1971. Effects of watering treatments on germination, survival and growth of Engelmann spruce: A greenhouse study. U.S.D.A. For. Serv. Res. Note RM-182.

Arnott, J.T. 1973. Germination and seedling establishment. Environ. Canada, C.F.S., Publ. No. 1339.

Auclair, A.N.D. 1983. The role of fire in lichen-dominated tundra and forest-tundra. *In:* The role of fire in northern circumpolar ecosystems. Ross W. Wein and David A. MacLean (Eds.). SCOPE 18, John Wiley and Sons, Chichester, England. pp. 235-256.

Barr, P.M. 1930. The effect of soil moisture on the establishment of spruce reproduction in British Columbia. Yale University, School of Forestry, Bulletin 26.

Beaufait, W.R. 1962. Procedures in prescribed burning for jack pine regeneration. Michigan College of Mining and Technology, Ford Forestry Centre, Technical Bulletin 9. L'Anse, Michigan.

Black, R.A. and L.C. Bliss. 1980. Reproductive ecology of *Picea mariana* (Mill.) BSP, at tree line near Inuvik, Northwest Territories, Canada. Ecol. Monogr. 50: 331-354.

Boerker, R.H. 1916. Ecological investigations upon the germination and early growth of forest trees. Ph.D. Dissertation, University of Nebraska. Lincoln, Nebraska.

Brown, G. 1973. Direct seeding in Ontario. Environ. Canada, C.F.S. Publ. No. 1339.

Brown, R.J.E. 1983. Effects of fire on the permafrost ground thermal regime. *In:* The role of fire in northern circumpolar ecosystems. Ross W. Wein and David A. MacLean (Eds.). SCOPE 18, John Wiley and Sons, Chichester, England. pp. 97-110.

Chrosciewicz, Z. 1970. Regeneration of jack pine by burning and seeding treatments on clear-cut sites in central Ontario. Environ. Canada, C.F.S., Info. Rep. O-X-138.

Chrosciewicz, Z. 1974. Evaluation of fire-produced seedbeds for jack pine regeneration in central Ontario. Can. J. For. Res. 4: 455-457.

Chrosciewicz, Z. 1976. Burning for black spruce regeneration on a lowland cutover site in southeastern Manitoba. Can. J. For. Res. 6: 179-186.

Day, R.J. and E.M. Harvey. 1981. Forest dynamics in the boreal mixedwood. Environ. Canada, C.F.S. Proc. O-P-9. pp. 29-41.

DeBano, L.F. 1969. Water movement in water-repellent soils. *In:* Water-repellent soils. L.F. DeBano and J. Letey (Eds.). Proc. Symposium on water-repellent soils. University of California, Riverside, California. pp. 61-89.

Dengler, A. 1925. Uber die Wirkung der Bedeckungstiefe auf das Auflaufen und die erste Entwicklung des Kiefersamens. Beifschrift für Forst-und Jagdwesen 57: 385-408.

El-Bayoumi, M.A. 1982. Simulation model of Acadian Forest succession (SMAFS). M.Sc. Thesis, University of New Brunswick, Fredericton, New Brunswick.

Flinn, M.A. and R.W. Wein. 1977. Depth of underground plant organs and theoretical survival during fire. Can. J. Bot. 55: 2550-2554.

Fowells, H.A. 1965. Silvics of forest trees of the United States. U.S.D.A., For. Serv., Handb. 271.

Frank, C.E. and R. Lee. 1966. Potential solar beam irradiation on slopes: tables for 30° to 50° latitude. U.S.D.A. For. Serv. Res. Pap. RM-18.

Frank, R.M. and L.O. Safford. 1970. Lack of viable seeds in the forest floor after clearcutting. J. For. 68: 776-778.

Fraser, J.W. 1970a. Cardinal temperatures for germination of white, red and jack pine seed. Environ. Canada, C.F.S. Info. Rep. PS-X-15.

Fraser, J.W. 1970b. Cardinal temperatures for germination of six provenances of black spruce seed. Environ. Canada, C.F.S. Info. Rep. PS-X-23.

Geiger, R. 1965. The climate near the ground. Harvard University Press, Cambridge, Massachusetts.

Haavisto, V.F. 1979. Some considerations for regenerating black spruce on peatlands in the northern clay forest section. Environ. Canada, C.F.S., Info. Rep. O-X-295.

Harper, J.L. 1977. Population Biology of Plants. Academic Press. London, England.

Heinselman, M.L. 1957. Silvical characteristics of black spruce. U.S.D.A. For. Serv. Lake States For. Expt. Stn. Pap. 45.

Hosie, R.C. 1969. Native trees of Canada. Canadian Forestry Service.

Isaac, L.A. 1938. Factors affecting establishment of Douglas-fir seedlings. U.S.D.A. Circ. 486.

Jeglum, J.K. 1979. Effects of some seedbed types and watering frequencies on germination and growth of black spruce: a greenhouse study. Environ. Canada, C.F.S. Info. Rep. O-X-292.

Jeglum, J.K. 1983. Changes in tree species composition in naturally regenerating strip clearcuts in shallow-soil upland black spruce. *In:* Resources and dynamics of the boreal zone. R.W. Wein, R.R. Riewe and I.R. Methven (Eds.). Association of Canadian Universities for Northern Studies. Ottawa, Ontario.

Johnson, J.A. 1974. Jack pine direct seeding on northern Michigan outwash plains. Michigan College of Mining and Technology, Ford Forestry Centre, Research Note 12. L'Anse, Michigan.

Johnston, W.F. 1971. Broadcast burning slash favors black spruce reproduction on organic soil in Minnesota. For. Chron. 47: 33-35.

Larsen, J.A. 1972. Vegetation and terrain (Environment): Canadian boreal forest and tundra. U.S. Army Research Office (Durham). Final Report. No. UW-G1128.

Leslie, A.P. 1953. Direct seeding as a means of reforestation. Pulp and Paper Mag. of Canada 54: 180-188.

McRae, D.J. 1979. Prescribed burning in jack pine logging slash: a review. Environ. Canada, C.F.S., Info. Rep. O-X-289.

McRae, D.J. 1980. Preliminary fuel consumption guidelines for prescribed burning in Ontario slash fuel complexes. Environ. Canada, C.F.S. Info. Rep. O-X-316.

Noble, D.L., C.B. Edminster and W.D. Shepperd. 1978. Effects of watering treatments on germination, survival and growth of Rocky Mountain Douglas-fir: A greenhouse study. U.S.D.A., For. Serv., Res. Note RM-357.

Place, I.C.M. 1955. The influence of seed-bed conditions on the regeneration of spruce and balsam fir. Canada, Dept. North. Affairs Nat. Res. For. Br. Bulletin 117.

Pruitt, W.O., Jr. 1978. Boreal Ecology. Studies in Biology No. 91. Institute of Biology. Edward Arnold, London, England.

Richardson, J. 1969. A study of germination and first-year development and survival of black spruce. M.Sc.F. Thesis, University of New Brunswick. Fredericton, New Brunswick.

Richardson, J. 1970. Broadcast seeding black spruce on a burned cutover. Environ. Canada, C.F.S. Publ. No. 1272.

Richardson, J. 1979. A comparison of methods of reforesting sites invaded by *Kalmia angustifolia*, using black spruce. Environ. Canada, C.F.S., Info. Rep. N-X-174.

Riley, L.F. 1980. The effect of seeding rate and seedbed availability on jack pine stocking and density in northeastern Ontario. Environ. Canada, C.F.S., Info. Rep. O-X-318.

Rouse, W.R. 1976. Microclimatic changes accompanying burning in subarctic lichen woodland. Arctic and Alpine Research 8: 357-376.

Rowe, P.B. 1947. Influence of woodland chaparral on water and soil in Central California. U.S.D.A., California Dept. Nat. Res. Div. For.

Rubner, K. 1927. Bedeckungstiefe und Kiemung des Fichtensamens. Forsti Centr. 49: 168-183.

Shepperd, W.D. and D.L. Noble. 1976. Germination, survival, and growth of lodgepole pine under simulated precipitation regimes: a greenhouse study. U.S.D.A. For. Serv. Res. Note RM-328.

Smithers, L.A. 1965. Direct seeding in eastern Canada. pp. 15-23. *In:* Direct seeding in the northeast, Symposium Proceedings. Amherst, Massachusetts.

Stoeckeler, J.H. and G.A. Limstrom. 1942. Ecological factors influencing reforestation in northern Wisconsin. Ecol. Monogr. 12: 191-212.

Tesche, M. 1975. Germination of conifer seeds under the conditions of simulated drought stress in polyethylene glycol (Lutrol). Biochem. Physiol. Pflanz BPP 167: 577-584.

Thomas, P.A. 1984. The influence of shelter and water regime on the establishment of conifers from seed on charred organic matter. Ph.D. Dissertation, University of New Brunswick. Fredericton, New Brunswick (In preparation).

Toumey, J.W. and E.J. Neethling. 1923. Some effects of cover over coniferous seedbeds in southern New England. Yale University, School of Forestry, Bulletin 9.

Uggla, E. 1973. Fire ecology in Swedish Forests. Tall Timbers Fire Ecol. Conf. 13: 171-189.

Van Wagner, C.E. 1974. Structure of the Canadian Forest Fire Weather Index. Environment Canada, Can. For. Serv., Publ. No. 1333.

Viro, P.J. 1974. Effects of forest fire on soil. *In:* Fire and Ecosystems. T.T. Kozlowski and C.E. Ahlgren (Eds.). Academic Press, New York. pp. 7-45.

Vyse, A.H. 1973. Economics of direct seeding versus other regeneration techniques. Environ. Canada, C.F.S. Publ. No. 1339.

Waldron, R.M. 1973. Direct seeding in Canada 1900-1972. Environ. Canada, C.F.S. Publ. No. 1339.

Wein, R.W. 1983. Fire behavior and ecological effects in organic terrain. pp. 81-95. *In:* The role of fire in northern circumpolar ecosystems. Ross W. Wein and David A. MacLean (Eds.). SCOPE 18, John Wiley, Chichester, England.

Winston, D.A. 1973. A comparative autecological study of black spruce, jack pine, lodgepole pine, and hybrid pine for direct seeding in northern Ontario. M.Sc. Thesis, Michigan State University. Michigan.

Zasada, J.C., L.A. Viereck and M.J. Foote. 1979. Black spruce seedfall and seedling establishment. *In:* U.S.D.A., For. Serv., Gen. Tech. Rep. PNW-90. pp. 42-50.

INFLUENCE OF HABITAT DIVERSITY ON THE ABUNDANCE AND DIVERSITY OF SMALL MAMMALS IN JACK PINE FORESTS IN ONTARIO

B.J. Naylor and J.F. Bendell
University of Toronto

Abstract: Rodent and shrew populations were studied in pure and mixed jack pine forests in northcentral Ontario to determine how habitat diversity influenced the abundance and diversity of small mammals. Mixed forests were floristically more diverse and physiognomically more complex. Mixed forests contained a greater diversity of small mammals, but approximately the same total abundance as pure pine forests. Similar overall density resulted from a greater individual abundance of species in pure pine forests.

Within forest types, changes in forest diversity with age apparently influenced rodent, but not shrew, diversity and abundance.

The heather vole and masked shrew accounted for 71 per cent of the total captures. Small mammals can be classified into three groups related to their apparent abundance in homogeneous, heterogeneous, or neither habitat type. Diversity and abundance of small mammals may influence their role in insect control, damage to forest regeneration, and nutrient cycling. Future research should quantify the value of small mammals to forest yield as predators of forest insects and as agents of accelerated nutrient cycling.

Introduction of the Forest Management Agreement in 1979 has led to a commitment to intensive forest management in boreal Ontario. Silvicultural techniques intended to maximize fibre production will include the conversion of diverse mixed forests to artificially regenerated coniferous monocultures. Forest managers must predict the ecological ramifications of this practice.

Pimlott (1969) emphasized the importance of habitat diversity to the production of forest wildlife. Miller and Getz (1977) observed a correlation between the diversity of small mammals and the total number of species of trees and shrubs in the forests of New England.

We were interested in determining how small mammals would be affected by a reduction in diversity of a typical boreal habitat accompanying intensive management. We therefore compared the diversity and abundance of small mammals in mixed (diverse) and pure (less diverse) jack pine (*Pinus banksiana*) forests. Jack pine is one of the most common species planted to monocultures in northcentral Ontario. A comparison between mixed and pure stands should reveal faunal changes resulting from a reduction in habitat diversity following artificial regeneration.

Small mammals were studied for a number of reasons. As a group, they are ecologically important since they comprise a significant proportion of the faunal biomass in forest communities (Hamilton and Cook 1940) and are an integral part of forest consumer food chains (Golley 1976). Furthermore, the bulk of small mammals inhabit the soil-litter interface where their activities may promote more rapid pedogenesis and nutrient cycling (Chew 1976; Burt 1977). Small mammals are also economically important to the forester. They may hinder reforestation through seed predation (Gashwiler 1970; Hawthorne 1980) and the girdling of planted stock (Smith 1962; Radvanyi 1975; Cleary *et al.* 1982). They may also be beneficial by controlling the population of some insect pests (Holling 1959; Platt and Blakely 1973). In addition, a census can be taken of the greater majority of the small mammal community using a standard sampling technique. This provides a measure of the diversity of rodents and shrews in each forest.

This preliminary investigation of small mammal response to monocultures is part of a long term study of the reciprocal interactions between forestry practices and wildlife at the University of Toronto.

Methods

Study Areas

Study areas were located (47°30′N, 81°40′W) in the Missinaibi-Cabonga section of the boreal forest region (Rowe 1972), approximately 30 km south of Gogama, Ontario. These sites were situated within an extensive sand flat region characterized by deep sands and gravels overlying felsic intrusive and metamorphic bedrock (Anon. 1981). Jack pine cover types dominate the forest community. Severe wildfires in the past, and the artificial regeneration of large clearcuts have created extensive blocks of monospecific forest. We were interested in the influence of the diversity of forests of different ages on small mammal populations. Therefore, forests were selected to provide a comparable range of ages within both the pure pine and mixed pine forest types (Table 1.) Pure pine forests (100 percent jack pine by basal area) were located on deep, fine to coarse sandy soils and exhibited a relatively continuous understory of blueberry (*Vaccinium angustifolium* and *V. myrtilloides*), sheep laurel (*Kalmia angustifolia*), and sweetfern (*Comptonia peregrina*). Tall shrub and herb layers were practically non-existent. A deep litter layer was composed of jack pine needles and foliage from ericaceous shrubs. Mixed forests (less than 75 percent jack pine by basal area) exhibited a well developed tall shrub layer of green alder (*Alnus crispa*), beaked hazel (*Corylus cornuta*), and pin cherry (*Prunus pensylvanica*). A patchy, discontinuous low shrub layer of bush honeysuckle (*Diervilla lonicera*), blueberry (*V. myrtilloides*), and wild sarsaparilla (*Aralia nudicaulis*) overtopped a well defined herb layer including bunchberry (*Cornus canadensis*), clintonia (*Clintonia borealis*), and Canada mayflower (*Maianthemum canadense*). Mixed stands were situated on slightly loamier sands of varying depth and had a litter layer composed of conifer needles and foliage from deciduous trees, shrubs, and herbs.

Within each forest, overstory composition (by basal area) was determined using a 20-point plotless timber cruise (Grosenbaugh 1952). The abundance (percentage cover) of understory species was visually estimated using twenty 4m^2 quadrats. Diversity values for overstory and understory vegetation were calculated using the Shannon-Weaver diversity index (H′) (Pielou 1966).

Sampling Techniques and Data Analysis

Rodents (Cricetids, Zapodids, and Scuirids) were censused using lines of standard 10 cm

Table 1. Characteristics of eight jack pine study forests near Gogama, Ontario. BURN and CUT represent mixed and pure recently disturbed sites, respectively.

	Cut	Burn	Young pure	Young mixed	Medium pure	Medium mixed	Old pure	Old mixed
Age (years)	2	1	20	21	36	41	58	56
Canopy height (m)	—	—	6.0	6.5	15.5	15.0	20.0	19.0
% Composition by basal area:†								
Pj	—	—	100	64	100	73	99	64
Other softwoods (Sb, Sw, B)	—	—	—	—	—	—	1	12
Hardwoods (Bw, Pot)	—	—	—	36	—	27	—	24
Diversity*								
Understory	1.04	1.56	0.84	1.29	0.66	1.59	1.04	1.61
Overstory	—	—	0.00	0.78	0.00	0.75	0.03	0.97

† Pj = *Pinus banksiana*, Sb = *Picea mariana*, Sw = *Picea glauca*, B = *Abies balsamea*, Bw = *Betula papyrifera*, Pot = *Populus tremuloides*.

* Diversity of vegetation calculated using Shannon-Weaver diversity index (H′).

Victor snap traps. Traplines consisted of 20 stations at 15 m intervals with three traps per station. Each of four parallel lines within a forest was run for three days and nights (72 hours) to produce a sampling effort of 720 trap nights (TN) per forest (1 TN = 1 trap set for 24 hours). Lines were placed 50 m apart and at least 50 m from a stand boundary. Traps were baited with a peanut butter-oatmeal mixture.

Shrews were censused using a grid of 20 plastic pitfall traps covering 10 ha within each forest. Pitfall pots were 15 cm in diameter and 18 cm deep, and contained 5 cm of a 30 per cent ethyl alcohol preservative mixture. Approximately 1120 pit nights (PN) were accumulated for each forest (1 PN = 1 pit in forest for 24 hours).

All sampling was conducted during June and July 1982. Species (nomenclature follows Banfield 1974), sex, weight, and reproductive condition were recorded for each specimen. We believe the use of both snap and pitfall traps enabled us to census all small mammals present with the exception of the eastern chipmunk (*Tamias striatus*), red squirrel (*Tamiasciurus hudsonius*), northern flying squirrel (*Glaucomys sabrina*), and star-nosed mole (*Condylura cristata*). These species were present but were either too large or spent too little time at the litter surface to be caught with our traps.

We also believe that our trapping intensity was adequate to sample the populations of the species we report on. In no forest did the number of traps limit the number of captures because: (1) less than 20 per cent of all snap traps were fired any night; (2) rarely were all three snap traps at a given station fired; and (3) pitfall traps continued to capture shrews even when they contained dead sorcids or voles.

Rodent and shrew data were analyzed separately because of the different methods of capture. Results are expressed as rates of capture (number of individuals per 100 TN or PN) because sampling effort was not uniform in all forests owing to unequal trap disturbance. Species richness (S) (Krebs 1972) and Brillouin's index (H) (Pielou 1966) are presented as two measures of small mammal diversity.

Some species of small mammals exhibit yearly fluctuations in density and may occur in a wider range of habitats when populations are high. This might affect the interpretation of results from a single season study such as this one. However, data collected in 1981 (Lattner, unpubl. report) from two of the eight forests suggest that numbers may have varied, but that trends in relationships between species observed in this study remained relatively consistent.

Results

Forest Diversity

Mixed stands appeared to provide a more diverse habitat for wildlife. They were floristically more diverse than pure pine forests of similar age (Table 1) and exhibited greater complexity in the physiognomy of both the vegetation and litter. Within each forest type, habitat diversity also varied with forest age. Lower diversity in young-medium age stands may reflect their aggrading nature (Bormann and Likens 1981).

Abundance of Small Mammals

A total of 446 small mammals representing nine species of rodents and four species of shrews was caught in the eight study forests.

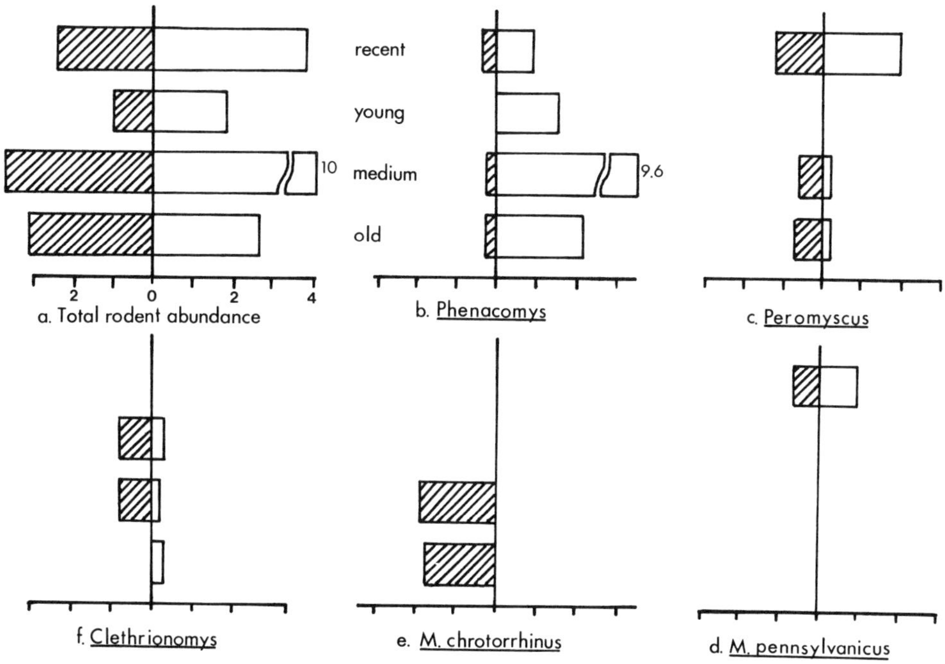

Figure 1. Abundance of rodents per 100 trap nights in four pure (open blocks) and four mixed (hatched blocks) jack pine forests ranging from one to two years (recently disturbed) to 60 years (old).

Pure forests exhibited a greater mean total abundance of rodents than mixed forests (Figure 1a), but this difference was not significant (pure: 4.5/100 TN, mixed: 2.5/100 TN $t = 0.999$ $P > 0.05$). Pure pine forests also displayed a greater variability in abundance between age classes (pure: CV = 83.29, mixed: CV = 45.98). Total abundance followed a similar age-related pattern in both cover types. Initial high densities on recently disturbed sites declined as the canopy closed in the young forest (20 years). This was followed by an increase to middle age (40 years) and a gradual decline with stand maturation (60 years).

The heather vole (*Phenacomys intermedius*) (Figure 1b) was caught in almost all ages in both forest types but was most numerous in pure pine forests, especially of middle age. The deer mouse (*Peromyscus maniculatus*) was most abundant on recently disturbed sites and in older mixed forests (Figure 1c). The meadow vole (*Microtus pennsylvanicus*) was caught only on recently disturbed sites (Figure 1d), while the rock vole (*Microtus chrotorrhinus*) was restricted to older mixed forests (Figure 1e). The red-backed vole (*Clethrionomys gapperi*) (Figure 1f) avoided recently disturbed sites and appeared to favour mixed forest. Its absence from the old mixed forest is puzzling. Other rodents caught, predominantly in the mixed forests in low numbers, included the least chipmunk (*Eutamias minimus*), the southern bog lemming (*Synaptomys cooperi*), and the woodland and meadow jumping mice (*Napaeozapus insignis* and *Zapus hudsonius*).

Total abundance of shrews appeared to increase with age in the mixed forests, but peaked and then declined after middle age in the pure pine forests (Figure 2a). While mean

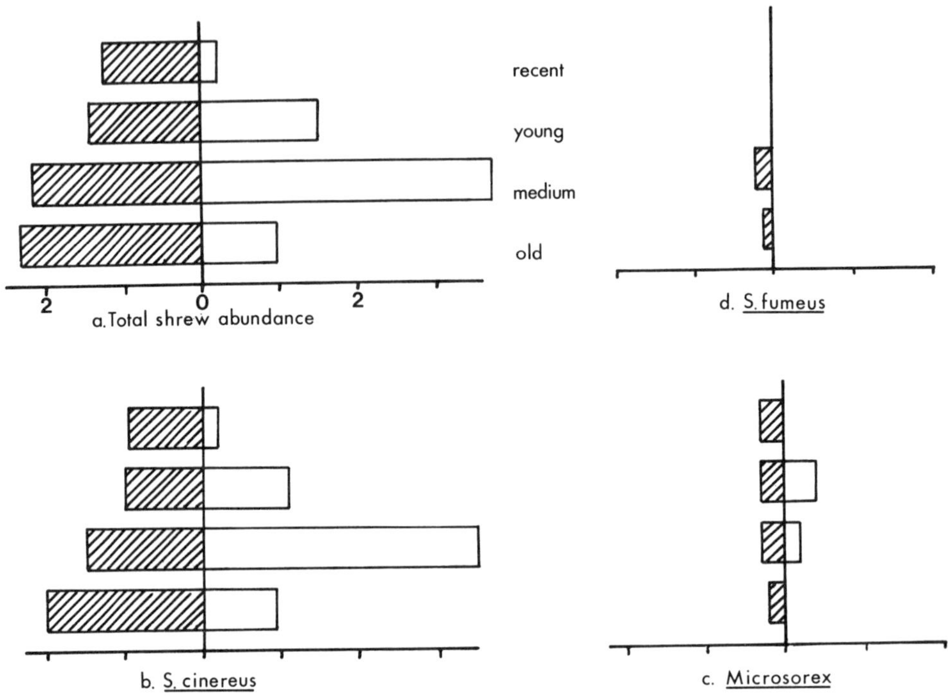

Figure 2. Abundance of shrews per 100 pit nights in four pure (open blocks) and four mixed (hatched blocks) jack pine forests ranging from one to two years (recently disturbed) to 60 years (old).

densities were similar (pure: 1.3/100 PN, mixed: 1.7/100 PN $t = 0.636$ $P > 0.05$), pure pine forests exhibited greater variability between age classes (pure: CV = 85.68, mixed: CV = 22.25).

The masked shrew (*Sorex cinereus*) was the most common species in both forest types and reached greatest density in the pure pine forest (Figure 2b). The pygmy shrew (*Microsorex hoyi*) preferred mixed forest of all age classes (Figure 2c), while the smoky shrew (*Sorex fumeus*) was caught only in the older mixed forests (Figure 2d). The short-tailed shrew (*Blarina brevicauda*) was trapped infrequently in mixed forests. Total abundance of rodents and shrews (Figures 1a and 2a) in the mixed forest is comprised of a mosaic of species with individual habitat preferences. Conversely, the majority of the total abundance in pure pine forests is accounted for by only two widespread species, the heather vole and masked shrew.

Diversity of Small Mammals

Mixed stands generally contained a greater number of species of shrews, rodents, and in total than pure pine stands of similar age (Figure 3). No species was found exclusively in pure pine forests; however, a number were caught only in mixed forests. As a rule, the diversity of rodents and shrews as measured by Brillouin's index was also greater in mixed forests than in pure pine forests of comparable age (Figure 4). Diversity of shrews was roughly similar in all ages of forest. However, the diversity of rodents was highest on recently disturbed sites and in older forests (Figures 3 and 4).

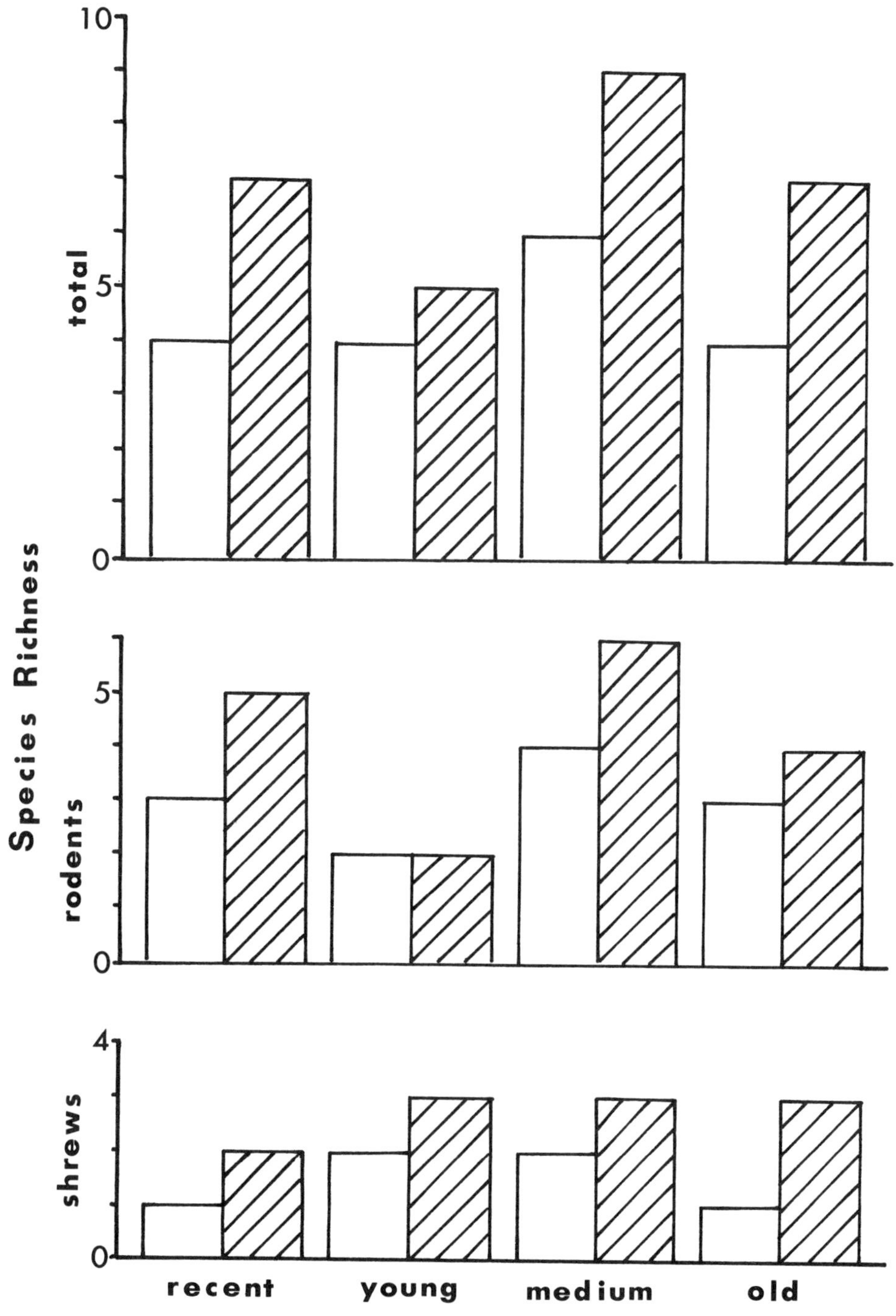

Figure 3. Number of species of small mammals (species richness) in four pure (open blocks) and four mixed (hatched blocks) jack pine forests that range from recently disturbed to old growth conditions near Gogama, Ontario.

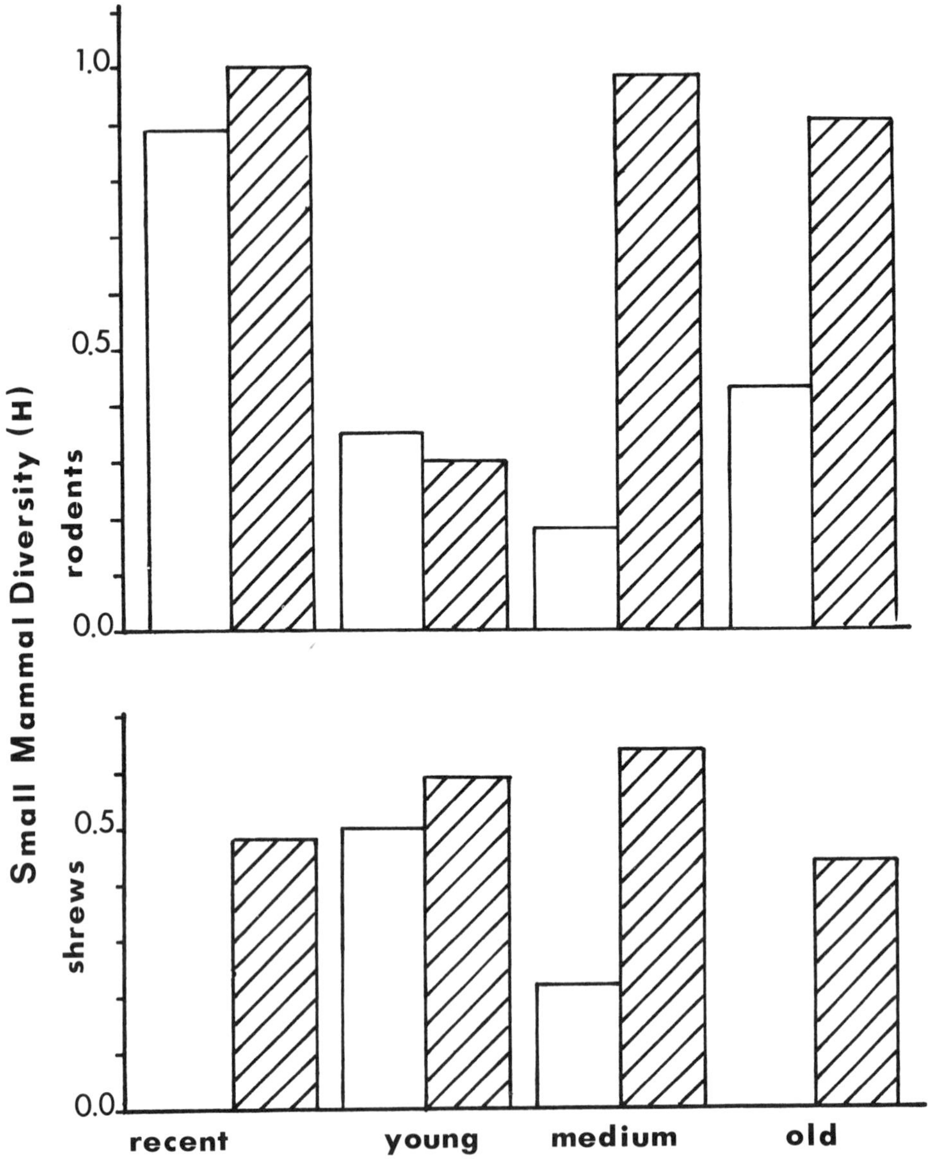

Figure 4. Diversity of small mammals (Brillouin's index, H) in four pure (open blocks) and four mixed (hatched blocks) jack pine forests near Gogama, Ontario.

Mixed forests contained a greater diversity of small mammals. However, species in pure pine forests reached a higher average individual abundance and this resulted in approximately an equal overall density in both forest types.

Discussion

Our study has identified the common small mammals of pure and mixed jack pine forests in northcentral Ontario. Species such as the heather vole, rock vole, and pygmy shrew are regarded by some as rare or uncommon over much of their ranges (Banfield 1974). Furthermore, the latter two species are not usually associated with jack pine cover types by other authors.

Forest diversity (mixed versus pure) exerted a stronger influence on the diversity, rather than the total abundance, of small mammals in these habitats. Reynolds (1980) found a similar result in populations of prairie small mammals in natural ungrazed fields of sagebrush (*Artemisia tridentata*) and sagebrush fields planted to a crested wheatgrass (*Agropyron cristatum*) monoculture. Conversely, Nagorsen and Peterson (1981) reported that both the diversity and density of woodland small mammals were lowest in the most homogeneous boreal forests they examined. Similar total abundance in habitats differing in diversity may be related to the fact that in our study, and in Reynolds' (1980), site types were relatively similar on all study areas.

Changes in forest diversity with age (Table 1) apparently influenced the diversity and abundance of small mammals. Lowest forest diversity and lowest rodent diversity and abundance occurred in young mixed and young-medium pure pine stands, and may represent a transition period from early successional species (deer mouse, meadow vole) on recently disturbed sites to woodland species (red-backed vole, rock vole, heather vole) in older forests. This change in the composition of small mammal communities during plant succession follows disturbance by fire (Ahlgren 1966, Krefting and Ahlgren 1974), clearcutting (Kirkland 1977), and strip-mining (Sly 1976). Furthermore, Atkeson and Johnson (1979) reported a similar decrease in rodent abundance during early to mid-succession in pine plantations in Georgia. However, shrews did not respond in the same manner as rodents. Since they are insectivorous, changes in floral diversity within habitat types may not influence their diversity or abundance to the same extent.

Forest diversity may also influence the seasonal abundance of small mammals. A number of authors (especially Elton 1958) suggest that species in simple communities are more subject to large fluctuations in population density. The greater variation in abundance between age classes in pure pine forests suggests such a phenomenon. Langley and Shure (1980) reported that populations of the cotton rat (*Sigmodon hisidus*) fluctuated more in plantations of loblolly pine (*Pinus taeda*) than in natural habitats.

Homogenized habitats (i.e., plantations) may influence small mammals in a number of ways. The diversity of small mammals may be primarily related to the diversity of food items (Miller and Getz 1977) and to structural attributes of the habitat (Grant *et al.* 1982). More diverse, mixed forests provide a greater variety of food types directly (number of edible plants and mast types) and indirectly (variety of host plants influences diversity of invertebrate prey species) (Miller and Getz 1977). Mixed forests also provide a more heterogeneous horizontal and vertical vegetation structure. Greater variety in food and habitat structure may lead to a larger number of niches, and hence, a higher diversity.

Homogeneous habitats (pure pine forests) offer fewer kinds of resources. This likely reduces species diversity, but also interspecific competition. Furthermore, monocultures tend to concentrate the limited kinds of resources present (Root 1973). Hence, species capable of exploiting these types of resources are confronted with an abundant supply and little competition, and so attain a greater individual abundance.

Small mammals responded to the range of habitat diversity in three main ways. The heather vole was most abundant in homogeneous pure pine forests. Others (Edwards

1955; Foster 1961; Banfield 1974) have noted this vole's preference for dry coniferous monocultures. Pure pine forests contained a relatively continuous mat of ericaceous shrubs. Since these plants comprise a significant proportion of the heather vole's diet (Foster 1961; Banfield 1974), concentration of this resource may be a factor in the high densities observed.

Other species (red-backed vole, rock vole, smoky shrew, pygmy shrew) occurred in greatest abundance in forests of high diversity. Many of these species typically inhabit heterogeneous mixed forests. The red-backed vole is most abundant in mixed boreal forests (Nagorsen and Peterson 1981; Kirkland and Schmidt 1982) and in habitats with a considerable structural complexity (Gunderson 1959; Miller and Getz 1977) and a diversity of seed- and fruit-producing plants (Ahlgren 1966). The rock vole most often inhabits rocky talus with a variety of mesophytic vegetation (Banfield 1974; Kirkland and Knipe 1979), mixed forests (Kirkland and Schmidt 1982), and forest-talus interfaces (Timm *et al.* 1977). Smoky shrews are typically found in mixed forests (Jackson 1961; Burt 1977; Richens 1974). The pygmy shrew has a wide habitat tolerance (Long 1972; Wrigley *et al.* 1979), but appears to prefer moist mixed boreal forests (Nagorsen and Peterson 1981).

The third group of species did not respond as clearly to the range of inter-habitat diversity we studied (masked shrew, deer mouse, meadow vole). The masked shrew was the most common insectivore in all ages of both forest types. Banfield (1974), Wrigley *et al.* (1979), and Nagorsen and Peterson (1981) have noted the ubiquitous nature of this shrew. Its local distribution appears to be relatively independent of vegetation structure or food supply (Getz 1961a). The deer mouse and meadow vole were abundant on the two recently disturbed sites regardless of site history. Both of these species are common inhabitants of early successional plant communities (Sims and Buckner 1973; Krefting and Ahlgren 1974; Fala 1975; Kirkland 1977; Martell and Radvanyi 1977). The deer mouse, as an omnivore, is able to exploit the variety of food resources on disturbed sites (Martell and MacAulay 1981). The meadow vole is more highly specialized and consumes grasses and sedges appearing after habitat disturbance (Getz 1961b). The distributions of the deer mouse and meadow vole may indicate that the recently disturbed sites are less different habitats than the succeeding pure pine and mixed forests.

The interaction between forest management practices and wildlife is of a reciprocal nature. Changes in the diversity and abundance of small mammals following intensive management practices may influence growth and yield of the subsequent managed forest in many ways, of which we note a few.

Holling (1959) suggested that small mammals may control some populations of insects as long as their functional and numerical responses keep insect numbers below some escape threshold. Monocultures result in a lower diversity of potentially insectivorous small mammals. However, the possibly greater interspecific competition in mixed forests may dampen the ability of each species to respond numerically to insect outbreaks (Turnbull and Chant 1961; Watt 1965). Hence, small mammals in pure forests may be better adapted to control insect outbreaks.

Granivorous and herbivorous small mammals may retard regeneration after timber harvesting (Smith 1962; Cleary *et al.* 1982). Burning prior to planting or direct seeding has been suggested as a technique to reduce the damaging effects of small mammals (Fala 1975). Higher densities of two potentially injurious cricetids (deer mouse, meadow vole) were observed on the recent clearcut than on the recent burn. This suggests that reduced habitat diversity (cut) may encourage some species to reach pest levels.

Herbivorous small mammals may also accelerate nutrient cycling in forests. For example, sheep laurel can cover up to 60 per cent of the ground in our pure pine forests. When fire is excluded from these forests, sheep laurel litter accumulates and decomposes at an uncommonly slow rate (Damman 1971). Site fertility and forest growth may be reduced. However, sheep laurel may be one of the major foods of the heather vole. Hence, this vole may release nutrients from plant tissues in the form of urine, feces, and carcasses that are readily transformed into states that permit absorption by vegetation.

Reduction of small mammal diversity in plantations may also have unknown effects on their predators (e.g., hawks, owls, furbearers) and competitors (e.g., ground feeding birds).

Summary

This preliminary investigation has suggested that a reduction in habitat diversity following artificial regeneration may influence the diversity and abundance of small mammals. The applicability of these results to other wildlife groups requires examination. Further study might examine the dynamics of small mammal populations in monocultures. In addition, the importance of small mammals to forest management needs to be quantified. Numerous studies focus on the detrimental influences of small mammals on forest growth. However, none has quantitatively examined the beneficial effects of small mammal predation of insects on forests yield. Modeling of insectivore-insect-forest growth relationships would provide estimates of the value of insectivores to forest management. Other studies could investigate the value of herbivorous small mammals to nutrient cycles. Study of the role of the heather vole in recycling nutrients immobilized in sheep laurel might be one approach.

Acknowledgements

The authors are indebted to Dr. S. Spires and M.T. Lattner for the use of some of their data. K.J. Szuba provided good criticism of an early draft of the manuscript and Dr. R.L. Peterson commented on our results. The study was funded by the Natural Sciences and Engineering Research Council of Canada and the Ontario Ministry of Natural Resources.

Literature Cited

Ahlgren, C.E. 1966. Small mammals and reforestation following prescribed burning. J. For. 64: 614-617.

Anonymous. 1981. Gogama district land use plan: background information. Ontario Ministry of Natural Resources.

Atkeson, T.D., and A.S. Johnson. 1979. Succession of small mammals on pine plantations in the Georgia piedmont. Amer. Midl. Nat. 103: 59-65.

Banfield, A.W.F. 1974. The mammals of Canada. University of Toronto Press, Toronto, Ontario.

Bormann, F.H., and G.E. Likens. 1981. Pattern and process in a forested ecosystem. Springer-Verlag, New York, NY.

Burt, W.H. 1977. Mammals of the Great Lakes region. University of Michigan Press, Ann Arbor, Michigan.

Chew, R.M. 1976. The impact of small mammals on ecosystem structure and function. Pymatuning Laboratory, Spec. Publ. No. 5, pp. 167-180.

Cleary, B.D., R.D. Greaves and R.K. Hermann. 1982. Regenerating Oregon's forests. Oregon State University, Ext. Ser., Cornwallis, Oregon.

Damman, A.W.H. 1971. Effect of vegetation changes on the fertility of a Newfoundland forest site. Ecol. Monogr. 41: 253-270.

Edwards, Y. 1955. The habitat preferences of the boreal phenacomys. Murrelet, 35: 1-4.

Elton, C. 1958. The ecology of invasions by animals and plants. Methuen, London.

Fala, R.A. 1975. Effects of prescribed burning on small mammal populations in a mixed-oak clearcut. J. For. 73: 586-587.

Foster, J.B. 1961. Life history of the phenacomys vole. J. Mammal. 42: 181-198.

Gashwiler, J.S. 1970. Further study of conifer seed survival in a western Oregon clearcut. Ecology, 51: 849-854.

Getz, L.L. 1961a. Factors influencing the local distribution of shrews. Amer. Midl. Nat. 65: 67-88.

Getz, L.L. 1961b. Factors influencing the local distribution of *Microtus* and *Synaptomys* in southern Michigan. Ecology, 42: 110-119.

Golley, F.B. 1976. Role of small mammals in ecosystems. Pymatuning Laboratory, Spec. Publ. No. 5, pp. 76-78.

Grant, W.E., E.C. Birney, N.R. French and D.M. Swift. 1982. Structure and productivity of grassland small mammal communities related to grazing induced changes in vegetative cover. J. Mammal. 63: 248-260.

Grosenbaugh, L.R. 1952. Plotless timber estimates — new, fast, easy. J. For. 50: 32-37.

Gunderson, H.L. 1959. Red-backed vole habitat studies in central Minnesota. J. Mammal. 40: 405-412.

Hamilton, W.J. Jr., and D.B. Cook. 1940. Small mammals and the forest. J. For. 38: 468-473.

Hawthorne, D.W. 1980. Wildlife damage and control techniques. *In* Wildlife management techniques manual. S.D. Schemnitz (Ed.). Wildl. Soc. Washington, DC. pp. 411-439.

Holling, C.S. 1959. The components of predation as revealed by a study of small mammal predation of the European pine sawfly. Can. Ent. 91: 293-320.

Jackson, H.H.T. 1961. Mammals of Wisconsin. University of Wisconsin Press, Madison, Wisconsin.

Kirkland, G.L., Jr. 1977. Responses of small mammals to the clearcutting of northern Appalachian forests. J. Mammal. 58: 600-609.

Kirkland, G.L., Jr. and C.M. Knipe. 1979. The rock vole (*Microtus chrotorrhinus*) as a transition zone species. Can. Field-Nat. 93: 319-321.

Kirkland, G.L., Jr. and D.F. Schmidt. 1982. Abundance, habitat, reproduction and morphology of forest-dwelling small mammals of Nova Scotia and southeastern New Brunswick. Can. Field-Nat. 96: 156-162.

Krebs, C.J. 1972. Ecology. Harper and Row. New York.

Krefting, L.W., and C.E. Ahlgren. 1974. Small mammals and vegetation changes after fire in a mixed conifer-hardwood forest. Ecology, 55: 1391-1398.

Langley, A.K., Jr. and D.J. Shure. 1980. The effects of loblolly pine plantations on small mammal populations. Amer. Midl. Nat. 103: 59-65.

Lattner, M.T. Unpublished. Notes on small mammal abundance in two coniferous monocultures. University of Toronto, Faculty of Forestry, Toronto, Ontario.

Long, C.A. 1972. Notes on habitat preference and reproduction in pygmy shrews, *Microsorex*. Can. Field-Nat. 86: 155-160.

Martell, A.M., and A.L. MacAulay. 1981. Food habits of deer mice (*Peromyscus maniculatus*) in northern Ontario. Can. Field-Nat. 95: 319-324.

Martell, A.M., and A. Radvanyi. 1977. Changes in small mammal populations after clearcutting of northern Ontario black spruce forest. Can. Field-Nat. 91: 41-46.

Miller, D.H., and L.L. Getz. 1977. Factors influencing local distribution and species diversity of forest small mammals in New England. Can. J. Zool. 55: 806-814.

Nagorsen, D.W., and R.L. Peterson. 1981. Distribution, abundance and species diversity of small mammals in Quetico Provincial Park, Ontario. Naturaliste Can. 108: 209-218.

Pielou, E.R. 1966. The measurement of diversity in different types of biological collections. J. Theoretical Biol. 13: 131-144.

Pimlott, D.H. 1969. The value of diversity. Trans. N. Amer. Wildl. and Natural Resources Conf. 34: 265-280.

Platt, W.J., and N.R. Blakley. 1973. Short-term effects of shrew predation upon invertebrate prey sets in prairie ecosystems. Proc. Iowa Acad. Sci. 80: 60-66.

Radvanyi, A. 1975. Harmful effects of small mammal populations on a tree plantation in southern Ontario. Can. Field-Nat. 89: 53-57.

Reynolds, T.D. 1980. Effects of some different land management practices on small mammal populations. J. Mammal. 61: 558-561.

Richens, V.B. 1974. Numbers and habitat affinities of small mammals in northeastern Maine. Can. Field-Nat. 88: 191-196.

Root, R.B. 1973. Organization of a plant-agthropod association in simple and diverse habitats: the fauna of collards (*Brassica oleracea*). Ecol. Monogr. 43: 95-124.

Rowe, J.S. 1972. Forest regions of Canada. Environ. Canada. C.F.S., Publ. No. 1300.

Sims, H.P., and C.H. Buckner. 1973. The effect of clear cutting and burning of *Pinus banksiana* forests on the populations of small mammals in southeastern Manitoba. Amer. Midl. Nat. 90: 228-231.

Sly, G.R. 1976. Small mammal succession on strip-mined land in Vigo County, Indiana. Amer. Midl. Nat. 95: 257-267.

Smith, D.M. 1962. The practice of silviculture (7th ed.). John Wiley and Sons. New York.

Timm, R.M., L.R. Heaney and D.D. Baird. 1977. Natural history of rock voles (*Microtus chrotorrhinus*) in Minnesota. Can. Field-Nat. 91: 177-181.

Turnbull, A.L., and D.A. Chant. 1961. The practice and theory of biological control. Can. J. Zool. 39: 697-753.

Watt, K.E.F. 1965. Community stability and the strategy of biological control. Can. Ent. 97: 887-895.

Wrigley, R.E., J.E. Dubois and H.W. Copland. 1979. Habitat, abundance, and distribution of six species of shrews in Manitoba. J. Mammal. 60: 505-520.

EARLY POSTFIRE EFFECTS ON SOME INVERTEBRATES, SMALL MAMMALS AND BIRDS IN NORTH-CENTRAL ONTARIO

S. Spires and J.F. Bendell
University of Toronto

Abstract: Animals were trapped or observed in burned and adjacent unburned forest during the first five weeks following an intense 380 ha ground and canopy fire in May 1981. The following general trends were found. Herbivores (e.g., Gapper's red-backed vole, snowshoe hare) were captured or observed in lower numbers in burned than in unburned forest. Granivores such as the deer mouse and white-throated sparrow were captured or observed in high numbers in both forests. Aerial insectivores (e.g., common nighthawk), ground insectivores (e.g., American robin, a tiger beetle) and bole insectivores (woodpeckers) were captured or observed in higher numbers in burned than in unburned forest. Foliage insectivores (mainly wood warblers) were observed in lower numbers in burned than in unburned forest. There were some exceptions to the above trends, for example the ground insectivorous masked shrew was captured in lower numbers in burned than in unburned forest. These exceptions may relate to requirements of the species involved for cover or moisture.

The effects of fire on birds and mammals have been reviewed by Scotter (1972), Bendell (1974) and Fox (1983). Ahlgren (1974) reviewed fire-effects on soil-dwelling invertebrates, and since then studies on sections of this group have been produced by Winter *et al.* (1980), Schaeffer (1980), Winter (1980), Schauermann (1980) and Richardson and Holliday (1982). Of these Scotter (1972), Fox (1983) and Richardson and Holliday (1982) deal specifically with the boreal region.

One documented effect is immediate fire-caused mortality, although birds and mammals (Bendell 1974), soil invertebrates (Ahlgren 1974) and even herbivorous insects (Gillon 1971; Gillon, Y. 1971) can to some extent evade fire either by sheltering in refuges or by flying over the fire front. The greatest and most long term effects of fire are generally thought to be the result of habitat change (Scotter 1972). Evidence for postfire succession in animal communities concurrent with vegetational succession has been presented by Erskine (1977) and Fox (1982, 1983). Changes in the importance of various foraging guilds appear to be associated with this succession (Sullivan 1980; Blake 1982; Fox 1983).

The present paper reports on birds, small mammals and ground-active invertebrates in burned and unburned forest within five weeks of an intense fire, before full development of herbaceous and shrubby regrowth. This appears to be the first such study in the boreal region of Ontario.

Materials and Methods

Study Area and Fire

The study area (approximately 47°30′N, 81°45′W) was located on an extensive sand-flat 15 to 22 km south of Gogama, Ontario. Most of this area was covered with moderately dense, second growth after logging and fire, mixed coniferous-deciduous forest. Dominant trees were the pioneer species jack pine (*Pinus banksiana*), trembling aspen (*Populus tremuloides*), and white birch (*Betula papyrifera*). This forest had a patchy understory including shrubs such as mountain maple (*Acer spicatum*), beaked hazel (*Corylus cornuta*), and pin cherry (*Prunus pensylvanicus*), and herbs such as *Aster* spp, *Clintonia borealis* and *Cornus canadensis*. The forest floor was covered with a variable litter layer interrupted by a few scattered logs and rocky ridges.

On May 23, 1981, a fire started in the southern part of the study area. At this time the forest was very dry (Fire Weather Index, 20). This resulted in a very intense ground and canopy fire (estimated: rate of spread, 27 m/min; total fuel consumption, 1.85 kg/m^2; intensity, 30,000 kW/m of fire front). The fire spread about 5 km in a northerly direction, burning 380 ha. Immediately after the fire, on May 25, there was no living vegetation of tree, shrub or herb left above ground level. The burned site was a black, open landscape with a few charred trees, fallen logs, stumps and organic matter. By July 2, 1981 (shortly after the end of the trapping period) a herb layer had developed into sparse patches of 20-30 per cent ground cover, including such species as fireweed (*Epilobium angustifolium*), pin cherry and *Aster* spp.

Data Collection

Immediately after the fire, on May 25, 1981, the burn was searched for a total of nine hours, noting birds heard or seen, other signs of animal life, and animal remains.

Subsequently, trap lines in burned forest (lines B1 and B2) and nearby unburned forest (lines U1 and U2) were run from June 11 to 17 (B1 and U1) or June 22 to 29 (B2 and U2). The weather was comparable in all census periods. B1 and B2 were approximately 500 and 800 m from unburned forest, respectively. U1 and U2 were approximately 500 m and 2 km from the burn, respectively (Figure 1).

Each line consisted of ten (U1), 15 (U2 and B2) or 20 (B1) stations 10 to 20 m apart. Trapping methods were as follows: three mouse snap traps 1 m apart, baited with peanut butter and oatmeal at each station; one rat snap trap at stations 5, 10 and 15 of lines B2 and U2; one pitfall trap 15 cm diameter by 18 cm deep containing a preserving 50:50 mixture of ethyl alcohol:ethylene glycol, at the first ten stations of each line. Small mammals were captured by snap and pitfall traps; invertebrates by pitfall traps. (There were always many more set traps than mammals captured.) Birds heard or seen while walking the trap lines were noted, except for obvious transients flying high overhead.

In addition snowshoe hares were censused from a vehicle along five contiguous sections of road (Figure 1). Four sections (Ua, Ub, Uc, Ud) passed through unburned forest, the other (Ba) passed through burned forest. Road length and number of trips were: Ua 2.7 km, 34 trips; Ub 2.2 km, 34 trips; Uc 1.9 km, 30 trips, Ud 4.0 km, 12 trips; Ba 1.9 km, 28 trips. Numbers of hares seen on each section of road were recorded.

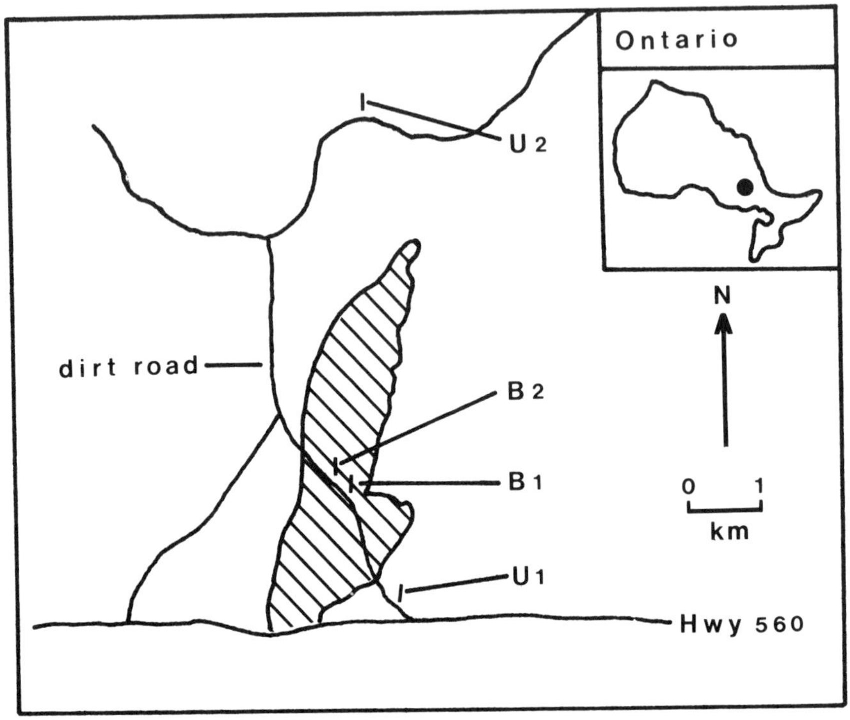

Figure 1. Location of May 23, 1981 burn (47°30′N, 81°45′W) showing trap lines in burned (hatched) and unburned areas.

With some of the above techniques there is a possibility of differences in trapping or observation efficiency in burned and unburned forest. Pitfall captures depend on an animal's absolute rate of horizontal movement (ground activity) as well as on its abundance. Reduced cover in the burn may possibly have increased ground activity and so increased captures. Similarly, observation of birds may have been easier in the burn because of reduced cover. In these cases the most reliable results were those showing greater numbers in unburned forest, despite the possible sampling bias in favour of the burn.

Results were broadly similar between burn lines and between unburned lines, and so were pooled to simplify data analysis.

Results

Initial Search of the Burn

The following species of bird were seen alive or heard in the burn on May 25, 1981: common nighthawk, white-throated sparrow, hermit thrush, American robin, belted kingfisher, northern raven, whip-poor-will, ovenbird, and either downy or hairy woodpecker. Signs of insect activity included 17 freshly-excavated ant hills, and boring dust produced by wood-boring beetles on one tree and one log. Remains of the following animals were found: three red squirrels, one snowshoe hare, one red-backed vole, bones of one unknown small mammal, a nest and three eggs of spruce grouse, an egg of ruffed grouse, 19 land snails (*Polygyra albolabris*), 100 to 200 small landsnails/m^2 in a wetland area.

Invertebrates

The three most abundantly captured groups of invertebrates were ants (Formicidae), beetles (Coleoptera) and spiders (Araneae) (Table 1). Ants were captured in appreciably higher numbers in burned than in unburned forest, while such differences were slight for beetles and spiders. Of the less common groups, the presence of caterpillars (larval Lepidoptera), snails (Gastropoda) and plant-bugs (Hemiptera) exhibited the greatest differences between burned and unburned forest, but numbers involved were very low.

Table 2 shows the beetle captures in more detail, with families and species ranked according to abundance. The tiger beetle (*Cicindela longilabris*), the ground beetle (*Agonum obsoletum*), and the weevils (*Hylobius* spp.) were captured in considerably higher numbers in burned than in unburned forest. The ground beetles (*Calathus ingratus, Agonum retractum* and *Agonum decentis*), and the rove beetles (*Staphylinidae*) were captured in considerably lower numbers in burned than in unburned forest.

The two closely related ground beetles (*Pterostichus pensylvanicus* and *Pterostichus adstrictus*) were unfortunately pooled during the early stages of identification for this study. Richardson and Holliday (1982) found more *P. pensylvanicus* on unburned sites and more *P. adstictus* on burned sites.

Table 1. Invertebrates captured by pitfall traps in burned (B) and unburned (U) forest during June, 1981. (260 pitfall days/forest)

	B	U	B-U	(B-U)%*
Formicidae	413	202	+211	51
Coleoptera	289	262	+27	9
Araneae	42	52	−10	19
Diptera	33	38	−5	13
Orthoptera	6	6	0	0
Diplopoda	8	4	+4	50
Lepidoptera (larvae)	0	5	−5	100
Gastropoda	0	5	−5	100
Hemiptera	0	3	−3	100
Total	791	577		

* (B-U)% = (B-U) as a percentage of the larger, B or U.

Table 2. Beetles captured by pitfall traps in burned (B) and unburned (U) forest during June, 1981. (260 pitfall days/forest)

	B	U	B-U	(B-U)%*
Carabidae				
Pterostichus pensylvanicus	125	90	+35	28
Pterostichus adstrictus				
Calathus ingratus	7	57	−50	88
Agonum retractum	1	29	−28	96
Agonum decentis	2	18	−16	89
Myas cyanescens	9	5	+4	44
Agonum obsoletum	13	0	+13	100
Pterostichus coracinus	3	4	−1	25
Scaphinotus bilobus	1	4	−3	75
Pterostichus punctatissimus	3	0	+3	100
Harpalus sp	2	1	+1	50
Carabus serratus	2	0	+2	100
Pterostichus lucublandus	2	0	+2	100
Cymindis cribricollis	2	0	+2	100
Sphaeroderus nitidicollis	0	1	−1	100
Curculionidae				
Hylobius spp	69	17	+52	75
Otiorhynchinae	1	2	−1	50
Staphylinidae	1	33	−32	97
Cicindelidae				
Cicindela longilabris	32	0	+32	100
Cicindela limbalis	1	0	+1	100
Elateridae	5	1	+4	80
Byrrhidae	4	0	+4	100
Tenebrionidae	2	0	+2	100
Scarabaeidae	1	0	+1	100
Silphidae	1	0	+1	100
Total	289	262		

*(B-U)% = (B-U) as a percentage of the larger, B or U

Small Mammals

Most small mammals feed on a variety of food items during the year. The crude categorization in Table 3 relates to main feeding habits during spring and summer, and is based principally on information in Banfield (1974). Species names are as in Banfield (1974).

Five species were captured in the burn, three granivores and two herbivores. Nine species were captured in unburned forest, three granivores, four herbivores and two insectivores. Gapper's red-backed vole (the most abundant herbivore) and the masked shrew (the most abundant insectivore) were both captured in appreciably lower numbers in burned than in unburned forest, despite greater trapping effort in the burn. The deer mouse (the most abundant granivore) was captured in high numbers in both burned and unburned forest.

Overall a higher proportion of males was captured in burned forest, while a higher proportion of females was captured in unburned forest. This difference is significant for the deer mouse, the most abundant species ($X^2 = 6.157$, $0.05 > p > 0.01$).

Road Census of Snowshoe Hares

On the basis of numbers seen on the road snowshoe hares appeared less abundant in burned than in unburned forest. Hares per trip per kilometre for the four unburned road sections were Ua 0.404, Ub 0.641, Uc 0.721 and Ud 0.668. These figures have a mean of 0.60 and 95 per cent confidence limits of 0.32 and 0.88. The value of 0.226 for the burned road section (Ba) falls well below this range.

Table 3. Small mammals captured by mouse or rat snap traps or pitfalls* in burned (B) and unburned (U) forest during June, 1981, grouped according to main food habit in spring and summer (see text)

	Burned			Unburned			B-U	(B-U)%†
Granivores	♂	♀	Σ‡	♂	♀	Σ		
deer mouse	12	1	13	5	4	10	+3	23
least chipmunk	0	4	4	0	1	1	+3	75
eastern chipmunk	0	0	0	0	4	4	−4	100
meadow jumping mouse	0	1	1	0	0	0	+1	100
Herbivores								
Gapper's red-backed vole	1	1	2	2	8	10	−8	80
woodland jumping mouse	0	1	1	0	3	3	−2	67
southern bog lemming	0	0	0	1	0	1	−1	100
northern flying squirrel	0	0	0	0	1	1	−1	100
Insectivores								
masked shrew	0	0	0	0	3	8	−8	100
pygmy shrew	0	0	0	0	1	1	−1	100
Total	13	8	21	8	25	39		

* 610 (mouse and rat snap) trap nights in burned and 417 (mouse and rat snap) trap nights in unburned forest, plus 260 pitfall days in both forests.

† (B-U)% = (B-U) as a percentage of the larger, B or U.

‡ Σ = total including unsexed specimens.

Table 4. Birds seen or heard in burned (B) and unburned (U) forest during June, 1981, grouped according to main food habit in spring and summer (see text)

	B	U	B-U	(B-U)%*
Granivores				
white-throated sparrow	19	13	+6	32
purple finch	4	4	0	0
evening grosbeak	6	0	+6	100
slate-coloured junco	3	0	+3	100
song sparrow	0	1	−1	100
Aerial insectivores				
common nighthawk	49	7	+42	86
whip-poor-will	7	2	+5	71
tree swallow	2	0	+2	100
eastern kingbird†	2	0	+2	100
olive-sided flycatcher†	2	0	+2	100
eastern phoebe†	0	1	−1	100
Ground insectivores				
hermit thrush	19	9	+10	53
Swainson's thrush	16	6	+10	62
American robin	19	0	+19	100
eastern bluebird	15	0	+15	100
common flicker	13	0	+13	100
ovenbird	0	9	−9	100
ruffed grouse‡	1	7	−6	86
chipping sparrow	6	1	+5	83
veery	0	5	−5	100
winter wren	1	1	0	0
wood thrush	0	1	−1	100
Bole insectivores				
hairy woodpecker	7	0	+7	100
downy woodpecker	3	0	+3	100
Foliage insectivores				
red-eyed vireo	4	8	−4	50
yellow-rumped warbler	4	5	−1	20
chestnut-sided warbler	1	7	−6	86
common yellowthroat	0	8	−8	100
Magnolia warbler	0	6	−6	100
golden-crowned kinglet	2	3	−1	33
black and white warbler	0	3	−3	100
black-throated blue warbler	0	3	−3	100
Nashville warbler	0	2	−2	100
ruby-crowned kinglet	1	1	0	0
Canada warbler	0	1	−1	100
mourning warbler	0	1	−1	100
scarlet tanager	0	1	−1	100
house wren	1	0	+1	100

* (B-U)% = (B-U) as a percentage of the larger, B or U.

† These species make short foraging flights from perches in the canopy or shrub layers.

‡ This species is an herbivore and insectivore.

Birds

In total 38 species of bird were heard or seen, 25 in burned forest and 27 in unburned forest (Table 4). Like small mammals, most species of bird feed on a variety of food items during the year. Categories in Table 4 relate to main feeding habit in spring and summer, are based primarily on information in Godfrey (1979), and resemble those of Blake (1982). Species names are as in the current Checklist of the American Ornithologists Union.

For granivores (B-U) is positive for three species, negative for one species and zero for one species. For insectivores corresponding values are: positive, 14 species; negative, 17 species; and zero, two species. Corresponding values for the sub-categories of insectivores are: (a) aerial insectivores, positive, five species; negative, one species; (b) ground insectivores, positive, six species; negative, four species, and zero, one species; (c) tree bole insectivores, positive, two species; (d) foliage insectivores, positive, one species; negative 12 species, and zero, one species. Bearing in mind the possible sampling bias in favour of the burn, these results at least suggest that the groups of birds most affected by the fire were the foliage insectivores (i.e., those gleaning insects from foliage of shrubs and canopy).

Discussion

The most obvious habitat change caused by the fire was destruction of above-ground vegetation and blackening of the landscape. This would affect many factors. In particular both food for herbivores, and cover would be drastically reduced. Results of the present study provide some evidence for an immediate reduction in herbivorous mammals (Gapper's red-backed vole, snowshoe hare) and herbivorous invertebrates (caterpillars, plant-bugs and snails). A similar abrupt decline in red-backed vole numbers has been reported after a forest fire (Krefting and Ahlgren 1974) and after clear-cutting and burning (Ahlgren 1966; Gashwiler 1970; Sims and Buckner 1973). This may be a direct response to food shortage, or may reflect this species' requirement for humid environments (Banfield 1974; Miller and Getz 1977). Snowshoe hares have also previously been found to abandon severely burned areas (Keith and Surrendi 1971). Burned areas are recolonized by both these species a few years after a fire, as vegetational regrowth develops (Gashwiler 1970; Krefting and Ahlgren 1974).

In addition to vegetation, litter and other debris on the soil surface were also destroyed by the fire. This may account for the probable lower abundance of the scavenging, litter-dwelling rove beetles and the ground beetles (*Calathus ingratus, Agonum retractum* and *A. decentis*) in burned as compared with an unburned forest. Richardson and Holliday (1982), 15 months after fire, caught fewer *C. ingratus* in burned than in unburned spruce sites, but caught similar numbers in burned and unburned aspen sites.

Fire is not solely a destructive agent. Freshly-burned habitat may provide adequate or optimal conditions for some animals. In general granivores did not appear as badly affected by the fire as did herbivores, suggesting a fairly good supply of seeds was available in the burn. In particular the deer mouse and white-throated sparrow appeared common in both burned and unburned forest. High abundances of deer mice in burned areas the year following fire have been reported previously, and have usually been attributed to the availability of seed (Ahlgren 1966; Gashwiler 1970; Sims and Buckner 1973; Krefting and Ahlgren 1974). Krefting and Ahlgren (1974) found it remained the most abundant small mammal on a burn for seven years following a fire. The deer mouse seems very responsive to short-term changes in food supply. Taitt (1981) found augmentation of food supply rapidly increased immigration and population levels and decreased territory size. This may account for the ability of the species to rapidly exploit freshly-burned habitat. On the other hand the low proportion of females captured in the burn in the present study suggests fire may somehow alter the sex ratio and possibly disrupt reproduction, at least temporarily. However, Sullivan (1980) found good survival and reproduction on a clear-cut four years after burning.

Freshly-burned habitat may also provide a good supply of insects for insectivores. This would in part be due to increased exposure resulting from destruction of vegetation and flushing effects, but may also involve attraction of some species of insect to the area. Attraction of wood-boring insects to forest fire is well known (Evans 1971). We saw evidence of wood-borer activity immediately after the fire. Additionally, the ground beetle (*Agonum obsoletum*) and the tiger beetle (*Cicindela longilabris*) were captured in considerably greater numbers in burned than in unburned forest. *A. obsoletum* belongs to a group of four ground beetles known to be attracted to burning wood, although the biological function of this behaviour is unknown (Lindroth 1969). Tiger beetles are very mobile predators, with a preference for open areas and a tendency to oviposit on sandy soil. It is likely that *C. longilabris* was also attracted to the burn. Ants and spiders were captured in as high or higher numbers in burned than in unburned forest, and evidence of ant activity was seen in the burn immediately after the fire. Total invertebrate captures were slightly higher in burned than in unburned forest. Although some species of insect appeared displaced by the fire, the freshly-burned habitat contained much insect life.

Broadly speaking, most aerial, ground and tree bole insectivorous birds appeared as or more abundantly in burned than in unburned forest, and this may in part be due to food availability. In contrast, foliage insectivores (principally wood-warblers — Parulidae) appeared displaced by the burn. This may be a result of presumed destruction of herbivorous insects by the fire. Blake (1982) found fewer foliage insectivores on burned than unburned sites one and a half to two years after fire. We would expect in later years as vegetation regrowth developed, that the herbivorous insect and so foliage insectivore populations would recover.

In contrast to the general trend for ground insectivore birds the insectivorous masked shrew was captured in high numbers in unburned forest but was not captured at all in the burn. This may be a result of the high moisture requirements of this species (Getz 1961; Banfield 1974).

In summary, most of the more marked faunal differences between burned and unburned forest in the present study may be related to fire-induced changes in food availability, cover and moisture. Some species seemed well able to exploit freshly-burned habitat, and this may in part reflect evolutionary adaptation to life in fire-prone areas.

This study was carried out for only a few weeks following fire. Observed faunal responses would involve changes in rates of immigration, emigration and death. Changes in rates of birth would probably not have had observable effects so shortly after fire. The results provide a rather blurred snapshot of a community in flux. Subsequent vegetational succession would probably be accompanied by further faunal changes (Erskine 1977; Fox 1982, 1983).

Acknowledgements

We thank Brian Naylor and Kandyd Szuba who helped search the burn and count hares, and Brian Stocks of the Canadian Forestry Service who estimated the rate of spread, total fuel consumption and intensity figures quoted in this paper. We also thank R.W. Wein and anonymous referees whose comments were instrumental in improving the manuscript. This study was supported by the Ontario Ministry of Natural Resources, Departments of Forestry and Wildlife; and the Natural Sciences and Engineering Research Council, Ottawa.

Literature Cited

Ahlgren, C.E. 1966. Small mammals and reforestation following prescribed burning. J. For. 64: 614-618.

Ahlgren, I.F. 1974. The effect of fire on soil organisms. *In:* Fire and Ecosystems. T.T. Kozlowski and C.E. Ahlgren (Eds.). Academic Press, New York. pp. 47-72.

Banfield, A.W.F. 1974. The mammals of Canada. University of Toronto Press, Toronto, Ontario.

Bendell, J.F. 1974. Effects of fire on birds and mammals. *In:* Fire and Ecosystems. T.T. Kozlowski and C.E. Ahlgren (Eds.). Academic Press, New York. pp. 73-138.

Blake, J.G. 1982. Influence of fire and logging on nonbreeding bird communities of ponderosa pine forests. J. Wildl. Manage. 46: 404-415.

Erskine, A.J. 1977. Birds in boreal Canada: communities, densities and adaptations. C.W.S. Rep. Ser. No. 41.

Evans, W.G. 1971. The attraction of insects to forest fires. Tall. Timbers Conf. on Ecol. Anim. Contr. by Habitat Manage. 3: 115-127.

Fox, B.J. 1982. Fire and mammalian secondary succession in an Australian coastal heath. Ecology 63: 1332-1341.

Fox, J.F. 1983. Post-fire succession of small-mammal and bird communities. *In:* The role of fire in northern circumpolar ecosystems. Ross W. Wein and David A. MacLean (Eds.). SCOPE 18. John Wiley and Sons, Chichester, England. pp. 155-180.

Gashwiler, J.S. 1970. Plant and mammal changes on a clearcut in west-central Oregon. Ecology 51: 1018-1026.

Getz, L.L. 1961. Factors influencing the local distribution of shrews. Amer. Midl. Nat. 65: 67-85.

Gillon, D. 1971. The effects of bush fire on the principal pentatomid bugs (Hemiptera) of an Ivory Coast Savanna. Tall Timbers Fire Ecol. Conf. 11: 377-417.

Gillon, Y. 1971. The effect of bush fire on the principal Acridid species of an Ivory Coast Savannah. Tall Timbers Fire Ecol. Conf. 11: 419-471.

Godfrey, W.E. 1979. The birds of Canada. University of Toronto Press, Toronto, Ontario.

Keith, L.B. and P.C. Surrendi. 1971. Effects of fire on a snowshoe hare population. J. Wildl. Manage. 35: 16-26.

Krefting, L.W. and C.E. Ahlgren. 1974. Small mammals and vegetation changes after fire in a mixed conifer-hardwood forest. Ecology 55: 1391-1398.

Lindroth, C.H. 1969. The ground beetles (Carabidae, excl. Cicindelinae) of Canada and Alaska. Opus Entomologica Supplementum, 14, 20, 29, 33, 34, 35.

Miller, D.H. and L.L. Getz. 1977. Factors influencing local distribution and species diversity of small mammals in New England. Can. J. Zool. 55: 806-814.

Richardson, R.J. and N.J. Holliday. 1982. Occurrence of carabid beetles (Coleoptera: Carabidae) in a boreal forest damaged by fire. Can. Entomol. 114: 509-514.

Schaefer, M. 1980. Sukzession von Arthropoden in verbrannten Kiefernforsten II. Spinnen (Araneida) und Weberknechte (Opilionida). Forstwiss. Centralbl. 99: 341-356.

Schauermann, J. 1980. Sukzession von Arthropoden in verbrannten Kiefernforsten IV. Moderkafer (Lathridiidae). Forstwiss. Centralbl. 99: 366-371.

Scotter, G.W. 1972. Fire as an ecological factor in boreal forest ecosystems of Canada. *In:* Fire in the Northern Environment. C.W. Slaughter, R.J. Barney and G.M. Hansen (Eds.). Pacific Northwest For. and Range Exp. Stn. Portland, Oregon. pp. 15-24.

Sims, P.H. and C.H. Buckner. 1973. The effect of clear cutting and burning of *Pinus banksiana* forests on the populations of small mammals in southeastern Manitoba. Amer. Midl. Nat. 90: 228-231.

Sullivan, T.P. 1980. Comparative demography of *Peromyscus maniculatus* and *Microtus oregoni* populations after logging and burning of coastal forest habitats. Can. J. Zool. 58: 2252-2259.

Taitt, M.J. 1981. The effect of extra food on rodent populations, I. deer mice (*Peromyscus maniculatus*). J. Anim. Ecol. 50: 111-124.

Winter, K. 1980. Sukzession von Arthropoden in verbrannten Kiefernforsten III. Laufkafer (Carabidae). Forstwiss. Centralbl. 99: 356-365.

Winter, K., J. Schauermann and M. Schaeffer. 1980. Sukzession von Arthropoden in verbrannten Kiefernforsten I. Methoden und Uberblick. Forstwiss. Centralbl. 99: 324-340.

FOOD POTENTIALS FROM AGRICULTURE, FISHERIES AND WILDLIFE IN THE CANADIAN BOREAL FOREST

Roderick R. Riewe, Margaret Boyle, Franklin Bustillos and Brian Reader
University of Manitoba

Abstract: There are arable lands in the boreal forest but economic constraints eliminate their usefulness. The greatest agricultural potential in the north is in the development of intensive gardens, greenhouses and the cultivation of edible native plants. The commercial fishery in the boreal forest is a highly subsidized industry which now exports its entire harvest south. If properly managed, the fisheries could provide a large share of the northern protein requirements. Acid rain and mercury pollution, however, could drastically reduce the biome's aquatic productivity. The production of wild meats could be greatly increased through habitat manipulation and game ranching if the hurdles of governmental controls were removed. Massive marketing campaigns for low value fish, fur bearer meat and wild ungulate meat could easily change the economic picture of the renewable resource industries in the north.

One of the major impacts of northern development has been an ever increasing dependence of northerners upon imported foodstuffs. This dependence led to rapid diet changes which have been deleterious to the health of the native peoples (Brown 1949; Rodahl 1950; Romer *et al*. 1981; Schaefer and Steckle 1980; Williamson 1971).

Now that the time of rapid resource development based on cheap energy supplies is gone, food prices in the northern stores have become prohibitive, primarily because of increased transportation costs. Hence a tremendous economic burden has been placed on the northern consumers and governments alike. Recently, there has been an increasing awareness that something must be done to decrease the northern dependence upon southern goods.

This paper explores the potential productivity of the boreal forest of Canada in terms of agriculture, fisheries and wildlife.

Agricultural Potential for Traditional Crops and Animals

The boreal forest in Canada extends over approximately 2,589,000 km^2. In a continuous belt south of the forest-tundra transition zone, the boreal forest extends from Newfoundland, west to the Rocky Mountains and northwestward to the Alaskan border. It is the most widespread forest region in Canada (Rowe 1972). Brown (1970) has reviewed the status of traditional European type agriculture in this region.

The agricultural capability of Canadian soils, including parts of the boreal forest, have been classified in the Canada Land Inventory (C.I.L.). In the C.I.L. system, seven classes and 13 sub-classes of mineral soils have been delimited. Unfortunately, the organic soils which compose a large percentage of the boreal forest soils have not been included in the

C.I.L. system to date. Organic soils offer limited potential for commercial agricultural production because they are often related to short frost-free periods, poor drainage, high fertilizer requirements and the presence of permafrost (Environment Canada 1972). Although these limitations occur, organic soils are often cultivated in order to enlarge fields in areas of Gray Luvisols. There are a few areas such as Carrot River, Saskatchewan and Edson, Alberta where entire farms are situated on organic soils (Hoyt *et al*. 1978).

Approximately two million hectares (22 per cent of the land area in Canada) has been surveyed by the C.I.L. Of this area, two per cent is classed as suitable for prime agricultural practices (classes 1 and 2), 12 per cent is suitable for field crops (classes 1 to 4), 22 per cent is capable of supporting some kind of agricultural activity (classes 1 to 6), and the remaining land has no capability for arable culture or permanent pasture (Shields and Nowland 1975).

There is a significant amount of land under cultivation in the region of the boreal forest, but it is only a small fraction of the total arable land in Canada. The greatest agricultural development in the boreal forest has taken place in the interior plains of the three prairie provinces and in northeastern British Columbia. This development has spread from earlier agricultural settlements in the parkland regions of the prairies. The total cultivated arable soils in the boreal forest region is estimated at 2.8 million hectares (Hoyt *et al*. 1978).

The area of potentially arable soils in this region is approximately 15.4 million hectares. These potentially arable lands are distributed as follows: 8.3 million hectares in the interior plains of Manitoba, Saskatchewan, Alberta and northeastern British Columbia, 5.3 million hectares in the clay belt of Ontario and Quebec, 1.3 million hectares in the Northwest Territories, 0.4 million hectares in the interior of British Columbia, and 0.06 million hectares in the Yukon. Undoubtedly, additional lands that have agricultural potential exist in small isolated pockets throughout the boreal forest especially in the form of deposition soils around lakes, river levees, and deltas (Environment Canada 1972; Hoyt *et al*. 1978; Mills and Smith 1981; Pettapiece and Lindsay 1981; Shields and Nowland 1975).

It has long been a popular belief that agriculture in the north is impossible. This idea is shared by southern communities, northern natives and white settlers alike. In the past, most of the native peoples of the boreal forest were hunters and gatherers not farmers. Settlements of whites in the boreal forest (with the exception of the Peace River District) were based on the fur trade and on timber and mineral resources. People in southern communities of Canada, United States and foreign countries mainly view the Canadian boreal forest in terms of recreation. These stigmatized notions which lasted through the years of Canadian development have limited agricultural inroads in the north.

Agriculture Canada carried out experimental crop production in the northern clay belt at Wabowden, Manitoba from 1955 to 1965. From 1956 to 1970, the University of Manitoba conducted field crop demonstrations in the Pasquia land settlement near La Pas to obtain information on the productivity of northern river delta soils. These northern agricultural studies demonstrated that under optimum management and average climatic conditions good to very high yields are possible for commercial crops such as spring wheat, oats, barley, rapeseed, potatoes, legumes and grasses. However, other commercial crops are limited by physical factors including the thermal regimes, soil types and structure, and land forms (Mills and Smith 1981). Although there are arable lands in the boreal forest there are enormous impediments to their commercial development.

Ideally, agriculture should be a net producer of energy but the North American style of extensive agriculture is often a net consumer of energy (Steinhart and Steinhart 1974). In individual farming operations, fuel, fertilizer, machinery production and maintenance make up the greatest amount of energy inputs. Tremendous energy is also consumed by the processing, transportation, distribution and preparation of agricultural products. These factors have important implications in the consideration of agricultural use of boreal soils. To achieve optimum crop production on boreal soils, great amounts of fertilizer, particularly nitrogen are required.

Potentially arable boreal soils also require large energy inputs for land clearing. If further development of farming occurs in the north, it will remain a considerable distance from large markets. The energy used for transportation of goods and services to the farm and produce from the farm to market becomes a very significant factor in energy budget considerations (Hoyt *et al.* 1978).

Although some traditional crops can be grown in the north, competition with the same crops produced more cheaply in southern agricultural areas raises the question of production viability. The small northern centres also are not large enough to support themselves in terms of farm services.

The severe limitation to agriculture in the boreal regions necessitates the implementation of high levels of farm management skills by farmers in order to run viable operations. The resources to acquire these skills are not readily available to the northerner. Unless the northern population centres reach such levels as to sustain reliable markets, southern style extensive farming practices will probably not continue to develop in the boreal forest region.

Cultivation of Northern Boreal Plants

From an ecological perspective, it is far more beneficial if future northern agricultural programs concentrate on innovative small scale intensive programs rather than on developing extensive southern style agricultural practices in the north. These programs could be established and funded initially through various federal and provincial agencies. Agricultural workshops could be held in various northern communities in order to transmit new information and skills to the northerners.

The greatest agricultural potential in the north today probably rests in the establishment of local intensive agricultural production programs, including home garden plots, solar greenhouses and the intensive propagation of native northern edible plants (Romer *et al.* 1981). It has been demonstrated that native species of edible plants transplanted and grown under simple ameliorative techniques increase their yields by four to twenty-five fold, over plants grown under natural conditions. However, before we proceed to produce native plants on a commercial scale it would be wise to reflect on the history of the wild rice industry in North America.

Historically, wild rice was an important staple crop of the Indian peoples inhabiting the boreal forests between the Great Lakes and the Prairies. Wild rice also became an important means of subsistence to the fur traders and settlers in this region. During this century, wild rice has become an expensive delicacy which in 1979 retailed for as much as $35/kg (Borys 1980). In 1979, the total production of wild rice harvested in North America was about 4 to 4.5 million kg of which Minnesota produced over 2.7 million kg and Ontario and Manitoba each produced about one million kg.

Until the 1960s, most of the North American wild rice harvest was obtained from natural stands. In Canada, 85 per cent of the production still comes from natural stands, the remaining 15 per cent is harvested from a few commercially developed rice paddies in southeastern Manitoba. In the United States the situation is just the reverse. There the crop is almost entirely produced from rice paddies (Rader 1970).

Benefit-cost analyses have shown that natural stands have higher benefit-cost ratios than paddy development (Lee 1976). This is due to the lower capital cost of natural stands which show positive returns under all assumptions. The capital costs for the same unit of exploitation in paddy development are about five times higher than for natural stands and the annual operating costs are two times greater. In Manitoba, it was demonstrated that in years of normal production, mechanically harvested wild rice from natural stands will not only be competitive but will be more profitable than paddy rice.

Traditionally, the Indians harvested wild rice by flailing the ripe grains off the emergent plants into their canoes. This technique harvested less than 10 per cent of the natural stand's productivity (Lithman 1973). Recently, numerous forms of innovative mechanical harvesters have been developed which are capable of harvesting from 10 to 25 per cent of a stand's production. Ecologically, however, the traditional harvest method is superior to these mechanical harvesters because it causes less damage to the rice plants.

As the wild rice industry shifts from traditional methods of harvest to mechanical harvesters to cultivated paddies, the Indian people lose control over the economy of the crop. Today, unfortunately, wild rice is no longer an important dietary item for the Indians because they cannot afford it. Wild rice is now a valuable commodity on the world market. About 70 per cent of the Canadian rice crop is exported to the United States. According to Lithman (1973) the Indians' diminishing control over wild rice is due to the Euro-Americans assuming ever increasing control through financial institutions (The Hudson's Bay Company in earlier days) and the major food industries. The Indian loss of control in the United States is partly due to their difficulty in financing paddy developments. It has been argued that the changes in the wild rice industry in effect contribute to further underdevelopment of the Indians.

Fisheries — Potential from Boreal Forest Waters

The annual Canadian inland fisheries production is about 40,000 tonnes of fish with a landed value of $24 million (Cauvin 1979). This fresh water harvest is about four per cent of the entire Canadian fisheries production (Fisheries and Environment Canada 1977). The commercial fish production of the boreal forest lakes is about half the entire Canadian freshwater harvest or about 20,000 tonnes.

In order to obtain a figure for the total fish harvested from boreal forest lakes we must add the harvest of the subsistence fisheries and the sport fisheries. Unfortunately, figures are not available for the entire biome, but two examples may help to illustrate their magnitude. The Cree subsistence fishery in the boreal forest lakes around James Bay amounts to 240 tonnes annually. In 1974-1975 this fish harvest represented about one-quarter of the total potential fresh food by weight and about half of the minimum adequate protein requirement for the Cree population (Berkes and Farkas 1978). In the planning region of Northwestern Ontario the sport fish harvest was recorded as 4,230 tonnes in 1970 while the commercial harvest is currently only 2,950 tonnes (Ontario Ministry of Natural Resources 1980).

In the boreal forest region as a whole, we would venture to say that the subsistence and sport fisheries harvest is as much or more than the commercial fisheries. The performance of the commercial fishermen is as variable as the geography of the boreal forest biome. Generally, however, it may be classified as poor. In 1977-1978, the mean commercial harvest for all enterprises operating in the jurisdiction of the Freshwater Fish Marketing Corporation was 2500 kg with a mean gross revenue of $2,063 (Thompson 1981).

This meager harvest is indicative of the low productivity of the lakes in this biome. There are numerous techniques for estimating productivity and fish yields. Regier (1978) has listed at least ten such approaches many of which require detailed biological data that unfortunately, are not available for most of Canada's north. Various authors have used these techniques to estimate the productivity of the northern lakes. Their values generally range from 0.4 to 2.5 kg/ha/year (Berkes 1981; Feit 1978; Fuller and Hubert 1981; Hartman 1981; Power and Le Jeune 1976).

The Morphoedaphic Index is one of the simpler techniques which requires information only on mean lake depth and total dissolved solids. This index has been tested on lakes in southern Canada and has proven reliable (Ryder *et al.* 1974). Berkes (1981), however, has shown that this index overestimates the productivity of northern lakes. He believes it may be corrected by adjusting the index to take into account possible temperature effects and the fact that most northern fisheries are based on what may be ecologically considered "climax" communities, consisting of slow growing, large individuals, and a low productivity:biomass ratio. By comparison to some tropical African lakes which produce up to 300 kg/ha/year our boreal forest lakes are truly poor producers.

Cauvin (personal communication, 1982) believes the present rate of harvest for the boreal forest biome could at least be doubled or perhaps quadrupled but that the fisheries operators on the northern lakes are economically marginal at best. Thompson (1981) paints the following gloomy picture of the freshwater fisheries

> ". . . a large proportion . . . of these fisheries have little or no potential for economic viability either in the short-run or the long-run. Their continued existence is entirely dependent on the myriad of public subsidies, the withdrawal of which would see the cessation of fishing activities. Conversely, the continuance of fishing will require perpetual public support. The subsidies cannot be justified on economic development grounds because the fisheries have no potential to be self-sufficient in the foreseeable future. Moreover, given present demand and supply trends, it appears that some of these fisheries may be actually becoming less self-sufficient. The continuance of subsidies to support the fishing industry, then must be justified on social grounds.".

While the biological potential of the boreal forest lakes may be underutilized, harvesting close to the theoretical yields is nearly impossible since the fishing effort is not distributed evenly throughout the north but rather tends to be concentrated in space and time. Harvesting the sparse productivity of these lakes from permanent settlements often creates over-fishing which results in changes in species composition and huge fossil fuel energy subsidies. For the Fort George Cree, Berkes (1981) estimated that the ratio of energy captured in fish to the energy expended to make the catch is 1:14 to 22.

In 1980-1981, the Manitoba government paid over a quarter of a million dollars to northern fishermen for freight assistance (Manitoba Department of Natural Resources 1981). In some remote parts of the province the government paid transportation subsidies of as much as \$2.20/kg for fish worth only \$1.10/kg on the market (Cauvin, personal communication, 1982). Since fisheries subsidies are merely forms of welfare, it may be desirable to increase social benefits while developing local markets in the northern communities. Greater efforts should also be made to market the low market value, or so-called ''trash'' fish such as sucker, carp and bowfin. A marketing campaign might increase the value of the freshwater fisheries and thereby increase the potential economic viability of the northern fisheries.

The major ills of the fisheries in the future may not be merely economic, but rather biological in nature. If acid rain continues to accumulate in the northern lakes, we may soon lose the few subsidized fisheries we possess. Mercury contamination is another serious problem affecting the future of northern fisheries. Mercury from industrial wastes is being deposited in many northern drainage systems. The rate at which mercury enters the aquatic food chains partly depends upon the water's acidity. The Canadian Shield which underlies much of the boreal forest is geologically susceptible to acid rain and hence to increased mercury accumulation. The flooding of reservoirs is also thought to cause an increase in mercury levels, presumably due to leaching (Bodaly and Hecky 1979; Bruce and Spencer 1979). It has also been hypothesized that high ground levels of mercury may be a general phenomenon in the north because of little topsoil, the long periods of snow cover and the mechanisms of trace metal migration from bedrock through snow (Eaton 1979).

The final fisheries problem that we wish to mention is that of allocation of the resource between the commercial, sports and subsistence sectors. These allocation problems are developing wherever there are two or more sectors competing for fish stocks. Northwestern Ontario is a good example of such a place. The present Ontario policy states that sports fishing has priority over commercial fishing. The native subsistence fishery is not specifically dealt with in the Strategic Land Use Plan (Ontario Ministry of Natural Resources 1980). This is creating an adversary situation between the Indians and the Government. Unless improved communications can be established and unless there are accurate harvest records kept by all sectors there cannot be equitable allocations and the fish stocks will surely continue to decline.

Hunting Potential of Boreal Species

Big game hunting provides an important protein contribution to many northerners. Moose occupy the full scope of the Canadian boreal forest, while caribou, elk, deer and bison are more localized.

Moose populations across Canada have been reported to range from 0.15 moose/km^2 to as many as 1.5 moose/km^2 (Cumming 1974; Lynch 1975; Mercer and Manuel 1974). Moose harvests vary from about one moose killed per 50 km^2 to about one moose per 10 km^2. Canadians annually harvest about 14 million kg of moose meat (Ritcey 1974). In Sweden harvests of one moose per 6.4 km^2 have been reported (Lykke 1974). If we Canadians managed our forests and moose as the Scandinavians do, we could probably increase our moose production by two to five fold. However, a decision to do so would be open to considerable social and economic debate.

Most wildlife management is actually people management through the setting of seasons and bag limits and restricting the use of harvesting techniques. The remainder of the techniques available to the wildlife manager primarily revolve around habitat manipulation.

It is a basic principal of wildlife management that biologically diverse habitats support larger wildlife populations than do simplified habitats. Therefore, wildlife utilize the forest edges or ecotones, far more than the simpler forest interior. Thus, the wildlife manager endeavours to improve the habitat by creating more "edge effect", and opening up the large continuous habitat.

Under some conditions the openings created by the pulp or timber operators are also beneficial to certain species of wildlife. Stelfox *et al.* (1973) reported that continuous mature forests on the Albertan foothills supported populations of ~ 0.2 to 0.8 ungulates/km^2; this included mule deer, elk and moose. After clear cutting operations, the cervids increased to a maximum of three per km^2. The logging operations consisted of cutting 2/3 of a 15 km^2 block by removing 116 ha strips and leaving intervening strips of mature forest 92 m wide. The logged strips were scarified to flatten unmerchantible timber and to expose the mineral soil. The cervids utilized the edge of the cut-over strips for feeding but were completely dependent upon the residual mature forest for shelter the year round. Immediately after cutting, the 16 ha strips were of least value to wildlife, but as plant succession advanced, forage and browse became readily available. Nine years after the logging the forage increased by four times and so did the three big game species. These cut over strips, however, provided no winter shelter for the ungulates for at least 20 years. Unfortunately, the remaining one-third of the forest was cut after only 12 years, which left the entire 15 km^2 devoid of big game.

All too often in Canada this is what the wildlife biologist faces. Rather than managing narrowly fluctuating wildlife populations the manager is constantly dealing with boom or bust conditions. He has little control over habitat manipulation since forest resources are manipulated by forest industries and agri-businesses for short term profit.

With today's forest technology and economics, clear cutting practices are all too often too large. Instead of small patches or strips being cut on a rotational basis, which would perpetuate the forest ecosystem, vast areas are stripped in order to secure a quick profit. Such large scale, clear cutting can cause numerous problems including erosion, river siltation, gross hydrological imbalances, destruction of wildlife habitat, poor forest regeneration, ecological instability, destruction of recreation potential and visual and economic poverty.

In order to change this situation the public and the governments must be made aware of the true value of wildlife. As Hartman (1981) pointed out, one of the most profound mistakes we frequently make is to misjudge the value of fish and wildlife. In 1978 he estimated that the annual income from fish and wildlife in the Yukon Territory was close to $15 million. This figure included consumptive uses but omitted the non-consumptive uses. With protection of the land base, these values, or some expanding value, may be realized annually. Perhaps if the true value of the fish and wildlife resources were appreciated, government decisions might reflect such knowledge in the future.

Game Ranching of Boreal Mammals

During the last decade, there has been considerable discussion regarding big game ranching in the Canadian boreal forest as a means of increasing the protein supply (Payne and Muir n.d.; Telfer and Scotter 1975). The protein content by weight of wild ungulate meat compares favourably with standard grade beef, but the fat (energy) content is considerably lower. Prime grade beef provides only half the protein of wild ungulate meat but has more than three times as much energy due to the high fat content.

In the 1970s the Manitoba government developed a moose ranch model and a proposal to establish an experimental moose ranch on a northern Indian reserve (Payne and Muir n.d.). This model suggested that four moose/km^2 would be maintained on good moose range by means of supplemental feed and by nurturing a quality habitat through controlled burning, scarification, fertilization and seeding. For an operation of any economic viability, the ranch would have required a minimum of 50 km^2 of boreal forest, for a herd of 200 animals. This hypothetical herd would have annually produced 13,000 kg of meat from 87 animals or 2.60 kg/ha. The possible advantages of moose ranching in Manitoba were considered to be increased meat and hide production, potential production of a local milk and cheese supply, the potential to develop draft animals, the employment of local people, increased tourism and relatively lower development cost compared with southern style agricultural practices in the north. Some of the disadvantages foreseen in the proposal included a loss of hunting opportunity, need for intensive predator control, the high cost of fencing and the unknown marketability of moose meat. This proposal, unfortunately, was never implemented. The Manitoba government, however, is now involved in licencing private elk and bison ranches.

It appears that elk ranching or elk in combination with moose and bison is a more viable operation than moose alone. At Elk Island National Park, it has been estimated that the land can support a combination of about seven bison, four and a half moose and four and a half elk per km^2 and that an annual harvest of 11.41 kg of ungulates per hectare can be obtained (Telfer and Scotter 1975). The 1973 costs of harvesting the ungulates from Elk Island, moving the carcasses to an abattoir, skinning, dressing, and hanging the meat was about 22¢/kg live weight. This is close to the cost of organized caribou hunting in the Northwest Territories which in 1972 provided meat at an average cost of 20¢/kg.

Even though game ranching is highly feasible, it has not been established in Canada on a large scale (except for the reindeer herd in the western Arctic, (Nasogaluak and Billingsley 1981)) because of legal and administrative problems. The major obstacles are the game laws which prohibit the sale of wild game across Canada except for bison and reindeer which are now considered to be domestic species. Unless regulations can be modified, the health of the animals, abattoir and storage facilities and government inspection and grading of meat may prove prohibitively expensive. There is also no doubt that strong opposition will be heard from cattlemen and perhaps even sportsmen to the establishment of game ranches in the boreal forest.

Trapping Potential of Boreal Species

Trapping, like the fresh water fishery, is an industry which provides many northern Canadians with a seasonal income. In 1980-1981 the industry generated over $119 million through the sale of raw furs (Statistics Canada 1982). Much greater revenues are naturally generated as spin-off benefits from support services and the fur industries. Trapping could produce additional income to the residents of the boreal forest if some of the highly edible furbearers were also utilized for their food value.

We have examined the Canadian fur records in an attempt to determine the average potential harvest of beaver, muskrat, lynx and black bear meat from the boreal forest. We calculated that more than 5×10^6 kg of meat is harvested annually. If the black bears which are killed as nuisance animals or shot by hunters are included in this figure, it would be raised to at least 6×10^6 kg. Although this meat is available in the bush there are several problems involved in getting it to the potential consumer. To begin with the trapper or hunter must find it worth his time to prepare meats as well as furs, and transport them to a market or centralized distribution and inspection plant. His willingness to do so ultimately relates directly to sufficient monetary rewards to warrant his extra labour.

Assuming the trapper is willing to bring these meats to the nearest settlement there still remains the problem of bringing this produce to the consumer. If it is to be sold in major population centers the transportation, storage, and handling costs may be prohibitively high and not competitive with the well established beef, pork and poultry markets. To be competitive with domestic meats it is likely that wild produce would have to be offered for sale either at a lower price to induce the public to try this food, or conversely, at a higher price as a gourmet item.

Initially only limited markets, such as native food restaurants, sportsman's dinners, and grocery stores dealing in exotic foods, would be available. Consumer preference, being solidly entrenched in the domestic meat markets, would require some time and perhaps an imaginative advertising and marketing campaign to be altered sufficiently for people to sample the new produce.

The present and probably the future market for wild meat is in the boreal communities. Development of this industry close to the harvest area would require lower transportation, handling and storage expenditures than a major urban market. A further advantage with this market is that many native people in these areas prefer wild meat to domestic meat which is not always in continuous supply, is more expensive, and often is perceived as providing less nutritional value than wild meat (Bissett 1974).

As with wild ungulate meat the greatest obstacles barring the creation of a fur bearer meat market are government laws and regulations. However, it may be possible to overcome these obstacles. Under section 5 (g) of the federal Meat Inspection Act the Governor in Council may exempt any person or meat product from the operation of all or any of the provisions of the Act providing that the generality of the provisions in the Act are not limited. By virtue of this section concessions enabling the operation of a wild meat industry could be granted.

The development of wild meat markets would be a positive step towards a more efficient use of our wildlife resources and could help to alleviate some of the boreal communities dependency on high priced southern-produced foods. Recently, Yukon Territory officials proposed that the development of a controlled wild meat industry along with a program of wildlife habitat improvement would be a more viable way to increase food production than the improvement of agriculture in the territory, as has been planned (Tempelman-Kluit 1982). With a wild meat industry in operation more revenue would accrue because both the meat and fur would be sold.

In the past, many attempts at increasing furbearer yields through wildlife management have been successful. Manitoba, in 1936, increased muskrat populations 40-fold over a period of four years by habitat manipulation on a 56,000 ha area in the Saskatchewan River delta. Royalties collected from the increase in pelt production more than repaid all expenditures incurred in the development. Saskatchewan, in a program of beaver habitat management, scored similar success, increasing the population from 25 to about 1200 after eight years of careful management (Cowan 1955).

Results such as these indicate that habitat management carried out near small communities throughout Canada's boreal forest could substantially increase fur and food production in this region. Even without intensive management, fur trapping could presently provide meat in sufficient quantities so that wild meat markets, adhering to sanitation and disease control standards, could be developed. Through consultation with federal and provincial public health officials and perhaps through minor amendments to the pertinent acts, meat from trapping could be inspected, approved, and sold commercially.

Acknowledgements

The authors are indebted to the numerous individuals who freely gave of their expertise during the preparation of this paper. Special thanks are due to Dr. Dennis Cauvin of Fisheries and Environment Canada for his opinions on the inland fisheries.

Literature Cited

Berkes, Fikret. 1981. Fisheries of the James Bay area and Northern Quebec: A case study in resource management. *In:* Renewable Resources and the Economy of the North. M.M.R. Freeman (Ed.). Association of Canadian Universities for Northern Studies, Ottawa, Ontario. pp. 143-160.

Berkes, F. and C.S. Farkas. 1978. Eastern James Bay Cree Indians: Changing patterns of wild food use and nutrition. Ecol. of Food and Nutrition 7: 155-172.

Bissett, D. 1974. Resource harvests-hunter-trappers in the Mackenzie Valley. Environmental-Social Committee, Northern Pipelines, Task Force on Northern Oil Development, Report No. 74-42.

Bodaly, R.A. and R.E. Hecky. 1979. Post-impoundment increases in fish mercury levels in the Southern Indian Lake reservoir, Manitoba. Fisheries and Environment Canada, Fisheries and Marine Service. MS Rep. No. 1531.

Borys, A. 1980. Wild rice: Manitoba's most historic crop. Conservation Comment. Jan.-Feb. Manitoba Department of Natural Resources.

Brown, M. 1949. Medical investigation of Southampton Island. Arctic 2: 70-71.

Brown, R.J.E. 1970. Agriculture. *In*: Permafrost in Canada. University of Toronto Press, Toronto, Ontario. Chapt. 8, pp. 180-207.

Bruce, W.J. and K.D. Spencer. 1979. Mercury levels in Labrador fish, 1977-1978. Department of Fisheries and Oceans, Canadian Industry Report of Fisheries and Aquatic Science No. 111.

Cauvin, D.M. 1979. Regulating access in Canada's inland fisheries. J. Fish. Res. Bd. Canada 36: 827-836.

Cowan, I.M. 1955. Wildlife conservation in Canada. J. Wildl. Manage. 19: 168-173.

Cumming, H.G. 1974. Annual yield, sex and age of moose in Ontario as indices to the effects of hunting. Le Naturaliste Canadien. 101: 539-558.

Eaton, R.D.P. 1979. High background mercury levels in northern latitudes. Arctic 32: 376-377.

Environment Canada. 1972. The Canada Land Inventory: Soil capability classification for agriculture. The Canada Land Inventory Report No. 2. Ottawa, Ontario.

Feit, H.A. 1978. Waswanipi realities and adaptations. Resource management and cognitive structure. Ph.D. Thesis, Dept. of Anthropology, McGill University, Montreal, Quebec.

Fisheries and Environment Canada. 1977. Annual statistical review of Canadian fisheries, Ottawa. Marketing Services Branch, Fisheries and Marine Service, Ottawa, Ontario.

Fuller, W.A. and B.A. Hubert. 1981. Fish, fur and game in the Northwest Territories: some problems of and prospects for, increased harvest. *In*: Renewable Resources and the Economy of the North. M.M.R. Freeman (Ed.). Association of Canadian Universities for Northern Studies, Ottawa, Ontario. pp. 12-29.

Hartman, G.F. 1981. Managing non-native renewable resource use in the North: rising expectations in unproductive ecosystems. *In*: Renewable Resources and the Economy of the North. M.M.R. Freeman (Ed.). Association of Canadian Universities for Northern Studies, Ottawa, Ontario. pp. 30-55.

Hoyt, P.B., W.A. Rice and A.M.F. Hennig. 1978. Utilization of northern Canadian soils for agriculture. *In*: The 11th International Congress of Soil Science, Symposium Session Papers, Vol. 3. University of Alberta, Edmonton, Alberta.

Lee, P.F. 1976. Northwestern Ontario wild rice management plan. Ontario Ministry of Natural Resources, Kenora, Ontario.

Lithman, Y.G. 1973. The capitalization of a traditional pursuit: the case of wild rice in Manitoba. *In*: Two Papers on Canadian Indians. Center for Settlement Studies, University of Manitoba, Winnipeg, Manitoba.

Lykke, J. 1974. Moose management in Norway and Sweden. Le Naturaliste Canadien. 101: 723-735.

Lynch, G.M. 1975. Moose populations and seasons in Alberta. Presented at the 11th Moose Workshop. Winnipeg, Manitoba.

Manitoba, Department of Natural Resources. 1981. Annual Report; year ending, March 31st, 1981. Department of Natural Resources, Winnipeg, Manitoba.

Mercer, W.E. and F. Manuel. 1974. Some aspects of moose management in Newfoundland. Le Naturaliste Canadien. 101: 657-671.

Mills, G.F. and R.E. Smith. 1981. Additional land for crop production in Manitoba. Presented at Professional Update "Agronomy of Cereal Crop Production". The 24th Annual Manitoba Soil Science Meeting.

Nasogaluak, W. and D. Billingsley. 1981. The reindeer industry in the Western Canadian Arctic: problems and potential. *In*: Renewable Resources and the Economy of the North. M.M.R. Freeman (Ed.). Association of Canadian Universities for Northern Studies, Ottawa, Ontario. pp. 86-95.

Ontario, Ministry of Natural Resources. 1980. Strategic Land Use Plan: Northwestern Ontario. Ministry of Natural Resources, Toronto, Ontario.

Payne, H. and T. Muir. n.d. Moose Ranching. Manitoba Department of Renewable Resources and Transportation Services, Winnipeg, Manitoba.

Pettapiece, W.W. and J.D. Lindsay. 1981. Soil resources and agricultural capability in Northwestern Canada. Agriculture and Forestry Bulletin, Vol. 4.

Power, G. et G. Le Jeune. 1976. Le potentiel de peche du Nouveau-Quebec. Cahiers de Geographie de Quebec 20: 409-428.

Rader, R. 1970. Marketing of wild rice. Northern Wisconsin Development Center, Wisconsin.

Regier, H.A. 1978. A balanced science of renewable resources. University of Washington Press, Seattle and London.

Ritcey, R.W. 1974. Moose harvesting programs in Canada. Le Naturaliste Canadien. 101: 631-642.

Rodahl, K. 1950. Arctic nutrition. Can. Geog. J. 40: 52-60.

Romer, M.J., W.R. Cummins and J. Svoboda. 1981. Is there a potential for Canadian northern agriculture? (A justification for research on northern native plants as potential foodcrops). *In*: Renewable Resources and the Economy of the North. M.M.R. Freeman (Ed.). Association of Canadian Universities for Northern Studies, Ottawa, Ontario. pp. 161-165.

Rowe, J.S. 1972. Forest regions of Canada. Environ. Canada, C.F.S., Publ. No. 1300.
Ryder, R.A., S.R. Kerr, K.H. Loftus and H.A. Regier. 1974. The morphoedaphic index, or fish yield estimator — review and evaluation. J. Fish. Res. Bd. Canada. 31: 663-688.
Schaefer, O. and J. Steckle. 1980. Dietary habits and nutritional base of native populations of the Northwest Territories. Science Advisory Board of the Northwest Territories, Yellowknife, Northwest Territories.
Shields, J.A. and J.L. Nowland. 1975. Additional land for crop production: Canada. *In*: The proceedings of the 30th Annual meeting of the Soil Conservation Society of America. (Land Use: Food and Living). San Antonio, Texas.
Statistics Canada. 1982. Fur production report 1980-1981. Statistics Canada, Ottawa, Ontario.
Steinhart, J.S. and C.E. Steinhart. 1974. Energy use in the U.S. food system. Science, 184: 307-316.
Stelfox, J.G., E.S. Telfer and G.M. Lynch. 1973. Effects of logging on wildlife. Fish and Game Sportsman Magazine, Fall.
Telfer, E.S. and G.W. Scotter. 1975. Potential for game ranching in boreal aspen forests of Western Canada. J. Range Manage. 28: 172-180.
Tempelman-Kluit, A. 1982. The struggle for the land. The Globe and Mail, Toronto, Ontario, April 10, p. 8.
Thompson, P.C. 1981. The economic performance of the commercial skiff fishery in Western Canada. Canadian Technical Report of Fisheries and Aquatic Sciences No. 1037. Government of Canada, Fisheries and Oceans.
Williamson, R.G. 1971. The Keewatin settlements. The Musk Ox. 8: 14-23.

THE BOREAL FOREST, CREE-OJIBWA FORAGING AND ADAPTIVE MANAGEMENT

Bruce Winterhalder
University of North Carolina at Chapel Hill

Abstract: By choice and partly by necessity, native peoples in the Canadian middle north retain much of their traditional economy, including widespread use of local renewable resources. Ecological studies of boreal forest ecology and Cree-Ojibwa foraging behaviors suggest that three types of understanding will facilitate analysis and management of the native boreal forest economy: i) populations of fauna fluctuate strikingly on several time scales, and they are widely scattered over a shifting mosaic of small-scale patches; ii) Cree-Ojibwa hunting and trapping strategies can be quite effective and sometimes are systematically patterned by a logic which accords generally with optimal foraging theory; and iii) the principle of "adaptive management" developed by Holling and his colleagues fits boreal ecology and Cree-Ojibwa foraging practices, and could well guide planning by boreal forest peoples.

In this paper I consider the continuing impact of development on a small community of Cree-Ojibwa speaking people living in the boreal forest of northern Ontario. I am especially concerned with the nature of their environment, their adaptations to that environment, and with planning concepts compatible with this human-ecological system. I write as an anthropologist who spent a year living in a small community with these people, studying their use of local resources (Winterhalder 1977). Like many anthropologists I am uncomfortable with development planning. Therefore, let me begin cautiously, with a definition and a carefully delineated context for my discussion.

First, I see development as a deliberate and organized form of a long-standing process: power, technology, knowledge and values are projected across political, socio-cultural or economic (class) boundaries with the effect of changing people's lives and surroundings. Development is pervasive and rarely neutral; either religion or construction may be helpful or harmful but not, commonly, inconsequential.

Second, residence of the Cree-Ojibwa within one of the affluent western industrial democracies makes them a somewhat anomalous example of peoples usually envisioned when development is discussed. Development is something done in Latin America, Africa or Asia. I will therefore address the comparison of the Cree-Ojibwa case to "underdevelopment" elsewhere, and will suggest that it is apt.

Third, I do not view my comments as recommendations to the Cree-Ojibwa people, nor to persons who have assumed responsibility "on their behalf." The Cree-Ojibwa have an extraordinary sense of themselves and their situation. I will try to make some important and useful observations, but I do not mean to mistake my possible insights for their interests.

I will proceed as follows: As background I will allude briefly to the historical development of the Cree. I will indicate how their present-day situation is and is not comparable to that of other third-world peoples. I will discuss briefly attitudes towards development assessment in Canada, and the place of local renewable resources in the present-day and future economy of northern peoples. This background will justify my definition of development, my sense of optimistic urgency with respect to this subject and the contemporary Canadian north, and my subsequent focus on renewable resources.

My main argument has three parts: i) the nature of the boreal environment and its description; ii) long-standing Cree-Ojibwa strategies for use of renewable boreal resources and their analysis; and iii) the compatibility of these ideas with the concept of ''adaptive management'' (Holling 1978).

Background

By the definition given above, the Cree-Ojibwa have undergone development throughout the historical period (Bishop 1974; Ray 1974; Rogers 1982). Commercial activity of the fur trade was first, augmented later by missionaries and finally by federal and provincial authorities. Development occurred in directions and at a pace set by external forces, whether corporate, ecclesiastical, or governmental (SCC 1977). There were transformations, but their impact ought not to be overstated. The early fur trade was built on an infrastructure of traders operating primarily as an extension of Indian skills and subsistence activities. Many of the goods traded were implements (e.g., firearms, snare wire, traps) that furthered the abilities of native peoples in their own lifestyles. Missionary activity was largely benign, in part because of Anglican monopoly and limited success. And, even if misconceived or badly implemented, government efforts (with the exception of treaties) were essentially the service industries of a liberal democratic society. The material culture of the Cree-Ojibwa has changed greatly, the nuclear family has grown in importance at the expense of extended family groups, and village life now prevails over life in the bush. But social and religious beliefs persist, as does the use of local resources in a native economy. Edward Rogers, an ethnographer with over 25 years experience in the region, puts it this way:

> ''. . . The fact remains that for all that the Indians of northern Ontario and adjacent Manitoba have had to endure at the hands of Euro-Canadians in their endeavors to assist the native people and in the face of changing environments, the Indians have not become assimilated into Western society. Fundamentally, they remain Ojibwa and Cree in spirit and in outlook. And, most likely they will continue to be themselves well into the foreseeable future'' (Rogers 1982).

Current Socio-economic Situation

Discussion of the Cree-Ojibwa in terms more commonly applied to third world peoples requires some justification. The Science Council of Canada has summarized the situation this way:

> ''. . . The political economy of Canada's north has most of the characteristics of an underdeveloped country: extremely high birth rates; declining mortality rates; relatively low incomes; extreme variations in income levels from the highest to the lowest; heavy reliance on imports; little local industry; preponderance of natural resource-based industry; a low level of political development, and so on . . .'' (Rea 1976, quoted in SCC 1977).

Historically, products originating in the north have generated wealth elsewhere:

> "... The profits from the fur trade and from whaling were earned in the markets of Europe and America and they generated secondary activity only in France, England and southern Canada. Only a fraction of the profits were returned to the Indians and Eskimos. The mining industry has also taken its profits out of the Northwest Territories, and the oil and gas industry will do the same ..." (Berger 1977).

Capital investment in the north buys equipment and materials in southern Canada (SCC 1977). Not least important, in their confrontations with southern culture and development proposals native peoples deliberately have sought guidance in the recognition that they have the political economy of a colonized people (Watkins 1977).

However, it is worth noting the differences as well. The Cree-Ojibwa live within a democratic state, scientifically sophisticated, affluent by world standards, and with an existing and highly developed industrial and agricultural base outside of their region. To a greater degree than elsewhere, the knowledge and funds exist to assist autonomous Cree-Ojibwa attempts to reach an accommodation with the dominant society, one that preserves their lifestyles and environment.

Development Assessment and Renewable Resources

The recent Canadian public experience with northern development has been dominated by one extended event: The Mackenzie Valley Pipeline Inquiry, also known as the Berger Inquiry (Berger 1977; Gamble 1978). Less visibly but concurrently, the Science Council of Canada drafted a set of reports on northern development and science (SCC 1977). Each of these reviews, and more recently Freeman (1981), has recorded the insistence of native peoples and the arguments of other experts that traditional resources such as game, fish and fur will continue to have a prominent role in the native economy of the north. Usher's (1981) conclusions are probably the most trenchant: continued use of these resources by Indians and Inuit is economically rational and a subsistence necessity because native peoples are and will continue to be marginal participants in the affluent industrial economy that surrounds them. In other words, the continuity of a native economy has gained widespread recognition.

To summarize, the Cree-Ojibwa and their boreal environment have been undergoing relatively slow endemic change and small-scale development by outsiders in the form of traders, churchmen and government agents for several hundred years. They have accommodated to some degree, but have remained Cree-Ojibwa — to a substantial degree in and of the boreal forest. In terms of population and economy their present situation is similar to that of third-world peoples. Like those people, they and the forest now face large-scale development projects initiated elsewhere and profitable elsewhere; like those people they continue to rely heavily on a local economy of renewable resources.

Environmental Analysis

I now turn to the central concerns of this paper. Indian peoples yet have and wish to retain much of their indigenous culture and economy. Successful pursuit of this resolve requires that all sides exercise a sound understanding of the renewable resources available to the Cree-Ojibwa, the effective ways of using them, and the management and planning methods compatible with the ecological practices of the Cree-Ojibwa and the nature of their boreal environment.

My first topic is shortcomings in standard environmental descriptions. Environmental assessments typically contain a description of the setting, an inventory of the natural resources and comments on how their numbers are likely to be affected by the proposed development. This type of characterization has its counterpart in the writings of ecological anthropologists, and the drawbacks are the same. The environmental descriptions in such studies are normative. Soils are described, climate is summarized by annual or monthly mean rainfall and temperature, vegetation is characterized by biome-level description, and a list of flora and fauna is provided. The environment is presented as a static background, portrayed by averaging statistics.

I want to emphasize (full discussion in Winterhalder 1980) that information in this form has limited utility. It is incompatible with theories for analyzing human adaptation; it misleads about the nature of ecosystems; and it provides little practical guidance in understanding the persistence or failure of human ecological systems as these are affected by local ecological events and processes.

The north is a particularly good example of a variable environment. Perhaps because of its vast size and lack of vistas, it suggests monotony. Its streams are sluggish; its vegetation dwarfed. But this appearance of homogeneity and timelessness is deceptive. As it affects the Cree-Ojibwa forager for example, it is a landscape vibrant with activity (see Winterhalder 1983a; papers in this volume), presenting a dramatic and, importantly, unusually well-documented, case of what is probably common to most natural environments: 1) dynamic patterns of spatial heterogeneity in the landscape and in dispersion of animal populations, and 2) recurrent fluctuations in the population density of flora and fauna. These fluctuations are more or less weakly synchronized because of the interlocking effects of food chains and overlapping responses to environmental factors such as climate and fire. They are more or less regular in their pattern. In its effect on a forager, this type of environment defies characterization and understanding in normative terms. Variability and range, rather than averages, are central to the functioning and persistence of these human populations, to the study of ecological adaptation, and to an attitude capable of effective management.

The Optimal Foraging Approach

My second concern is the analysis of foraging, the gathering or hunting of non-produced resources. I am interested in the rationality of this kind of productive activity. Microeconomics can illuminate the behavior of a firm operating in a market, given basic assumptions like profit maximization, knowledge of market conditions and the like. If we take seriously the proposition that the Cree-Ojibwa wish to continue a lifestyle which includes using renewable game resources of the boreal forest, then we have to come to a comparable understanding of the rationality of foraging. One set of concepts which can contribute to this goal, optimal foraging strategy models, guided my study of Cree-Ojibwa ecology (Winterhalder 1977; 1981a,b).

This work had a narrow focus, the attempt to analyze systematically the foraging behaviors of a Cree-Ojibwa, harvesting food-producing resources found in the boreal forest. I used theory from evolutionary ecology (Schoener 1971; Pyke *et al.* 1977; Winterhalder 1981a). It is based on the observation that individuals who improve their performance of basic life tasks relative to time and energy investment will be more adapted. In this case the optimization assumption is as follows: human foragers will develop behaviors which allow them to achieve the highest possible net rate of energy capture while foraging. The last part of this statement, 'while foraging', is worth emphasis. It circumvents the assumption that hunter-gatherers attempt simply to maximize resource use. The optimization principle is fully compatible with behaviors which attempt to minimize the time required to obtain a fixed and perhaps rather limited quantity of resources. The models which give an operational form to this principle are fairly simple; they are general and meant to be realistic but not quantitatively precise (Levins 1966).

The Models

Most organisms select their diet from a variety of potential food sources. The diet breadth model (MacArthur and Pianka 1966; Pyke *et al.* 1977) attempts to specify which and how many of those resources an optimal forager will take in different circumstances. The forager is assumed to randomly encounter prey. He or she decides which and how many of the encountered types to pursue. Resources are ranked by their net value relative to handling and pursuit costs. As more of the types encountered are pursued, average search cost per unit pursued decreases; average pursuit and handling costs, on the other hand, increase because the diet is expanding to include those types less desirable or more difficult to capture. These opposing trends can be used to identify an optimal diet, that set of prey which will give the maximum net rate of energy intake.

Various factors with a potential influence on foraging can be carried into this model by observing how they would affect prey ranking, search costs, or pursuit costs. Searching foragers, those spending relatively large amounts of foraging time looking for prey, will be dietary generalists. Pursuing foragers, those spending relatively more time in pursuit, will be specialists. Increased abundance of high ranked prey types reduces search time and leads to specialization. Behaviors or technologies which improve pursuit efficiency should cause the diet to expand; those which increase the efficiency of searching should cause it to contract. Finally, a prey type not worth harvesting when rare is not worth harvesting even if it suddenly becomes abundant. In effect, whether or not an item is included in the optimal diet is independent of its own abundance and depends only on its rank and the abundance of items of higher rank.

Foraging movement is the subject of another of these models, the marginal value theorem (Charnov 1976; Charnov *et al.* 1976). It considers a patchy or heterogeneous habitat, and predicts when a forager should leave the patch under current exploitation in order to search out another, unharvested one. There are travel costs between patches, food is encountered and harvested within patches, and within each patch the rate of food intake is "depressed" as foraging occurs. Prey are depleted, dispersed, become alerted and therefore wary or evasive, or become simply more difficult to locate as the easier to find items are taken. In these circumstances, the optimal forager leaves the patch under exploitation when the marginal capture rate drops to the average capture rate for the habitat. The organism that remains beyond this time has an immediate rate of harvest which is below what it could expect if it moves on.

The Horn model (Horn 1968; Wilmsen 1973) considers settlement pattern in relation to resource distribution and density. It assesses the geometric distance from a foraging group to food locations. If resources are predictably and evenly dispersed, then the more efficient pattern is regular dispersion of small forager social units. In contrast, if the available resources are clumped together and move unpredictably throughout a large range, then the optimal strategy is aggregation of the foraging population at the center of that range. The ability of humans to directly exchange information about resource location enhances the value of central place aggregation.

A related model indicates that moderately dense, predictable and evenly distributed resources create conditions which lead to the evolution of territoriality (Brown 1964; Dyson-Hudson and Smith 1978). Territories arise because the benefits of exclusive use outweigh the costs of defense. Once a territory is established it creates circumstances more favorable to long-term conservation because the forager is assured of the delayed benefit of short-term restraint.

Refuging (Hamilton and Watt 1970) or central place (Orians and Pearson 1979) foragers are defined by a pattern in which the forager makes an outward trip, forages, and then makes a return trip to a home base with captured prey. In this case, diet choices must consider the costs of prey search and pursuit, and also radial travel to and from the foraging locations (Orians and Pearson 1979). Although details of the optimal solutions vary with modifications in the assumptions, a general pattern is evident: as radial travel time increases the optimal diet breadth progressively is restricted to prey species of higher value, and the optimal load for the return trip increases in size.

With central place foraging one expects some depletion of prey close to the settlement. Faced with such depletion, the optimal forager will increase radial travel distance whenever the cost is offset by the greater richness of the more distant locales. Thus, if prey and habitat are used optimally, local depletion is to some degree self-correcting. The lower the travel costs, the greater will be the evenness in use of successive habitats surrounding a settlement.

Finally, a paper by Morrison (1978) makes a related point about central place foraging. In circumstances like those affecting the Cree-Ojibwa, the optimal search path of a forager attempting to locate randomly distributed prey is one in which the predator makes few turns. An unexpected corollary of this pattern indicates that this search pathway is not one expected always to find the prey nearest to the central place (Morrison 1978, p. 931). This reduces the impact of localized depletion around a settlement. As with the previous models, travel costs have an important impact on this strategy.

Cree Foraging

In my fieldwork (Winterhalder 1981b, 1983b) I traveled with Cree foragers, noting time and energy investments, foraging tactics and choices, knowledge of animal behavior, and the game captured. These trips, and others recorded in less detail through conversation with returning hunters, were mapped on airphotos in order to measure distances and placement relative to the vegetation mosaic. I intentionally worked with proficient hunters. In this village snowmobiles were used to travel to hunting areas, but not to pursue game, a practice prohibited by game regulations and impeded by the deep, soft snows of the bush. Small outboards were used for travel during open water periods.

In analyzing diet breadth I originally predicted that the low density and high dispersion of boreal forest resources would mean relatively high search times and consequently generalized hunting strategies. In fact, Cree-Ojibwa foragers often set out for only one or a few species. With snowmobiles, search efficiency, subsidized with fossil fuels, was high enough to account for the narrow diet observed. I had not anticipated the effects of transportation on search velocity. However, 15 years ago, snowmobiles and outboards were not in widespread use, and the prediction of more generalized hunting is supported by informants' statements, early ethnographies, and by commentary by northern anthropologists (Winterhalder 1981b; 1983b). The optimal diet breadth model seems to give us insight into the causes of changing Cree-Ojibwa foraging practices.

Qualitative observations of Cree foraging also give support to the marginal value theorem. When muskrat were shot as they sat on the new ice lining river coves, only the more readily available animals were taken in each location before the forager moved on to the next population concentration. Foragers say that hare snarelines in a particular place generally are abandoned for more favorable spots before they cease to be productive (for other instances see Winterhalder 1983b).

With respect to settlement, an original pattern of dispersed family hunting groups is widely documented for the eastern boreal forests, as is the recent change to more permanent and larger settlements (Rogers 1963; additional references in Winterhalder 1983b) with economic and government inducements. In an environment of fairly evenly dispersed resources the original pattern corresponds to that predicted by the Horn model. However, the shift to nucleated settlements occurred roughly coincident with the introduction of bush planes and outboards (snowmobiles came later), which gave the Cree-Ojibwa continued access to large areas of surrounding land, including traditional hunting and trapping ranges. The Horn model accurately characterized the original situation; the loss of foraging efficiency with settlement was somewhat offset by improved transportation, if at a high monetary cost (Berkes 1981a).

I do not have the data to closely assess the predictions concerning radial game depletion, or search path and turning. In the small community (about 100 people) that I studied, localized game depletion was not perceived as a problem, and the distribution of moose kills, for which I have evidence, did not suggest that it was. Search pathways regularly took people far from the village and made few turns.

Overall, in the instances cited here (Winterhalder 1981b), Cree-Ojibwa foraging corresponds rather well to the predictions of optimal foraging models. This suggests that the models provide insight into the rationality of foraging behavior, and a guide to efficient ecological practice. Exceptions, such as aggregated settlement pattern, can be explained by alternative causes and their effects investigated.

Evolutionary ecology theory illuminates foraging, but the practice of foraging and especially its dynamics require detailed information on the variables actually incorporated into the model and decisions. Here the type of dynamic environmental analysis mentioned above is essential.

The boreal forest is vast, but only a small portion of it is productive for the forager. Species used by the Cree are associated with vegetation habitats that make up only 28 per cent of the landscape; if the large areas of open water not used for fishing are excluded, then only about 10 per cent is used intensively. The favorable locations are small patches, randomly distributed, and by virtue of disturbance and succession, always shifting. Mobility is essential, and considerable experience goes into acquiring familiarity with a hunting area. Each trip reappraises that information as once favorable areas become less alluring to game through succession, and as disturbances such as fire create newly attractive habitats.

The temporal dynamics are equally important. Fluctuations of population density by season or longer periods are dramatic. These density changes plus seasonal alterations in visibility, snow and water conditions, and other factors which affect encounter rates give a highly dynamic quality to the search component of the diet breadth model. Similar environmental conditions impart flux to pursuit costs, and hence the ranking of prey species. Taken together, those spatial and temporal factors directly influencing foraging decisions are unstable and nonrecurrent on the scale of a forager's lifetime. The environmental elements of foraging are always novel, even if the guiding strategies persist.

It is impossible in a paper to portray the skill of a proficient Cree-Ojibwa forager. While the optimal foraging rules themselves may be rather simple, adroit application is demanding. The effective environment changes rapidly, in multiple ways, and with significant degrees of unpredictability. Among the major adaptive skills of a Cree-Ojibwa forager are: observational sensitivity to the state of the ecosystem and its likely changes; the ability to evaluate simultaneously and selectively many environmental factors which will affect foraging decisions; and flexible responses. Foraging is the application of simple rules in a complex and skill-demanding setting.

One commonly hears that western technology has disrupted the lifeways of the Cree-Ojibwa or other hunters; that a flannel shirt from Sears and a Winchester shotgun inevitably compromise this subsistence pattern. There may be cases in which this is true, but the Cree-Ojibwa have adopted Euro-Canadian technologies and other developments without altering their underlying strategies of harvesting forest resources. Optimal foraging models help to demonstrate the continuing proficiency and rationale of their hunting and trapping efforts even, for instance, as they take fewer species. The latter is not evidence of "disengagement" from their forest habitat; rather it is a predictable response of an optimal forager living in a certain kind of new technological context.

Foraging Management

Foraging models help to identify aspects of foraging behavior which are systematic and efficient. These should not be impeded if effective use of renewable game resources is to be encouraged. And, the models can aid in understanding the impact of effective game exploitation on the environment. However, use of this information requires that we overcome western myths about socio-cultural evolution, progress and our hunting "heritage", and the western experience of hunting as a sport (a leisure activity, with qualities enhanced precisely because they are the opposite of occupational pursuits).

Usher (1981) has provided a penetrating analysis of the first problem, that of the prevailing attitudes toward hunters. Western attitudes suggest that foraging production is an anachronism, to be preserved only as a romantic lifestyle preference or cultural legacy. Hunters are predators, and not very good ones, being limited by privation, natural processes and immature technologies. This view allows neither for efficient use nor effective management by foragers of their resources. If given a technological edge, these hunters will presumably overcome the dominance of nature and destroy the resource base. All of these conceptions deny that foraging is an important socio-economic activity, with rational as well as meaningful virtues. All are flawed.

Foraging strategy analysis is one aspect of an alternative and more enlightening view. It highlights the systematic nature of foraging, and it does so with clear implications for management. For instance, foraging harvest is most efficient when the hunter takes that set of game identified by the optimal diet breadth model. Game regulation categories which contravene these groupings will impede effective use of renewable resources and thwart skilled foraging practices. Separate management and regulations for categories like "big game", and "waterfowl" might be an example. In given conditions the optimal diet may include some species from each of these groups. Or, to give a more specific example, since the harvest of an organism depends on its "ranking," innovations in species-specific capture technologies should be a more effective way of controlling or directing harvest toward an under-utilized species than would be increasing its population.

The marginal value theorem suggests that the effective hunter moves from patch to patch without fully exploiting and hence depleting each location. Habitat heterogeneity not only enhances game diversity and productivity (Rowe and Scotter 1973), but reduces forager impact in given locations. Practices such as fire control or lumbering which reduce habitat patchiness will have a negative impact in this respect, and may also reduce the resilience of the human ecological system (Roff 1974; May 1975; Smith 1972).

The Horn model shows that growth of Cree-Ojibwa settlements decreases foraging efficiency. This same pattern risks uneven use of resources, with depletion adjacent to the populated area and underuse in the hinterlands. Effective policy here would encourage the proliferation of small settlements. Central place models and those relating to optimal search pathways suggest that only the most valuable game will entice foragers far from their villages, unless radial transportation costs are low. If even harvest of resources in the bush and near to settlements is desired (Fuller and Hubert 1981; Berkes 1981a), then use of snowmobiles and bush planes should be encouraged, perhaps through subsidy.

Ecological models suggest that conservation is more likely if conditions allow the conserver to realize the benefits of short-term restraint. Usher (1981) and Berkes (1981b) have also suggested that restriction of native hunters to traditional hunting ranges would allow customary conservation practices to operate more effectively.

Finally, good hunting conditions are unpredictable and ephemeral. Hence an effective hunter requires considerable flexibility in the timing of his or her work. Subsidiary activities such as wage labor should be organized so as to provide for this. And, although the boreal forest is vast, only a limited proportion scattered in small patches is good foraging habitat at any given time. The location of these patches is always in flux, due to factors like fire. Large areas contain small, scattered opportunities which are elusive unless the forager has experience within a region. This provides another argument for continuity of association between foragers and home ranges, and it highlights the need that these ranges be large.

What about the impact of optimal foraging patterns on the game itself? Several of these models identify circumstances in which game conservation is consistent with maximally effective exploitation. This may seem to be a paradox, but in a patchy environment the most efficient forager leaves some animals, a potential breeding population, in each patch. An optimal forager from a central place does not necessarily devastate resources close to home and ignore those further away. In the absence of population-wide regulations, resources used by one or few foragers within a territory are more likely to be "managed" for long-term harvest. Conversely, in an environment without strong spatial heterogeneity, the diet breadth model may suggest a circumstance where more deliberate practices might be needed. Resources which are ranked highly will be pursued whenever encountered. Moose or beaver might be examples. Depletion is more likely in this instance, and deliberate conservation more necessary. Similarly, costly radial travel to distant locations will lead the optimal forager to concentrate on game with the highest value, to the neglect of other species.

In effect, ecologically sound practice requires quite different perceptions of and regulations for subsistence and sport hunters. Sport hunting rules operate by impeding the effectiveness of the predator. This is compatible with the goals of sport hunting and game management, but probably not appropriate for the native population in the north (Wilkinson 1981). In some instances, subsistence hunting rules should be designed to maximize the effectiveness of the forager. This is compatible with the goal of subsistence hunting. But more important and what I want to stress, there are circumstances in which it can be compatible with management of game with only minimal guidance. To those for whom hunting is an occupation, controls designed for recreation management frustrate effectiveness, and will impede retaining a lifestyle based on renewable resources. A Cree-Ojibwa informant, chafing under diligent observation of restrictions on the use of snowmobiles to pursue game, made this comment (see also Usher 1981):

> ". . . We do not tell all of those bureaucrats in Toronto that they have to write everything with pencils, that they cannot use typewriters to do their work. Why can't we use snowmobiles to do our work?".

Berkes (1981b) has stated a similar position, calling it management by self-regulation and design. Design works by "establishing criteria to govern . . . [a] process so that the desired result will occur more or less automatically [and] without further human intervention", as opposed to regulation, which works through rules imposed from the outside and close policing. Berkes argues that the design approach will find greater acceptance in the native communities and hence has greater likelihood of being effective. Such a system would work through customary law and the traditional adaptive practices of the Cree, "keeping in mind that these practices may be under considerable stress at a given point in time."

Usher (1981) has made observations which complement these. He suggests that the hunting economy has continuity and rationality; its essential qualities persist. Usher takes concepts of political economy and mode of production to identify some of those qualities. I have used optimal foraging theory to identify other attributes of foraging rationality. Our conclusion is the same: technology and material culture may change, but as a mode of production and a strategy of foraging, the native economy persists:

> "What is important about wildlife to native people is not simply that it is of cultural or aesthetic importance, but that it is the obvious, and indeed the only possible basis, in the Arctic and subarctic, for the maintenance of the social relations that characterize the traditional mode [of production]" (Usher 1981).

To summarize, optimal foraging strategy models can illuminate the economic and ecological rationality of foraging adaptations. The models discussed here generally were supported by field studies (see also Smith 1980). This is one approach challenging notions which are dismissive of the continuity and importance of foraging economies; Usher's political economy analysis is another. The models show the systematic and predictable nature of foraging decisions even in an environment as complex and unstable as the boreal forest. Hence, their use is important to the effective management of foraging economies.

And, they illuminate the circumstances in which conservation is incidental to efficient exploitation, in which management can occur by internal design rather than external regulation.

Adaptive Management

The boreal resource environment is dynamic and unpredictable; non-normative conceptualizations are needed to properly characterize its effective qualities. Optimal foraging models can help identify decision rules that the Cree seem to follow in their efficient use of those resources, and can suggest specific management practices. There is need for a comprehensive statement about ecological management, a philosophy or theory for implementing these insights. This brings me to the concept of adaptive management (Holling 1978; Walters and Hilborn 1978).

Ecosystem Behavior

Until recently, images of natural ecosystems have been dominated by analogies based on community succession or organismal development. A biotic community was viewed as a homeostatic, equilibrium-preserving assemblage of species with a composition and physiognomy determined by stable edaphic and climatic factors. If disturbed by some external agency, this community would quickly pass through predictable states to regain its former equilibrium, the "climax". Further, disturbance was the exception; environments were ahistorical. More recent views note that succession does occur, but viewed over the long term an ecosystem is more often characterized by change than by equilibrium, and by internal as well as external disturbances. Many of the changes appear to be unpredictable in their timing and magnitude; others unexpected. The boreal forest is a good example, in its post-glacial climatic and biogeographic history, in the flux of its vegetation mosaic, and in the fluctuations of its fauna (see Holling 1973; Botkin and Sobel 1975, for other examples).

Holling (1973) and also Botkin and Sobel (1975) have developed two concepts to characterize this newer view of ecosystems. Stability refers to recovery of an equilibrium state after a disturbance. The less the fluctuation, the more rapid the return, then the greater the stability. The lynx-hare population cycles would be an example of a highly unstable system.

Resilience, on the other hand, is the ability of a system to absorb change or undergo a perturbation and yet retain the same basic relationships or configuration. It assesses the likelihood of system persistence. The lynx-hare cycle is highly resilient. Holling (1973) argues that clearer understanding and better management of ecosystem behavior will result if we shift from emphasis on stability to an emphasis on resilience.

Management

This change of view has important implications for studies of human adaptation and for the practice of environmental assessment and management. If ecosystems were globally stable they would be relatively easy to characterize, highly predictable and quite forgiving of stress and misunderstanding. The alternative implies not only that most natural systems (including those dominated by humans) tend to fluctuate unpredictably, but that they contain uncertainties and the capacity always to generate the unexpected. Incremental quantitative change can give way to qualitative transformation abruptly. These and other concepts suggest a rather thorough reformulation of environmental assessment and management practice (Holling 1978). The central question becomes how to manage when much is unknown, some things are uncertain, and the unexpected must be acknowledged as possible. Holling (1978) provides a series of explicit, practical answers to this question.

The history and characteristics of the boreal forest show it to be an environment like that envisioned in Holling's views on ecosystem uncertainty, resilience and stability. Similarly, Cree-Ojibwa foraging strategies and their use of skill and observation to meet the vicissitudes of their situation with adaptive flexibility will appear familiar to the student of Holling's views. Because their effective environment is one of uncertainty and novelty, the Cree have developed knowledge, tactics and foraging strategies which readily take into account instability and the unexpected. Cree foragers have learned and practice much of what the "adaptive manager" would propose.

This adaptive resilience extends beyond strategies for immediate harvest of renewable resources. Berkes (1981b) has described how the land tenure and ethno-ecological practices of the Cree have weathered the successive perturbations of the fur trade, the introduction of technology, growth, economic development and road construction, and although changed, have survived intact. In the face of development and other instabilities, the system has proven remarkably resilient: ". . . good resource use practices develop over time among peoples who are dependent on a particular resource. Despite various perturbations, the persistence of such resource use practices have been extensively documented. . ." (Berkes 1981b).

Recognition of how Cree-Ojibwa subsistence adaptations fit adaptive management should result not only in good ecology, but effective planning as well. Usher (1981) has listed multiple advantages of designing northern game-management on the basis of native custom and practice, "of modifying the 'idiom of management' (La Rusic 1979) in favour of something that native harvesters could recognize as congruent with their own mode of thought." Berkes' similar argument was cited earlier.

To summarize, advances in theoretical ecology and in understanding of ecosystems have converged in the idea of adaptive management. The use and development of renewable resources in the north is an ideal situation in which to apply the principles of adaptive management, in part because some of the "design" is already in place in the foraging mode of production and foraging strategies of native peoples.

Conclusions

> ". . . The development of the whole renewable resource sector — including strengthening of the native economy — would enable the native people to enter the industrial system without becoming completely dependent on it. An economy based on modernization of hunting, fishing and trapping, on efficient game and fisheries management, on small-scale enterprise, and on the orderly development of gas and oil resources over a period of years — this is not retreat into the past; rather, it is a rational program for northern development based on the ideals and aspirations of northern native peoples" (Berger 1977).

The Cree-Ojibwa have experienced slow development over several hundred years. Their response has been flexible and selective; they have resisted to the degree that much of their culture is yet intact. Their current political ability to influence the future schedule and route of their development is problematic, but tenaciously defended. There is evidence that renewable resources will continue to play a significant role in the native economy. Hence the protection and management of that economy is of great importance.

More specifically, these circumstances require: (1) an accurate appraisal of the short- and long-term dynamics of the boreal environment, its resources, and especially their spatial heterogeneity and temporal instability; (2) models which can systematically illuminate the Cree-Ojibwa foraging economy. Cree-Ojibwa hunting and trapping practices are intricate, always changing, and superficially without pattern. It takes an assumption of rationality and an appropriate body of ideas — here optimal foraging models — to find the coherence. Management policies may follow to a large extent from design already in the adaptations; and (3) a practical set of planning principles which can combine these observations on environmental properties and adaptive rationality.

The Cree-Ojibwa know these things tacitly. Ecosystem analysis and adaptation study are recognizing them explicitly and systematically. If planning can adapt, then forager, forest, and manager will gain the advantage of greater congruence.

Acknowledgements

I appreciate the guidance, comments and assistance of Sara DeGraff, Carol Goland, Eric Smith, John Yellen, and especially the people of Muskrat Dam Lake.

Literature Cited

Berger, T.R. 1977. Northern frontier — northern homeland. The Report of the Mackenzie Valley Pipeline Inquiry, Volume I. Minister of Supply and Services, Ottawa, Ontario.

Berkes, F. 1981a. Fisheries of the James Bay area and northern Quebec: A case study in resource management. *In*: First International Symposium on Renewable Resources and the Economy of the North. M.M.R. Freeman (Ed.). Association of Canadian Universities for Northern Studies, Ottawa, Ontario. pp. 143-160.

Berkes, F. 1981b. The role of self-regulation in living resources management in the north. *In*: First International Symposium on Renewable Resources and the Economy of the North. M.M.R. Freeman (Ed.). Association of Canadian Universities for Northern Studies, Ottawa, Ontario. pp. 166-178.

Berkes, F. and C.S. Farkas. 1978. Eastern James Bay Cree Indians: Changing patterns of wild food use and nutrition. Ecology of Food and Nutrition 7: 155-172.

Bishop, C.A. 1974. The Northern Ojibwa and the fur trade. Holt, Rinehart, and Winston of Canada, Toronto, Ontario.

Botkin, D.B. and M.J. Sobel. 1975. Stability in time-varying ecosystems. Amer. Nat. 109: 625-646.

Brown, J.L. 1964. The evolution of diversity in avian territorial systems. Wilson Bull. 76: 160-169.

Charnov, E.L. 1976. Optimal foraging, the marginal value theorem. Theor. Pop. Biol. 9: 129-136.

Charnov, E.L., G. Orians and K. Hyatt. 1976. Ecological implications of resource depression. Amer. Nat. 110: 247-259.

Colwell, R.K. 1974. Predictability, constancy and contingency of periodic phenomena. Ecology 55: 1148-1153.

Dyson-Hudson, R. and E.A. Smith. 1978. Human territoriality: An ecological reassessment. Amer. Anthro. 80: 21-41.

Feit, H.A. 1973. The ethno-ecology of the Waswanipi Cree; or how hunters can manage their resources. *In*: Cultural Ecology. B. Cox (Ed.). McClelland and Stewart, Toronto, Ontario. pp. 115-125.

Freeman, M.M.R. (Ed.). 1981. Proc. First International Symposium on Renewable Resources and the Economy of the North. Association of Canadian Universities for Northern Studies, Ottawa, Ontario.

Fuller, W.A. and B.A. Hubert. 1981. Fish, fur and game in the Northwest Territories: Some problems of, and prospects for, increased harvests. *In*: First International Symposium on Renewable Resources and the Economy of the North. M.M.R. Freeman (Ed.). Association of Canadian Universities for Northern Studies, Ottawa, Ontario. pp. 12-29.

Gamble, D.J. 1978. The Berger inquiry: An impact assessment process. Science 199: 946-952.

Hamilton, W.J., III and K.E.F. Watt. 1970. Refuging. Ann. Rev. of Ecology and Sys. 1: 263-286.

Hare, F.K. and J.C. Ritchie. 1972. Boreal bioclimates. Geog. Rev. 62: 333-365.

Holling, C.S. 1973. Resilience and stability of ecological systems. Ann. Rev. of Ecology Sys. 4: 1-23.

Holling, C.S. 1978. Adaptive Environmental Assessment and Management. John Wiley and Sons, Chichester, England.

Horn, H.S. 1968. The adaptive significance of colonial nesting in the Brewers blackbird *(Euphagus cyanocephalus)*. Ecology 49: 682-694.

MacArthur, R.F. and E.R. Pianka. 1966. On optimal use of a patchy environment. Amer. Nat. 100: 603-609.

May, R.M. 1975. Stability in ecosystems: Some comments. *In*: Unifying Concepts in Ecology. W.H. van Dobben and R.H. Lowe-McConnell (Eds.). Dr. W. Junk B.V. Publisher, The Hague, The Netherlands. pp. 161-168.

Morrison, D.W. 1978. On the optimal searching strategy for refuging predators. Amer. Nat. 112: 925-934.

Orians, G.H. and N.P. Pearson. 1979. On the theory of central place foraging. *In*: Analysis of Ecological Systems. D.J. Horn, G.R. Stairs and R.D. Mitchell (Eds.). University of Ohio Press, Columbus, Ohio. pp. 155-177.

Pianka, E.R. 1978. Evolutionary Ecology. Second ed. Harper and Row, New York.

Pruitt, W.O., Jr. 1958. Qali, a taiga snow formation of ecological importance. Ecology 39: 169-172.

Pruitt, W.O., Jr. 1978. Boreal Ecology. Edward Arnold, London, England.

Pyke, G.H., H.R. Pulliam and E.R. Charnov. 1977. Optimal foraging: A selective review of theory and tests. Quart. Rev. Biol. 52: 137-154.

Ray, A.J. 1974. Indians in the Fur Trade: Their Role as Trappers, Hunters and Middlemen in the Lands Southwest of Hudson Bay. University of Toronto Press, Toronto, Ontario.

Roff, D.A. 1974. Spatial heterogeneity and the persistence of populations. Oecologica, 15: 245-258.

Rogers, E.S. 1963. Changing settlement patterns of the Cree-Ojibwa of northern Ontario. Southwest J. Anthro. 19: 64-88.

Rogers, E.S. 1982. Cultural adaptations of the northern Ojibwa of the boreal forest, 1670-1980. *In*: Boreal Forest Adaptation: The Algonkians of Northern Ontario. A.T. Steegman, Jr. (Ed.). Plenum, New York. In press.

Rowe, J.S. and G.W. Scotter. 1973. Fire in the boreal forest. Quat. Res. 3: 444-464.

Schoener, T.W. 1971. Theory of feeding strategies. Ann. Rev. of Ecology and Sys. 2: 369-404.

SCC. 1977. Northward Looking: A Strategy and a Science Policy for Northern Development. Science Council of Canada Report No. 26, Ottawa, Ontario.

Smith, E.A. 1979. Human adaptation and energetic efficiency. Human Ecology 7: 53-74.

Smith, E.A. 1980. Evolutionary ecology and the analysis of human foraging behavior: An Inuit example from the east coast of Hudson Bay. Ph.D. Dissertation, Cornell University, Ithaca, New York.

Smith, F.E. 1972. Spatial heterogeneity, stability and diversity in ecosystems. Trans. Conn. Acad. Sci. 44: 309-335.

Southwood, T.R.E. 1977. Habitat, the templet for ecological strategies? J. Anim. Ecol. 46: 337-365.

Usher, P.J. 1981. Sustenance or recreation? The future of native wildlife harvesting in northern Canada. *In*: First International Symposium on Renewable Resources and the Economy of the North. M.M.R. Freeman (Ed.). Association of Canadian Universities for Northern Studies, Ottawa, Ontario. pp. 56-71.

Watkins, M. 1977. Aboriginal people and staple production: A commentary on the Berger Report. West. Can. J. Anthr. 7: 83-94.

Wilkinson, P.F. 1981. Some thoughts towards a policy for the rational use of renewable resources in the Canadian north. *In*: First International Symposium on Renewable Resources and the Economy of the North. M.M.R. Freeman (Ed.). Association of Canadian Universities for Northern Studies, Ottawa, Ontario. pp. 73-85.

Wilmsen, E.N. 1973. Interaction, spacing behavior, and the organization of hunting bands. J. Anthr. Res. 29: 1-31.

Winterhalder, B. 1977. Foraging strategy adaptations of the boreal forest Cree: An evaluation of theory and models from evolutionary ecology. Ph.D. Dissertation, Cornell University, Ithaca, New York.

Winterhalder, B. 1980. Environmental analysis in human evolution and adaptation research. Human Ecology 8: 135-170.

Winterhalder, B. 1981a. Optimal foraging strategies and hunter-gatherer research in anthropology: Theory and models. *In*: Hunter-Gatherer Foraging Strategies: Ethnographic and Archeological Analyses. B. Winterhalder and E.A. Smith (Eds.). University of Chicago Press, Chicago, Illinois. pp. 13-35.

Winterhalder, B. 1981b. Foraging strategies in the boreal forest: An analysis of Cree hunting and gathering. *In*: Hunter-Gatherer Foraging Strategies: Ethnographic and Archeological Analyses. B. Winterhalder and E.A. Smith (Eds.). University of Chicago Press, Chicago, Illinois. pp. 66-98.

Winterhalder, B. 1983a. History and ecology of the boreal zone in Ontario. *In*: Boreal Forest Adaptations: The Algonkians of Northern Ontario. A.T. Steegmann, Jr. (Ed.). Plenum, New York. pp. 9-54.

Winterhalder, B. 1983b. Boreal foraging strategies. *In*: Boreal Forest Adaptations: The Algonkians of Northern Ontario. A.T. Steegmann, Jr. (Ed.). Plenum, New York. pp. 201-241.

QUANTIFYING THE HARVEST OF NATIVE SUBSISTENCE FISHERIES

Fikret Berkes
Brock University

Abstract: A number of questionnaire-based studies have been carried out in recent years to quantify native subsistence fisheries, along with other animal resource harvests, in the Canadian north. A major problem in quantifying fish catches is that harvesters usually have difficulty remembering the actual numbers of these relatively small and numerous prey animals caught over a yearly cycle. The present study is based on actual field checks of fish harvests of native fishermen over a six-year period in one native community, Chisasibi, in James Bay, northern Quebec. The data so obtained are compared with the results of questionnaire studies conducted under the provisions of the James Bay Agreement.

The mean annual per capita fish harvest of eleven family groups was 60 kg (1975-1981) as compared to questionnaire study results in which the community-wide mean annual per capita harvest ranged from 26 kg in 1976-1977 to 82 kg in 1974-1975. On the basis of these results and other studies elsewhere in the north, the following approximations are offered. A "non-traditional" family living in a relatively large community is likely to obtain 10-20 kg fish per capita per year. A "traditional" family in the same size community is likely to obtain 50-100 kg fish per capita per year. A "traditional" family in a small community is likely to obtain 100-200 kg per capita per year. To improve the methodology of studies on native subsistence it is suggested that questionnaire surveys be combined with field work. This would help identify and confirm the species, the fishing areas, and produce data which would help verify the overall harvest estimate.

The study of resource harvesting by northern native peoples is important for several reasons. In terms of theory, the relation between hunters and animals is of interest to anthropology and ecology (e.g. Winterhalder and Smith 1981). For applied natural and social sciences, harvesting studies are important for native land and resource claims issues (Feit 1980), fisheries management (Berkes 1977), wildlife management (Feit 1973), and environmental and social impact studies of northern development projects (Berger 1977; Tester 1979; Berkes 1982).

The plan of this paper is first to provide an overview of native harvest and land use studies and to place native subsistence fisheries in perspective. Secondly, the paper presents the results of a field study on native subsistence fisheries which is discussed in some detail to provide a context for more general observations on subsistence fisheries. Thirdly, the paper includes a review of some of the questionnaire-based studies of native subsistence fisheries. Lastly, the paper offers some practical guidelines on ways in which the methodology of subsistence fishery studies may be improved.

While much of the material presented here is on the boreal zone, many studies are cited from other areas. This provides comparative information and helps emphasize the similarities rather than the differences between the boreal zone and the zones further to the north.

An Overview of Native Harvest and Land Use Studies

Until the 1970s, the two main sources of information on the use of fish and wildlife by northern native peoples came from "area economic surveys" (for a summary, Lotz 1976) and from investigations by anthropologists and ethnologists (e.g., Balikci 1968). Since about 1970, two new approaches have become common. The first is the inventory of land use, as carried out, for example, by Freeman and colleagues (Freeman 1976). This approach results in maps of resource use areas, in some cases, species by species; it is particularly relevant to establishing and arguing land claims. But it yields no information on the numbers of animals harvested.

The second approach seeks an inventory of harvests by species, as done for example in the James Bay and northern Quebec area (NHR 1982). This kind of study yields quantitative information on harvest by species and by year for a given community. Results of such studies have been used in establishing native peoples' share of the regional harvests of wildlife and fish, for example, in implementing the allocation provisions of the James Bay and Northern Quebec Agreement. Ideally, the two approaches should be combined to provide information on harvests as well as on their geographical distribution. However, the data base for such a study becomes difficult to manage, and to carry out the study properly requires enormous research time. At least, this is the experience in the few studies in which both approaches were simultaneously used, such as in Weinstein's (1976) work in the Cree village of Chisasibi and in the work of Kemp *et al*. (1977) in the Resolute area. The emergence of the land use and harvest approaches in the 1970s is not accidental but coincides with the advent of native land and resource claims and the push to develop northern resources. Almost all of these studies serve a purpose relating to land claims, agreements, or the impact assessment of development projects.

The role of the native people in many of these studies is an ambiguous one. On the one hand, there is pressure on native groups to prove that they do use the land and its animal resources. On the other hand, if their (considerable) harvests are quantified, they may be accused of over-hunting or, at least, governments may move to control their harvesting activities.

Given this situation, is the information provided by native hunters/trappers/fishermen really reliable? Or are the responses manipulated to serve political purposes? If so, would native harvesters over-state or under-state their kills? It is important, therefore, to have some means of checking questionnaire responses on harvests. In the case of the Northwest Territories, for example, much information is available from the Fish and Wildlife Service and the Royal Canadian Mounted Police for selected animal groups. Berger (1977), for example, relied more heavily on such information than on harvesting studies carried out by non-government groups. Fur records kept by the various provincial governments similarly serve the purpose of providing a check on questionnaire results of harvesting studies.

To improve the reliability of questionnaire responses, internal checks have been incorporated into the design of some harvesting studies. Other means have also been used to improve the accuracy of responses. For example, in some investigations, hunters were given diaries to keep a record of their kills (NHR 1982; Kemp *et al*. 1977). The experience with studies in the James Bay area shows that harvesting diaries or harvesting calendars are of real value in "getting around problems of recall, and preferences of informants not to quantify recall, at least for those informants prepared to use the diaries"

(H. Feit, personal communication). Thus, the "state of the art" in the early 1980s permits a reasonably reliable inventory of native peoples' wildlife harvests which can be independently verified, at least for some animal groups.

The reliability of native harvesting data is not the same for all animal groups. While the information on big game, fur animals and probably also waterfowl harvests are reasonably good in the various studies that have been conducted over the years, information on small game and fish harvests tend to be more questionable.

Berger (1977), for example, commented on the lack of data on subsistence fisheries. In studies on the subsistence economy of the Beaver people of northeastern British Columbia, no data were reported on fish because ". . . their numbers are too variable and imprecise. . ." (Brody 1981). Similarly, in a detailed study of resource harvesting in the Eskimo Point area, NWT, McEachern (1978) points out that ". . . data on (fish) catch size must be treated with caution; the survey results are highly variable within the sample. It may be that the questions posed were not easy for the fishermen to fully understand or to answer".

Many researchers have commented on the very high value northern native people place on truthfulness (Freeman 1976; Brody 1981). Some have also emphasized, however, that many native cultures, perhaps the Inuit in particular, ". . . until recently did not stress numerical exactness . . ." (Freeman 1979). As Tester (1979) puts it:

> "Concern for . . . how many fish are caught and any concern for recalling such numbers is understandably absent. Consequently, questions as to how many fish were caught and where and when are referred to as 'Kabloona questions'.".

In the case of James Bay area harvest studies, those conducting the questionnaire ran into similar difficulties. While the Cree had relatively little problem in providing numerical answers regarding big game, waterfowl and fur animals, they themselves pointed out that the numbers for fish (and small game) were usually only approximations. It became important, therefore, to try to verify independently the numbers of fish as reported to the Native Harvesting Research study (NHR 1982). This was done by using field data from a study that had been carried out since 1975 in one of the communities covered by NHR (1982). It should be pointed out that the field study in question (Berkes 1977; 1979; 1982) was not designed in the first place to provide backup for the 1975-1979 questionnaire/ diary investigation of NHR (1982). Nevertheless, it was possible to evaluate the reliability of the reported harvest levels of fish in Fort George (Chisasibi) by comparing them with the data gathered by direct observation of subsistence fisheries of that community.

A Field Study on Native Subsistence

The area and methods of study: Called Fort George until 1980 at which time it was relocated and renamed Chisasibi, the community in question is within the area covered by the James Bay and Northern Quebec Agreement. The harvest study was carried out under the provisions of the Agreement and represents the joint efforts of native groups, the governments and the crown corporations which are the co-signatories of the Agreement. There have been a series of NHR reports, covering different years, but only the final NHR report is cited here as NHR (1982). There is a separate series of reports on the harvests of the Inuit, but the final report of the Inuit NHR had not been completed at the time of writing.

A complicating factor in the NHR results for fish is the low harvest values in the years following 1975 and the adverse publicity regarding mercury in fish. There was field evidence to suggest that the low fish harvest reported in 1976-1977 for example is related to the 'mercury scare' of 1975-1976. For this reason, the pre-scare year of 1974-1975 is considered to provide more representative data on the Chisasibi subsistence harvests of fish.

Table 1 is a summary of the field data being compared to the NHR study results. Data base of the field study consists of 584 net checks, yielding 8,944 fish totalling 3,094 kg, and belonging to 27 different fishermen, between the spring of 1975 and the spring of 1979 in Fort George. Since the field study has continued after the NHR study was completed, this report also includes data from late 1979, 1980 and 1981, based on another 207 net sets belonging to an additional ten fishermen (37 fishermen in all). The number of fish actually recorded in the field constitute 1.5 per cent to 3.4 per cent of the total community catch as recorded by the NHR study for the five years of the investigation (Table 2). The sample locations are shown in Figure 1.

In evaluating the reliability of the reported harvests of fish, a three-pronged approach is taken. First, the field data are organized in such a way as to permit an estimate of the total community harvest, which can then be compared with the NHR results. Secondly, the annual harvests for eleven family groups (covering the years 1975-1976 to 1980-1981) are estimated on the basis of both field data and of interview information on the number of days fishing. These harvest figures are then converted into per capita harvests. The units used are "kg round weight per person", for every man, woman and child, regardless of age. The choice of these units minimizes inferences and extrapolations, and makes comparisons possible with other published data in the literature. This provides the third approach, comparing per capita harvest levels from the field study with those from the NHR study and other information across the north.

Results: Table 3 is a summary of the numbers of fish harvested per net-check (in the case of gillnets) or per day (in the case of seining or angling) in different years and areas/seasons. Therefore Table 3 shows the year-to-year variation in fishing success, and the rather large variations by season and area. The mean harvest levels are calculated by season and area. There were no samples for inland areas, 1975-1979, but supplementary data were available from 1980-1981: 88 fish (N = 12 net-checks) for a mean catch of 7.3 fish. Table 4 parallels Table 3, but is based on weight units. Here the 1980-1981 supplementary data for inland areas consists of 140 kg (N = 12 net-checks) for a mean of 11.7 kg.

Table 5 is constructed using the mean values in Tables 3 and 4 and the supplementary data from inland areas. The "soft data" in Table 5 are the estimates of the number of participants in the fishery and the number of net-checks (or days of fishing) by each participant. These numbers, based on interviews, general community knowledge and educated guesses, are given as ranges. There are two exceptions: the number of participants in the First Rapids fishery and the Fort George area fall/winter fishery were taken from Band Council records and are considered reliable.

The community harvest by numbers, as estimated by Table 5, ranges from a low of 56,128 to a high of 179,306. The variation is about three-fold, and stems from both the method of estimation of the first two columns of Table 5 and from the real year-to-year

Table 1. Summary of the field data being compared with the NHR diary/questionnaire studies, 1975-1979

Year	Season[1]	Number of net checks	Number of fish[2]	Weight of fish, kg[3]	Number of fishermen[4]
1979	Spring	26	430	115.1	3
1978/1979	Winter	32	425	109.1	1
1978	Summer/fall	50	1313	418.7	4
1978	Spring	6	133	33.4	1
1977/1978	Winter	38	776	269.2	3
1977	Summer/fall	43	722	213.2	3
1977	Spring	—	—	—	—
1976/1977	Winter	42	981	221.3	2
1976	Summer/fall	135	1237	653.0	6
1976	Spring	71	927	346.5	3
1975/1976	Winter	11	280	63.7	2
1975	Summer/fall	73	826	306.7	6
1975	Spring	57	894	343.9	10
Total		584	8944	3093.8	27

[1] Time periods used in the NHR studies were, counting backwards:
(1) from break-up to June 30, i.e. spring
(2) from freeze-up to break-up, i.e. winter
(3) from July 1st to freeze-up, i.e. summer/fall.

[2] All species caught, except sculpins, *Myoxocephalus*, which were thrown out.

[3] Whole (or round) weights.

[4] Many of the fishermen participated in more than one season of the field study.

Table 2. Sampling intensity, NHR study vs. field data for numbers of fish harvested

Year	NHR total	Field study	% of total
1974-1975 (time period 1 only)	57,714	894	1.5
1975-1976	116,680	2,033	1.7
1976-1977	64,649	2,218	3.4
1977-1978	62,243	1,631	2.5
1978-1979	102,762	2,168	2.1
Mean sampling intensity			2.2

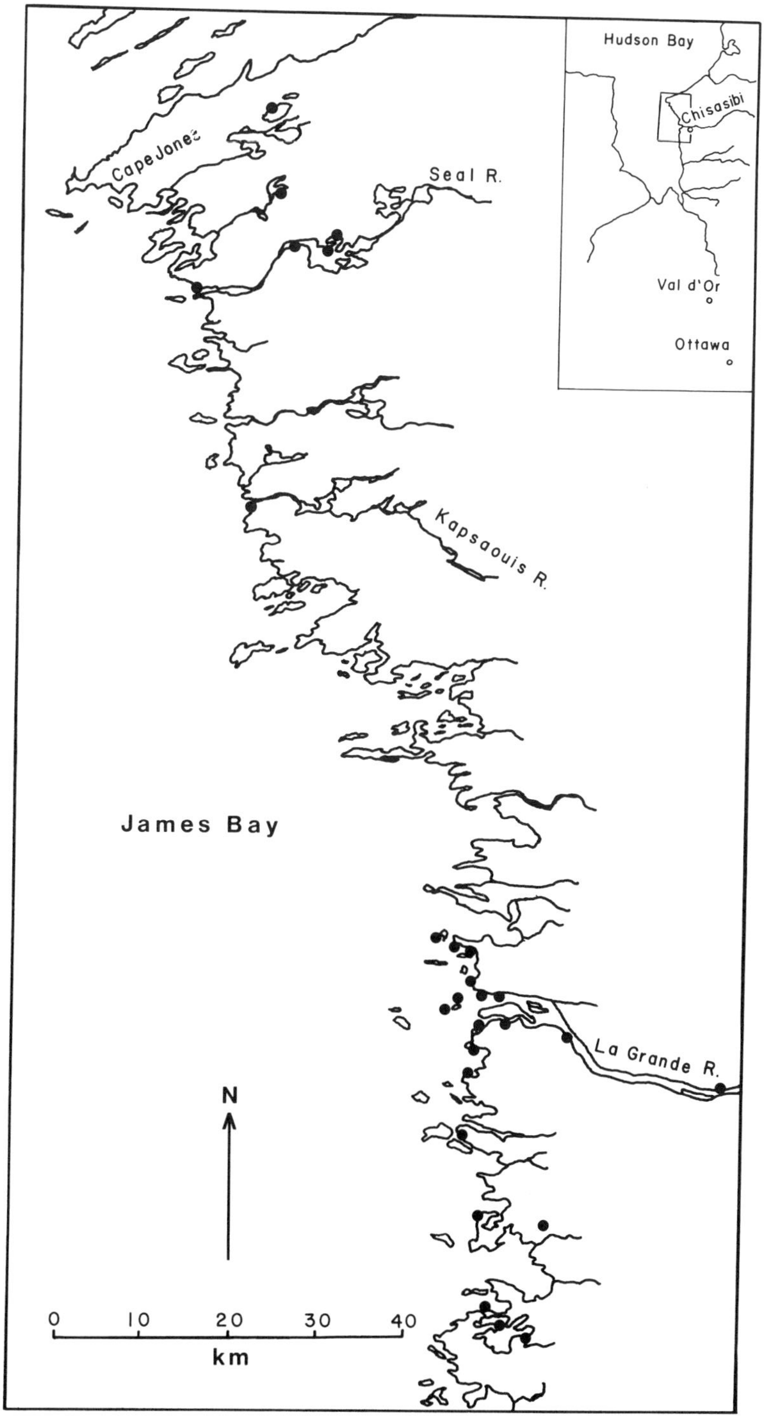

Figure 1. The Fort George area and the sampling stations (closed circles); there were an additional four inland field study stations, all in the LG-2/LG-3 area along the roads.

Table 3. Calculation of mean harvests in the field study is based on the numbers of fish by year and season/area. Number of net-checks is given in brackets.

Year	Fort George area spring/summer	F. George area First Rapids	F. George area fall/winter	North coast	South coast	Angling
1975	1,090 (90)	141 – 1 day	535 (25)	114 (7)	120 (14)	900 est. – 2 days – 20 people
1976	1,264 (137)	—	1,507 (55)	374 (56)	—	—
1977	—	256 – 1 day	920 (53)	217 (15)	—	—
1978	133 (6)	260 – 1 day	1,478 (72)	—	—	—
1979	430 (26)	—	—	—	—	—
Mean harvests	15/net-check	219/day	21.5/net-check	12.5/net-check	8.5/net-check	22.5/day

Table 4. The calculation of mean harvests in the field study based on whole weights (kg) by year and season/area. Number of net-checks is given in brackets.

Year	Fort George area spring/summer	F. George area First Rapids	F. George area fall/winter	North coast	South coast	Angling
1975	379 (90)	54 – 1 day	124 (25)	66 (7)	92 (14)	450 est. – 2 days – 20 people
1976	417 (137)	—	354 (55)	450 (56)	—	—
1977	—	90 – 1 day	237 (53)	—	—	—
1978	33 (6)	91 – 1 day	436 (72)	155 (15)	—	—
1979	115 (26)	—	—	—	—	—
Mean harvests	4.3/net-check	78.3/day	5.5/net-check	9.2/net-check	6.6/net-check	11.2/day

Table 5. Estimate of the total community catch at Fort George, partially based on field data.

	No. of participants (estimate)	No. of net-checks per participant (estimate)	Mean no. of fish per net-check (field data)	Mean yield per net-check, kg (field data)	Total harvest number of fish	Total harvest kg (weight)
1. Fort George Area						
(a) spring/summer	20	10	15	4.3	3,000	860
	30	20			9,000	2,580
(b) First Rapids, August	57	1 day	219/day	78.3/day	12,483	4,463
		2 days			24,966	8,926
(c) Fall/winter	43	10	21.5	5.5	9,245	2,365
		20			18,490	4,730
2. North coast	20	40	12.5	9.2	10,000	7,360
	40	60			30,000	22,080
3. South coast	15	40	8.5	6.6	5,100	3,960
	30	60			15,300	11,880
4. Inland lakes	25	40	7.3	11.7	7,300	11,700
	100	60			43,800	70,200
5. Angling (various	200	2 days	22.5/day	11.2/day	9,000	4,480
locations)	300	5 days			33,750	16,800
Totals: Low estimate					56,128	35,188
High estimate					179,306	137,196

Table 6. Annual total and per capita harvests of fish for several family groups in Fort George

		Fish harvest, round weight, kg				
Family group	Year	Monitored	%	Additional harvests[1]	Annual total	Per capita[2]
A	1975 - 1976	485	45	598	1,083	90
B	1975 - 1976	92	10	800	892	89
C	1976 - 1977	605	54	513	1,118	86
D	1976 - 1977	97	10	900	997	91
E	1977 - 1978	421	44	536	957	74
F	1977 - 1978	34	36	60	94	13
G	1978 - 1979	452	69	202	654	47
H	1979 - 1980	114	56	90	204	15
I	1979 - 1980	67	51	64	131	19
J	1980 - 1981 (6 mo.)	37	65	20	2 × 57	8
K	1980 - 1981 (6 mo.)	215	26	600	2 × 815	125
overall means			33.3			60

[1] Estimated by interview, from number of days fishing at a given location and the known catch per unit of effort at that location and season (Table 4).

[2] Total catch divided by the number of people who normally share the catch; this could be more than one family unit and includes children.

variation in the harvest. For comparison, questionnaire/diary studies produced the following figures:

1978-1979 – 102,762 fish (the last year of NHR)
1976-1977 – 64,649 fish (the low year of NHR)
1974-1975 – 173,349 fish (the high year of NHR)
1973-1974 – 171,720 fish (Weinstein 1976).

Table 6 shows the family harvests and per capita harvests of several groups (some groups are covered more than once) over several years. These groups and their monitored harvests comprise about 85 per cent of the total field data summarized earlier in Table 1. There is a wide range in the per capita harvest, but most data seem to fall into two distinct groups: those harvesting 10-20 kg/capita/year and those harvesting 50-100 kg/capita/year.

Discussion: The range of estimates in Table 5 is very similar to the range of results of the NHR study. But of course, Table 5 has the advantage of hindsight. Perhaps the point to emphasize here is that NHR results and their year-to-year variations can indeed be explained by using field data on the catch per unit of effort and a set of reasonable assumptions on the fishing effort.

Those parameters which are estimated (i.e., the fishing effort) are probably on the conservative side. For example, only 25 to 100 fishermen are assumed to participate in inland lake fisheries, whereas the number in some years could easily exceed 200. Further, the assumed fishing effort per participant is 40 to 60 net-checks per year. This means only 20 to 30 days of fishing (with two nets) per year. Whereas, a good number of families may be fishing more than that many days per season.

The harvests summarized in Table 6 may be taken with a relatively high degree of confidence. The percentage of the annual catches actually monitored in the field averaged 33.3 per cent ranging as high as 69 per cent; the cutoff point for inclusion was 10 per cent. The age composition of the fishermen in the sample is similar to that of NHR respondents. However, only 27 per cent of the fishermen in the sample were 'intensive' hunters (those who were on the income security program for hunters and who were harvesting for at least 120 days of the year). This is lower than the Fort George average, which was 48 per cent in 1975-1979, according to H. Feit (personal communication). Thus, there is a conservative bias in the field study. In fact, none of the fishermen in the sample can be called really 'traditional'. Even those who were 'intensive', returned to the community on the average of once a week. (For the researcher, the logistics of sampling the catch of a family which stays in the bush for long periods is difficult.)

With the above consideration, the average per capita harvests in Table 6 may be compared with the NHR estimates and Table 5 estimates. Note that in Table 5, the conversion factor is 650 g whole weight per fish, which is also applied to the NHR data and to Weinstein's (1976) data. To obtain per capita estimates from Table 5, a population figure of 1,690 is used (the 1978-1979 average). For the NHR data, the appropriate registered population figures for the year are used. The following were the per capita harvests in Fort George:

1978-1979; 40 kg (the last year of NHR)
1976-1977; 26 kg (the low year of NHR)
1974-1975; 75 kg (the high year of NHR)
1973-1974; 70 kg (Weinstein 1976)

1975-1981; 20 kg the low estimate (Table 5)
1975-1981; 80 kg the high estimate (Table 5)
1975-1981; 60 kg mean of selected family groups (Table 6).

Note that the data for selected family groups in Table 6, which comprises 85 per cent of all field data, do not show the rather large fluctuations in the community harvest as revealed by the NHR study. In fact, the monitored per capita harvests in 1976-1977 (the lower year of NHR) are as large as those in any other year (Table 6; see also Table 2). Hence, the field study did not pick out the low or high years in the total community harvest. The reason for this is that the field study used the services of whoever happened to be fishing. Thus, it appears that these low and high years do not so much depend on the fluctuation of effort of individual fishermen. Rather, the high and low years may depend on the number of participants in the fishery in a given season and year.

Conclusions: The mean annual community harvest estimated on the basis of field data is similar to the NHR questionnaire/diary estimates for 1975-1979. The large year-to-year fluctuation in harvests is explained mainly in terms of the number of participants in a given year. The annual per capita harvest levels of eleven family groups (with 69 per cent to 10 per cent field monitoring of their harvests) was 60 kg (1975-1981). This is well within the range of NHR data, 26 kg (1976-1977) to 82 kg (1974-1975). There are two lines of evidence that the NHR figures are too low: (a) comparing per capita harvest by years, the families followed in the field consistently harvested more than the NHR community averages, and (b) the field sample had proportionately fewer (27 per cent) 'intensive' harvesters than the community as a whole (48 per cent).

A comparison of the species composition of the field harvest with the species composition reported to NHR indicates that the lesser-valued fish such as suckers (used as dog food and trap bait) are systematically under-reported by the Cree. On the basis of the above considerations, it is concluded that the questionnaire/diary study underestimated the actual harvest. The level of under-reporting is not likely to be as much as 50 per cent, but may well be around 30 per cent.

Per Capita Subsistence Harvests in the North: A Literature Review

To place Fort George catches in perspective, Table 7 attempts to compare the NHR figures from James Bay and northern Quebec and other published data on subsistence fisheries elsewhere in the north. The information summarized in Table 7 does not constitute an exhaustive literature review. There are no doubt other sources of data on native wildlife harvests. With rare exceptions, native harvesting data do not seem to appear in the primary literature but rather in the "grey literature" which is not so readily accessible even to the specialist.

Thus, Table 7 represents a first attempt to cull from the information base on native subsistence fisheries a synthesis of some of the more accessible studies. To do this, information was summarized in terms of per capita harvest or consumption of fish, kg whole (round) weight per year. This unit was chosen to minimize subjective interpretation of data, and in some cases this choice of units necessitated going back to the reported raw data on harvests and on the population size. The table also specifies whether the authors expressed their findings as harvested weight or consumed weight. The difference between the two would presumably represent the loss of potential human food due to wastage and due to the use of fish for dog food and for trap bait.

Table 7. A summary of subsistence (domestic) fish harvests in Canada's north

Location	Year	Population	Total fish, kg	Per capita fish, kg/yr	Harvested or consumed	Reference
James Bay Cree, Quebec	1974/75	6,267	320,000[1]	51	harvested	NHR (1982)
Northern Quebec Inuit	1978	3,981	233,000[1,5]	59	harvested	JBNQNHRC (1982)
Churchill R. study area, Saskatchewan	1972	3,474	31,462	9	consumed	Stabler *et al*. (1975)
Eskimo Point, NWT	1976/77	834	52,727[2]	63	harvested	McEachern (1978)
MacKenzie Valley, NWT	1972	7,485	374,954	50	harvested	Bisset (1974)
Aklavik, NWT	1973	677	31,112[3]	46	consumed	Jessop *et al*. (1974)
Arctic Red River, NWT	1973	96	58,848[4]	613	harvested	Jessop *et al*. (1974)
Lac La Martre, NWT	1972	160	45,018[3]	281	consumed	Bond (1973)
Old Crow, Yukon	1973	142	13,638[4]	96	harvested	Stager (1974)
Baker Lake, NWT	1976	900	157,400[4]	175	harvested	Stager (1977)
Baker Lake, NWT	1977	900	65,000	72	consumed	IDS (1978)
Resolute area, NWT	1976	179	8,100[4]	45	harvested	Kemp *et al*. (1977)
Clyde River, Grise Fiord and Pond Inlet, NWT	1979	1,185	19,145	16	harvested	Finley & Miller (1980)

[1] Converted to whole weights, edible weight = 0.70 whole weight.
[2] This is the "low" estimate. The "mid-range" estimate is 82,586 kg; consumed weight = 65 per cent of harvested weight.
[3] Another 102,490 kg estimated to go into dog food in Aklavik; 54,736 kg in Lac La Martre.
[4] Includes dog food, no separate estimate given for human consumption alone.
[5] The 1976 reported harvest of 606,500 kg gives 176 kg per capita. Includes small commercial catches of salmon and char (L. Brooke, personal communication).

The overall Cree and Inuit harvests from the NHR study area, on a per capita basis, tend to fall in the mid-range of available data. The remarkable spread of figures requires some explanation, even though no attempt is made here to provide a systematic critique of individual studies. The figure of 9 kg per capita consumed weight for the Rocky Cree people in six communities of the Churchill River study area (Stabler *et al*. 1975) is comparable to the low group in Fort George (10 to 20 kg harvested weight). Nevertheless, other data from northern Saskatchewan indicate higher harvests. Jarvenpa (1980) found an average per capita consumption level of 112 kg for a group of 26 Chipewyan people of Patuanak. The other low figure, 16 kg for the three northerly Inuit communities, comes from an area where, simply, there is not much fish available.

In the case of the apparently very high per capita harvest of Arctic Red River, the investigators actually monitored the landing of about 6,200 kg over a period of only three and one half months. This alone would give a harvest level of 65 kg per capita, and the total annual harvest is obviously higher. This abnormally high harvest may be due to the possible utilization of the Arctic Red River catches by other communities (Inuvik?); this point is not clear in Jessop *et al*. (1974).

In addition to these community-wide estimates, there is also one regional estimate in the literature. The summary report prepared for the NWT Science Advisory Board (SAB 1980) suggests that a conservative estimate for the volume of the annual subsistence harvest would be about 2½ million pounds or 1,136,360 kg. This gives 52 kg per capita if only the native population in the NWT (22,000 in 1975) is considered.

Fish harvests summarized in Table 7 may be evaluated on the basis that a non-traditional Fort George family obtains 10 to 20 kg fish per capita per year. A more 'traditional' but largely community-based family appears to obtain about 50 to 100 kg per capita per year. It would not be unexpected, therefore, if families spending longer periods in the bush obtained considerably more.

Fort George is a relatively large native community, and thanks to the James Bay hydro project, wage-labour opportunities through the late 1970s were reasonably good. In the case of smaller native communities and/or those with fewer wage-labour opportunities than in Fort George, fish harvests are likely to be higher than those in Fort George. Thus, the following may serve as first order approximation for estimating annual per capita fish harvests in northern native communities: 'non-traditional' family in a relatively large community, 10 to 20 kg; 'traditional' family in a relatively large community, 50 to 100 kg; 'traditional' family in a small community, 100 to 200 kg.

Designing Studies of Native Subsistence Fisheries

In designing a questionnaire that would provide defensible data on subsistence harvesting, there are certain considerations that have to be taken into account. Based on the James Bay experience, the NHR committee recommended "that a number of specific measures be incorporated into research design to assure scientifically and professionally acceptable level of reliability" (NHR 1982). These were as follows:

(1) there should be a maximum of participation by native people, and interviewing should be conducted in the native languages;

(2) research instruments (questionnaires and diaries) should be adapted to local concepts of animals, seasonality, geographical location and social organization, and should be pre-tested;

(3) explanations of the research and its objectives should be communicated to the communities before interviewing and before distribution of diaries;

(4) internal checks on response consistency should be made by questioning harvesting time intensity, differences between animals caught and pelts sold, and catches by group and individual;

(5) comparisons should be made of harvest results with data available from earlier studies, and from provincial fur sales statistics;

(6) comparisons should be made of harvest results with estimates made by tallymen and community councils;

(7) spot checks should be conducted by research staff and NHR committee members.

More specifically regarding the investigation of subsistence fisheries, certain kinds of background information would be necessary to design an appropriate questionnaire. The following are some of the major considerations.

Native harvesters are selective in the kinds of species they harvest. Thus, background biological work may be misleading as regards to which species are more available. In the Baker Lake area, for example, arctic char *(Salvelinus alpinus)* comprised only 5.6 per cent of the biological samples but 42 per cent of the native domestic harvest (Tester 1979). In the Fort George area, coregonids (whitefish, *Coregonus clupeaformis*, and cisco, *C. artedii*) comprised 59 per cent of the biological samples but 78 per cent of the native harvest (Berkes 1979). Further, there may be identification problems in translation. For example, the James Bay Cree did not make a distinction in English between cisco and whitefish, even though these species certainly have two distinct names in Cree (Berkes and MacKenzie 1978). Thus, the reported Cree harvest of whitefish included cisco; indeed it also included the round whitefish (*Prosopium cylindraceum*) and landlocked cisco (to biologists also *C. artedii*) which to the Cree is a distinct fish with a distinct name (Berkes and MacKenzie 1978).

If the questionnaire study involves a sample of the harvesters, it is important to stratify the sample so that the fishermen who are interviewed provide a representative sample of those who harvest many and those who harvest few. For example, in Clyde River with 90 hunters, 52 per cent of the total char harvest was taken by 4.4 per cent of the fishermen; in fact 41 per cent of it was taken by one fisherman (Finley and Miller 1980). This problem is likely to be more acute in large communities with a mix of different ethnic native groups, such as Baker Lake and Mistassini. Some groups in the community may rely much more heavily on fishing than others.

A knowledge of the fishing areas used may be important. Among the Inuit, different family groups have distinct favourite fishing areas. Among the Indians, some groups may be obtaining all of their fish from their traditional hunting/trapping areas, which may or may not coincide with their government-registered traplines. Other groups may be obtaining a relatively greater portion of their catch from communal fishing areas. The choice of areas, in turn, will be related to the fishing methods used: gillnets, set lines, fish traps or seines, and hence to the seasonal concentration of species harvested.

It is important to note whether sled dogs are used in a community. Even if sled dogs are rare, significant numbers of fish may still be fed to pets or hunting dogs and used as trap bait.

For many native groups, fish are considered a staple resource. They are harvested throughout the year, and as a matter of course even during the more important seasonal events such as the goose hunt in James Bay. A casual interview will not alert the researcher that most harvesters would have a couple of nets out while hunting geese. Except in a few places such as the Pacific coast, fishing is not a particularly high-profile and prestigious activity. For the boreal zone native fishermen, perhaps only sturgeon would be considered important enough that the numbers caught would be noted and remembered. Similarly, large lake trout were treated with reverence in Mistassini, at least in the old days. Most species would command less respect. Freeman (1976) writes: "When asked to mark sculpin fishing areas many respondents declined, or with an amused disdain marked only a few places, despite the fact that very many of the men have eaten and enjoyed this abundant marine fish". For such fish, questions about numbers would not elicit very good answers. Perhaps what the researcher needs to do is to find some convenient and locally meaningful unit to quantify the harvest of small fish. In Fort George, before the people were subjected to year-after-year of questionnaire studies, such a convenient unit would have been a "tubfull". However, native respondents do adapt to the kind of questions they are likely to receive. One old Fort George fisherman, carrying his tubfull of cisco to the smoking tent, stopped long enough to give the researcher a broad smile, "ninety-seven", he said. Freeman (1976, 1979) provides some pointers on how to put questions in the correct cultural context.

With these general considerations in mind, the greatest accuracy in a subsistence fishery survey could probably be achieved by combining a field study approach with a questionnaire approach. A questionnaire study alone would leave lingering doubts about the data; a field study alone would not work. It would be impossible or nearly so to field-record completely a subsistence fishery, especially where the fishermen are eating the fish about as fast as they catch them. A researcher on native subsistence fisheries, coming into a new community and wishing to optimize his time, may proceed in three phases, as follows.

1. The preparatory phase — The researcher should, of course, be familiar with any land use or harvest reports previously done for that area. If none are available, the researcher should be prepared to spend a few days visiting fishermen, informally over a mug of tea in the kitchen, to get a feeling for the areas used for fishing, seasons and methods. He/she should become familiar with place names and should carry a good fish reference book with pictures to ascertain species identifications.

2. The interview phase — To be able to verify the results of the harvesting questionnaire study, the researcher should trace the seasonal fishing cycle of selected informants, recording names of locations, the number of days fished per location and season, and the major species taken. Based on the James Bay experience, this is relatively easy to do, and the number of days of fishing can be readily converted into numbers and weights of fish, once some field data are available.

3. The field verification phase — Native fishermen are likely to appreciate that the researcher is willing to go fishing with them, sharing the expenses, and is willing to learn by doing, which is a culturally more acceptable way of obtaining information than asking

(often ill-conceived) questions about harvests. Data on the catch per unit of effort in a given area and season provide the necessary information to quantify the harvests of selected groups, and thence to verify the community-wide harvest reported in questionnaires. It was found in James Bay area studies (Berkes 1979) that individual variation in fishing success at a given time and place falls within the relatively narrow range of ± 33 per cent in three cases out of four.

The field study also makes it possible to verify the species, the land use pattern, and to collect biological data at very low cost. Collecting data on the fish obtained by subsistence fishermen is both feasible and cost-effective, as shown in the James Bay area studies (Berkes 1977, 1982). Paying the fishermen for transportation and for the privilege of examining their catch makes it possible to trim costs on the equipment that would otherwise have to be purchased, transported, maintained and stored.

The general approach outlined here would work only with the cooperation of the local population. Needless to say, harvesting studies and subsistence fishery studies can only be carried out if the local population perceives these studies as being in some way beneficial to them in protecting what they view as their land and resources. In a wider perspective, there is general and worldwide interest, for example, in Food and Agricultural Organization (FAO) circles in subsistence and artisanal fisheries. Through FAO, self-administered questionnaire approaches, similar to the harvesting diary approach used in many of the studies cited here, have been developed. It would be naive to assume that such questionnaire studies can be widely applied in the Canadian north, as long as the native people perceive that their harvesting rights are being threatened by the dominant industrial society, and that any research over which they themselves have no control represents intrusion that may work to compromise their own interests.

Acknowledgements

I thank the Chisasibi Band Council, the Cree Regional Authority and many fishermen for cooperating with this study. I thank H.A. Feit, L. Brooke and other members of the James Bay and Northern Quebec Native Harvesting Research Committee. The comparison between the field study and the harvest study was undertaken for the Harvesting Research Committee, and appears as Appendix 10 in NHR (1982). The field study was supported by the Social Sciences and Humanities Research Council of Canada.

Literature Cited

Balikci, A. 1968. The Netsilik Eskimos: adaptive processes. *In*: Man the hunter. R. Lee and I. DeVore (Eds.). Aldine, Chicago, Illinois. pp. 78-82.

Berger, T.R. 1977. Northern frontier, northern homeland. The report of the Mackenzie Valley pipeline inquiry. Volume Two. Supply and Services Canada, Ottawa, Ontario.

Berkes, F. 1977. Fishery resource use in a subarctic Indian community. Human Ecology, 5: 289-307.

Berkes, F. 1979. An investigation of Cree Indian domestic fisheries in northern Quebec. Arctic, 32: 46-70.

Berkes, F. 1982. Preliminary impacts of the James Bay hydroelectric project, Quebec, on estuarine fish and fisheries. Arctic, 35: 524-530.

Berkes, F. and M. MacKenzie. 1978. Cree fish names from eastern James Bay, Quebec. Arctic, 31: 489-495.

Bisset, D. 1974. Resource harvests — Hunter-trappers in the Mackenzie Valley. Task Force on Northern Oil Development Report No. 74-42. Information Canada.

Bond, W.A. 1973. An investigation of the commercial fishery at Lac La Martre. Tech. Rep. Series No. 73-5. Fisheries and Marine Service, Department of the Environment, Winnipeg, Manitoba.

Brody, H. 1981. Maps and dreams. Indians and the British Columbia frontier. Douglas and McIntyre, Vancouver/Toronto.

Feit, H.A. 1973. The ethno-ecology of the Waswanipi Cree; or how hunters can manage their resources. *In*: Cultural ecology. B. Cox. (Ed.). McClelland and Stewart. Toronto, Ontario. pp. 115-125.

Feit, H.A. 1980. Negotiating recognition of aboriginal rights: History, strategies and reactions to the James Bay and Northern Quebec Agreement. Can. J. Anthr. 1: 159-172.

Finley, K.J. and G.W. Miller. 1980. Wildlife harvest statistics from Clyde River, Grise Fiord and Pond Inlet, 1979. LGL/Petro-Canada Explorations, Calgary, Alberta.

Freeman, M.M.R. (Ed.). 1976. Inuit land use and occupancy project. Volume Two. Supply and Services Canada, Ottawa, Ontario.

Freeman, M.M.R. 1979. Traditional land users as a legitimate source of environmental expertize. *In*: The Canadian National Parks: Today and tomorrow, Volume One, J.G. Nelson *et al*. (Eds.). University of Waterloo. Waterloo, Ontario. pp. 345-361.

IDS. 1978. Effects of exploration and development in the Baker Lake area. Prepared by Interdisciplinary Systems Ltd., Winnipeg, for DINA. (Executive summary reproduced in Musk-Ox 21: 62-66, 1978).

JBNQNHRC. 1982. Research to establish present levels of native harvesting. Harvests by the Inuit of Northern Quebec. Phase II (Years 1977 and 1978). James Bay and Northern Quebec Native Harvesting Research Committee. Montreal, Quebec.

Jarvenpa, R. 1980. The trappers of Patuanak: Toward a spatial ecology of modern hunters. National Museum of Man Mercury Series, Canadian Ethnology Service. Paper No. 67.

Jessop, C.S., K.T.J. Chang-Kue, J.W. Lilley and R.J. Percy. 1974. A further evaluation of the fish resources of the MacKenzie River Valley as related to pipeline development. Task Force on Northern Oil Development Report No. 74-7. Information Canada.

Kemp, W.B., G. Wenzel, N. Jensen and E. Val. 1977. The communities of Resolute and Kuvinaluk: A social and economic baseline study. The Resolute Community Council, Kuvinaluk Community and the Polar Gas Project.

Lotz, J. 1976. Area economic surveys: critique and assessment. pp. 23-30. *In*: Inuit land use and occupancy project, Volume Two. M.M.R. Freeman (Ed.). Supply and Services Canada, Ottawa, Ontario.

McEachern, J. 1978. A survey of resource harvesting, Eskimo Point, NWT, 1975-77. Quest Socio-Economic Consultants Inc., Delta, B.C., for the Polar Gas Project.

NHR. 1982. The wealth of the land. Wildlife harvests by the James Bay Cree, 1972-73 to 1978-79. James Bay and Northern Quebec Native Harvesting Research Committee, Quebec City, Quebec.

SAB. 1980. Fish, fur and game in the Northwest Territories. Science Advisory Board of the Northwest Territories, Yellowknife, Northwest Territories.

Stabler, J.C. *et al*. 1975. Churchill River study (Missinipe Probe) Socio-Economic (Saskatchewan). Final Report No. 25.

Stager, J.K. 1974. Old Crow, Y.T., and the proposed northern gas pipeline. Task Force on Northern Oil Development. Report No. 74-21. Information Canada.

Stager, J. 1977. Baker Lake, NWT. A background report of its social and economic

development. The Baker Lake Settlement Council and the Polar Gas Project.
Tester, F.J. (Ed.). 1979. The potential social, economic and environmental impacts of the proposed Polar Gas pipeline on the District of Keewatin, Northwest Territories. Volume Two, Analysis. Department Indian and Northern Affairs, Ottawa, Ontario.
Weinstein, M.S. 1976. What the land provides. Report of the Fort George Resource Use Study. Grand Council of the Crees (of Quebec), Montreal, Quebec.
Winterhalder, B. and E.A. Smith, (Eds.). 1981. Hunter-gatherer foraging strategies. University of Chicago Press, Chicago, Illinois.

HORSE PRODUCTION FOR NORTHERN AREAS

William L. Pringle
Agriculture Canada

Abstract: It is estimated that there are approximately 700 thousand horses in Canada. Almost without exception they will all be exported as human food. Canada exported 16 million kilograms of horse meat in 1980 which grossed about 25 million dollars. This represented about 70,000 horses slaughtered and was 5 times the amount reported for 1967. The horse can utilize lower quality forage than the ruminant animal and so is able to survive and thrive in our northern areas where beef cattle would perish. The use of land in northern Canada for raising beef is limited by the length of the grazing season. In most areas it is necessary to provide up to 3 tonnes of hay per cow to bring it safely through the winter. Horses can paw through 50 cm of snow for their sustenance. They are therefore able to utilize much of the aftermath from seed fields as well as some of the sedgegrass mixtures from wet and inaccessible areas. A horse operation in northern British Columbia running 450 head has shown that horses wintered easily even in the deep snow of 1981-1982. In most years only mares in foal and weanlings require supplemental feed for about 60 days from February on. Horses have a great flexibility in marketing, do not need grain for finishing but do need maturity and fat cover. One of the greatest advantages of a large scale horse production operation is that it is energy efficient as well as labor efficient compared to cattle production. To date Canada has very little research underway as to breeding or feeding of horses for the meat trade.

This paper will present and discuss some of the theories and practical aspects about utilizing northern forage resources for horses. It is designed to show that horse raising for the meat industry will become an accepted and profitable agricultural enterprise or form of land use over much of the fringe areas of northern Canada in the foreseeable future.

Anon. (1979) reported that in the first six months of 1978 the United States shipped abroad 26.5 million kilograms of horsemeat as compared with 30.1 million kilograms of beef and 33.9 million kilograms of pork. They outline the horse meat business as follows: ''Horses are bought at small horse sales in rural areas then trucked to the packing plants where they are penned until they are slaughtered. Dressed carcasses are then trucked to international airports where they are flown to foreign distributors. The entire trip from sale to European table usually takes less than a week''.

Canada exported 15.4 million kilograms of horse meat in 1978, which grossed 24.6 million dollars; one-half of the meat was sold to France (Jonas 1980). In 1979, 15.8 million kilograms were exported (Matthews 1980). This represents a three-fold increase in sales since 1967. There are three major and four minor packing plants in Canada processing horse meat; two are in Alberta, one each in Manitoba and Ontario and three in Quebec. These plants handle about 70,000 animals annually or approximately ten per cent of the total horse population in Canada. The Canadian attitudes regarding the slaughter of horses for export are obviously changing.

With such large numbers of horses being removed from the population we can see that if exports are to expand a production industry must be established and nurtured. This would provide a supply of meat horses and at the same time help to preserve the light horse industry from pressures of the marketplace.

Through the western Canadian north it is a common sight to see horses surviving the winter either on snow-covered hayfields or on the edge of forests where they receive some protection. In some cases they are fed some hay to supplement the meagre supply they can dig from beneath the snow. The horse in northern areas is a survivor and can even thrive where cattle would perish. Can their potential for producing under adverse climatic conditions be exploited so that a new industry can be established where horse production will become an ongoing and integral part of ranching operations?

French and Soviet Horse Productions

There is a scarcity of literature on horse production for meat. The majority of the research has been conducted in the U.S.S.R., France, Belgium and Italy. These papers deal mainly with subjects such as production numbers, market development, breeding of meat horses, meat quality and horse nutrition. The research station at Theix in France specializes in all aspects of horses for the meat trade. Their research projects include topics such as feeding and management of a herd of breeding mares, grazing poor hill pastures, over-wintering, foal fattening at different ages, and carcass composition. Martin-Rosset *et al.* (1982), reported that France consumes 100,000 tonnes of horse meat annually which is 1.8 kg/person/year with about eighty per cent being imported. France has 400,000 horses, half of which include draught types currently used to produce horse meat. The reproduction rate of mares is low (40 to 45 per cent) because the fertility rate obtained with the traditional mating system is limited to fifty-nine per cent and foal mortality is from 13 to 20 per cent.

Half of the horse meat produced in France is supplied by heavy horses, particularly colts. Research in the past ten years has shown that colts can be fattened in feedlots or on forage in all types of production systems which include intensive (slaughter age: 12 months), semi-intensive (slaughter age: 18-24 months), or extensive (slaughter age: 30 months) systems.

Andreyev (1971) has described horse breeding for meat in Yakutskaya, A.S.S.R. He indicated that horse herding like reindeer herding is based on natural food resources and has long been one of the main branches of the economy. These animals permit wider use of food resources of distant regions and require little labour for care or feeding. Most of the horses remain on pasture throughout the year and receive additional fodder only under deep snow conditions.

During 1961-1966 in Yakutskaya, A.S.S.R., horse meat contributed 20.4 per cent of the total meat produced. Some state farms produced over 40 per cent of the total meat production from horses. An average cost of producing horse meat on the Verkhoyanskiy State Farm during that period was 63.8 roubles per centner (i.e., per 100 kg) while the cost of producing beef was 149 roubles per centner. Foals of Yakut horses have weight increases of 1.0 to 1.2 kg/day before weaning during the summer months. This drops to 30 to 80 g/day during the winter.

Herds consist of up to 11 mares for one stallion; the most common herd structure is about 65 per cent mares, 62 per cent of which will have foals at any one time. In the Kolyma basin swampy land around lakes which cannot be used for hay cutting or grazing cattle is ideal for horses. Tussocky swamps in the forest tundra unsuitable for reindeer also make excellent horse pasture. Managers claim that in localities with light snowfall and continental climate the Yakut horse is superior but where there is heavy snowfall and a less continental climate the horse bred from the Kamchatka area is preferred. They suggest that other areas suitable for horse herding would be found in Alaska and northern Canada.

Potential In Canada

The use of land in northern Canada for ranching or for beef cattle production is limited by the shortness of the grazing season. In order to assure that beef cows are brought safely through the winter, it is necessary to provide up to 3 tonne of hay per animal. The northern rancher, then, is always going to be at a disadvantage in comparison to livestock producers in more favourable climatic areas.

Many parts of northern Canada are capable of growing vast quantities of forage. This forage resource is mainly a sedge grass mixture which is found on the flat or boggy areas along some of our northern rivers and annually yields two to three tonnes of hay per hectare. Most of this forage will never be utilized for hay or summer grazing. A great deal of it occurs on wet areas where haying is impossible or where no livestock are available for its utilization. This forage resource when cured on the stem ranges from four to eight per cent crude protein (CP) (Pringle 1976). Alberta Agriculture (1977) feed test laboratory reports mean values of 8.6 per cent crude protein with a range of 2.4 to 17.0 per cent crude protein for 139 samples of sedge grass. The analyses of these samples were very similar to that recorded for brome grass hay. Since this forage will be grazed year long, the critical time for nutrients will be in the winter and at that time the bulk of the material is less than five per cent CP.

In addition, large amounts of cereal straw, aftermath from grass seed fields of fescue and brome and very often poor quality hay goes to waste. What is needed is a means of utilizing much of this wasted forage in an energy efficient manner.

Under conditions of ample cured forage, relatively low snowfall (30 to 50 cm) and extensive areas of open meadows the horse has many advantages over the beef cow as a converter of vegetation into meat. First, it can utilize relatively poor quality vegetation; second, it can paw through snow to reach its feed; and third, it does not require a finishing ration of grain in order to turn off an acceptable carcass. The horse is gregarious and can be herded relatively easily. Because it has a strong herd instinct and appears to have superior intelligence, compared to most animals, it is capable of withstanding predation. It can also adapt to a number of ecological niches. Proof of its ability to survive in northern conditions is readily seen in the Peace River area where many horses over-winter on fescue fields and stubble fields with little or no supplementary feed. One rancher near Fort Smith, Northwest Territories who rents horses in the summer stated that when his "season" is finished, he chases his horses out onto the native meadows and does not see them again until spring breakup. Literally hundreds of horses belonging to game outfitters survive all year on the ranges in the Yukon Territory. This also occurs in the valley of the Kechika, Muskwa and Sikanni rivers and other areas near Fort Nelson where hundreds of "pack horses" range all year long.

How is it that the non-ruminant, (horse, zebra, elephant) is able to survive on forage that is too low in nutrients to support a ruminant (cow, sheep, deer)? Bell (1971) and Hintz *et al*. (1977) compare the digestive systems of both types of animals. Put in simple terms the horse has no restraining mechanism imposing a limit on the rate of passage of material through its gut. Compared to the cow (ruminant) the horse can digest forage twice as fast as the cow. However, the horse using caecal fermentation is only 2/3 as efficient as the ruminant in extracting protein from its feed. Since the non-ruminant processes twice as much feed then its rate of assimilation is four thirds that of the ruminant and so can support itself on a diet that is too low in protein to support a ruminant.

An *in vivo* trial using a ruminally fistulated cow and a cecally fistulated mare found that the digestible dry matter intake for three common grass hays averaged 0.85 kg/100 kg body weight for the cow and 1.28 kg/100 kg body weight for the horse. Turnover time for all feeds moving through the gut was seven hours for the mare compared with twenty-four hours for the cow (Johnson *et al*. 1982). Thus this information supports the contention that mares are adapted to higher intake rates with lower rates of digestion than cows. This strategy is beneficial if sufficient quantities of forage are available especially if they have readily digestible components.

A Case Study from British Columbia

In April 1980, the Bavarian South Peace Stock Farm near Dawson Creek, in northern British Columbia purchased 48 two-year old horses with an average weight of 300 kg. In August and October of the same year an additional 400 head of between 270 and 320 kg yearlings and two year olds were purchased at meat prices and brought to the farm. In 1980, 25 of the originally purchased mares were bred to a grade Belgian stallion and in May 1981, 19 foals were born. These along with a few others that had come from mares that were bred at purchase were weaned in December and averaged 173 kg for 28 animals. They were wintered with the saddle horses. By April 1982 they averaged 250 kg for an average daily gain of 0.52 kg. Of the 400 horses bought in late 1981, 38 head were shipped to slaughter on February 1, 1982; they averaged 400 kg and the heavier animals brought good prices. Of the original 400 horses, about half were mares and these were bred in 1981 and therefore foaled in 1982. In the first year only the weaker animals and mares in foal were fed for about 60 days. In 1982, because of deep snow the whole herd was fed along with the cattle on hay, aspen shavings and a grain mixture. It seemed that the horses ate only enough to give them strength to go out and paw through 75 cm of snow for the fescue. As for management, 30 mares are put with a stallion. Animals are branded in the autumn and wormed both in the spring and in the autumn using an injectable vermifuge.

This innovative venture by the Bavarian South Peace Stock Farm involves buying young horses for growing and fattening and raising its own second generation animals in a breeding and production operation.

It is recognized now that the industry prefers at least three-year old horses for meat. This represents an extra year of feeding, compared to a cattle operation. However, the reduced amount of management and winter feed should balance the economics of a large scale horse production unit with traditional beef cattle production.

The Bavarian South Peace Stock Farm has thousands of hectares in fescue seed production. This crop must be rejuvenated every three to four years and this is less necessary if the crop is judiciously grazed after seed harvest. In addition, fescue grows well in the autumn, retains nutrients well into the winter, and so makes ideal horse range.

The Bavarian herd was started with a great many small pony-type horses since the small horse does not command the higher prices. It is necessary to breed to produce larger animals and so Belgian stallions are being used. It has been found that when snow is over 60 cm deep the animals require supplementation on pasture. Mares in foal and weaned foals require supplement for about 60 days after January. Weaning presents no problems, winter feeding helps to gentle the horses, stallions easily sort out their mare groups, male animals do not require castrating if they are to be marketed as three year olds, horses range well in with cattle, and the foaling percentage has been higher than expected. It was pointed out that labour for the horse herd is minimal. Predation from wolves and bears has not occurred. Some unexpected advantages of horse ranching have been noted. Certain animals can be sold as saddle stock or bucking horse stock. There is marketing flexibility in that animals can be held for any length of time without deterioration of quality. Two of the greatest advantages of a large scale horse operation are labour efficiency and greater low quality feed utilization. At this time in history, these two factors are of paramount importance for any farming operation.

Feedlot Trials

A feedlot trial was carried out at Lakeside Feeders in Brooks, Alberta to obtain an initial assessment of the practicality of feedlot handling of horses (Grimson 1982). Ninety head of both sexes of foals with an average weight of 230 kg and an age of about eight months old were used. Three diets, 70 per cent alfalfa cubes and 30 per cent oats, 30 per cent alfalfa cubes and 70 per cent oats, and 30 per cent alfalfa cubes and 70 per cent barley, were fed for 20 weeks. The mean body weight at slaughter was 403 kg. Average gain was 0.96 kg/head/day with no significant differences among treatments.

In a second trial (Lakeside Research 1982) 90 horses were fed 70 per cent, 85 per cent, and full feed on a 15.6 per cent crude protein ration consisting mainly of alfalfa cubes and flaked oats. Up to 112 days each increase in consumption gave a corresponding increase in average gain of 0.43, 0.54 and 0.79 kg/day for 70, 85 per cent and full feed, respectively. The horses were then all put on full feed for another 84 day period. During that time the 70 per cent animals attained growth equal to the 85 per cent fed animal group. However, over the combined periods the full feed horses had significantly greater gain than the other two treatments. Full feed gave 0.92, 85 per cent gave 0.80 and 70 per cent gave 0.77 kg/head/day. Costs in dollars/kg gain were $1.46, $1.53 and $1.50, respectively. Of interest was the fact that females had significantly greater rates of gain, better exterior fat cover as well as marbling in the meat.

Conclusions

Horses for meat represent a much larger part of the agricultural economy than most people realize. Exports valued at 25 million dollars/year are significant. Evidence is present to indicate horse raising can compete with other forms of livestock production and because it is less labour and energy intensive it has advantages over beef cattle raising in cold or northern areas. There is a place for horse production on a large scale so as to assure a supply for an expanding export market. I think we will continue to have some ups and downs in the horse meat market but over the past five years sales overseas have been increasing. As long as we have the animals or exploit our potential to supply them we will continue to increase our exports.

It is obvious that there are a great many personal preferences and unknown factors about the horse meat industry. This large industry at present is dependent upon the recreational horse maintaining its place as a part of the social structure of our society. Market requirements for carcass quality (age, type and condition) are not widely known. Countries such as France, Belgium, U.S.S.R. and Italy are researching many aspects of horse production for human meat consumption whereas to date, Canada has not provided resources for such studies. If the industry is to survive and flourish ranchers need to know how long it takes to grow animals that will produce a prime carcass. What breeds are best? What are the efficiencies of gain for various forages? What are the mineral and other feed supplement requirements? What diseases and other disorders can be expected if animals are raised in large numbers? We can learn from the literature and by observing operations such as the Bavarian South Peace Stock Farm but I think it is time the agricultural researchers accepted this new challenge and became involved in horse management projects.

Literature Cited

Alberta Agriculture. 1977. Average analyses of Alberta feeds. 1965-1975. Agricultural Soil and Feed Testing Laboratory.

Andreyev, V.N. 1971. Horse herding for meat in Yakutskaya, A.S.S.R. The Polar Record, 15: 931-933.

Anonymous 1979. Selling by the Pound, Special Report, Equus 17: 41-48; 68-69.

Bell, R.A.V. 1971. A grazing ecosystem in the Serengeti. Sci. Amer. 225: 86-93.

Grimson, R.E. 1982. Feedlot performance of young horses. Unpublished presentation to Expert Committee on Animal Nutrition. Edmonton, Alberta.

Hintz, H.F. 1977. Digestive physiology. Chapter 7. Nutrition of the horse, part 3. *In*: The horse. J.W. Evans, A. Barton, H.F. Hintz and L.D. Van Vlack (Eds.). W.H. Freeman and Co., San Francisco, California. pp. 241-258.

Johnson, D.E., M.M. Borman and L.R. Rittenhouse. 1982. Intake, apparent utilization and rate of digestion in mares and cows. Proceedings, Western Section American Society of Animal Science, 33: 294-296.

Jonas, Marg. 1980. The trail ends here. Cattlemen 43: 64-66.

Lakeside Research. 1982. Feedlot performance of weanling foals: II. Effects of feeding level on growth rate, feed efficiency and carcass quality. A Research Report prepared for Horse Industry Branch, Alberta Agriculture.

Martin-Rosset, W., M. Jussiaux, C. Trillaud-Geyl and J. Agabriel. 1982. Horse meat production in France: Husbandry and production systems. Proceeding European Association for Animal Production 33rd Annual Meeting, Leningrad, U.S.S.R.

Matthews, B. 1980. Leading a horse to slaughter and profit. Macleans, July 14, 1980.

Pringle, W.L. 1976. An assessment of the Agricultural Potential of the Slave River Lowlands, Northwest Territories, Canada. A report by Northern Research Group, Beaverlodge, Alberta.

BIOCONVERSION OF PEAT: UTILIZATION OF PEAT EXTRACTS AS A FERMENTATION SUBSTRATE

Antonio M. Martin
Memorial University of Newfoundland

Abstract: Peat and some derived products (e.g., peat extracts) contain the nutrients required to support the growth of microbial populations. This paper deals with the submerged production of edible mushroom mycelium utilizing peat hydrolysates as the substrate source. *Sphagnum* peat acid extracts have been obtained and utilized in submerged fermentation processes for producing fungal mycelium from the species *Morchella esculenta* and *Agaricus campestris*. The adaptation of *Agaricus campestris* to growth on non-supplemented peat extracts was found superior to the results with *Morchella esculenta*. The effects of several operating variables on the production of *Agaricus campestris* mycelium are discussed. The results are expressed by means of the biomass concentrations obtained under different fermentation conditions.

Peat, the organic soil of peatlands, consists mainly of organic residues of dead plants and microorganisms, and offers a variety of opportunities for the development of products of interest to the biochemical industry. Several countries with extensive peatlands, including Canada, are interested in the efficient utilization of this resource. Peat and some derived products (e.g., peat extracts) are rich in carbohydrates, minerals and other substances (McLoughlin and Küster 1972). Those nutrients are required to support the growth of microbial populations. Peat extracts or hydrolysates have been tested in the past as culture media for microorganisms (LeDuy 1979).

Worldwide resource shortages have become evident, heightening the importance of finding new resources, or more efficiently exploiting the existing ones. A cheap, plentiful supply of nutrients for fermentation processes is one of the primary goals of biotechnology; the importance of the industrial utilization of microorganisms in the production of a wide variety of products, including protein and fuels, is rapidly increasing.

Peat deposits cover large areas of North America, but until now have received little consideration as an industrial raw material. Peat markets generally have been restricted to horticulture and fuel utilization only.

In Europe, the situation has been different. Work done in Finland, Ireland, the Soviet Union and other countries have contributed to a better understanding of the possibilities of peat as a feedstock for the chemical industry.

The production of microbial biomass has special economic importance to boreal zones and other northern regions such as the province of Newfoundland, where the agricultural production is limited by climatic factors. Any attempt to produce an inexpensive protein source for animal feeding using the resources of the area will have a positive impact on the local economy. In those areas, the development of livestock with commercial aims necessarily implies the importation, at high prices, of the basic components for the nutrition of the animals, specifically protein sources. Therefore, the production of single cell protein for animal feeding utilizing the local peat resources could be an attractive idea. Canada has one of the world's largest peat reservoirs, with much of the peat deposits located in boreal areas.

Studies of the influence of peat and peat extracts on the growth and metabolism of microorganisms show that the peat with a low value as feed is the most appropriate for microbial utilization (Küster 1968) and that the chemical composition of peat provides a stimulation in the metabolism of yeast (Evodokimova 1976). With respect to biomass production for food, yeasts and specifically *Candida utilis* yeast, appeared to give the best results (McLoughlin 1971; Quierzy *et al*. 1979) because *Candida utilis* is able to metabolize both the hexoses and pentoses found in peat extracts. An unidentified acidophilic fungus has also been cultivated in peat hydrolysate (Boa and LeDuy 1981).

Among the various methods employed to remove the fermentable substrates from peat, dilute acid autoclaved extracts have been found to be the most suitable (McLoughlin and Küster 1972). The acid-catalyzed peat extracts are similar to those obtained after the so-called wet carbonization process utilized to dewater peat and to increase the heating value of peat for the production of briquettes. This process results in large quantities of waste liquid by-products. One of the problems associated with this procedure is the disposal of the effluent (McLoughlin 1971). Thus, it is reasonable to try to use a microbial process with two aims with respect to the commercial utilization of peat as a feed source. The aims include the recovery of materials as nutrients for fermentation processes that can produce valuable by-products, and the biological treatment of the waste effluent.

Submerged culture production of mushroom mycelium is relatively recent as compared with the older process of producing fruiting bodies (or sporophores) in manure or in compost beds. Mushroom mycelium has potential as a food or a feed supplement. A mushroom-flavored product could be used in dehydrated and formulated foods such as soups, sauces, etc. Several substrates have been employed in the submerged growth of mushroom mycelium (Litchfield 1967a; Kosaric *et al*. 1973; Kosaric and Miyata 1981) and studies have been conducted in assessing their nutritional and flavor properties (Litchfield 1967b; LeDuy *et al*. 1974). Although peat is an important component in the traditional cultivation of commercial mushrooms, no attempts have been made in the past to utilize peat extracts as a nutrient source for the submerged culture of edible fungi. Recently, work has begun at Memorial University of Newfoundland on the submerged growth of the mushroom mycelium utilizing peat extracts as a substrate source (Martin 1982a; Martin 1983).

Methods

Two species of mushroom have been utilized. *Agaricus campestris* NRRL 2334 (American Type Culture Collection) is practically the only mushroom species cultivated and eaten in North America. *Morchella esculenta* NRRL 2603 (American Type Culture Collection) is a member of the better known morel mushrooms, recognized for their pleasant flavor. Both species were previously adapted to growth in peat extracts by successive transfers (Martin 1982b, c).

Ground *Sphagnum* peat moss was mixed with 1.5 per cent (v/v)H_2SO_4 and autoclaved at 15 psig (121°C) for two hours. The final total carbohydrate (TCH) concentration was determined by the peat:H_2SO_4 solution ratio. The extracts were obtained by pressing the autoclaved product in a Carver Laboratory Press Model C (F.S. Carver Inc., Wisconsin) followed by filtration through Whatman No. 1 filter paper.

Blended pure culture mycelium were assepticaly inoculated to 100 mL sterile peat extracts in 250 mL shaker flasks and incubated in a Gyrotory water bath shaker (Model G76, New Brunswick Sci. Co., Inc.). The initial pH of the media was adjusted by addition of 15N NaOH. Several values of inoculum ratio, temperature, initial pH, fermentation time and agitation were utilized in different processes. Some preliminary batch fermentations have been done in a 2L bench fermentor (Model G76, New Brunswick Sci. Co., Inc.)

Total carbohydrate concentration in peat extracts and in the fermented media were determined by the anthrone reagent method without correction for color interference (LeDuy *et al*. 1973).

After fermentation the culture was filtered through oven dried Whatman No. 1 filter paper. The filter paper and the mycelium were oven dried (105°C to constant weight), and the mycelium (biomass) dry weight determined.

Results and Discussions

The effects of the different variables analyzed on the biomass concentration of *Agaricus campestris* in shake flask fermentations are shown in Figures 1 to 3. Results obtained with *Morchella esculenta* have been presented elsewhere (Martin 1982d). Further analysis of the growth of *Agaricus campestris* on peat extracts, including information on the biomass yields obtained, can also be found in a previous presentation (Martin 1982e). Figures 1 and 2 show the highest biomass concentrations were obtained at 4 per cent (v/v) inoculum concentration. The fact that the biomass concentration decreases, in general, with increased inoculum concentration, could suggest the production of inhibitory substances in previous steps of the cultivation of *Agaricus campestris*. This was also observed in the previous adaptation of this species to peat extracts media (Martin 1982b). The highest biomass concentrations were obtained at 24°C (Figure 1).

A pH of 6.0 was found to be the optimum initial value for *Agaricus campestris* growing in a peat hydrolysate medium (Figure 2). Peat hydrolysate with 25 g/L of TCH concentration was utilized to evaluate the effect of fermentation time and agitation rate on the biomass concentration of *Agaricus campestris*. Figure 3 shows the biomass concentration increased from four to six days, and afterwards it decreased. This suggests an autolitic process occurs after six days of fermentation, probably due to a lack of some nutrient. Although it is possible not all the carbohydrates present are assimilable by the mushroom, it is our idea that one or more nutrients, such as nitrogen or phosphorus, become the growth limiting factor at this point. Further research will explore the use of the supplemented peat hydrolysates in order to overcome this limitation to the biomass growth.

The optimum agitation found was 150 rpm (Figure 3). Although higher agitation rates would imply better mixing and mass transfer in the medium, it is possible that some physiological damage to the biomass also occurs.

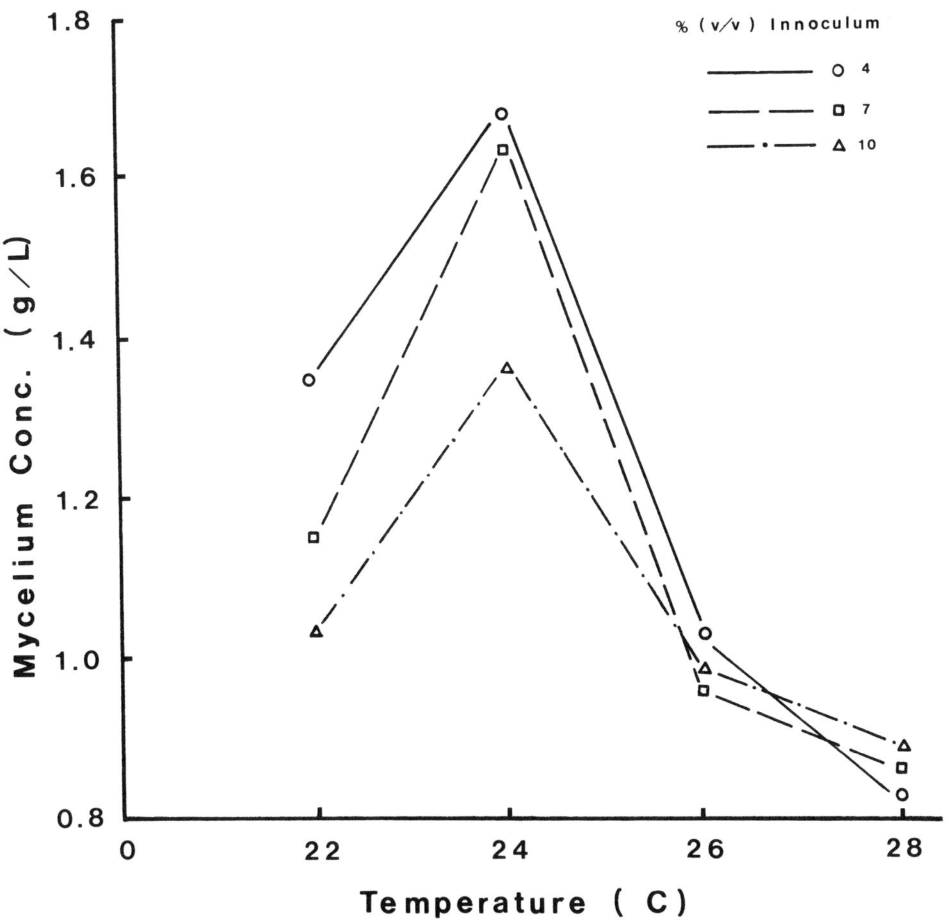

Figure 1. Effect of temperature and inoculum ratio on the final dry mycelium concentration of *Agaricus campestris* in peat extract medium at initial pH 6.0, 150 RPM, 17.7 g/L initial TCH concentration. Fermentation time is 6 days.

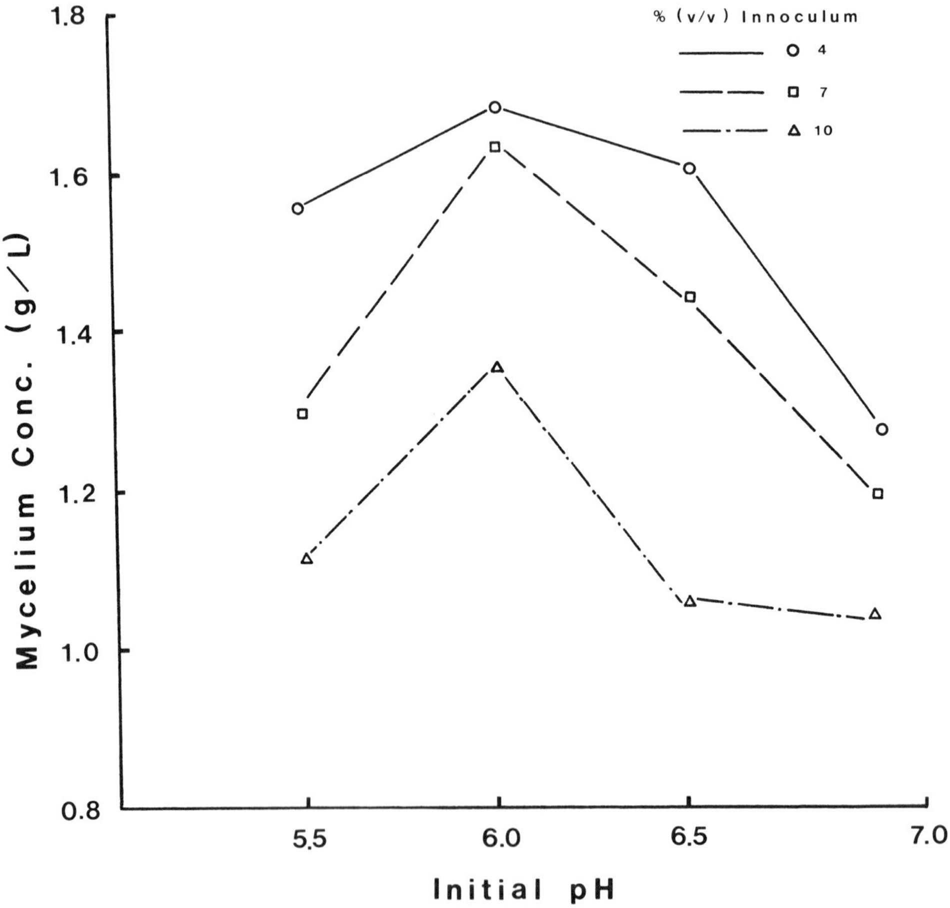

Figure 2. Effect of initial pH and inoculum ratio on the final dry mycelium concentration of *Agaricus campestris* in peat extracts medium at 24°C, 150 RPM, 17.7 g/L initial TCH concentration. Fermentation time is 6 days.

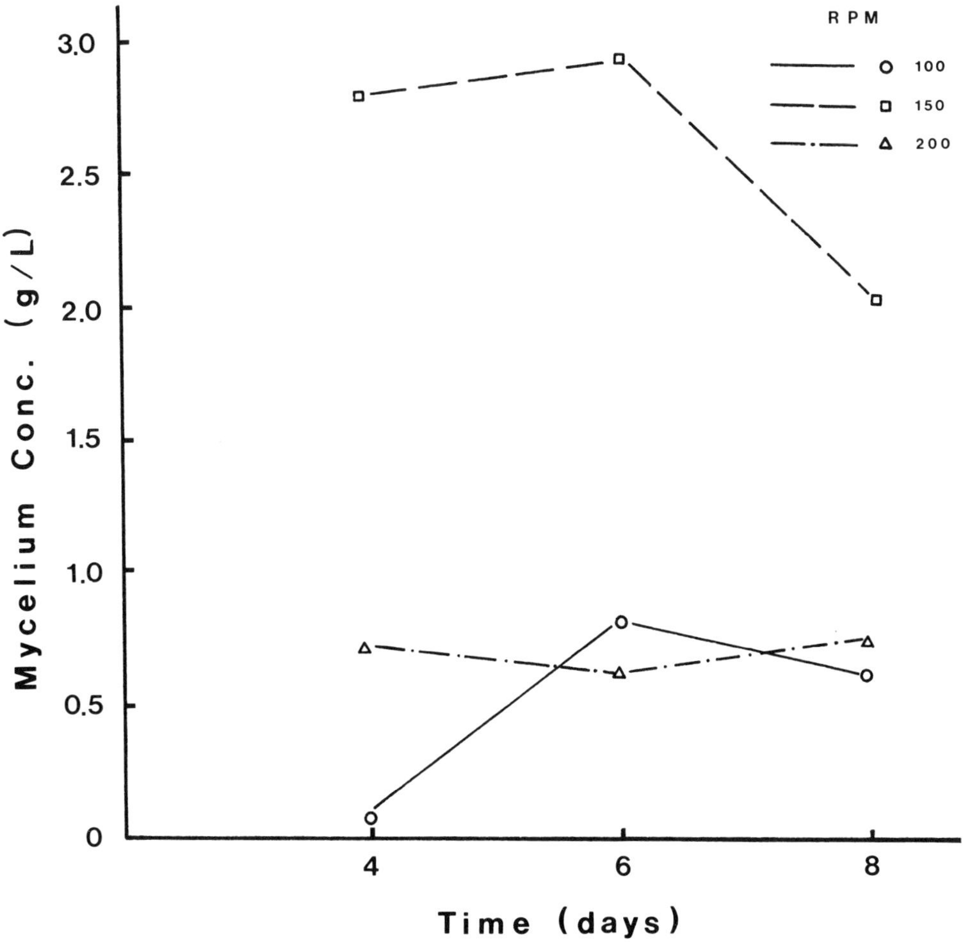

Figure 3. Effect of fermentation time and agitation on the final dry mycelium concentration of *Agaricus campestris* in peat extract medium at pH 6.0, 24°C, 4 per cent (v/v) inoculum ratio. Initial TCH concentration is 25 g/L.

Conclusions

Fermentable substrates have been extracted from peat, and the mushrooms *Agaricus campestris* NRRL 2334 and *Morchella esculenta* NRRL 2603 were adapted and cultivated, for the first time, utilizing peat hydrolysate as the only substrate source. Peat, which is often utilized as casing soil in the traditional production of the *Agaricus campestris* mushroom sporophore (fruit body) can also be employed as a source of nutrients in the submerged production of mushroom biomass. *Agaricus campestris* shows a better adaptation and growth than *Morchella esculenta* when cultivated in non-supplemented peat extracts. Further studies with nutrient-supplemented peat extracts are necessary to evaluate completely the potential of these species to growth in peat based culture media.

The aforementioned results are examples of the potential utilization of peat as fermentation substrate. There are other scientific and socioeconomic aspects as well.

• The design of an industrial process for the integrated utilization of peat (production of fuel plus microbial biomass such as single cell protein (S.C.P.) or other fermentation product) can be an important technological consequence of the studies being carried out in the bioconversion of peat in fermentation processes.

• The chemical composition of peat is complex and varies with location and other factors. It is necessary to evaluate the different types of peat as fermentation substrates. This work has not yet been carried out in a systematic way.

• The exploitation of an abundant Canadian resource such as peat, can be made more attractive by the integrated utilization of all the components of this resource. In the case of peat, one probable utilization is as fuel. The use of peat as a fermentation substrate would not compete with its utilization as fuel, because the hydrolysate effluent from wet carbonization can be utilized as the substrate source. In addition, this use contributes to eliminating a potential waste and pollution problem.

Acknowledgements

Assistance of V. Bailey, Department of Biochemistry, Memorial University of Newfoundland, is appreciated. This work was supported in part by grants from the Natural Sciences and Engineering Research Council of Canada and the New Technology Employment Programs, Government of Canada.

Literature Cited

Boa, J.M. and A. LeDuy. 1981. Acidophilic fungus SCP from peat hydrolysate. Paper presented at the A.I.Ch.E. 74th Annual Meeting. New Orleans, Louisiana.

Evodokimova, G.A. 1976. Investigation of chemical composition and biological effect on living organisms of peat chemical treatment products. *In*: Proceedings of the 5th International Peat Congress. Wydawinictwa Czasopism Techiczhych Nat., Warzawa, 3: 23-32.

Kosaric, N., A. LeDuy and J.E. Zajic. 1973. Submerged culture growth of edible mushrooms on waste sulphite liquors. Can. J. Chem. Eng. 51: 186-190.

Kosaric, N. and N. Miyata. 1981. Growth of morel mushroom mycelium in cheese whey. J. Dairy Res. 148: 149-160.

Küster, E. 1968. Influence of peat and peat extracts on the growth and metabolism of microorganisms. *In*: Transactions of the Second International Peat Congress, Leningrad. R.A. Robertson (Ed.). 2: 945-951.

LeDuy, A. 1979. Single cell protein from peat hydrolysates. Process Biochem. 14: 5-7.

LeDuy, A., N. Kosaric, and J.E. Zajic. 1973. Correction factor for anthrone carbohydrate in coloured wastewater samples. *In:* Proceedings of the 10th Canadian Water Pollution Research Symposium. T.C. Hutchinson (Ed.). University of Toronto, Toronto, Ontario. pp. 126-131.

LeDuy, A., N. Kosaric, and J.E. Zajic. 1974. Morel mushroom mycelium growth in waste sulphite liquors as source of protein and flavoring. Can. Inst. Food Sci. and Tech. J. 7: 44-50.

Litchfield, J.H. 1967a. Submerged production of mushroom mycelium. *In*: Microbial Technology. H.J. Peppler (Ed.). Reinhold Publishing Corp., Amsterdam and London. pp. 107-144.

Litchfield, J.H. 1967b. Morel mushroom mycelium as a food flavoring material. Biotech. Bioeng. 9: 289-304.

Martin, A.M. 1982a. Submerged growth of *Morchella esculenta* in peat hydrolysates. Biotech. Letters, 4: 13-18.

Martin, A.M. 1982b. Adaptation of *Agaricus campestris* NRRL 2334 to a peat extract medium. Paper presented at the 1982 Annual Meeting of the Canadian Society of Microbiologists. Quebec City, Quebec.

Martin, A.M. 1982c. Adaptation of *Morchella esculenta* to a peat extract medium. Paper presented at the 13th International Congress of Microbiology. Boston, Massachusetts.

Martin, A.M. 1982d. Submerged production of edible mushroom mycelium. Paper presented at the 25th Annual Conference of the Canadian Institute of Food Science and Technology. Montreal, Quebec.

Martin, A.M. 1982e. Submerged production of *Agaricus campestris* mycelium in peat extracts. Paper presented at the 1982 Annual Meeting of the Institute of Food Technology. Las Vegas, Nevada.

Martin, A.M. 1983. Submerged production of *Agaricus campestris* mycelium in peat extracts. J. Food Sci. 48: 207-208.

McLoughlin, A. 1971. Preliminary investigations on industrial pollutants as substrate for microorganism. Water Res. 5: 1117-1119.

McLoughlin, A.J. and E. Küster. 1972. Utilization of peat. 1.—Methods of obtaining fermentable substrates. Proc. Roy. Irish Acad., Section B. 72: 1-7.

Quierzy, P., N. Therien and A. LeDuy. 1979. Production of *Candida utilis* protein from peat extracts. Biotech and Bioeng. 21: 1175-1190.

ENERGY PRODUCTION IN THE BOREAL FOREST ZONE

Ralph P. Overend
Division of Energy
National Research Council of Canada
NRCC 21270

Abstract: The boreal forest zone effectively extends from Inuvik (NWT), to St. John's, Newfoundland in a broad arc sometimes as much as 800 km in depth. It is of interest to energy planners and producers because it has the potential to supply energy in its many forms to the consumers both within Canada as well as in export markets. The major energy consumers in the zone are the energy industries themselves along with the minerals extraction and refining industries. As a consequence the region has one of the highest energy consumption/capita indices in the world; however, there are also many 'remote' communities of native peoples who have extremely low per capita commercial energy consumption mainly as a consequence of the high cost of electrical energy supplies. Energy projects will be a major stimulus to the development of the boreal forest zone with oil, oil sands and gas predominating west of Hudsons Bay and hydro-electric projects mainly in Quebec and Labrador.

The 1980s ushered in a period of considerable uncertainty with respect to the Giga-Dollar energy projects in the zone. Not only projects based on fossil fuels such as gas, oil, oil sands and uranium, but also those based on renewable hydro electricity production, have been cancelled or deferred in the light of current economic conditions.

The energy production potential of the oil sands was projected at 150,000 cubic m/day of syncrude by the year 2000. This corresponds to almost 50 per cent of the anticipated domestic crude requirement in that year. The role of the zone in making up the shortfall in traditional crude production is clear: the reason for the deferment of projects was primarily the high investment costs required; approximately half a million dollars is required for a production rate of a cubic metre per day. Similar large investments in the tens of billions are required for gas and oil pipelines and hydroelectric developments.

The boreal forest zone is a source of minerals, energy, water and forest resources for the industrial economy of Canada. Gold, silver, uranium, copper, zinc, lead, iron, nickel, and cobalt are all extracted, concentrated or refined within the region.

Fossil fuels are to be found in large quantities within the boreal forest zone (BFZ), with current oil and gas production coming largely from Alberta and Saskatchewan. Oil sands and heavy oils, the successors to conventional oil production, exist entirely within the BFZ. Quantitatively, they might contain as much energy as all of the world's light conventional crude oil resources. Coal reserves in the form of sub-bituminous coals and lignite are very extensive and promise to yield significant energy supplies in the long term.

Renewable resources are a significant component of development with hydro-electric production in Labrador, Quebec and Manitoba already contributing a large part of the electricity supply. The remaining unexploited hydro resource is primarily in this region. The wind regime is such that a high potential exists in only a few areas which await the provision of high voltage transmission lines before they can be economically exploited on a large scale. The forest resource although already exploited along the southern margin, could sustain through a combination of forest stand conversion, logging residue, mill residues and the merchantable surplus, a large renewable carbon based fuel supply.

The soils of much of the boreal forest zone are themselves composed of organic materials, described variously as muskeg and peat. Because of the large areal extent it is estimated that the energy content is extremely large, although to date, there are no technologies available for its exploitation under the special climatic conditions that prevail over much of the zone.

The Fossil Fuels

Coal

Lignites exist at Onakawana and at Wapaweka and both deposits have been reviewed periodically for their potential commercial exploitation. Other coals are found in the Yukon and the Northwest Territories as well as in Newfoundland. The most extensive coal resources are however in Alberta; the plains and foothills coal deposits lie on the southern edge of the boreal forest zone (Figure 1). The size of the deposits and the energy content are given in Table 1.

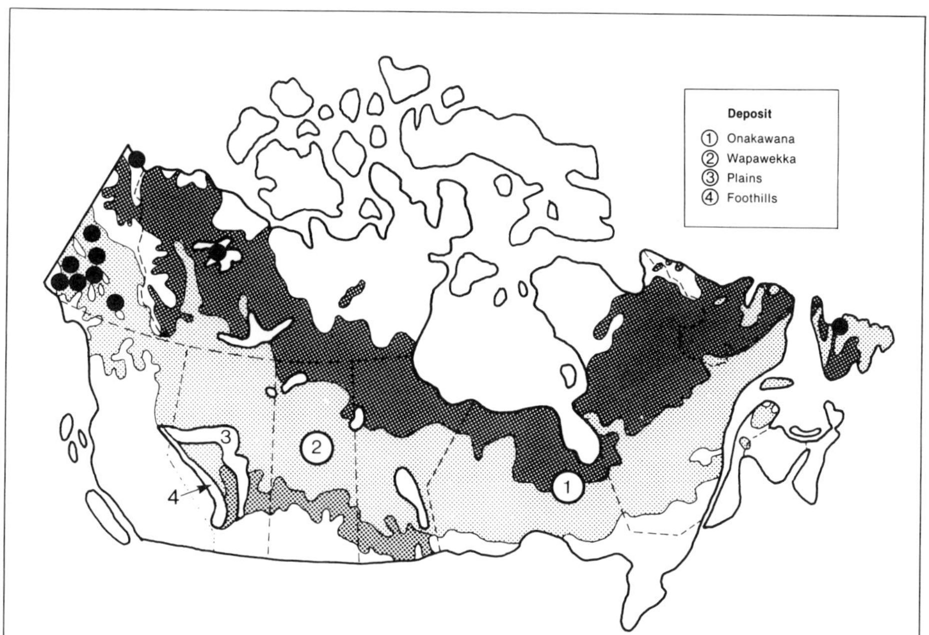

Figure 1. Location of major coal deposits in the boreal forest zone (adapted from Bielenstein *et al*. 1980).

Table 1. Estimated size and energy content of coal deposits in Canada (adapted from Bielenstein *et al*. 1980)

Deposit	Type	Quantity / Mt	Energy / EJ
Onakawana	Lignite	218	3.5
Wapaweka	Lignite	—	—
Plains	Sub-bituminous	102,000	2040
Foothills	Bituminous	7,700	192

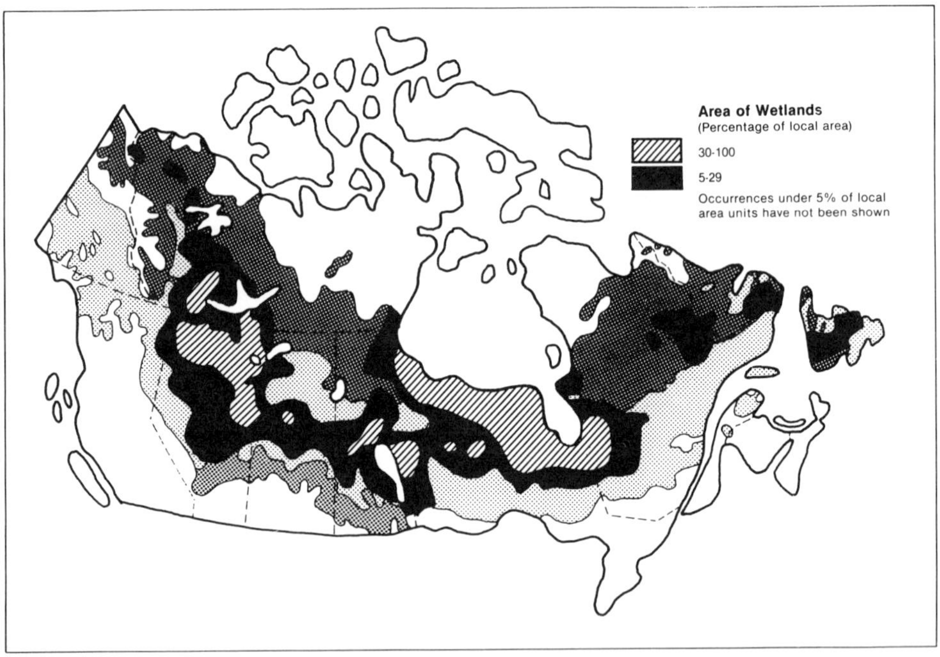

Figure 2. Location of wetlands in the boreal forest zone (adapted from Radforth and Bramner 1977).

To put the scale of these mainly thermal coal deposits in perspective it is useful to recognize that 57 TWh were generated from about 30 Mt of thermal coal in 1980 (Anon. 1981a).

Peat

The peat resource is not known with any certainty. Current research is targeted towards a first level inventory of the total Canadian resource by 1985. Using the data on the distribution of wetlands and of muskeg such as that published by the National Research Council Associate Committee on Geotechnical Research (Radforth and Bramner 1977), it is possible to estimate the peat content of the boreal forest zone at around 100 Gt on an anhydrous basis. This estimate derived from the areal extent and a presumed depth of peat $\geqslant 1$ m corresponds to an energy content of 2000 EJ (Figure 2). While there is at present no technology available to exploit this young coal resource, the extent (quantitatively similar

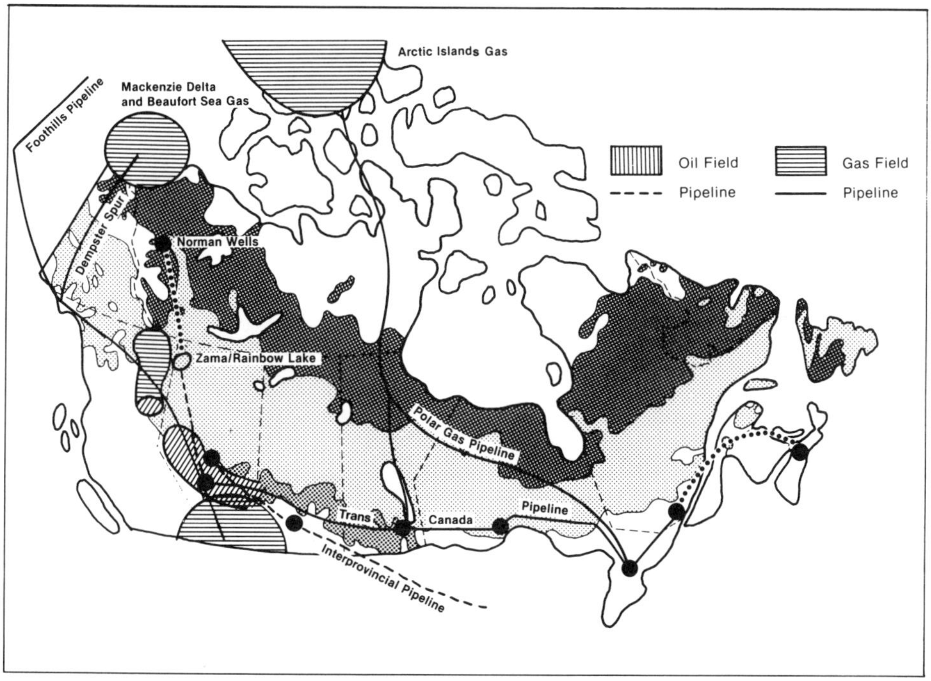

Figure 3. Location of major oil and gas production fields and pipelines.

to coal) makes it an interesting option particularly for provinces within the boreal forest that lack conventional fuel resources.

Conventional Oil and Gas

In the main, the production of oil and gas in Canada takes place in British Columbia, Alberta and Saskatchewan (Figure 3). Once found, reservoirs of oil and gas can be linked to an extensive system of pipelines and exploited. Individual wells have a finite capacity and their output inevitably declines with time. In the case of crude oil, the current production rate is expected to decline from 200,000 m^3/day to 25,000 m^3/day by the year 2000 (Anon. 1981b). To sustain both oil and gas production, a very high level of exploration is required each year. A measure of this activity in the boreal forest is the cutting of extensive seismic lines through the forest "the felling of trees for seismic lines exceeds the commercial forest industry harvest in Alberta" according to apocryphal account. While the oil search has been relatively unsuccessful, that for gas has been able to build an economic reserve such that the current delivery rate of about 10^8 m^3/a could be sustained for the next 20 years (Anon. 1981b). Further additions to the gas supply are foreseen from the frontier areas: the Beaufort Sea, the Mackenzie Delta and the Arctic islands. The impact of these will be felt in the boreal forest zone on account of the necessary pipelines required to deliver the gas to southern Canada and export markets. The building of pipelines is contingent upon adequate reserves being found and upon gas prices that will justify the large multi-billion dollar expenditures. Two major proposed pipelines are the foothills section of the Alaska Gas Pipeline with its associated Dempster

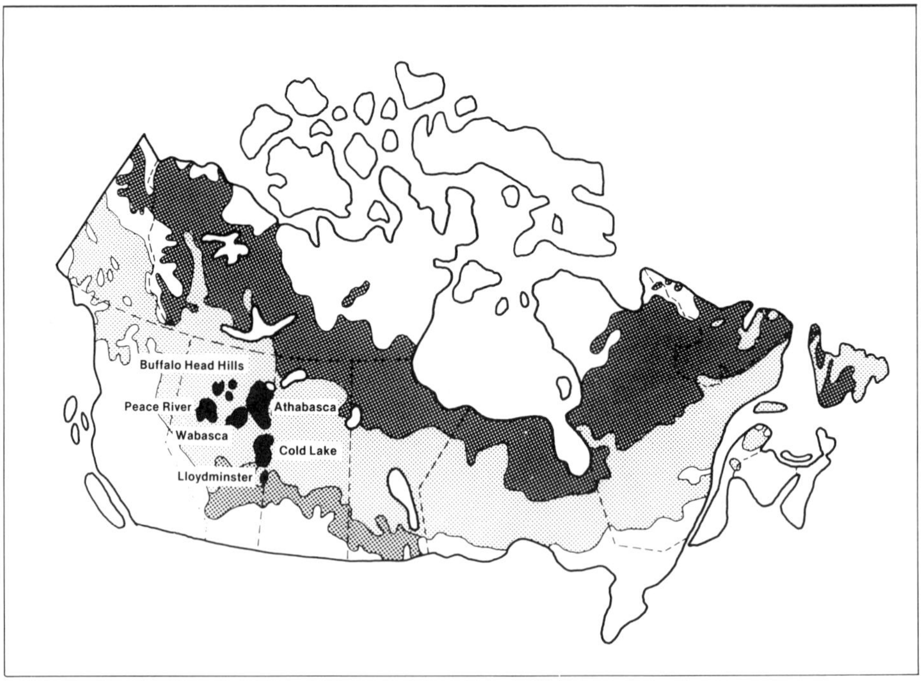

Figure 4. Location of bitumen-bearing oil sands and heavy oil deposits.

spur connection with the Mackenzie Delta, and the Polar Gas Line from the Arctic to Ontario.

The integrated oil production anticipated is around 750 10^6 m^3 with a total energy content of 30 EJ, currently the production rate is 2.6 EJ/year with a decline to 0.33 EJ/year anticipated by the year 2000.

The gas reserve in the region is estimated at 66 EJ of marketable gas with an ultimate potential of 145 EJ (Anon. 1981b). The current production rate is about 4 EJ/a.

Oil Sands and Heavy Oils

The bitumen-bearing oil sand and the heavy oil deposits are situated in the northeastern part of Alberta and run into the western part of Saskatchewan (Figure 4). The deposits are one of the largest of those containing petroleum-like hydrocarbons; the energy content exceeds that of the OPEC oil reserves. Regrettably the ease with which the oil is obtained from the bitumens is not at all comparable with that of conventional crude oil recovery. Recent news items have detailed the abandonment or deferral of several large scale schemes for syncrude recovery as a consequence of the cost and technical difficulty of operations. Table 2 details the nature of the resources and their gross energy content (Anon. 1977).

The current operations of Great Canadian Oil Sands and Syncrude are surface mining of the bitumen bearing oil sand and then upgrading of the bitumen to a syncrude which is

Table 2. Estimates of heavy oils and oil sands

Deposit	Areal extent km²	Pay thickness metre	Volume in situ Giga m³	Energy content Exa Joule
Heavy oils				
Cold Lake A	7384	16	19	716
Cold Lake B	2630	12	5	200
Cold Lake C	2873	5	2	85
Crude bitumens				
Athabasca				
Mineable	1983	30	12	449
In-situ	21286	21	88	3357
Buffalo Head	643	2	0.16	6
Peace River	6499	14	12	455
Wabasca A	5431	8	8	291
Wabasca B	6965	6	6	231
Totals	55596	115	152	5792

then shipped to conventional oil refineries for processing into end-products such as gasoline. The surface mining technique involves a considerable amount of preparation prior to recovery of the sand. The overburden removal consists of: tree clearing, overburden drainage, stream diversion, muskeg removal, dyke construction and overburden stripping. Once exposed, the oil-sand is excavated, transported to the plant, and the bitumen is extracted and upgraded to a synthetic crude (syncrude). The residual sand and tailings are then ponded and ultimately the land is reclaimed. The recovery factor for surface mining is about 50 per cent to the bitumen and the upgrading to syncrude is about 75 to 80 per cent efficient. Overall, the final efficiency is comparable to that of conventional oil recovery.

A much more extensive deposit is made accessible if in-situ techniques are utilized. Drilling, heating by steam or fire flood followed by solvents can be used to recover both the bitumens and the heavy oils. The recovery of bitumen is less than that of surface mining.

It is estimated that at least 6×10^9 m³ are recoverable by today's surface and in-situ technologies and that this corresponds to about 230 EJ in energy units.

Inexhaustible Fuels

Uranium

In 1980 there were seven primary uranium producers with the majority of the production outside of the boreal forest zone in Ontario. A considerable expansion of activity is underway in northern Saskatchewan though a number of the ore bodies are past their midpoint of exploitation, with one mine closing this year (Figure 5).

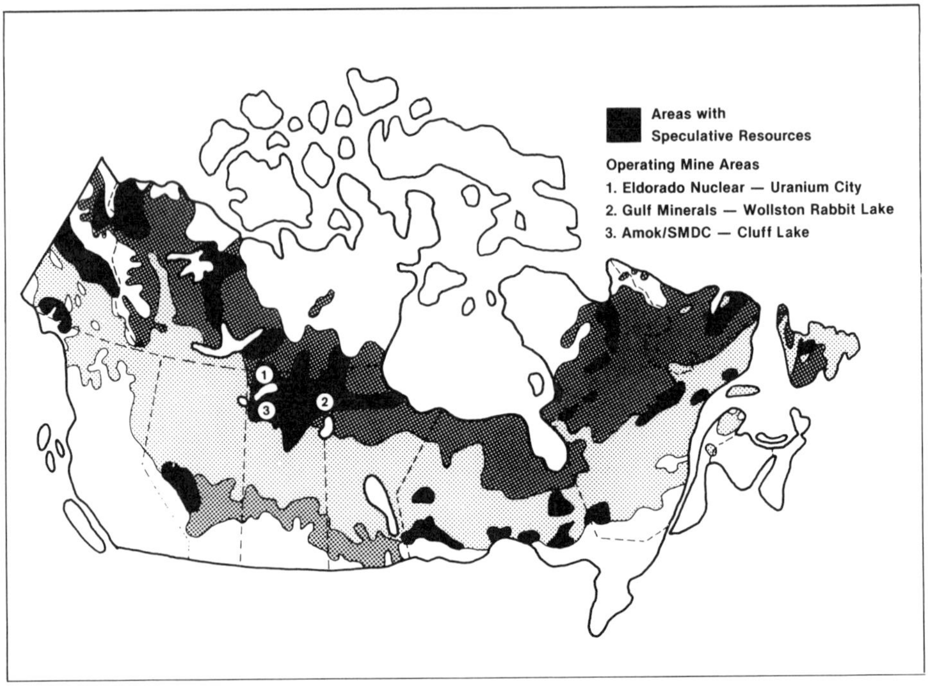

Figure 5. Potential locations of uranium resources and operating mine areas.

Canada is a leading exporter of uranium and is currently the second largest producer after the U.S.A. It is estimated (Anon. 1981c) that about 12 per cent of the world's low cost, reasonably assured resource base is in Canada. With increased exploration, this position is likely to be maintained in the face of increasing world demand.

The 1980 estimates of uranium in Canada (Anon. 1981c) are in three categories: measured, indicated, and inferred. The quantities are respectively 73, 185, and 315 kt at prices of up to $200/kg U. Approximately one-third of this material is in the boreal forest zone.

The energy content of uranium is a function of the nuclear fuel cycle in which it is "burnt". Table 3 lists the 95 per cent load factor fuel consumption of the various nuclear fuel cycles.

If the inferred resource of 100 kt is burnt in a CANDU system then the approximate electrical output would be 625 GW•year or 20 EJ. For reference, the 1980 electrical generation in Canada from all sources was 42 GW•year of which only 9.8 per cent was nuclear (Anon. 1981a). Clearly, the adoption of an advanced fuel cycle with reprocessing would extend the Canadian resource to such an extent that it could only be considered to be inexhaustible. This is of little consequence to the boreal forest zone where the deposits are expected to have mine lifetimes of around 20 years.

Table 3. Uranium consumption in electricity generation

	1000 MW_e reactor fuel consumption (95% load factor)		
Reactor	Fuel cycle	Fuel consumption (t/a)	
		U	Th
CANDU	Natural U once-through	160	–
	Pu/U	80	–
	Pu/Th	44	1
	Th	–	2
LWR	Enriched U once-through	220	–
	Uranium Recycle	190	–
	Pu/U	155	–
Breeder	Pu/U	2	–

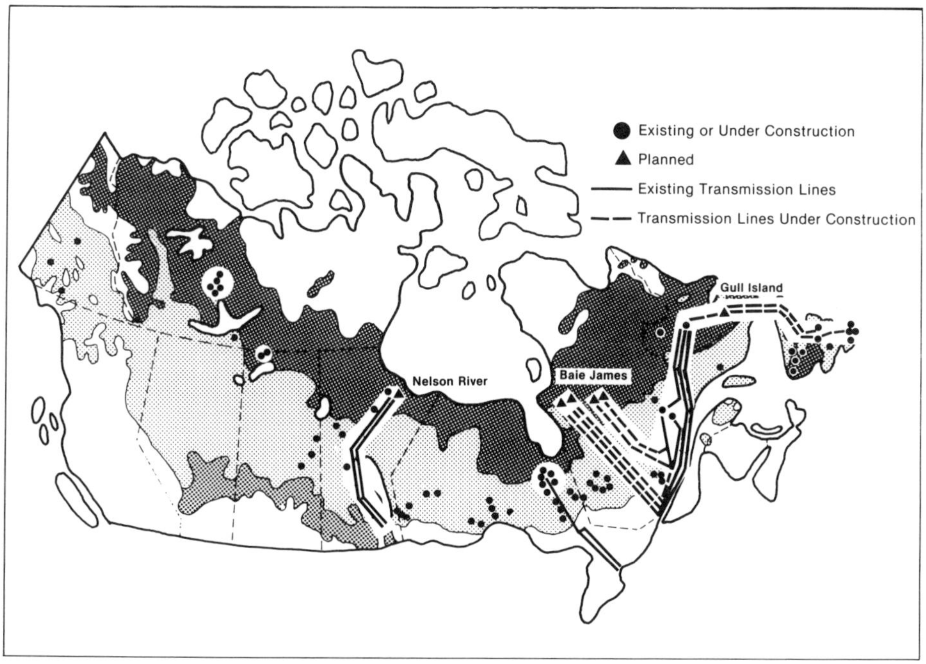

Figure 6. Location of major hydro-electric power producing areas and energy corridors.

The Renewable Fuels

Hydro-Electricity

Hydro-electricity is a major contributor to the electrical supply of Canada. The provinces of Newfoundland and Labrador, Quebec and Manitoba derive almost all of their hydro-power from river systems within the boreal forest zone, whilst Ontario and British Columbia mainly utilize other river systems (Figure 6). The major watersheds are generally far from centres of load demand and a characteristic of the development of hydro-electricity in the boreal forest zone is the requirement for extensive high voltage transmission lines, which run for thousands of kilometres.

The scale of hydro-electric development is dictated by the availability and reliability of river flow, the capital cost of civil engineering for dams and river diversions, as well as the economic cost of transmission. As a consequence, new hydro developments are likely to be of two scales; the Giga Watt such as at James Bay and the so-called micro and mini hydro where kilowatt and megawatt supplies are developed adjacent to a load. The early industrialization of the boreal forest zone by the forest products industries, mining and minerals processing was facilitated not only by the raw material base but also by the ready availability of hydro-power as can be seen by inspection of the locations of the towns. In Quebec, Ontario and British Columbia about 70 per cent of the installed industrial generating capacity is based on hydro (Anon. 1981a).

Large scale generation is extremely capital intensive as is illustrated by the James Bay project. Based on a watershed known as the La Grande complex, located almost 1500 km north of Montreal, the drainage area of 176,000 km^2 covers almost one-tenth of Quebec. It includes 174 dykes, eight dams, 1,000 km of roadway, the diversion of four rivers, the creation of six lakes and the installation of 37 generators and is a measure of the logistics of large scale hydro-power. The capital cost for the 10.3 GW capacity of the first phase is around $1,400/kW. The annual production from the LG2, LG3 and LG4 stations scheduled for completion in 1984 is anticipated to be 62.2 TWh. The second phase of La Grande will bring the total capacity to 13.6 GW and the annual production to 80.5 TWh from a total of 63 generating units (Anon. 1980).

Projected expansion of the Quebec system would encompass the Nottoway, Broadback and Rupert rivers of the NBR complex to the south of James Bay and the Grande Baleine to the north. The transmission system involves five parallel 735 kV AC lines in two corridors, the western corridor is now complete while the eastern corridor will be completed in 1984. The lines and their associated sub-stations cost about a billion dollars and are a major engineering feat in themselves.

Other major hydro-electric projects are detailed later in the paper in the summary list of major capital expenditures in the boreal forest zone.

Wind

Wind energy has been investigated at a number of scales, these can usefully be classified on the basis of the size range and expected application as is indicated in Table 4 (Chappell 1981).

Note that unit sizes differ by two orders of magnitude between each category. Only the megawatt-scale installations are likely to provide a measureable contribution to provincial or national energy supply systems. Moreover, as wind energy and other energy resources are non-uniformly distributed amongst Canada's provinces, the likely contribution of aeroelectricity varies greatly throughout the nation.

Other considerations affecting the projected aeroelectricity contribution in Canada are:

- Windmills cannot capture all the wind energy. The theoretical maximum is 59.3 per cent, but other factors reduce this to a realizable value of about 40 per cent.

Table 4. Applications for aeroelectric VAWT systems

Application	Scale
Special purpose (communications) (navigation aids)	– 1 kW (DC) single-unit installations – on-demand power (requires storage) – reliability (remote, no back-up) – expensive: $10,000/kW
Remote community (northern Canada) (isolated areas)	– 100 kW (AC) single-unit installations – local network – no storage – coupled to (diesel) base – fuel saving mode operation – $1500/kW (15¢/kWh)
National energy (utilities)	– 10,000/kW (AC) multi-unit installation – grid coupled – no storage – fuel saving plus capacity credit – $500/kW (5¢/kWh)

- Maximum penetration in any provincial grid is expected to be about 20 per cent in order to maintain current standards of stability without buffer storage.
- Windmills must be reasonably close to main electric grids to keep transmission costs low.
- Windmills cannot be too close together. A conservative assumption is that the windmill area should be less than 1 / 1000 of the ground surface area allocated to the windmill.

The annual energy output from a given size of windmill is very dependent on the wind regime at the given site. To give a guide to regional wind intensities the map in Figure 7 has been produced to show the expected average annual power output. In general the wind energy prospects are inversely correlated with the existence of forest cover! The highest wind energy potentials are in the marginal northern part of the boreal forest zone and along the shores of the St. Lawrence.

When the electrical output of large wind turbines is evaluated at the busbar, the cost of the energy produced is competitive with that from large scale hydro and nuclear plants. The imputed energy value is primarily because of the fuel saver aspect of intermittent wind energy production. It has been established that the installed capacity of wind energy in large grid systems can be as much as 20 per cent of the total before special considerations of energy storage become necessary. With the extreme northern location of much of the boreal forest zone wind energy potential the exploitation awaits the installation of the necessary transmission lines which would only be economically justified if associated with large scale hydro-electric developments. Provided that these are installed it is anticipated that wind farms comprised of between 20 and 100 multi-megawatt units could be in place by the year 2000.

The current status of wind turbine developments is that several multi-MW units are already operational in the U.S.A. and Europe. In Canada Hydro-Québec and NRCC are jointly developing a 4000 m^2 unit – project EOLE –which would give power outputs of 3 to 4 MW in the better wind regimes at a cost of 3-5¢/kWh at the busbar.

Although units of smaller size produce much more expensive power, it is likely that they

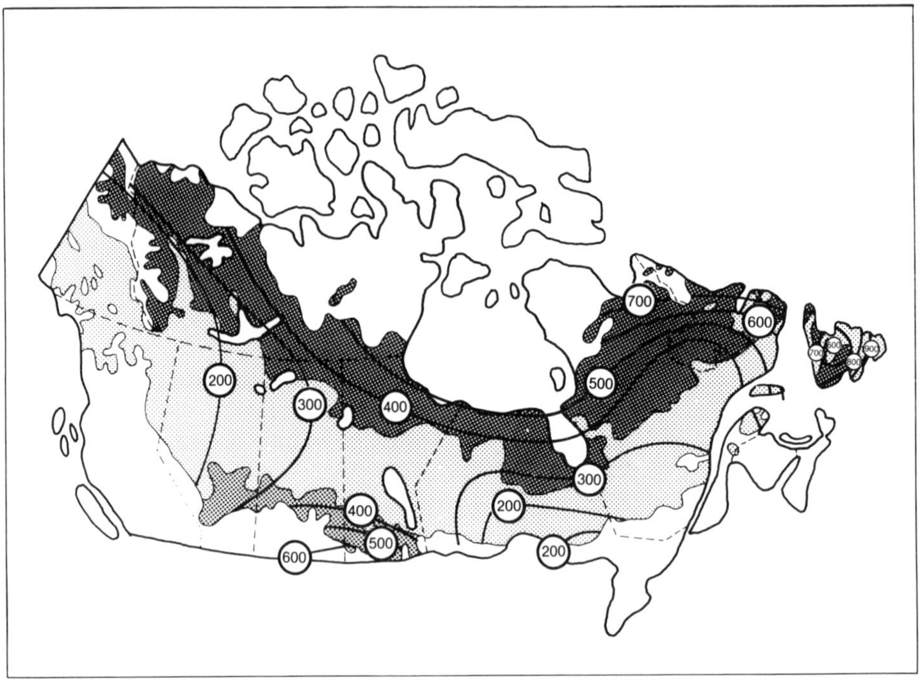

Figure 7. Annual average wind power density at 50 m (W/m^2) in the boreal forest zone.

will be cost competitive in replacing expensive diesel generation in appropriate areas of the boreal forest zone such as remote communities.

Geothermal

In the western region of the boreal forest zone the potential for geothermal energy based on hydrothermal deposits with temperatures ≥80°C is estimated to be very significant. Figure 8 shows the regions that have temperature gradients that are likely to be economically interesting for low-temperature thermal applications such as space and greenhouse heating. Because of the high initial capital costs of drilling wells recent studies have shown that economic exploitation is dependent on significant loads ≥5 MW thermal and high load factors (Anon. 1982a).

Forest Biomass

The forestry energy potential of the boreal forest zone was first identified as a potential large scale energy supply during the 1970s when much of the zone was described as surplus forest. In the meantime, the use of self-generated residues in the forest industries has increased due to the price signals for traditional hydrocarbon fuels changing significantly since 1973. The current wood energy contribution to the primary supply is about four per cent. The forest industry 'self-use' is expected to at least double by the 1990s. The available resource has been under intensive study as part of the ENFOR program of the Canadian Forestry Service, and the current indications are, that there is a large and well distributed resource which would be able to supply one EJ of secondary

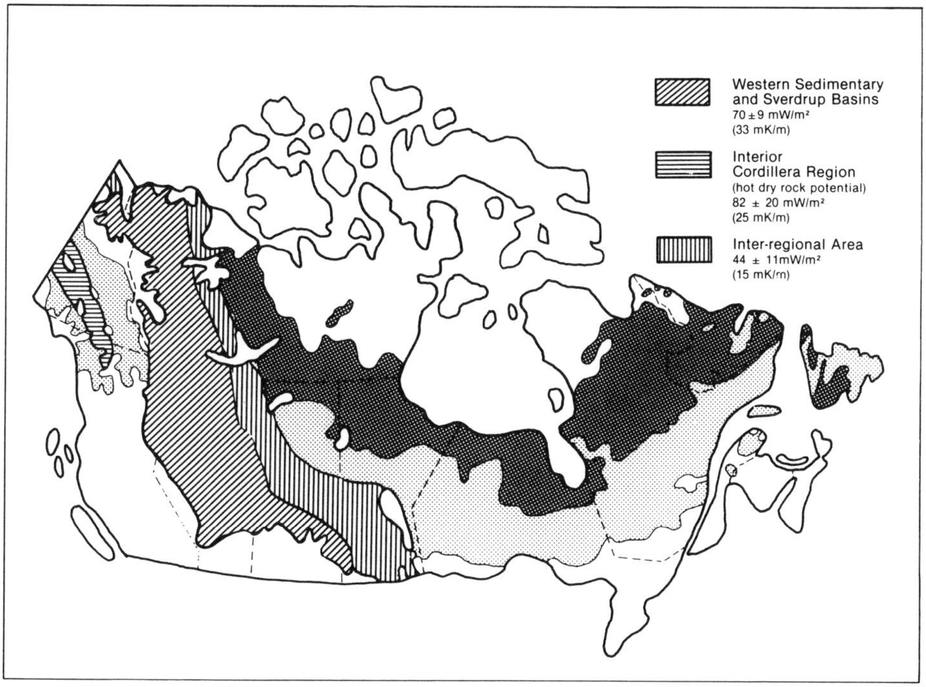

Figure 8. Locations of temperature gradients that are of economic interest for geothermal energy production.

energy assuming a conversion efficiency of around 55 per cent for the 100 million dry tonne of material available.

The current estimate of forest biomass availability (Anon. 1982b) is shown in Table 5. The estimate of delivered costs are also shown in Table 6, both for the year 2000. It can be seen that the majority of the biomass is available for less than \$60/t or \$3/GJ as a feedstock delivered to a plant gate. Established combustion technologies and emerging gasification technologies could be used on a large scale to manufacture electricity by cogeneration at forest industry sites, to provide transportation fuels such as methanol or ethanol or even to produce synthetic natural gas which could then be pipelined to southern markets.

The current barrier to more extensive use of forest biomass in electricity generation is the extremely low price the Canadian utilities are prepared to offer private cogeneration facilities. In the U.S.A. where legislation (PURPA) has been enacted which gives the private utility the marginal as against the average cost of generation of power in the utility, such applications are expanding rapidly.

The projected costs of methanol production (at the plant gate) from forest biomass at around the 1000 t/day output level are similar to those currently projected for syncrudes derived from the Alberta bitumens (when compared on an energy equivalent basis). Unlike syncrude which can be put into the existing refinery structure, methanol requires new distribution facilities and some vehicle modifications in order to be implemented on a large scale. The methanol option is, however, under close scrutiny as it offers 'hydrogen

Table 5. Forest biomass supply (kt) by province and biomass category for the year 2000. It is assumed that Gross Energy Content 100 kt = 2 PJ

Year	Province	Merchant-able surplus	Mill residues	Logging residues	Salvage	Stand conversion	Planta-tions	Total
2000	British Columbia	0	11,976	7,679	3,088	2,400	600	25,743
	Alberta	5,109	439	1,164	938	900	600	9,150
	Saskatchewan	1,570	254	768	1,859	450	600	5,501
	Manitoba	1,576	187	486	1,096	450	600	4,395
	Ontario	7,736	2,013	6,841	9,705	3,600	1,872	31,767
	Quebec	7,115	3,017	8,795	7,215	1,800	2,340	30,282
	Atlantic	2,035	1,182	4,129	3,899	1,200	1,260	13,705
	Total	25,141	19,068	29,862	27,800	10,800	7,872	120,543

Table 6. Forest biomass supply potential at different price levels for the year 2000.

	Delivered cost estimate ($1981/ODt)							
Province	1-15	16-30	31-45	46-60	61-75	76-100	Over 100	Total kt
British Columbia	8,984	3,355	3,803	3,687	899	406	1,521	22,655
Alberta	330	225	5,273	2,210	174	0	0	8,212
Saskatchewan	191	139	2,124	1,073	115	0	0	3,642
Manitoba	140	96	2,014	977	73	0	0	3,300
Ontario	1,511	1,187	12,174	6,165	1,024	0	0	22,061
Quebec	2,263	1,634	10,207	5,306	1,319	0	0	20,729
Atlantic	886	709	4,709	2,882	619	0	0	9,805
Total	14,305	7,345	40,304	22,300	4,223	406	1,521	90,404
Total Energy/EJ	0.29	0.15	0.81	0.45	0.085	0.008	0.03	1.81

economy' like interchangeability of carbon sources since methanol can be produced economically from many carbon sources such as natural gas, coal and peat. There are also some interesting hybrid options which could combine coal, peat or biomass derived carbon with electrolytic hydrogen derived from hydro and wind or nuclear.

From the viewpoint of boreal forest zone development between 30 and 40 plants having an average of 1200 t/day output would be required to equal the output of a single Albertan oil-sands plant on an energy basis (≈ 1 PJ/day output). The capital investment is estimated at around 8 Gigadollars (1982) which would be geographically well distributed as would be the employment impacts. With wise management of the resource, the renewable energy would maintain stable communities for generations.

Remote Communities

A characteristic of the boreal forest zone is that transportation and distribution links are very meagre. As a result there are many communities that are not grid connected for electricity supply. These communities have access to electrical power, although often on an intermittent basis and nearly always at costs that are very much higher than those in the cities in the south. Often the consumer price is subsidized for some low quantity of energy (e.g., 320 kWh/month (Anon. 1981d)) at rates similar to southern cities and then priced at the cost of generation which can be as high as 20¢/kWh.

This reflects not only the high cost of diesel fuel but also the very high cost of shipping oil to the remote community by seasonal ice road, waterway, or even by air. Diesel fuel costs for the NWPC result in electricity production costs of 6-10¢/kWh (Anon. 1982c). The Department of Indian and Northern Affairs has diesel fuel costs that range from 20¢/litre to 60¢/litre (Anon. 1982d). With a typical heat rate of 2 kWh/litre for the fuel alone, electricity cost is between 10 and 30¢/kWh. The situation is similar for mining and other industrial operations remote from the grid.

Almost by definition, many of the remote communities in the boreal forest zone have access to forest biomass, in fact the majority of space heating requirements are satisfied with wood. Other energy sources that are available, although on a very site dependent basis, include wind, geothermal and small scale hydro.

On account of the small scale of most remote community electrical generation plants – installed capacities are in the range of 20 to 1200 kW – the capital cost of alternatives to inexpensive (to purchase) diesel generation has to be offset by the fuel cost differential on a life cycle basis if diesel is to be replaced. Only recently, have the increases in diesel cost been such that alternatives are being demonstrated. Small scale hydro is increasingly being utilized and a demonstration of wind assisted diesel generation is underway in Sudbury. Experiments and demonstrations of wood gasification to promote gaseous fuel substitution in existing or modified diesel engines are also underway.

Discussion

The energy potential of the boreal forest zone is truly immense as can be seen in the summary in Table 7. While some of the resources are truly finite such as those of conventional oil, gas and uranium, others such as heavy oils, bitumens, coal and peat are so extensive that, relative to foreseeable rates of exploitation they should be viewed as essentially inexhaustible. Renewable resources are also extensive in the boreal forest zone

Table 7. Energy in the boreal forest zone

Source	Resource EJ	Reserve EJ	Probable rate of exploitation EJ/year 1982	2000
		Non-renewables		
Coal	2200	500	0.5	2
Peat	2000	–	–	?
Conventional				
Oil	80	30	2.6	0.3
Gas	145	66	4	4
Heavy oil and bitumen	5800	230	0.4	2.5
Uranium	?	(100 kT) 20 EJ	0.13	0.4
Geothermal	11000	1000	–	0.02
		Renewables*		
Hydro	5	0.5	0.5	2
Wind	10	2	0.001	0.020
Bioenergy	10	2	0.35	1

* The resource and reserve concept is not adequate for renewable energy sources which constitute a *flux* of energy rather than a store. The figures quoted are annual fluxes.

and in some instances offer an interesting challenge to the development of fossil fuel options.

The impacts of energy developments in the boreal forest zone are very hard to quantify. Two measures might be to describe the investment levels required and try to establish the land use impacts.

Capital Investment

To put the investment levels into some perspective it is worth recognizing that the current retail value of energy in the Canadian economy is about $\$90 \times 10^9$/annum. This is derived from the product of primary energy use and consumer costs of energy. Electricity at 3.5¢/kWh and heating oil at 30¢/litre correspond to almost $10/GJ while the total primary energy consumption is in the neighbourhood of 9 EJ.

Table 8 is a list of energy investments proposed for the boreal forest zone which has been compiled from newspaper clippings as well as from the report of the Major Projects Task Force (Anon. 1981e). Not all of the dollar values have been reported on a consistent basis, some are in constant dollars of some unspecified year, while others are in current dollars of the year of expenditure. It is therefore not entirely accurate to add the dollar total and perform manipulations with the figures. If one does, the projected 1980-2000 investments total 150 G$ with approximately half of the investment in hydro-electric projects, one-third in oil-sands and heavy oils, one-fifth in transmission systems and oil and gas pipelines.

It should be noted that some of the projects in the list of Table 8 have recently either been deferred or cancelled. Nonetheless, these projects or similar substitutes will eventually be

Table 8. Major capital projects in the boreal forest zone

Province	Class	Description	Cost 10^9 $	Year	Capacity comments
Ont.	Biomass	Various Waste to Energy	3.2	80-90	– Mainly P&P
B.C.	Gas process	Fort St. John	0.25		
NWT	Gas process	Parsons Lake	0.3	85-90	
NWT	Gas process	Niglintgak-Mackenzie Delta	0.2	85-90	
YNWT	Gas transm'n	Dempster Lateral	2.5	85-90	34 $10^6 m^3/d$
Atl	Generation	Muskrat Falls	3.2	81-86	600 MW
Alta	Generation	Peace River	1.8	85 —	1500 MW
Man	Generation	First Rapids	0.5	95- +	210 MW
Ont	Generation	Onakawana Lignite	1.0	86-92	1000 MW
Man	Generation	Waskwatim-Thompson River	0.54	90-98	280 MW
Que	Generation	Manic 5 expansion	0.9	82-85	1000 MW increment
Atl	Generation	Gull Island	4.5	80-95	1700 MW inc Trans
Que	Generation	NBR I & II	30.2	81-99	8500 MW
Que	Generation	Grande Baleine	7.7	81-90	—
Man	Generation	Limestone	1.54	87-93	1100 MW
Man	Generation	Conowapa – Lower Nelson	2.7	95- +	1330 MW
Yuk	Generation	Five Fingers Hydro	0.1	81-83	
Man	Generation	Manason	0.6	90-98	119 MW
Que	Generation	La Grande I & II	10.8	80-92	3000 MW
Sask	Heavy oil	Lloydminster	1.75	80-99	inc Alta
Sask	Heavy oil	Lloydminster up-grader	1.0	81-86	4.8 $10^3 m^3/d$
Que	Mining	Aluminum – 3 Projects	1.3	80-86	—

Atl	Mining	Aluminum smelter	0.5	83-90	180 kt/a
Alta	Mining	Judy Creek Coal	0.4	82-84	2.3 Mt/a
NWT	Oil	Norman Wells	0.4	81-84	4 $10^3 m^3/d$
Alta	Oil-sands	GCOS expansion	0.2	—-82	2.1 $10^3 m^3/d$ increment
Alta	Oil-sands	In-situ program	0.1	84-88	
Alta	Oil-sands	Sundalta-Gulf	10	86-90	
Alta	Oil-sands	Fort McMurray Mining	10	85-90	21 $10^3 m^3/d$
Alta	Oil-sands	Peace River in-situ	5	85-92	
Alta	Oil-sands	Alsands Project	7.2	80-87	22 $10^3 m^3/d$
Alta	Oil-sands	Syncrude expansion	1.5	84-86	9.5 $10^3 m^3/d$ increment
Alta	Oil-sands	Cold Lake in-situ	10	81-99	22 $10^3 m^3/d$
Mult	Pipelines	TQ&M	2.3	81-84	—
Mult	Pipelines	Cochin Loop	0.13	82	—
Mult	Pipelines	Alaska Gas (Foothills)	8.4	82-87	56 $10^6 m^3/d$ 2035 km
Mult	Pipelines	Polar Gas (Arctic Gas)	12.3	89-94	38 $10^6 m^3/d$ 4800 km
Mult	Pipelines	Norman Wells Oil	0.37	81-83	— 870 km
Mult	Pipelines	Trans Canada Looping	0.56	80-84	— 11 Compressors
Alta	Tert-oil	Judy Creek CO2 Flood	0.5	82-85	
Mult	Transmiss'n	Western Power Grid	0.62	—-87	—
Man	Transmiss'n	HVDC Conversion	0.45	81-89	
Que	Transmiss'n	James Bay HV system	1.86	81-90	— 315 kV
NWT	Transport'n	Barge System Mackenzie	0.15		Beaufort Sea
Sask	Uranium	Cluff Lake	0.17	79-81	1.5 kt/a
Sask	Uranium	McLean Lake	0.3		
Sask	Uranium	Warman-UF6 Plant	0.1		
Sask	Uranium	Midwest Lake	0.3	83-86	5.4 kt/a
Sask	Uranium	Key Lake	0.5	81-83	

Table 9. Energy intensity of different energy resources

Energy source	Energy/ha	Area/km^2 for 1 EJ
Crude oil wells (Alta)	22 PJ	0.46
Athabasca Oil Sands Exploitation by mining	2.2 PJ	4.5
Lignite coal Shaunavon (Sask.)	380 TJ	26
Peatlands in eastern Canada	50 TJ	200
Solar energy incident on Ottawa in one year	46 TJ/year	217
Hydro-electric schemes – James Bay (Que)	16.5 GJ/year	606,000
Forest – mature 500 Mg/ha	9.3 TJ	1,080
– annual rate of accumulation		
5 Mg/ha	9.3 GJ/year	108,000/year
30 Mg/ha	560 GJ/year	17,900/year

constructed so that Table 8 can be considered to be a representative guide to the magnitude and sequence of, if not the timing of, energy investments in the boreal forest zone.

Land Use Impacts

The non-renewable energy sources have surprisingly small land use requirements. The local impacts of oil-sands mining are very intensive, but are limited to a small area compared to the forest harvest area required for equal output. As indicated in Table 9, even peat which is the accumulation of 7000 years or so of carbon has a much smaller impact than the James Bay hydro project. These energy/unit area indices are also a guide to the social and environmental impacts. High energy intensity implies an extremely large transient construction work force with town site construction that far exceeds the operating work force requirements. Even the operational plant staffing is likely to far exceed the native population in the region. Many of these concerns were raised during the Berger Commission (Berger 1977) and if energy development on the scale of that indicated in Table 8 is undertaken then continuing attention will have to be paid to the social impacts of these developments.

Environmental impacts for new operations are the subject of intensive study and it is recognized that a significant cost component of new developments has to be the investment in technologies and strategies that minimize and ameliorate potential disruption.

Literature Cited

Anonymous. 1977. Oil sands and heavy oils: the prospects. Dept. Energy, Mines and Resources, Rept. EP 77-2. Ottawa, Ontario.

Anonymous. 1980. Electric power in Canada – 1979. Dept. Energy, Mines and Resources (DSS Cat. No. N23-14180-7).

Anonymous. 1981a. Electric power in Canada – 1980. Dept. Energy, Mines and Resources (DSS Cat. No. M23-14 (81 – 6E)) Ottawa, Ontario.

Anonymous. 1981b. Canadian energy supply and demand – 1980-2000. National Energy Board (DSS Cat. No. NE23-15/1981E). Ottawa, Ontario.

Anonymous. 1981c. Uranium in Canada – 1980. Assessment of supply and

requirements. Dept. Energy, Mines and Resources, Rept. EP 81-3. Ottawa, Ontario.
Anonymous. 1981d. The social, environmental and resource impact of wood gasification in isolated northern communities – Part I. Dept. Environment, ENFOR Project C-8(2). Ottawa, Ontario.
Anonymous. 1981e. Major Canadian projects, major Canadian opportunities. Ministry of Industry, Trade and Commerce. Ottawa, Ontario.
Anonymous. 1982a. An overview of geothermal heat extraction technology. National Research Council. Report OSX81-00087. Ottawa, Ontario.
Anonymous. 1982b. Availability and cost of forest biomass in Canada. Dept. Energy, Mines and Resources, ENFOR Program. Report OSQ81-00085. Ottawa, Ontario.
Anonymous. 1982c. Northern Canada Power Commission 33rd Ann. Review for the year ended 31st March, 1981. Northern Canada Power Commission. Edmonton, Alberta.
Anonymous. 1982d. Feasibility study for wood gasification and small scale hydro generation for Indian reserves. Phase I – National Screening. Dept. Indian and Northern Affairs (File 10026). Ottawa, Ontario.
Bielenstein, H.U., L.P. Chrismas, B.A. Latour and T.E. Tibbetts. 1980. Coal resources and reserves of Canada. Dept. Energy, Mines and Resources, Rept. ER 79-9. Ottawa, Ontario.
Berger, T.R. 1977. Northern frontier, northern homeland. The report of the Mackenzie Valley pipeline inquiry. Volume Two. Department of Supply and Services. Ottawa, Ontario.
Chappell, M.S. 1981. The Canadian wind energy research and development program: Overview and current studies. National Research Council, Rept. TM-WE-005. Ottawa, Ontario.
Radforth, N.W. and C.O. Bramner (Eds.). 1977. Muskeg and the northern environment in Canada. University of Toronto Press. Toronto, Ontario.

ENERGY PRODUCTION FROM PEAT AND WOOD RESOURCES IN NORTHERN FINLAND

Eino Siuruainen
University of Oulu

Abstract: From World War II until the early 1970s, the heating energy produced traditionally by fire wood, was largely replaced by imported fossil fuels, even in the forest-dominated rural communities in northern Finland. This trend reached a peak in the middle 1970s when the international energy crisis encouraged the development of more domestic energy reserves. Parallel with the increase in nuclear power production, was an increase in heating energy and electricity from peat and wood chips in power stations located in both cities and rural centres. Research was required to determine total regional peat and wood reserves, size of local deposits, deposit availability, transportation systems required, and finally, the regional impact of this utilization on human activity and nature conservation. This paper summarizes recent field studies led by the Department of Geography, University of Oulu. Detailed research was based on map sheets (1:20,000) and other statistics were received from a detailed questionnaire sent to all 74 communal authorities, and numerous companies producing domestic energy. The study was initiated in 1980 and will be completed in 1985. The total amount of peat in northern Finland is approximately 66 billion cubic metres of which less than three per cent is being used, or planned for use in the near future. Annual wood biomass harvest will be approximately 5.5 million cubic metres in the same geographic area. Less than one-tenth of the annual growth is utilized at present. Although these percentage use values are low, the present harvest equals 200,000 tons of crude oil with a value of 80 million marks. In 1981, the production and use of these energy sources employed approximately 1600 man years with a salary value of 63 million marks and with taxes for local communities of 11 million marks.

Finland is situated between Sweden and the Soviet Union reaching from 60°N to 70°N latitude. The northernmost part of the country is bounded by Norway not far from the Arctic Ocean. The total area of the country is 337,009 km^2. The average altitude is only 152 m, and most parts of the country are less than 200 m above sea level. Lakes and river systems are common throughout the country and cover some nine per cent of the total area. The country belongs to the boreal coniferous forest zone with large areas of wilderness, mires and wastelands, and only 4.7 million inhabitants. Most of the labour force is occupied in secondary and tertiary sectors, a continuously diminishing part staying in agriculture and forestry. Especially since World War II, the economy and social affairs have modernized to a degree that Finland now ranks among the most highly developed countries in the world. Between the latitudes 64°N−70°N, northern Finland is relatively low (0-200 m) in its southern and south-western parts, but in the north and east some summits reach 600-1000 m. The relative height varies greatly from only 0-200 m in the western part, 20-200 m in the east and generally more than 200 m, in the north.

Typical glaciated landforms are gentle sloping moraine hills and sandy eskers surrounded by mires consisting of layers of moss or sedge peat. Most of northern Finland belongs to the boreal forest zone, but in the north there are also some features of Arctic tundra at altitudes of 600-1000 m.

The climate of northern Finland is relatively harsh with a cold long winter and a cool short summer. The average annual temperature varies between +2° and −2°C. The average temperature in January is −12° to −14°C and in July is +12°C to +14°C. The average annual precipitation is some 500-700 mm, most of which is received as snow. The permanent, 60-90 cm deep snowcover lasts from late October until early May, or in the northernmost parts, even to mid-June. The annual evaporation is one-half of the total precipitation and when combined with low relief creates favourable conditions for mires dominated by mosses, sedges and various willows.

Some 30-40 per cent, and locally even more than 60 per cent, of the total land area of northern Finland is covered by various types of bogs (Figure 1). Some 38 per cent of this mire area is open treeless bogs (*neva*), approximately 42 per cent is small pine-dominated swampy tracts (*räme*) and 18 per cent spruce-dominated peatbogs (*korpi*). The average percentage of productive forest land is approximately 50-60 per cent of the total land area (Figure 2). The two conifers in this area are the Scotch pine (*Pinus silvestris*) and the spruce (*Picea abies*). The main deciduous trees are birch (*Betula pubescens, B. verrucosa,* and in the north *B. tortuosa*), aspen (*Populus tremula*) and alder (*Alnus incana, A. glutinosa*). The most common bushes are numerous varieties of willows (*Salix* spp.). The virgin nature of northern Finland has been much changed, especially in this century. The boom of lumber felling took place just after World War II and lasted until the late 1960s. The main reasons for this large-scale change were the expanding pulp industry, which rebuilt the almost totally destroyed infrastructure of the area, and the heavy war losses to the Soviet Union. Hundreds of square kilometres of forest on private and public land were clearcut. Some of these areas were fired and most of them replanted or reseeded just after the felling operations. Much valuable productive forest land has been gained by draining wet mires. It has been estimated that by the late 1970s some 550,000 km of drainage ditch had been constructed in this area. During the intensive resettlement period between 1940 and 1960 an enormous area of bog and forest land was occupied for farm-related purposes.

Northern Finland consists of two provinces with nearly one-half of the total land area but less than one-sixth of the population. The present settlement pattern is based on rapidly growing urban centres surrounded by sparsely settled rural villages located mainly on lake shores, riverbanks and along main roads (Figure 3). During the last 20 years many young people moved to centres for employment in expanding sectors of industry and service. This movement has been weakening rural areas, leaving more and more houses, even rural villages, unoccupied. Modern techniques and automation have released people from primary production but now development also decreases employment in industry and other urban occupations. This development trend has resulted in growing economic and social problems.

During the last few years attempts to solve these growing problems led to a regional development policy which supports, for instance, domestic energy reserve use for heating energy and for generating electricity. To replace imported fossil fuels by domestic energy resources have been regarded as an important way to get new jobs for people who formerly worked in agriculture, forestry or in the industrial sector. The multiple effects of this policy will affect the rural as well as the urban areas.

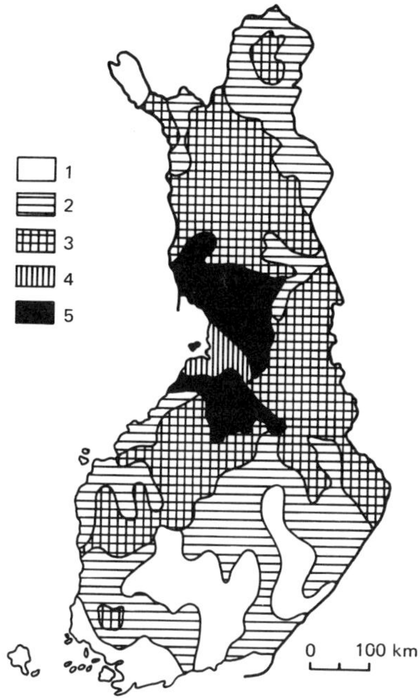

Figure 1. Percentage of bog of the total land area in Finland: (1) -10, (2) 11-30, (3) 31-50, (4) 51-60, (5) 61- (Ilvessalo 1961).

Figure 2. Percentage of productive forest land of the total land area in Finland (1) -20, (2) 21-40, (3) 41-60, (4) 61-80, (5) 80- (Kuusela 1976).

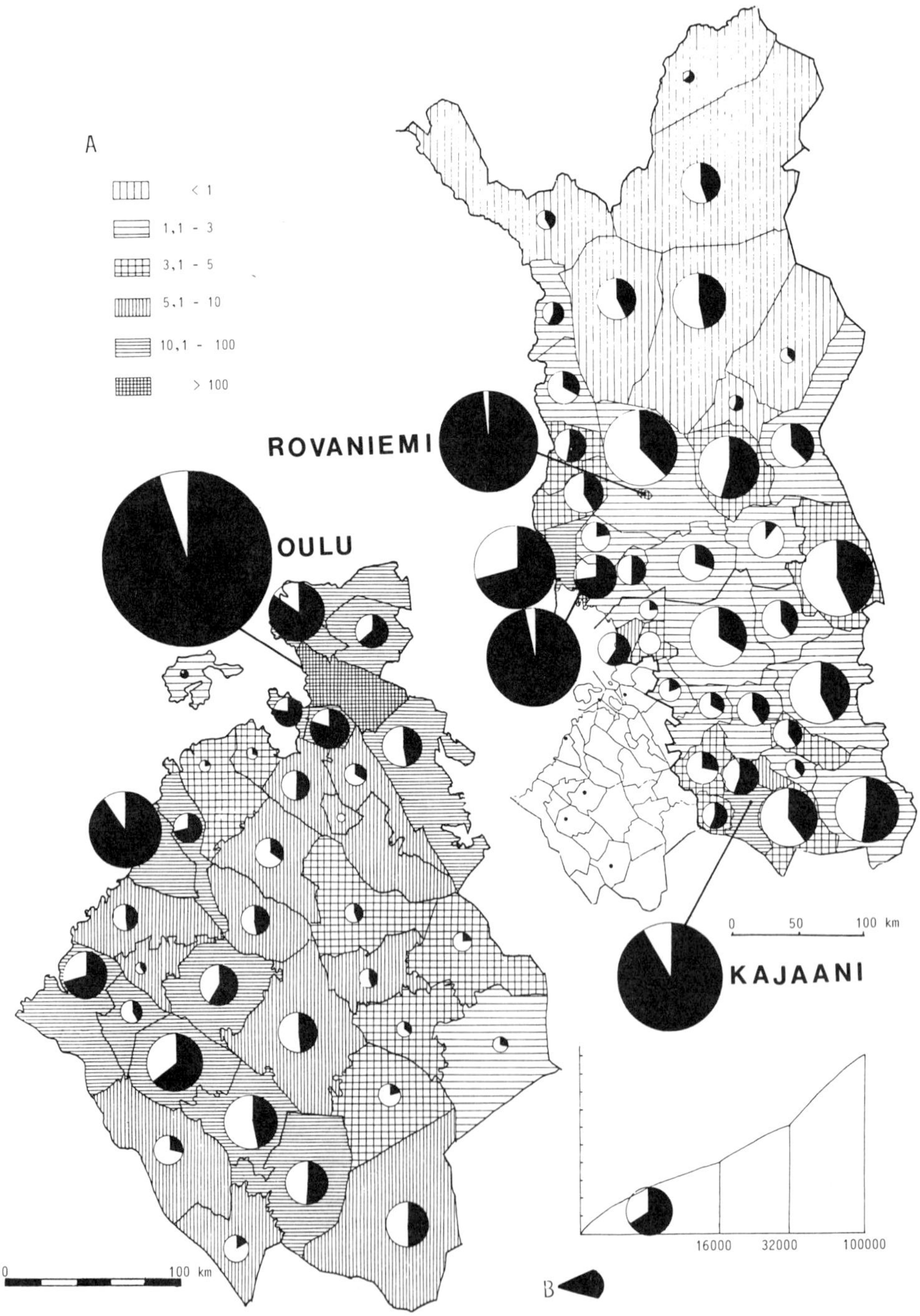

Figure 3. Population density inhabitants/km^2 (A) and total population with proportion living in built-up areas (B) in northern Finland in 1980 (Tarakkamäki *et al.* 1982).

Problems and Methodological Questions Related to Domestic Energy Reserve Use

When making plans to reallocate the forms of energy, while keeping in mind the national economy, the regional impacts on local people and environment, and multiple effects on each level, one has to solve numerous problems such as:

- What are the local energy resources, what techniques have been used before, and what are the potential total reserves for each type of use?

- Where are the deposits located when compared with present railway and road systems and principal consumption centres?

- What are the economics of deposit development, and what kind of harvesting techniques are required?

- Are the final forms of energy-heating fuel, electricity or even liquid fuels?

- Are the mires or forests owned by the state, industrial companies, or individuals? Do the owners want to sell or rent out the deposits, or would they prefer to wait for the next 20 years for a more serious energy crisis which would increase the value of their resource?

- What will be the national energy policy? Do public authorities support the production of domestic energy forms in areas with serious unemployment problems, even though nuclear power is the least expensive way to meet the national energy needs?

Many other problems of great importance must be solved before industrial units or heating systems are set up in cities, in rural centres or even on one-family farms. For the present, we have enough knowledge to initially answer these questions, but in the long run we may need more methodological skills and general knowledge on this complex problem. For instance, what is the general opinion about the further utilization of water energy resources? Should a fifth nuclear power station be built in this country? What shall we do with large mires from which peat layers have been taken for energy purposes? Should they be planted to forest, seeded for agriculture or modified to produce biomass energy?

During the last few years much research work has been done in Finland on domestic energy resources. For example, valuable large-scale work has been done by the Geological Research Institute of Finland, the Fuel Centre of the State, the Ministry of Commerce and Industry, numerous research institutes and various scientific departments of universities. Energy wood reserves are inventoried during the periodic field inventories of Finnish forests. Much research activity in this field has concentrated on the most important regions around the biggest urban areas and on the most profitable production areas; for instance, on bogs more than two m in depth. Considerable research resources have been invested in the development of processing technology. These results have been encouraging during the last few years. To make further energy plans we may need more methodological research and general knowledge on regional peat reserves and annual forest growth useable for energy purposes. In each region one has to optimize more closely present natural resources and needs of human activity with economic and social perspectives.

Present Deposits of Domestic Energy Resources

This paper is a brief presentation of the results of a recent interdisciplinary research project in the Department of Geography, University of Oulu, Finland, concerning peat and firewood reserves and their present use in northern Finland. This part of the project concentrated on the topic of peat and wood energy reserves based on map sheet (1:20,000) analyses and on results received from a detailed questionnaire sent to all the 74 public communal authorities and to numerous companies involved in the production or consumption of domestic energy resources.

Methodologically the map analysis is based on a chorological point matrix formation on 100 km^2. In the centre of each square kilometre the landforms, dominant vegetation types, such as open treeless mire, pine- and spruce-dominated peatlands, and productive forest types, have been analysed from 1:20,000 basic map sheets and aerial photographs. This research has also been supported by results from practical fieldwork done sporadically in the research area. At each point of the matrix some important human factors have also been inventoried. For example, the settlement pattern, the distances to the nearest road or railway, and the distance to the nearest urban centre (Figure 4), and the potential location of future users are noted. The number of basic map sheets covering the research area is 1759; this means the total number of analysed points is about 175,900. Using this enormous data bank, an atlas of domestic peat and wood energy reserves was completed including the numerous maps and statistics. This volume was printed in 1982-1983 and the present paper includes a sample of maps and statistics from that volume. Table 1 summarizes the peat and wood energy resources of northern Finland.

The total amount of mire in Finland is some 10.5 million hectares or 30 per cent of the total land area of the country. In Figure 2 we see that regionally the percentage is as high as 60 per cent. Of this enormous area in northern Finland some 38 per cent is still open treeless bogs (*neva*), 42 per cent mainly pine-dominated swampy tracts (*räme*), and 18 per cent spruce-dominated backwoods (*korpi*). These types of mires provide some 66 million m^3 peat and a moderate amount of energy wood reserves. All mire types contain great amounts of old lignous material, such as tree trunks and stumps, but this is just as useful for energy purposes. Potential peat reserves have been calculated using generalizations from each type of bog in the research area, and the average depth of each type measured in fieldwork from a wider area. Forest energy has been estimated in each area using generalities of each forest type and the measured average annual growth. The author is aware of the numerous sources of error and the general level of results of this study, but this type of general analysis gives necessary information on regional possibilities to domestic energy forms. This analysis may help to concentrate future research activity on those potential areas where there are as yet no plans to utilize local energy reserves.

The estimated depth of open treeless mire in northern Finland is some 1.5 m, and the estimated amount of peat is 33.6 billion m^3. This type of mire is most common around the city of Oulu, and in northernmost Lapland (Figure 5). On the base of the analysis some 39 per cent of that peat is relatively easily available to transportation networks, some 23 per cent is located between 2-4 km distance from roads, and the rest (38 per cent) is located at a distance of more than 4 km from roads and railways.

Forested bogs are more shallow than open peatlands (Figure 6). The estimated mean depth of pine-dominated swampy tracts is 1.0 m, and that for spruce-dominated wetlands, only 0.65 m. In this study the depth of 0.9 m layer is used as the multiplier to get the total peat

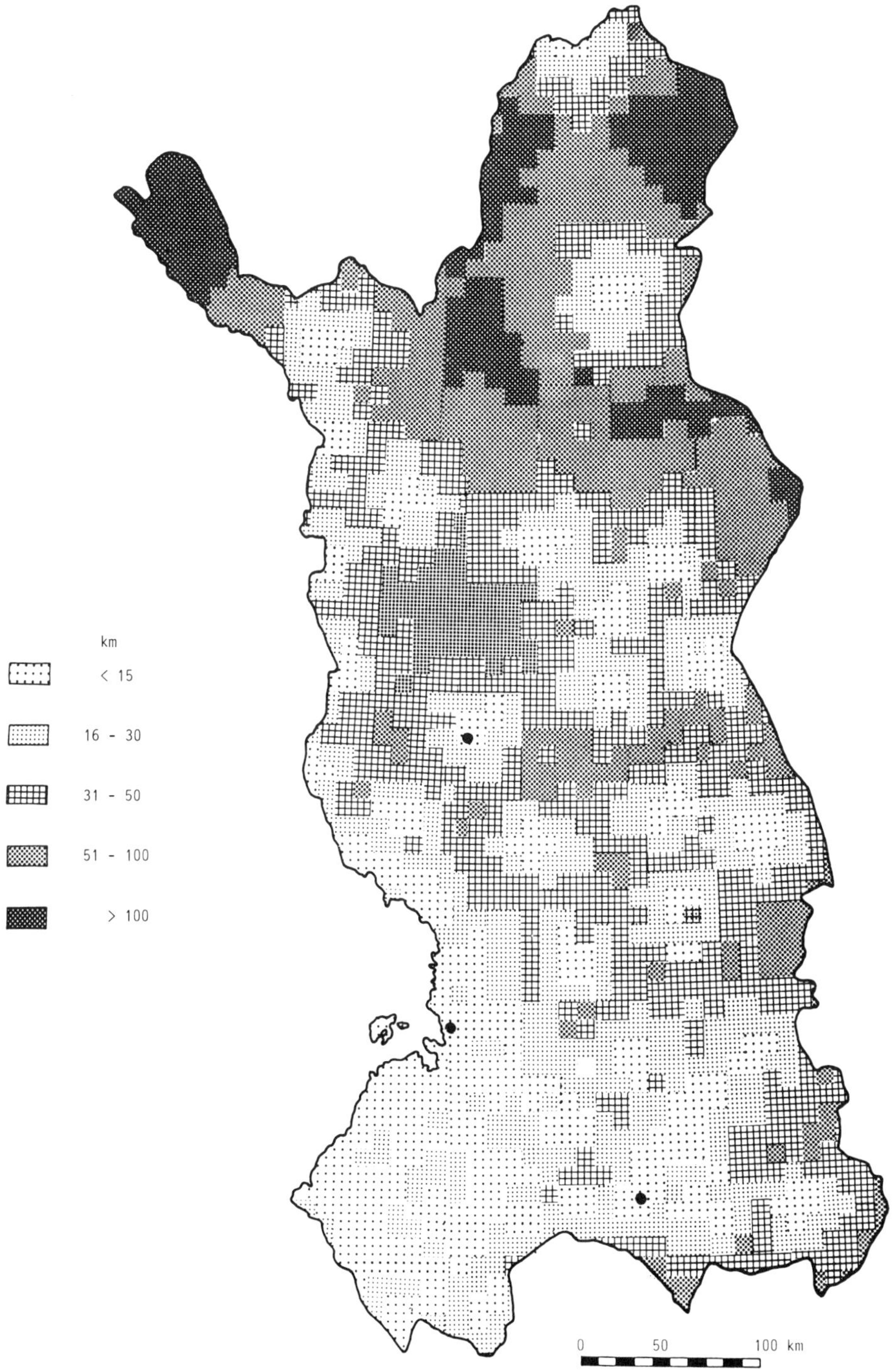

Figure 4. Average distance from each basic map sheet to nearest urban centre in northern Finland (Aitto-oja and Siuruainen 1982).

Table 1. Estimated peat (10^9 m^3) and wood (10^6 $m^3.y^{-1}$) energy resources in northern Finland.

Accessibility (distance from road systems)	Peat				Wood					
	Open mire	Forested mire	Total	%	Forested mire	Conifer. forest	Mixed forest	Mountain birch forests	Total	%
<2 km	13.0	18.7	31.7	48.2	1027	1605	265	32	2929	53.0
2-4 km	7.8	8.0	15.8	24.2	441	594	112	24	1171	21.2
>4 km	12.8	5.4	18.2	27.6	297	658	241	233	1429	25.8
Total	33.6	32.1	65.7	100.0	1765	2857	681	289	5529	100.0

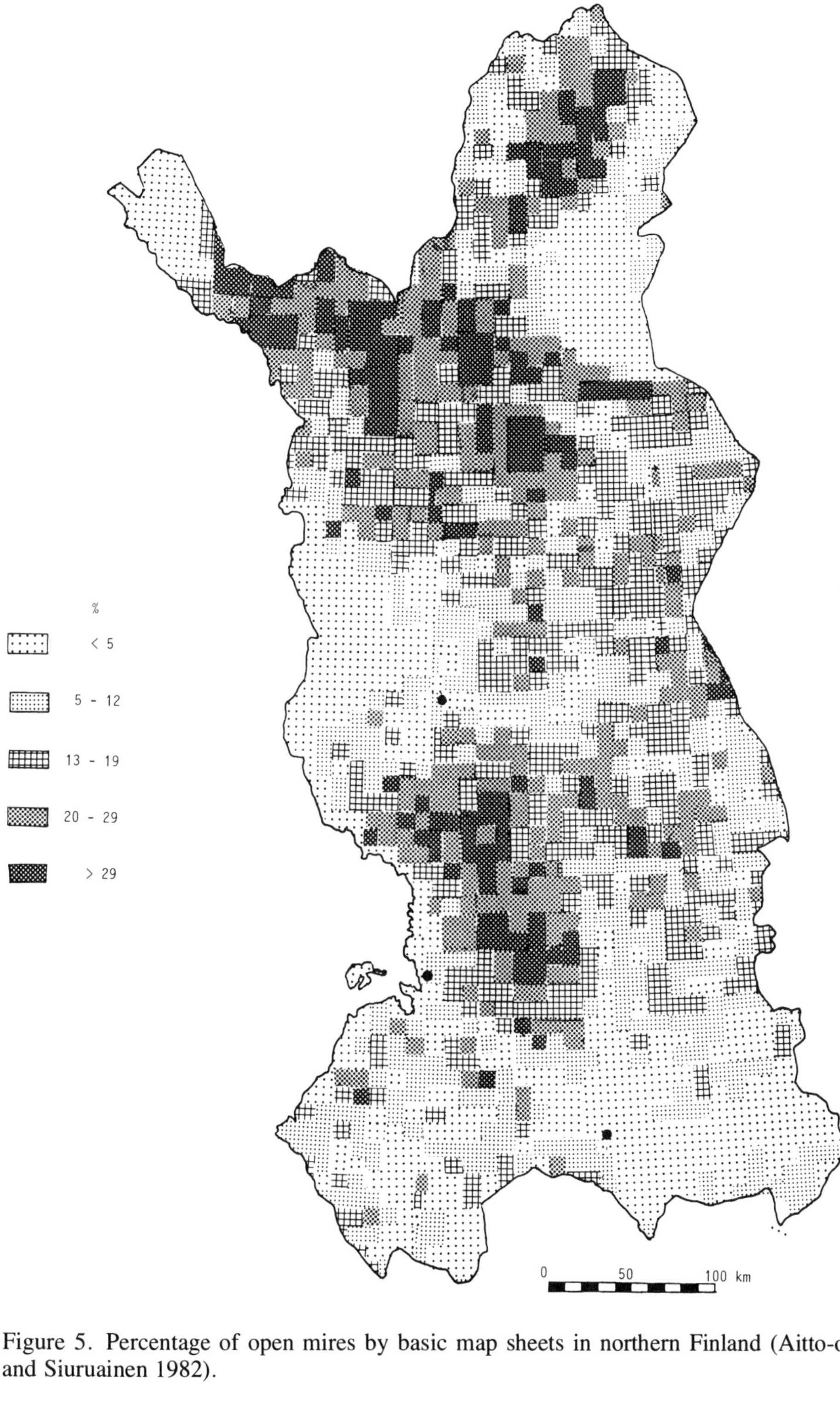

Figure 5. Percentage of open mires by basic map sheets in northern Finland (Aitto-oja and Siuruainen 1982).

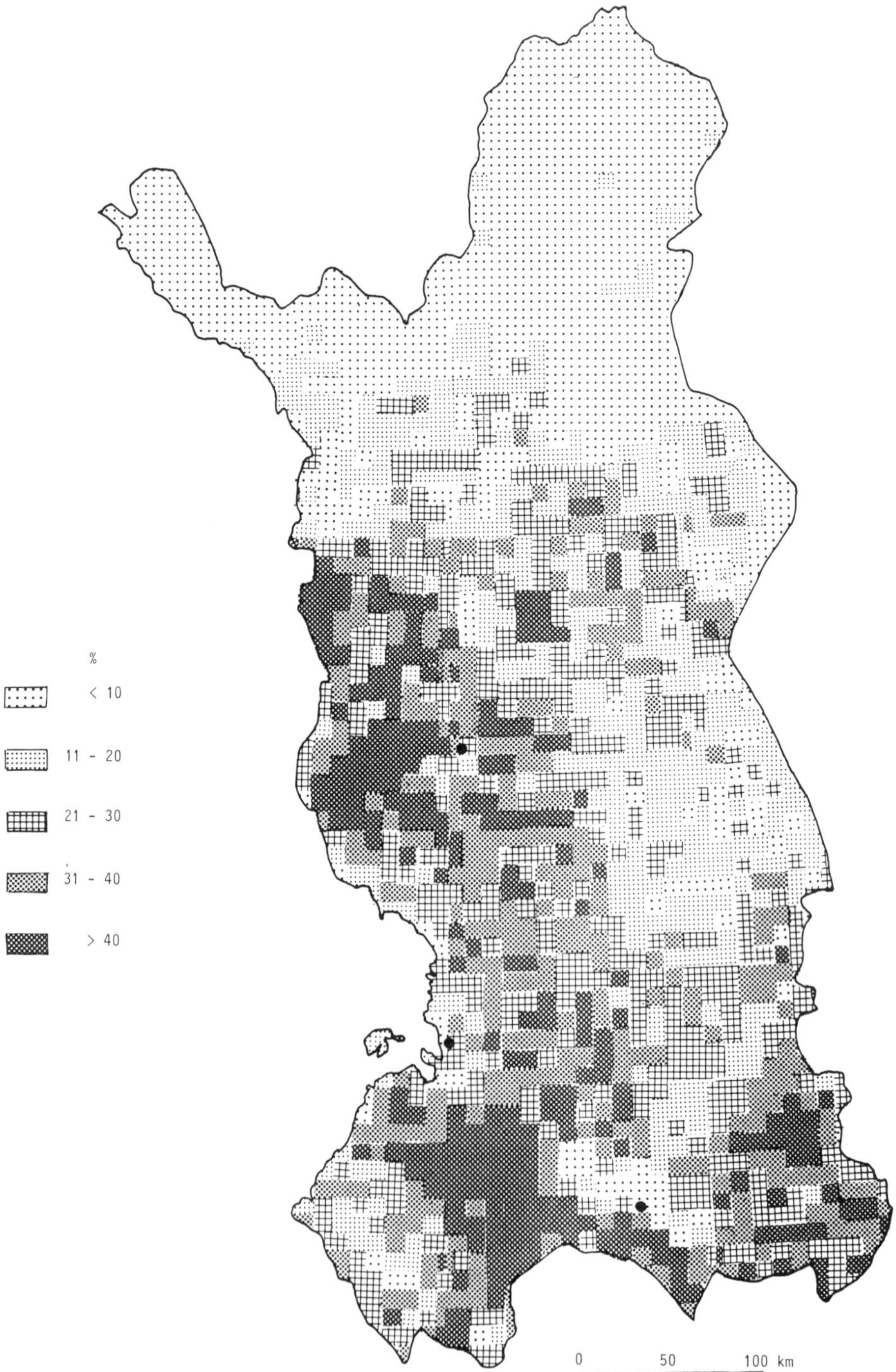

Figure 6. Forested mires as a proportion of total surface area in northern Finland (Aitto-oja and Siuruainen 1982).

layer deposits in these mire types. The total amount of peat in these categories is 32.1 billion m^3. Some 58 per cent of these reserves are located closer than 2 km from traffic systems. The estimated amount of all peat reserves in northern Finland is 65.7 billion m^3, of which more than one-half is easily accessible to roads and rail, 24 per cent is at the distance of 2-4 km, and the rest is more difficult to transport to traffic systems. From Figure 7 we see that the cumulative total amount of peat reserves around the main centres of Oulu, Rovaniemi, and Kajaani are very great. The largest reserves seem to be about 50 km from the centre of Oulu, but most of the reserves are within 30 to 50 km. The reserves around the two other cities are less great, but the availability of the forest-dominated mire types is better than in the case of Oulu.

The percentage of productive forest area in Finland varies between 40 to 60 per cent, but in northernmost parts of the country it is less than 40 per cent. Coniferous forests are the most common types in northern Finland (Figure 8). In the coniferous forests of the research area more than one-half is pine-dominated, one-third is spruce-dominated and the rest is mixed forest. Based on numerous field data, it is agreed that the annual growth of timber usable for energy purposes is some 0.5 m^3 per hectare. The total area of coniferous forests in northern Finland is 57,142 km^2, which, perhaps, produces 2,857 million m^3 of energy material per year. Because of the highly developed forest road system some 56 per cent of this material is situated less than 2 km from road accessibility.

Mixedwood and birch forests total some 7.7 per cent of the land area, which annually produces 618,000 m^3 of energy wood. At the Arctic treeline the birch (*Betula tortuosa*) is an important local energy resource. The annual growth of that species is estimated to be some 0.289 million m^3, and as much as 80 per cent of this material is situated more than 4 km from the highway and forest road system. The use of this subarctic forest zone is, as a rule, limited because of its low capacity to reproduce.

Figure 9 suggests that wood energy resources are similar around each main centre in the research area. Kajaani and Rovaniemi are situated in forested areas, while around Oulu, there is more open mires and arable land. Figure 9 shows that there is much wood energy within 40 km of the towns.

When regionally summarizing the above results, Figure 10 suggests that the energy reserves in the eastern part of northern Finland consist mainly of coniferous forests, while in the north, mixedwood and mountain birch forests dominate. Large areas of the north are also covered with forested mires which are a potential source of domestic energy resources. Open mires with peat layers are the main deposits of domestic energy forms around the city of Oulu and in the western part of Lapland province.

Use of Domestic Energy Resources and Effects on Regional Development

The change of energy forms from fossil fuel to domestic peat and wood sources, and even to nuclear power, has been moderately slow. The total energy consumption in Finland has been decreasing slightly, but domestic energy is increasing in importance. The greatest changes have taken place in urban heating energy systems and in industry. Some 9.2 million m^3 of energy peat was produced in 1980, of which 2.8 million was produced in northern Finland (Figure 11). In the research area most of this amount was used in community heating power stations, and some 0.5 million m^3 in industry. Only 0.26 million m^3 of wood chips were used, primarily in industry. The annual use of traditional firewood was some 0.16 million m^3 in the same period (Figure 12).

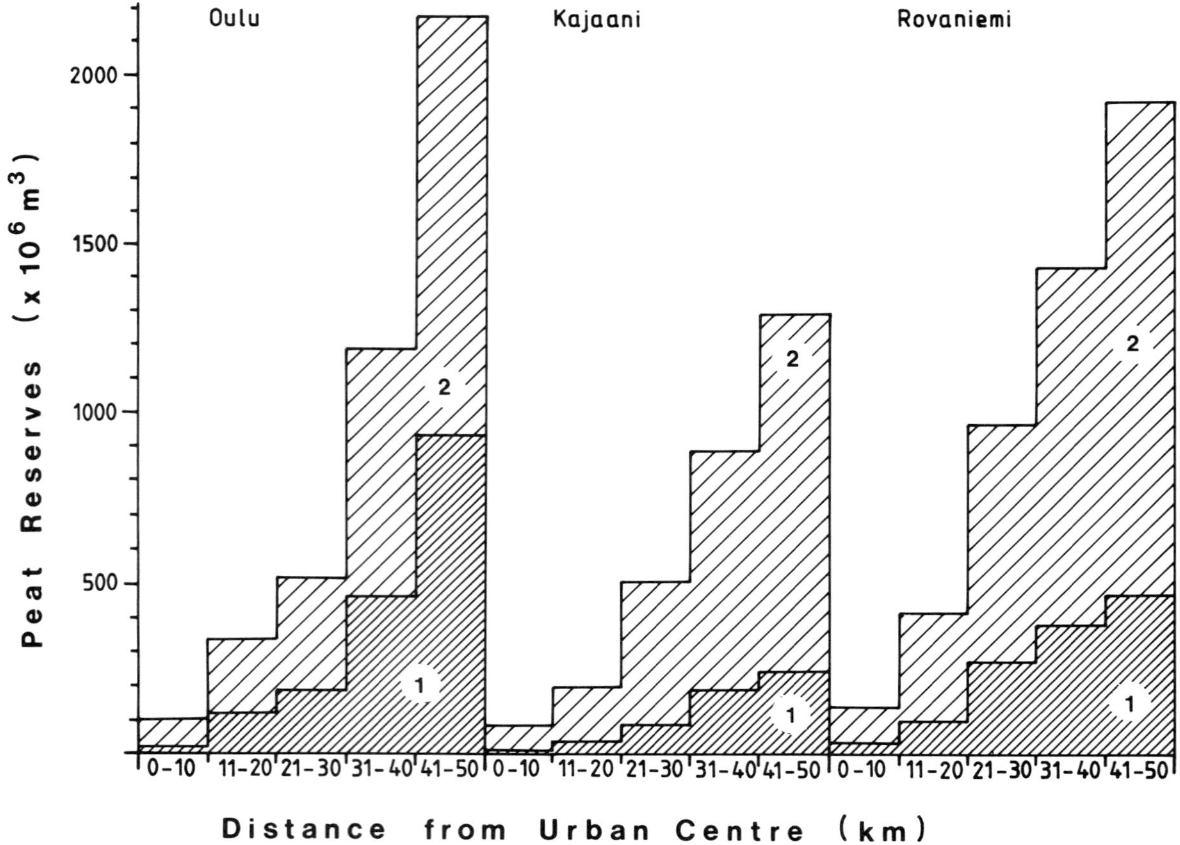

Figure 7. Cumulative peat reserves in the areas around Oulu, Kajaani and Rovaniemi by distance from the centre: (1) peat reserves of treeless mires, (2) peat reserves of forested mires (Aitto-oja and Siuruainen 1982).

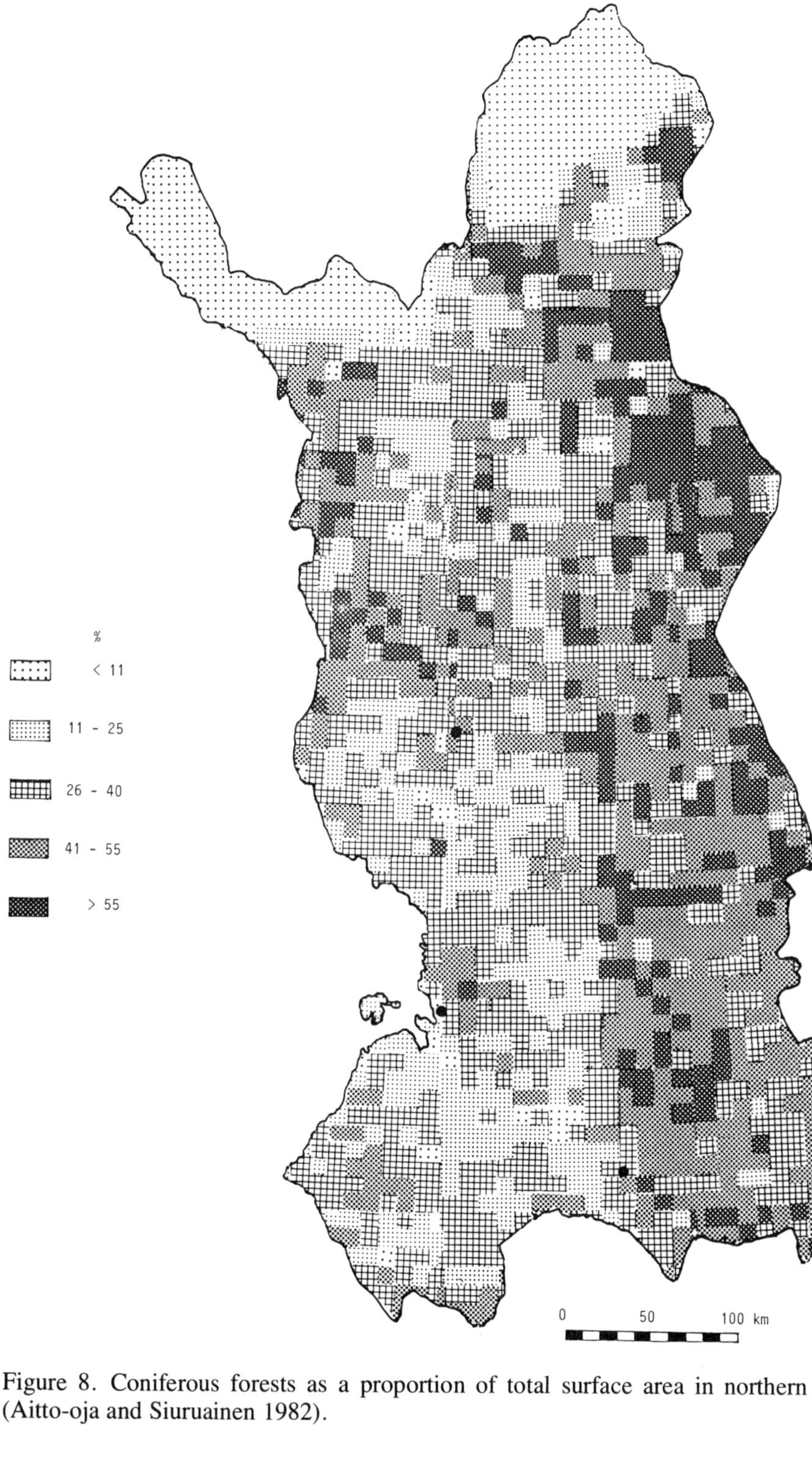

Figure 8. Coniferous forests as a proportion of total surface area in northern Finland (Aitto-oja and Siuruainen 1982).

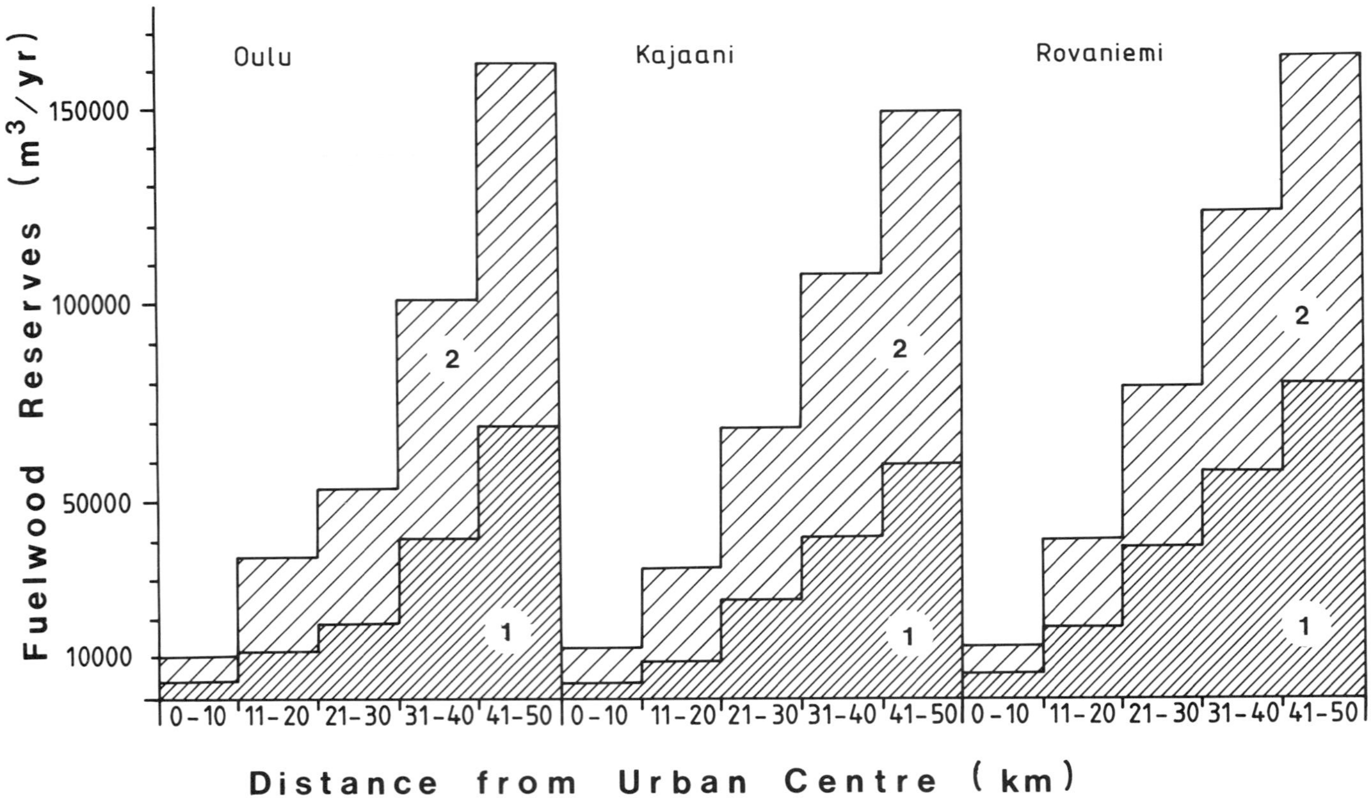

Figure 9. Cumulative firewood reserves in the areas around Oulu, Kajaani and Rovaniemi, by distance from the centre: (1) reserves of firewood in forested mires, (2) reserves of firewood in coniferous and mixed forests (Aitto-oja and Siuruainen 1982).

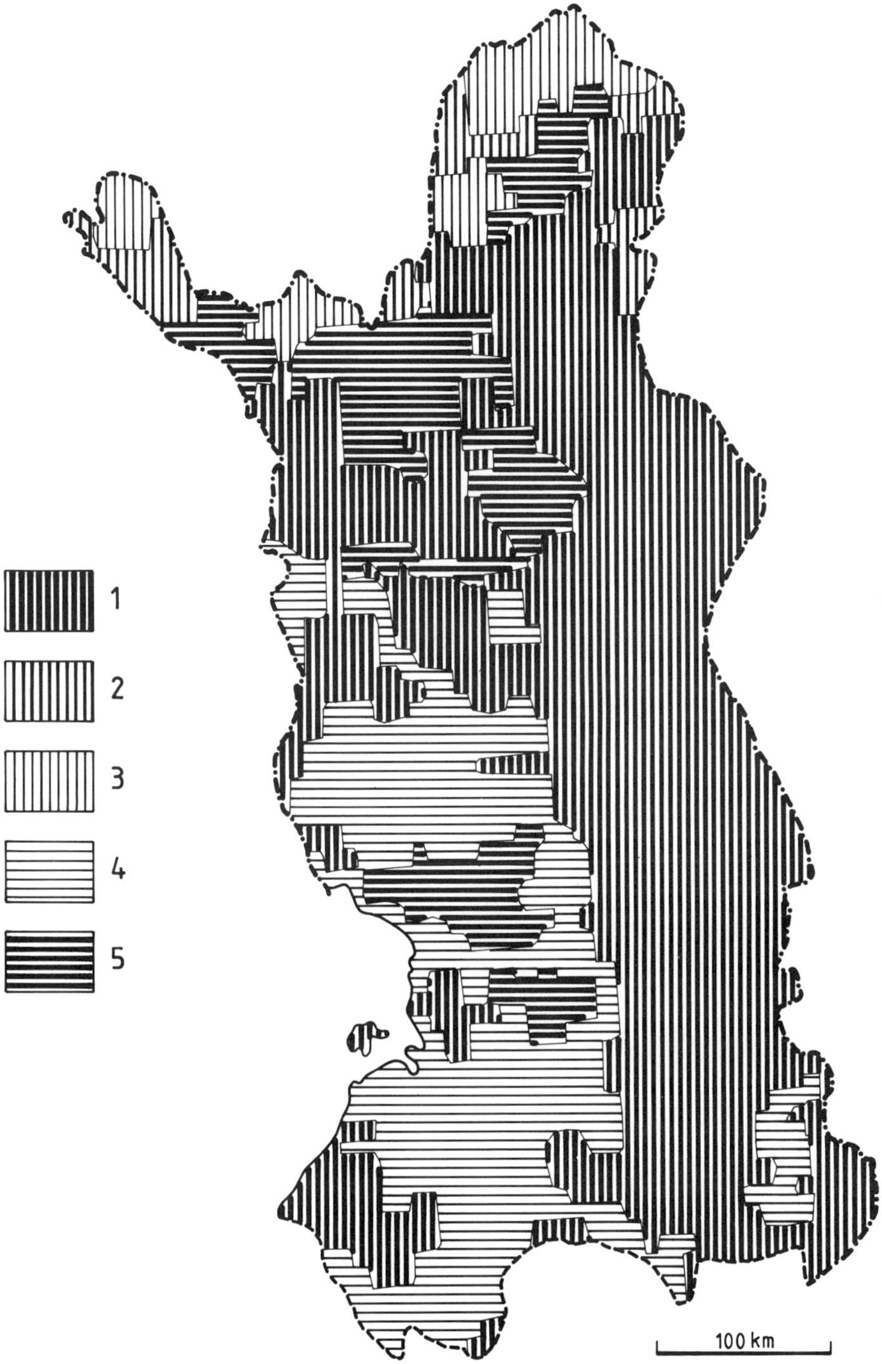

Figure 10. Areas of northern Finland supplying sources of fuel: (1) coniferous forests, (2) mixed forests, (3) mountain birch forests, (4) forested mires, (5) treeless mires (Aitto-oja and Siuruainen 1982).

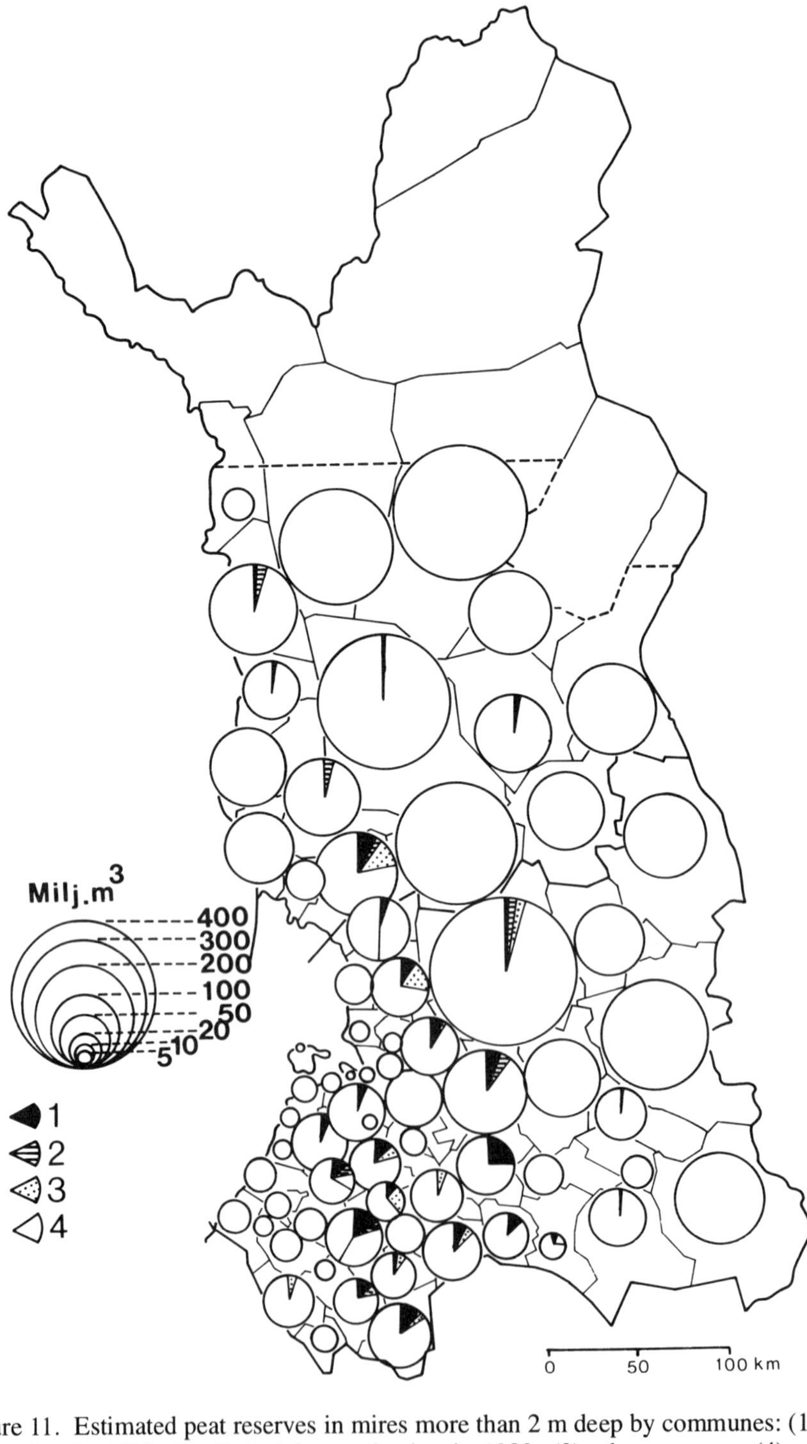

Figure 11. Estimated peat reserves in mires more than 2 m deep by communes: (1) under production in 1980, (2) ditched for production in 1980, (3) other reserves, (4) potential available peat reserves (Tarakkamäki *et al*. 1982).

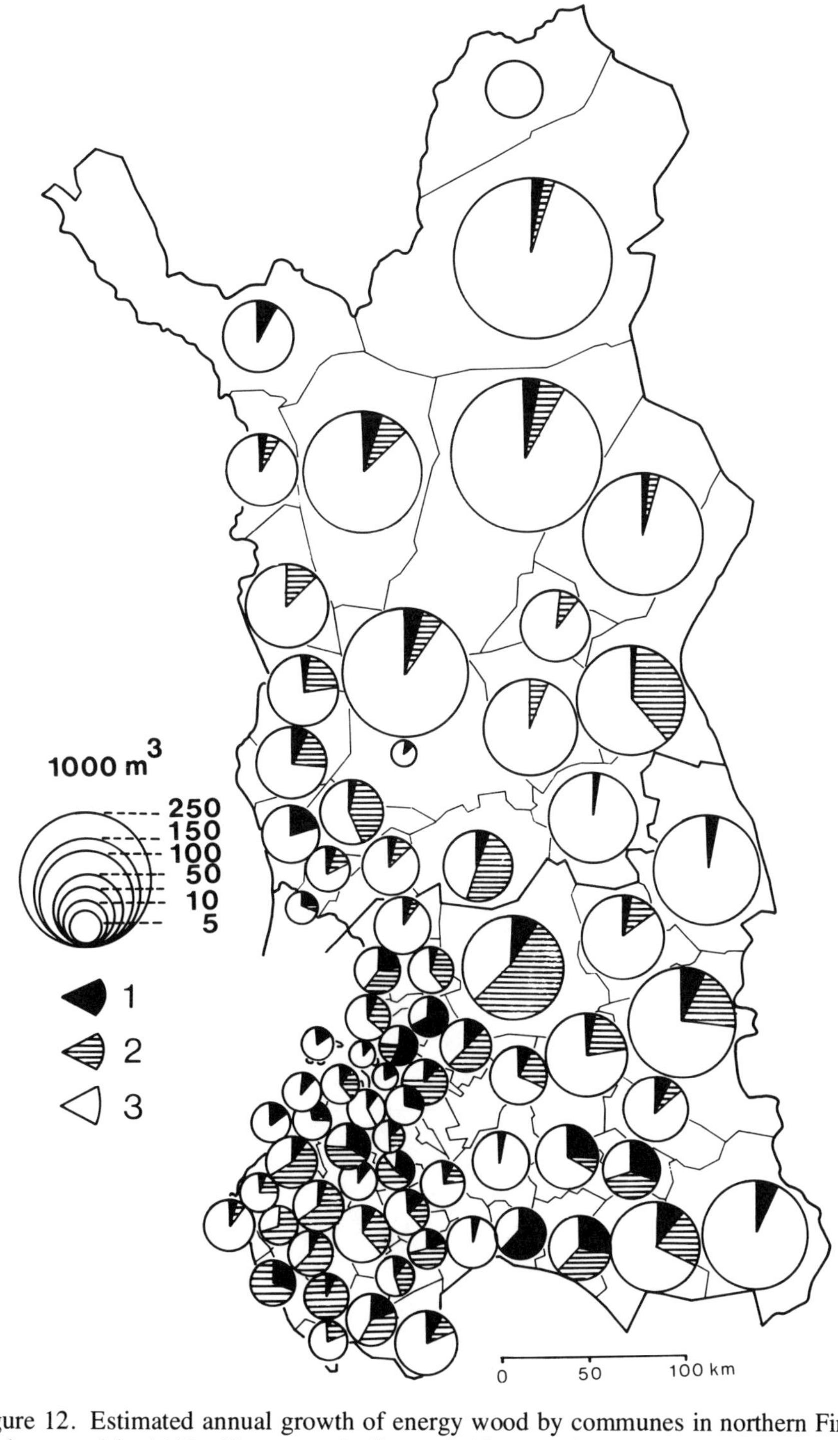

Figure 12. Estimated annual growth of energy wood by communes in northern Finland: (1) harvested in 1980, (2) economically available annual growth, (3) present available reverves (Tarakkamäki *et al*. 1982).

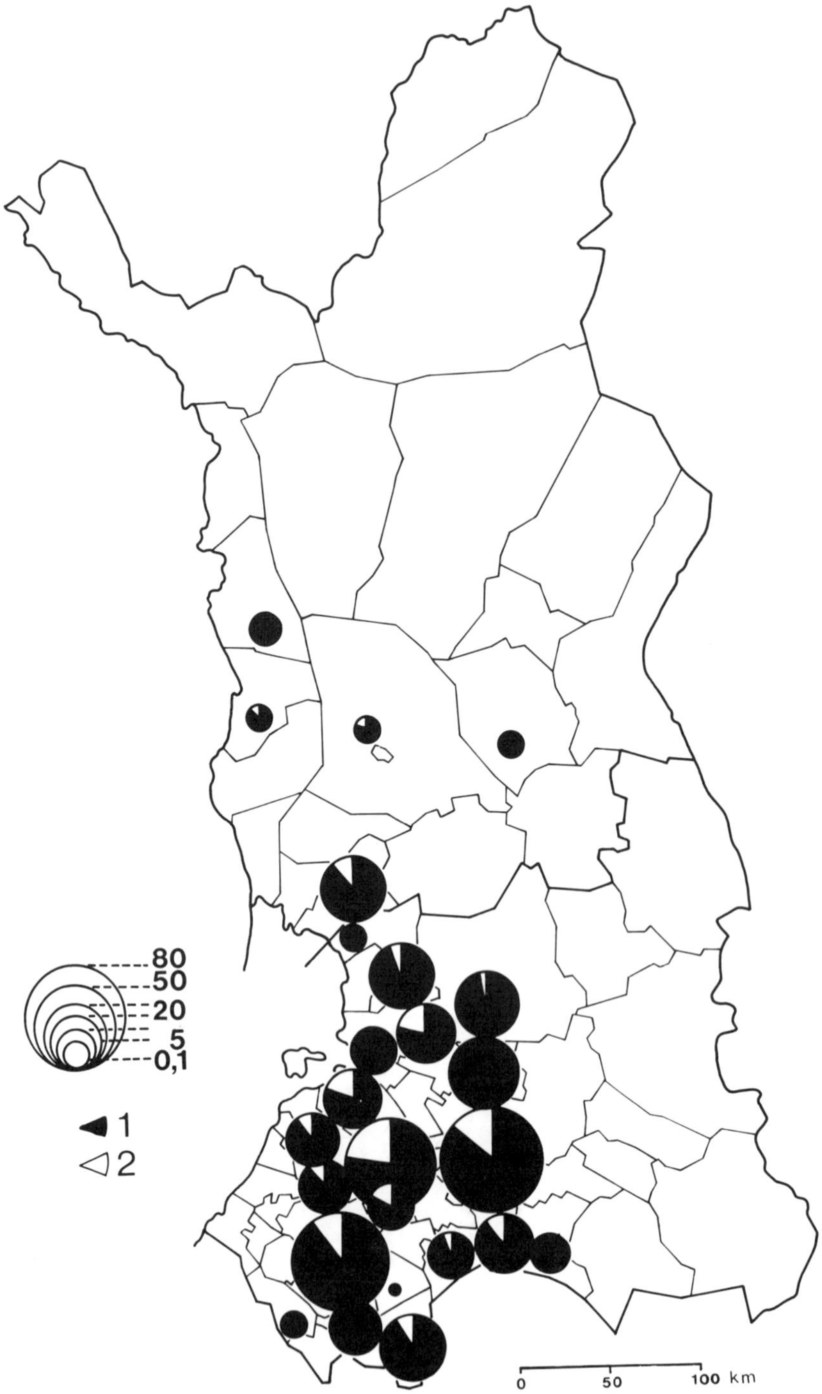

Figure 13. Man labour years involved in peat production by communes in northern Finland: (1) production, (2) transportation, both in 1981 (Tarakkamäki *et al*. 1982).

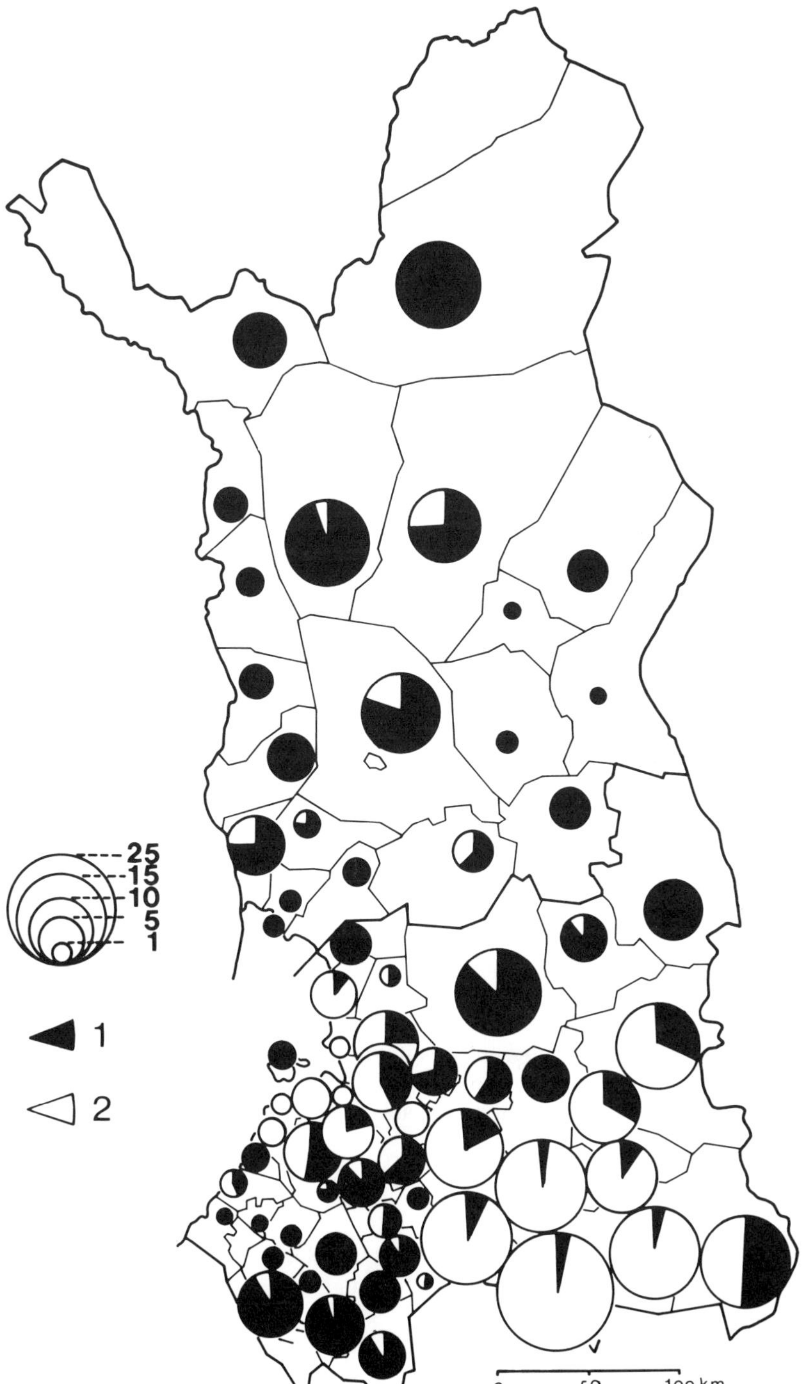

Figure 14. Man labour years involved in energy wood production by communes in 1981: (1) firewood, (2) wood chips (Tarakkamäki *et al.* 1982).

At present there are only two main energy production companies in northern Finland. The state-run company Fuel Centre of Finland, "VaPo", produces some 70 per cent of the total energy on state owned or privately owned and periodically rented mires. One initially private, but later communally-run company, "Turveruukki Ltd", produces peat for the biggest power station at Oulu. This enterprise also has branch plants in Canada. Some peat for local use is produced by smaller enterprises and local farmers. In 1980 peat was produced from 8,900 ha and in 1981 an additional 4,400 ha were drained. The harvesting rights for the year 1981 covered some 23,000 ha; so nearly 40,000 ha now support this activity (see Figure 11). Most mires used for energy production are treeless, at least 50 ha in size, and the distance to the site of processing is generally less than 50 km. Peat depth is some 1.5 to 2.0 m, which means the area can support peat harvesting for about 15 years. Numerous farms have sold or rented their peat reserves to these companies, and new uses for these abandoned fields are planned.

Although the use of domestic energy forms is still in its first stages, it has had important effects, especially on rural areas of northern Finland. The activity totalled an employment of 1,567 man years in 1980 (Figures 13 and 14). That employment effect consisted of the following sources: 609 years in peat production and transportation, 120 years at power stations, 195 years to construct power stations, pipelines and other heating facilities, and the remaining 290 years to plan and to construct peat harvesting and processing machinery. The highest amount of employment was reached in summer, when local farmers received work for themselves and farm tractors.

The estimated salary totalled some 61 million marks, and the communal taxes reached 10.7 million marks in 1980. The harvested domestic energy compensated for some 200,000 tons of crude oil. This was equivalent to 80 million marks in the same period. It has been estimated that the use of peat will increase to 4 million m^3 by the year 1985 and to 10 million m^3 in 1990. The planned use of wood energy resources for the next years suggests a doubling of the harvest until the year 1985. This prediction gives positive prospects for the development of rural employment, and it estimated that in 1990 some 2,300 man years will be involved in the production and transportation of peat.

Although we now have new perspectives of domestic energy resources in northern Finland, we are still faced with numerous unanswered questions. Much depends on international and national policies. What will be the availability and the price of imported fossil fuels? What will be the technical development of harvesting and processing of domestic energy resources? At present Finland is among the first countries investing in this activity. This has resulted in strong scientific and practical cooperation between Finland and other circumpolar countries, notably other Nordic countries, Canada, and the Soviet Union.

Literature Cited

Aitto-oja, Jouni and Eino Siuruainen. 1982. Pohjois-Suomen turve-ja puupolttoainevarat sekä niiden sijainti ja saavutettavuus kartta-analyysien perusteella. Kotimaisen polttoaineen tutkimusprojekti, Oulun yliopisto, maantieteen laitos, Oulun työvoimapiiri, No. 9, Oulu.

Ilvessalo, Yrjö. 1960. Soita prosentteina maa-alasta. Suomen kartasto – Map of Finland, 11/15, Helsinki.

Keskitalo, Jorma. 1982. Reserves of peat and ligneous material in the mires of Siikajoki valley and the Oulujärvi area. Publicationes Instituti Geographici Universitatis Ouluensis. No. 75.

Kuusela, Kullervo. 1976. Metsämaan osuus koko maa-alasta. Suomen kartasto – Map of Finland, vihko 234, 3c. Helsinki.

Tarakkamäki, Jaakko, Jukka Matero and Aimo Vitikka (1982). Turve-ja puupolttoaineiden hyväksikäyttö ja sen työllisyysvaikutukset Pohjois-Suomessa. Nordia. Sarja B, No 1/1982.

Vitikka, Aimo, Tuomo Huhtala and Eino Siuruainen. 1981. Kotimaisten kiinteiden polttoaineiden tuotanto ja käyttö Pohjois-Suomessa vuonna 1980. Oulun yliopisto, Pohjois-Suomen tutkimuslaitos, Sarja C. No. 39.

ENERGY FROM FOREST BIOMASS FOR REMOTE COMMUNITIES

C.R. Silversides
National Research Council of Canada

Abstract: The potential for forest biomass as a feedstock for energy, in whatever form, has been discussed at length in relation to existing forest industries in Canada. To move beyond this physically, to the north, outside of the economic transport range of the forest industries, to northern communities beyond the electric grid, to small settlements with high cost energy but relatively low demand requires rethinking of the problems involved.

Wood has been a source of energy to mankind since pre-recorded history. It offers Canadians in the north a means to free themselves from the increasing costs of fossil fuels at the expense of some lack of convenience but with a resulting increase in local employment and the retaining of energy costs within the community. New technologies in the harvesting, transformation and final conversion of biomass make it increasingly attractive both physically and economically to remote communities.

To date most consideration of forest biomass and peat as sources of energy has taken place in the framework of high rates of energy demand and supply. Existing forest industries (pulp and paper, sawtimber) have been considered as both producers and users of biomass energy feedstocks.

The potential for forest biomass as a feedstock for energy, in whatever form, has been discussed at length in relation to existing forest industries in Canada (Silversides 1981a, 1982). To move beyond this physically to the north, outside of the economic transport range of the forest industries, to northern communities beyond the electric grid, to small settlements with high cost energy but relatively low demand requires rethinking of the subject.

The development of the large scale use of forest biomass for energy could be the best that could happen to implement intensive forest management in Canada. Biomass for energy has no specification and all cellulosic material, limbs, twigs, bark as well as fire- and insect-killed timber is included. This means that all trees and portions of trees which are presently unusable for conventional forest products can be used for energy and are thus given a value. As a result it is foreseen that our present junk forests and so-called silvicultural slums could be converted to higher value merchantable stands with much of the cost recovered from the biomass removed.

In the northern boreal forest where there are limited forest resources but also limited demands to date, there has been little effort to manage the forest in any manner to improve the growing stock and to ensure its continuity. The growth rates of the tree species tend to be low and regeneration, both natural and artificial is not conducive to maximum energy production.

Table 1. Mean annual increment for areas of the boreal forest presently north of the forestry operations in Canada. The reference codes relate to those in Figures 1 and 2. The information is adapted from Bickerstaff *et al*. (1981)

Province or territory	Reference* code	Access and/or stand operability**	Mean annual increment m^3/ha
Yukon	B26a	2	0.6
	B26b	2	0.7
	B26c	2	0.6
	B26d	2	0.6
Northwest Territories	B23a	2	0.8
	B23b	3	0.5
	B27	3	0.4
	B32	3	0.2
	B33	3	0.2
British Columbia	K	2	1.3
	L	2	1.4
Alberta	B18b	2	0.6
	B23a	2	0.8
Saskatchewan	B22a	2	1.0
	B22b	2	0.7
	B27	3	0.4
Manitoba	B22a	2	1.3
	B27	3	0.4
Ontario	B5	3	0.6
	B22a	2	1.4
	B32	3	0.2
Québec	B1b	2	1.3
	B6	2	0.8
	B13a	3	0.3
	B13b	3	0.3
	B31	3	0.5
	B32	3	0.2

* See boreal forest location in Figures 1 and 2.
** 2 = marginal but will be operable in future;
3 = unsuitable for economic commercial use.

Table 1 is derived from a study of the potential biomass energy production and use undertaken by the Canadian Forestry Service (Bickerstaff *et al*. 1981). The reference codes chosen are for those areas, still treed, which are north of the presently occupied boreal forest with good access and stand operability characteristics (Figures 1 and 2). Access has not been defined in this report but operability classes reflect differences in the economic availability of wood fibre between sections. Class 1 is defined as currently operable, Class 2 mostly operable in the future, and Class 3 unsuitable for economic

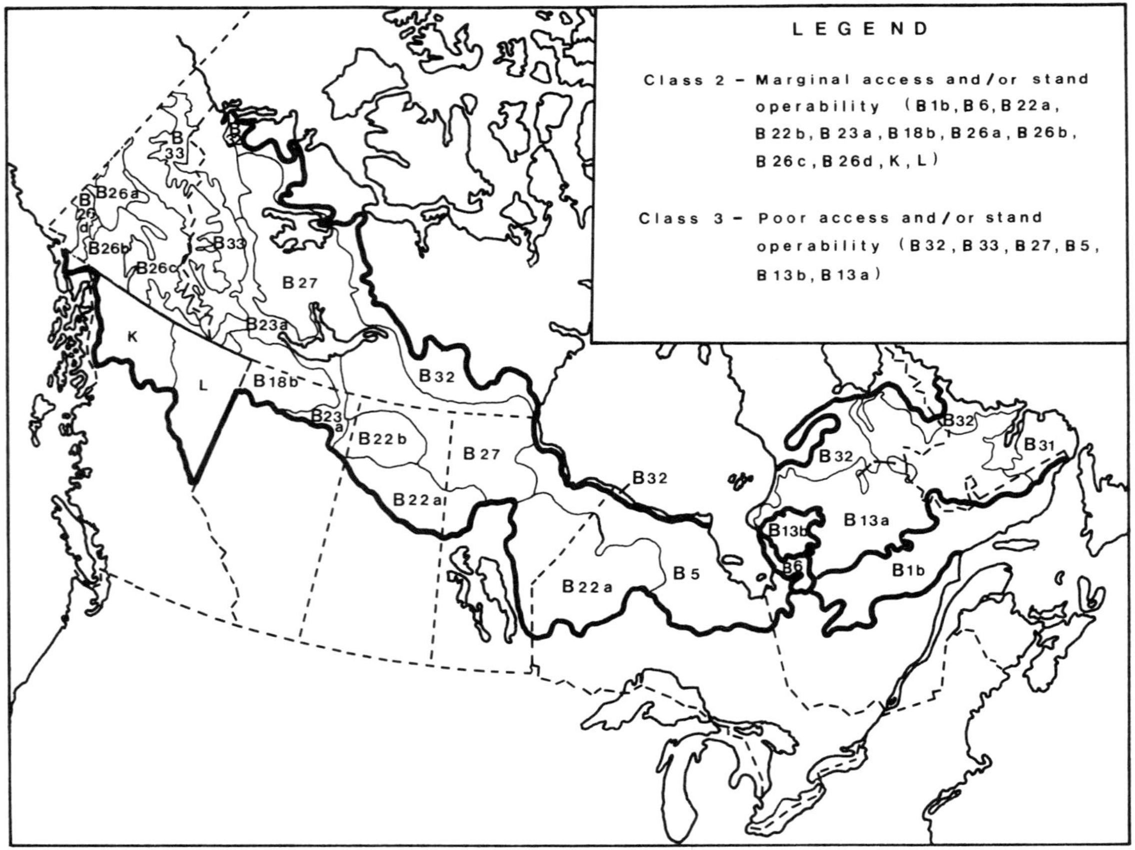

Figure 1. Operability of forest sections (adapted from Bickerstaff *et al*. 1981).

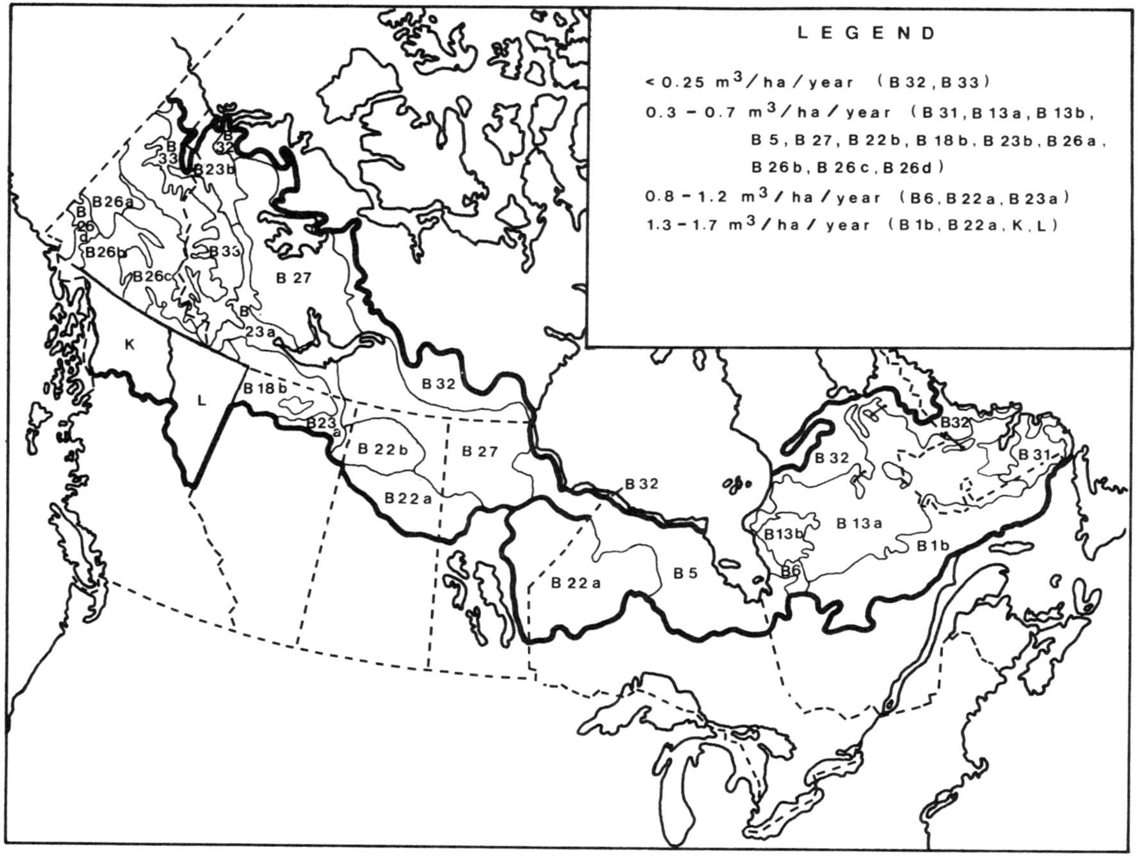

Figure 2. Mean annual increment per hectare on productive forest land within forest sections (Bickerstaff *et al*. 1981).

commercial use. Table 1 shows some of the area considered to be of current economic value. It also shows that the mean annual increment of the forest in these regions varies from 0.6 – 1.4 m^3/ha/year for Class 2 to 0.2 – 0.6 m^3/ha/year for Class 3 (Figure 2). Forest lands with growth rates at this low level offer little prospect for management on any intensive level for conventional forest products but do offer potential for production of forest biomass. A study of standing crop density in the northern boreal region (Section B18a and B19a, Figure 1) showed exceptionally high figures for certain shrub communities and dense stands of conifer saplings (Peterson *et al*. 1981). It has been shown that certain shrubby stands may have standing crop densities up to ten times as great as the 1.0 to 1.5 kg/m^3 that is considered the normal range for fully closed stands of mature forests (Kira and Shidei 1967).

Table 2 adapted from Peterson *et al*. (1981), shows some examples of apparent upper limits of above-ground standing crop for unmanaged 3, 4 or 5 year old stands of vegetation types in comparison with the highest density classes of fully stocked aspen stands of the same age (Bella and DeFranceschi 1980).

It must be stressed that the maxima given in Table 2 should not be extrapolated to larger areas. Their value is to indicate the upper limits of yields that could be encountered in localized areas.

Many small communities within the tree line have similar energy supply problems as those of developing countries (Overend and Silversides 1981). Long supply lines and low capital availability are not restricted to what is often called the third world. Over 50 MW of electrical generations in Canada is in small generating sets using diesel fuel shipped in over long distances.

In remote power systems, diesel fuel represents a cost of up to $3.40 per KWH generated. Current federal-provincial pricing agreements indicate that the price of fuel alone may reach as high as $7.00 per KWH by 1990.

In some remote settlements using diesel electric power No. 2 diesel fuel costs $0.94/litre. It has been estimated that locally harvested biomass fuel equivalent could be produced for one-quarter of the cost of such fuel. The energy demand will be relatively small and the supply will be local, dependent mostly on labor intensive harvesting. As much of the production will be for domestic consumption, it may be carried on outside the money economy. Under such circumstances E.F. Schumacher's basic assumption that technology must be adapted to the situation in which it will be used, rather than vice versa, should hold true.

The basic requirements for energy in the situation envisaged is low temperature heat, high temperature heat, fuel technologies, mechanical energy and electrical energy.

Forest biomass as an energy feedstock is very versatile as it can be converted into energy in solid, liquid, or gaseous form. For the scale of application envisaged in this paper the liquid form as methanol or ethanol has been ruled out. This form of energy conversion is still energy inefficient and capital intensive. The economy of scale requires large conversion plants. InterGroup Consulting Economists Limited (1977) considered 300 oven dry tonnes/day of product or 550,000 m^3/year of biomass feedstock an economic scale.

Table 2. Examples of apparent upper limits of above-ground standing crop for unmanaged, 3, 4 and 5 year old stands compared to fully stocked aspen stands of the same age

Vegetation type	Standing crop kg/ha at age 3	at age 4	at age 5
Comparative baseline for fully stocked aspen (Bella and DeFranceschi 1980)	8,166	9,335	10,971
Some apparent upper limits (Peterson *et al*. 1981)			
Balsam poplar	12,000	?	?
Balsam poplar − willow mixture	13,500	?	?
Aspen − hazel mixture	14,000	?	?
Willow	10,800	?	?
Aspen − alder mixture	16,900	18,100	?
Aspen	11,300	12,500	29,600
Red-osier dogwood	?	?	12,500
Mountain maple	?	?	16,200

Wood in different forms can be burned directly to generate heat for space heating, for generating hot water to use for heating or process, to develop mechanical power through the production of steam or electric power by means of electric generators. Wood can be pyrolized in special combustors to produce a gas which can be used to power internal combustion engines. Such engines can be used to supply mechanical energy directly or can be used to power electric generators.

In consideration of the increased use of forest biomass the technology to be used should be low cost, labor intensive, relatively small scale, and have a low-level skill requirement. A technology can be all of the above and still be technically first rate. However, if a technology is considered to be primative or second rate it will be almost impossible to introduce, particularly if it is to replace a convenient energy intensive fuel like oil.

If one looks at the progression in the use of fuels in Canada it began with wood, moved to coal, to oil and to gas. The changes were to a large degree a matter of convenience, of heat value per weight or volume (Table 3) as much as cost. Today, in considering a move back to wood, convenience will still play a large part in winning acceptance. For this reason the burning of wood in chip form rather than solid wood should be given careful consideration.

The production of chips from forest biomass of the size and quality encountered in the north should prove to be more productive and economical than the production of solid wood. In harvesting small diameter stands whole tree chipping may produce up to 100 per cent more biomass per unit area than if the same material is harvested for cordwood. Whole tree chipping is more economical than the production of solid wood. Fewer operations are required and man day productivity is greatly increased. It is true that cordwood can be produced with only an axe and bowsaw while chips require power and a

Table 3. Approximate heat values for various fuels (adapted from Stock 1981)

Fuel type	Heat value	
	Unit weight MJ/Kg	Unit volume MJ/m³
Fossil fuels		
Gasoline	44	32,000
Fuel oil	44	39,000
Kerosine	45	36,000
Bituminous coal	33	42,900
Peat	20	18,200
Biomass fuels		
Wood	18	14,400
Charcoal	28	11,000
Wood gas	—	11

fairly substantial capital investment in chipping equipment. However, such investment can usually be justified on both cost and energy input/output basis.

Chipping of forest biomass material will permit the utilization of otherwise unusable material such as alder, willow and birch thickets which can be highly productive and self reproducing through coppicing. In Finland, for example, at the Arctic Circle such natural coppicing stands produce 2-3.5 oven dry tonnes/ha/year (Pohjonen 1982). At present brush stands are harvested manually but several harvesters specifically designed for coppice harvesting are under development.

If wood is to be used in direct combustion on a small scale it should be properly prepared to obtain the greatest return on its use. Drying is important in the direct combustion of woods since air-dried wood yields approximately 80 per cent of the energy value of oven dried wood, while green wood yields only 50 per cent. The evaporation of moisture prior to combustion consumes about 1000 Btu's per pound of moisture. Moisture content is one of the few properties of wood which can be manipulated by man, under the circumstances, and this should be exploited as much as possible. Moisture content in trees varies seasonally. Species such as aspen and white birch show very marked seasonal changes. They are driest at the end of summer, just before the leaf fall, then they increase in moisture content during the autumn, lose moisture during the first months of the year and have a very high moisture content at leaf opening. Conifers show comparatively small changes. The trees have low moisture content in April-May and again in August-September. They are high in moisture in November-December and again in June-July.

Sour felling, or felling trees and leaving them with foliage intact to permit transpiration, to draw moisture out of the bole, is one means of reducing moisture. Trees felled in this manner require two to three weeks to dry and if left on the ground for a longer period will pick up moisture. When leaves have become so dry and shrivelled that they break off it is a sign that the bole has dried to the minimum of about 50 per cent. Experiments have shown that the butt of the tree is the first portion to dry and the leaves are the last. Therefore when the leaves become dry the trunk will also be dry (McIntosh 1949).

If wood is produced in stove lengths and is split and piled under cover it will air-dry over a period of up to nine months. Splitting increases the rate of drying as the drying process to a large extent is dependent upon surface area exposed (Miller 1951). As described, sour felling can speed up this process considerably.

The reversion back to burning solid wood in space heaters or furnaces offers problems and some inconvenience to those accustomed to heating with fuel oil. The care and maintenance of such heating appliances creates problems not only of production and storage of fuel but also of heat output control and creosote build-up in chimneys. The burning of wood in chip or chunk form in household furnaces is growing. The production of wood in this form permits it to be metered into the combustion chamber and the heat supply can be regulated almost as well as with an oil burner (Schneider 1980).

Most commercial disc chippers produce a flat ''pulp quality'' chip which is not the ideal shape for burning. A better chip shape is a cube as it produces an open granular fuel with air spaces present. These air spaces promote complete combustion by assuring a good oxygen supply throughout the fuel. Such cubic chips are easy to convey and handle. The Finnish ''Kopo'' chipper produces such a chip (Silpala 1982). It utilizes a large conical screw cutter rather than a disc with inserted knives as is used to produce pulp chips.

If the timber available is large and wood is produced in a solid form consideration should be given to mechanizing the bucking and splitting operations. Firewood processors are on the market which can take tree length or log length material and reduce it to short quartered bolts, suitable for burning in space heaters or furnaces. Such units only require one operator and can produce up to six face cords per hour (Cord King of Canada 1981).

One concept, that is now being exploited and accepted, which has considerable potential for space heating is the combustion of wood and the storage of the heat in a large insulated water tank. From this storage tank either using hot water directly, or hot air by means of a simple heat exchanger, a building may be heated. Wood up to 40 per cent moisture content is burned in a refactory combustion chamber at 650-1100°C under a forced draft of hot air so that all tars (creosote) and wood gases are consumed, which eliminates many of the problems associated with burning wood. Depending upon the outside temperature, efficiency of insulation and size of heat storage tank, the intervals between firings may be hours or even days.

Where available forest stands are too small for the production of chunkwood or cordwood and facilities permit the burning of chips or small material, the economic harvesting of this small material in the form of brush, small trees, or coppice, requires thought and planning. Brush saws may be used or conventional powersaws can be readily modified to work efficiently with brush size material. The brush can be piled manually, loaded and hauled mechanically to a central location for chipping. Another approach might be to use a highly mobile chipper and chip and load the material as it is felled. There can be as many variations in these harvesting methods as there are in collecting agricultural field crops. Different methods can be specific to conditions of the crop, capital and/or labour availability. Harvesters designed specifically for this brush crop and rough terrain exist in prototype in Canada and Scandinavia. Such equipment should, in the near future, make the utilization of this form of biomass for energy socially and economically acceptable.

This paper has been concerned with the production and direct combustion of forest biomass. There are many other approaches to the production of energy from forest biomass. These include the densification of biomass into fuel pellets to make handling more convenient. These approaches are considered to be technically and economically outside the possibility of application in remote northern communities at this time.

Literature Cited

Bella, I.E. and J.P. DeFranceschi. 1980. Biomass Productivity of Young Aspen Stands in Western Canada. Environ. Canada, C.F.S., Inf. Rep. NOR-X-219.

Bickerstaff A., W.L. Wallace and F. Evert. 1981. Growth of the forest in Canada – Part 2 – A quantitative description of the land base and the mean annual increment. Petawawa National Forestry Institute. Chalk River, Ontario.

Cord King of Canada. 1981. 31 Arthur Street, Ottawa, Ontario, K1R 7B8.

Intergroup Consulting Economists Limited. 1977. Liquid fuels from renewable resources.

Jetstream. 1981. Hampton Technologies Corporation, Box 2277, Charlottetown, Prince Edward Island, Canada, C1A 8B9.

Kira, T. and T. Shidei. 1967. Primary Production and Turnover of Organic Matter in Different Forest Ecosystems of the Western Pacific. Jap. J. Ecol. 17 :70-87.

McIntosh, D.C. 1949. Observations of Trees Felled and Left with Foliage Intact. Project 129, Mimeo 0-141, Forest Products Laboratories. Ottawa, Ontario.

Miller, R.H.P. 1951. Characteristics of wood as fuel. Wood Fuel – Report of Conference at Philadelphia, Pa.

Overend R. and C.R. Silversides. 1980. Energy from forest biomass, A Canadian perspective. NRCC. 18064. Ottawa, Ontario.

Peterson, E.B., V. Levson and R. Kabzems. 1981. Upper limits of standing crop density and growth rates for wood species in the prairie provinces. Western Ecological Services Limited, Edmonton, Alberta.

Pohjonen, V. 1982. Wood as an energy source in Finland. Pera-Project 1981. The Finnish Forest Research Institute, Helsinki, Finland.

Schneider, M.H. 1980. A wood-chip-fired hot-air home heating system. Solwest 80 Joint Solar Conference, University of British Columbia, Vancouver, British Columbia.

Silpala, Mauno. 1982. Letter. Kopo Inc., 2000 Rhode Island Ave. N., Minneapolis, Minnesota 55427.

Silversides, C.R. 1981a. Forest biomass for energy and for forest industries. Pulp and Paper Canada, October.

Silversides, C.R. 1981b. Mill and forest residues as fuel. Pulp and Paper Canada, March.

Silversides, C.R. 1982. Energy from forest biomass – its effect on forest management practices in Canada. Biomass, 2: 29-41.

Stock, F. 1981. Renewable energy in the third world. Petroleum Economist, July.

ENVIRONMENTAL IMPACT OF HYDROELECTRIC DEVELOPMENT IN NORTHERN ONTARIO: A Cost-Benefit Approach

R.A. Muller and P.J. George
McMaster University

Abstract: This paper reports the preliminary results of a cost-benefit analysis of ten proposed hydroelectric projects in northern Ontario currently under active consideration by Ontario Hydro, and on exploratory estimates of the environmental costs likely to be associated with the projects. The partial net benefits from hydroelectric expansion of the Lower Mattagami River complex and construction of the Moose River Basin new sites are estimated to be of the order of $125 million if coal-fired generating capacity remains in use, and about $1000 million if excess coal capacity is unavailable. Together with a preliminary estimate of maximum environmental costs in the Moose River system of approximately $22.5 million, the cost-benefit analysis suggests that the projects will generate sufficient partial net benefits to more than compensate for the environmental costs incurred. The tentative conclusions are, then, that the social net benefits from hydroelectric development in northern Ontario will be large, and that the projects would be a 'socially profitable' investment. However, these findings will need to be refined as more complete data on local social impacts and environmental consequences become available.

This paper is a preliminary report on our attempt to assess the probable economic and environmental impacts of ten potential hydroelectric developments in northern Ontario which are currently under active consideration by Ontario Hydro. We briefly describe the proposed projects, and then apply conventional cost-benefit techniques to evaluate the partial net benefits (excluding environmental and social costs) of developing these ten sites. Finally, we provide a short discussion and make rough estimates of the environmental costs likely to be associated with these projects. Our tentative conclusions are that the potential net benefits of hydroelectric development of these sites in northern Ontario are substantial and in most cases greatly outweigh the probable environmental effects identified at present. Ongoing research on local social impacts and environmental consequences will undoubtedly add to the completeness of our study and the validity of our conclusions.

The TASO Research Program

McMaster University has established, with financial support from the Donner Canadian Foundation, a multidisciplinary research program for Technology Assessment in Subarctic Ontario (TASO). TASO is undertaking long-term research in anticipation of the social, economic, and environmental consequences of major resource investment projects expected to occur in northern Ontario over the next several years. Although assessment of the Onakawana lignite deposits and the construction of the Detour Lake gold mine come readily to mind, the development of hydroelectric energy resources in several northern river basins undoubtedly would be the most wide-reaching in its consequences. As a result, TASO members have chosen initially to concentrate on the impact of potential hydroelectric projects in the Moose-Mattagami-Abitibi Basin and the Little Jackfish River, since these sites are under active consideration by Ontario Hydro (Ontario Royal Commission on Northern Environment 1978; Ontario Hydro 1980, 1981). The TASO program is motivated by the fact that research on economic, environmental and social impacts of the Baie James project in Quebec was only undertaken after the fact, without an overall strategy, and has been far from adequate. We contend that, if an appropriate research strategy had been developed and implemented in the 1960s, many of the problems associated with Baie James could have been anticipated and possibly alleviated or avoided. We propose, then, to develop and implement a research strategy for northern Ontario before further construction of hydroelectric and other large resource investment projects begins, in the hope of providing findings and recommendations useful to all parties interested in northern development.

Potential Hydroelectric Development in Northern Ontario

A hydroelectric development program for Ontario, including proposed site extensions and new plant construction in northern Ontario, was approved by the Board of Ontario Hydro in July, 1978, and revised in 1980 (Ontario Hydro 1980). In the earlier version of the program, seventeen projects in northern Ontario were identified, with peaking capacity of 2000 MW and average capacity of 550 MW. A target increase in hydroelectric generating capacity of 2000 MW was incorporated in the provincial energy security goals announced by the Minister of Energy in 1979 (Ontario Ministry of Energy 1979). Ontario Hydro's (1980) review revised the number of northern Ontario sites to sixteen. These sites are under active consideration by Ontario Hydro: for some, feasibility studies are underway or have already been completed, and at least two sites are currently undergoing environmental assessment.

Existing hydroelectric generating capacity in northern Ontario, operated by both Ontario Hydro and others, is depicted in Figure 1, and potential developments and plant extensions are presented in Figure 2. Of special interest to this paper are the ten sites that we have grouped for analytical purposes as:

- Little Jackfish River — two sites (Upper, Lower);
- Lower Mattagami River — four site extensions (Little Long, Smoky Falls, Harmon, Kipling);
- Moose River Basin New Sites — four sites (Grand Rapids, Nine Mile Rapids, Grey Goose Island, Renison).

The other six sites include Maynard Falls, Long Sault Rapids, Patten Post, Ear Falls, Ragged Chute, and Matabitchuan, but they have been recommended for 'much lower priority' than the first ten sites (Ontario Hydro 1981). We understand from Ontario Hydro officials that no proposals are currently being investigated by Ontario Hydro for the Albany, Attawapiskat, Winisk, or Severn Rivers, and in fact development has been delayed indefinitely on the Albany and the Winisk.

The ten sites itemized above have been found by Ontario Hydro to be 'cost effective' based on a comparison of their lifetime costs, including capital, operating and maintenance costs, with the alternatives of nuclear and thermal generating stations. At this stage of their analysis Ontario Hydro has taken no account of socioeconomic and environmental considerations (Ontario Hydro 1981). According to their assessment, the

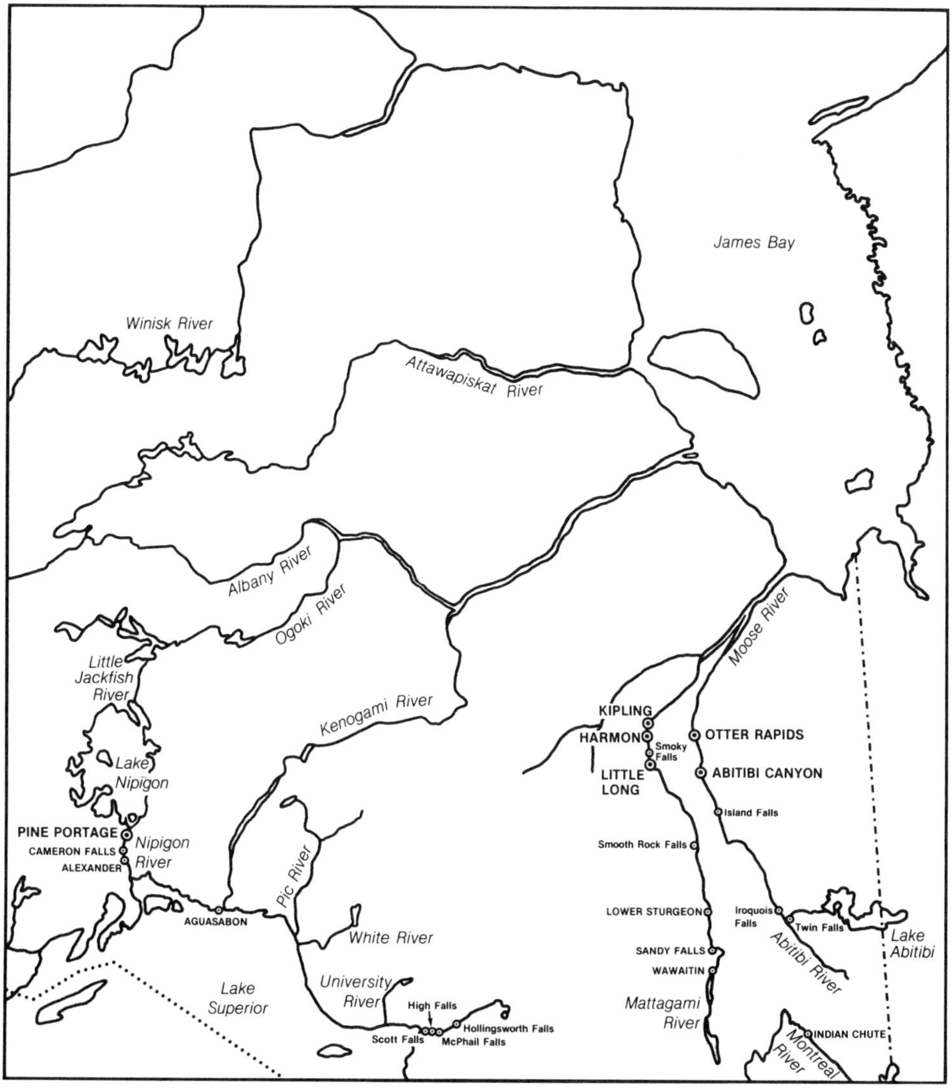

Figure 1. Existing hydroelectric generating stations (adapted from Ontario Hydro reports).

most cost-effective projects are those involving extensions of existing plants on the Lower Mattagami River and new construction at the Grand Rapids site. However, the private cost-effectiveness criterion employed by Ontario Hydro is not as appropriate for evaluating investment projects from society's point of view as other techniques, notably cost-benefit analysis. In the following section we report on our cost-benefit analysis for these projects.

A Preliminary Cost-Benefit Analysis

In this section we present a preliminary cost-benefit analysis of the potential hydroelectric projects on the Moose River and its tributaries, and the Little Jackfish River. Cost-benefit

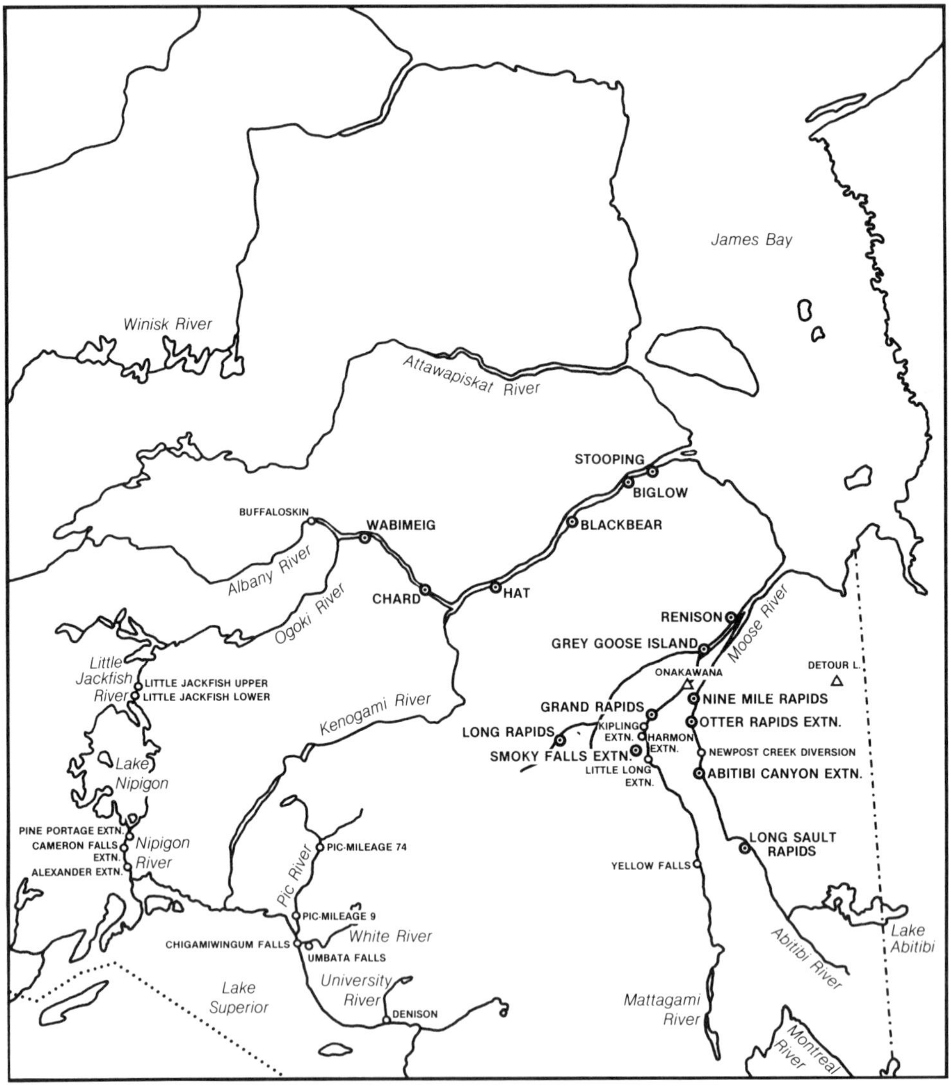

Figure 2. Potential hydroelectric sites and plant extensions (adapted from Ontario Hydro reports). In this map, Ontario Hydro indicates potential sites on the Albany River, but the development of these sites has been delayed indefinitely by the Government of Ontario (RCNE 1978).

analysis is a systematic effort to identify and place a value on all the advantages (benefits) and disadvantages (costs) of a project when compared with a well-specified alternative. Cost-benefit analysis is based on the concept of a potential Paretian improvement: if the value of all the benefits associated with a project exceeds the value of the costs, those gaining from the project could, in principle, fully compensate the losers and still be better off than before the project was undertaken (Mishan 1976).

Cost-benefit analysis offers a valuable framework for organizing the assessment of a major project and its environmental effects. Conventional environmental assessments, such as those required by the Ontario Government, stress the identification of environmental consequences and the need to consider alternatives (Ontario Ministry of the Environment 1982). However, they do not provide a weighting scheme which allows comparison of alternatives or separation of important from trivial consequences. In cost-benefit analysis such a weighting scheme is provided by dollar values. In every case we seek to identify the sum which would be acceptable as compensation to those suffering losses. While this may be difficult in some circumstances, the effort to identify and evaluate these losses provides a valuable clarification of the complex nature of the decisions involved in assessing major projects.

In the present case, our strategy will be first to estimate the value to society of undertaking the hydroelectric projects if there were no environmental consequences of building new dams. We will call these the partial net benefits. Then we will consider some potential environmental impacts and their values. The social net benefits of the hydroelectric projects are the difference between the partial net benefits and net environmental costs. Since this is a preliminary study, we must emphasize that we are seeking only to identify the orders of magnitude of the various costs and benefits. This is worthwhile first in judging whether or not the projects may be socially viable and secondly, in identifying those environmental consequences which are of the greatest potential importance. Moreover, we note that many local social impacts and environmental consequences have yet to be identified and measured. This will be one of the ongoing tasks of the TASO research program. For example, an important benefit of construction projects may be the employment of unemployed resources, particularly labour, and an important cost may be the infrastructure expenses required to support construction work forces and to deal with social problems generated by a large influx of workers. As these types of social and environmental consequences are determined, they will be incorporated into our cost-benefit analysis. In fact, some of them may prove so important that they will warrant particular consideration in separate studies.

Net Benefits Before Externalities

Ontario Hydro (1981) has prepared a cost assessment of their proposed hydroelectric development program in which ten of sixteen sites are identified as less costly than a nuclear alternative. In this section, we discuss Ontario Hydro's cost assessment and compare it to two alternatives. We conclude that partial net benefits are most appropriately measured in an explicit cost minimization framework. Application of a simple linear programming model shows that the Lower Mattagami redevelopment and the new sites on the Moose River and its tributaries would yield partial net benefits in the year 2000 of about $125 million, on the assumption that excess coal-fired generating capacity remains available to the system. However, if existing coal-fired generation has been retired, the net benefits rise to the order of $1000 million. We stress that these estimates are subject to important qualifications.

Ontario Hydro's estimates: Ontario Hydro has assessed the cost of the hydroelectric projects by comparing the discounted capital and operating costs of each to an equivalent nuclear project. A 60-year time horizon is used. Since hydroelectric generating capacity is utilized much less fully than nuclear, a credit is given for the excess energy generated by the latter. The results are reported in Table 1. All three groups of projects exhibit potential cost savings relative to the nuclear alternatives when discounted at eight per cent. At higher rates of discount, the nuclear projects dominate since the present value of the replacement nuclear station required after 30 years is greatly reduced (Ontario Hydro 1981).

Table 1. Ontario Hydro's estimates (adapted from Ontario Hydro 1981). The original data, reported in 1995 dollars, were calculated on the basis of escalated costs of construction and operation, and then discounted to 1995 using a nominal interest rate of 8 per cent. We have deflated to 1981 dollars by the implied cost index of 290 (derived from Tables 7 and 8 of the Ontario Hydro report, pp. 21, 22).

Project	Discounted cost (1981 M$)	Nuclear alternative (1981 M$)	Net benefit (1981 M$)
Little Jackfish River	211	242	31
Lower Mattagami River Redevelopment	228	383	155
Moose River Basin New Sites	840	1071	231

These net benefit estimates are not fully satisfactory for our purposes for a number of reasons. First, they are based on escalated costs of fuel and construction over a 60-year time horizon using projections which are not reported in the Ontario Hydro document. Consequently, important assumptions concerning the prices of coal and uranium relative to each other and to the general price level are hidden. Similarly, excess energy from nuclear projects is valued at the 'system value of energy', the definition of which is not clear.

Secondly, the prices relevant to Ontario Hydro for accounting purposes are not always those relevant to a social cost-benefit analysis. For example, Ontario Hydro pays water rentals to the Ontario Government. These payments are relevant to the corporation but, in the absence of alternative uses of the water, do not reflect social opportunity costs. Similarly, Ontario Hydro purchases coal and uranium under long-term contracts which may not reflect their true opportunity costs which are world prices. Moreover, the current price paid for uranium may not adequately reflect the social costs of monitoring and managing mine tailings and the potential damage to the health of mine workers. Finally, although Ontario Hydro has included estimates of the future cost of decommissioning nuclear generating stations, the source of these estimates is not reported and it is unclear whether the data represent reasonable estimates of the marginal costs of spent fuel disposal as well as the cost of dismantling and sterilizing the contaminated site.

Thirdly, since hydroelectric and nuclear generation have quite different operating characteristics, an optimal expansion program may include building both. The cost savings relative to a nuclear plant may therefore be a poor estimate of the net value of a hydroelectric project to the power system as a whole.

Fourthly, the Ontario Hydro report takes as the opportunity cost of capital an assumed nominal interest rate (eight per cent in the base case) plus an unexplained net income requirement expressed as a percentage of capital costs. Cost-benefit analysis is usually performed using real discount rates reflecting the social opportunity cost of funds. Harberger (1974) recommends that this be close to the marginal rate of return before tax in the private sector and the Canadian Treasury Board (1976) recommends the use of real discount rates of ten per cent with sensitivity tests at five and 15 per cent.

Alternative net benefit calculations: Some of the difficulties with the Ontario Hydro estimates can be resolved by reworking their data using alternative approaches to net benefit estimation. For the sake of comparison, we include our estimates of net benefits from the Little Jackfish projects, although our estimates of environmental damages discussed below relate only to the Moose River Basin. We first report on net benefits calculated according to a procedure recommended by Harberger (1974). In this procedure, a clear distinction is drawn between peak and off-peak energy. Energy produced by hydroelectric stations during peak periods should be valued at the full cost of constructing and operating a nuclear station to meet this demand. Thus the shadow price per kilowatt hour of peak-period energy should equal the fuel cost of nuclear generated electricity plus the annualized capital and operating cost of the nuclear capacity sufficient to produce this energy. Energy produced in off-peak periods should be valued at the marginal cost of production from other sources, in this case the fuel cost of coal-fired thermal generation.

Table 2 presents the results of recalculating the net benefits of hydroelectric projects in this manner. The relative prices of coal and uranium are as assumed by Ontario Hydro for 1995. No trends in relative prices beyond then are assumed. The capital and operating costs for hydroelectric and nuclear generation are those reported by Ontario Hydro, after excluding water rentals from operating costs (Ontario Hydro 1981). It is assumed that the thermal peak is 2920 hours per year; that is, that the first 2920 hours of annual nuclear generation should be considered a contribution to peak load (Harberger 1974).

Table 2. Net benefits estimates using the Harberger procedure (adapted from Ontario Hydro 1981). Discounted gross benefits are calculated as the discounted capital and operating costs of providing the same amount of peak-period energy from a nuclear generating station. However, a nuclear station would produce additional energy at off-peak periods, displacing energy from coal-fired stations. The foregone fuel saving is the fuel-related cost of the displaced energy net of nuclear fuel costs. Prices are assumed by Ontario Hydro to 1995 with no trend beyond that date. The thermal peak is assumed to be 2920 hours per year.

Project and discount rate	Discounted gross benefits (1981 M$)	Discounted capital and operating costs (1981 M$)	Discounted foregone fuel saving (1981 M$)	Discounted net benefits (1981 M$)
Little Jackfish River				
5%	304.6	179.6	303.2	−178.2
10%	240.2	175.7	167.3	−102.8
15%	220.7	174.3	116.9	− 70.4
Lower Mattagami River Redevelopment				
5%	393.6	199.4	391.7	−197.6
10%	310.3	195.3	216.1	−101.1
15%	285.2	193.8	151.1	− 59.7
Moose River Basin New Sites				
5%	1,316.6	727.9	1,310.5	−721.8
10%	1,038.0	711.0	722.9	−395.8
15%	954.1	704.7	505.4	−256.0

Table 2 shows that the value of the energy produced by the hydroelectric projects exceeds their cost by a wide margin. However, the discounted value of the off-peak energy produced by the nuclear alternative is so great that the net benefit from the hydroelectric projects is negative when compared to the nuclear alternative.

Sensitivity tests suggest that the assumption of a shorter thermal peak increases the relative advantage of nuclear stations (at low discount rates) due to the increased value of the off-peak fuel savings. On the other hand, reducing the assumed price of coal by 50 per cent is sufficient to generate positive net benefits of the order of $200 million for the three projects combined.

These results confirm the critical importance of the assumed mix of generation types and the assumed time path of fuel prices in estimating the net benefits of the hydroelectric projects. Clearly, if there were no coal-fired generation for the nuclear off-peak energy to displace, the value of foregone fuel savings would be zero and the hydroelectric projects would show large net benefits. This suggests that the hydroelectric projects should be evaluated in a framework which explicitly accounts for the expected future generation mix.

A linear programming formulation: In principle, the net benefits imputed to a hydroelectric project should be the reduction it provides in the total cost of generating and distributing electricity. (This assumes that the price of electricity and hence total quantity demanded is not affected by the choice of project. If electricity prices are reduced, allowance must be made for the increase in consumer surplus.) These net benefits can be estimated by assuming that Ontario Hydro will act so as to minimize the total cost of providing electricity subject to a number of constraints including those on the availability of hydroelectric generating sites. The net benefits of a specific hydroelectric project are then the reduction in minimum cost which occurs when Ontario Hydro is allowed to take this project into account in planning its system. Mathematically, this may be accomplished by expressing Ontario Hydro's problem as a linear program. This approach is well known and has several advantages (Turvey and Anderson 1977). By placing the evaluation of the hydroelectric project in an explicit cost-minimization context, it is clear what the term 'net benefits' is intended to convey. It is easy to incorporate special constraints, such as those associated with the inability of nuclear plants to closely follow changing loads. The analysis of the dual problem guides the analyst in selecting shadow prices and the problem formulation makes it clear that hydroelectric and nuclear alternatives are not mutually exclusive.

A simple linear programming problem is sufficient to capture the main elements relevant to the selection of hydroelectric projects. We assume that Ontario Hydro wishes to minimize the total discounted costs of meeting a stationary load equal to that expected in the year 2000 or, equivalently, that it wishes to minimize current operating costs plus the annualized capital costs of meeting the annual load. (We adopt the medium growth scenario of the Porter Commission which called for a system load of 203.3 TW.hr.; see RCEPP 1980, Vol. 1, p. 107.) It is assumed that the load duration curve is approximated by the three period distribution shown as Figure 3. This approximation has a short peak period of 876 hours per year, an intermediate period of 3942 hours per year and a low period, again of 3942 hours per year. The system is allowed to expand using any combination of nuclear, coal-fired and hydroelectric generation. Each of the three main project groups — the Little Jackfish, the Lower Mattagami River redevelopment, and the Moose River Basin new sites — enters as a separate potential project. In one series of calculations the program is allowed to utilize existing thermal capacity at zero cost. In a

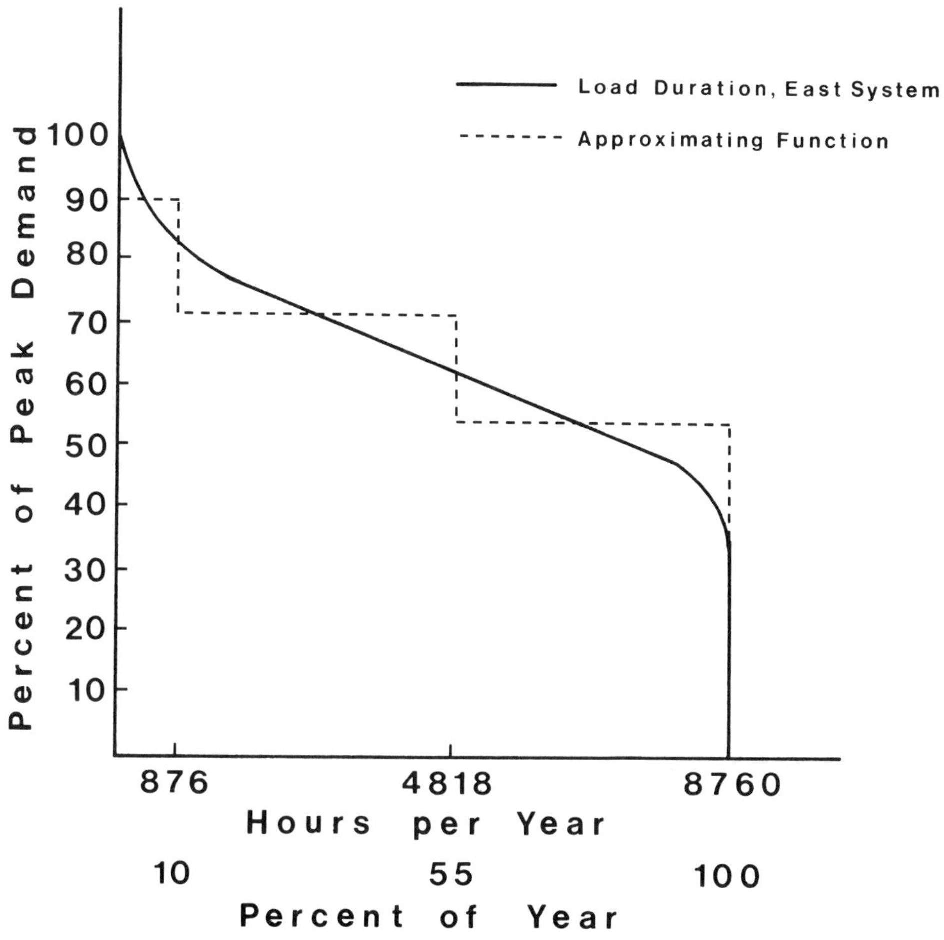

Figure 3. Load duration curves for Ontario Hydro (adapted from RCEPP 1980).

second series, the existing capacity is assumed to have been retired and replacements must be built at full cost. In both, the inability of nuclear plants to follow loads is modelled by constraining the power developed at peak periods not to exceed the power at intermediate periods and by preventing nuclear power at off-periods from falling below 50 per cent of the intermediate load (our interpretation of the comments in RCEPP 1980).

The results of this exercise are reported in Table 3. At real discount rates of five per cent, all three hydroelectric projects enter the optimal solution. In the presence of excess coal-fired capacity the net benefits are small, about $135 million. However, when existing thermal capacity is valued at replacement cost, the net benefits increase to almost $1.2 billion. At higher rates of discount, no hydroelectric plants are built while excess coal capacity remains, but the net benefits remain substantial when the coal-fired capacity is due to be replaced. We conclude that there will be a strong incentive to develop these sites when existing coal-fired capacity is retired, if not before.

Table 3. Net benefit estimates using the linear programming model. Data are from Appendix 1 which is available from the authors on request. Net benefits are calculated as the reduction in minimum cost of meeting an annual load of 203.3 TW year/year for a 60-year time horizon (assuming the approximating load distribution curve of Figure 3 and the costs of hydraulic, nuclear and coal-fired generation reported in Ontario Hydro 1981) which occurs when the hydroelectric project is available to the system.

	Net benefits	
	Existing thermal capacity valued at	
Project/ real discount rate	Zero (1981 M$)	Replacement cost (1981 M$)
Little Jackfish		
5%	7.6	85.6
10%	0	7.3
15%	0	0
Lower Mattagami		
5%	44.3	533.5
10%	0	307.4
15%	0	242.5
Moose River New Sites		
5%	83.1	566.9
10%	0	180.2
15%	0	73.9
Total		
5%	135.0	1,186.1
10%	0	494.9
15%	0	316.4

Although these linear programming estimates have a clear theoretical basis, they must be interpreted with great caution. One obvious difficulty occurs because the 'lumpy' characteristic of these investment projects has been ignored. In particular, nuclear power units in Ontario are expected to come in increments of 850 MW and the hydroelectric projects are not infinitely divisible. These characteristics can be incorporated into a mixed integer programming problem (Rowse 1978). However, the ultimate desirability of the hydroelectric projects should not be affected since the optimal linear program does not call for a fractional amount of any project. More important is the lack of detail on special constraints affecting the operation of the system, such as the apparent difficulty of operating the generating stations at Niagara Falls on a peaking basis. However, constraining the alternatives to the new hydroelectric projects is likely only to increase the cost of the alternatives and the net benefits of the hydroelectric projects. Another important limitation is the static nature of the analysis. A multi-period model would predict the optimal retirement date of existing coal-fired generating plants and allow a more precise estimate of net benefits.

An even more serious caveat concerns the assumed load duration curve. The high net benefit of the hydroelectric projects when existing thermal capacity is valued at replacement cost occurs because of the high value of peak-period energy once coal-fired capacity is retired. If peak energy demand can be successfully reduced by time of day and seasonal pricing schemes, the net benefits of the hydroelectric projects will be reduced

(RCEPP 1980). Of course, the net benefits can be re-estimated for any assumed shape of the load duration curve.

Finally, no allowance has been made for changes in exports and imports of power by Ontario Hydro, a factor which might prove to significantly alter the desired expansion of the system. Consideration of these and other demand factors would require a shift from the objective of cost minimization to a broader approach of profit maximization.

Despite these caveats, we consider the programming approach to be the most satisfactory method of evaluating the net benefits of hydroelectric projects, not least because it allows all of the above objections to be considered explicitly in the analysis. We believe that the net benefits imputed to hydroelectric generation are best measured by the reduction in total system cost which they permit. This is most conveniently done using linear programming or a variant. Regardless of the method chosen, net benefits are subject to important uncertainties such as the mix of generation capacity and the price of conventional and nuclear fuels. Nevertheless, adopting a five per cent real discount rate and using Ontario Hydro's own cost estimates, it appears that net benefits in the year 2000 of the order of $135 million are available from developing the Little Jackfish, the Moose River new sites and the Lower Mattagami complex, and that these would rise to approximately $1200 million once coal-fired capacity is due for replacement. These estimates of large benefits suggest that there will be strong pressures to develop these hydroelectric sites in the future.

Evaluation of Environmental Impacts

In this section we attempt to provide reasonable estimates of the importance of some environmental changes which might be caused by additional hydroelectric development in the Moose River and its tributaries. Our purpose is to compare these estimates with the expected partial net benefits (evaluated before externalities) discussed in the previous section. Should the expected environmental costs exceed the partial net benefits, then according to the conventional cost-benefit criterion, the hydroelectric projects should not be undertaken. If environmental costs are much less, the projects should be undertaken, but it may be desirable to ensure that full or partial compensation is paid to losers. Our method is to identify the major potential environmental impacts and the groups affected by them. We then attempt to estimate the amount of money which would need to be paid to fully compensate those suffering losses as a result of the project or which would need to be spent to eliminate the environmental damage.

Possible impacts: In principle, the construction and operation of new hydroelectric generating stations in the Moose River Basin might have a number of undesirable consequences. The flooding of headponds upstream from the dams could drown valuable timber and animal habitat and eliminate the potential for future harvests. The peaking operation of generating capacity can lead to large fluctuations in downstream water conditions, with prolonged daily low water periods. This might affect river travel and wildlife and fish habitats. Increased downstream fluctuations might prevent the riverbank soil from stabilizing and lead to increased erosion and downstream siltation. This would harm fish habitats and might increase the difficulty of river travel. Altered stream flow patterns could affect the timing of break-up and freeze-up, with attendant social disruption. Finally, the cumulative effect of diverting or reducing fresh water flows into James Bay in a number of rivers could be an alteration in the salinity of the Bay, an event which might affect the biology of the west coast, which is an important pathway for migratory birds. Our knowledge about the probability of these occurrences is limited at present, and it is a major purpose of the overall TASO research program to contribute to building up a data base on environmental as well as social impacts. However, for the purpose of this paper, it was necessary to make a preliminary judgment about the possible importance of each identifiable effect.

Our impression is that the effects of flooding would not be severe, since no large reservoirs are to be created. Bodaly *et al.* (1982) discuss the deleterious effects of large-scale flooding in similar circumstances for the Churchill-Nelson River diversion. It is unlikely that any reservoir would overtop the immediate banks of the rivers. This means that the total area inundated will be small. It is true that the general area is under timber licence (RCNE 1978, Maps Graphs Charts Ontario North of 50°) but the long rotation period and difficulty of access to riverbanks suggest that the value in forestry of the drowned land will be slight. This point remains to be verified. Similarly, we have no specific information concerning the possible alterations in break-up or freeze-up, nor about the possible effects on the salinity of James Bay. These topics are under investigation by other TASO members.

The major local environmental concerns are focused on the potential effect of hydroelectric projects on the water levels in the Moose River and on the possible extent of increased siltation downstream from the dams. In this paper we make no judgment as to the probability of these occurrences, but rather ask what would be the maximum loss should they occur.

Uses of the river: The Moose River is important both for local transportation between Moose Factory and Moosonee and for upstream travel. Although the river is tidal at Moosonee, it is possible that low river flows would exacerbate the already low water conditions at low tide. In the extreme case, freight canoe and barge service between Moosonee and Moose Factory might be cut off during low water periods. Similarly, low water conditions now prevent easy travel by freight canoe more than a few miles upriver from Moosonee. Reduced flows might effectively prevent river travel between the dam site at Renison, for example, and Moosonee, as is now the case in the Abitibi River below Otter Rapids.

There is no single reliable source of information about the frequency of river use. The Ministry of Natural Resources is currently compiling information on land use patterns in the development of a regional land use plan. These data will be most valuable when they become available. For our present purpose of estimating the magnitude of environmental costs, however, it will be necessary to proceed with less precise estimates of the important numbers.

Local water transportation between Moose Factory and Moosonee is used by local residents and tourists. The usual fare for a one-way freight canoe trip is $3.00. Crude calculations indicate that crossings by local residents should be of the order of 20,000 per year, while tourists might make 7,000 crossings per year. (For local residents, assuming 4 boat passages per hour with an average of three passengers each, 10 hours per day, 170 days per year, yields 20,400 crossings. For tourists, 50 round trips per day, 70 days per year, yields 7,000 crossings.) We estimate the annual market value of these crossings to be $81,000. However, the loss of water transportation between the communities during low water periods is properly measured by the minimum amount users would voluntarily accept in compensation. A probable upper limit to this amount is the cost of building a bridge across the Moose River. This would represent an annualized cost of the order of \$1,000,000 or a discounted value of about \$20 million. (At a real discount rate of 5 per cent and a 30-year lifetime, a \$10 million bridge has an annualized cost of \$610,782 per year. Adding maintenance of less than 5 per cent of original cost yields an approximate annual cost of \$1 million.) In fact, a dredging program to maintain a channel between Moosonee and Moose Factory might be an acceptable lower cost alternative and, if so, our estimates based on bridge construction will be biased upwards. The federal government has already dredged the channel at least once. (Personal communication, Dave Mann, Ontario Ministry of Northern Affairs, August 24, 1982.) It should be noted that while most users of water transportation could be presumed to prefer a bridge crossing, the operators of freight canoe water taxis might experience a loss of rents. To the extent these come from local residents this would be an intra-community transfer, whereas the rents collected on the tourist trade would represent a net loss to the community. However, a bridge would have major offsetting gains to the community, especially during break-up and freeze-up when no river travel is possible.

The main upstream users of the Moose River are commercial fishermen, local residents on hunting or fishing trips, guides and their fishing or hunting parties, and recreational canoeists. We consider each in turn.

There is only one commercial fisherman on the Moose River. Current yields are estimated by a local official of the Ministry of Natural Resources to be of the order of 1000 kilograms per year, with a market value of approximately \$4,000. (The quantity estimates are from John Weir, Ministry of Natural Resources, Moosonee. The approximate value represents a selling price of \$4 per kilogram.) If this fishery were completely eliminated, the loss would be equal to the profits earned on the fishing operation, which we might value at \$2,000 per year, plus the loss in consumer surplus by purchasers. Since close substitutes are available through imported protein foods, this loss should not be too great, say at most \$2 per kilogram of fish. Thus the maximum value for the surplus at risk in the commercial fishery is about \$4,000 per year, or \$80,000 capitalized at five per cent.

Local residents use the river for hunting, fishing and recreation, but the extent of their use is difficult to determine. According to one informant, about ten per cent of the Moose Factory band live entirely ''off the land'' (William Small, Assistant Band Manager, Moose Factory, June 24, 1982). Another suggests that there are about 60 full-time trappers within a 100 km radius of Moosonee, perhaps 30 per cent of whom would be directly affected by hydroelectric development on the Moose River (Chief Dave Fletcher, Moose Factory Band, June 24, 1982). On the other hand, as many as 50 or 60 parties might be camping upstream of Moosonee on a long weekend (William Small, June 24, 1982). Local residents often meet a portion of their protein requirements through fishing in this way (Tom Archibald, June 24, 1982). Local officials at the Ministry of Natural Resources consider that the total fish harvest might be about ten times greater than the commercial catch (John Weir, Moosonee, telephone interview, July 16, 1982). This would include the catch of guided fishermen (see below). In addition to fishing, the river is used for some hunting expeditions. (For example, the important goose hunt mainly takes place on the coast, but in some seasons the geese travel inland in search of fresh water. Travel to the hunt has been hampered by low water under such conditions according to Chief Dave Fletcher, June 24, 1982.)

Accepting the Ministry's estimate of the total fish harvest by residents, the non-commercial fish catch might amount to 9,000 kilograms per year with an annual market value of approximately $36,000. As a measure of the rent received by local residents from the fishery this figure is biased upwards because it includes the outfitting costs of fishing expeditions and downwardly biased because it ignores the satisfaction derived from fishing and camping activity. It suggests, however, that losses from total elimination of fishing on the Moose River might be of the order of $800,000 ($40,000 per year capitalized at a 5 per cent real interest rate).

The Moose River is also used by recreational fishermen, some of whom employ local residents as guides. We estimate that there are fewer than 10 local residents regularly employed as guides on the river. During the summer, possibly as many as seven parties per week with two to four men per party might go up river on guided fishing trips. The usual charge is $50 per day plus gasoline and food, but this charge is sometimes applied on a per person basis and sometimes on a per party basis. A crude estimate of income from this source is $21,000 per year (seven parties per week for ten weeks, three days per party at $100 per party per day), from which must be deducted depreciation, maintenance, and capital cost of boats and motors as well as the opportunity cost of the guides' time in order to obtain a measure of the economic rent derived from guiding. Allowing $5,000 per year for the capital items, we obtain a rent of $16,000 per year provided the opportunity cost of the guides' time is entered at zero (the capital cost of seven boats and motors, each valued at $5,000 is about $4,300 per year amortized over ten years).

In addition to the loss of rent to the guides, a loss of fishing opportunities on the Moose River would cause a loss of consumer surplus to recreational fishermen. An upper estimate of this would be the additional cost of reaching equivalent fishing. On the assumption that equivalent fishing would be available at Fort Albany, an upper limit on the foregone consumer surplus to the fishermen would be the return air fare to Albany from Moosonee, estimated at $200 per fisherman. This would amount to $42,000 for 210 fishermen per year. Notice that the fishermen might create economic rents for guides in the alternative location. If so, the rents lost by Moosonee guides should be offset against rents gained by others.

The final group of users to be considered are recreational canoeists. The Moose River system is the most accessible far-northern river system in Ontario. Rough estimates are that perhaps 100 canoeists come down each of the Abitibi, the Mattagami and the Missinaibi rivers each year (Steve Anderson, Ministry of Natural Resources, Moosonee, telephone conversation, July 16, 1982). It is difficult to evaluate the loss which would be suffered by these people, if more hydroelectric dams were built on the Abitibi, Mattagami and Moose Rivers. The Missinaibi trippers would be largely unaffected, but more might wish to end their trip at Moose River Crossing. Travel on the Abitibi would be made easier above Nine Mile Rapids, but more difficult downstream of that point, and travel on the Mattagami might be made more difficult below Grand Rapids. Much of the wilderness experience would be spoiled by paddling on headponds rather than free flowing rivers. The consumer surplus lost by these canoeists is bounded above by the cost of reaching equivalent canoeing elsewhere. Some of the Abitibi and Mattagami canoeists might consider the Missinaibi an equivalent opportunity; others might only consider a fly out trip down the Albany or Harricanaw rivers as equivalent. These would be much more expensive than Moose River trips due to the need to charter aircraft. We allow $25,000 per year as a maximum estimate of the loss to canoeists from the reduction in opportunities to travel on the Abitibi and the Mattagami Rivers (roughly the cost of flying 100 canoeists out of Fort Albany, calculated at $1,000 per round trip for an Otter Aircraft carrying two canoes and four passengers).

Table 4. Maximum estimate of environmental costs.

	Annual 000$	Discounted at 5 per cent 000$
Local transportation	1,000	20,000
Commercial fishing	4	80
Local fishing and hunting	40	800
Rents to guides[a]	16	320
Guided fishermen	42	840
Recreational canoeists	25	500
Total environmental costs	1,127	22,540

[a]These may be offset by increased rents to guides elsewhere.

Estimated environmental costs: We reiterate that these data are to be considered useful only as indications of the order of magnitude of the environmental costs associated with hydroelectric power development on the Moose River and its tributaries. The estimates are to be considered upper bounds, because the loss estimates assume total loss of fishing and transportation on the Moose River whereas the most probable outcome is lesser damage, and because we have chosen upper bounds on most losses.

Table 4 indicates that by far the most costly environmental consequence of the hydroelectric projects would be the loss of local transportation between Moose River and Moose Factory. One substitute, a bridge, would cost about $1 million per year, although it would entail a number of other benefits that should in principle be netted out. The remaining costs are divided about equally between losses to fishermen, hunters and guides, and those incurred by recreational fishermen and canoeists. In each case, the annual losses are of the order of $60,000 per year. It should be remembered that lost rents to guides may be offset by increased rents to guides in other communities. Total environmental costs are capitalized at $22.5 million, using a five per cent real discount rate.

Conclusions

In this paper, we have reported on the preliminary results of our cost-benefit analysis of ten proposed hydroelectric projects in northern Ontario, including our first tentative steps towards an appraisal of the environmental costs associated with the projects.

We have estimated the partial net benefits from hydroelectric expansion of the Lower Mattagami River complex and construction of the Moose River Basin new sites to be of the order of $125 million if coal-fired generating capacity remains in use, and about $1000 million if excess coal capacity is unavailable. Given our preliminary estimate of maximum environmental costs in the Moose River system of approximately $22.5 million, our analysis suggests that the projects will generate sufficient partial net benefits to more than compensate for the environmental costs incurred. Our tentative conclusions are, then, that the social net benefits from hydroelectric development in northern Ontario will be large, and that the projects would be a 'socially profitable' investment. However, these findings will need to be refined as more complete data on social impacts and environmental consequences become available to the TASO research program.

Acknowledgements

The authors are respectively Associate Professor and Professor of Economics, and principal investigators in the Research Program for Technology Assessment in Subarctic Ontario (TASO) at McMaster University. The financial support of the Donner Canadian Foundation and the helpful comments of F.T. Denton, R.J. Preston, and J.R. Williams are gratefully acknowledged. An appendix describing the linear programming problem employed in this paper is available from the authors on request.

Literature Cited

Bodaly, R.A., *et al*. 1982. Ecological Effects of Hydroelectric Development in Northern Manitoba, Canada: The Churchill-Nelson River Diversion. Canada Department of Fisheries and Oceans, Freshwater Institute, Winnipeg, Manitoba.

Canada Treasury Board. 1976. Benefit-Cost Analysis Guide. Ottawa, Ontario.

Harberger, A.C. 1974. Project evaluation: Collected papers. Markham Press, Chicago, Illinois.

Mishan, E.J. 1976. Elements of cost-benefit analysis. Second Edition. George Allen and Unwin, London, England.

Muller, R.A. and P.J. George. 1982. Evaluating the environmental impact of hydroelectric development in northern Ontario: A cost-benefit approach. TASO Research Report No. 1. McMaster University, Hamilton, Ontario.

Ontario Hydro. 1980. Hydroelectric development program, 1980 review. Report No. 213-306. Toronto, Ontario.

Ontario Hydro. 1981. Cost assessment of the hydraulic development program. Report No. 600 SP. Toronto, Ontario.

Ontario Ministry of Energy. 1979. Energy security for the eighties: A policy for Ontario. Toronto, Ontario.

Ontario Ministry of the Environment. 1982. General guidelines for the preparation of environmental assessments. Revised. Toronto, Ontario.

Ontario Royal Commission on Electric Power Planning. 1980. Report. Toronto, Ontario.

Ontario Royal Commission on the Northern Environment. 1978. Interim report and recommendations. Toronto, Ontario.

Rowse, J. 1978. Toward optimal capacity expansion for an electric utility: The case of Saskatchewan Power. Can. J. Econ. 11: 447-469.

Turvey, R., and D. Anderson. 1977. Electricity economics: Essays and case studies. Johns Hopkins University Press, Baltimore, Maryland.

EFFECT OF THE CHURCHILL RIVER DIVERSION ON WATERFOWL USE OF THE LOWER CHURCHILL RIVER, MANITOBA

P.N. Boothroyd
Canadian Wildlife Service

Abstract: The effect of reduced flows, resulting from Manitoba Hydro's Churchill River diversion project, on waterfowl use of the lower Churchill River was investigated in 1978 and 1979. Densities of waterfowl using habitat on and away from the river in 1978 and 1979 are compared to pre-diversion waterfowl densities in the study area and regional population trends. Altered flow regimes appear to have indirectly resulted in reduced use of the Churchill River by breeding ducks and geese. Low nesting effort by Canada geese was apparent on the river in 1979. It is suggested that this may be attributed to reduction in availability of attractive goose breeding habitat.

Introduction

In 1973, construction work began on a coffer dam at Missi Falls, the outlet of Southern Indian Lake in northern Manitoba. This project was part of a scheme by Manitoba Hydro to divert water from the Churchill River to the Nelson River for the purposes of generating electricity. The completed control structure has resulted in a drastic reduction in flows of the Churchill River. Long-term average June flows for the river were 1,052 m^3/s. Average flows in June 1982 were 33 m^3/s.

The lower Churchill River is important as reserve waterfowl breeding habitat when dry or drought conditions exist in prairie and parkland areas further south (Cowan 1975). The river is also an important breeding area for the western prairie population of Canada geese. In anticipation of the potential adverse effects of the diversion on waterfowl, data were collected in 1969 by Raveling (1977) to document the importance of the river for Canada geese. In 1974, further pre-project data were collected on a wider group of waterfowl (Cowan 1975). Raveling (1977) predicted that lowered water levels would adversely affect Canada geese due to: enlargement or connection to shore of nesting islands on the river, promotion of plant succession on islands and feeding areas from sedge (*Carex* spp.) and willow (*Salix* spp.) to black spruce (*Picea mariana*) because of lack of flooding, and alteration of creek drainages and their pond water levels. Cowan (1975) anticipated that geese may not be attracted to or may not recognize the enlarged configurations of traditional nesting islands or newly-exposed rocky or gravelly islands barren of vegetation. A decrease in goose production was expected due to reduced reproductive efforts and higher predation losses on traditional nesting islands connected to the mainland.

Several studies have documented the effects of elevated water levels or flooding of nesting habitat on Canada geese (Craighead and Craighead 1946; Gibson and Buss 1972; Allen *et al.* 1978); however, little work has been done on the effects of drawdown on Canada geese and other waterfowl populations. This paper documents the initial results of a study being conducted by the Canadian Wildlife Service to monitor the effects of the diversion on waterfowl populations and waterfowl habitat of the Churchill River. It was felt that the results of the study would be valuable in assessing the potential impacts of future hydroelectric development proposals in northern Canada.

Study Area

The study area includes the portion of the Churchill River from Missi Falls to Mountain Rapids and includes Partridge Breast, Northern Indian, Fidler and Billard lakes (Figure 1). Within the study area, the river flows over granites and granitic gneisses of the Precambrian Shield (Study Board 1975). The river channel is controlled by the underlying bedrock while in upland areas the overburden is predominantly thick (up to 30 m) glacial till (Newbury 1973).

The study area lies within the closed coniferous forest physiographic region, as described by Ritchie (1962), dominated by black spruce. Located within a subarctic climatic region, the area experiences short, cool summers and cold, snowy winters. Average temperatures in January and July are −28°C and 13°C, respectively. Average annual snow fall is 102 to 114 cm with total annual precipitation averaging 43 cm at Southern Indian Lake (Study Board 1975). Permafrost is prevalent throughout the study area (Newbury 1973).

Methods

Thirty-two transects, established along shorelines of the Churchill River between Missi Falls and Fidler Lake by Cowan (1975), were surveyed for breeding and non-breeding waterfowl using standard aerial survey techniques. The surveys were conducted on June 5 and 6 in 1978 and June 10 in 1979. In addition, observations of breeding and non-breeding Canada geese were recorded along the river between Fidler Lake and Mountain Rapids on June 6, 1978 and June 12, 1979 (Churchill River Reconnaissance Survey). In order to compare densities in areas unaffected by the diversion away from the river, another survey was flown of the habitat north and south of the river on June 7, 1978 and June 12, 1979 (Interior Reconnaissance Survey). The route selected generally followed creeks, rivers, lake shorelines and wetlands, located from aerial photographs, which appeared suitable for nesting Canada geese. All surveys were flown in a Cessna 180 at an airspeed of 135 to 160 km/hour at an altitude of 30 to 60 m above ground level. Two observers recorded waterfowl observed within 0.2 km of each side of the aircraft. Survey routes are shown in Figure 2.

To compare levels of nesting effort by Canada geese before and after diversion, the perimeters of 46 islands in Northern Indian Lake were surveyed for active goose nests. The survey was conducted on June 15 and 16, 1979 using a Bell 206B helicopter flying 20 to 40 m above ground level.

The effect of lowered water levels on the use of the lower Churchill River by ducks and geese was investigated by comparing the data collected with pre-project data of Raveling (1977) and Cowan (1975). The basic approach underlying these methods was to construct 95 per cent confidence intervals for differences between breeding and non-breeding densities before and after diversion.

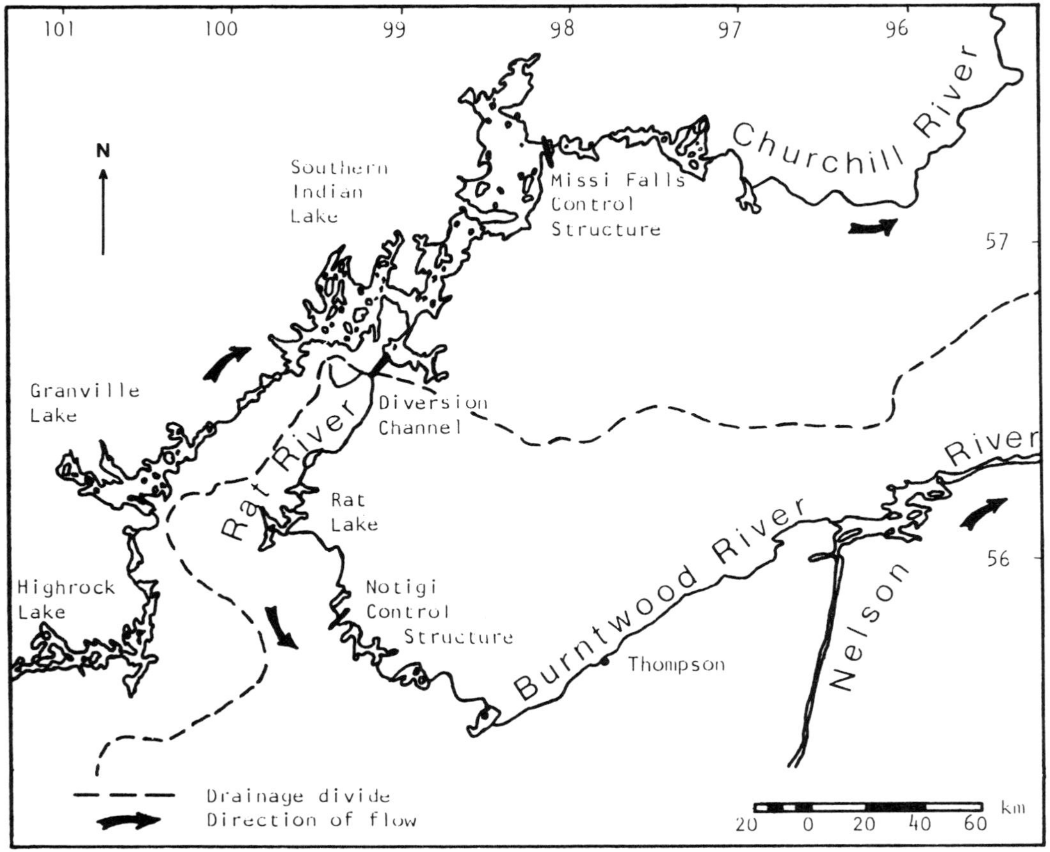

Figure 1. The diversion scheme.

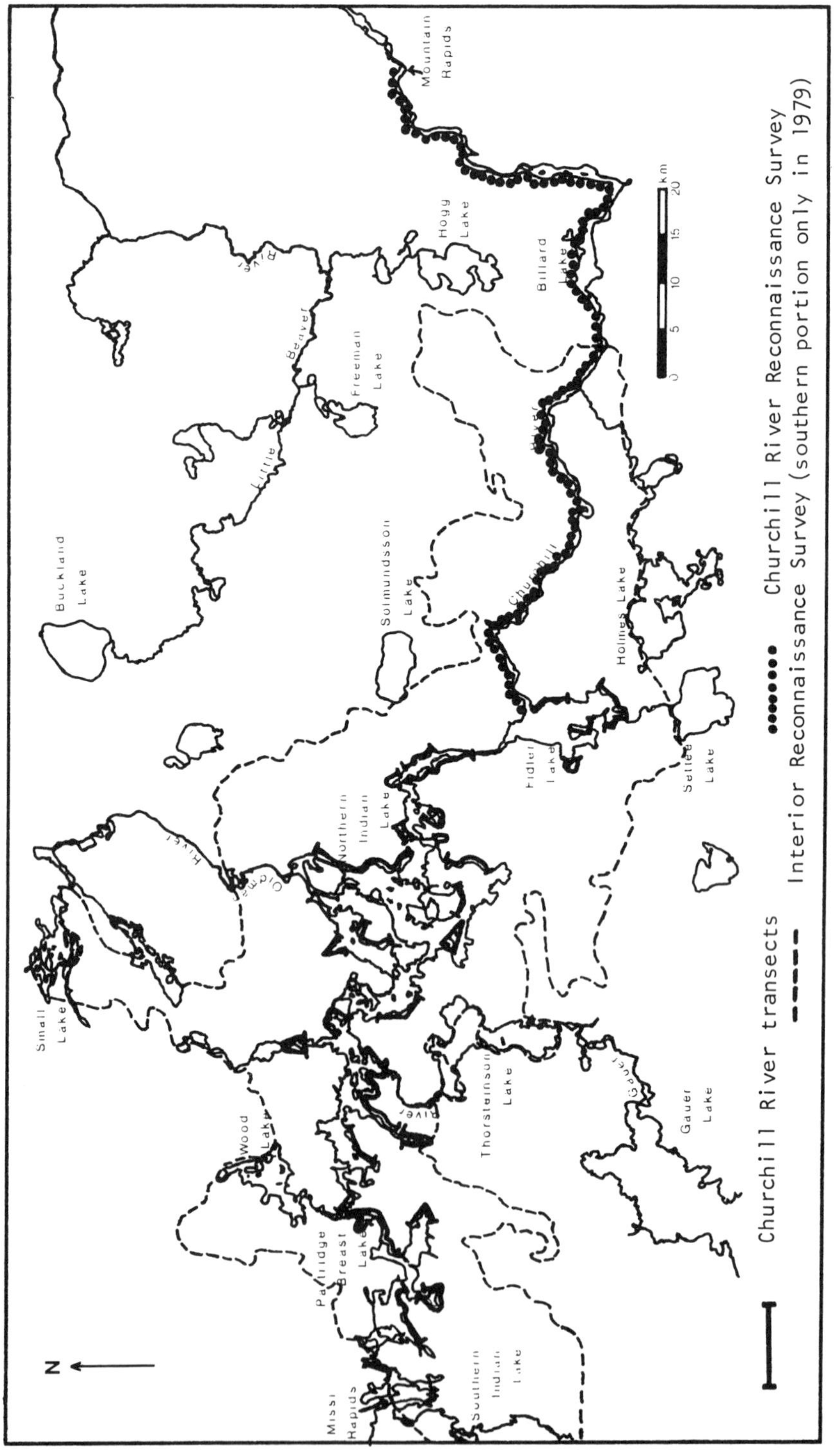

Figure 2. Aerial survey routes.

The data collected in each of the four years (1969, 1974, 1978, 1979) have been used to determine the densities of waterfowl using the Churchill River and adjacent area and therefore describe how the waterfowl are distributed spatially. The spatial distribution in any year can be approximated by the Poisson distribution. In addition, because the sample sizes (numbers of km^2 surveyed) are fairly large, the Central Limit Theorem applies and the data will tend to also exhibit a normal distribution. These facts allow inferences to be made concerning the significance of observed differences between waterfowl densities before and after diversion.

Ninety-five per cent confidence interval end points can be calculated for the differences between mean numbers of waterfowl observed per km^2 before and after diversion using:

$$(\overline{y_1} - \overline{y_2}) \pm z_{0.025} \sqrt{\frac{\sigma_1^2}{n_1} + \frac{\sigma_2^2}{n_2}}$$

where $\overline{y_1}$ = mean number of waterfowl observed per km^2 after diversion;
$\overline{y_2}$ = mean number of waterfowl observed per km^2 before diversion;
$z_{0.025}$ = 1.96;
σ_1 = variance associated with y_1;
σ_2 = variance associated with y_2;
n_1 = number of km^2 surveyed to obtain y_1;
n_2 = number of km^2 surveyed to obtain y_2.

For the Poisson probability function, the mean is equal to the variance. Therefore, the 95 per cent confidence interval end points can be approximated by:

$$(\overline{y_1} - \overline{y_2}) \pm z_{0.025} \sqrt{\frac{\overline{y_1}}{n_1} + \frac{\overline{y_2}}{n_2}}$$

For example, we wish to know whether the density of breeding geese observed away from the Churchill River in 1979 was significantly different from the density in 1978. Therefore:

$\overline{y_1} = 0.20 \qquad n_1 = 82$

$\overline{y_2} = 0.25 \qquad n_2 = 190$

95 per cent C.I. end points:

$$(0.20 - 0.25) \pm 1.96 \sqrt{\frac{0.20}{82} + \frac{0.25}{190}} = 0.05 \pm 0.12$$

Therefore, the end points are -0.17 and 0.07. Since this confidence interval overlaps zero, the difference in densities is not significant.

Results and Discussion

The effect of the diversion on use of the river by breeding dabblers, divers and Canada geese was investigated by comparing the 1978 and 1979 transect data to data collected in the same manner in 1974 by Cowan (1975). Dabblers, divers and geese were all more numerous in 1978 than in 1974 (Table 1). However, the apparent increase in dabbler breeders was not significant (Figure 3). In 1979, the situation had reversed with breeding dabblers and divers now less numerous than before diversion (1974) (Table 1, Figure 3). Geese were still present in marginally higher densities than before diversion but numbers were down from 1978.

Table 1. Densities of waterfowl breeding pairs along 32 transects on the Churchill River before diversion (1974) and after diversion (1978 and 1979).

	Waterfowl breeding pair density (pr/km²)		
Year	Dabblers	Divers	Canada geese
1974[a]	1.16	1.29	0.31
1978	1.33	2.61	1.00
1979	0.58	0.73	0.69

[a]Data from Cowan (1975).

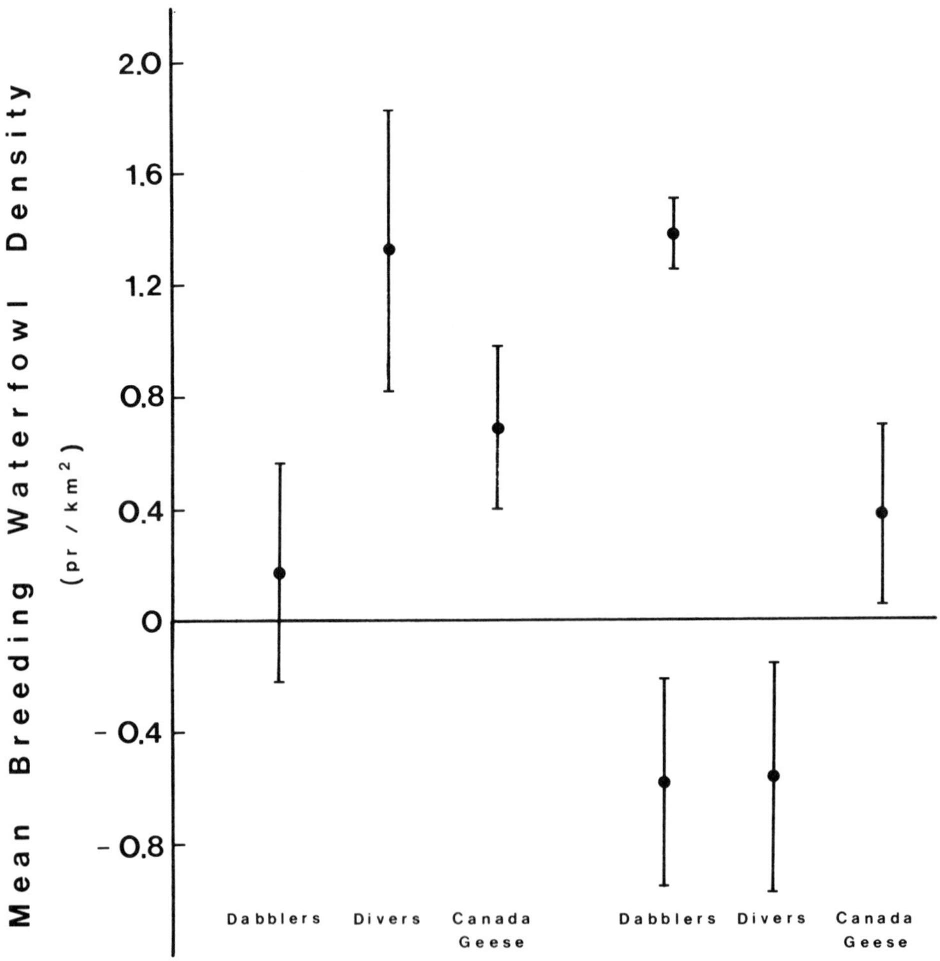

Figure 3. Differences in mean breeding pair waterfowl densities on the Churchill River, 1974-1978 (left), and 1974-1979 (right). Vertical line is 95 per cent confidence level.

The significance of the trends observed on the Churchill River is evident when we compare regional population trends established from data collected by the U.S. Fish and Wildlife Service during waterfowl breeding pair surveys of northern Manitoba. Data were obtained for the area currently described as stratum 24 (approximately 54° to 58° Lat. and 95° to 102° Long.) which overlaps almost the entire study area. Breeding pair counts obtained from aerial surveys have been adjusted to arrive at population estimates for each species (USFWS 1977). Results of straight-line transects flown over boreal forest habitat are of course fairly crude and must be interpreted cautiously. For example, Malecki (1976) discusses the influence of observer visibility on results of aerial surveys flown in forested areas. Nevertheless, population fluctuations exhibited by the data are useful for comparative purposes.

Figure 4 shows estimated dabbler and diver population trends for stratum 24 (or equivalent geographic area) for the period 1955-1979. In the stratum as a whole, dabbler and diver populations both experienced a substantial increase from 1974 to 1978, followed by a slight decrease in 1979. The apparent general increase in density of diving ducks using the Churchill River for breeding in 1978 following diversion is, therefore, most likely a reflection of the increase in diver populations throughout the stratum. However, the reduction in breeding density on the Churchill River in 1979 to levels less than those observed in 1974 does not reflect a general situation experienced in stratum 24 as a whole. Instead, the reduced use of the river by breeding ducks has likely resulted from reduced flows of the diversion.

Data are also available on Canada goose breeding pair densities in stratum 24 during the same period (Figure 4). Goose breeders were at a very low level in the stratum in 1974 but increased substantially reaching a level in 1979 not experienced since 1968. The reduction in breeding pairs from 1978 to 1979 observed on the Churchill River did not occur in the stratum as a whole. This implies that factors influencing the reduction were confined to the river itself. The most likely explanation is that the reduced flows resulting from the diversion have reduced the attractiveness of the Churchill River for breeding geese through alterations in river morphology and shoreline habitat.

The apparent reduction in use of the Churchill River by breeding geese concurred with the results of the survey of island nesting conducted in Northern Indian Lake in 1979. Only nine nests were found on five islands compared to the 28 nests on 26 islands found by Raveling (1977) in 1969 (Figure 5).

When this study was being designed, the possibility was foreseen that the geese, finding the altered shoreline habitat of the Churchill River unattractive, might move to more favourable habitat away from the river which was not affected by the diversion. The purpose of the Interior Reconnaissance Surveys was to determine whether habitat away from the river was in fact being used to a greater extent by breeding geese following diversion. Table 2 shows densities of breeding and non-breeding Canada geese on and away from the river in 1969, 1978 and 1979. The density of breeding pairs on the Churchill River was greater in 1978 than in 1969 (Figure 6). Breeders away from the river were also slightly more numerous. However, the increase was greater on the river than off. The density of non-breeding geese was marginally greater in 1978 whereas non-breeding density off the river remained unchanged. The increase in breeding pairs on the river again corresponds to the increase in breeders that occurred during the same period in stratum 24 as a whole.

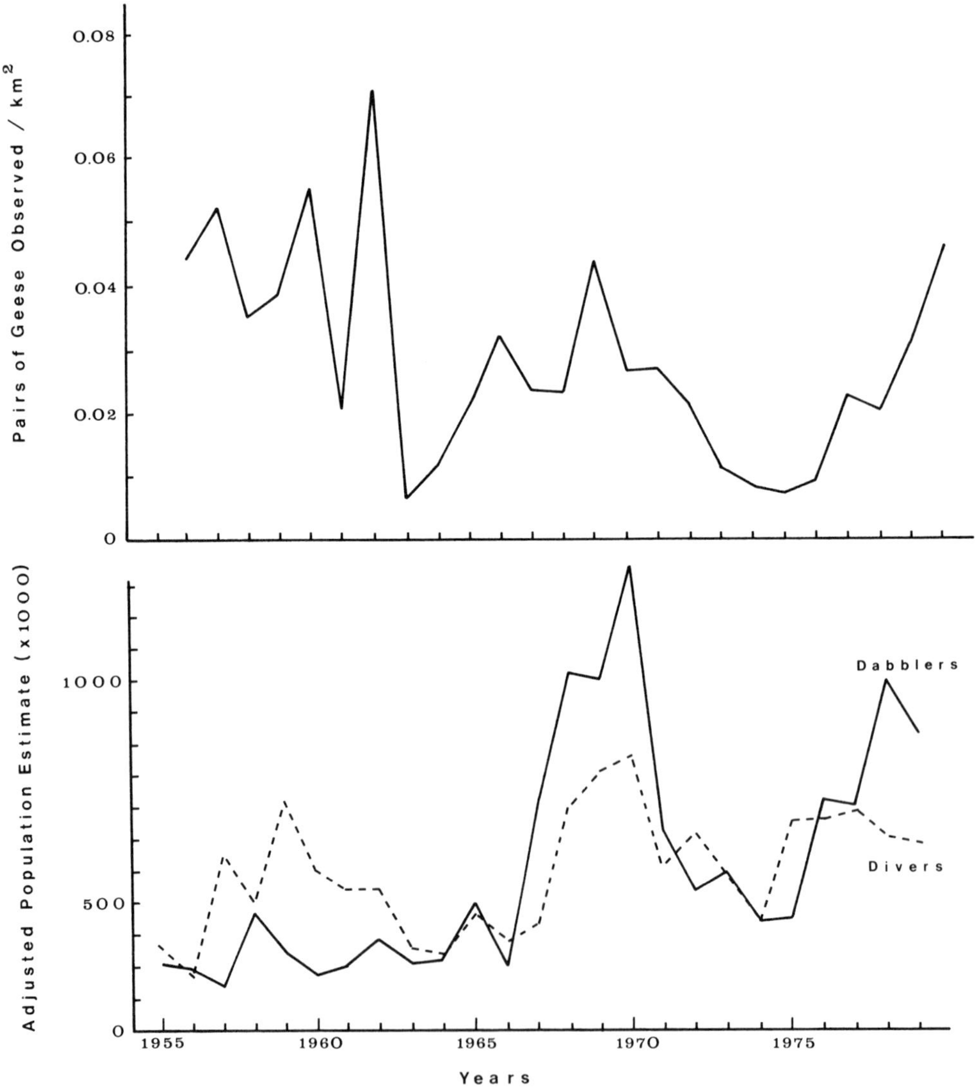

Figure 4. Breeding Canada goose densities (upper) and dabbler and diver populations (lower) in stratum 24, 1955-1979.

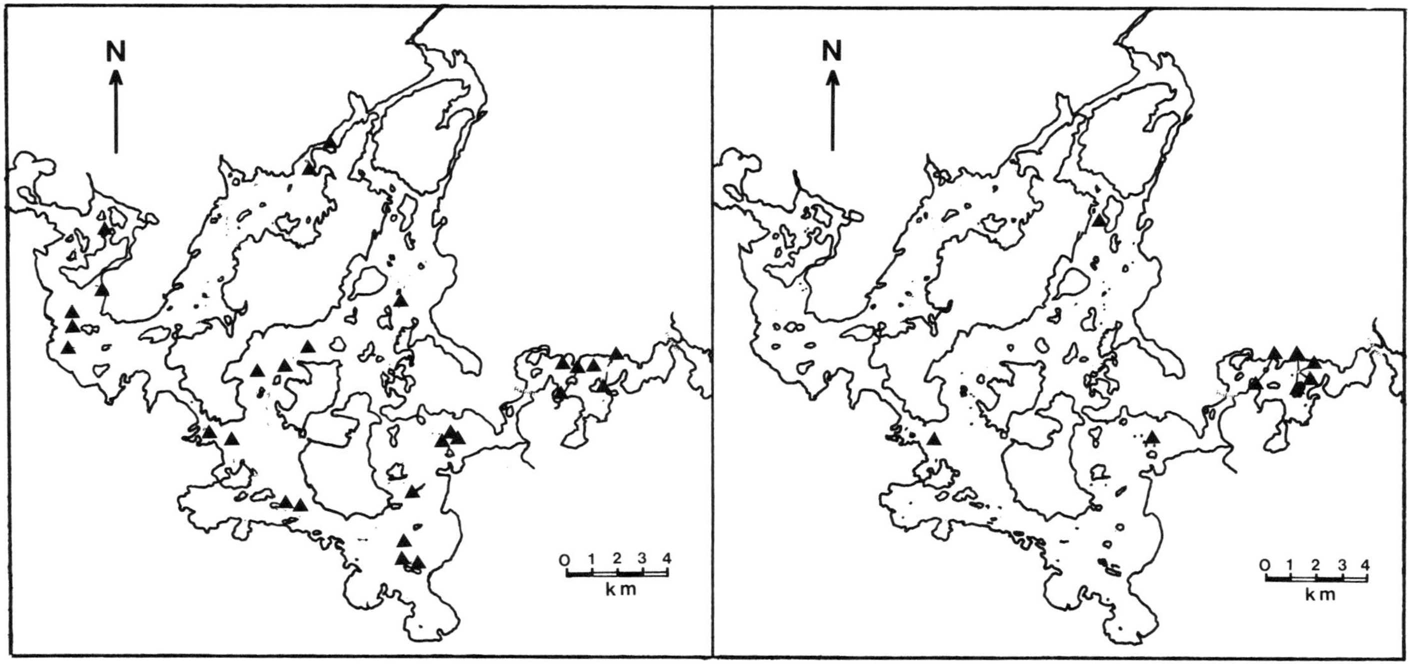

Figure 5. Locations of island goose nest sites in 1969 (pre-diversion) (left) and in 1979 (post-diversion) (right).

Table 2. Densities of breeding and non-breeding Canada geese on and away from the Churchill River before diversion (1969) and after diversion (1978 and 1979)

Year	Goose density	
	On river	Off river
1969[a]	0.57[b]	0.13
	(1.25)[c]	(0.24)
1978	1.46	0.25
	(1.61)	(0.31)
1979	0.81	0.20
	(1.31)	(0.42)

[a] Data from Raveling (1977).
[b] Breeders (pr/km²).
[c] Non-breeders (birds/km²).

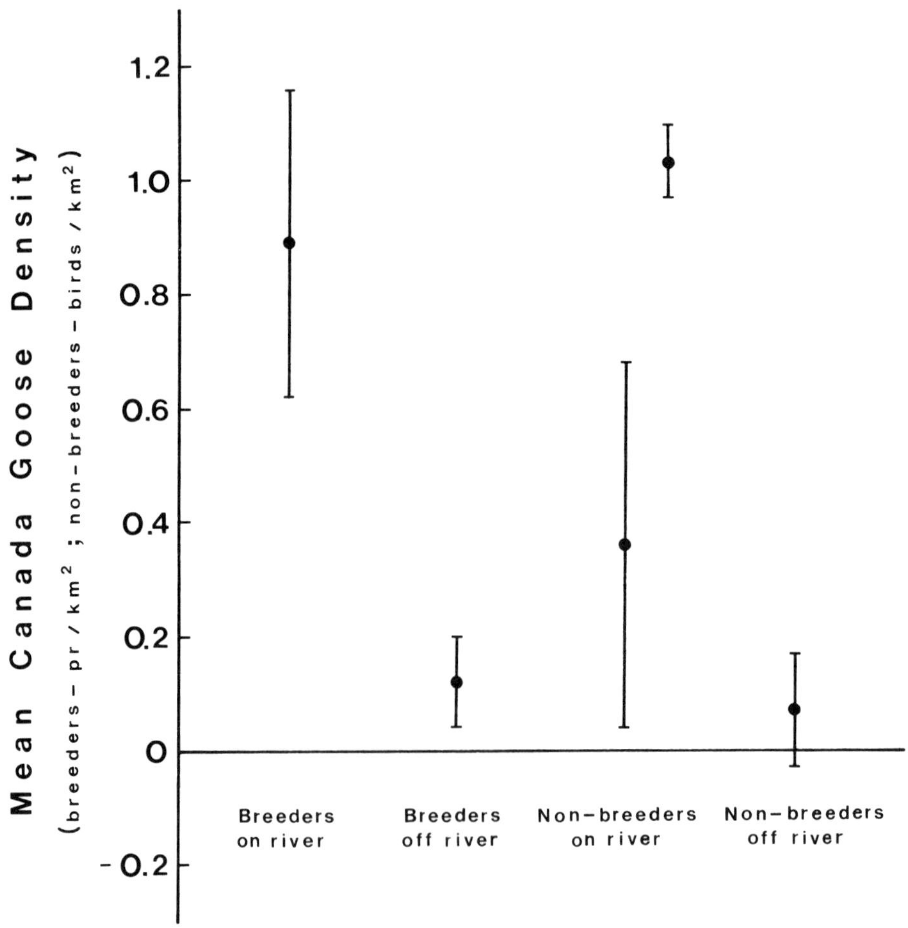

Figure 6. Differences in mean Canada goose densities, 1969-1978. Vertical lines represent 95 per cent confidence intervals.

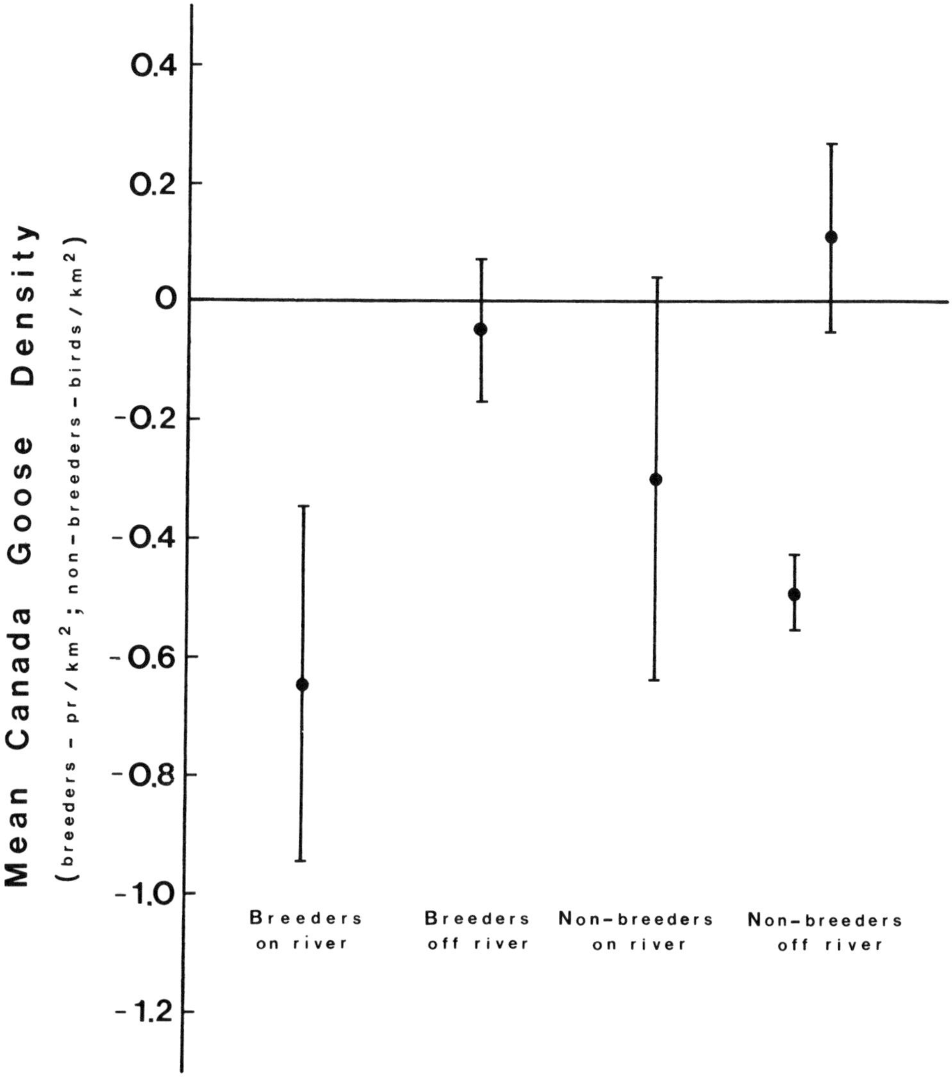

Figure 7. Differences in mean Canada goose densities, 1978-1979. Vertical lines represent 95 per cent confidence intervals.

In 1979, breeding geese were significantly less numerous on the river than in 1978 (Figure 7). The density of non-breeders on the river essentially did not change from 1978 to 1979. There was no significant change in breeders or non-breeders away from the river. Therefore, the 1978 and 1979 data do not show that breeding geese had shifted their breeding efforts from the Churchill River to habitat undisturbed by the diversion.

Figure 8 shows that, from 1969 to 1979, breeders increased only marginally on the river. The density of non-breeders away from the river was slightly higher in 1979 than in 1969 but there was no significant change in breeders away from the river or non-breeders on the river.

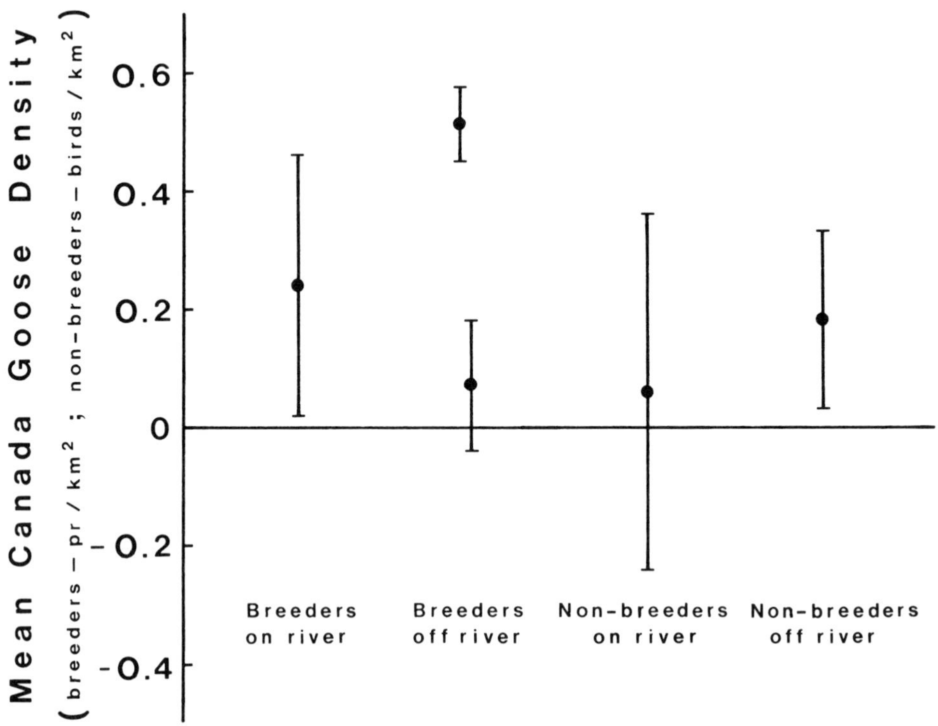

Figure 8. Differences in mean Canada goose densities, 1969-1979. Vertical lines represent 95 per cent confidence intervals.

The fact that breeding pair density on the Churchill River was virtually the same before (1969) and after (1979) diversion, and yet nesting effort in Northern Indian Lake was much lower after diversion, implies that only a small proportion of the paired birds were actually nesting. Copulations by geese occur before the mated pairs arrive at the nesting areas (Barry 1962, 1967; Hanson 1965; Ryder 1967). However, if certain environmental stimuli, such as appropriate nesting conditions, are lacking, ovulation can be inhibited (Barry 1962; Wood 1964). Continuing appropriate behaviour between members of a mated pair is necessary for follicle maturation to occur. The behaviour of the male at the potential nest site and stimulus of nest-building encourage the ovarian follicles to grow and produce estrogen leading to oviduct development, ovulation and egg laying (Murton and Westwood 1977). Absence of such stimulation through lack of appropriate mate behaviour or unsuitable nesting conditions can result in follicle atresia and abandonment of nesting.

Barry (1962) found atretic follicles in non-nesting brant and geese caused by a delayed season on the nesting grounds. Spring arrived later than normal in northern Manitoba in 1979. It is possible that the delayed spring resulted in a high incidence of atresia in mated geese of the lower Churchill River. It is also quite possible that atresia occurred as a result of the unattractive appearance of post-diversion shoreline habitat. Whether the apparent low nesting effort in 1979 can in fact be attributed to atresia is speculative since this possibility was not investigated during the present study.

Table 3. Some physical characteristics of Canada goose island nest sites in Northern Indian Lake, 1979

Nest	Height above water level (m)	Distance from lakeshore to nest (m)	Distance from lakeshore to nearest vegetation (m)
1	3.1	32.2	12.5
2	4.6	51.7	33.7
3	2.6	33.7	33.7
4	3.6	40.1	29.4
5	2.9	74.2	47.2
6	3.1	75.6	57.6
7	1.6	36.1	36.1
8	3.2	43.1	14.0
9	2.1	32.9	32.9
Mean	3.0	46.6	33.0

Certainly, 'negative' environmental stimuli, such as large distances from open water to nesting cover and the absence of shoreline emergent vegetation and shallow marsh feeding areas, were present in the study area. Distances from the water's edge to island nests in Northern Indian Lake in 1979 varied from a minimum of 32.2 m to a maximum of 75.6 m (Table 3). Nest sites were elevated above the lake level as much as 4.6 m. These values are considerably higher than those usually recorded (Craighead and Craighead 1946; Kossack 1950; Cooper 1978). Suitable nesting cover was distant from the shoreline and, in several cases, large boulders exposed by the reduced water levels separated the nesting cover from the shoreline.

Conclusions

Surveys were conducted in 1978 and 1979 to determine the effect of reduced flows, resulting from Manitoba Hydro's Churchill River diversion project, on waterfowl use of the lower Churchill River. Lowered water levels created extensive areas of exposed mudflats and rock shoreline and caused accelerated drainage and drying-out of marsh areas along the river and changes in general river morphology.

Diver and Canada goose breeding pairs were more numerous on the Churchill River in 1978 than in 1974. This pattern paralleled regional population trends for ducks and Canada geese in stratum 24 which encompasses the study area. Densities of breeding dabblers and divers on the river dropped in 1979 to levels less than in 1974 (before diversion). This reduction could not be attributed to a regional phenomenon occurring throughout stratum 24. Therefore, it was felt that the apparent reduction in breeding ducks had resulted from habitat changes caused by low post-diversion flows. In 1979, Canada goose pairs were still present in marginally higher densities than before diversion but numbers were down from 1978. Conversely, numbers of breeding geese continued to increase in stratum 24. This discrepancy could be explained by changes in river morphology and habitat which rendered the Churchill River less attractive for breeding geese.

Canada goose breeders were slightly more numerous on the river in 1979 than in 1969. No change in the numbers of breeders away from the river, or in the number of non-breeding geese on the river, had occurred during this period. However, goose nesting effort in Northern Indian Lake in 1979 was significantly reduced from the 1969 level. It was postulated that certain environmental stimuli, required for maturation of ovarian follicles and subsequent egg-laying to occur, had been suppressed by drastic changes in the physical appearance of shoreline habitat. Follicle atresia and termination of nesting effort by mated pairs would have resulted. This explanation would account for the observed low nesting effort even though breeding pair densities were comparable before and after diversion.

Conclusions drawn from the 1978 and 1979 results are only tentative. Further data are required to confirm the observed trend of reduced use of the river by breeding waterfowl.

Literature Cited

Allen, G.T., S.E. Fast, B.J. Langstaff, D.W. Tomrdle and B.L. Troutman. 1978. Census of Canada geese on the Palouse River, Washington, during the spring of 1977. Murrelet 59: 96-100.

Barry, T.W. 1962. The effect of late seasons on Atlantic brant reproduction. J. Wildl. Manage. 26: 19-27.

Barry, T.W. 1967. The geese of the Anderson River Delta, Northwest Territories. Ph.D. Dissertation. University of Alberta, Edmonton, Alberta.

Cooper, J.A. 1978. The history and breeding biology of the Canada geese of Marshy Point, Manitoba. Wildl. Monogr. 61.

Cowan, W.F. 1975. The impact of reduced flows on waterfowl of the lower Churchill River. Preliminary assessment. Can. Wildl. Serv. Unpubl. Rep.

Craighead, F.C., Jr. and J.J. Craighead. 1946. Nesting Canada geese on the Upper Snake River. J. Wildl. Manage. 13: 51-64.

Gibson, L.W. and I.O. Buss. 1972. Reactions of Canada geese to reservoir impoundment on the Snake River in Washington. Northwest Sci. 46: 301-318.

Hanson, H.C. 1965. The giant Canada goose. Southern Illinois University Press, Carbondale, Illinois.

Kossack, C.W. 1950. Breeding habits of the Canada goose under refuge conditions. Amer. Midl. Nat. 43: 627-649.

Malecki, R.A. 1976. The breeding biology of the eastern prairie population of Canada geese. Ph.D. Dissertation. University of Missouri, Columbia, Missouri.

Murton, R.K. and N.J. Westwood. 1977. Avian breeding cycles. Clarendon Press, Oxford.

Newbury, R.W. 1973. Physical impact study. Interim report. 1972. Lake Winnipeg, Churchill and Nelson Rivers Study Board. Winnipeg, Manitoba.

Raveling, D.G. 1977. Canada geese of the Churchill River basin in north-central Manitoba. J. Wildl. Manage. 41: 35-47.

Ritchie, J.C. 1962. A geobotanical survey of northern Manitoba. Arctic Inst. N. Amer. Tech. Paper. 9.

Ryder, J.P. 1967. The breeding biology of Ross' goose in the Perry River region, Northwest Territories. Can. Wildl. Serv., Rep. Ser. 3.

Study Board, Lake Winnipeg, Churchill and Nelson Rivers. 1975. Summary Report. Winnipeg, Manitoba.

U.S. Fish and Wildlife Service. 1977. Standard operating procedures for aerial waterfowl breeding ground population and habitat surveys. U.S. Fish Wildl. Serv. and Can. Wildl. Serv. Unpubl. Rep.

Wood, J.S. 1964. Normal development and causes of reproductive failure in Canada geese. J. Wildl. Manage. 28: 197-208.

ELEMENTAL SULPHUR DUST DEPOSITION ON SOILS AND VEGETATION OF LODGEPOLE PINE STANDS IN WEST-CENTRAL ALBERTA

D.G. Maynard, P.A. Addison and K.A. Kennedy
Northern Forest Research Centre
Canadian Forestry Service
Edmonton, Alberta, Canada

Abstract: The influence of the addition of elemental sulphur dust on forest vegetation and soils in west-central Alberta was studied using four sites from a larger study on biomonitoring of sulphur effects. The two sites close to the sulphur dust sources (<200 m) had significantly higher total and water soluble soil sulphur contents in all horizons to a depth of 20 cm. Associated soil changes included both the lowering of pH and a reduction in the total concentration of Mg, K and Mn in the LFH horizon. The effect of the S deposition on the vegetation was the almost complete elimination of the mosses and a reduction in the herb cover. Research is continuing on the controlling factors in the oxidation of elemental S in the soil, the effects of the S dust on soil processes such as nutrient leaching, and the response of vegetation to the changes in soil chemistry.

The natural gas processing industry is the largest anthropogenic sulphur source in western Canada. Almost half of the natural gas found in Alberta (the major gas and oil producing province in Canada) is sour; that is, it contains significant quantities (up to 30 per cent) of H_2S. There are 109 sour gas processing plants scattered throughout the province and collectively they processed 6.5×10^{10} m^3 of gas in 1981 (Energy Resources Conservation Board, Report 82-D, 1982). The gas industry as a whole is very efficient in sulphur recovery (97 per cent) as compared with other sulphur releasing industries in western Canada such as oil sands mining and processing (65 to 85 per cent) or nonferrous smelting (80 per cent).

High efficiencies in the extraction of S from natural gas results in large amounts of the element S being produced. Before 1979, most of the recovered S was poured into blocks which accumulated millions of tonnes. An improvement in the S market initiated break-up or melting of accumulated blocks. The methods used for handling S resulted in the production of elemental S dust and although largely confined to the immediate vicinity of the plants (i.e., <1 km), contributed significant quantities to local soils.

In 1981, we initiated a study in consultation with Canterra Energy Company Ltd. (formerly Aquitaine Company of Canada) and Gulf Canada to assess the impact of S emissions from two sour gas processing plants on the forest system.

The specific objective of this paper is to describe preliminary results as to the magnitude of the impact of the deposition of elemental S particulates on forest soils and vegetation.

Study Area

The study area (Figure 1) is located approximately 45 km southwest of Rocky Mountain House, Alberta. The lower elevation forest type is dominated by *Pinus contorta* var.

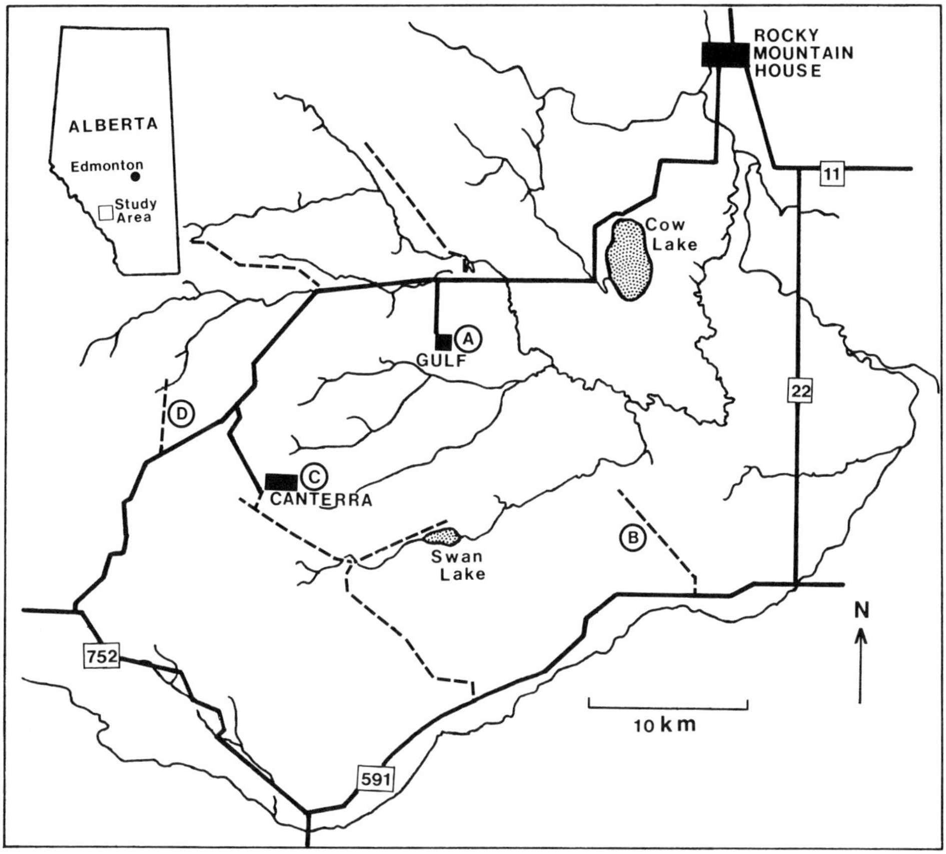

Figure 1. Location of the Gulf and Canterra gas plants and study sites (A, B, C, D) within the study area.

latifolia and the soils are mainly Brunisolic or Podzolic Gray Luvisols with subdominants of Orthic Gray Luvisols and Eluviated Dystric Brunisols. These soils have developed on a glacial till parent material that has a 20 to 30 cm aeolian veneer.

Sites A and C are located respectively 50 and 200 m from the Gulf and Canterra sulphur blocks (Figure 1) and sites B and D represent controls which are at least 9 km from the source. The sites were selected so that the vegetation was as similar as possible. All sites are characterized by 25 m *Pinus contorta* with *Vaccinium vitis-idaea* var. *minus, V. myrtilloides, Cornus canadensis, Pleurozium schreberi,* and *Hylocomium splendens*. Variation in vegetation does occur however, and appears to be related to elevation and stand age. It was on this basis that control sites (B and D) were selected to match the two sites that are in close proximity to the sulphur blocks. The stands at sites A and B are young (49 and 53 years) and are at low elevations (1200 and 1160 m). They are of a mixed wood composition with *Populus tremuloides* and *Picea glauca* in the tree layer, *Alnus crispa* and *Ledum groenlandicum* in the shrub stratum and the herbs *Pyrola asarifolia, Viola renifolia* and *Arnica cordifolia*.

The stands at sites C and D are older (75 years) and at higher elevations (1400 and

1420 m) than sites A and B. The species at site D include *Abies balsamea, Rubus pedatus, Antennaria neglecta, Pyrola secunda,* and *Arnica cordifolia*.

Materials and Methods

At each site, a 20 × 20 m plot was established within which each tree greater than 1 m in height was recorded as to its species, location, vitality and diameter. Ten *Pinus contorta* (>10 cm DBH) were selected and age was determined on increment cores that passed through the trees' centre. Lower stratum vegetation description of each site was carried out by estimating the cover of each plant species in each of 20 randomly selected and permanently marked quadrants (1 × 1 m).

Ten *Pinus contorta* trees on the perimeter of each plot were selected and felled. Height and diameter (at 30 cm height) were recorded as well as the leader length interval for each of the past five years. The needles and stems of each of the past five years from each of the five youngest lateral branches in excess of five years old were collected. Needles were counted and stem length was measured. Dry weight of both stems and needles were recorded.

Five soil pits were dug at each site and both the description and classification were carried out according to the system of soil classification for Canada (Canada Soil Survey Committee 1978). Samples from each of the four horizons in the top 30 cm of each soil pit were collected. Moist subsamples were extracted with water or ammonium acetate (pH 4.8), at 1:25 moist soil to solution ratio. The pH of the horizons was determined on a saturated soil paste (Richards 1954) also using moist subsamples. The remaining portion of the samples was air-dried and ground to pass through a 100 mesh sieve and digested with $HNO_3 - HClO_3 - HF - HCl$ (modified from McQuaker *et al*. 1979). All solutions were analysed for Ca, Mg, K, Al, Fe, Mn, S and P by inductively coupled argon plasma atomic emission spectrometry (ICAP-AES).

Results and Discussion

The parent material of the four sites consists of a slightly calcareous glacial till overlain by a 20 cm aeolian veneer. The soils of sites A, B and D were classified as Brunisolic Gray Luvisols generally well to moderately well drained. Site C is imperfectly drained with gleying at 6-21 cm and was classified as a gleyed Podzolic Gray Luvisol. The surface texture of the surface mineral horizons ranged from silt loam to clay loam. All four sites had a 3 to 6 cm organic litter layer overlying the mineral soil.

The total soil sulphur data (Table 1) indicated that a large amount of elemental S had been deposited to soils in close proximity to the blocks (sites A and C). The total soil S values were much higher (56,950 and 36,710 $\mu g\ S\ g^{-1}$ soil; sites A and C, respectively) compared with the respective control soils (841 and 1231 $\mu g\ S\ g^{-1}$ soil; sites B and D). The water soluble soil S concentration of all four horizons at site A and C were also considerably higher than those at sites B and D (Table 1). The majority of total S in all horizons at the control sites appeared to be either organic or adsorbed since the water soluble S was negligible. At sites A and C, however 20 to 35 per cent of the total S in the LFH horizon and greater than 60 per cent in the mineral horizon (total S of the Ae_2 horizon of site A was only 40 per cent water soluble S) was water soluble.

Table 1. Total soil sulphur, water soluble sulphur concentrations and pH of the four surface horizons at selected sites at varying distances from sulphur dust sources

Site		pH[a]	Soil sulphur[b] ($\mu g.g^{-1}$) Total	Water soluble
A	(.05 km)			
	LFH	2.5(2.4-2.7)	56950 ± 10210	10550 ± 6370
	Ael	4.1(3.7-4.9)	225 ± 56	186 ± 83
	Bf	5.3(4.9-5.4)	258 ± 53	154 ± 74
	Ae2	5.1(4.9-5.5)	104 ± 33	37 ± 32
B	(9 km)			
	LFH	5.2(4.4-5.5)	841 ± 220	29 ± 4
	Ae	4.9(4.0-5.4)	90 ± 12	3 ± 2
	Bf	5.5(5.0-6.0)	100 ± 13	4 ± 6
	2Ae	5.2(5.1-5.5)	67 ± 12	1 ± 1
C	(0.2 km)			
	LFH	3.0(1.6-5.4)	36710 ± 9780	13110 ± 10160
	Ahe	4.1(3.8-4.5)	349 ± 100	479 ± 206
	2ABgj	4.5(4.4-4.6)	280 ± 103	300 ± 154
	2Bfgj	4.5(4.4-4.7)	207 ± 72	197 ± 125
D	(18 km)			
	LFH	4.5(4.3-4.7)	1231 ± 142	31 ± 17
	Ahe	4.7(4.5-5.1)	107 ± 39	0
	Bf	5.4(5.2-5.7)	113 ± 19	2 ± 3
	2Bm	5.5(5.2-6.0)	102 ± 42	0

[a] Values are average pH and the range.
[b] Values are means ± 95 per cent confidence limits.

Decreases of approximately two pH units in the LFH horizons of sites A and C were observed (Table 1). Smaller decreases in the pH (approximately 0.5 pH unit) of the Ae horizon of both sites A and C and the two lower horizons of site C were also observed. Increases in the water soluble S content and a decrease in the pH of the LFH horizon indicate the oxidation of elemental S, probably by microbial activity.

Total soil analysis for the various horizons at the four sites are given in Table 2. The total Ca concentrations were considerably higher in the LFH of sites A and C compared with B and D. The high levels of Ca (sites A and C) are the result of lime ($CaCO_3$) applications of approximately 2.5 tonnes ha^{-1} applied by aircraft in 1979 and 1980. The high concentrations of Ca in the LFH horizon along with the high concentrations of S suggest that $CaSO_4$ (gypsum) could be an important constituent of the LFH horizon at sites A and C. Total Mg, K, Mn, Fe and Al concentrations in the LFH horizons of sites A and C are lower than at their respective control sites. Studies in the literature suggest the addition of sulphate in the form of acid rain can have a potential effect on leaching (Cole and Johnson 1977; Mollitor and Raynal 1982) and ion translocation processes (Johnson and Cole 1980; Hovland *et al.* 1980). In addition, the high concentrations of Ca in the LFH could be important in the leaching of other cations from the LFH horizon. Excessive Ca in the LFH, as a result of liming, could result in the replacement of other cations on the exchange sites by Ca, subjecting them to leaching. No conclusive evidence is available however, and decreases in the nutrient content of the LFH horizon also may be a result of

Table 2. Total soil calcium, magnesium, potassium, manganese, iron and aluminum of the four surface horizons at selected sites at varying distances from sulphur dust sources. Values are means ± 95 per cent confidence limits.

	Total soil concentrations ($\mu g.g^{-1}$)					
Site	Ca	Mg	K	Mn	Fe	Al
A (0.05 km)						
LFH	11488 ± 6227	722 ± 434	1856 ± 639	649 ± 996	3532 ± 2321	7985 ± 4580
Ael	3276 ± 493	1801 ± 38	8615 ± 609	282 ± 183	7131 ± 1301	29867 ± 1881
Bf	2772 ± 429	2160 ± 202	7494 ± 442	183 ± 47	11934 ± 1568	36227 ± 4105
Ae2	1954 ± 95	2398 ± 227	8632 ± 209	80 ± 11	9440 ± 965	27019 ± 1742
B (9.0 km)						
LFH	6525 ± 2466	1759 ± 404	3003 ± 622	3145 ± 4733	5370 ± 2490	11831 ± 3842
Ae	2717 ± 245	1467 ± 198	8222 ± 561	264 ± 269	5929 ± 1003	24440 ± 1992
Bf	2242 ± 468	1738 ± 59	7180 ± 300	179 ± 58	9659 ± 882	28222 ± 7267
2Ae	2108 ± 333	2592 ± 1084	8187 ± 1116	127 ± 28	10108 ± 1032	26128 ± 5982
C (0.2 km)						
LFH	18714 ± 24557	577 ± 229	1035 ± 370	143 ± 81	2913 ± 1602	5711 ± 2979
Ahe	3058 ± 730	2168 ± 248	8172 ± 554	254 ± 230	8702 ± 1441	31904 ± 2752
2ABgj	1952 ± 899	2344 ± 391	7175 ± 344	209 ± 124	11404 ± 2002	32104 ± 6021
2Bfgj	1539 ± 151	2823 ± 305	7698 ± 272	130 ± 22	13953 ± 2852	30040 ± 2078
D (18 km)						
LFH	4192 ± 989	931 ± 215	1963 ± 95	706 ± 378	3669 ± 2946	6180 ± 2973
Ahe	1794 ± 377	1622 ± 567	6605 ± 598	172 ± 103	9278 ± 3056	25821 ± 5072
Bf	2085 ± 1766	1722 ± 182	6791 ± 1905	173 ± 92	11464 ± 6006	27127 ± 6502
2Bm	920 ± 270	2053 ± 117	6454 ± 394	119 ± 36	14721 ± 1965	25443 ± 3135

Table 3. Species number and cover in selected *Pinus contorta* stands in the vicinity of sulphur dust sources

Site	Stand Age (years)	Distance from source (km)	Number of species			Per cent cover		
			Shrubs	Herbs	Mosses	Shrubs	Herbs	Mosses
A	49	.05	7	11	4	5	10	1
B	53	9.0	8	8	6	6	7	16
C	76	0.2	7	10	1	7	4	0
D	75	18.0	3	11	7	3	7	18

Table 4. Growth of *Pinus contorta* at selected sites at varying distances from sulphur dust sources. Values are means ± 95 per cent confidence limits

Site	Tree		Terminal growth $cm \cdot year^{-1}$	Lateral branch	
	Age (years)	Height (m)		Growth $cm \cdot year^{-1}$	Needle production $g \cdot cm^{-1}$
A	49	17.5 ± 1.3	26.5 ± 7.3	7.8 ± 1.0	0.86 ± 0.08
B	53	15.3 ± 2.2	21.7 ± 7.6	5.5 ± 0.8	0.80 ± 0.06
C	76	20.3 ± 1.3	13.3 ± 2.9	5.3 ± 0.6	0.83 ± 0.10
D	75	17.9 ± 3.9	8.3 ± 4.3	3.5 ± 0.5	0.64 ± 0.11

a reduction in litter fall. Thus while the data (Table 2) indicated a decrease in the total concentration of Mg, K, Mn, Fe and Al, in the LFH horizon of the sites in close proximity to the S block the reasons for the decrease are largely unknown and are currently under investigation.

The effect of the elemental S dusting on the vegetation was dramatic particularly on the mosses (Table 3). Both the number of mosses and their cover were reduced as was the cover at site C as compared with site D. Site A showed a similar reduction of cover when compared with site B but there did not appear to be any effect on either the composition or cover of herbaceous species. The difference between the young (A and B) and old (C and D) sites in the response of the herbs appears to be related to a combination of the magnitude of deposition and the duration of exposure.

There did not appear to be any response of woody plants to the soil changes described above (Tables 3 and 4). None of the growth measurements taken on *Pinus contorta* gave any indication of an effect of elemental S addition. In fact, site D appeared to be much less productive with respect to tree growth than any of the other sites studied (Table 4). Also there were fewer species and a smaller cover of shrubs at site D than at site C (Table 3).

In general, even though there has been an almost total elimination of mosses at sites that have had significant elemental S deposition, there is no measurable response of woody species. It appears that before tree and shrub species are affected, soil chemical changes will need to be more intense or of longer duration.

Acknowledgements

This research project was funded by a contract between Canadian Forestry Service, Canterra Energy Ltd. and Gulf Canada Ltd. Technical assistance was provided by J. Ridgway, S. L'Hirondelle, F. Radford, D. Carnochan and F. Theriault.

Literature Cited

Canada Soil Survey Committee. 1978. The Canadian system of soil classification. Res. Branch, Can. Dep. Agric. Publ. 1646. Ottawa, Ontario.

Cole, D.W. and D.W. Johnson. 1977. Atmospheric sulfate additions and cation leaching in a Douglas Fir ecosystem. Water Resource Res. 13: 313-317.

Energy Resources Conservation Board. 1982. Sour gas processing in Alberta. A review of evidence presented at the recent ERCB hearings respecting the impacts and surveillance of sour gas plants. ERCB Report 82-D. Energy Resources Conservation Board, Calgary, Alberta.

Hovland, J., G. Abrahamsen and G. Ogner. 1980. Effects of artificial acid rain on decomposition of spruce needles and on mobilisation and leaching of elements. Plant and Soil, 56: 365-378.

Johnson, D.W. and D.W. Cole. 1980. Anion mobility in soils: Relevance to nutrient transport from forest ecosystems. Environ. International 3: 79-90.

McQuaker, N.R., D.F. Brown and P.D. Kluckner. 1979. Digestion of environmental materials for analysis by inductively coupled plasma-atomic emission spectrometry. Analytical Chem. 51: 1082-1084.

Mollitor, A.V. and D.J. Raynal. 1982. Acid precipitation and ionic movements in Adirondack forest soils. J. Soil Sci. Soc. Am. 46: 137-141.

Richards, L.A. (Ed.). 1954. Diagnosis and improvement of saline and alkali soils. U.S. Salinity Laboratory, U.S.D.A., Handbook 60.

FALLOUT AND NATURAL RADIOACTIVITY IN THE CANADIAN NORTHERN ENVIRONMENT

E.A.H. Benson, T.T. Lei, J. Svoboda and H.W. Taylor
University of Toronto

Abstract: Twenty years after the moratorium on atmospheric nuclear weapons testing (1963) a persistent radioactive burden of ^{137}Cs and other fission products can be detected on vegetation, and in peat and soils of the boreal and arctic zone in Canada. Although the levels of radioactive fallout were originally higher in southern Canada, at present the main zone of contamination in vegetation was identified to be between 60-70 latitude. All species tested displayed a bell-shape distribution of ^{137}Cs burden with dry habitat lichens showing the highest contamination. This paper will review the present status of our knowledge of this burden based on the literature, and a study of ^{137}Cs conducted by the authors along the latitudinal gradient from Riding Mountain National Park (50°N) to Alert, Ellesmere Island (82°N). It will also present information based on a recent survey of naturally occurring radioisotopes due to uranium mineralization areas in the Keewatin District, Northwest Territories. The objective of this project is to provide baseline data on the naturally occurring radioactivity in selected and arbitrary environments of the Canadian northern mainland.

The distribution and retention of long-lived fission products associated with fallout, particularly ^{90}Sr and ^{137}Cs, has been extensively studied in temperate regions during the 1960s when concern over human exposure to these radionuclides was widespread. During the same time period extensive work was carried out in Alaska and in Scandinavia in connection with the bioaccumulation of ^{90}Sr and ^{137}Cs in the short northern food chain (Hanson 1966, 1971, 1980; Hanson and Palmer 1965; Hanson and Eberhardt 1969; Blanchard and Kearney 1967; Martin and Koranda 1971; Svensson and Lidén 1965). Much less documentation exists for the Canadian north, although federal government agencies have continued to monitor levels of ^{137}Cs (Mohindra 1967) in selected sites. Deposition of ^{137}Cs and ^{90}Sr in the northern latitudes was only 25 to 50 per cent of the deposition measured in temperate regions (Collins 1963; Feely 1976). However concentrations of these fission products reported in native people whose diet largely consisted of caribou and reindeer were some 100 to 1000 times higher than in vegetation (Hanson 1980; Koranda and Martin 1971; Blanchard and Kearney 1967).

As part of our on-going research on the distribution and persistence of ^{137}Cs in vegetation and soils of the Canadian north, measurements have been taken in the boreal forest region along the southwest coast or Great Slave Lake, Northwest Territories. More extensive sampling of soils and vegetation has been carried out north of 60°N, particularly in the Keewatin District of the Northwest Territories and in the arctic islands.

Methods

Top soils (0-5 cm) and dominant plant species were collected from each site in sufficient volume to meet the requirements of the analytical procedure (Svoboda and Taylor 1979). The gamma ray counting system consisted of a high resolution (1.7 keV at 1332 keV) Ge

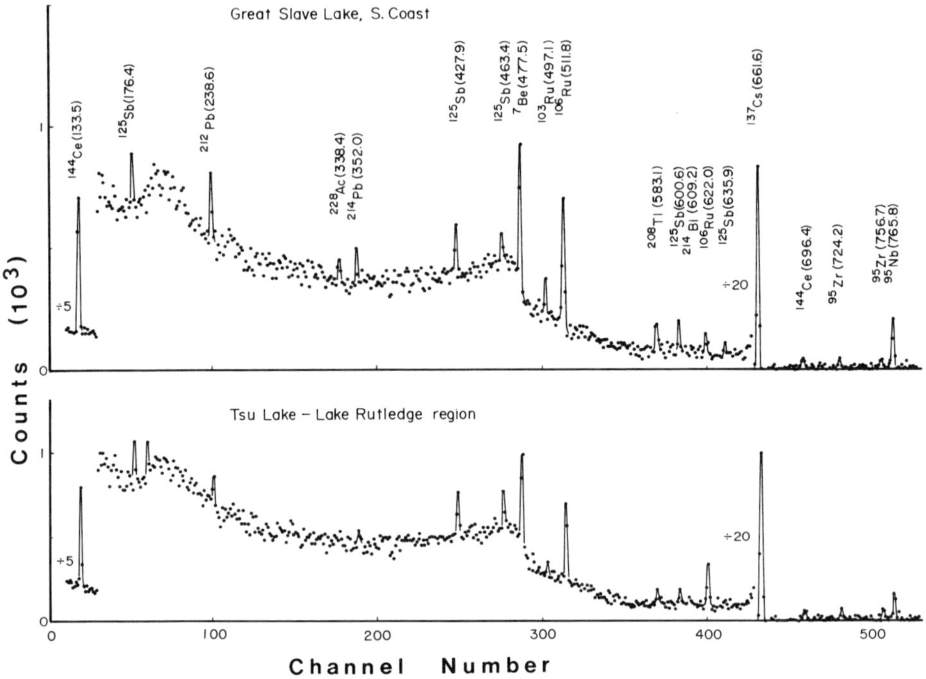

Figure 1. Gamma ray spectrum of composite vegetation collected on a transect along the south shore of Great Slave Lake, Northwest Territories (upper) and on a transect south of Rutledge Lake, Northwest Territories (lower). Both transects were sampled in August 1978.

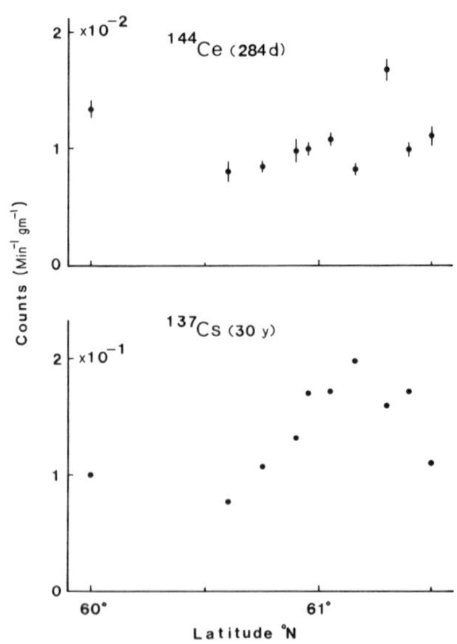

Figure 2. Distribution of ^{141}Ce (35 day) (upper) and ^{137}Cs (30 year) (lower) in vegetation along a Tsu Lake-Rutledge Lake transect at approximately 60°00′-61°30′N.

counter, a linear amplifier and either a 1024-channel or a 4096-channel pulse-height analyser. A tape recorder and a line printer were used for read out purposes. The gamma ray spectrum produced by the system consists of a number of sharp lines (one for each gamma ray) superimposed on a continuum which is produced through Compton scattering in the Ge counter. Since each sample is only weakly radioactive, a counting time of 12-24 hours is required to produce a spectrum with an acceptable statistical validity. The area of the gamma ray line corresponding to the strongest gamma ray emitted by a radioisotope is taken as a measure of the abundance of the isotope in the sample. For inter comparison of samples, areas are normalized to 1000 min of counting and 1 g of sample. ^{137}Cs is detected through a single gamma ray with energy 662 keV; ^{226}Ra is monitored by a 186 keV line, ^{214}Bi through a 609 keV line, ^{228}Th by means of a 583 keV line.

Results and Discussion

Fallout Radioactivity

Typical gamma ray spectra obtained with the Ge spectrometer for lichen species collected in the region of the boreal forest southwest of Great Slave Lake are shown in Figure 1. The line due to the 662 keV gamma ray emitted by the fission product ^{137}Cs (30.1 year) dominates the spectra. Short-lived radionuclides such as ^{144}Ce (284.4 day) ^{125}Sb (2.73 year) ^{7}Be (53.3 day) ^{103}Ru (39.6 day) ^{106}Ru (396 day) ^{95}Zr (65.5 day) and ^{95}Nb (35.1 day) contribute to the spectra also, although their associated activities represent only a small fraction of that due to ^{137}Cs. (Note that the ^{137}Cs peak is divided by 20.) The contributions to the spectra from the short-lived fission products will be essentially zero after a period of about 1-2 years. Naturally-occurring isotopes such as ^{228}Th (1.91 year) and ^{226}Ra (1600 year) are also present in lichens collected from the boreal forest and may indicate uptake of these radionuclides from the substrate. These spectra were obtained a few months after a Chinese nuclear test and as a consequence, the gamma rays from short-lived isotopes are readily observed. The spectra normally encountered contain a fallout contribution from ^{137}Cs only.

Figure 2 shows the variation with distance of the concentration of ^{144}Ce and ^{137}Cs along a transect which is approximately parallel to the south shore of Great Slave Lake for a distance of 170 km. The concentrations of these isotopes vary no more than $\pm$25 per cent for ^{144}Ce and $\pm$45 per cent for ^{137}Cs. The upper panel of the figure shows that fallout is deposited relatively uniformly over short distances. The increase of the ^{137}Cs content in lichens along the transect may be due to local variation. Our general observations on latitudinal distribution of radionuclides, however, indicate that the noted increase of $\sim$ 50 per cent in ^{137}Cs might be due to latitudinal effect (1.5° change) along the Fort Smith – Lake Rutledge transect. This is in accord with our findings within the Churchill, Manitoba – Ellesmere Island corridor (Figure 3), described below.

For many years, lichens have been recognized as ideal indicators of airborne contaminants. The persistence and longevity of these primitive plants, and their dependence on airborne nutrients, accounts to a great extent for their ability to intercept and retain long-lived radionuclides. It occurred to us that a systematic inventory of ^{137}Cs in the Canadian north could make a valuable contribution to the base line data in uranium prospecting and sites exploration prior to development of northern uranium reserves, and to our knowledge of the long term fate of fallout ^{137}Cs in the Arctic.

The latitudinal distribution of ^{137}Cs in top soil and vegetation was determined using the

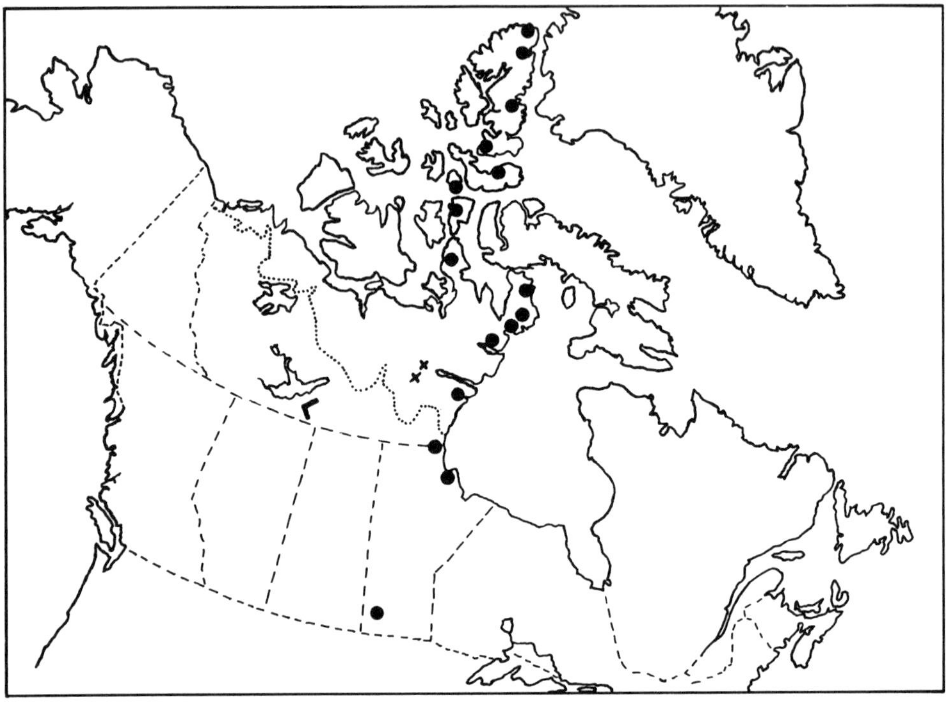

Figure 3. Sampling corridor extending from 50°N to 82°N. Individual sites are shown as full circles. Crosses indicate the uranium outcroppings sampling sites. The darkened areas represent two sampling transects near the south shore of Great Slave Lake. The boreal forest-tundra transition is shown as a dotted line.

sampling sites between 58°N and 82°N, which are shown in Figure 3. At each site dominant lichen and vascular species were collected and analyzed for ^{137}Cs in the Toronto laboratory. The results shown in Figure 4 indicate that the latitudinal distributions of ^{137}Cs in the fruticose lichen *Alectoria nigricans* and *Polytrichum juniperinum* had well defined maxima at latitude of 60 to 70°N. The dashed curves of the figure indicate the trends of the observational points. They were determined by the least squares fitting of a Gaussian distribution on a continuum of arbitrary slope. All distributions were treated in this manner, however, the dashed curves should not be interpreted literally. The bell shaped distribution was approximately representative of all lichens measured in the study. Arctic vascular species growing in the same habitat tended to have less ^{137}Cs activity than lichens and exhibited maximum ^{137}Cs activities per unit area at the more southerly regions of the latitudinal transect. As further examples, the latitudinal distributions for *Silene acaulis* and the total vegetation are shown in Figure 5. Despite the higher initial deposition of ^{137}Cs at the southern end of the transect, it is apparent that the vegetation has retained proportionately less here than the same species and plant communities sampled in the mid-Arctic. In the sub and low Arctic, the vegetation has been 'cleansed' from the originally deposited ^{137}Cs and the radionuclide can now be identified prevalently in soils and in sediments. This is even more true for the temperate environments which have a more vigorous biological turnover. The highly conservative nature of arctic plant communities allows intercepted ^{137}Cs to remain locked into thalli of the slow growing persistent lichens and the shoots of vascular plants with less opportunity for movement of the radionuclide into the abiotic components of the ecosystem. The slow growth rate,

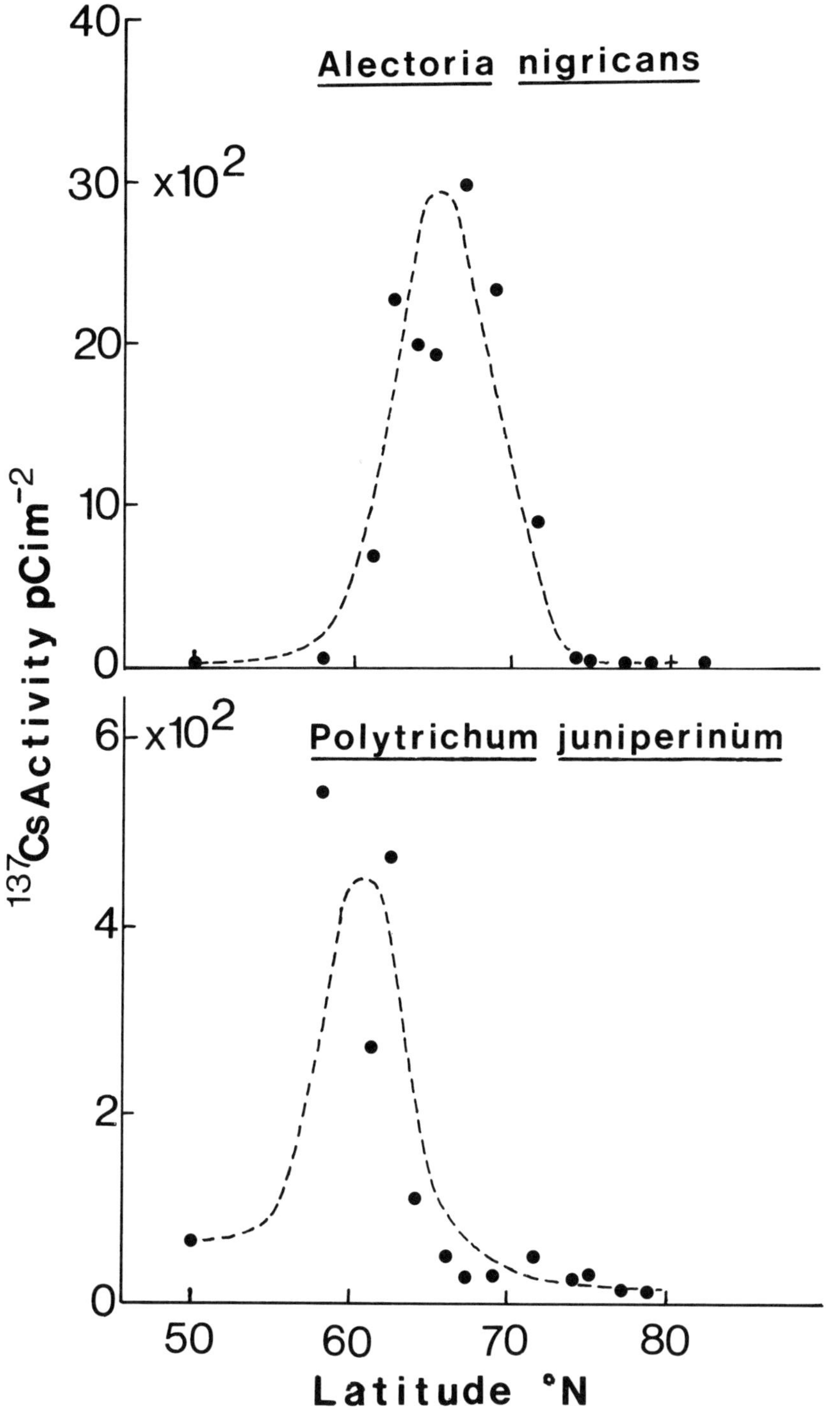

Figure 4. Latitudinal distribution of ^{137}Cs in the lichen *Alectoria nigricans* and the bryophyte *Polytrichum juniperinum* (units are picocuries per m^2).

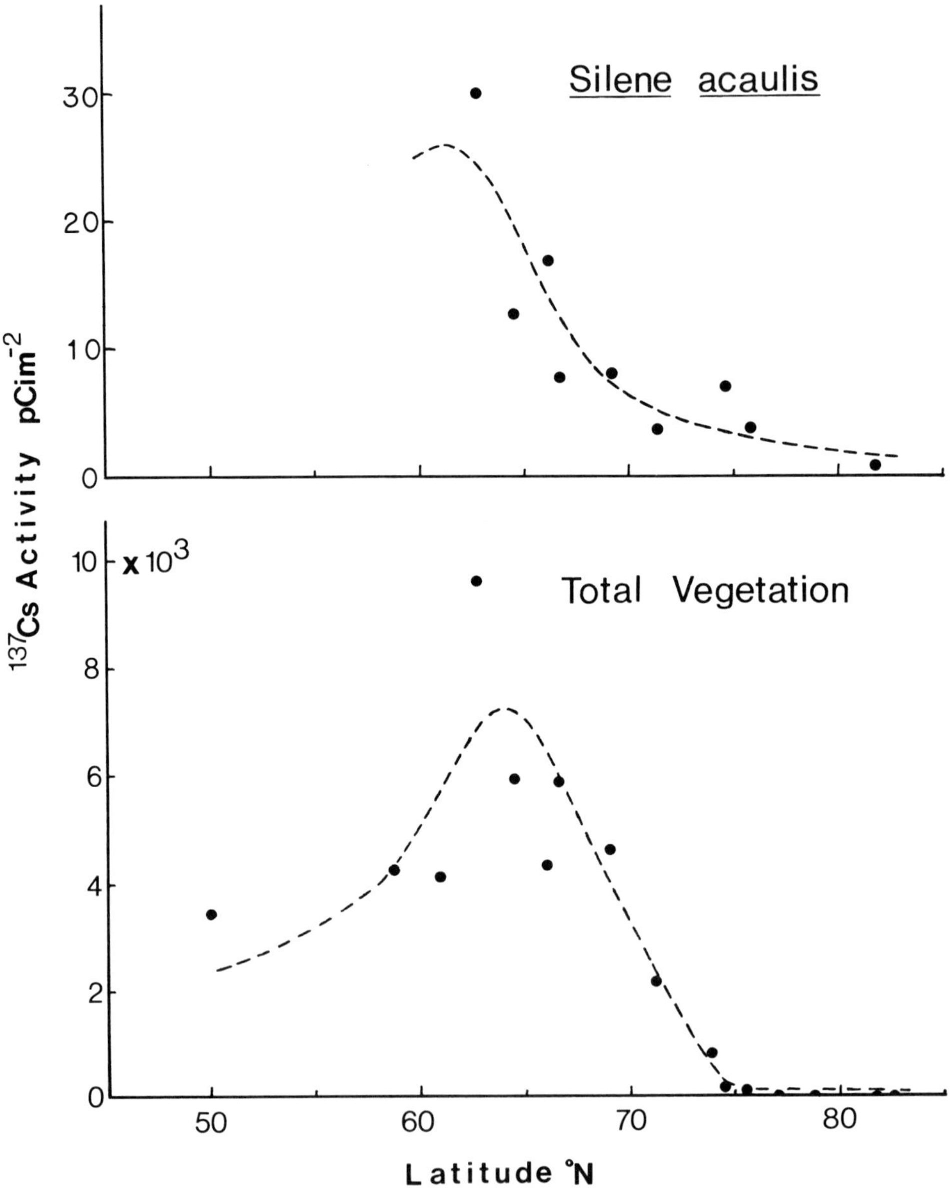

Figure 5. Latitudinal distribution of ^{137}Cs in the cushion plant, *Silene acaulis* and the total dry habitat vegetation extending from 50° N to 82° N.

cushion-like form, large surface area and absence of a cuticle are important factors which enhance the ability of lichens to intercept and retain fallout ^{137}Cs (Benson *et al*. 1983).

North of 70° N much less ^{137}Cs was deposited in comparison with lower latitudes. On the contrary, in large areas of the high Arctic the paucity of vegetation coverage greatly

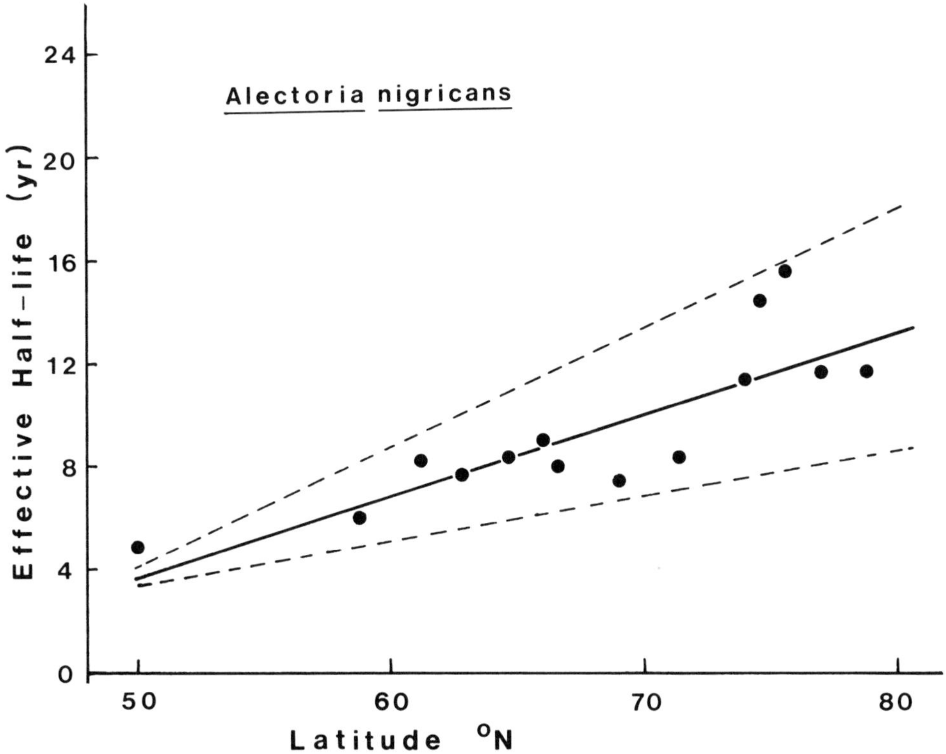

Figure 6. Plot of the effective half-life (T_{eff}) versus latitude for the lichen, *Alectoria nigricans*. The solid line in the figure represents the result of a regression analysis to the data points assuming a linear relationship, between T_{eff} and latitude. The dotted lines give an estimate of the experimental errors.

reduced the possibility of interception of airborne fallout by plants. As a result, most of the ^{137}Cs was deposited on bare ground and persists in the soil.

Feely (1976) has discussed the amount and distribution of the accumulative fission products in northern latitudes from nuclear explosions. If the assumption is made that most of the ^{137}Cs was deposited during the period 1955-1963 (i.e., that the amount of ^{137}Cs from post − 1963 atmospheric tests is negligible) then one can use the latitudinal distribution of ^{137}Cs for a species, such as *Alectoria nigricans*, to estimate the effective half-life (T_{eff}) of ^{137}Cs in the species. The effective half-life of a radioisotope is the time period during which its original quantity is reduced to half due to a combined effect of its radioactive decay and physical removal. (Taylor *et al*. 1983). Figure 6 shows the effective half-life vs. latitude regression for *Alectoria nigricans*. The solid line was obtained by a least-squares fitting of the data to the equation of a straight line. The points are subject to fairly severe fluctuations. The dashed lines indicate the variation in the fitting if the errors (~50 per cent) on the parameters are taken into consideration. The T_{eff} measurements for a typical arctic vascular plant such as *Dryas integrifolia* are lower than the result obtained for the lichen *Alectoria nigricans*.

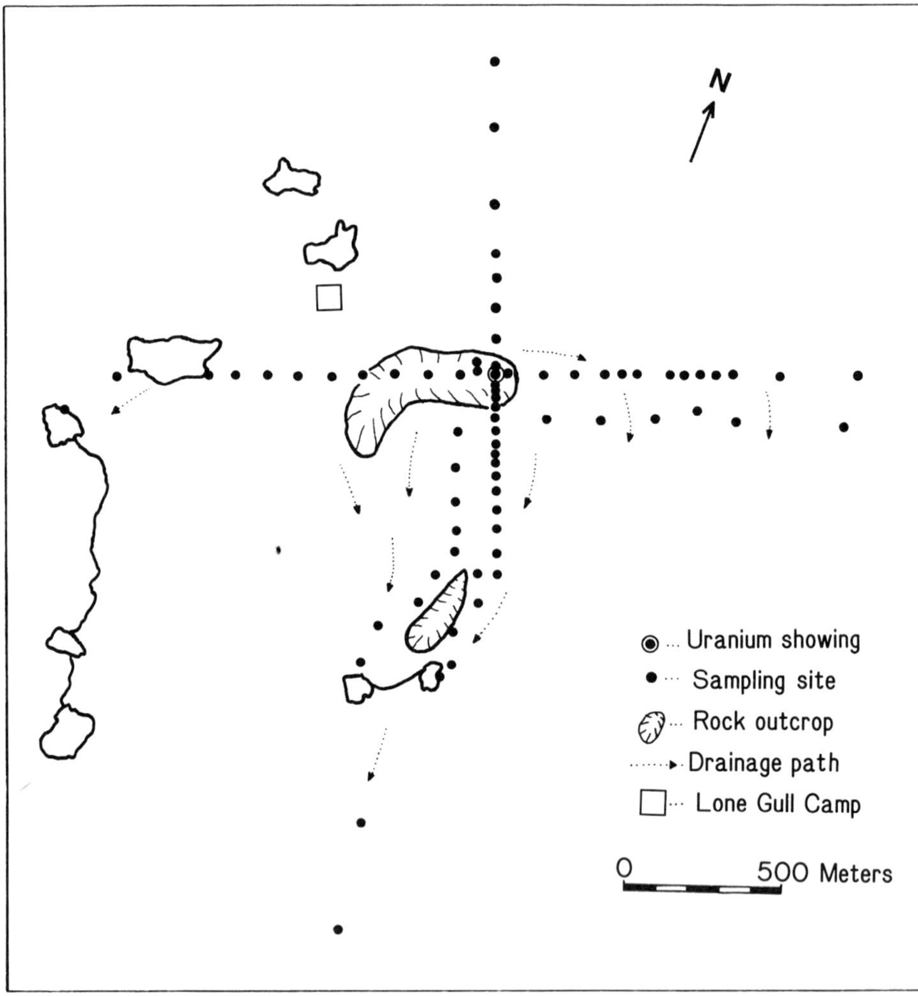

Figure 7. Sampling matrix at the Lone Gull (Urangesellschaft Camp) site 64°27′N, 97°39′W, Keewatin District, Northwest Territories. The uranium mineralization "showing" is at the center of the sampling grid.

Naturally-occurring radionuclides

Evidence of radioactivity in vegetation, peat and mineral substrate due to uranium mineralization outcroppings in the central regions of the Keewatin District is presently being investigated by the authors. Two uranium mineralization sites have been sampled: one, which has a visible outcropping is located at Lone Gull Camp (64°27′N, 97°39′W) which is operated by Urangesellschaft Canada. The second is located at Nogash Camp (62°37′N, 98°40′W) and was operated by Pan Ocean Oil Ltd. Figure 7 shows the sampling matrix that was set up at the Lone Gull site during the summer of 1981. Figure 8 shows the distribution of sampling sites near the Nogash Camp. Sampling points represented in both figures were chosen to emphasize the mineralization zone but were extended into the surrounding area also to include the drainage path associated with the

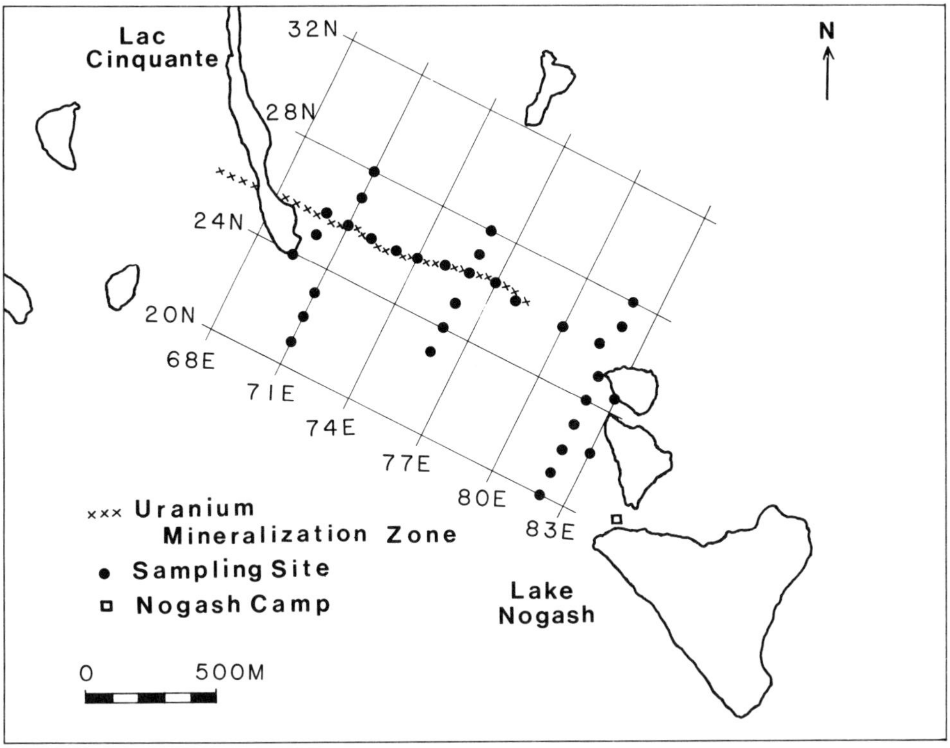

Figure 8. Sampling matrix at the Nogash Camp site (Pan Ocean Oil Ltd.), 62°37′N, 98°40′W.

zone. Gamma ray analyses of the samples collected from the mineralization outcroppings have low concentrations of the following: ^{214}Bi which is a daughter product of ^{238}U and may thereby be an indicator of the abundance of ^{238}U-^{226}Ra in the sample, and ^{137}Cs, present in arctic environments as an anthropogenic contaminant. Figure 9 shows that the uranium activity in the vegetation (top panel) is high close to the uranium outcropping (showing), and maintains elevated activities at distances up to 600 m from the showing. This suggests that plants are accumulating uranium from the substrate. The concentrations of ^{225}Ra, ^{214}Pb, ^{214}Bi, ^{137}Cs and ^{40}K in plant species sampled in the vicinity of the uranium ore outcropping are given in Table 1. The ^{40}K concentration is essentially independent of the plant species but the concentrations of other isotopes show strong variations among the species. The preliminary data of Table 1 show considerable variation in the uptake of radium among the various species with *Carex* and moss species accumulating the largest burdens. On the other hand, lichens are the most heavily contaminated plants as far as fallout ^{137}Cs is concerned.

More pronounced fluctuations of the radionuclide burdens are observed for the peat and soils of the southern sampling sites (centre and bottom panels). These fluctuations reflect the topography and drainage pattern of the immediate area. The peaks observed in the peat samples likely represent local depressions that act as sinks for the radionuclides transported by drainage water from the showing and the upstream watershed. The corresponding results for the Nogash Camp site are shown in Figure 10. In this case, the absence of a distinct uranium ore outcrop results in a less dramatic variation in counting

Table 1. The relative intensity of the signature gamma ray of selected radionuclides in the plant taxa growing in the vicinity of the uranium outcropping. Intensity values are expressed in counts/1000 min./g (dry weight)

Species	^{226}Ra (186 keV)	^{214}Pb (352 keV)	^{214}Bi (609 keV)	^{137}Cs (662 keV)	^{40}K (1460 keV)
Lichens	58.4±2.6	47.5±1.9	25.0±2.2	54.2±1.3	1.4±1.3
Mosses	299.0±3.6	262.8±2.4	152.0±2.2	27.5±1.1	0.9±1.0
Grasses	175.0±2.5	113.2±1.6	61.6±1.7	3.9±0.7	1.8±0.9
Carex spp.	340.7±6.8	499.3±5.7	290.2±5.4	13.4±2.3	2.7±2.2
Salix spp.	145.7±2.3	118.8±1.5	68.1±1.5	3.9±0.7	1.6±0.8
Betula glandulosa	114.6±2.1	151.6±1.7	88.9±1.7	4.1±0.7	1.4±0.8
Oxytropis maydelliana	104.1±2.6	209.4±1.7	123.1±1.6	9.3±0.7	2.5±0.8
Arctostaphylos rubra	128.8±2.8	65.0±1.9	39.3±1.9	8.8±1.1	1.8±1.2
Vaccinium uliginosum	40.0±2.1	25.4±1.1	14.0±1.3	5.9±0.7	3.1±0.9
Vaccinium Vitis-idaea	163.9±3.4	108.2±2.2	61.9±2.3	6.3±1.0	3.0±1.5
Ledum decumbens	192.6±2.2	128.5±1.5	73.0±1.4	6.8±0.6	1.7±0.7
Empetrum nigrum	64.8±1.5	48.4±1.0	26.9±1.1	9.4±0.6	0.9±0.6

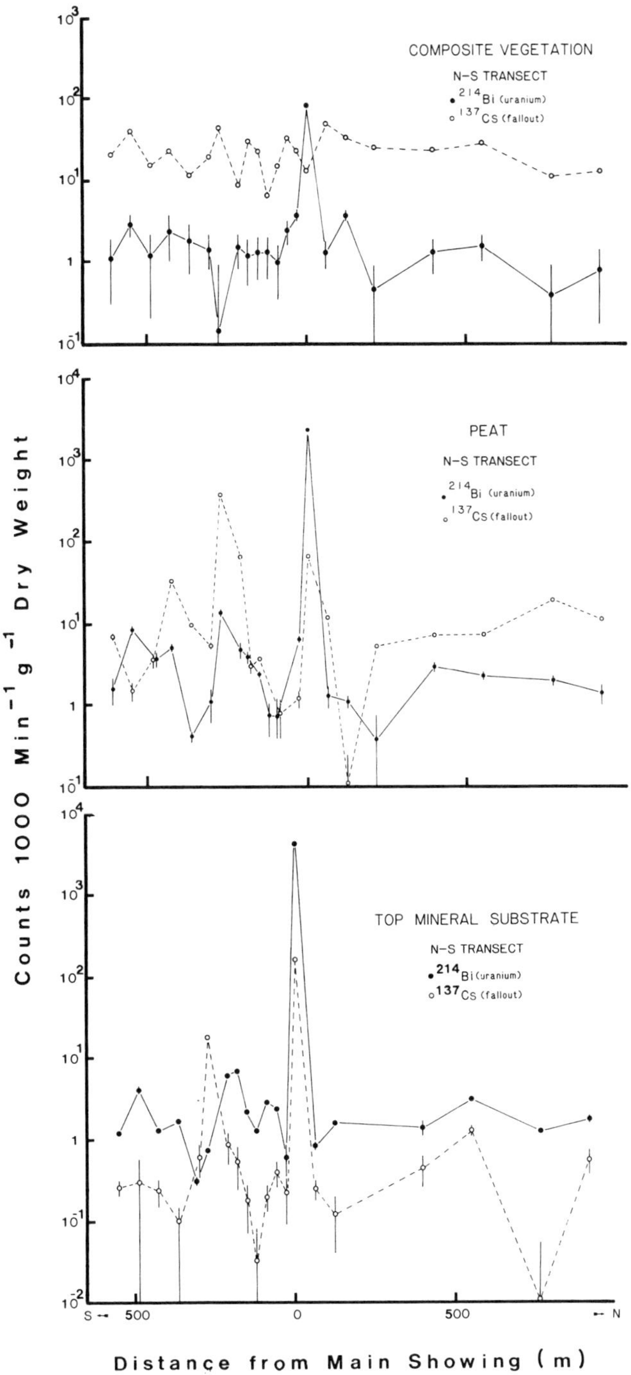

Figure 9. Concentration of ^{238}U-^{226}Ra indicator ^{214}Bi and fallout indicator ^{137}Cs in the composite vegetation, peat and top mineral substrate along the N-S transect at Lone Gull Camp site, shown in Figure 7.

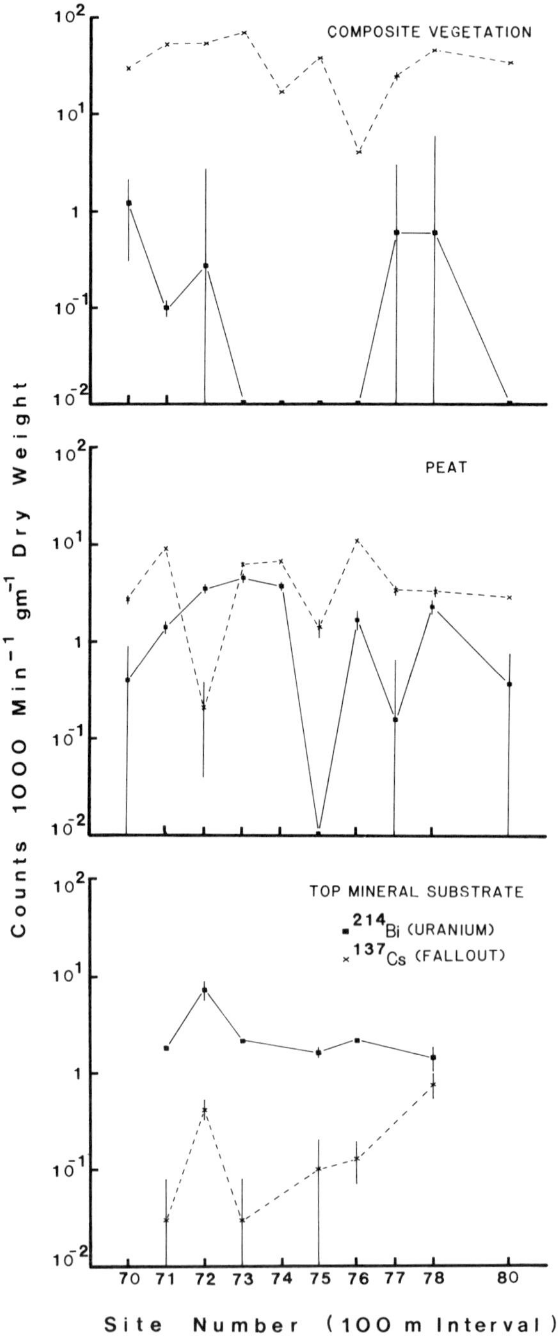

Figure 10. Concentration of ^{214}Bi and ^{137}Cs in the composite vegetation, peat and top mineral substrate along the transect which runs parallel to the mineralization zone near the Nogash Camp site shown in Figure 8.

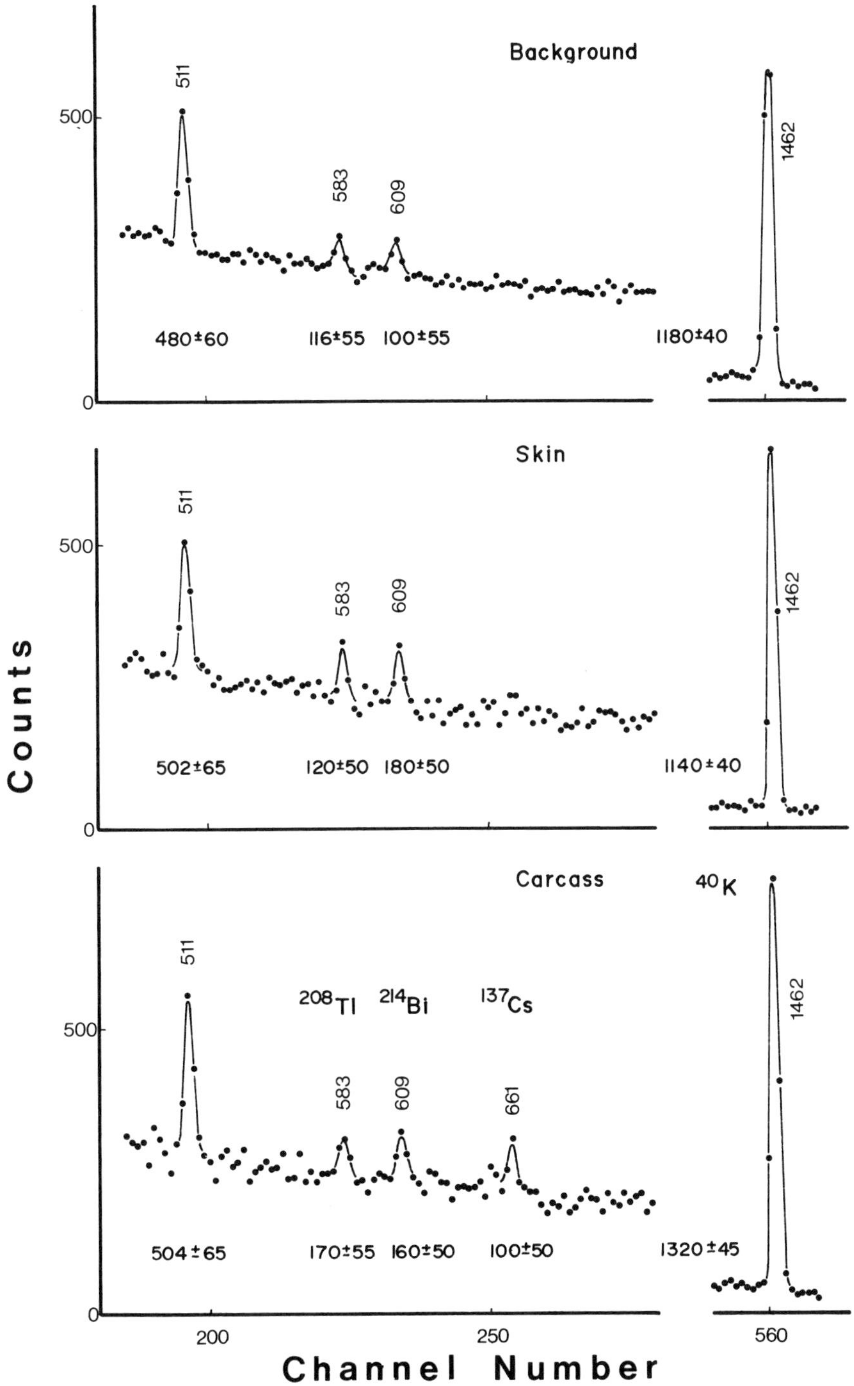

Figure 11. Gamma ray spectrum of fallout ^{137}Cs and naturally-occurring radioisotopes (^{208}Tl, ^{214}Bi and ^{40}K) in the skin and carcass of arctic ground squirrel (*Citellus parryi*) burrowing in the immediate vicinity of the radioactive outcropping as compared to laboratory background.

rate along the transect shown in Figure 10. The behavior of ^{137}Cs is the same for both sites, however.

As an ancillary experiment to our pilot investigation of 1981, one Arctic ground squirrel (*Citellus parryi*) burrowing in the immediate vicinity of the radioactive outcropping, was sacrificed, dissected, dried and analysed for ^{214}Bi, ^{208}Tl (a thorium "indicator") and ^{137}Cs. Measurements of the skin and carcass showed levels of ^{214}Bi and ^{208}Tl not significantly different from background levels (Figure 11). However, the ^{137}Cs activity was slightly elevated above the background which indicates of some bio-accumulation of the fallout radioisotope due to consumption of contaminated vegetation. Although no definite conclusion can be made on the basis of one analysis, it nevertheless appears that uranium daughter products do not concentrate readily in *C. parryi* tissue, which support other evidence for the rather localized effects of the outcropping. More study is necessary in order to endorse these preliminary findings. Intensive sampling of vegetation, peat and soils of the mineralization outcroppings and arbitrarily chosen localities was completed during the summer of 1982. The new analyses should clarify the trends mentioned briefly here.

Even though global fallout deposition is greatly reduced today compared to the late 1950s and early 1960s, there is a need to monitor radionuclide deposition and retention in northern environments. Our observations show that conservative growth rates and slow biological turnover, limit the effectiveness of natural dissipative processes which have been mainly responsible for the dilution of ^{137}Cs and other long-lived radionuclides in temperate environments. It is of some importance that present levels of radionuclides present in the Canadian north be systematically monitored in order that prudent decisions can be made in the future concerning the development of the north within the boundaries of health protection and environmental integrity.

Acknowledgements

The authors wish to thank the Arctic Land Use Research (ALUR, Northern Environmental Protection Branch of the Department of Northern and Indian Affairs) and the Natural Sciences and Engineering Research Council for financial support of this work.

Literature Cited

Benson, H.E.A., J. Svoboda and H.W. Taylor. 1983. The latitudinal inventory of ^{137}Cs in vegetation and top soil in Northern Canada 1980. In preparation.

Blanchard, R.L. and J.W. Kearney. 1967. Natural radioactivity and ^{137}Cs in Alaskan caribou and reindeer samples. Envir. Sci. and Tech. 11: 932-939.

Collins, W.R., Jr. 1963. Cs-137/Sr-90 Ratio in Fallout in the United States. *In:* Fallout Program Quarterly Summary Report, AEC Report HASL-132. New York. pp. 247-248.

Feely, H.W. 1976. Worldwide deposition of strontium-90 through 1975. *In:* Health and Safety Laboratory Quarterly Report, HASL-308. New York. pp. 121-137.

Hanson, W.C. 1980. Transuranic elements in arctic tundra ecosystems. *In:* Transuranic elements in the Environment. W.C. Hanson (Ed.). U.S. Dept. of Energy Rep. TIC-22800. Springfield, Virginia.

Hanson, W.C. 1971. Fallout radionuclide distribution in lichen communities near Thule. Arctic, 24: 269-2176.

Hanson, W.C. 1966. Fallout radionuclides in Alaskan food chains. Amer. J. Veterinary Res. 116: 359-366.

Hanson, W.C. and L.L. Eberhardt. 1969. Effective half-times of radionuclides in Alaskan lichens and Eskimos. *In:* Symp. radioecology, Proc. 2nd Natl. Radioecology Symp. D.J. Nelson and F.C. Evans (Eds.). Ann Arbor, Michigan. (U.S. Atomic Energy Commission Rep. CONF. 670503).

Hanson, W.C. and H.E. Palmer. 1965. Seasonal cycle of ^{137}Cs in some Alaskan natives and animals. Health Physics, 11: 1401-1406.

Koranda, J.J. and J.R. Martin. 1971. Gamma-emitting radionuclides in Alaskan environments 1967-1970. *In:* Radionuclides in Ecosystems. D.J. Nelson (Ed.). Proc. of 3rd Natl. Symp. on Radioecology. Vol. 1. pp. 81-107.

Martin, J.R. and J.J. Koranda. 1971. Recent measurements of cesium-137 residence time in Alaskan vegetation. *In:* Radionuclides in Ecosystems. D.J. Nelson (Ed.). Proc. of 3rd Natl. Symp. on Radioecology. Vol. 1, May 1971. pp. 108-122.

Mohindra, V.K. 1967. Cesium-137 in the Canadian north. Acta Radiologiae Therapy Physics Biology, 6: 481-490.

Svensson, G.K. and K. Lidén. 1965. The quantitative accumulation of ^{95}Zr + ^{95}Nb and ^{140}Ba + ^{140}La in carpets of forest moss — a field study. Health Physics, 11: 1033-1042.

Svoboda, J. and H.W. Taylor. 1979. Persistence of Cs-137 in Arctic lichens, *Dryas integrifolia* and lake sediments. Arctic and Alpine Research, 11: 95-108.

Taylor, H.W., H.E.A. Benson and J. Svoboda. 1983. Effective half-life of fallout ^{137}Cs in vegetation in the Canadian Arctic. (In preparation)

PROVINCIAL MINISTRIES OF NORTHERN AFFAIRS: A COMPARATIVE ANALYSIS

Geoffrey R. Weller
Lakehead University

Abstract: The paper presents a comparative analysis of the northern ministries of primarily Ontario, Manitoba and Saskatchewan as well as of the Secretariat des Activités Gouvernementales en Milieu Amerindien et Inuit (S.A.G.M.A.I.) in Quebec. The paper begins with an analysis of the old provincial north and the reasons for the formation of the northern ministries. Then a comparison is made of the policy objectives and consequent range of responsibilities of the agencies that were formed through the four kinds of bureaucratic responses to the changes in the north during and after World War II. This involves, among other things, a comparison of staffing in terms of size, background and location. The next section of the paper deals with the relations of the northern agencies with both vertical and horizontal ministries within each province and with the relevant federal agencies. This will be followed by an analysis of the relations of the agencies with regional pressure groups. The paper concludes with a look at the policy outputs and problems faced by these relatively new structures and makes some suggestions as to what the future might hold.

The land between the Yukon and the Northwest Territories and the populated south, the Canadian provincial north, is a truly forgotten area. It has little or no place in the public consciousness. It has not been, until recently, very important to governments or bureaucracies either at the federal or the provincial level. Moreover there has never been any general agreement on the territorial extent of the provincial north. The definition used here is thus necessarily stipulative. I have chosen to use political boundaries, largely those established by the provinces themselves. This is partly because of the simplicity of this approach and partly because the paper deals with political, local government and bureaucratic matters for which political boundaries have real meaning.

In Ontario, Manitoba and Saskatchewan the north is defined, for the purposes of this paper, by the boundaries set by the statutes establishing the departments of northern affairs in those provinces[1]. In Alberta, the north is taken to be the area for which the Northern Alberta Development Council is responsible[2]. In Quebec, the north is taken to be the area covered by the James Bay and Northern Quebec Agreement[3], while in Newfoundland it is taken to be Labrador. There is no truly satisfactory way of precisely

1. See Ontario, The Ministry of Northern Affairs Act, 1977 (1st Session, 31st Legislature, Ontario, 26 Elizabeth II, 1977); Saskatchewan, The Department of Northern Saskatchewan Act, 1972; Manitoba, The Northern Affairs Act, 1974 (S.M. 1974, c56 – cap. N100).

2. See Northern Alberta Development Council, Five Year Review of Activities, April 1, 1973 – March 31, 1978, Edmonton: Department of Business Development and Tourism, 1978, especially the map on page 13.

3. Quebec, The James Bay and Northern Quebec Agreement, Quebec City: Editeur Officiel du Quebec, 1976.

determining the north in British Columbia for although the province is organised into regional governments, there is no clear agreement on which of these are truly northern and which are not. It should be noted here that northern Ontario has always had a much larger population and been more developed than the other provincial 'norths'. These days it is the 'remote' north of Ontario that is comparable to the other provincial norths. The remote north is normally regarded as being the area north of the Canadian National Railway line, namely the area that is now of concern to the Fahlgren Commission[4].

Thus the definition of the provincial north used here deliberately avoids using concepts of the 'north' or 'mid north' that do not take account of political reality. I am thus avoiding the definition of the 'mid north' given by the Canadian Council on Rural Development[5], or the definition of 'Mid-Canada' given by Richard Rohmer and Acres-Bechtel[6]. More especially I wish to avoid the complicated definitions of north and 'nordicity' suggested by Hamelin[7]. I am also avoiding the biological 'boreal forest' definition[8].

The paper will proceed by analysing the nature of what can be termed the old provincial north and then the nature of the changes that have occurred in recent decades. This will be followed by a discussion of the bureaucratic responses these changes engendered. After an analysis of the responsibilities and priorities of the new bureaucratic structures, the paper will conclude with an assessment of their problems and future prospects.

The Old Provincial North

Until very recently the provincial norths were social and economic backwaters. Their populations were extremely small numbering only a few thousand or tens of thousands. Moreover these few people lived in tiny scattered pockets along the major transportation routes or in extremely remote areas. The scattered nature of the population weakened regional consciousness and cohesiveness. In addition these small populations were very variegated racially and ethnically. The provincial norths contained a much higher proportion of natives to non-natives than was the case in the south and the native population was divided between status Indians, non-status Indians, Metis and, in some cases, Inuit. The white population of the provincial norths was also ethnically very mixed.

The economies of the old provincial norths were based on the traditional pursuits of hunting, fishing and trapping. There was some tourism but this was a seasonal and low wage industry that did not employ large numbers of people. The income from the traditional pursuits was insignificant and a great deal of poverty, even extreme poverty,

---

4. See Ontario, Royal Commission on the Northern Environment, Terms of Reference, Objectives, Work Program (2nd edition, July 1979), Toronto: Royal Commission on the Northern Environment, 1979, especially pp. 3-5.

5. Canadian Council on Rural Development, A Development Strategy for the Mid North of Canada, Ottawa: CCRD, 1976, especially the map on page 15.

6. See, for example, Richard Rohmer, The Green North: Mid Canada, Toronto: Maclean-Hunter, 1970 and Acres Research and Planning Ltd., Mid Canada Development Corridor: A Concept, Toronto: Acres Research and Planning Ltd., 1969.

7. Louis-Edmond Hamelin, Canadian Nordicity, Montreal: Harvest House, 1978.

8. As used by, for example, the International Symposium on the Dynamics of Boreal Forest Ecosystems, Thunder Bay, August 1982.

resulted. Most of the population became dependent on government handouts. In essence the old provincial north could not be described as being part of the normal wage economy of the country.

The general conditions of life in the old provincial north of the traditional economy were not good. Apart from the difficulties presented by a harsh climate, the populations remained isolated, often extremely so, as a consequence of large distances combined with very poor transportation and communications facilities. Poverty, remoteness and neglect by government resulted in truncated or non-existent local government structures and poor servicing in all respects. Most communities in the remoter areas had no clean water supply and no roads let alone sewers or proper garbage disposal facilities. Moreover health, educational and other social services were very badly neglected in comparison with southern regions.

The federal government's concern for the provincial north really only began during World War II, but continued throughout the cold war era, largely as a result of the need for the construction and maintenance of defence sites of various types across the north. This was followed by a general concern for the basic health and welfare of northern residents which flowed largely from the federal government's responsibility for status Indians and its decision to treat Inuit in the same manner as status Indians. This led to a much more visible federal rather than provincial government presence in the provincial norths. The Medical Services Branch nursing station, the federal schools, the Department of Transport and Department of National Defence establishments and the Royal Canadian Mounted Police posts were all symbolic of this. In many ways the provincial norths, perhaps especially northern Quebec, were more like extensions of the Northwest Territories than parts of their respective provinces[9].

The provincial governments rarely showed any political or bureaucratic interest in their northern regions until very recently. This is reflected in the poverty of bureaucratic agencies intended to deal specifically with the north before the 1970s. When an interest was expressed it largely related to a concern for the development of natural resources, both renewable and non-renewable. This concern was essentially economic and not social and was related largely to the potential exploitation of resources for the needs of the south and not the needs of the north. This is clearly revealed when the precursors of the present provincial northern agencies are surveyed.

Some provincial concern for northern Ontario had been expressed in bureaucratic structures ever since 1911 when Premier Whitney established a Northern Development Branch within the Department of Lands and Forests. The purpose of this branch was, essentially, to encourage and aid settlement. In 1926, the Branch became a full Department of Northern Development but had no separate minister. In 1937, the Department was merged with the Department of Highways. Only in 1970 was the north again given distinctive bureaucratic recognition. In that year, the Department of Mines became the Department of Mines and Northern Affairs and twenty-four information offices were established throughout the north. These offices were taken over in 1972 by the newly created Ministry of Natural Resources[10].

---

9. See Hamelin *op. cit.*, Chapter 4.

10. See M. Zaslow, The Opening of the Canadian North 1870-1914, Toronto: McClelland and Stewart, 1971; H. V. Nelles, The Politics of Development: Forests, Mines and Hydro-electric Power in Ontario, 1849-1941, Toronto: MacMillan, 1974, and G. Brock, The Province of Northern Ontario, Cobalt: Highway Book Shop, 1978.

The Manitoba Northern Affairs Act passed by the legislature in 1966 established the Office of the Commissioner of Northern Affairs. This remained a very small unit even though it was expanded somewhat in the 1960s. The Commissioner was located in Winnipeg, a director in Thompson and two co-ordinators in Lac du Bonnet and The Pas. The Commissioner approach was dubbed a 'colonial' operation by the opposition New Democratic Party[11]. It was established largely because of pressure for basic services from northern residents and because there were problems within the Conservative Party concerning the lack of attention paid to the north, especially with regard to economic development[12].

Quebec did not possess a northern agency until 1962 when the Direction General du Nouveau Quebec (D.G.N.Q.) was established by René Lévesque, then the Minister of Natural Resources. The main purpose of the D.G.N.Q. was to establish a Quebec presence in an area largely dominated by the federal government, but beyond that its emphasis was on economic development. René Lévesque was apparently shocked upon becoming minister to discover that there had been so little provincial interest in the area that it was essentially English speaking and influenced mainly by the federal government. The D.G.N.Q. began as quite a large organisation under Eric Gourdeau with line responsibilities for service delivery. Apart from conflicts with the federal authorities, the D.G.N.Q. came into conflict with the traditional line departments of the government of Quebec, especially Health and Natural Resources. The James Bay and Northern Quebec Agreement of 1975 really pre-empted most of the economic and many of the other roles of the D.G.N.Q. and in 1977 it disappeared to be replaced, in effect, by the Secretariat des Activités Gouvernementales en Milieu Amerindien et Inuit (S.A.G.M.A.I.) which, as the name suggests, has a relatively restricted mandate[13].

The other provinces showed even less interest in their northern regions. A great deal of thought was given to the creation of a Ministry of Northern Affairs in Saskatchewan beginning in 1947 and eventually a northern affairs branch was created within the Department of Natural Resources but it was not very large or influential[14]. Discontent in northern Saskatchewan led the Liberals under Premier Thatcher to establish a Task Force on Northern Affairs in the early seventies. The Liberals were voted out of office before they had a chance to implement any changes but when the New Democratic Party succeeded them they established the Department of Northern Saskatchewan (D.N.S.). Newfoundland created a separate Department of Labrador Affairs in 1966 when it was felt there should be special treatment for the region[15]. The department was abolished in 1973

11. Interview with the Hon. Ron McBryde, 17th August 1979.

12. Pressure to upgrade the office of Commissioner of Northern Affairs into a full Department of Northern Affairs was applied by northern representatives as was reported in Manitoba, Royal Commission of Inquiry into Northern Transportation, Report, Winnipeg: Queen's Printer, 1969, p. 448. The Royal Commission recommended (p. 449) that consideration be given to such a change.

13. See Hamelin, *op. cit.*, Chapter 6.

14. See, for example, Saskatchewan, Administrative Management Brief Dealing with Government Services Performed in the Northland, Regina: Budget Bureau, 19th February 1947; Saskatchewan, Report on the Organization for a Single Agency to Provide Government Services in the North, Regina: Budget Bureau, 9th June 1961.

15. See Newfoundland, Royal Commission on Labrador, Report, St. Johns: Queen's Printer, 1974, pp. 1296-7.

when it was decided that each of the line departments should deal with Labrador separately[16].

This brief survey of the precursors of the present provincial northern agencies reveals not only the spotty interest in the north but the great mixture of agencies experimented with.

The Changing Provincial North

In the post war period, the natural resources of the provincial north began to be exploited to a significant degree and by the 1970s resulted, finally, in the provincial governments showing a much greater interest in their northern regions. Natural resource exploitation, principally mining, forestry and power generation produced a new, second, economy alongside the traditional economy. Most of the developments have been on a large scale and many of them have been gigantic and they have slowly spread further and further north.

Although most of the development of the provincial norths has been performed by very capital intensive industries, they have still wrought considerable changes on the northern population. While the total size of the northern populations remains small in relation to the southern parts of the provinces, it has increased considerably in the post war period. Moreover, it has radically changed in racial mix and, thereby, deeply divided the provincial norths. Towns, some of them quite sizeable, have developed around the resource and power development sites and they are largely populated by whites. The traditional populations of the north have, by and large, been excluded from the workforce of the newer economy and, therefore, do not normally reside in these towns. The division can be clearly seen in most of the provincial norths but a good example of the phenomenon is northern Manitoba where all of the larger towns, with the exception of Flin Flon, have been founded since 1945 and the most recently created resource site, Thompson (founded in 1960), has become the largest community in Manitoba's north[17].

The development of the newer economy created pressures which in the 1970s led to new bureaucratic initiatives in connection with the provincial norths. The first of these pressures was the sheer economic importance of the new resource based economies of the provincial norths which generated an enormous quantity of goods and services most of which was exported and were thus of great importance to the provincial, and indeed national, economies[18]. A further pressure was that the Indians and Metis became resentful of the major resource developments because they have often robbed them of their traditional means of livelihood while there was no really serious attempt to incorporate them into the new economy[19]. As a consequence they became much more politically assertive.

---

16. See Newfoundland, White Paper on the Organization of the Public Service of Newfoundland and Labrador, St. Johns: Queen's Printer, 1972, p. 15, and Newfoundland, Department of Labrador Affairs (Repeal) Act, St. Johns: Queen's Printer, May 1, 1973.

17. For a description of the two economies in Northern Manitoba see Hickling-Johnston, Manitoba Northlands Transportation Study: General Appendix, Toronto: Hickling-Johnston, 1975, Chapters 1 & 2.

18. The value of production from northern Ontario alone has been estimated at $5 billion annually. See Hon. Leo Bernier, "Remarks to the Federation of Northern Ontario Municipalities," Parry Sound, 29th April 1977.

19. See, for example, the Treaty 3 brief to the Ontario Royal Commission on the Northern Environment in North of 50°, No. 2, 21st December 1977, page 16.

The new industries created serious environmental problems and this also created pressure for new bureaucratic initiatives. The forest industry has produced a number of horror stories such as the mercury pollution of the English-Wabigoon river system in northwestern Ontario[20] and has also been accused of treating the forests as if they were a non-renewable resource[21]. The mining and power generation industries have also come in for criticism. Provincial governments have felt the need to make some kind of response to the pressures placed upon them by environmental groups, nature organisations and the tourist industry.

Provincial governments also became increasingly pressured by local governments in their northern regions. The smaller communities were largely left out of the new developments and they began to pressure for basic services, at least a minimal level of self-government and a fair share of the economic benefits resulting from the development of the resource sector[22]. The larger, newer, resource based communities began to pressure for economic diversification. This, they believed, would create more jobs, stem the out-migration of the young, provide a broader and more stable tax base and the possibility of more and better municipal services. The pressure has largely been brought by municipal governments and chambers of commerce[23].

Because of the problems just mentioned, there has been a growing amount of political discontent in the provincial norths in recent years. Political discontent has taken many forms. In many of the provincial norths, it has led either to a rejection of the governing party, as it has in the past in Manitoba, or to very strong undercurrents of protest voting, as has traditionally been the case in northern Ontario[24]. It has also taken rather more dramatic forms which, while often not taken seriously in the south (or often in the north for that matter), are perhaps indicative of underlying frustrations. These forms are the calls in Labrador for either union with Quebec or a separate status and the repeated calls for a separate province in northern Ontario. While these northern regions have very few representatives in the provincial legislatures few political parties are happy to alienate any group of voters. Much of the increased interest in the provincial norths by provincial governments might, therefore, be explained by a desire either not to lose northern support or to regain it. Indeed nearly all of the provincial ministers of northern affairs have cited political reasons as among the more important for the establishing of their portfolios[25].

---

20. See Warner Troyer, No Safe Place, Toronto, and G. Hutchison and D. Wallace, Grassy Narrows, Toronto: 1977.

21. See K. Lie, "The Plight of Ontario's Northern Forests," Alternatives, Vol. 7, Autumn 1978, pp. 17-25; and K. W. Hearndon, "Growing the Second Forest in Ontario — A Political and Professional Challenge." Speech given to the O.P.F.A. Seminar for MPP's on the Status of Ontario's Forests, Lakehead University, Thunder Bay, August 30-31st, 1976.

22. See, for example, Northern Association of Community Councils Inc., Resolution on Local Autonomy, 1979 Annual General Conference and various submissions to the Ontario Royal Commission on the Environment such as those by the mayor of Dryden in Ontario, Royal Commission on the Northern Environment, North of 50°, Issue No. 1, 29th November 1977, p. 18.

23. See, for example, Sudbury and District Chamber of Commerce, A Profile in Failure, Sudbury: Sudbury and District Chamber of Commerce, 1976.

24. See, for example, G. Brock, The Province of Northern Ontario, Cobalt: Highway Book Shop, 1978; Ed Diebel, "A New Province for Northern Ontario," Boreal, No. 6, no date, pp. 3-12; J. Aitken, "The Scary Things I Hear About Canada," Weekend Magazine, 6th May 1981.

25. Interviews with Mr. Ron McBryde, 17th August 1979; Mr Leo Bernier, 20th November 1978; Mr. Ted Bowerman, 29th August 1979.

The Bureaucratic Response

The problems caused by the changes in the provincial norths in the post war era led to four kinds of bureaucratic response in the 1970s. The first of these was what might be termed a non-decision or non-response, namely a decision not to establish any kind of separate northern agency or ministry. This was the case in British Columbia. Those who believe there should not be a separate agency argue there should be equal treatment of all residents of a province wherever they may happen to reside and not an institutionalisation of separate and unequal treatment. It is also argued that line departments are likely to be well co-ordinated and united in relation to their own functions whereas a ministry set up on geographic criteria would have difficulty co-ordinating widely varying line activities. Thus, if there is inequitability in service provision by region, the answer is to encourage or command established line ministries to extend their services adequately.

Against these arguments are those which hold that equal treatment has obviously not been meted out to the provincial norths in the past. Line departments have simply not had the same unity of purpose or even interest in delivering their services in northern regions as in southern regions. Northern regions have consistently been given a lower priority in most policy areas and in most provinces. It is also argued there is less local input into service provision in the north and thus there has often been a poor match to circumstances and little 'customer' satisfaction. In addition it is argued that the delivery of services is centralised in the south and this has resulted in little co-ordination among line ministries concerning their northern services and few unified attempts to tackle northern problems[26].

The government of British Columbia has apparently been convinced by the arguments against the need for a separate northern agency or ministry. As a British Columbia document states, "British Columbia does not have a specific ministry for northern development or Indian affairs nor a co-ordinating agency or a policy framework for such policies"[27]. British Columbia is represented at the Interprovincial Conference of Ministers Responsible for Northern Development by the Ministry of Industry and Small Business Development.

The second type of bureaucratic response was the creation of special agencies dealing with the provincial north but which have relatively limited mandates. This was the case in Alberta, Quebec and Newfoundland. The limited nature of the mandates reveals the relatively limited or low level of concern felt for the north and the fact that there is not extensive co-ordination of northern affairs in these provinces.

The relevant agency in Alberta is the Northern Alberta Development Council. This organisation has a relatively limited focus in that it is primarily concerned with the promotion of economic development[28]. The relevant agency in Quebec is the Secretariat

26. Many of these arguments for and against a single northern agency can be seen in a memorandum from A.W. Johnson, Secretary of the Treasury Board (Saskatchewan) to all members of the Treasury Board entitled Organization of a Northern Bureaucracy, 21st October 1959.

27. British Columbia, Manpower Services in Northern British Columbia with Special Reference to the Transition from Traditional to Wage Economy. Paper presented to the Third Annual Interprovincial Conference of Ministers with Responsibility for Northern Development, Thompson, Manitoba, September 9-11th, 1980, p. 7.

28. As can be seen in Northern Alberta Development Council, Five Year Review of Activities, April 1, 1973 – March 31, 1978, Edmonton: Alberta Business Development and Tourism, 1978 and the annual reports of the Council.

des Activités Gouvernementales en Milieu Amerindien et Inuit (S.A.G.M.A.I.). This agency is primarily responsible, as its name implies, for matters affecting the indigenous population[29]. The James Bay and Northern Quebec Corporation has the major responsibility for economic development in northern Quebec. Thus Quebec has two separate agencies dealing with the north, both somewhat limited in their roles. The relevant agency in Newfoundland is the Department of Rural, Agricultural and Northern Development. This department has branches in Labrador and is mainly concerned with economic development[30].

The third type of bureaucratic response was the creation, in the 1970s, of Ministries of Northern Affairs in several of the provinces. These ministries were of three different types. The first type of northern ministry could be described as a limited ministry because it is only responsible for some of the major policy areas affecting the north. The Department of Northern Affairs in Manitoba is of this type. The Department's mandate is limited largely to two areas. One is local government development in the more remote parts of the province. The larger cities and towns in the north are explicitly excluded from its mandate. The other is the co-ordination of federal-provincial and other agreements affecting the north such as the Manitoba Northlands Agreement, the special Agriculture Rehabilitation and Development Act agreement and the Nelson hydro-electric project[31].

The second type of northern ministry could be called a full service ministry. The Department of Northern Saskatchewan (D.N.S.) in the days of the N.D.P. government was of this type. Nearly everything that happened or was delivered in northern Saskatchewan used to be the responsibility of the D.N.S. As a consequence the department was accused of being a 'state within a state'[32]. This approach has long been associated with the N.D.P. and a number of advantages are normally cited. A single agency with full responsibility would or should evince, it is argued, a greater awareness of the total needs of the north and be able to develop a comprehensive, interrelated programme to meet these needs. Moreover such an agency would possess delivery capacity and could, therefore, more easily co-ordinate the delivery of those services decided upon. In addition it is argued that such an agency would more easily be able to obtain community input and match delivery to local desires. Finally it is argued that a single voice for the north both within the bureaucracy and the cabinet would be likely to increase the awareness of the north and an emphasis on solving its problems[33].

29. As can be seen by consulting the magazine Rencontre published by S.A.G.M.A.I. Volume 1, Number 1, appeared in the Fall of 1979.

30. This can be seen in Newfoundland, Department of Rural Agricultural and Northern Development, The Northern Development Branch, a paper presented to the Third Annual Interprovincial Conference of Ministers with Responsibility for Northern Development, Thompson, Manitoba, September 9-11th, 1980.

31. Manitoba, Department of Northern Affairs, Responsibilities of the Minister, Department of Northern Affairs, Renewable Resources and Transportation Services. A paper prepared for the meeting of Provincial Ministers Responsible for Northern Development, Fort McMurray, Alberta, 27-29th September 1978.

32. See statements attributed to George McCleod, PC critic of the DNS in the Toronto Globe and Mail, 19th February 1980.

33. See Saskatchewan, Administrative Management Brief Dealing with Government Services in the Northland, Regina: Budget Bureau, 19th February 1947, p. 14.

It was, however, recognised that there might be difficulties with this approach. Not the least of these being that while it would be a relatively small department it would have a vast range of responsibilities and would, therefore, be unlikely to be able to develop the functional expertise in each area that would be equivalent to that of a line ministry covering the whole province. Another difficulty would be that of maintaining province wide standards. While rough equivalency might be obtained, there was bound to be a divergence in detail even if only because delivery could be better matched to local desires. A fear of the Conservative opposition was that the Department of Northern Saskatchewan, in becoming the "ministry of everything" in the north, and a ministry with a vested interest in developing the region, would eventually develop an attitude in the northern population that government is responsible for everything thereby destroying local initiative[34].

The third type of northern ministry is what can be called a co-ordinating ministry. The Ontario Ministry of Northern Affairs is of this sort and is, in effect a compromise between the Manitoba and the old Saskatchewan approach for, despite significant delivery capacity in the area of transportation, its main role is really that of co-ordinating the activities of line ministries and federal-provincial agreements in the north[35].

The fourth type of bureaucratic response, and perhaps the most indicative of the growing level of interest in the provincial north, was the formation of the Interprovincial Conference of Ministers Responsible for Northern Development. The first meeting of this group took place in Fort McMurray in 1978 and has been held in a different northern locale in each subsequent year. So far the meetings really only constitute a useful way of sharing information and ideas.

Responsibilities, Priorities and Relations with Regional Groups

As one would expect given the widely differing structures of the northern agencies, the range of their responsibilities as well as their range of priorities vary widely. Moreover, they have changed over time in each jurisdiction. Very broadly, however, they have five major areas of responsibility.

Firstly, nearly all of the agencies are responsible for economic development in one form or another. Until very recently most of the mining and forestry developments were largely left to the operation of market forces although there were some examples of encouragement by government agencies. In the past decade or two there has been much greater governmental encouragement of economic development. This has developed in, essentially, two stages. The first stage was exemplified by a great deal of emphasis upon infrastructure development as a means of promoting economic development. In most provinces this has taken the form of help with water supply and sewer systems, the building of access roads for the forest industry, townsite development for the mining industry and road and transportation improvements generally[36]. Another major thrust has

34. See Toronto Globe and Mail, 19th February 1980.

35. See Ontario, Ministry of Northern Affairs, Brief to the Royal Commission on the Northern Environment, Toronto: Ministry of Northern Affairs, 1979.

36. See for example British Columbia, Alberta and Saskatchewan, Economic Development Strategies for Canada's Provincial North, Joint Paper prepared for the Interprovincial Meeting of Ministers Responsible for Northern Development, Thunder Bay, 5-7th September 1979, and Ontario, Ontario's Approach to Northern Development, Paper presented to the Interprovincial Meeting of Ministers Responsible for Northern Development, Thunder Bay, September 5-7th, 1979.

been upon the delivery of incentive grants to the major industries already established in the northern regions[37]. In short, this first stage basically involved a 'more of the same' approach. The second stage is exemplified by a slow change to an emphasis upon job creation and the type of economic development that is likely to broaden the economic base of the regions. It also includes a new awareness of the need to include native northerners in northern development and a new awareness of environmental and social problems in the north[38].

The economic development policies undertaken by the northern agencies have been criticised quite severely. One of the major criticisms is that they have not been pursued in accordance with overall plans that are likely to lead to the development of reasonably integrated, broad and balanced economies that would include most of the residents of the north. Another criticism is that far from there being real plans the northern ministries operate either on an *ad hoc* basis or conduct plain porkbarrelling for political motives under the guise of economic development[39]. Yet another criticism is that the northern agencies only have limited responsibility in this area and all kinds of other agencies, provincial and federal, are also involved[40].

Secondly, local government development is a responsibility of nearly all the northern ministries and agencies although they differ in the degree of importance attached to this role[41]. Only the Department of Northern Saskatchewan is responsible for all of the northern region. In Manitoba the Department of Northern Affairs deals only with the unorganised part of the northern region because the organised cities and towns are excluded from its mandate. The Ministry of Northern Affairs in Ontario is responsible for

37. See, for example, Ontario, Statement by Ontario Treasurer, Frank S. Miller on Ontario's Program to Assist the Province's Pulp and Paper Industry, Toronto: Ministry of the Treasury, 31st January 1979 and Canada, Principal Elements of a National Development Policy in Support of Canadian Forest Industries: Background Notes, Ottawa: Regional Economic Expansion, February 1979.

38. See the various papers presented on the topic of transition from a traditional to a wage economy at the Third Annual Interprovincial Conference of Ministers with Responsibility for Northern Development, Thompson, Manitoba, September 9-11th, 1980. See especially Saskatchewan, Development in Northern Saskatchewan: Transition from a Domestic to a Wage Economy, Paper presented to the Third Annual Interprovincial Conference of Ministers with Responsibility for Northern Development, Thompson, Manitoba, September 9-11th, 1980, p. 1.

39. See, for example, Ontario New Democratic Party, Briefing to the NDP Caucus on the Performance of the Ministry of Northern Affairs and N.O.D.C., Sault Ste. Marie: Ontario New Democratic Party, 29th November 1978; Kathryn Fournier and Gilbert Heroux, "Lesson from Quebec: How Ontario Could Really Involve People in Decision Making," Noract, No. 11, Spring 1981, pp. 12-15; article by Stan Oziewicz entitled "Big Leo: Playing Santa with pots of money and standing up against the big bad south," Toronto Globe and Mail, 12th September 1979.

40. The complexity is indicated in The Northern Development Research Group, A Selected Overview of Ontario's Public Decisional Framework of Northern Primary Resource Development, Toronto: Northern Development Research Group, 1980.

41. For a detailed discussion of local government problems in the Canadian provincial north see G.R. Weller, "Local Government in the Canadian Provincial North," Canadian Public Administration, Vol. 24, No. 1, Spring 1981, pp. 44-72.

local government in the unorganised areas but also has extensive contacts with the organised municipalities throughout northern Ontario. In all cases the provincial agencies have relatively few dealings with Indian reserves as they come under federal jurisdiction.

The approach of the northern ministries to their responsibility for local government development can be said to have developed in two broad stages. The first stage involved a response to the pressure for basic services from the northern communities. This is not at all surprising as the need was great and urgent. The smaller more remote northern communities, and reserves, usually had and still have no tax assessment base and were thus thoroughly dependent upon government as a source of revenue. Government expenditures have not been exactly lavish in the north with the result that few of the communities had water or sewer systems let alone other community services such as fire protection or recreational facilities. The northern ministries and agencies thus faced a bad situation and one in need of rapid remedy so it is not surprising that they responded in what might be described as a paternalistic manner by largely assessing needs centrally and then trying to meet them. Although the situation in most provincial norths is far from ideal, the situation has markedly improved as a result of the new ministries and agencies.

The second stage involved a gradual change of emphasis to attempt to develop a greater degree of local self government. The enhancement of local self government was, in fact, always an objective of the northern ministries but it had to be delayed until the urgently needed infrastructure was in place and the necessary skills had been developed among northern residents. There has been an almost bewildering variety of attempts to foster local self government in the provincial norths. Some provinces have been content with promoting rudimentary forms of local self-government bodies. This is the case in Ontario where Local Service Boards (L.S.B.), which are hardly local government units in any accepted sense, have been established on a purely voluntary basis. Other provinces have set about installing organisations of a much more sophisticated nature. Saskatchewan, for example, developed a whole hierarchy of local government units ranging from Local Community Authorities (L.C.A.) to Local Advisory Councils (L.A.C.) to a Northern Municipal Council (N.M.C.) which covers the whole northern region. It is still too early to judge which approach is likely to be the more successful in the long run. While the Ontario approach may be cautious, it was developed in close association with local residents and it was they, not the Ministry of Northern Affairs, who advocated the caution because they feared a loss of the voluntary spirit if they moved too rapidly. The Saskatchewan approach relatively rapidly built up a desire, if not full ability, on the part of local government bodies to control a wide range of functions. The Department of Northern Saskatchewan, right up until the recent Conservative victory, was not exactly keen to divest itself of its responsibilities and fulfill its stated goal of doing itself out of a job.

Thirdly, nearly all of the northern ministries and agencies have some degree of responsibility for native peoples but the situation is, however, very confused. While the federal government is responsible for status Indians on reserves, there are many federal-provincial agreements and much overlap between the two levels. Within the provincial governments the location of responsibility varies considerably. In Saskatchewan the D.N.S. exercises that provinces responsibilities for native peoples in the north. In Manitoba and Ontario the situation is not so clear cut. In Manitoba, the Department of Northern Affairs has *de facto* responsibility for many Indians and Metis in its part of the north but no clear mandate. In Ontario, the Ministry of Northern Affairs has many dealings with Indians and Metis but shares the responsibility, primarily with the Native Community Branch of the Ministry of Culture and Recreation. In Quebec S.A.G.M.A.I. has responsibility but the line ministries have a strong presence.

The progression of policy in relation to native peoples has been much the same as with local governments in the more remote northern areas. This is hardly surprising because many of the reserves and non-reserve native communities suffer from the same problems, namely a startling lack of even the most basic services and a very high degree of paternalism exercised by governments[42]. The first stage in the development of policy was to try and improve the level of basic community services. While many improvements have been made, progress could not exactly be described as rapid nor the results entirely satisfactory. The second stage consisted of attempts to increase the degree of local self government and the level of employment opportunities for native people. While local self government still remains more an ideal than actuality in many places at least the direction of policy is in the direction desired by most native northerners. Attempts to expand employment opportunities by developing the economies of the reserves and the smaller non-reserve communities and by trying to ensure that native peoples also acquired economic benefits and spin-offs from the resource economy have had very mixed results. In the case of money intended for servicing and economic development many of the native communities and groups would like the governments to simply hand it over to them so they could manage matters themselves. This is especially true of the Metis in Manitoba and Saskatchewan. These two groups are clearly worried that the northern ministries will reduce their influence over their people if such a bureaucracy provides the services and development. This has resulted in the northern ministries being the focal point of Metis protest and numerous bitter confrontations[43].

There has been a great deal of conflict between the two senior levels of government concerning which level should have responsibility for native peoples in the future. The federal government clearly wants to shift the responsibility to the provinces but the provinces, while not totally opposed, are leery of the prospect. Unfortunately the first 'experiment' in a transfer of authority has not gone at all well. This is the transfer of responsibility to the Province of Quebec which has so upset the James Bay Cree[44]. In fact the relations between S.A.G.M.A.I. and native peoples has long been marked by conflict. This may well be because the native population feels it is the pawn in a nationalistic power struggle being pursued by the province.

The fourth major area of responsibility of the provincial northern ministries and agencies is that of co-ordination. This co-ordination is of two types. Firstly, the co-ordination of federal provincial activities that affect the provincial norths and secondly, the co-ordination of the operations of provincial ministries and agencies where these impinge on the north. In most of the provinces under discussion, the northern ministries main

42. See, for example, The National Indian Socio-Economic Development Committee, To Have What is One's Own, Ottawa: The National Indian Socio-Economic Development Committee, 1979.

43. See, for example, Winnipeg Tribune, 17th August 1979 and 21st August 1979 concerning the occupation of the Manitoba legislature.

44. See I.E. LaRusic et al, Negotiating a Way of Life: Initial Cree Experience with the Administrative Structure Arising from the James Bay Agreement, Montreal: SSDCC Inc., 1979.

contact with the federal government is via the economic development agreements signed between the Department of Regional Economic Expansion and the provincial governments. In Manitoba, for instance, there is a joint federal-provincial committee overseeing the Manitoba Northland Agreement. It is sometimes argued that the provincial ministries are often not so much co-ordinators but merely conduits for federal monies directed at the provincial north[45]. While this may be partly true, it would appear that, by and large, the relationships between the two levels of government in relation to the operation of these agreements is reasonably amicable.

The co-ordination of the activities of other provincial ministries by ministries of northern affairs has not always been particularly easy or amicable in any of the provinces under discussion. Many ministries resented and resisted the creation of full ministries in Manitoba, Saskatchewan and Ontario. In part this was because they did not like to see another bureaucratic entity come into existence with a strong claim to special and large sums. It was also partly because they did not wish to have the extra work and annoyance of another department scrutinising some of their activities and suggesting changes. In some cases, as with the D.N.S. in Saskatchewan in its early years, the northern ministries helped regional groups protest the actions of other ministries, such as highways. This did not endear the northern ministry to other ministries. In addition the northern ministries, because of the nature of some of the people they initially hired, were seen as hotbeds of radicalism. One tendency of the other established ministries was to try and dump either their major problem areas, in terms of responsibilities, or unwanted personnel on the new ministries of northern affairs[46]. This again was an approach hardly likely to create harmonious relations.

The chief rival ministry at the provincial level is frequently the ministry of natural resources. These are older well established ministries that often still have clear responsibilities in the north. The Ministry of Natural Resources in Ontario, for example, controls all timber and mineral resources and is both the chief custodian and developer of these resources. Moreover, these ministries often conduct their own planning, usually fairly independently of other provincial agencies. In Ontario, for example, while the Ministry of Northern Affairs was made responsible for what were supposed to be the general plans for the northeast and northwest, namely the Design for Development plans[47], these became defunct and the Ministry of Natural Resources Strategic Land Use Plans have become very influential and often quite controversial[48]. The new northern ministries and agencies also sometimes have rivals in the form of other non-departmental agencies and commissions with responsibility for the north. In Ontario, for example, there is the Northern Ontario Development Corporation which is responsible to the Ministry of Industry and Tourism and the Royal Commission on the Northern Environment (the Hartt/Fahlgren Commission). These agencies operate with little or no contact with the

45. Interview with Bud Wildman, MPP Algoma, NDP Northern Affairs critic in the Ontario legislature.

46. Interviews with Ted Bowerman, Ron McBryde and Leo Bernier, *op. cit.*

47. Ontario, Department of Treasury and Economics, Design for Development: Northwestern Ontario Region, Toronto: Queen's Printer, 1970; and Ontario, Ministry of Treasury, Economics and Intergovernmental Affairs, Design for Development: Northeastern Ontario Regional Strategy, Toronto: Queen's Printer, 1977.

48. See, for example, Ministry of Natural Resources, Strategic Land Use Plan: Northwestern Ontario, Toronto: Ministry of Natural Resources, 1980.

Ministry of Northern Affairs although their responsibilities clearly overlap to some degree.

As was earlier indicated, the Ontario Ministry of Northern Affairs was deliberately established as a co-ordinating ministry. This co-ordination has two aspects in connection with relations with other provincial ministries and agencies. The first area of co-ordination consists of it having the role of "lead" ministry in the north. This consists of it being the first point of contact with government for most northern residents via its network of northern affairs offices. These offices act as clearing houses where people or groups are put into contact with the appropriate ministry. It also consists of the ministry being the co-ordinator of several ministries where these activities all impinge on a particular area or problem. The second area of co-ordination consists of the M.N.A. setting the priorities in a particular policy area, such as road construction, applying funds at its own disposal to meet these priorities but with the details of the work, such as the engineering studies and the actual construction, being performed by the ministry with the appropriate expertise, in this example the Ministry of Transportation and Communications[49]. In these particular and pragmatic areas of operation, the relationships of the ministry with the other agencies seems to have become reasonably amicable.

The fifth major area of responsibility of the provincial northern ministries and agencies are party political. All of the agencies were established with political priorities in mind but often these were not openly stated. The political priorities of the northern ministries are essentially twofold. Firstly, they were established to stave off, minimise or capitalise upon northern discontent. In Ontario, for example, the Progressive Conservatives were worried by the inroads being made into their northern support by the New Democratic Party which had long been the second party in the region. Mr. Bernier, the Minister of Northern Affairs, claims that his ministry was in part intended to counter this and indeed did so because it was partially responsible, in the election immediately following its founding, for regaining three northern seats from the New Democratic Party[50]. Secondly, the northern ministries were, in part, established to assert full provincial control over the northern territories or rather establish provincial sovereignty. Most provincial governments were traditionally quite willing to tolerate a strong federal presence in their northern regions. This has, however, changed and in recent years there has been quite a lot of federal-provincial competition.

Problems and Prospects

The new ministries and northern agencies have had a rather large number of problems during their relatively short existence and their long term future is in considerable doubt in some provinces. The problems of the new agencies began at the point of their formation because they were often regarded by other parts of the provincial bureaucracies as what might be termed convenient bureaucratic 'dumping grounds.' This attitude is partly the result of bureaucratic snobbery in that they are not glamour ministries and assignments to them might mean being located away from the main centres in each province. It is also partly true in that it is clear that those ministries which lost part of their role to the newly created northern agencies did try and transfer people they wanted to get rid of or who were not particularly productive in their then present settings. These attitudes made it difficult

49. See Hon. Leo Bernier *op. cit.*

50. Interview with the Hon. Leo Bernier, 28th November 1978.

for the new ministries to establish credibility within their respective provincial bureaucratic structures.

The new ministries had difficulty developing new ideas and approaches in the short run because of the fact that a significant proportion of their initial personnel were carry overs from previous smaller northern related agencies or the northern related sections of the various line ministries. For instance, the senior personnel in the Department of Northern Saskatchewan were initially mainly ex Department of Natural Resources employees who had had experience in the north. Similarly the senior S.A.G.M.A.I. officials were initially ex D.G.N.Q. or Department of Natural Resources employees and a significant number of the senior personnel in the Ontario Ministry of Northern Affairs were former planners who had responsibility for northern Ontario within the Ministry of Treasury, Economics and Intergovernmental Affairs. This situation inevitably meant that there was a carry over of previous thinking and attitudes towards the north and northerners in the new ministries and agencies. This, in turn, meant the new ministries had difficulty in getting a coherent operation underway because of the many and varied backgrounds of the personnel which was itself a source of much internal conflict. The variety of backgrounds and attitudes was magnified by the wide range of functions that had to be performed.

In addition the new ministries and agencies had some difficulty obtaining good staff. Most aspiring individuals were not attracted by unglamorous relatively low priority departments working for which involved lots of uncomfortable travel in poor conditions and remoteness from not only the centres of power but also of culture, education and recreation. Those who were attracted to these ministries were often those whom the charitable called "idealists" and others "left wing rabble rousing academics." Indeed many of those attracted initially, especially in Saskatchewan and Manitoba, were interested in "social animation" and were concerned not so much to promote the delivery of services but to agitate among the recipients of the service to promote change via conflict.

The ministries of northern affairs in Saskatchewan and Manitoba were established by N.D.P. governments who were, if not entirely sympathetic to the left wing interpretation of the ills and the cures for the provincial norths, at least disinclined to prevent the appointment of such people — especially since they were often the only ones particularly keen to be employed in the north. The result was that the conflict between the advocates of 'hard' and the advocates of 'soft' services was most clearly evident in these two provinces.

Much of the early emphasis of the D.N.A. in Manitoba was on the delivery of 'soft' services. There were quite a lot of ex 'academics' on the staff at first and they established social orientation courses to pursue 'social animation.' In addition the Minister, Mr. McBryde, was keen on establishing a large extension programme. The D.N.A. quite quickly began to acquire a reputation in the north for stirring up trouble. Local government groups, for instance, were not at all keen on having D.N.A. employees promoting activist groups as opposition forces that would stand up to them. The D.N.A. employees were, of course, applying their conflict theory of social progress. Premier Schreyer reportedly did not much care for many of the 'soft' service programmes or the manner of their application but he seemed unwilling or simply not firm enough to do much about it[51]. The left wing approach tended to be concentrated in the Policy and Planning

51. Interview with Mr. Bob Mcleverty, Mayor of Thicket Portage.

Branch and the internal problems were eventually only solved by a large scale set of firings from that branch. When the Progressive Conservatives became the governing party they, essentially, did away with the soft services side of the department which had, in their eyes, become too severely tainted by socialists to be worth preserving. This reaction, though understandable from their point of view, was perhaps regrettable because without the ideological overtones many of the soft services were useful and some of them, such as extension services, employed a relatively large number of native northerners.

The same early emphasis can be seen in Saskatchewan with the D.N.S. Again the Policy and Planning Branch of this Department was initially staffed by 'idealists' who supported a confrontational approach which even involved them supporting native protestors and supporting a petition for the removal of the minister, Ted Bowerman. Many within the D.N.S. thought the activities of this group discredited the D.N.S. in the eyes of the other government departments thus making it very difficult for them to operate successfully within the provincial bureaucratic structure. As in Manitoba the upshot of the internal conflict was a large scale set of firings especially within the Policy and Planning Branch. There was, however, an interesting twist in Saskatchewan compared with Manitoba for in 1980 one of those fired by Bowerman, Jerry Hammersmith, became the Minister of the D.N.S. and another of those fired, Brian Hill, returned to the department as a senior civil servant.

The new ministries did not hire many northern residents, let alone northern native residents, and this did nothing to increase their credibility in the north. All kinds of apparently sound arguments were used to explain this situation, such as the lack of educational preparation among northerners, but apart from a few weakly applied affirmative action programmes, the fact remains that no concerted effort in this regard was made by any of the new ministries and agencies. Part of the explanation for this may be that some of the ministries were conceived of as being of relevance only to the white north and that many native northerners did not want to work for agencies they often regarded as either a sop or exploitative. Some of the more left wing ideologues were opposed to native employment in the ministries, using such apparently right wing phrases as 'they aren't ready for it', because they were worried it would lead to co-option to 'the system' and a lessening of actual or potential conflict. In some cases native northerners had to face racism, prejudice and job protectionism. Thus native northerners contemplating employment with a northern ministry or agency had to face not only the possibility of prejudice, but the possibility that they would be manipulated by ideologues, and the certainty they would be regarded with some suspicion by their own people. It is not surprising, therefore, that in Manitoba, for instance, the few native northerners who were hired initially were quickly lost.

The ministries were, and to some degree still are, plagued by deep conflict internally between those who thought primary emphasis should be placed upon the delivery of 'hard' services and those who thought it should be placed upon the delivery of 'soft' services. The conflict could be described as one between the 'sewer and water' types and the 'social development' types. This conflict was heightened by the fact that most of the sewer and water types were older and were transferees from the line ministries whereas the social development types were younger, university educated, left wing and appointed directly to the new ministry.

The new ministries have also had to face the problem of deciding upon the issue of whether or not to decentralise their operations and to locate their headquarters in the north.

The dilemma was generally seen as one of balancing the need to create credibility in the north by having a physical presence there with the need to maintain a strong presence in the provincial capital so as not to be isolated from the centre of bureaucratic and political decision making. The ministries have dealt with this problem differently.

In Saskatchewan, it was decided to locate the department in the north. A big new headquarters was constructed in La Ronge which became, in effect, the northern capital. In Manitoba, the Deputy Minister of Northern Affairs was originally located in Winnipeg but was moved to Thompson in 1978 and into a new provincial government office building. It is perhaps significant that Thompson was in the riding of the then Minister of Northern Affairs (Mr. Ken McMaster). Some time later the Deputy Minister was moved back to Winnipeg because it was argued Thompson was an awkward location as the interplay of ministries took place in Winnipeg and the M.N.A. could not afford to be shut out of the process. The move was, however, a blow to many northerners for they thought it represented a downgrading of government interest in the north. In Ontario, the Ministry of Northern Affairs has its headquarters in Toronto but is otherwise quite decentralised. Assistant Deputy Ministers and regional offices are located in Sault Ste. Marie and Kenora while district offices have been established in Thunder Bay and Sudbury with area offices in 29 other locations. There was political controversy concerning the location of the regional offices. The two major cities, Thunder Bay and Sudbury, felt slighted and accusations were made to the effect that Sault Ste. Marie and Kenora were chosen because they were strong Progressive Conservative seats, with Kenora being on the Minister's, Mr. Leo Bernier's, home turf[52]. The decentralised approach was intended to give the M.N.A. a strong and very visible presence in the north and this has, indeed, been achieved.

Some of the northern ministries and agencies have suffered from what might be termed "bureaucratisation" in that they have, once established, developed a momentum of their own increasingly divorced from the expressed interests and basic needs that they were established to fulfill. Perhaps nowhere was this problem more evident than in Saskatchewan which is not surprising because of its wide range of functions. Local commentators asserted that under the N.D.P. government the D.N.S. bureaucracy became "vast and bewildering" and that it became too uncoordinated or specialised with the result that sometimes three or four D.N.S. officials would separately pass through one small community on a given day[53]. Moreover, the Northern Association of Community Councils argued that the D.N.S. budget was being spent more on staff, especially senior staff, than on services and that they would like to see this emphasis reversed. They also commented that the department was becoming so big it was in danger of becoming uncontrollable. The D.N.S. was also accused of having the goal of promoting local autonomy but in fact resenting initiative when shown by the newly promoted local government units because it meant things were not always done the way the D.N.S. wanted[54]. The Minister of Northern Affairs in Manitoba, though smaller than the D.N.S., was expanding and may have developed into something closer to the Saskatchewan full

52. See, for example, the Thunder Bay Chronicle Journal, 16th April 1977.

53. Murray Dobbin, "DNS: A Major Reform Betrayed," Next Year Country, Vol. 1, No. 6, February-March 1974, pp. 32-37.

54. See "Yew Knocks Bureaucracy DNS has brought to the North," The Saskatchewan Indian, June 1974, p. 11.

service ministry approach. Political forces, however, stopped the further bureaucratisation of both departments in both cases and that brings us to the final problem to be considered.

The biggest problem for the northern ministries and agencies is that their very creation, and certainly their survival is linked very closely to that of particular political parties. In Manitoba and Saskatchewan the very idea of a northern ministry is seen as a fundamentally N.D.P. idea. In Ontario it is seen as a Conservative idea and is opposed by the N.D.P. and Liberals. It is thus not surprising that the Ministry of Northern Affairs in Manitoba had its responsibilities and manpower seriously reduced (from over 300 to just over 100) when Premier Lyon and the Conservatives gained power or that Grant Devine and the newly elected Conservatives in Saskatchewan are in the process of, in effect, dismantling the Department of Northern Saskatchewan. By next year over 80 per cent of the D.N.S. responsibilities will be transferred back to line ministries. The D.N.S. will then only be responsible for economic development and municipal services in the north.

A SOCIAL SCIENCE VIEWPOINT OF RESEARCH IN THE BOREAL REGION

Paul F. Wilkinson
McGill University

Abstract: Social problems, such as few material benefits, high unemployment, high cost of living, higher than average rates of addiction to drugs and alcohol and greater feelings of powerlessness in the face of government and business are not unique to the small boreal forest communities and not necessarily unique to natives. Racism, prejudice, double standards and patronizing viewpoints are part of the realities to be faced by scientists who wish to act out their social responsibility. Solutions to these problems will not come only from fast processing of land claims but will require long-term and expensive methods as well. Indeed, land claim settlements may aggravate rather than solve existing problems if natives feel forced to continue to rely on outside experts rather than quickly developing their own expertise and obtaining a clear self-image.

I am not a social scientist, and I know little about the boreal forest. Despite this, I am presenting my thoughts in the guise of a social scientist and about the boreal forest. I do so for two principal reasons: first, because I have for the past seven years worked for a small group of Indians, who live in a small mining town near the northern limit of the boreal forest in Québec; and, secondly, because I do not believe that the societies and cultures of the boreal forest, which is essentially an ecological or phytogeographic concept, are unique with respect to the problems that they face today.

Let me elaborate briefly on each of the preceding points. I do not believe that it is disadvantageous that I am not a social scientist, or for that matter a natural scientist, most of all because I do not believe that the principal requirement of the peoples and communities of the boreal forest today is so much for scientific studies as for social, economic, and political initiatives. In this respect, the situation pertaining to the natural sciences, about which I know much less, strikes me as being quite different, and the need for increased research in the natural sciences in the boreal forest is obvious even to the layman.

Further, not being a social scientist is not disadvantageous because our federal, provincial, and territorial governments are usually markedly unresponsive to the efforts of scientists to provoke change of any kind. Seven years of experience in the twilight world that exists on the fringes of government, universities, native organizations, trade unions, and northern mining communities lead me to endorse in full the view expressed recently by Professor Bruce Doern that ". . . Although scientific and technological variables are important aspects of decision processes within the federal government, one is forced to conclude that scientific and technological controversies are, on balance, viewed by decision-makers as peripheral to the formal decision process in comparison with competing political, economic, legal and other elements." I would go even further than Doern, however, and suggest that the formal decision-making process that he describes is in many instances peripheral to the real decision-making process, and that this latter totally precludes scientific and technological factors.

Let me mention here a story that I would like to think is apocryphal, but that I fear is true. Several meetings between biologists and native fishermen familiar with the northern regions of one of our provinces led to a unanimous recommendation to the government in question that the sport-fishing season for brook trout be closed at an earlier date than provided for in the regulations and that the bag-limit for that species be lowered. In due course, a reply came back from the government refusing the proposed changes, purportedly on the grounds that the scientific evidence advanced in support of them was inconclusive. In fact, however, knowledgeable persons reported that the decision had been made personally by the responsible minister, quite simply because he enjoyed fishing for brook trout and did not want to diminish his pleasure by lowering the bag-limit.

The preceding observation is in no way a revelation to Canadian academics. I have been struck by the willingness in many universities to acknowledge that academics must be socially responsive, especially to the demands of minority groups, and that they must be political or devious enough to sell their views to politicians in the harsh competition that exists with lobbyists from business. In consequence, we saw in the Canadian universities in the early 1970s the emergence, with the full approval of the universities, of a group of social and natural scientists who were establishing a middle ground between the pursuit of knowledge and its application. Sadly, however, the universities that had encouraged their younger faculty members in this direction with the tacit implication that they would subsequently be judged for professional advancement on the basis of criteria other than research, publication, and teaching, often failed to live up to their promise. I raise this point now, because it is relevant to a matter that I shall discuss later, namely an emerging estrangement between universities and many social groups, especially natives, including those of the boreal forest.

I am not unduly discouraged by the fact that I have little experience in or knowledge of the boreal forest, because my limited reading and experience suggest to me that there are few, if any, social problems that are unique to the boreal forest. To be sure, the concept of the boreal forest has meaning outside the natural sciences: the archaeologist, ethnologist, and linguist, for example, find the peoples of the boreal forest to share a variety of features that differentiate them to varying degrees from their neighbours to the north and south. On balance, however, it seems to me that, from a sociological viewpoint, the communities of Canada's boreal forest are similar in most major respects to communities of comparable size in other areas of rural Canada.

My experience in this area has been acquired in an unsystematic way as a result of working for seven years on a variety of problems, and my presentation will therefore be equally unsystematic. I shall try to mention briefly a rather random selection of matters that have captured my interest or attention or that have caused me particular sadness or dismay. I am concerned principally with Indians, because I have worked principally for and with Indians. More specifically, I am going to discuss one small group of Indians, because they are the people whom I know best. But the Indians that I know live in the same town as a much larger number of whites, and I get the impression that they share many of the same problems: an unwanted alienation from the principal material benefits of our society, high levels of unemployment, high cost of living, higher-than-average rates of addiction to drugs and alcohol, a feeling of powerlessness in the face of government and business.

One of the things that we have been doing with the Indians is building an entire new village. There is a shortage of skilled manpower among the Indians, and we are obliged therefore to seek trained carpenters, in particular, among the non-Indian segment of the population, a task greatly facilitated by the current high levels of unemployment brought on by the depressed demand for iron ore. A few months ago, therefore, I interviewed forty men who had been laid off by "The Company" with little or no prospect of recall, even though many of them were skilled workers and had in some cases over twenty years of seniority! I found this experience especially depressing because I appreciated for the first time that the assistance that these workers can count upon from their governments and other representatives is negligible compared to the programmes that are available to Indians.

Thus, while I shall restrict my comments to one group of Indians, I do not wish in any way to downplay the social, economic, and cultural problems that face small, non-native communities in the boreal forest as a result of the phenomenon of boom-and-bust economies documented by students such as John Bradbury and his colleagues at McGill University.

The Social Responsibility of Scientists

I have referred to the emergence of a relatively new class of researchers, to which I belong, practitioners of what is known by the ugly and ungrammatical name of "advocacy research." I think that we have been in some ways too successful, or it may be that some academics are too naive in that they always take what we have to say at face value, forgetting that we are advocates and politicians, highly paid by special-interest groups to advance their interests.

When I write that persons such as myself, the practitioners of so-called advocacy research, have been too successful in a way that risks being in the long term prejudicial to the interests that we are seeking to protect or to advance, I mean that we seem often to have conveyed persuasively the notion that all scientists have an equal and identical social responsibility, in particular, towards the minority groups of Canada's rural areas. A moment's thought is enough to show that this is nonsense. The social concerns and responsibilities of scientists in different fields vary enormously, and in some fields they are at best tenuous and intangible. Biologists studying large mammals, fish, or waterfowl, for example, have a clear responsibility to inform local populations of that research in advance, to allow them the possibility of participating in it, and to inform them of their results. The same does not hold true for the entomologist, the ice-physicist, or the meteorologist, for example, unless their work impinges directly or indirectly on the activities of the native populations.

What I fear now is a backlash on the part of scientists, a refusal to accept the legitimate requests of native groups to be informed of and involved in research projects affecting the lands and animals upon which they depend, because of earlier, unrealistic, and excessive demands for consultation about and participation in research projects on the part of certain native groups or their representatives. The present volume includes presentations by researchers from most branches of the natural and social sciences. Obviously, I urge researchers to pay due attention to the legitimate social and economic concerns of the peoples of the boreal forest, among whom they work; equally, however, I urge them to be realistic in assessing the nature and extent of that social responsibility and the manner in which it should be discharged. If researchers have doubts about any particular instance, they should discuss them with the local natives, who are usually sympathetic and helpful.

The Extent of Prejudice

While I try to be honest about the role of the advocates of special social groups and interests and am ready to concede our occasional tendency to exaggerate for the sake of our clients and causes, I think that such behaviour is principally a reaction against the racism and prejudice that exist at every level from the small communities of the boreal forest, through trade unions, professional associations and universities, to every level of government. Such prejudice and racism are difficult or impossible to document, although I suspect that every researcher who has spent even a short time in a small, mixed-race community in the boreal forest was made quickly aware of them from comments of taxi-drivers, barmen, police officers, and store-keepers, for example, or quite simply from observing what went on around him with the interest and acuity that typify the good scientist. Within my own experience, the systematic and officially condoned brutalizing of Indians by policemen in one northern mining town, which ceased only after repeated complaints had been lodged with the appropriate authorities, is ample proof of the unreasoning hostility that exists.

The racism that exists in a more official way and among people whom one might expect to be better educated, better informed, and more sensitive, can be illustrated by reference to a document published in October 1981 by the Association de pêcheurs sportifs de saumons du Québec Inc. on behalf of some twenty local, national, and international organizations, including the prestigious International Atlantic Salmon Foundation. For me, this document bears all the classic hallmarks of racism: unnecessarily provocative language, xenophobia, inexcusable distortions and omissions of facts. As evidence of unnecessarily provocative language, let me cite only the following: ". . . la thèse soutenue par les Amérindiens à l'origine de leurs revendications, soit la pêche de "subsistance," est une absurdité grotesque qui constitue une insulte grave pour l'intelligence même médiocre d'un profane." There are, perhaps, legitimate reasons for doubting, contesting, or not wishing to accept the arguments of the Indians concerning their right to fish salmon, but nothing is to be gained by describing these arguments as a "grotesque absurdity."

As evidence of the xenophobia demonstrated by these so-called experts on and defenders of the salmon, we may take note of their view that salmon spawned in the rivers of Québec, but which travel through the waters of Newfoundland, through international waters, and which pass long and critical portions of their life-cycle in the waters of Greenland are in some peculiar sense "Québécois" and should not be fished by Greenlanders or Newfoundlanders!

I shall cite only one example of the incomplete and misleading presentation of facts, namely the claim that Indians are exempted from provincial and federal income tax with respect to income earned on reserves. This is, of course, true, but it is misleading to state this fact without noting simultaneously that most reserves offer virtually no opportunities for employment because of their size, location, or for other reasons, so that the number of Indians for whom the exemption from income tax has any practical meaning is very small indeed. The high levels of federal and provincial income tax today are a sensitive issue for all citizens, and misinformation of this type from individuals and organizations that should know better can only aggravate latent tensions between non-natives and natives. More to the point, what is the connection between exemption from income tax and the right to fish salmon?

As for evidence of a double standard, this report, when evaluating claims by Indians concerning the importance to their economy of salmon, insists that depriving Indians of their right to fish salmon would in no way threaten the subsistence of anyone in Québec. A few pages later, however, the same document describes commercial salmon fishing as the only means of gaining a living or of economic independence for certain non-Indians! Neither claim is documented, and no explanation other than racism seems to explain this assertion with respect to Indians. I shall deal a little later with the broader phenomenon that this represents; namely, the repeated claim that Indians today do little or no harvesting, a claim belied by a growing body of evidence.

Similar prejudices exist also among academics and in professional associations. I know first-hand of one eminent biologist who refuses flatly to have anything to do with the native people of northern Canada for no other reason than that they are natives, and of another biologist whose outspoken comments about the native peoples made the views of many members of the John Birch Society seem positively liberal. Discrimination among academics takes many forms, and it is particularly easy for scientists to conceal it from themselves and from others on the grounds that they are concerned only with professional qualification, experience, appreciation of western scientific tradition, etc. The speech made by Kenneth A. Brynaert, Executive Vice-President of the Canadian Wildlife Federation, to the Annual Meeting of the Canadian Society of Environmental Biologists in January is a good example of ill-informed racism.

Brynaert starts from the observation that "we are only now beginning to recognize the extent to which agriculture, forestry, industry, mining, petroleum and urban development have served to modify and degrade the environment." Not surprisingly, he does not accuse the native peoples of being responsible for this degradation, but he does devote most of the rest of his speech to a gratuitous, ill-informed, inflammatory, and unnecessary attack on natives and native rights, instead of advancing solutions to the problems caused by non-natives:

> "Regardless of what rights the native people may have, I am certain they do not include the right to abuse, waste or deplete resources."
>
> "Native people have not demonstrated to society that they possess either the will or the capability to assume the responsibility for management of wildlife . . . Although they claim rights, they have consistently failed to fulfill the obligations and responsibilities inherent in the exercise of those rights."

Unfortunately, politicians rarely see any political capital in publicly opposing the type of racism that I have just described. The native peoples of the boreal forest are too few and too scattered to possess enough political power to merit the full attention and protection of their elected representatives, whereas sport fishermen and hunters, for example, are numerous, vocal, relatively wealthy, urban residents, to whom politicians are traditionally especially responsive.

The Obverse of the Coin of Racism

The obverse of the coin of racism is, of course, the image that many natives and many apologists for the native "cause" present of the Indians and Inuit as the blameless victims of changes totally beyond their control. I have met a few Indians who are as prejudiced, as racist, as unprincipled, and as uncaring for their environment and their fellow-men as are many non-Indians. I am familiar also with native persons who have learned to exploit to the full the white man's system and to play upon our sympathies and our desire to right past injustices, real or imagined.

I have been very struck by the huge differences that seem to exist between the groups of Indians and the individual Indian described by anthropologists and my perception of the Indians with whom or for whom I have worked for the past seven years. *Maps and Dreams*, for example, Hugh Brody's latest book, seems to me to attempt to convey the impression of a group of Indians who are, individually and collectively, orders of magnitude more perfect than any Indians or whites that I have ever known. Peter Usher's discussions of the potential role of customary law in wildlife management strike me as being predicated on a degree of interest and leisure that are rare or absent among the Indians that I know.

It may be that Usher and Brody were more fortunate than I in the groups with or for whom they worked. It may be that the Indians with whom I worked are atypical, and indeed they are a group that have been described by virtually all who have known them for 150 years as being particularly ''difficult.'' It may be that my perceptions and sensibilities are less sensitive or more prejudiced than those of a trained anthropologist or sociologist. Somehow, I doubt these explanations, however, for the varying personalities among the Indians that I know and the problems that arise in their community are strikingly similar to those that I encounter in the small village in New York State where I live or that I have glimpsed in Greek villages or Israeli kibbutzim.

One important factor must surely be the nature of our differing relationships with the native groups with which we are associated. Many academics writing about Indians and Inuit today, have assisted the peoples they study in preparing for or in negotiating land-claims settlements, and they are naturally anxious, therefore, to present their 'clients' in the most favourable and flattering light possible. It is certainly true that during the time that I was negotiating a land-claims settlement for the Indians I represent, I was careful to present publicly and even privately a selective, flattering, and incomplete view of my clients. Indeed, during the three years that negotiations lasted, my contacts with my clients, the questions that I asked them, the parts of their communal and individual lives that I chose to see, and the individual and collective views that they presented to me because they seemed most relevant to our common concerns, were such as to emphasize certain more positive features of their lives, and to conceal or ignore others.

It was only during the four years that have passed since the signing of the agreement in question and during which the nature and extent of my contacts with the same Indians have multiplied and diversified, that other features, less attractive, of life on the reserve have gradually been revealed to me.

It is normal and natural for all parties attempting to resolve a problem by means of negotiation to formulate and advance positions that they know to be extreme. What disturbs me is that, on the one hand, so many persons on the native side seem to have become convinced by their own rhetoric, while so many persons on the other side are convinced that everything good said by or about natives is automatically false!

The Extent of Subsistence Harvesting

It is a favourite argument of developers and of persons inimical to the interests of native persons and of others who use or value the renewable resources of the boreal forest to claim that these persons in fact no longer rely to any great extent on wildlife, or that wildlife no longer plays an important role in their spiritual and cultural lives.

I have mentioned already the claim by the Association des pêcheurs sportifs de saumons du Québec Inc. to the effect that "since the last World War, few Amerindians of the south of the Province of Québec have practised hunting and fishing activities different from those of non-Amerindians" (my translation). Another excellent example is the testimony by expert witnesses for the proponents of the James Bay Hydro-Electric Project. Justice Malouf rejected the contention that the Cree Indians of the boreal forest zone of Québec no longer depended economically or culturally on harvesting to an important degree, and later studies have confirmed the correctness of his decision.

Section 24 of the James Bay and Northern Québec Agreement provided, among other matters, for what are known as guaranteed levels of harvesting for the beneficiaries of that agreement. The guaranteed levels are to be based on "present" levels, which are themselves to be negotiated principally on the basis of a study of the numbers of animals harvested by the Cree between 1972 and 1979. The final report on this study will be published shortly, but the preliminary reports so far available indicate that the Cree of the boreal forest of Québec harvest up to one million kilograms of bush food annually, coming from more than thirty-two species of fish, mammals, and birds, and representing 0.33 kilograms of potential food per Cree per day. Furthermore, this study found that, for 1978/1979, "The hunters constitute nearly all the men who are physically able to conduct harvesting activities" and that ". . . Cree hunters spend an average of about seven months a year hunting." Harvey Feit and Adrian Tanner, among many others, have demonstrated the continuing importance for the spiritual and cultural life of the Cree of Québec of harvesting itself and of the products of that activity.

The Nature of the "Problem"

It is correct, but misleading and incomplete, to describe the problems that many of Canada's native peoples face as being principally those of so-called native "rights," whether these be broken treaties or lands usurped without permission, without compensation, or without treaties or agreements. I state that this is misleading because for many it seems to imply the corollary that the problems will be solved quite simply by respecting past treaties or by negotiating new land-claims settlements, a view that seems in fact to be the current policy of the federal government. Obviously, past treaties must be respected, and cessions of rights must be obtained where they are required and where the holders of those rights are willing to cede them. But the signing of agreements and the renewal of treaties is at most a first step, indeed a step that often does not begin to address the basic problems or, I would argue, that may even aggravate them!

I am continually struck by what I can only conceive of as being a remarkable inability or unwillingness on the part of many persons — natives and non-natives alike — to recognize or to admit, at least publicly, the real nature of the problem, which I see as being a prerequisite for formulating workable and acceptable solutions.

Once again, it is possible that the only group that I know well is unique in Canada in the nature of its problems, but I doubt this. I find it more plausible that governments do not wish to recognize the real nature of the problems because they would then have to admit that the solutions, if there are any, will be very long-term, and very expensive — not the neat, rapid signing of a treaty, with all the attendant political capital to be reaped from publicity photographs of smiling ministers and Indian chiefs, but the long, dull, mundane, politically unrewarding and unrewarded efforts to assist physically, morally, and socially disadvantaged and depressed persons define for themselves a new collective and individual relationship to society and then acquire the skills to attain that position.

When I say that land claims settlements may aggravate rather than solve existing problems, I am very serious. The two land-claims settlements that have been signed recently, the James Bay and Northern Québec Agreement (JBNQA) and the Northeastern Québec Agreement (NEQA), are immensely long and complex documents, devised and negotiated by lawyers and administrators, and conceived to appeal to the most unashamedly and unredeemedly bureaucratic and western mind. To implement these documents successfully requires advanced training and experience in law, accounting, administration, and a host of other esoteric disciplines. Paradoxically, if the native persons for whose benefit and self-improvement these agreements were supposedly negotiated possessed the skills required to negotiate the agreements, it is highly unlikely that they would have ever needed the agreements in the first place, because they would have possessed all the skills that they might need to command status, respect and high salaries in our society or in their own!

At the risk of great over-simplification, what I have experienced personally may be summarized as follows: a small native group, suffering from an extreme degree of alienation from its own past and from the life of the dominant society, puts its trust in a small group of its own members and of white experts to negotiate a land-claims settlement. The native group assumes that the resulting agreement will in some near-miraculous way solve all its problems over-night, including its dependence on external skills and personnel. The native group comes to realize in the three or four years following the signing of the agreement that one of its principal effects is to highlight its members' lack of the skills required to implement it, to increase their sense of impotence and inability to control and shape their own lives and destinies, and to extend the nature and degree of their dependence on outsiders.

The preface to the James Bay and Northern Québec Agreement contains a sincere expression of the hope that that Agreement will put its beneficiaries ". . . on a new and more dignified footing, as a collectivity, than they have known in the past." My fear, and my growing conviction, is that, in most cases, land-claims settlements as we have experienced them in recent years, will have the opposite effect to that which is intended and that they will increase the sense of insufficiency and inferiority vis-à-vis our society of the vast majority of their beneficiaries.

Let me repeat in this context my conviction that land-claims settlements are primarily a mechanism for removing a certain class of potential legal obstacles or impediments to the exploration and exploitation of the resources, especially the non-renewable resources, of Canada's north. They are only secondarily, in my opinion and experience, a means of ensuring that Canada's native peoples ". . . obtain satisfaction and from this blossom socially, culturally and economically," as the federal government apparently intends.

Having asserted that the "problem" is not principally a legal one of the recognition of rights, I am obliged to try and say what I think it is. Obviously, there is no single problem; there are probably many communities, native and non-native, that do not have any particular problem at all; among those that do (and, sadly, one seems to get the impression that they are the majority), the problems may well be varied. Once again, then, I am reporting only what I have seen and experienced, and I leave it to the reader to determine whether any useful and valid generalizations can be made.

The root of the difficulties that I have perceived, which is particularly clear among young persons, is in part the lack of a clear self-image. I believe that this problem is aggravated by the fact that many native leaders and non-natives working for or sympathetic to native persons indulge constantly in powerful, emotive, and poetic rhetoric advocating an image of an Indian that is at variance with the real desires of many or most young Indians and that is probably no longer viable. These young persons are then hesitant to express or try to live up to their own self-image, which is not usually that of the independent, land-based, philosophical, natural conservationist, 'noble savage'. The dichotomy and the suffering that it causes to so many young persons is visible in the difference between their words and their actions, and is occasionally verbalized in moments of affection, intimacy, or intoxication.

My observation is that the situation is especially difficult for young men, perhaps because they have suffered proportionately more than young women, as a result of decades or even centuries of contact with white culture and economy. Women have at least maintained their traditional responsibilities for bearing and raising children and for other domestic functions.

Conclusions

As noted at the outset, I have touched upon a few of the many topics that have engaged my interest over the past years, so that my remarks have been highly personal and have in no way purported to constitute a 'state-of-the-art' review.

Looking at the list of papers in this volume, I cannot help but wonder whether my views are included more because it was thought necessary or expedient to invite a social scientist, especially one who works with native peoples, than because the majority of researchers in scientific fields are interested in what a social scientist has to say.

In either case, my message, I hope, is a reassuring one, for I suggest that researchers forge ahead with scientific deliberations and studies with a due respect for the social implications of what they are saying and doing, but without any of that sense of paranoia that permeates too much of science in Canada today.

As sensitive, informed, and intelligent individuals, most scientists will sense when research requires consultation with local populations. Researchers do not need an anthropologist any more than the average Indian or Inuit family needs one.

As for the small minority among researchers who do not care what local populations think of their activities, a battalion of anthropologists at their side day and night would not change them.

Acknowledgement

I wish to dedicate these thoughts to the memory of John Jerome Teal Jr., domesticator of the musk ox, explorer, scientist, and friend.

COMMUNITY PLANNING WITH THE NORTHERN RESIDENT

Alison Gill
Brandon University

Abstract: Many communities located in the boreal forest region are associated with resource extraction. Such northern resource communities differ demographically and socially from communities in the south, and the needs of residents vary accordingly. High rates of labor turnover characterize many of the communities creating serious economic and social problems. One partial solution to this problem is to design communities which produce higher levels of residential satisfaction. To achieve this goal the attitudes and preferences of residents must be incorporated into the planning process. A case study of two northern Manitoba communities, Thompson and Leaf Rapids, is presented to illustrate methodological procedures for eliciting residents'attitudes. An adapted form of repertory grid technique is employed. Selected results of the study are examined and the design implications are discussed.

The focus of this paper is northern resource communities, although some aspects have application in the broader context of all northern settlement, and planning in general. Frequent reference is made to the differences in perception that exist between researchers and managers in resource planning. The problem is compounded in the field of community development because a third component, the resident, is introduced. Improving the lines of communication between the planner and the resident is of paramount importance if northern resource communities are to succeed in meeting the specific needs of the northern inhabitant. The underlying theme of this paper is the need for planners to move away from the egocentric approach, so frequently adopted, and incorporate input from residents into the planning process.

In this paper I will review some aspects of my work on two single-resource communities in northern Manitoba: Thompson and Leaf Rapids (Gill 1982). The technique I employed for eliciting residents' responses may, from an applied perspective, seem somewhat esoteric. I am not, however, necessarily advocating that this technique should be employed, I am simply presenting it as one example of how, through the application of psychometric methods, a better understanding of residents' attitudes can be acquired and utilised as the basis for more appropriate planning decisions. First, I will present background to the problem, then, after discussing the technique employed, I will give some examples of the research results and their relevance to planning. I will conclude with some general comments on the need for ''behavioral'' input into design.

Background

The high rates of labor turnover which characterize many northern resource communities are indicative of dissatisfaction with the ''quality of life'' in these towns (Siemens 1973; Riffel 1975; Lauder 1977). This dissatisfaction may be related to the job, the company, the environment, or the individual, (or any combination of these factors) (Cawsey and Richardson 1975).

My research focuses on the relationship between the community and the individual. In an attempt to resolve problems between the environment and the individual, the first step is to establish the nature of the relationship. Since the early 1960s a branch of psychology, most commonly referred to as environmental psychology, has seen this relationship as one in which man's response to the environment is mediated by the mind's cognitive processes. (I should note that I am using the term ''environment'' in its broadest sense to include all aspects of both the physical (natural and man-made) and human or social environments.) Behavior is therefore not a direct response to the environment but instead to a distortion of ''reality''. It is the nature of these ''filters'' or distortions that create the differences in perception between the planners and the residents. In order to plan appropriately the planner must view the environment not through his own ''filter system'', but through those of the resident. In an attempt to resolve problems between the environment and the individual, the first step is to establish the nature of the relationship. The question a researcher must ask is ''What are the sources of dissatisfaction or satisfaction with the community?''

Initially, many northern resource towns were designed as carbon copies of southern communities. Thompson, Manitoba, located 750 km north of Winnipeg, is a good example of such a town. Built after INCO's discovery of nickel in the area in 1959, it grew to be the third largest city in Manitoba, with a peak population in 1978 of around 21,000. The population has since declined to around 14,500 as the result of economic recession. This boom and bust syndrome is a fact of life in all single-resource communities, and a fact which, in many instances, makes the viability of investing large amounts of capital on townsite development questionable. The design of Thompson by the Metropolitan Planning Commission of Winnipeg follows the ''neighbourhood principle'' and evokes frequent comparison to a Winnipeg suburb. The term ''wilderness suburb'' has been widely applied to such design (Porteous 1979).

Planners during the 1970s viewed such design as unsuited to the northern environment, and sought to plan communities that were congruent with the climatic and ecological conditions of the north. The result was communities such as Leaf Rapids, Manitoba. This town was built in 1973 to support the copper and zinc mining operation of Sherritt Gordon. The innovative design includes an architectural-award winning town centre which houses under one roof, nearly all retail, educational, and service facilities. A second factor which made Leaf Rapids unique was the joint agreement between the provincial government and Sherritt Gordon in the town's establishment. These two factors have made Leaf Rapids a show-piece for northern design. However, although the progress of the town has to some extent been monitored, residents' evaluations do not appear to have been elicited to ascertain if the planning has lived up to its rather lofty aspirations (Clarke 1978). In fact, there are indications from previous research that people prefer the familiar, and would rather live in a suburban-type environment (Siemens 1976). The original plan to have high-density housing in Leaf Rapids was, in fact, adapted to one with lower-densities as the result of objections by the residents of Lynn Lake, who were transferring to Leaf Rapids.

One should bear in mind a point made by Siemens (1976) who states that,

> ''By and large, the designers of new mining communities are not planning for northerners but, in large part, for professionals and para-professionals who have become accustomed to a southern suburban life-style and who wish to continue living in the manner with which they are familiar''.

Nevertheless, *are* the attitudes, needs, values and preferences of people who choose to live in northern resource towns identical to those of their southern counterparts? On the basis of my own research and personal experience I feel the answer to that question is "no". Even among northern residents various subgroups can be identified on the basis of attitude. For example, there are those whose primary concern is making money, which can then be converted into the required status items (e.g., snowmobiles, stereo equipment, motor-boats and four-wheel drive vehicles) for that group. Another group is oriented to outdoor wilderness-type activity (e.g., fishing, cross-country skiing and camping).

The specific aim of my study was to ascertain what components of place evaluation are most significant to the northern resident. This is based on the theoretical assumption that relative to the assessment of any object, a person employs only a limited number of dimensions. For example, in evaluating a place as a potential migration destination, a person might assess only the job opportunities and the cost of housing as being significant to the migration decision. Once these aspects are identified, they can then be utilised to assist in planning priorities.

Research Methodology

Existing attitude studies are often non-quantitative and frequently conducted by the company employer, thus tending to overemphasise the responses of male employees and management (Porteous 1976). Researchers in environmental design have since the late 1960s, looked to psychology for appropriate research techniques. Attitude studies have been criticised because they often include bias introduced by the researcher. In the case of work in northern resource towns, (where due to distorted information channels images are often stereotyped), I felt it important to attempt to overcome this potential bias.

The method I selected, the repertory grid technique (Kelly 1955), is potentially useful in the context of environmental design as it ensures the relevancy of the variables to the resident. The technique was initially developed in the field of clinical psychology and is based on the theory of personal constructs. Basically this states that a person organises information according to scientific principles in which objects and events are organised along personally established constructs. A construct is seen as a bipolar scale along which a person, in a comparative fashion, identifies one object as being different from another.

Most frequently researchers elicit the nature of constructs by asking respondents to state which of three objects (or elements) presented to them is different from the other two (Fransella and Bannister 1977). The construct is then identified by asking the reason for this distinction. I used a simplified version of this technique by presenting only pairs (or dyads) of elements. I asked a sample of residents to state in what way their community differed from each of 17 other places of varying size and function throughout Canada which were presented to them as elements. In either case (using triads or dyads) one ends up with a set of adjectives which represent a person's construct system with reference to a particular set of elements (in this instance places). The discussion with the respondent continues to determine the opposite end of the bipolar scale. In some cases this is obvious, for example, if a person in assessing the difference between Thompson and Winnipeg says that "Winnipeg has better shopping", then the bipolar construct can fairly readily be ascertained as "good-bad shopping facilities". Other responses are sometimes more difficult to identify. For example, several times I encountered the description of a community as "rough". After some discussion with the respondent the best opposite that could be found was "civilised". On average each individual appears to use about five or six different constructs in place evaluation, although there is considerable variation. Obviously this procedure is time consuming, although my technique of presenting only pairs did reduce the complexity of the procedure without a loss of validity. This procedure represents only the preliminary stage of investigation and is conducted to ensure that relevant aspects are included for evaluation.

The next stage is to have respondents rate their community according to the scale of adjectives they have provided. At this stage the technique as employed by psychologists was once again modified. Whereas psychologists deal with the individual (so that each person can employ his own constructs), larger scale environmental enquiry calls for an aggregation of responses. This diminishes the sensitivity of the technique but nevertheless does provide a set of scales which are much more relevant to the resident than one provided by the researcher. In my work, I took the most frequently mentioned constructs and ended up with a set of 46 bipolar adjectival scales which I then asked a much larger sample of 400 residents in the two communities to rate on a seven-point scale.

Analysis and Results

As my analysis was rather extensive, I will restrict this part of my discussion to some examples, and indicate their significance in a planning context. The responses to the rating scales were subjected to statistical reduction using principal components analysis. This procedure organises responses into groupings which can then be interpreted on the basis of their composite components. In the analysis, several different variables were examined to determine their influence on place evaluation, these included comparisons between responses of residents in Thompson and Leaf Rapids; between residents of differing residence periods, and of different sex and marital status. For example, in comparing the responses of short-term residents (less than five years residence), and longer-term residents, the components analysis reveals that short-term residents place greater emphasis on urban amenities and physical environment than do longer-term residents. Residents of over five years consider social aspects of the environment to be more important, and also appear to emphasise the "northern" aspects of the environment. What do these results mean in practical term? Certainly such attitude studies can only offer preliminary direction for planners. However, the greater importance of the physical environment, amenities, etc. to shorter-term residents may suggest the need for some alternative to the formal community structures (such as town councils and school boards) as these tend to be largely composed of longer term residents.

A further example of the results, which reveals differences in response between Thompson and Leaf Rapids residents, indicates a greater awareness and appreciation of the community's natural environment by Leaf Rapids residents. This is obviously a positive response to the careful ecological planning. However, despite this, the overall level of community satisfaction does not appear to be greater as a result of this. The most significant components relate to other community factors such as the social environment and amenity provision. This is seen largely as a function of size, and the implications seem to be that if a town is too small to support an adequate level of amenities, then satisfaction will be low despite careful planning. It also implies that attempts should be made to maximise the input into social planning for small communities.

A comparison of the differences between male and female responses confirms existing research (Jackson and Poushinsky 1971; Siemens 1973; Cawsey and Richardson 1975) that indicates that women have lower levels of residential satisfaction. The main components of place evaluation which are revealed for women (the majority of whom were married) in Thompson and Leaf Rapids are associated with urban amenities and the physical and recreational environments. The components extracted for men reflect several of the same dimensions but also reveal a concern for the social environment. It would seem that living in northern communities tends to focus concern on the needs of family life. The examples that I have presented offer some idea of the type of information that can be obtained using available psychometric techniques.

Conclusions

Recognition of the need for attitudinal research as a basis for design is not new, for example, in the early 1970s much of the research by the Centre for Settlement Studies at the University of Manitoba reflected this (Jackson and Poushinsky 1971; Matthiasson 1971; Siemens 1973; Riffel 1975). However, as far as I can determine little has been done in the context of northern settlement. One reason for this relates to the slow process of changing the training and attitudes of planners and designers to create an awareness of the importance of "behavioral" input. The Environmental Design Research Association (EDRA), an inter-disciplinary organisation based in Washington, DC, has been operating since the late 1960s and is making some headway in that area. In Canada, Canada Mortgage and Housing Corporation (CMHC), which provided much of the funding for the now defunct Centre for Settlement Studies, and also for my own research in the north, has shown an interest in such work, especially in the context of northern settlement. One of the reasons that there appears to be little action is that attitudinal studies are frequently time consuming, the measurement techniques are problematic, and the findings are often nebulous and not prone to generalisations. Nevertheless, these problems do not negate the need for such studies. I believe that this work is not necessarily the task of the planner, (although he or she should be aware of the value of it), but rather is best suited to trained social scientists in universities or private organisations who can contribute as consultants.

Although this paper has focused solely on northern resource towns the need for this type of residential input into planning is equally valid, and probably more essential in the context of native communities where cultural differences create an even greater gap between planner and user. Finally, I must add, that I have outlined only one preliminary phase of resident input into the planning process, obviously public participation is also desirable in a more direct manner at later stages.

Acknowledgements

The author gratefully acknowledges the receipt of a Canada Mortgage and Housing Graduate Scholarship to support this research.

Literature Cited

Cawsey, T. and P. Richardson. 1975. A Study of Labour Turnover in the Canadian Mineral Industries. Prepared for the Mining Association of Canada, School of Business Administration, University of Western Ontario, London, Ontario.

Clarke, W.F. 1978. Leaf Rapids − Year 1 + 4: A re-appraisal and Cost Analysis, Sheritt Gordon Mines Ltd. (Mimeographed).

Fransella, F. and D. Bannister. 1977. A Manual for Repertory Grid Technique. Academic Press, London, England.

Gill, A. 1982. Residents' Images of Northern Canadian Resource Communities. Ph.D. Thesis, University of Manitoba, Winnipeg, Manitoba.

Jackson, J.W. and N.W. Poushinsky. 1971. Migration to Northern Mining Communities: Structural and Social-Psychological Dimensions. Centre for Settlement Studies, Research Report 8, University of Manitoba, Winnipeg, Manitoba.

Kelly, G.A. 1955. The Psychology of Personal Constructs. Norton, New York.

Lauder, K. 1977. Planning for Quality of Life in New Resource Communities. M.A. Thesis, University of Waterloo, Waterloo, Ontario.

Matthiasson, J.S. 1971. Resident mobility in resource frontier communities: An examination of selected factors. *In*: Two Studies of Fort McMurray. Centre for Settlement Studies, Research Report 6, University of Manitoba, Winnipeg, Manitoba.

Porteous, J.D. 1976. Quality of life in British Columbia company towns: Residents' attitudes. New Communities in Canada, Special Issue, Contact, J. Urban and Environ. Affairs, 8: 332-346.

Porteous, J.D. 1979. Quality of life in resource towns: Problems and proposals. *In*: The Viability and Livability of Small Urban Centres. B. Sadler and F. Jankunis (Eds.). Environment Council of Alberta, Edmonton, Alberta. pp. 43-52.

Riffel, J.A. 1975. Quality of life in resource towns. Centre for Settlement Studies. University of Manitoba, Winnipeg, Manitoba.

Siemens, L.B. 1973. Single-enterprise community studies in northern Canada. Centre for Settlement Studies, Occasional Paper 7, University of Manitoba, Winnipeg, Manitoba.

Siemens, L.B. 1976. Single-enterprise communities on Canada's resource frontiers. New Communities in Canada, Special Issue, Contact, J. Urban and Environ. Affairs, 8: 277-297.

NATIVE CANADIAN INTERPRETERS MULTIPLE ROLE FUNCTIONS IN HEALTH CARE DELIVERY

William W. Koolage, Lorna W. Guse, and Joseph M. Kaufert
University of Manitoba

Abstract: Native Canadian interpreters in two Manitoba hospitals perform multiple functions in the delivery of health care to northern native clients. They serve as translators in clinician/native patient interaction, culture-brokers in mediating between two different cultural systems, patient advocates in maintaining patients' rights, and compassionate visitors for native patients who are far from home and family.

During preliminary investigations into the problems experienced by northern native patients evacuated to southern Euro-Canadian dominated hospitals, we were impressed by the vital roles played by native Canadian interpreters as translators, culture-brokers, patient advocates, and compassionate visitors for native patients who were far from home and family. We also became aware of the problems faced by the interpreters due to their multiple role functions, the expectations of other health personnel, and the needs of their native clients. In an effort to develop teaching materials for health care personnel (and social science students), we began to look more closely at the ways in which interpreters worked to alleviate cultural and linguistic barriers between health care practitioners and native clients.

One of the problems we have found is that misunderstandings arise between health care practitioners and interpreters when the former view the latter's role as simply that of a 'pragmatic translator' — that is, someone who simply translates one technical language into another. Many health care workers become frustrated when they think the translators are censoring or inadequately translating a patient's reply. Since the clinicians do not know Dene, Inuktitut, Algonkian (Cree, Saulteau), or Siouan, they have no way of knowing that the interpreters are establishing rapport, putting the patient at ease, or finding culturally appropriate analogies for complex western scientific terminology and concepts.

In their role as culture-brokers, the interpreters act as intermediaries between different cultural (e.g., native and Euro-Canadian) and subcultural (e.g., remote northern communities and urban southern hospitals) systems. This mediation involves sensitizing patients to the hospital milieu, explaining procedures and treatments, and obtaining meaningful informed consent. In patient education, the interpreters must use their knowledge of health care procedures, human physiology, and their own cultural background to make education culturally relevant. For example, Mrs. Pauline Steiman, a native nurse, uses the analogy of mixing gas and oil in the right proportions for an outboard motor in order to make diabetes education meaningful to a northern native woman.

The interpreters also try to sensitize practitioners to the cultural and environmental background of the native patient. This information is especially important where treatment regimens appropriate in the south are unsuitable to northern communities and lifestyles.

Interpreters act as intermediaries between clinicians and patients and their families when treatment agendas are different. One example of how an interpreter acted as culture-broker/mediator concerns a man who was brought to a Winnipeg hospital for treatment. He was in a coma and his family insisted that a traditional medicine man be brought in from outside the city. The interpreter, a registered nurse, was caught between the physician-directed treatment regimen in the intensive care unit and the family's insistence on a four-day treatment regimen by the medicine man. Through her understanding of both cultures and their ''healing'' systems, the interpreter mediated the situation so that both treatment regimens were accommodated. The patient recovered and both his physician and his family were satisfied.

As patient advocates, the interpreters are often faced with conflicts between their role of health system employee/practitioner and their patients' rights. This is especially true when meaningful consent for a procedure has not been obtained and the procedure runs counter to the patient's cultural and/or individual beliefs and values. We have noted this conflict in several instances of the implantation of pacemakers to which many elderly natives strenuously object.

As compassionate visitors, the interpreters attempt to greet all native monolingual patients on their arrival and to visit with them during their hospitalization. While most of their work is with obstetric/gynecology patients and on the children's wards, they also attempt to visit all native patients in the hospital. While visiting, the interpreters will provide assistance with the particular problems experienced by native patients. Northern patients especially comment on the loneliness of hospitalization and their appreciation for visitors, their problems in obtaining news from home, and their difficulties in obtaining information concerning their length of stay in hospital.

In Manitoba, efforts have been made since 1973 to increase the number of full-time interpreters in the health care system. There are now eight full-time personnel at the Health Sciences Centre and St. Boniface Hospital. Hopefully, with better sensitivity to the numerous roles performed by native Canadian interpreters, we can all gain from this vital resource in health care delivery to northern native clients.

Literature Cited

Anderson, R.B.W. 1976. Perspectives on the role of interpreter. *In*: Translation: Applications and Research. R.W. Brislin (Ed.). Halstead Press, New York.

Brislin, R.W. 1976. Translation: Applications and research. Halstead Press, New York.

Clatworthy, S.J. 1980. The demographic composition and economic circumstances of Winnipeg's native population. Institute of Urban Studies Publication No. 70: 1-57. Winnipeg, Manitoba.

Guse, L. 1981. Impact of regionalization on the native maternity patient. Paper presented to the Nursing Association of the American College of Obstetricians and Gynecologists, Winnipeg, Manitoba.

Guse, L. 1982. Health care delivery: Structural and perceptual features of medical evacuation of native maternity patients. University College Ethnology Papers, No. 2.

Guse, L. and W. Koolage. 1979. Medical anthropology, social science and nursing: A bibliography. University of Manitoba Anthropology Papers 22: 1-99.

Kaufert, J.M. 1979. Patterns of primary care and secondary service use among the population of northern native communities. Winnipeg Clinic Foundation, Winnipeg, Manitoba.

Kaufert, J. 1979-1981. Data base: Taped interviews with 40 physicians, nurses and interpreters. Unpublished tapes.

Kaufert, J., W. Koolage and J. Conner. 1981a. Cultural practices in therapeutic interaction. Instructional Media, Faculty of Medicine, University of Manitoba, Winnipeg, Manitoba.

Kaufert, J., W. Koolage and J. Conner. 1981b. The role of native interpreters and the urban health care delivery system. Instructional Media, Faculty of Medicine, University of Manitoba, Winnipeg, Manitoba.

Koolage, W. 1968. The Chipewyan of Camp 10. pp. 61-127. *In*: Ethnographic survey of Churchill. John J. Honigmann (Ed.). Institute for Social and Economic Research. Chapel Hill, North Carolina.

Koolage, W. 1976. Differential adaptations of Arthapaskans and other native ethnic groups to a Canadian northern town. Arctic Anthropology, 13: 70-83.

Koolage, W., and J. Kaufert. 1982. Health workers see ways culture affects communication. Bulletin, University of Manitoba, 16: 1-2.

McCaskill, D. 1980. The urbanization of indians in Winnipeg, Toronto, Edmonton, and Vancouver: A Comparative Analysis. Culture, 1: 83-89.

Moore, Esther. 1982. Native services programme, St. Boniface Hospital. Paper presented at the Ninth Annual Meeting, Canadian Ethnology Society, Vancouver, British Columbia.

Paine, R. (Ed.). 1971. Patrons and brokers in the east Arctic. Institute of Social and Economic papers No. 2. Memorial University of Newfoundland, St. John's, Newfoundland.

Ray, M.D. 1976. Canadian urban trends: Metropolitan perspective, Vol. 2. Ministry of Supply and Services Canada, Ottawa, Ontario.

Smith, Margaret. 1982. Services to native patients, Health Sciences Centre. Paper presented at the Ninth Annual Meeting, Canadian Ethnology Society, Vancouver, British Columbia.

FOREST RESEARCH IN THE TAIGA OF ALASKA: ISSUES FOR INTEGRATED RESEARCH

C.T. Dyrness, R. Norum, C.W. Slaughter, V. Van Ballenberghe, L.A. Viereck, R.A. Werner and J. Zasada
USDA Forest Service

Abstract: High-latitude forests and associated renewable resources are being subjected to increasing pressures for extraction, 'development' and overt management in Alaska and Canada. We outline several integrated research programs underway in the discontinuous-permafrost taiga of central Alaska. Willow Island studies — The river bottom-land forests of central Alaska are highly productive of wood fiber, and comprise an important wildlife habitat and aesthetic (tourism) resource. The research plan focusses on the consequences of cutting and site preparation approaches for regeneration rates, site productivity, and floodplain succession processes and strategies. Headwaters catchment management — The role of taiga uplands in providing freshwater resources, aquatic habitat and forest products is under study in a 104-km^2 research watershed. The scope provides for analysis of the influences of selected resource management practices (timber harvest, prescribed fire, accelerated biomass production) on the catchment/stream system. Wildfire — A study of the way upland forests respond to fire was begun at the Washington Creek Fire Study Area in 1978. Foraging ecology of moose in Denali National Park — Relationships between forage quality and population dynamics of moose in a subarctic ecosystem is the main focus of this study. Management of white spruce stands to prevent losses of fiber production caused by bark beetles — The development of silvicultural treatments that promote vigorous tree growth on selected sites is the ultimate objective of this work.

These examples of current Forest Service integrated research are indicative of the breadth of questions and problems confronting resource managers in the north. Environmental issues which must be faced in the taiga are not susceptible to simplistic, single-discipline attack; developing the understanding which is requisite to rational planning, management, and development demands continuance and intensification of multi-discipline, integrated long-term research in northern ecosystems.

The Institute of Northern Forestry at Fairbanks, Alaska, is one of three forestry research facilities maintained by the USDA Forest Service in Alaska; other research units are at Juneau and Anchorage. The mission of the Fairbanks Laboratory is to develop an understanding of the ecology of forest ecosystems of interior Alaska for use in land-use planning and management. Twelve research scientists use a multi-disciplinary approach to research as a method of studying ecosystems which have closely related components. Disciplines of staff scientists include silviculture, forest genetics, forest ecology, forest entomology, fire science, forest hydrology, soil science, and wildlife ecology.

Until recently forestry uses of interior Alaska taiga were limited to harvesting for local sawmills, fuel wood, and house logs, but today the region stands on the threshold of greatly accelerated forest usage. A large part of this potential for increased use of forest resources can be attributed to the transfer of large tracts of land from public to private ownership under provisions of the Alaska Native Claims Settlement Act of 1972 and

Alaska National Interest Lands Act of 1980. Naturally, these new owners are most anxious to learn what economic benefits they may derive from the land. Public agencies must be in a position to recommend sound land use practices if wise decisions are to be made. Unfortunately, present understanding of the many and varied facets of taiga soil–water–plant–animal systems is often inadequate to suggest environmentally sound management strategies.

We outline several integrated research studies underway in the discontinuous-permafrost taiga of central Alaska. All studies are in environmental settings selected because they are subject to immediate and long-term ''people pressures'', amenable to rational management for sustained production of selected renewable resources, and poorly understood from both a scientific and resource management viewpoint.

Regeneration of Floodplain White Spruce Stands Following Logging

The floodplain forests of interior Alaska represent an important potential source of wood and fiber products for local use and export, as well as providing important wildlife habitat. These forests were exploited during the search for gold and the settlement period of the early 1900s but have received little attention as a forest resource in the past 50 years. In recent years, however, an increased interest in interior Alaska forests as a source of fuel and lumber has resulted in increased demand to utilize the floodplain forests. In spite of the extensive harvesting of floodplain forests in Alaska in the early 1900s, there is no documented information on management of this resource in Alaska. There has, however, been research on regeneration of white spruce (*Picea glauca*) stands on floodplains in the southern Yukon Territory of Canada (Gardner 1980) and on upland sites in Alaska (Zasada and Grigal 1978). Knowledge of the primary successional sequence on floodplain sites is relevant to their management for forest products.

The bare silt bars formed by deposition of river-borne sediments are initially colonized by various willows (*Salix spp.*), alder (*Alnus tenuifolia*) and balsam poplar (*Populus balsamifera*). The willows and alder dominate these sites for a period of 15 to 20 years. Then, balsam poplar becomes the dominant species as it overtops the willows and alder. For the next 60 to 80 years, balsam poplar continues to dominate. During this time, the white spruce understory develops and eventually overtops the balsam poplar. White spruce can invade these sites rapidly or become established gradually over a period of years. This process will depend on such variables as seed crop periodicity, proximity to seed sources, flooding, silt deposition, and other variables. The length of time which white spruce dominates these sites is not well documented. Several generations of white spruce may occur on these sites before black spruce (*Picea mariana*) invades and becomes the dominant species. Important changes in organic layer depth, soil temperature regime, and permafrost occur in association with the changing overstory structure (Viereck 1970; Van Cleve and Viereck 1981).

At present the species in greatest commercial demand is white spruce. The earlier successional stages, however, hold great potential for fiber and energy biomass production as markets and demand develop. University of Alaska staff are currently studying the potential for short rotation, intensive culture production on these and other sites in interior Alaska. In order to develop forest management alternatives for white spruce on floodplain sites and to determine the impact of harvesting on the white spruce component of this ecosystem, an interdisciplinary study is being conducted on Willow Island, a 235 ha island in the Tanana River about 25 km southwest of Fairbanks.

Willow Island is an ideal setting for the study of floodplain white spruce because it contains a number of different-aged river terraces with two different soil series. This combination supports eight different white spruce associations, from a woodland type with less than 25 per cent canopy cover to dense closed stands with over 75 per cent cover. In addition, small areas have early successional stages of willow and young balsam poplar. The younger terraces of the island have permafrost-free soils formed recently from river alluvium and are classified as Typic Cryofluvents. The soils of the older terraces of the island are underlain by permafrost with an active layer of one-half to one metre and are classified as Pergelic Cryaquepts. The open stands of white spruce have a well developed alder layer of both *Alnus tenuifolia* and *A. crispa* and a nearly continuous moss mat of predominantly *Hylocomium splendens*. Lingenberry (*Vaccinium vitis-idaea*) is an important component of the dwarf shrub–herb layer in some stands but is absent or of low cover in others. The closed stands have a sparse alder layer and little development of a dwarf shrub–herb layer but have a luxuriant feathermoss cover. Some tamarack (*Larix laricina*) and balsam poplar occur in the closed stands.

Tree ages in the stands to be harvested range from 100 to 350 years. Stands are generally even-aged with some examples of multiple-aged stands. A small area in the center of the island, occupied by a mixed black and white spruce woodland and with a low shrub layer dominated by *Ledum groenlandicum*, may have passed through two or more generations of white spruce before developing its present stand characteristics. No evidence of previous fires exists in any of the spruce stands on the island.

Pre-logging treatment work has included inventory of vegetation, soils, and small mammals. Selected stands have been intensively studied and sampled in order to thoroughly document conditions before treatment. This work has included the installation and inventory of permanent understory vegetation and tree layer plots. Standard tree measurements have been made (density, volume, basal area, diameter, age, and height), and a number of stands have been stem-mapped and vertical stand profiles constructed.

Two silvicultural treatments will be applied: clearcutting and shelterwood. Treated units will vary in size from 2.4 to four ha (six to ten acres). Two shelterwood densities will be applied: 37 to 62 trees/ha (15 to 25 trees/acre) and 86 to 111 trees/ha (35 to 45 trees/acre). Post-harvest site preparation treatments will consist of mechanical scarification and burning. Regeneration treatments will consist of natural seeding, artificial seeding, and planting of containerized seedlings.

After harvesting, the course of tree regeneration will be followed on all units. In some instances individual seedlings will be permanently marked and their development followed each year. In addition, seedfall within the harvested units will be monitored. Response of shrubs, herbs, mosses, and lichens to the silvicultural treatments will be followed on the established permanent plots.

Headwaters Catchment Management

Two facets of watershed management research are of high priority in central Alaska. These are: (1) precipitation–runoff relations in the taiga, and (2) stream system quality and site productivity in relation to landscape stability and sediment production. Both areas include concern with the natural, undisturbed state and with consequences of resource management and landscape perturbation.

A principal characteristic of the northern taiga is the paucity of quantitative information concerning all aspects of watershed relationships, whether precipitation inputs, snowpack distribution and ablation, influences of vegetation on disposition of precipitation and timing of streamflow, stream system physical–chemical–biological relationships, rates of sediment production and transport, site productivity alteration due to soil movement, or management and rehabilitation practices (Helmers and Cushwa 1973).

The focal point of the Institute of Northern Forestry watershed studies is the Caribou–Poker Creeks Research Watershed (CPCRW). The 104 km^2 watershed is in the Yukon–Tanana Uplands of central Alaska, 48 km north of Fairbanks. The Caribou Creek basin is reserved for low-impact process and observational studies, while the Poker Creek basin is designated for experimental landscape manipulation research. This division corresponds to the 'representative' and 'experimental' watershed research categories recommended by the United Nations Education, Scientific and Cultural Organization (Toebes and Ouryvaev 1970).

The setting is a region of uplifted Precambrian rock (Birch Creek schist) overlain by a thin mantle of loess originating from glacial outwash plains north of the Alaska Range. Elevations range from 210 to 826 m. The watershed is partially underlain by permafrost, with south-facing slopes generally free from perennially frozen soils. Valleys and north-facing slopes have perennially frozen ground. The natural vegetation reflects typical subarctic patterns: north slopes are dominated by black spruce underlain by shrubs and thick moss/lichen ground cover, while south-facing slopes support vigorous stands of aspen (*Populus tremuloides*), birch (*Betula papyrifera*), and lesser amounts of white spruce and black spruce. Alder is present on north and south slopes. Valley floors are generally poorly drained and support dense stands of willow, dwarf birch (*Betula glandulosa*), and assorted shrubs and forbs.

The entire catchment is free from recent (about the past 30–40 years) human activities, with no evidence of recent prospecting, mining, logging, or other disturbances. The area has a typical fire-influenced vegetation pattern, with charcoal present in many soil profiles; charred stumps are found in most sectors of the watershed. There are traces of early prospecting and logging activities, probably dating from the 1905–1930 era.

Reports now available, specific to CPCRW, concern geology, stream and bedrock geochemistry, soils (Rieger *et al.* 1972), vegetation distribution, and selected vegetation community analysis (Troth *et al.* 1976). A map defining effective solar radiation input for the entire basin has been prepared. Baseline climatic and hydrologic data have been acquired since late 1969; several tabulations are available (see Lotspeich and Slaughter 1981). Other completed work concerns local site/vegetation relationships, hydrologic modelling and analysis, and mesoclimates of slopes. Stream quality research has included both chemical and biological parameters in the first- through fourth-order stream continuum (Hilgert 1979), with initial results pointing up the need for more detailed work in this area.

A number of studies in the watershed have related to the role of soil moisture in the active layer and documented moisture dynamics of the soil mantle in a black spruce stand (Kane *et al.* 1978). Other studies have investigated the role of organic soils in subarctic hydrology (Slaughter and Kane 1979), and erosion and soil movement associated with vegetation removal on experimental plots (Aldrich and Johnson 1979).

The watershed provides an ideal opportunity to study the role of aufeis ('icings' or 'nalyds') in catchment hydrology (Slaughter 1982). Aufeis may involve as much as 40 per cent of the total winter streamflow of a headwaters stream in temporary solid state storage. Water movement in winter between soil and snowpack has been measured as has the snowpack areal distribution and snow drift formation in an upland catchment. Still, many questions concerning the hydrologic and ecologic roles of seasonal snow in the taiga remain unresolved.

Work on developing the catchment as a viable research site in terms of facilities and access has continued. A 20 km vehicular trail network linking study sites has been completed with major assistance from the Alaska Division of Lands, the Bureau of Land Management, INF, and USA CRREL. Among the positive results of these efforts are recent designation of CPCRW as an Experimental Ecological Reserve by The Institute of Ecology (TIE), and emergence of the research area as a focal point for circumpolar cooperative programs in high-latitude watershed research (Slaughter and Santeford 1975).

Among new (1982) Institute of Northern Forestry catchment related research initiatives are: (1) study of the influence of vegetation clearing along a first-order stream on stream physical and chemical quality, aquatic invertebrates, and primary productivity. The baseline measurement and analysis were begun in 1982, with treatment (clearing) scheduled for summer 1983 and, (2) hydrologic calibration of a permafrost-free, south-facing 2.5 km^2 watershed in preparation for study of the effects of timber harvest on stream quality and productivity. The measurement program was initiated in 1981, with timber harvest projected for 1984–1985.

Fire Effects

Wildfire has always been an important element in Alaska taiga ecosystems. Between 1940 and 1970 an average of 400 000 ha burned annually in interior Alaska. Fire control efforts have reduced this to an average of 240 000 ha annually in the 1970s. Under the natural fire regime, some forest types may burn as frequently as every 26 years whereas other types have much longer fire cycles (Yarie 1981). Tree stands older than 170 years are rare except on islands in the floodplain, at treeline, or within moist drainages.

For the past decade the Institute of Northern Forestry has been involved in research on the effects of wildfire on vegetation and soils (Viereck and Dyrness 1979; Viereck and Schandelmeier 1980). Controlled experimental burning in the black spruce/feathermoss forests of interior Alaska began in 1976 (Viereck *et al.* 1979). Although those early efforts were a modest start, they gave guidance for planning the work that followed. Four small units were burned under rather severe, dry fire conditions. Three were burned when both the surface fuels and forest floor were dry. In the other fire, the surface fuels were very dry, but the forest floor was moist. The results indicated that the fire intensity was not nearly as influential as the moisture content of the forest floor in determining the degree of organic layer consumption and consequent plant responses.

Taking guidance from the 1976 fires and from work done in other places, an expanded study was designed to determine what prefire variables can be used to predict potential fire results in upland black spruce forests. Seven study units, ranging in size from two to about 3.5 ha, were prepared for careful measurement, inventory, and documentation in 1977. All seven units were burned in 1978 over a considerable range of fire conditions, giving a useful array of treatments.

A coordinated interdisciplinary effort was mounted to evaluate fuels, vegetation, soil physical and nutrient status, and fire behavior. Distribution of trees, shrubs, forbs, mosses, and lichens was quantified. The recognized importance of the forest floor led to extensive measurements of the depth of the forest floor, both before and after burning, along with classification of burn severity. Measurements of virtually all transects, quadrats, and plots were repeated after burning, and some additional studies undertaken. An intensive study of the artificial regeneration of trees and tall shrubs was completed, as well as a thorough study monitoring the response of two species of *Vaccinium* to different severities of forest floor burn. The chronological changes and shifts in plant abundance, composition, and patterns, as well as shifts in soil physical and chemical properties, remain under close scrutiny.

Fire behavior variables were measured during the fires and have since been used as part of a successful attempt to adapt the national fire behavior forecasting system to black spruce forests in Alaska.

Forest floors after burning were comprised of a small-scale mosaic of five classes: unburned, scorched, lightly burned (surface blackened but not burned to any depth), moderately burned (about one-half the organic layer consumed), and heavily burned (organic materials entirely consumed). Percentages of the seven units in moderately and heavily burned condition ranged from 11 to 77 per cent and percentage decrease in average forest floor thickness varied from 27 to 63.

For the first three years after fire, biomass production was greater on heavily burned than on lightly burned sites (58 g/m^2 per year versus 37 g/m^2 per year on an annual basis). Heavily burned sites were completely dominated by the invading species *Epilobium angustifolium*, *Ceratodon purpureus*, and *Marchantia polymorpha*, whereas lightly burned plots were occupied by sprouting species such as *Calamagrostis canadensis*, *Vaccinium uliginosum*, and *Ledum groenlandicum*. In addition, tree seedlings were almost completely restricted to heavily burned locations.

Post-fire soil analyses also showed some striking correlations with forest floor burning classes. Soil pH and amounts of total P and available P in the forest floor increased significantly as a result of burning. In all cases, increases reached a maximum in moderately and heavily burned areas. Total N in the forest floor increased significantly in moderately burned areas, but decreased slightly in heavily burned areas. Total N and total P showed smaller increases in the surface mineral soil as a result of burning. Supplies of available P in the mineral soil increased almost fourfold in moderately burned and over sixteenfold in heavily burned areas.

Moose Habitat Ecology

In 1980 wildlife scientists at the Institute of Northern Forestry began a long term study of moose habitat ecology in Denali National Park. This work was initiated to better understand the kinds of upland plant communities that moose select in summer, to determine how moose exploit such communities, to determine how these communities contribute to the seasonal energetics of moose, and to develop management prescriptions for creating, preserving, and enhancing these communities as integral components of the environment that moose require. Denali National Park was chosen because it contains shrub-dominated, treeline communities typical of moose summer range across a broad area of Alaska. In addition, the park supports a dense population of moose, it has road access, park moose are habituated to humans, and information is available on park moose from other studies extending back to the 1930s.

Research to date has focused on four field objectives as follows: (1) to determine the location and size of seasonal home ranges of moose, (2) to determine seasonal food habits, (3) to determine time/activity budgets of bulls and cows and estimate rates of forage consumption, and (4) to determine vegetation quantity and quality in plant communities heavily utilized by moose. Radiotelemetry and direct observation of moose serve as the primary field techniques. Prolonged daylight from mid-May through early August facilitates observation around the clock.

Preliminary results suggest that summer home ranges of moose average about 27 km^2, but certain individuals move straight-line distances as great as 64 km. Adult males use shrub dominated, high elevation habitats more than females with calves. The latter prefer spruce dominated habitats with relatively sparse understories. One shrub species, *Salix planifolia* spp. *pulchra*, comprises about half the summer diet of both males and females. Time of foraging peaks in early June at about 10 hours per day. By late July, this is reduced to about 60 per cent of early June levels. By early September, adult bulls stop feeding and do not begin again until about 3.5 weeks later, just prior to the peak of the rut. Bulls gradually increase foraging time in October and by early November are apparently ingesting food at a maximum rate governed by processing time in the rumen.

Rates of forage consumption and defecation have been estimated by direct observations. These, along with the results of nutrient analyses of simulated moose bites, will yield estimates of digestible energy that, in turn, can be related to the quantitative characteristics of different plant communities selected by moose. Forage intake as a function of plant species availability will be used as an index to species preference. We are trying to better understand how quality influences forage and habitat selection by assuming that energy is the most limiting 'nutrient' and gathering data to determine how intake of digestible energy is related to vegetation biomass. Eventually, these findings should allow land managers to produce the arrangement of plant communities that moose find optimal.

Two aspects of this moose research provide the opportunity for multifunctional interactions among scientists. First, plant ecologists are assisting in the vegetation studies by quantitatively sampling certain plant communities. This will provide a basic description of these communities and will facilitate estimates of vegetation biomass and production of moose forage. Second, research entomologists are assisting with studies of external parasites of moose that may be important drains on reserves of moose energy, and are involved in studies of insects consuming various willow species. The impact of these consumers as potential competitors of moose will be evaluated.

Spruce Beetle Studies

The spruce beetle, *Dendroctonus rufipennis*, is the most destructive forest pest in Alaska (Werner *et al*. 1977). From 1978 to 1981, 140 000 ha of white spruce, *Picea glauca*, have been infested in the Cook Inlet area of south central Alaska. Recurring infestations have resulted in the decimation of mature white spruce stands (Werner and Holsten 1983). Many stands appear to be succeeding toward subclimax or climax timber types, and some contain white spruce as a minor component.

Studies on spruce beetle behavior and ecology have been underway for six years. Major emphasis includes the development of a hazard rating model for tree and stand susceptibility to beetle attack, the determination of the relationship between rapid beetle population increases and one- and two-year life cycle beetles, and the field testing of several chemicals for preventive and remedial control of beetles in high value trees and stands.

Results indicate that trees with less than average radial growth during the last five years were highly susceptible to beetle attack. Contrary to accepted theory, trees with larger than average diameter are not necessarily high risk trees unless radial growth is less than average. A rudimentary stand hazard rating model was developed (Hard *et al*. 1983). Recommendations for silvicultural treatment of stands to prevent beetle attack are that stands with basal areas of 45 to 50 m^2/ha be reduced by 11.5 to 23 m^2/ha. Trees with more than four mm cumulative radial growth during the last five years should be retained since they are least susceptible to beetle attack.

The rapid spread of spruce beetle populations throughout southcentral Alaska during the past four years was not readily understood because spruce beetles normally have a two-year-life cycle in Alaska. One-year-life cycle beetles were found in 1978. One-year populations only developed on the south sides of standing trees and the south and top sides of felled trees which were warmer because of direct contact with solar radiation. Inner bark temperature was the environmental factor that regulated the rate of beetle development. The one-year-life cycle populations partially explain the rapid dispersion of infestations.

High value trees in campgrounds, roadside rest areas, and residential areas should be protected with insecticide treatment. Sumithion at a 2 per cent concentration and permethrin at a 0.25 per cent concentration protected uninfested white spruce trees from beetle attack for 16 months. Sumithion at a two per cent concentration killed 90 per cent of emerged and 66 per cent of non-emerged beetles when applied to infested trees. Sumithion is being recommended for registration for use as a preventive and remedial insecticide treatment for spruce beetles in Alaska.

Beetle-killed standing white spruce trees dead for as long as 50 years produced high quality bleached and unbleached pulp (Werner *et al*. 1983). Pulp yield and paper strength properties remained extremely high regardless of how long trees were dead. Harvest times of beetle-killed spruce can be extended without undue loss of wood pulp fiber. This information will enable the forest resource manager to adequately plan harvest times when market facilities are available, foreign demand for export develops, and areas now inaccessible are open.

Conclusions

The five studies discussed briefly in this paper are a sample of the types of research being carried on at the Institute of Northern Forestry. Other current or recent projects include vegetation classification and description, forest genetics, host–hardwood defoliator relationships, small mammal ecology, and the structure and functioning of a black spruce ecosystem. In the five studies described, a multifunctional approach has proved useful in developing basic ecological information that will be needed for developing management guidelines as more intensive forest land use occurs in interior Alaska.

Literature Cited

Aldrich, J.W. and R.A. Johnson, 1979. Surface erosion and sedimentation associated with forest land use in Interior Alaska. Report IWR-99, Inst. of Water Resour., University of Alaska, Fairbanks, Alaska.

Gardner, A.C. 1980. Regeneration problems and options for white spruce on river floodplains in the Yukon Territory. *In*: Forest regeneration at high latitudes. M. Murray and R.M. Van Veldhuizen (Eds.). U.S.D.A, For. Serv., Gen. Tech. Rep. PNW-107. pp. 19-24.

Hard, J.S., R.A. Werner and E.H. Holsten. 1983. White spruce susceptibility to attack during the early years of a spruce beetle outbreak in Alaska. Can. J. For. Res. 13: 678-684.

Helmers, A.E. and C.T. Cushwa. 1973. Research opportunities and needs in the taiga of Alaska. U.S.D.A., For. Serv., Gen. Tech. Rep. PNW-2.

Hilgert, J.W. 1979. Upland stream quality in the Caribou−Poker Creeks watershed. *In*: Proceedings, 30th Alaska Science Conference, Alaska Division, AAAS, and Alaska Section, American Chemical Society, Fairbanks, Alaska.

Kane, D.L., R.D. Seifert and G.S. Taylor. 1978. Hydrologic properties of subarctic organic soils. Report IWR-99, Inst. Water Resour., University of Alaska, Fairbanks, Alaska.

Lotspeich, F.B. and C.W. Slaughter. 1981. Preliminary results of a study on the structure and functioning of a taiga research watershed. Report IWR-101, Inst. Water Resour., University of Alaska, Fairbanks, Alaska.

Rieger, S., C.E. Furbush, D.B. Schoephorster, H. Summerfield, Jr. and L.C. Geiger. 1972. Soils of the Caribou−Poker Creeks Research Watershed, interior Alaska. Soil Conserv. Serv., U.S.D.A. and USACRREL. Tech. Rep. 236; USACRREL, Hanover, New Hampshire.

Slaughter, C.W. 1982. Occurrence and recurrence of aufeis in an upland taiga catchment. *In*: The Roger J.E. Brown Memorial Volume. Proc. Fourth Canadian Permafrost Conference, Calgary, Alberta. National Research Council of Canada, Ottawa, Ontario. pp. 182-188.

Slaughter, C.W. and Kane, D.L. 1979. Hydrologic role of shallow organic soils in cold climates. *In*: Canadian Hydrology Symposium 79 — Cold Climate Hydrology Proceedings. National Research Council of Canada, Ottawa, Ontario. pp. 380-389.

Slaughter, C.W. and H.S. Santeford. 1975. High-latitude research basins as settings for circumpolar environmental studies. *In*: Proceedings, Circumpolar Conference on Northern Ecology. National Research Council of Canada, Ottawa, Ontario. pp. IV-57-68.

Toebes, C. and V. Ouryvaev. 1970. Representative and experimental basins: An international guide for research at practice studies and reports in hydrology 4. UNESCO. Paris, France.

Troth, J.L., F.J. Deneke and L.M. Brown. 1976. Upland aspen/birch and black spruce stands and their litter and soil properties in interior Alaska. For. Sci. 22: 33-44.

Van Cleve, K. and L.A. Viereck. 1981. Forest succession in relation to nutrient cycling in the boreal forest of Alaska. *In*: Forest succession — concepts and application. D.C. West, H.H. Shugart and D.B. Botkin (Eds.). Springer-Verlag, New York. pp. 185-211.

Viereck, L.A. 1970. Forest succession and soil development adjacent to the Chena River in interior Alaska. Arctic and Alpine Research 2: 1-26.

Viereck, L.A. and C.T. Dyrness (Tech. Eds.). 1979. Ecological effects of Wickersham dome fire near Fairbanks, Alaska. U.S.D.A., For. Serv., Gen. Tech. Rep. PNW-90.

Viereck, L.A., Joan Foote, C.T. Dyrness, Keith Van Cleve, Douglas Kane and Richard Seifert. 1979. Preliminary results of experimental fires in the black spruce type of interior Alaska. U.S.D.A., For. Serv., Res. Note PNW-332.

Viereck, L.A. and L.A. Schandelmeier. 1980. Effects of fire in Alaska and adjacent Canada — a literature review. U.S.D.I. BLM-Alaska Tech. Rep. 6.

Werner, R.A., R.H. Baker and P.A. Rush. 1977. The spruce beetle in white spruce forests of Alaska. U.S.D.A., For. Serv., Tech. Pap. PNW-61.

Werner, R.A. and E.H. Holsten. 1983. Mortality of white spruce during a spruce beetle outbreak on the Kenai Peninsula in Alaska. Can. J. For. Res. 13: 96-101.

Werner, R.A., E.E. Elert and E.H. Holsten. 1983. Evaluation of beetle-killed white spruce for pulp and paper. Can. J. For. Res. 13: 246-250.

Yarie, J. 1981. Forest fire cycles and life tables: a case study from interior Alaska. Can. J. For. Res. 11: 554-562.

Zasada, J.C. and D.F. Grigal. 1978. The effects of silvicultural system and seedbed preparation on natural regeneration of white spruce and associated species in interior Alaska. *In*: Proc. Fifth North American Forest Biology Workshop, Gainesville, Florida. pp. 213-220.

FIRE IMPACT MANAGEMENT IN THE BOREAL FOREST REGION OF CANADA

David L. Martell
University of Toronto

Abstract: Increasing awareness that fire is a natural component of boreal forest ecosystems and escalating suppression costs are fostering growing opposition to traditional fire exclusion policies. However, since many people live and work in the boreal forest region of Canada, threats to their social and economic well-being make fire managers understandably reluctant to alter their exclusion stance. The author advocates the adoption of fire impact management policies whereby decisions concerning the suppression of wildfires and the use of prescribed fire are based on sound social, economic and ecological principles, and discusses some of the practical problems associated with the development and implementation of such policies.

Most Canadian forest fire management agencies presently operate under what can best be described as fire exclusion policies[1]. Fire managers strive to minimize area burned, subject to somewhat flexible (although increasingly rigid) constraints, on fire control resource availability and use[2]. These policies were initially developed in response to forest fire related disasters[3]. The current widespread reliance upon fire exclusion policies resulted from the prevailing attitudes of society in general, and forest resource users in particular, concerning values at risk in forested areas.

Enhanced communications and fire suppression resource mobility now enable fire managers to quickly deploy powerful but costly suppression forces. Modern fire management agencies can expend suppression funds much faster than has ever been the case in the past. But, there is growing evidence that in the boreal forest region of Canada, fire exclusion is neither economically nor ecologically sound (see for example, Stocks *et al.* (1981) and Rowe and Scotter (1973)). It has become obvious that we must abandon fire exclusion in favour of fire impact management policies, whereby decisions concerning the use of prescribed fire, and the suppression of wildfires, are influenced by their potential social, economic and ecological impacts.

Such non-exclusion fire management policies are not as novel as they might at first appear. The Government of Canada implemented a non-exclusion priority zoning system

1. Many provincial fire management agencies have delineated "Outside Intensive Protection (OIP)" areas that receive minimal protection due to their perceived low values at risk. Fires in OIP areas are usually not attacked unless they appear to pose a threat to public safety or property.

2. Most forest fire management agencies can avail themselves of "Extra (or Emergency) Fire Fighting (EFF)" funds that are used to fund operations associated with heavy fire loads, when costs usually exceed budget levels. In the past, EFF funds were viewed as being almost limitless, and very little fiscal control was exerted by legislators or managers not directly involved in suppression activities. Recent years have witnessed an increase in centralized control of EFF expenditures.

3. For historical accounts of some Canadian forest fire related disasters see McClement (1969).

in the Northwest Territories in 1967 (Murphy *et al.* 1980). The United States National Park Service began to revise its fire exclusion policy in 1968, primarily as a result of the 1963 Leopold report (Kilgore 1975). The United States Forest Service has since 1974, used a designated control (DESCON) policy in the Frances Marion National Forest in South Carolina. Under DESCON, some benign wildfires are not suppressed unless they threaten to burn beyond designated control areas (Devet 1976). The United States Forest Service began a widespread revision of its exclusion (i.e., 10:00 A.M.[4]) policy in 1978 (Dahl *et al.* 1981). As noted in footnote 1, many Canadian fire management agencies already have OIP areas. In 1982, the Ontario Ministry of Natural Resources approved the adoption of a non-exclusion fire management policy that will be applied throughout the province of Ontario[5]. Parks Canada is presently in the early stages of developing non-exclusion fire policies for selected parks (Lohnes 1981).

Canadian fire managers are experiencing considerable pressure to abandon fire exclusion in favour of fire impact management. At the same time, forest resource users continue to demand that their investments receive the protection they think they deserve. The change from exclusion to fire impact management must be very carefully planned and implemented, or disastrous consequences could result.

The purpose of this paper is to briefly outline some of the important practical issues that must be adequately addressed in order to ensure the development and implementation of sound non-exclusion alternatives to traditional fire exclusion practices.

Forest Fire Management Objectives

Barney (1975) and many others have pointed out that the role of a forest fire management agency is to provide a service, namely fire management, to land managers. The fire manager's ultimate objective is to satisfy society's desires, as expressed by land managers and others, in terms of public safety, economics, and ecology. In other words, fire managers must decide how to maximize the net benefit that results from the use of the fire management resources that society places at their disposal.

In its simplest form, fire exclusion is based on the implicit assumption that every unit of burned area is equally undesirable[6]. The fire manager's objective is therefore quite simple (although difficult to achieve); namely, to use the fire control resources at his or her disposal to minimize the extent of fire on the forest area within his or her jurisdiction. If we replace that simple objective with the more complex objective of maximizing the net (difficult to assess) benefit that results from the use of fire management resources, we complicate the fire management task dramatically[7].

---

4. The United States Forest Service "10:00 a.m." policy specified that suppression efforts should be aggressive enough to ensure that the fire is brought under control by 10:00 a.m. on the day following the day the fire was reported.

5. The new policy which is described in Ontario Ministry of Natural Resources (1982), is a non-exclusion policy which specifies that every forest fire in Ontario will receive a response that is governed by the predicted behaviour of the fire, its potential impact on people, property and values, and the estimated cost of the response.

6. It should be noted however, that fire managers recognize that fire damage varies over time and space, and they deploy their scarce resources accordingly.

7. In technical terms, we replace a relatively simple signle attribute utility function, (i.e., area burned), with a very complex, difficult to quantify, multi-attribute utility function.

Implementing Forest Fire Management Policy Changes

Forest fire managers are being forced to revise their practices, partly in response to a complex mixture of demands from diverse constituencies. Biologists emphasize the natural role of fire in boreal forest ecosystems, and advocate the re-introduction of fire into the environment. Silviculturists who are becoming increasingly conscious of site treatment costs, urge fire managers to expand their prescribed burning activities. Government policy analysts question the cost effectiveness of traditional fire exclusion practices, and demand rigorous quantitative assessments of fire management expenditures. Forest industries see valuable timber stands destroyed by fire, and urge governments to enhance their fire suppression capabilities.

It is clear that fire management agencies must abandon fire exclusion in favour of fire impact management, and some land management agencies have already begun to implement policy changes. But the transition and the ultimate goal cannot be achieved as easily as many critics imply. Some of the many complex issues that fire managers and their clients must resolve are as follows.

Fire Impact Assessment

Forest fire management agencies honed their exclusion skills in response to the generally accepted belief that the net impact of fire is destructive. One of the most appealing attributes of exclusion is that it reduces uncertainty (i.e., once a fire is extinguished, you no longer have to worry about it). This is an important benefit when one considers the fact that some unextinguished fires might well burn ''until the snow flies'', and meteorologists are unable to reliably forecast weather more than a few days in advance. It would therefore be socially irresponsible for fire managers to replace the relative certainty of exclusion with the uncertainty inherent in fire impact management, until they are capable of predicting the ultimate consequences of their actions.

The true social cost of large fires that rage through populated areas of the boreal forest has not been carefully assessed. History demonstrates that forest fires threaten people's lives, although current surveillance, communications, transportation and fire management technologies have all but eliminated this threat. Although many people enjoy the boreal forest ecosystems that have been shaped by fire, fires destroy homes and cottages, and severely disrupt peoples lives by causing evacuations and forest closures.

The development and implementation of sound fire impact management practices awaits the development of reliable operational fire damage appraisal systems that can be used to evaluate the economic impact of fire management actions. Simple fire damage appraisal schemes based on stumpage losses alone, fail to properly account for even the timber losses that result from fire. Van Wagner (1979) has illustrated why we must look beyond the stand that has burned, and assess the impact of a fire on the entire forest, (e.g., timber supplies, harvesting schedules and harvesting costs). Furthermore, although timber values are no doubt paramount throughout most of the boreal forest where fire exclusion objectives are now actively pursued, it is essential that non-timber economic impacts also be assessed.

Although fire exclusion practices are now being questioned primarily on economic grounds, it was the 'naturalness' of fire that precipitated the early calls for non-exclusion fire management practices. Forest ecologists and others have gathered tremendous amounts of data which describe the ecological impact of fire in boreal forest ecosystems (see for example, Rowe and Scotter (1973)). But much of this data is anecdotal in nature, and ecological effects are seldom related to important fire behaviour parameters (e.g., intensity). Consequently, there is little information that fire managers can use to quickly assess the ecological impact of their actions.

In summary, it is unreasonable to expect fire managers to practice fire impact management unless we provide them with the means of quickly assessing the social, economic, and ecological impacts of their actions. Fire managers have traditionally been expected to develop damage appraisal schemes in order to justify their budget requests. Since fire management is but a service to land users, fire management clients must bear some responsibility for the development of fire damage appraisal systems. Until such time as reliable and comprehensive fire damage appraisal systems are available, uncertainty concerning impact will understandably encourage risk averse fire managers to act in favour of fire exclusion.

Operational Fire Management Objectives

Burned area and fire management expenditures alone, are no longer adequate measures of fire management performance. Although natural fire regimes may well be suited to some designated 'wilderness' areas, it is clear that society cannot afford to allow fires to ''roam'' unchecked throughout the boreal forest. Neither can we afford the cost of attempting to exclude all fire from forested land, even portions that are devoted primarily to timber production.

The ultimate objective of a fire management organization is to use the resources at its disposal so as to best satisfy the fire management desires of society. This abstract objective must be translated into precise operational objectives and performance measures, so that fire managers can use the type of impact assessment schemes discussed above, to evaluate alternatives and assess their performance. Land managers and others must integrate the diverse expectations of different segments of society and develop operational objectives that fire managers can act upon.

Fire managers play two important roles in land management; one operational (i.e., to suppress, ignite, and otherwise manage the impact of fire in the forest, so as to satisfy land management objectives in a cost effective manner), and another which is advisory (i.e., to advise land managers with respect to the impact of fire in the forest). In some cases it may prove necessary to initiate organizational changes so that fire managers can properly fulfill their obligations to their clients.

Public Involvement

Fire exclusion practices evolved partly in response to an assumption that the net impact of every fire is destructive. Since the Crown owns most of the land in the boreal forest region of Canada, it is essential that the public be consulted if non-exclusion alternatives are considered. Murphy *et al*. (1980) stress the need to look beyond simple economic measures, and consider the potential impact of fire on people's lifestyles. Public consultation will help ensure that fire management practices do satisfy society's needs. Furthermore, an abrupt, poorly explained switch in fire management policy might well undermine the credibility of fire management agencies, diminish essential public involvement in the prevention and reporting of fires, and precipitate an increase in the use of 'personally prescribed fire'.

Legal Implications

When a forest fire management agency operates under a policy of fire exclusion, a wildfire that causes excessive damage is often viewed as an 'act of God'. The implementation of fire impact management policies will no doubt result in some good decisions leading to bad outcomes (e.g., a sound decision is made not to aggressively attack a particular fire which subsequently causes enormous damage, when with the benefit of hindsight, it should have been aggressively attacked). Such an incident may be viewed as an 'act of government', and will sometimes result in litigation. The fire management implications of forest management agreements between governments and industries (e.g., fire control obligations, prescribed burning activities, and the impact of burned area on the determination of annual allowable cut) must also be carefully assessed. For an interesting discussion of some legal aspects of forest fire management, see Spaeth (1979). It is essential that everyone involved in the development and implementation of non-exclusion fire management policies fully understand and assess the legal implications of policy alternatives.

Decision Support Systems

Forest fire managers are presently not properly equipped to administer non-exclusion fire management policies. Exclusion policies simplify fire management decision-making. Under such policies, fire managers do not have to decide which fires to extinguish; they simply decide upon suppression alternatives[8]. If we expect fire managers to successfully cope with the heightened complexity of fire impact management decision-making, they must be provided with powerful computer-based decision support systems to enhance their skills[9]. Such a system would transform massive amounts of fire related data into meaningful information. It could be used by fire managers to help evaluate policy alternatives, the strategic deployment of resources, and tactics for dealing with individual fires.

Some of the requisite characteristics of a forest fire management decision support system have been discussed by Barrows (1974) (fire intelligence system), and Kessell (1979) (gradient modeling). Maloney and Potter (1974) addressed many important development and implementation issues associated with fire management decision support systems, and provided valuable insight concerning some of the many complex scientific, technical and management problems that must be resolved. Kourtz (1981) has been developing and field testing an experimental computer-based decision support system in the Outaouais region of Quebec.

Examples of the types of information and capabilities that should be incorporated in a fire management decision support system include:

- fuel and topography information;
- observed and forecast weather information;
- current and projected fire management resource status information;

---

8. Fire managers are accustomed to assigning priorities to fires, and postponing action on low priority fires, when heavy fire loads result in suppression resource requirements that exceed supplies.

9. For a discussion of decision support systems see Keen and Morton (1978).

- values at risk information;
- fire occurrence and behaviour prediction models;
- fire management resource allocation models;
- fire impact prediction models;
- prescribed burn planning models, and
- procedures to facilitate the assessment of past performance.

The development of such decision support systems will require substantial amounts of money, and the skills of many specialists including; land managers, fire managers, fire behaviourists, economists, ecologists, operational researchers, and computer scientists.

Gradual Change

While it is becoming increasingly obvious that fire exclusion must be abandoned in favour of fire impact management, it is not clear how such a major shift in policy should be implemented. Although some American fire management agencies were required to quickly deal with the significant impact that fire exclusion has had on selected ecosystems in a relatively short period of time, the boreal forest is a comparatively stable ecosystem that does not yet appear to have been radically altered by past suppression activities. Canadian land managers can therefore afford the luxury of changing their fire exclusion practices gradually.

Detailed recommendations concerning the change from attempted exclusion to fire impact management are beyond the scope of this paper. Each agency must develop a detailed schedule of policy and procedure changes that will bring about the desired transition in a responsible manner. The following general steps might be incorporated into such transition schedules[10].

Step 1: Increased use of prescribed fire and rigidly supervised use of discretionary initial attack analysis and escaped fire situation analysis in low value areas[11].

Step 2: Extension of escaped fire situation analysis to all areas.

Step 3: Extension of discretionary initial attack analysis to all areas.

Fire management clients have come to expect effective fire exclusion, and forest fire management agencies have evolved into powerful exclusion forces. Only gradual change will enable clients to properly incorporate fire impact management into their land management programs, and allow fire managers to acquire the resources and expertise required to provide their new service in a cost effective manner.

---

10. Throughout the transition and beyond, fire management agencies must maintain strong suppression (i.e., exclusion) capabilities.

11. The term ''escaped fire situation analysis'' is used to denote the formal process of evaluating suppression alternatives when initial attack has failed (Hafterson 1979). I have used the term 'discretionary initial attack analysis' to denote a similar process applied to fires that are candidates for initial attack.

Conclusions

The adoption of fire impact management policies will not lead to the intentional burning of large areas in the boreal forest region of Canada. In the near future, fire managers will no doubt be directed to continue to aggressively attack most fires that occur in areas that were formerly designated as 'intensive protection' zones. Eventually, intentionally burned areas will increase as a consequence of fire impact management practices. However, social and economic considerations will continue to dominate ecological considerations. We simply cannot afford to allow fire to play its ''natural role'' throughout the boreal forest region of Canada[12].

Canadians can no longer afford the monetary or ecological consequences of fire exclusion throughout the boreal forest. But historical and recent events vividly illustrate the enormous harm that might result if fire is improperly managed. Although the time for a revision of traditional attempted fire exclusion practices has arrived, those involved in the transition to fire impact management must proceed slowly and with extreme caution.

Acknowledgements

The author wishes to thank R.J. Drysdale, R.V. Brady, D.L. Booth, D. Boychuk, J.M. Fullerton, T.J. Lynham, G.D. Puttock, and J.D. Walker, for their helpful comments on earlier versions of this paper. This work was supported in part by the Natural Sciences and Engineering Research Council of Canada.

Literature Cited

Barney, R.J. 1975. Fire management: a definition. J. For. 73: 498-519.

Barrows, J.S. 1974. Forest fire management — for ecology and people. Proceedings, Let the Forests Burn? College of Forestry and Natural Resources, Colorado State University, Fort Collins, Colorado. pp. 1-9.

Dahl, D.W., S. Pyne, E.V. Anderson and T. Crow. 1981. Fire and public policy. Proceedings, Fire Regimes and Ecosystem Properties. U.S.D.A. For. Serv. Gen. Tech. Rep. WO-26. pp. 557-569.

Devet, D.D. 1976. Descon: Utilizing benign wildfires to achieve land management objectives. Tall Timbers Fire Ecol. Conf. 14: 33-43.

Hafterson, J. 1979. Escaped fire situation analysis. Proceedings, Intermountain Fire Council and SAF Fire Working Group Symposium — Fire Control in the 80's. Missoula, Montana. pp. 220-226.

Keen, P.G.W. and M.S. Scott Morton. 1978. Decision support systems: an organizational perspective. Addison – Wesley Publishing Company Inc., Reading, Massachusetts.

Kessell, S.R. 1979. Gradient modeling: resource and fire management. Springer-Verlag, New York.

---

12. Brady (personal communication, 1983) has observed that the adoption of fire impact management policies will ead fire managers to implement revised forms (i.e., based on revised values), of the site specific, value determined strategies that characterized the early days of fire control, when suppression efforts were concentrated in high value areas (i.e., near towns, villages, transportation corridors and timber harvesting operations). Thus fire managers who gradually expanded their 'sphere of exclusion influence' to almost the entire forest, will once again focus on site specific fire impact, and tailor their response to individual fires.

Kilgore, B.M. 1975. Restoring fire to national park wilderness. Amer. For. 81: 16-19; 57-59.

Kourtz, P.H. 1981. Computer applications to daily forest fire management decision-making in southwestern Quebec. Proceedings, Intermountain Fire Council Symposium. Computer Modelling: It's Application in Fire Management. Salt Lake City, Utah. pp. 86-105.

Lohnes, D. 1981. Toward a new fire policy. Parks Canada, Natural Resources Bulletin. Ottawa, Ontario.

Maloney, J.E. and M.U. Potter. 1974. The fire management system — the fire management centre — preliminary results and design concepts. Environ. Canada, C.F.S., Inf. Rep. FF-X-48.

McClement, F. 1969. The flaming forests. McClelland and Stewart Ltd., Toronto, Ontario.

Murphy, P.J., S.R. Hughes and J.S. Mactavish. 1980. Forest fire management in the Northwest Territories. Indian and Northern Affairs Canada, Ottawa, Ontario.

Ontario Ministry of Natural Resources. 1982. Forest fire management policy. Rep. AFMC #126. Aviation and Fire Management Centre, Ontario Ministry of Natural Resources, Sault Ste. Marie, Ontario.

Rowe, J.S. and G.W. Scotter. 1973. Fire in the boreal forest. Quat. Res. 3: 444-464.

Spaeth, G.L. 1979. Legal constraints of firefighting. Proceedings, Intermountain Fire Council and SAF Fire Working Group Symposium — Fire Control in the 80's. Missoula, Montana.

Stocks, B.J., C.E. Van Wagner, W.R. Clark and D.E. Dubé. 1981. The 1980 forest fire season in west-central Canada — social, economic, and environmental impacts. Environ. Canada, Ottawa, Ontario.

Van Wagner, C.E. 1979. The economic impact of individual fires on the whole forest. For. Chron. 55: 47-50.

A RESOURCE MANAGEMENT VIEWPOINT OF BOREAL FOREST ECOSYSTEMS

G. Baskerville
University of New Brunswick

Abstract: Forest management is defined in terms of the four basic tools, scheduling, distribution, renewal and protection. Emphasis is placed on the role of forecasting dynamic development in management decision making. More research is needed to relax the constraint to management enforced by the need to forecast development of many different types of stands, on different sites and in response to different kinds of interventions. Particularly, there is a need for research towards better ways to integrate geographically spread individual stand development patterns into a whole forest forecast.

Management consists of controlling biological dynamics over a forest area in such a manner that it is possible to take a continuous flow of various materials or uses from the area without endangering the rate of production of these materials and uses by the resource. Basically, management is control of the amount and timing of availability of the materials and uses. This control is gained by regulating the dynamics of the resource. Management only indirectly addresses present problems. Management is undertaken in order to ensure future continuity of the forest uses, and to improve the future situation. With a few exceptions, the benefits of forest management actions taken today will not be captured until some forty to fifty years from now.

This paper is an introduction to forest management. The reader is strongly encouraged to consult references such as Davis (1966), Duerr *et al*. (1975), Raiffa (1970) and Reed (1978) for more detailed information.

The Four Basic Actions

Management of a forest, or for that matter of any renewable resource, requires the integration of four basic forms of action. These are the means by which forest dynamics are regulated; scheduling the harvest, distributing the harvest, renewal of the resource, and protection of the resource.

Scheduling the harvest consists of fixing the amount of harvest, or use, that can be taken and the location of that harvest for each year of the planning period. Normally the time horizon in forest management planning is of the order of 25 to 40 years. For example, if a series of stands in specific geographic locations can be identified to yield the necessary harvest for each of sixty years into the future, there can be sustainability over that period. By that time the first area (or some other) must be available for harvest again, so that it can be rescheduled for the 61st year. The key here is that the actual locations where each annual harvest will be taken during the planning period are identified. Alternatively, one could view it as the scheduling of each hectare with respect to the pattern of development of the forest on it and the harvesting of that forest.

As each annual harvest is made there will be an array of raw materials of various qualities available for distribution to users. In the exploitive stages of forest use this distribution is not viewed as a particularly important matter, however, as forest use diversifies the quality of raw material becomes crucial to survival of the forest-based use. It is then necessary to get the best possible distribution or allocation, of the available raw material and uses each year.

Renewal, or silviculture, consists of those activities such as planting, spacing, weeding and thinning that are used to speed the rate of growth of a forest stand or renewal for a non-consumptive use. There is a rather widespread belief that these actions themselves constitute 'forest management'. It is worth emphasizing that renewal is but one of the four major elements of forest management.

Protection from unscheduled harvest by fire, insects or disease is essential if the management benefits of the first three actions are to be achieved. In the obvious context, a stand scheduled for harvest in 1990 must be protected from fire and insects until then. If it is not protected, and the stand is lost, it will be necessary to revise the schedule. Where there is a reserve of 'unused' forest this does not pose a problem. However, on a management area where all the stands are scheduled in order to maintain sustainability, the loss of a portion of the forest means that a future rupture in sustainability is certain.

Linkage of Actions

The four tasks that together comprise forest management, namely scheduling, distribution, renewal, and protection are intricately linked together by geographic location and by time. Each of these actions is carried out on a specific location at a specific time. A specific stand may be (i) scheduled for harvesting in 1982, to (ii) yield 200 m^3 ha^{-1} of pulpwood and 100 m^3 ha^{-1} of sawlogs, with (iii) the area to be scarified in 1983, (iv) planted to 2-2 stock in 1984, (v) weeded as necessary between 1986 and 1990, and (vi) protected as necessary so that the area can be, (vii) scheduled for harvest in the year 2024 to yield 300 m^3 ha^{-1} of wood. The point here is that these management actions are local in nature, specific to place and time, and they are distributed over wide areas, usually affecting a very small portion of the total forest in any one year.

Management planning, or strategic planning as it is called in the jargon of the '80s, consists of integrating the four basic actions of scheduling, distribution, renewal and protection, over a geographic management area and across time. These four tools, or tactics, are the means by which forest dynamics are controlled. Management planning therefore attempts to use these tools in the right amount, in the right places, in the right years, so that over a long period of time the desired management goals for the forest are obtained. No one of these tools constitutes management in itself, and none of them is an instant cure for any management problem. Management is achieved by the integration of all four kinds of action over the entire management area and over a time horizon of at least 40 years.

Forest management decisions centre on the use of the four basic tools. Since the tools are used in specific locations there is a tendency to consider 'management' at the level of a specific location. In fact, management concerns the relationship of the specific actions (at specific places and times) to development of the whole forest. Thus the year-to-year mix of the four basic actions in location, and in amount, is determined with respect to the total forest and the goals for its development. To comprehend resource management it is essential to grasp the distinction between these two levels of scale: the stand level at which individual management actions are taken, and the forest level at which the integrated effects of the stand level actions are planned and evaluated.

Forecasting and the Scale Problem

Decisions with respect to all four of the basic actions of management require forecasts of future development of parts of the forest system. In fact, all decisions are essentially a choice between two or more alternative forecasts of the future. For example, the decision choice is not really between planting 2-0 stock and planting 2-2 stock, but rather between the two probable future development patterns that would result from these two courses of action in a specific situation. Because of the two distinct scales involved, forest management decisions require two distinct types of forecasts. First, the choices of action in a particular stand must be characterized by a forecast of the dynamic development in that stand that would likely follow the implementation of each alternative action under consideration. Second, to choose an appropriate mix of actions, it is necessary to integrate combinations of stand level forecasts for all stands, whether subjected to action or not, in the whole forest into a whole forest picture to develop a forecast of the probable response at the forest level.

Management decisions aim at choosing a combination of scheduling, distribution, renewal and protective actions which, when applied at specific locations with specific intensity and at specific times, will give a development of the whole forest consistent with the management goals. That is, the set of actions for which the combined forecasts yield the desired future for the forest. In this context you will see that it is unlikely that anyone will manage "the boreal forest". A forest as extensive as the boreal is managed by parts, that is in units of a million or so hectares where a manager can exercise control over the scheduling, distribution, renewal, and protection. Further, management is only undertaken where there is some benefit in doing so. In general, management is undertaken when the flow of benefits to man is in some way imperiled, ie., the resource-based economy is imperiled. In this case it becomes economically reasonable to expend the dollars and manpower required to execute the four management actions.

Management does not attempt to get a return on the dollars and manpower of management effort on a hectare-by-hectare basis. At current interest rates standard discounting practices would prohibit all activities except exploitation. Management is the integration of the four activities over the whole forest, and over a long period of time, and therefore the costs of management are assessed against the whole forest and against the industry based on it. By conventional economic theory it may not pay to plant a hectare of forest where the end point of the calculation is the harvest, in the distant future, of that hectare alone. However, it can pay to plant many hectares of forest where the end point in the calculation is the several hundred million dollar investment in a pulp mill. Particularly when the mortgage on the mill lasts forty to fifty years.

Forecasts play a contentious role at the management, or whole forest, level. No chief executive officer wants to spend money on management actions if these actions are not essential to protect the long term viability of his total industrial operation. Similarly, no chief executive officer wants to face the necessity of reporting to his shareholders, and bankers, that failure to take timely forest management action to secure a sustainable supply is limiting opportunities for mill expansion or forcing reductions in mill capacity. Good forecasts are essential to the timely planning and execution of forest management and to the industry based on the resource.

The essential feature here is that the forecasts necessary to the management of scheduling, distribution, renewal, and protection are interlinked. For example, it is not possible to forecast scheduling without considering renewal and protection. Because of this linkage there is relatively little room to trade off one of these forms of action against another. Planting trees by itself cannot save a forest if there is no controlled scheduling of the harvest. If the forest is all harvested, within say 30 years, it is not possible to plant enough trees to prevent a rupture of forest productivity 30 years hence since it takes longer than that to grow planted trees to useable size. While the actions are linked, it should be obvious that there is a continuum of intensity of effort in which the four actions are applied. The mix of the four actions, and the intensity with which each is utilized depends on the current management goals for the forest area. That is, the effort expended in the four actions is determined by what is necessary to achieve the goals.

Forecasting Dynamics of Development

Let us now address the issue of how dynamics in boreal forest systems impose constraints on forest management. Clearly, a key dynamic is the economic one. Although economics is not a dynamic of the boreal forest, it is necessary to cast the biological dynamics in a context that permits economic evaluation. In broadest terms, biological dynamics of the forest impose a constraint to management simply because the spatial variability of the dynamics leads to a failure to comprehend their impact, and hence to a failure to recognize the need for management control with sufficient lead time to allow for the slow response of the system to the management action. The difference between a managed and an unmanaged forest could be characterized as, in an unmanaged situation we let the future happen to us, whereas in the managed situation we attempt to choose the future we will endure. The manager must first comprehend the dynamics to realize the need for action. He then compares reasonably possible futures. Having chosen a 'future', the four management actions are then used to regulate the dynamics of the forest so that the desired 'future' unfolds.

The importance of forest dynamics in decision making can be illustrated in terms of the four management actions in turn. With respect to scheduling, the key questions are "when will this stand be available?", "how long will it continue to be available?", "what is the likely successional response following the harvest?", "when can this stand be rescheduled?". Forecasts are necessary in order to address the issues of availability and whether or not renewal action is required.

The development of an existing stand depends on its composition, structure and site. The successional pattern following a harvest depends on the biological interactions among the forest that was cut, the site conditions, and the method of harvest used. Scheduling, therefore, requires a forecast of the development of each stand up until it is harvested and of succession following harvest. The need for interim protection action or post-harvest renewal action is then addressed in the context of the total forest. If the amount and timing of availability of wood from the whole forest is acceptable, given the individual forecasts of pre- and post-cut development, there is no need to treat further. On the other hand, if the amount of wood that will be available in the future will be reduced, and/or if the timing of availability will be delayed, the maintenance of current harvesting levels requires that treatment be undertaken so that the harvest can be taken as planned and the area can be rescheduled at a time appropriate to the maintenance of sustainability.

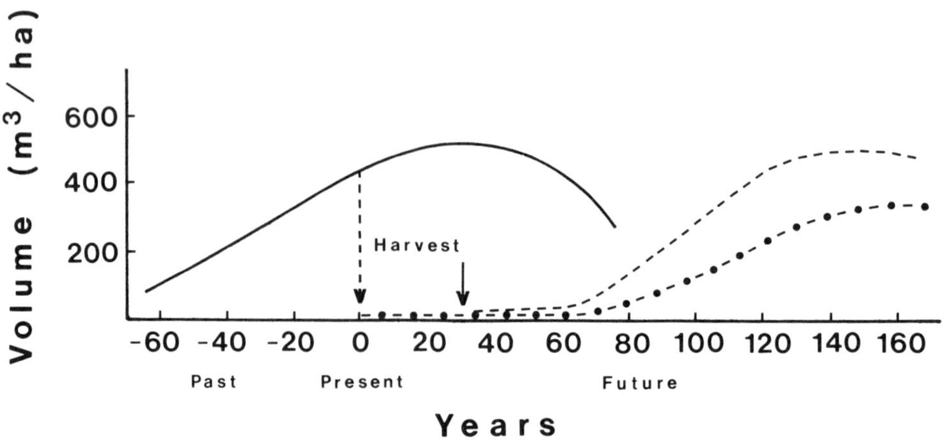

Figure 1. The past development and probable future of a single stand (solid line), and the probable pattern of development following logging the stand now (dashed line), or 30 years from (dash dot line) now, using the same logging method.

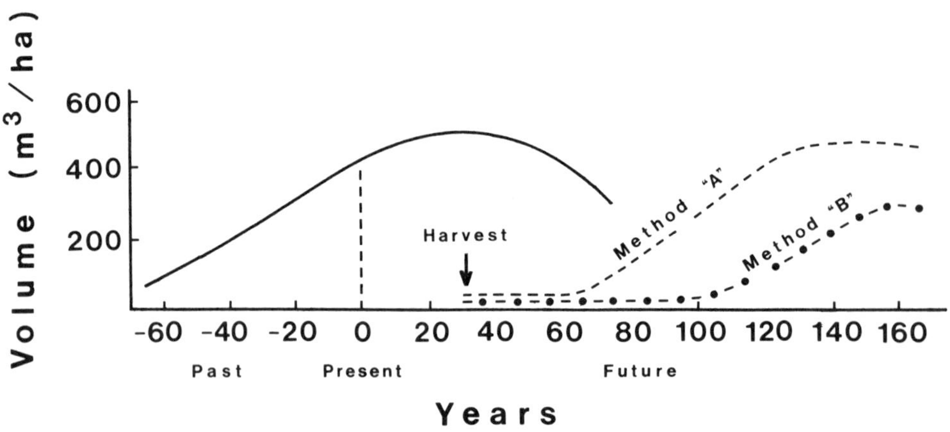

Figure 2. The past development and probable future of a single stand (solid line), and the probable pattern of development following logging by method A (dashed line) and method B (dash dot line).

The stand level dynamics are characterized in Figure 1 which illustrates some reasonably possible futures with respect to a single indicator, stand volume, in a single stand. The forecast shows that the stand will be available for harvesting, with slight increases in volume, for the next 30 years given normal protection. Obviously, if it is not protected from fire and insects, there is a risk that the stand will not maintain its current volume over that period. If the stand is harvested this year the successional pattern following the harvest is forecast to result in a usable stand being available for rescheduling in about 140 years. On the other hand, if the stand is not harvested until 30 years from now its post harvest development forecast will permit rescheduling, for a substantially higher volume, about 120 years from now. The logging system and the site are the same in both cases, the difference resulting from the development of the stand, particularly its advance growth, over the next 30 years.

The method of harvesting also powerfully influences the post cut dynamics. Two different methods of harvest at the same stage of development of the stand can lead to very different post-harvest successional patterns. The diagram in Figure 2 suggests that the forecast for one method permits rescheduling in 150 years while the forecast for the second method of harvesting permits rescheduling for a higher volume in about 110 years, both without any renewal effort.

The point here is that for a single stand there are a number of possible dynamic reactions depending on the level of protection afforded until it is cut, on the stage of development of the stand when it is cut, and how it is cut. The choice of when and how to cut requires a comparison of forecasts of these reaction patterns. The forecasts in the figures are for a single indicator, but the manager will also want information on other indicators such as tree size, species mix, and deer habitat capability.

Achieving the set of stand level forecasts of dynamic responses is the first step. The forest is made up of hundreds of thousands of stands, each with its own peculiar site and compositional characteristics and each at a particular stage of development. To characterize the forest, it is necessary to composite the stand forecasts and develop a set of indicators which show how the whole forest will probably develop given particular sets of actions. The forest level forecast is primarily a problem of addressing spatial patterns. Average forest level responses are exceedingly dangerous in management and their use in the past has lead to many 'local' problems.

Determining the distribution of raw material qualities that can be obtained from a stand that is being cut this year is easy. However, management also requires a forecast of what products will be available from the stands that are scheduled for cutting, 1, 2, 3, 4, 5, 6, 7 . . ., years hence. This requires a forecast of the structural development of each stand that is scheduled for harvest over the period of interest.

If the stand in Figure 1 is harvested now it will yield a volume of 450 m^3 ha^{-1} all of which is in pulpwood material. If the stand is left to develop and harvested in 30 years time, the volume is forecast to increase to 550 m^3 ha^{-1}. However, at that time it is forecast that 300 m^3 ha^{-1} is in trees usable as sawlogs and 250 m^3 ha^{-1} is in pulpwood material. The relative needs of pulpwood and sawlog quality material over time force an examination of the dynamics of change in size class distribution in the stand. Again, this will lead naturally to the question – "is treatment needed?" Also again, this can only be determined by comparing the forecasts of the future results of treatment and no treatment in the context of the total forest, and deciding which result is most consistent with management goals.

The action of renewal is the one with which you will no doubt be most familiar. The key questions are 'is it needed' and 'how much?' Assuming that analyses of post-cut forecasts in the total forest context have indicated that renewal action is needed, the issues become 'how much area to plant?', 'where?' and 'with what technique?' Will the area simply be planted, will it be scarified and planted, will it be burned and planted, and, in any case, with what species? For each alternative form of action it is necessary to forecast the probable development on the specific site so that decisions can be made with respect to the amount and type of action consistent with the management goals for the whole forest.

Perhaps the easiest way to illustrate this forecasting with respect to silvicultural treatment is the case of weeding plantations with herbicides. A forecast is made of the development of a plantation on a specific site if it is given appropriate weeding in the early years,

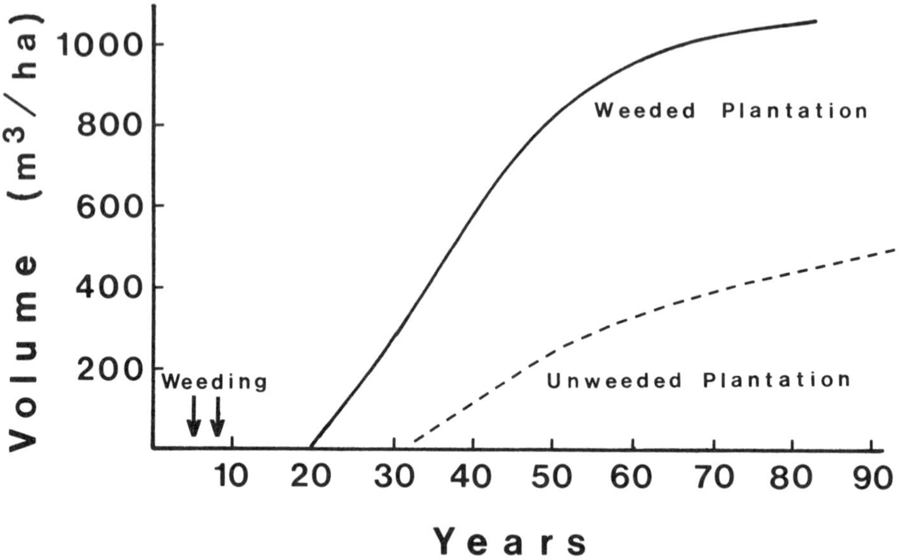

Figure 3. The probable development of a plantation on a given site if it is weeded in the early years (solid line) and if it is not weeded (dashed line).

usually once or twice in the first five or six years in the life of the plantation. A second forecast is made of the probable development of the same plantation if weeding is not undertaken (Figure 3). These forecasts are then compared in the whole forest context to the required amount and timing of availability of wood, and an appropriate choice is made. The use of herbicides is based on forecasts of the probable development of plantations with and without early weeding and how these forecasts relate to whole forest goals.

As with the other actions, decisions with respective protection require forecasts. In simplest terms the question is 'what losses can be accepted?', or 'what protection is needed?' Again, an analysis of the total forest is necessary to determine what particular losses of growth, or of growing stock, and what delays in stand development, are tolerable with respect to the required availability of wood. Further, there is always a choice between various forms of protective action. For example, with respect to protection from spruce budworm, such possibilities exist as doing nothing, using conventional aerial spraying of chemical insecticides, and spraying with BT. A decision must be made with respect to both how much of the forest is protected, and in what manner. This is done by developing forecasts of the probable development of the forest following each of the possible patterns of action in the whole array. One particularly interesting aspect of this problem has to do with the dynamic relationship of budworm populations and the host forest. This insect and the boreal forest have evolved together for a very long period of time and have developed some exquisitely neat dynamics. Building a forest that will be less susceptible to budworm is simple in concept. The four actions of management are used so that, over a period of time, the structure of the forest is changed such that the budworm/forest dynamics that have prevailed over evolutionary time have been disrupted. The problem, of course, is to discover the budworm/forest dynamic that can be challenged on the necessary spatial and time scales. There is no point in spending 50 years changing a forest unless a biological forecast indicates the changed forest will be less habitable for the bug.

The point of this discourse is to impress upon you the fact that managing a forest requires extensive use of an understanding of forest dynamics. Put bluntly, it is not possible to make management decisions about a forest without involving, in one way or another, an understanding of stand and forest dynamics.

The Constraint

Forest management is accomplished by a series of decisions which, year by year, over long periods of time integrate, on-the-ground, the four management actions of scheduling, distribution, renewal and protection. The goal is to organize the particular on-the-ground actions, year by year, so that over the planning period the whole forest develops in the desired manner. Management decisions determine the use of these actions at a particular place and a particular time, so that at that same particular place a particular kind of forest will be available at some particular future time. I want to reinforce the idea that a decision is a choice between two or more forecasts. There is a powerful tendency for loose discussion of management decisions to focus on local actions, often as if the choice were between one local action and inaction. Actions are simply a means to an end, and the choice is made for the means which is most likely to produce the desired end. That is, the choice is dependent on a comparison of forecasts. A real constraint to management then, is the ability to forecast the dynamic development of stands and forests. In this respect, it is fair to say that there are very few wrong choices in resource management. Analysis of 'bad' decisions, that is decisions which are subsequently found to be bad, almost always show that the difficulty lay in bad forecasts. Management decisions are made by reasonable people using the best available information. Objective analysis of 'bad' decisions will reveal that, given the information available at the time of decision, the choice was reasonable. The forecasts of what would follow from the chosen actions were inadequate, and consequently the decision turned out to be 'bad'. I distinguish this situation from that where management decision making is challenged based on a local observation. In this latter case the observer has not placed his 'need for action' in a management, that is whole forest, context and his comment on management is meaningless.

There is a long history in forestry of forecasting stand and forest development. These forecasts range from intuition based on extensive observation of the natural system, to elegant models based on rather fundamental understanding of how the system operates. Although there is currently research interest in this area, there is still not enough work, and it is very poorly focussed. Many years ago when I was expounding on the virtues of my research program to a woodlands' manager, he brought me up short with the statement that 'you scientists are always telling me what happened on your 0.01 ha plot over the past ten years, while what I need to know is what is going to happen on my 100,000 ha forest over the next forty years'.

Forecasting problems center on scale and direction. Science commonly addresses dynamics at a scale not usable in management decision making. Science has a tendency to look backwards (so that it can be factual) rather than forewards in time.

The scale at which information is presented for decision making is absolutely crucial. For information to be useful in decisions it must be presented at a scale usable in the planning and execution of scheduling, distribution, renewal, and protection. If these actions are planned and executed at a scale of 100 to 500 ha then the forecast of dynamic response must be made available at that same scale. Individual tree data and whole forest averages are worse than useless to the management decision maker. Since they do not relate to specific locations at the scale of management they cannot be used directly in decision making, and their gratuitous presentation by scientists almost always leads to confusion and resentment on the part of the decision maker.

Much better ways of systematically forecasting the stand dynamics that occur at one place over periods of time as long as 50 to 100 years are needed. Further, much better means of integrating these spatially disaggregated forecasts into development of a whole forest over similar periods are needed. There is no visible research in Canada that addresses the management level concerns relating to spatially variable dynamics.

The Need for Theories

The fundamental issue here for scientist researchers is that by definition all forecasting lies in the realm of theory. The future does not repeat the past in dynamic systems such as a boreal forest system under management. Management is regulation of the dynamics of development, and it is unrealistic to expect that, as the mix of actions is changed, the dynamic development will be similar to that observed in the past with no intervention. Hence the desperate need for a theory of stand development that can realistically encompass response to a variety of actions, and a need for a theory of forest development that is sufficiently comprehensive to grasp the spatial variation of the possible actions carried out in the forest.

No one suggests that these theories/forecasts be presented as fact. No decision maker believes a scientist who suggests he can do this. Decision makers, if nothing else, know all about the uncertainty of the future and are accustomed to dealing with uncertainty in forecasts. What the decision maker needs is the best possible application of science in developing the forecasts of the reasonably possible futures from which he as a manager chooses his management actions. A particular frustration to those of us interested in management has been the absence of research into ways of putting all of the stand forecasts together so that an array of scheduling, distribution, renewal and protective actions can be designed to reach a desired forest development over time. There has been much work published on optimization of actions in administration and economic journals, but virtually all of this is done with forecasts that do not recognize the biologically dynamic responses. Indeed, many of our really 'bad' resource decisions are traceable to elegant mathematical decision frameworks that used really dumb forecasts of biological dynamics. This area urgently requires the attention of biological scientists in addition to the current mathematical orientation.

Forecasting the reasonably probable futures that will result from management intervention in boreal forest systems is the essence of management. These considerations of dynamics enter into all decisions respecting management. Sometimes the considerations of dynamics have been explicit, sometimes implicit, sometimes the decision maker has not even realized he has made an assumption or postulate about dynamic response, but beneath every resource management decision lies a group of forecasts about dynamic responses. Using forecasts based on scientifically sound theories would render the forecasts more explicit, and thereby make it possible to evaluate the theories by carefully monitoring and comparing the outcomes. Science has not served resource management very well. By failing to provide usable theories of dynamics at the temporal and spatial scales used in management, science has left the integration of biological dynamics to the manager. Disciplinary blinkers have been a major problem in this respect. Scientists have not integrated their observations into usable theories, and they have been quick to pass value judgements on the managers when the necessary integration is made to the best of the managers abilities.

Scientific reporting, particularly in the refereed journals has completely missed the boat in failing to address resource management at the appropriate geographic and time scales. The literature contains many papers on the scale of 'the floristic characteristics of the boreal forest' (or whatever). No manager believes such material. He knows the forest is variable and that what is a 'natural' pattern in one place does not hold in another, particularly not in a managed system. In any event, no one will ever take management action at the scale of the whole boreal forest and consequently these papers are useless to management. The literature contains many papers that describe in detail what one branch or one tree, or one hectare looks like, or occasionally, what one tree or one hectare looked like over the past five years. This scale of observation has some potential value, if these various past incidents in successional history can be integrated into a theory of future dynamic development. Seldom in the biological literature does one find work at the scale used in management decision making and more importantly, seldom does one find comprehensive theories of dynamic development. This latter is changing as some useful models of succession are evolving although as yet few of these embody a useful temporal/spatial scale.

Conclusions

The absence of a rigorous approach to the development of a usable understanding of dynamics in boreal forest systems is a serious constraint to management of these systems. We can perhaps most conveniently gain this understanding by observing and experimenting with undisturbed systems. However, the bottom line is that managed, means disturbed. The understanding of dynamics that is required is one that comprehends dynamic reactions to the perturbations caused by the four kinds of management actions. In short, we need good, testable, theories of dynamic development. These theories can be used to develop safer forecasts and the theories can be checked by comparing the actual result with the forecast. This should lead to systematic revision of these theories to make them more comprehensive.

Improving the level of science in the forecasting of stand and forest dynamics will not render the future certain. It can, however, reduce the level of uncertainty and it can help to protect decision makers from sequences of decisions that lead to dead ends. The basic constraint to management resulting from the dynamics of the boreal forest is the lack of a systematic approach to the development of a scientific understanding of dynamics at the stand and forest levels.

Literature Cited

Davis, K.P. 1966. Forest management: regulation and valuation. McGraw-Hill, Toronto, Ontario.

Duerr, W.A., D.E. Tecguarden, S. Guttenberg and N.B. Christiansen. 1975. Forest resource management: Decision-making principles and cases. O.S.U. Bookstores Inc. Corvallis, Oregon.

Raiffa, H. 1970. Decision analysis: introductory lectures on choices under uncertainty. Addison-Wesley, Don Mills, Ontario.

Reed, F.L.C. 1978. Forest management in Canada. Environment Canada, Canadian Forestry Service. Info. Rep. FMRX-102.